Advances in
Human Aspects of
Road and Rail Transportation

Advances in Human Factors and Ergonomics Series

Series Editors

Gavriel Salvendy
Professor Emeritus
School of Industrial Engineering
Purdue University

Chair Professor & Head
Dept. of Industrial Engineering
Tsinghua Univ., P.R. China

Waldemar Karwowski
Professor & Chair
Industrial Engineering and
Management Systems
University of Central Florida
Orlando, Florida, U.S.A.

3rd International Conference on Applied Human Factors and Ergonomics (AHFE) 2010

Advances in Applied Digital Human Modeling
Vincent G. Duffy

Advances in Cognitive Ergonomics
David Kaber and Guy Boy

Advances in Cross-Cultural Decision Making
Dylan D. Schmorrow and Denise M. Nicholson

Advances in Ergonomics Modeling and Usability Evaluation
Halimahtun Khalid, Alan Hedge, and Tareq Z. Ahram

Advances in Human Factors and Ergonomics in Healthcare
Vincent G. Duffy

Advances in Human Factors, Ergonomics, and Safety in Manufacturing and Service Industries
Waldemar Karwowski and Gavriel Salvendy

Advances in Occupational, Social, and Organizational Ergonomics
Peter Vink and Jussi Kantola

Advances in Understanding Human Performance: Neuroergonomics, Human Factors Design, and Special Populations
Tadeusz Marek, Waldemar Karwowski, and Valerie Rice

4th International Conference on Applied Human Factors and Ergonomics (AHFE) 2012

Advances in Affective and Pleasurable Design
Yong Gu Ji

Advances in Applied Human Modeling and Simulation
Vincent G. Duffy

Advances in Cognitive Engineering and Neuroergonomics
Kay M. Stanney and Kelly S. Hale

Advances in Design for Cross-Cultural Activities Part I
Dylan D. Schmorrow and Denise M. Nicholson

Advances in Design for Cross-Cultural Activities Part II
Denise M. Nicholson and Dylan D. Schmorrow

Advances in Ergonomics in Manufacturing
Stefan Trzcielinski and Waldemar Karwowski

Advances in Human Aspects of Aviation
Steven J. Landry

Advances in Human Aspects of Healthcare
Vincent G. Duffy

Advances in Human Aspects of Road and Rail Transportation
Neville A. Stanton

Advances in Human Factors and Ergonomics, 2012-14 Volume Set:
Proceedings of the 4th AHFE Conference 21-25 July 2012
Gavriel Salvendy and Waldemar Karwowski

Advances in the Human Side of Service Engineering
James C. Spohrer and Louis E. Freund

Advances in Physical Ergonomics and Safety
Tareq Z. Ahram and Waldemar Karwowski

Advances in Social and Organizational Factors
Peter Vink

Advances in Usability Evaluation Part I
Marcelo M. Soares and Francesco Rebelo

Advances in Usability Evaluation Part II
Francesco Rebelo and Marcelo M. Soares

Advances in
Human Aspects of
Road and Rail Transportation

Edited by
Neville A. Stanton

CRC Press
Taylor & Francis Group
Boca Raton London New York

CRC Press is an imprint of the
Taylor & Francis Group, an **informa** business

CRC Press
Taylor & Francis Group
6000 Broken Sound Parkway NW, Suite 300
Boca Raton, FL 33487-2742

© 2013 by Taylor & Francis Group, LLC
CRC Press is an imprint of Taylor & Francis Group, an Informa business

No claim to original U.S. Government works

Version Date: 20120529

International Standard Book Number: 978-1-4398-7123-2 (Hardback)

Visit the Taylor & Francis Web site at
http://www.taylorandfrancis.com

and the CRC Press Web site at
http://www.crcpress.com

Table of Contents

viii

Preface

Human Factors and Ergonomics have made a considerable contribution to the research, design, development, operation and analysis of both road and rail vehicles and their complementary infrastructure. This book presents recent advances in the Human Factors aspects of Road and Rail Transportation systems. These advances include accident analysis, automation of vehicles, comfort, distraction of drivers (understanding of distraction and how to avoid it), environmental concerns, in-vehicle systems design, intelligent transport systems, methodological developments, new systems and technology, observational and case studies, safety, situation awareness, skill development and training, warnings and workload. The book is divided into 8 sections, as shown below:

    I.   Design of driver - vehicle systems interaction (22 chapters)
    II.   Distraction of the driver (4 chapters)
    III.   Environmental impact of vehicles (5 chapters)
    IV.   Integrated transport systems (5 chapters)
    V.   International differences in drivers (2 chapters)
    VI.   Investigations into safety and accidents (2 chapters)
    VII.   Methods and Standards (18 chapters)
    VIII.   Studies of locomotives, drivers and rail systems (9 chapters)

I am grateful to the Scientific Advisory Board which has helped elicit the contributions and develop the themes in the book. These people are academic leaders in their respective fields, and their help is very much appreciated, especially as they gave their time freely to the project.

K. Bengler, Germany	D. Kaber, USA
G. Burnett, UK	J. Krems, Germany
P. Chapman, UK	F. Mars, France
F. Chen, Sweden	D. McAvoy, USA
L. Dorn, UK	R. Risser, Austria
I. Glendon, Australia	P. Salmon, Australia
J. Groeger, Ireland	S. Sharples, UK
R. Happee, Netherlands	G. Walker, Scotland
S. Jamson, UK	K. Young, Australia

This book will be of interest and use to transportation professionals who work in the road and rail domains as it reflects some of the latest Human Factors and Ergonomics thinking and practice. It should also be of interest to students and researchers in these fields, to help stimulate research questions and ideas. It is my hope that the ideas and studies reported within this book will help to produce safer, more efficient and effective transportation systems in the future.

March 2012

Neville A. Stanton
University of Southampton
United Kingdom

Editor

Section I

Design of Driver–Vehicle Systems Interaction

CHAPTER 1

Evaluation of a Contact Analog Head-Up Display for Highly Automated Driving

Daniel Damböck, Thomas Weißgerber, Martin Kienle and Klaus Bengler

Technische Universität München – Institute of Ergonomics
Boltzmannstraße 15, 85747 Garching, Germany
damboeck@lfe.mw.tum.de

ABSTRACT

The automotive industry is working on assistance systems to increase safety and comfort of today's vehicles. In the course of this development combined with increasingly capable sensors, assistance systems become more and more powerful. This whole development leads to a change in the role of the human, from the actual driver of the car to a supervisor of the automation state. The following article describes a display concept to inform the driver about the current status and future actions of the automation system. The concept is based on a contact analog Head-Up Display. By giving visual feedback the driver's comprehension of the automation actions is increased, and therefore the driver-automation-cooperation is improved. Correspondingly the presented experiment reveals significantly decreased reaction-times by visual feedback regarding automation failure.

Keywords: highly automated driving, driver assistance, cooperative control, contact analog, Head-Up Display

1   INTRODUCTION

Today a lot of traffic-accidents happen due to human failure. Since quite a while the automotive industry is working on assistance systems that help to avoid or at least reduce this kind of accidents. Such systems not only assist the driver in his task

4

driving the vehicle. In fact, the evolution of these systems could lead to a complete removal of the driver as a factor in the driving task. Given the fact mentioned in the beginning, that human failure is one of the main reasons for traffic-accidents, this approach seems to be a promising idea. But it clearly has to be said that this approach can only work as long as the assistance system is working without error and has a perfect representation of the surrounding environment in every possible situation. This means the driver wouldn't need to take over at any time. As this demand is impossible to achieve by any technical system a paradigm of vehicle driving has to be found in which both, the automation system and the human can collaborate mutually compensating the insufficiencies and failures of the respective partner. By such a cooperative way of vehicle driving the driver shall be kept active and in the loop. Thus automation effects can be reduced and the driver is available to take over vehicle control when necessary.

Figure 1 shows one possible occurrence of a cooperative human-automation-vehicle interaction as it has been developed in the nationally funded project H-Mode (Flemisch et al., 2003). Based on the initial driving task, both, human and automation establish a mental representation of the environment and try to influence the vehicle movement. By doing so, they have to communicate with each other to negotiate the best possible solution for the current driving situation (Damböck et al., 2011).

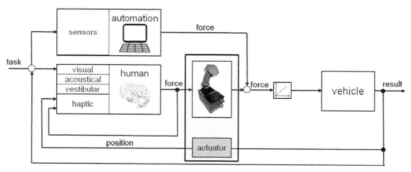

Figure 1: Schematic illustration of a cooperative human-automation-vehicle-system

One of the essential features of this idea of cooperative vehicle-driving is a bi-directional haptic coupling with continuous and/or discrete communication between driver and automation (Flemisch et al., 2008). In order to provide the driver with a haptic feedback of automation recommendations (Schieben et al., 2008), active control elements (e.g. Active Sidestick or Active Steering Wheel) are used as a basis (Kienle et al., 2009). Within the project H-Mode the idea of haptic communication has been described and its benefit has been shown in several experiments (e.g. Damböck et al., 2009, Flemisch and Schieben 2010). Although the haptic communication between driver and automation system can be a solid basis for a cooperative human-machine-system, it is inevitable to complete the haptic signals with visual information. It is necessary that vital information is transferred between human and machine, considering the actual and future status of the driving task and future intentions of the automation about driving maneuvers.

This approach is based on Bubb (2001) who claimed that the haptic channel should be used to show the driver what to do, while the visual channel is used to explain why. By doing this the driver's comprehension of the automation actions can be increased, and therefore the driver-automation-cooperation is improved.

An ideal way to use the visual channel in order to inform the driver about the intentions of the automation and therefore the future behavior of the vehicle is the application of a contact analog Head-Up Display (caHUD). Contact analog content is spatially related to the outside world. The displayed image refers to some object in the background, and the driver perceives a sort of fusion between real world and virtual object. The information displayed in the windscreen merges with the real environment in the sense of augmentation and thus a contextualization of the displayed information can be established (Poitschke et al., 2008). The potential of this technique is pointed out for example by Bergmeier (2009) in the context of information representation regarding driving at night and by Schneid (2008) in combination with driving with Adaptive Cruise Control (ACC). While both of them used display-prototypes not quite applicable for every days use, Israel (2010) could show that it is possible to realize this highly innovative display concept even regarding the high requirements of serial automotive application.

The following study aimed at the potential of a contact analog HUD in combination with highly automated vehicle driving. Especially the understanding of the automation and reactions to automation failures were expected to improve by a preliminary visual announcement of automation behavior.

2   METHOD

2.1   Experimental Setting

Apparatus: The experiment took place in the static driving simulator at the institute of ergonomics, Technische Universität München. The driving simulator consisted of a mockup with a BMW convertible and six projection screens. Three of them were used to generate the front view creating a 180 degree field of view. The rear view was realized by one projection screen for every rearview mirror (Figure 2 and Figure 3 left). The driving simulation software SILAB™ directly received all commands from the control elements and provided the vehicle and driving dynamics simulation. Furthermore it generated the information for the haptic feedback at the active steering wheel, including actual steering torque and automation feedback. Data sampling and logging facilities were also provided by the SILAB software.

Automation: The driver was supported in longitudinal direction by an Adaptive Cruise Control (ACC), which was complemented by a Traffic Sign Recognition. For the duration of the experiment the time gap to the vehicle in front was fixed at 1.5 seconds and could not be manipulated by the driver.

In lateral direction a very restrictive Lane Keeping Assistant System (LKAS) actively kept the vehicle in the lane. Even small deviations from the center of the lane were corrected by the system. The driver got haptic feedback via the active steering wheel. The LKAS was supplemented by a Lane Change Support (LCS).

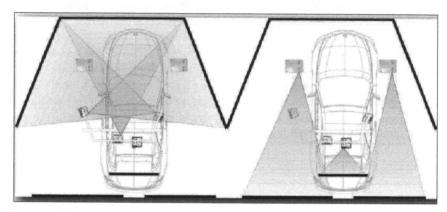

Figure 2: Schematic topview of the projection screens (left: front view, right: rear view) at the static driving simulator.

Lane changes could either be initiated by overpowering the force on the active steering wheel or by using the indicator. Once initiated, lane changes were carried out by the automation itself. The automation contained no Lane Change Assistant (LCA). That means the automation had no sensorial recognition to the neighboring lane (e.g. blind spot) and the driver had to self-reliantly secure each lane change. The automation kept perfectly to speed limits. In case of a reduced speed limit, the necessary deceleration was induced in front of the traffic sign. The deceleration began in 100 m distance ahead to the traffic sign to ensure that the vehicle sticks to the speed limit when reaching the sign. In situations where the allowed speed was increased, the necessary acceleration was initiated right after passing the according sign. It was possible to activate and deactivate the automation by pressing a button on the direction-indicator control. Pressing the brake also deactivated the system.

Contact Analog Head-Up Display (caHUD): Figure 3 shows an example for the driver's perspective during the simulation including caHUD information presentation. Three information items were included in the prototypical contact analog display concept in order to give feedback regarding the current and future automation state. The "trajectory" chosen by the automation was visualized as a dashed line. It was always located in the middle of the lane and followed the course of the road. It marked the lane, which was detected and used for regulation by the automation. A horizontal and upwards open oriented "clamp" beneath the front vehicle signaled the driver, that an object was detected and could be regulated upon if necessary. Additionally a quadratic frame was implemented to highlight detected speed limits at a distance of 100m. The symbols of the caHUD were directly implemented in the SILAB scenery in order to ensure a better synchronization and to enable a reliable manipulation during the experiment compared to an additional projection via beamer.

Test Track: Due to the fact that high vehicle automation will be most probable in highway scenarios, the test track was implemented as a three-lane freeway. The track with a total duration of about 25 minutes was divided into single sections with a length of one to two kilometers. Within these sections various maneuvers had to be completed by the test subjects. The separate sections were randomly arranged for each subject to reduce the influence of learning effects and to avoid that subjects remember the time and sequence of automation failures. During each drive there were two types of maneuvers to complete: lane changes (18 times) and speed changes (31 times). Speed changes were actively initiated and carried out by the automation, lane changes had to be initiated by the driver as described above. During the whole track a front vehicle had to be followed. In the course of the test drive subjects were confronted with three different kinds of automation malfunctions: speed limit failures, lane keeping failures and distance failures.

Figure 3: Static driving simulator (left). Scenery and caHUD symbols from the driver's perspective (right): a) Trajectory, b) Clamp and c) Frame

The speed limit failure occurred twice per run – once at a necessary deceleration from 120 to 100 km/h and once from 100 to 80 km/h. In these situations the automation failed to detect the speed limit sign. Therefore, the vehicle did not decelerate autonomously. The driver had to detect the failure and intervene by either pressing the brake pedal or deactivating the automation via the on/off button. Both operations were indications for the drivers' awareness of the failure. In the run with active caHUD all signs recognized by the automation were marked by the frame (Figure 3 right c). Accordingly not recognized signs were not marked by the frame. So the driver was precociously informed about the automation failure and should have been able to react adequately. In the run without caHUD no visual information was given whether signs were recognized or not.

The lane change failure also occurred twice per run. In this situation the automation misleadingly detected either the left or the right lane and switched to the neighboring lane. As the automation applied a torque on the steering wheel, the driver got a haptic feedback about the automation's lane change intention. In case of the lane change to the right, no nearby traffic participant occupied the target-lane. The lane change to the left was constricted by a following vehicle on the target lane which was too close to actually allow the maneuver. Driving with caHUD, the displayed trajectory bounced to the incorrectly detected lane and provided the driver visual information about the intended lane change of the automation.

In the third failure situation (distance failure) the automation suddenly failed to detect the vehicle in front, which led to an acceleration of the own vehicle and finally to a rear-end collision - in case the driver did not react. In the condition with caHUD the driver was visually informed about the missed detection by a sudden absence of the clamp. The third situation is not part of this paper.

2.2 Experimental Design and Participants

The experiment was set up using a within-subjects design with the visual display as independent variable and reaction times and different kinds of driving data (e. g. lateral deviation) as dependent variables. The workload was assessed using the NASA-TLX questionnaire. Eye-Tracking Data had been gathered and analyzed too (e.g. Weißgerber et al. 2012), but will not be subject to this paper as well as the results of the NASA-TLX. For the assessment and the statistical analysis of the dependent variables t-tests and chi^2-tests were applied where appropriate. The relevant level of significance was determined by $\alpha = .05$.

24 subjects (5 female and 19 male) with an average age of 31.6 years (range from 19 to 66 years) participated in the experiment. They had an average driving experience of 13.4 years (range from 2 to 47 years). Only four subjects had a yearly mileage of less than 5000 km. All subjects had normal or corrected-to-normal visual acuity.

2.3 Experimental Procedure

Most of the subjects never participated in a driving simulator study before. Due to this fact the experiment started with a training session of about 10 minutes to make the subjects familiar with the simulator environment and to introduce them to the specification of the automation as well as the information items on the caHUD. Subsequent to the training, two trial runs were performed in permuted order. The runs differed in the allocation of visual information (with caHUD and without caHUD). Both runs had to be performed with the same automation and on the same randomly arranged sections. All subjects were asked to strictly stick to speed limits and to survey the vehicle guidance of the automation at all times. Every run was accompanied by the NASA TLX.

3 RESULTS

3.1 Speed Limit Failure

As mentioned above, two speed limit failures occurred during the test run. Regarding the speed limit change from 120 to 100 km/h, Figure 4 (left) shows the number of subjects that did not take over vehicle control from the automation although the allowed speed was exceeded by 20 km/h. The number of test persons without reaction is nearly equal in both conditions (13 persons with caHUD, 14 persons without caHUD, χ^2 (1) = .048, p = .771). So at first glance there seems to be no improvement caused by the additional visual information.

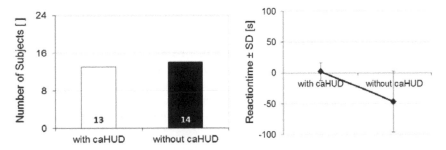

Figure 4: Left: Number of test persons who did not intervene during the automation failure (speed limit 120 to 100). Right: Point of intervention of those subjects who reacted to the automation failure in both conditions (N=6). The relevant traffic sign was located at meter 0; a negative value corresponds to a later reaction.

Regarding the subjects who did react to the failure in both conditions however, the intervention point - related to the position of the sign - is shown in Figure 4 (right). The reaction of the drivers with caHUD in average occurred 2 meters prior to the sign, compared to 47 meters after the sign for drivers without caHUD. This difference is significant ($t(5)$=-2.655, p=.045).

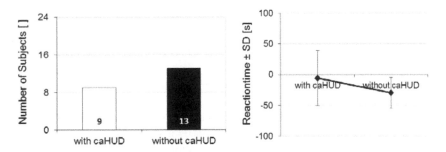

Figure 5: Left: Number of test persons who did not intervene during the automation failure (speed limit 100 to 80). Right: Point of intervention of those subjects who reacted to the automation failure in both conditions (N=10). The relevant traffic sign was located at meter 0 ; a negative value corresponds to a later reaction.

The results of the second speed limit failure (100 to 80 km/h) are shown in figure 5. Here again no difference can be found in the number of subjects that noticed the failure and reacted accordingly ($\chi^2(1)$ = 1.342, p = .247). Nine subjects missed the speed limit in the condition with caHUD, 13 subjects missed it while driving without caHUD. Regarding the point of intervention no significant influence is revealed (t(9)=-1.619, p=.140) but a tendency to a later reaction without visual feedback can be seen. With caHUD the reaction of the drivers in average occurred 6 meters after the sign compared to 30 meters without caHUD.

3.2 Lane Change Failure

Additional to the speed limit failure, unintended lane changes were initiated by the automation at two points during the test run. Figure 6 shows the data corresponding to the lane changes to the left. Due to technical problems in this scenario only 18 subjects could be taken into account. As the target lane was occupied in this scenario no subject allowed the automation to fully change the lane. In the condition without caHUD one subject crossed the lane borders with the edge of the car before intervention. With caHUD no crossing was recorded.

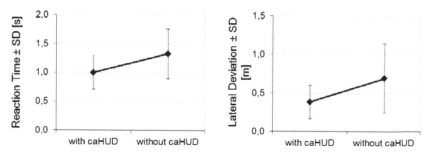

Figure 6: Results of the lane change to the left (target lane occupied).
Left: Reaction time from the start of the unintended lane change until intervention of the subjects.
Right: Maximal lateral deviation caused by the unintended lane change.

Both, reaction time (t(17)=-2.399, p=.028) and maximal lateral deviation induced by the automation (t(17)=2.411, p=.027), significantly decrease in the condition with additional visual information. In this critical scenario the caHUD allowed the subjects a faster reaction due to the visualization of the future behavior of the car.

Regarding the unintended lane change to the right, subjects showed a completely different behavior. 7 subjects driving with caHUD and 4 subjects driving without caHUD allowed a full lane change by the automation. As in this scenario the target lane was not occupied by another vehicle, the maneuver was possible and not critical. The fact that more subjects actually changed lane with caHUD is probably based on a better understanding of the situation due to the visual feedback. The displayed trajectory showed the new target lane and therefore the driver knew what the automation had planned. He had been able to gather information if the maneuver was possible and to approve it if he wanted to. Due to the fact that a number of subjects allowed the unintended lane change to the right, only 14 subjects intervened in both conditions and therefore were considered for the statistical comparison. The reaction time (t(13)=-1.425, p=.178) and the maximal lateral deviation (t(13)=-.617, p=.548) reveals no significant influence of whether the caHUD was active or not, although there is a tendency to faster reaction times with an active visual display (Figure 7).

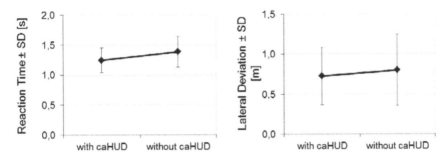

Figure 7: Results for the lane change to the right (target lane not occupied).
Left: Reaction time from the start of the unintended lane change until intervention of the subjects.
Right: Maximal lateral deviation caused by the unintended lane change.

4   SUMMARY AND DISCUSSION

In the long run completely autonomous cars are not very likely to be on the market due to technical restrictions (e.g. sensor range) as well as the Vienna Convention which states that the driver must be kept in the loop and active. Nevertheless the increasing number and quality of driver assistance systems enables the feasibility of a highly cooperative vehicle. One potential implementation of this idea follows the H-Mode interaction paradigm as a parallel-simultaneous interaction concept with active control elements for a haptic communication between driver and automation. In order to increase the driver's understanding of the automation and its future intentions, the haptic communication needs to be complemented by visual information. The application of a contact analog HUD could be a possible solution for this kind of future driver-automation-vehicle interaction. As the presented experiment and the according results show, an additional visual display can help to understand the automation and react faster to automation failures. Especially in critical situations a benefit in terms of faster reactions can be found, whereas in non-critical situations the understanding of the situation comes to the fore. In the course of the future development of the project H-Mode the idea of contact analog visualization will be further prosecuted and evaluated. Especially the displayed trajectory is subject to ongoing enhancement.

ACKNOWLEDGMENTS

The authors would like to acknowledge the contribution of the German Aerospace Center (DLR) especially Prof. Flemisch as a reliable partner within the

12

project H-Mode. In addition the German Research Foundation (DFG) has to be especially mentioned as there would not be a project without their kind assistance.

REFERENCES

Bergmeier, U. 2009. Kontaktanalog markierendes Nachtsichtsystem - Entwicklung und experimentelle Absicherung. Dissertation an der Technischen Universität München.

Bubb, H. 2001. Haptik im Kraftfahrzeug. In T. Jürgensohn, K.-P. Timpe, & H.-P. Willumeit (Eds.), Kraftfahrzeugführung (ISBN 3-540-42012-6, pp. 155–176). Berlin: Springer-Verlag.

Damböck, D., M. Kienle, F. O. Flemisch, J. Kelsch, M. Heesen, A. Schieben, K. Bengler 2009. Vom Assistierten zum Hochautomatisiertem Fahren - Zwischenbericht aus den Projekten DFG-H-Mode und EU-HAVEit. VDI-Congress „Fahrer im 21. Jahrhundert".

Damböck, D., M. Kienle, K. Bengler and H. Bubb. "The H-Metaphor as an example for cooperative Vehicle driving." Paper presented at the 14th Conference on Human Computer Interaction, Orlando, Florida, 2011.

Flemisch, F. O., Adams, C. A., Conway, S. R., Goodrich, K. H., Palmer, M. T., Schutte, P. C. 2003. The H-Metaphor as a Guideline for Vehicle Automation and Interaction (No. NASA/TM-2003-212672). Hampton: NASA. Langley Research Center.

Flemisch, F. O., J. Kelsch, C. Löper, A. Schieben, J. Schindler and M. Heesen, 2008. Cooperative control and active Interfaces for vehicle assistance and automation. FISITA World Congress.

Flemisch, F. O. and A. Schieben 2010. Highly automated vehicles for intelligent transport: Validation of preliminary design of HAVEit systems by simulator tests. Deliverable to the EU-commission D33.3.

Israel, B., M. Seitz, B. Senner and H. Bubb. Kontaktanaloge Anzeigen für ACC - im Zielkonflikt zwischen Simulation und Ablenkung. Paper zur Tagung aktive Sicherheit, Garching, 2010.

Kienle, M., D. Damböck, J. Kelsch, F. O. Flemisch, K. Bengler 2009. Towards an H-mode for highly automated vehicles driving with side sticks. Automotive User Interfaces 2009. Essen. Germany.

Poitschke T., M. Ablaßmeier, G. Rigoll, S. Bardins, S. Kohlbecher, and E. Schneider. Contact-analog Information Representation in an Automotive Head-Up Display. In: Proc. of the 2008 symposium on Eye tracking research & application ETRA 08, Savannah, Georgia, USA, pp. 119–122. ACM Press, NY, 2008. March 2008.

Schieben, A., D. Damböck, J. Kelsch, H. Rausch, F.O. Flemisch 2008. Haptisches Feedback im Spektrum von Fahrerassistenz und Automation. 3. Tagung Aktive Sicherheit durch Fahrerassistenz.

Schneid, M. 2008. Entwicklung und Erprobung eines kontaktanalogen Head-up-Displays im Kraftfahrzeug. München. Dissertation an der Technischen Universität München.

Weißgerber, T., D. Damböck, M. Kienle, K. Bengler 2012. Erprobung einer kontaktanalogen Anzeige für Fahrerassistenzsysteme beim hochautomatisierten Fahren. In: 5. Tagung Fahrerassistenz. Schwerpunkt Vernetzung. München, 15.-16. Mai 2012. TÜV SÜD Akademie GmbH.

CHAPTER 2

Optimized Combination of Operating Modalities and Menu Tasks for the Interaction between Driver and Infotainment-System using a Touchpad with Haptic Feedback

Andreas Blattner, Klaus Bengler, Werner Hamberger

Institute of Ergonomics
Technische Universität München
85747 Garching, Boltzmannstraße 15, Germany

HMI Development
AUDI AG
85045 Ingolstadt, Germany

ABSTRACT

A touchpad with adaptive haptic feedback is specified in the context of this contribution. This innovative control element offers several operating modalities in order to handle the various menu tasks of modern car-infotainment-systems. A test in a static driving simulator has been conducted to compare the different operating modalities of the touchpad for the appearing menu tasks with each other and to identify their optimized combination.

Keywords: haptic, touchpad, automotive, HMI, operating modality, menu task

1   INTRODUCTION

The quantity of features in modern car-infotainment-systems is steadily increasing. There are more and more functions integrated into the infotainment-system, which the driver must be able to handle with a minimum of distraction using a limited number of control elements even while driving. The various car-manufacturers in the world try to solve this problem with different types of control elements like touchscreen, joystick, voice control, rotary push button, etc. A new approach to actualize a preferably facile and intuitive interaction between driver and infotainment-system is the usage of a touchpad, like the Audi AG applies it in the Audi A8.

According to Hamberger (2010) an in-car-touchpad offers several potentials. It is familiar to the users because of the accustomed usage with computer touchpads and enables handwriting recognition. Robustness, optics and the ease of use are additional positive arguments. The results of an experiment in a driving simulator occupy, that a touchpad reduces lane deviation compared to a rotary push button and a touchscreen. In case of a text entry task a touchpad decreases gaze diversion times in comparison to a rotary push button (Bechstedt et al., 2005). In addition to the mentioned results customers prefer the usage of a touchpad compared to a touchscreen (Hamberger, 2010).

In such ambitious dual-task-situations, i.e. the interaction between driver and infotainment-system while driving, interferences occur, if tasks draw upon the same mental resources (Wickens, 1984; Wickens et. al, 2004). In accordance with Bubb (1992) it is moreover beneficial for interacting with technical systems to give a redundant feedback via multiple sensory channels. Therefore it is reasonable to unload the mainly used visual sensory channel in this special dual-task by using a touchpad with an additional haptic feedback. Spies et al. (2009b) and Peters et al. (2010) have developed an automotive touchpad with haptic feedback based upon braille-technology, in order to map display-elements, like buttons and sliders for instance, as sensible and operable elements onto the touchpad. In a driving-simulator experiment this touchpad with haptic feedback was compared to a conventional touchpad. The results were that a touchpad with haptic feedback decreases time for interaction and improves driving-performance in a dual-task-situation significantly (Spies et al., 2009b; Spies et al., 2010).

2   THE TOUCHPAD WITH HAPTIC FEEDBACK

Based upon the findings of Spies et al. (2009b), Peters et al. (2010) and Hamberger (2010) the development of a new touchpad with haptic feedback was accomplished. The functional principle of this haptic touchpad (see figure 1) is explained consecutively.

The displayed contents (e.g. buttons) in the menu screen can be imaged on the touchpad as sensible and operable elements by deploying the appropriate pins.

Figure 1: Prototype of the touchpad with haptic feedback

Thus the user is able to feel and press every elevated element on the touchpad representing concrete menu content, for instance the selectable button "Navigation". This principle is shown by the cross section of the touchpad with haptic feedback in figure 2.

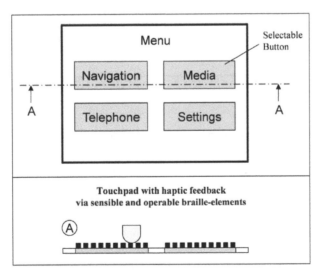

Figure 2: Technical realization of the touchpad with haptic feedback

A visual feedback maintains the haptic feedback of the touchpad by highlighting the currently touched graphical widget on the screen and thus demonstrates the user the effective finger position on the touchpad.

The next section describes different operating modalities, which are offered by the touchpad with haptic feedback, in order to operate the menu functions of car-infotainment-systems.

3   OPERATING MODALITIES

In addition to the aforementioned feedback and normal operations using one finger this innovative touchpad contains the following special operating modalities that are already common and state of the art for conventional touchpads:

1. Handwriting recognition
2. Multitouch gestures
3. Sensible sliders

The integration and the functional areas of these three operating modalities on the touchpad surface are shown in figure 3.

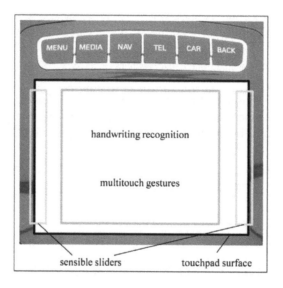

Figure 3: Functional areas of the operating modalities on the touchpad surface

The first two operating modalities, handwriting recognition and multitouch gestures, are placed in the central area of the touchpad surface, where users have enough space to write letters (e.g. for entering a destination) or to make a zooming gesture (e.g. for enlarging a map) for example. For the third operating modality, the sensible and also relative sliders, users have to put their finger onto the right or left edge of the touchpad surface, where they can move their finger up and down, guided by the sensible edges of the touchpad. Thus users can scroll in lists (e.g. list of last destinations) or adjust concrete values (e.g. loudness of the navigation system) for instance.

In order to decrease the distraction of the driver from the main driving task and to enable a preferably easy interaction with the infotainment-system, these operating modalities have to be combined with the various menu tasks, which are specified in the next section, in a preferably optimal way.

4   MENU TASKS

In order to deal with the mentioned increasing quantity of functions in modern infotainment-systems the various car manufacturers apply different control elements and also varied menu structures – like a hierarchical structure or a task-oriented structure (Spies et al., 2009a), to name a few. In accordance with Peters et al. (2010) most of these different menu structures of current car-infotainment-systems consist of the following standard menu tasks:

1. Selection task, <= 9 elements (e.g. menu selection)
2. Selection task, > 9 elements (e.g. list selection)
3. Move/adjust, one-dimensional (e.g. bass)
4. Move/adjust, two-dimensional (e.g. navigation map)
5. Free interaction task (e.g. selection of points of interest in a map)

In the special case of using the described touchpad with haptic feedback as control element of the car-infotainment-system, the normal touchpad operation modality, i.e. using one finger, is only available for finding and selecting elevated menu elements. But conflicts will occur, if users have to operate further tasks on one menu screen – like for example "zooming the navigation map" in order to select a point of interest on the map using one finger.

Investigations on menu tasks, which create conflicts for handling the above mentioned standard menu tasks (Peters et al., 2010) using the touchpad with haptic feedback, have mainly resulted in the following three tasks: scrolling in lists, zooming maps or pictures and adjusting concrete values. For using these further menu tasks it is necessary to offer users the special operation modalities (handwriting recognition, multitouch gestures and sensible sliders) of the touchpad with haptic feedback.

Hence there is need for research to figure out the optimized combination of the various operating modalities using the touchpad with haptic feedback and the three specified menu tasks in a dual-task situation. The next section describes the method of a usability test, which was conducted in order to compare the mentioned operation modalities in combination with the described menu tasks and to clarify the following hypotheses:

1. There is a difference in operation times using the various operation modalities for the three menu tasks.
2. The gaze behaviour while performing the three menu tasks differs for the various operating modalities.
3. Using the various operation modalities for the three menu tasks induces a different driving performance.

5   USABILITIY TEST

A usability test in a static driving simulator has been conducted with 30 test persons using a within subject design. The subjects had to perform the following six interaction tasks, representing the above mentioned further menu tasks. They were

18

driving on an endless track and had to follow a preceding car in a constant distance without using any technical devices, like e.g. cruise control:

1. Scrolling in lists:
   a. Entering a given destination using the last destinations
   b. Calling a specified contact member of the telephone book
2. Zooming maps or pictures:
   a. Zooming into the navigation map
   b. Zooming out of a contact picture
3. Adjusting concrete values:
   a. Reducing the volume of the navigation system
   b. Increasing the tone of the park distance control

These interaction tasks were integrated in a specially implemented infotainment-system and the test persons handled the six different tasks using the aforementioned operation modalities of the touchpad with haptic feedback alternately while driving. During the test time for completing the particular interaction tasks, data of driving performance and information about gaze deviation dialing the various tasks while driving were measured. After the experiment the subjects were additionally asked both to indicate which operation modality they prefer for each menu task and to reply several questions evaluating the usage of the touchpad with haptic feedback.

6   CONCLUSION

As a conclusion it can be said that the described operation modalities of the touchpad with haptic feedback offer solutions for handling all the menu tasks of a modern car-infotainment-system. The knowledge of the optimized combination of the operation modalities and the menu tasks is very important in order to create a special menu system for the touchpad with haptic feedback and to guarantee an easy usage and a minimum of distraction from the main driving task.

Thus the results of this experiment provide the basis for the further development of the touchpad with haptic feedback and the menu structure of the specially implemented car-infotainment-system, in order to assure a preferably facile and intuitive interaction between driver and infotainment-system even while driving.

REFERENCES

Bechstedt, U., Bengler, K. & Thüring, M. (2005). Randbedingungen für die Entwicklung eines idealen Nutzermodells mit Hilfe von GOMS für die Eingabe von alphanumerischen Zeichen im Fahrzeug. *6. Berliner Werkstatt MMS – „Zustandserkennung und Systemgestaltung".* Berlin.

Bubb, H. (1992). Menschliche Zuverlässigkeit, Definitionen - Zusammenhänge - Bewertung. Landsberg/Lech, EcoMed-Verlag.

Hamberger, W. (2010). MMI Touch – new technologies for new control concepts. *IQPC – Automotive Cockpit HMI 2010.* Stuttgart.

Peters, A., Spies, R., Toussaint, C., Fuxen, D. & Hamberger, W. (2010). Haptisches Touchpad zur Infotainmentbedienung. i-com: Vol. 9, No. 01, pp. 58-61.

19

Spies, R., Blattner, A., Horna, G., Bubb, H. & Hamberger, W. (2009a). Entwicklung einer aufgabenorientierten Menüstruktur zur Infotainmentbedienung während der Fahrt. In A. Lichtenstein, Ch. Stößel and C. Clemens, C. (Eds.), *Der Mensch im Mittelpunkt technischer Systeme*. VDI-Fortschritt-Berichte Reihe 22 (29), S. 411-416, Düsseldorf, VDI-Verlag.

Spies, R., Peters, A., Toussaint, C. & Bubb, H. (2009b). Touchpad mit adaptiv haptisch veränderlicher Oberfläche zur Fahrzeuginfotainmentbedienung. In H. Brau, S. Diefenbach, M. Hassenzahl, K. Kohler, F. Koller, M. Peissner, K. Petrovic, M. Thielsch, D. Ullrich, and D. Zimmermann (Eds.), *Usability Professionals 2009*. Stuttgart, Fraunhofer Verlag.

Spies, R., Hamberger, W., Blattner, A., Bubb, H. & Bengler, K. (2010). Adaptive Haptic Touchpad for Infotainment Interaction in Cars – How Many Information is the Driver Able to Feel?. *AHFE International – Applied Human Factors and Ergonomics Conference 2010*. Oxford, Wiley-Blackwell.

Wickens, C.D. (1984). Attention, Time-Sharing and Workload. In C.D. Wickens (Ed.), *Engineering psychology and human performance*. Columbus u.a., Merril.

Wickens, C.D., Lee, J., Liu, Y., & Becker, S.G. (2004). Cognition. In C.D. Wickens, J. Lee, Y. Liu and S.G. Becker (Eds.), *An introduction to human factors engineering*. Upper Saddle River, NJ, Pearson Prentice Hall.

CHAPTER 3

A Review of HMI Issues Experienced by Early Adopters of Low Carbon Vehicles

Andree Woodcock[1], Tom Wellings[2], Jacqueline Binnersley[1]

[1]Integrated Transport and Logistics
Department of Industrial Design
Coventry School of Art and Design, Coventry University, UK
A.Woodcock@coventry.ac.uk

[2]Formerly Warwick Manufacturing Group, Warwick University, |UK
Now, Human Machine Interface, Accommodation and Usability,
Jaguar Land Rover. UK
twellin2@jaguarlandrover.com

ABSTRACT

Low carbon vehicles offer a means of reducing carbon emissions and thereby helping sustainability. However, their rapid introduction, accompanied by new IT capabilities (for example in wireless and sensor technology and cloud computing) has meant that there has been little opportunity to fully consider the nature of the eco-driving task, the added demands it places on drivers, the HMI requirements and the development of a road infrastructure that can support anxiety free driving. This research considered the experiences of early adopters through a review of blogs, news articles, reviews by motoring journalists and reports of trials, and interviews with those who had taken part in a nationwide UK trial.

Common themes emerged relating to range anxiety, problems with charging, feedback from the vehicles and the adaptations drivers need to make to their driving descriptions in order to use the vehicles effectively. The most serious concern for inexperienced drivers was whether the range of their vehicle would be sufficient for their needs. More experienced drivers learnt to plan journeys to take account of the limited range. Reliable feedback about the range and charging status helped drivers to adapt their behaviour and use their vehicles effectively

Keywords: HMI, low carbon vehicles, early adopters, sustainability

1   INTRODUCTION

It is widely accepted that the continued use of fossil fuels is unsustainable due to the environmental impact of CO^2 emissions on global warming. In 2007, the International Panel on Climate Change (IPCC) reported that an 80% cut in greenhouse gases was needed from developed countries to limit the future damage. European politicians have committed to achieve this by 2050. At present, transport represents a third of total CO2 emissions in the UK. Over the last decade, this has increased faster than any other sector. It is evident that the target will not be achieved unless there are significant reductions in emissions from transport (RAE, 2010). In response to this challenge, the future will see improved efficiency in new cars, increased take up of new model hybrids, vehicle charging infrastructure initiatives, and introduction of early market ultra-low carbon vehicles.

In response to commercial and legislative factors, Original Equipment Manufacturers (OEMs) are developing low carbon vehicles and including technology and driver assists which increase eco-friendly driving and reduce energy costs in other ways (e.g. by using vehicles to store energy and transfer it back to the grid). As people become more eco-aware, the road infrastructure and design of new vehicles improves, and eco-friendly cars become less expensive there will be a steady rise in interest and uptake of these vehicles.

The estimated market for fully electric vehicles will reach 2.6 million by 2015 and Deloitte (2010) has estimated that by 2020, electric vehicles will account to 3.1 percent of U.S. automotive market. The UK Committee on Climate Change (2009) expects dramatic improvements in the carbon efficiency of cars and measures to cut the growth in traffic, potentially including road pricing. It recommends that there should be 1.7m electric cars, with 3.9m drivers trained in fuel-efficient techniques by 2020.

Taken together, these trends create three types of challenges for drivers of electric vehicles in relation to 1) the driving task, 2) the manner of driving and 3) the interpretation of information presented in the car and on other devices. These influence the user experience of the electric car and safety issues. In the following sections each of these challenges will be considered in turn.

1.1 The Driving Task

Traditionally the driving task has been characterised in terms of three tasks, navigation, control and hazard identification (e.g. Stanton et al, 2001). The emphasis has been largely on maintaining the safe and efficient progress of the car towards the destination. The requirements of eco-driving (see below) will see an enlargement of the control function to embrace ecologically efficient driving.

In the UK, with a mature carbon fuel based traffic infrastructure, the driver does not have to be over concerned about the state of the vehicle prior to departure (e.g. in terms of the amount of fuel, tyre pressure), as frequent opportunities are provided to refuel or correct minor issues. On the other hand, the driver of an alternatively fuelled vehicle has additional tasks concerned with assessing the amount of fuel available, the distances this will allow him/her to travel, and deciding the charging strategy. In most countries the road infrastructure does not provide many opportunities for the owners of alternatively powered vehicles to refuel, or to refuel quickly. This means that the preparatory phase of the driving task is extended; for electric vehicles which require long charge times, this preparatory phase can be extended by up to 8 hours. Additionally, with few recharging points available, the early adopter has to plan (and possibly book) where the car can be recharged for the return journey. At the time of writing the UK had 700-750 charging points, with two third of these located in London (Vaughn, 2011). The research reported in this paper highlights the need to recognise these new areas associated with the driving task.

1.2 Eco-driving

Eco-driving is an adapted driving style which contributes to reduced fuel consumption and is not restricted to driving low carbon vehicles. It comprises the following elements: educating novice drivers, re-educating licensed drivers; fuel saving in-car devices; managing tyre pressures; and purchasing behaviour (Treatise, 2005.). Strategies include shifting to a higher gear as soon as possible, anticipating the road conditions, maintaining a steady speed in as high a gear as possible, driving slower than normal, switching off air conditioning, removing excess weight, switching off rather than idling the car, and keeping tyres at the optimum air pressure. Some low carbon vehicles have HMI that can assist with eco-driving through the provision of feedback on driving performance (e.g. through econometers, shift indicators, rev(olution) counters and speed limiters).

1.3 HMI and Information Presentation

Typically, driver related automotive HMI consists of the following four elements (just-auto, 2010)
1. Primary driving controls (steering wheel, pedals, shifter). Instrument cluster
2. Displays / controls associated with supporting the primary driving task (e.g. Navigation)
3. Supplementary displays associated with the primary driving task (e.g. head-up display, blind spot monitoring)

Market assessments for LCVs show a trend towards standardizing telematics and associated HMI. By 2015, Frost and Sullivan (2009) forecast, that more than 80% of electric vehicles sold globally will have standard telematics features (up from approximately 50% in 2010). Future instrument clusters will not consist of two primary dials (speedometer and tachometer) but an array of reconfigurable, digital displays informing the driver of the vehicle's state, their driving performance, the performance of others, current and future road conditions, points of interest, and availability of charging points. Non standard means of presenting information will be available; displays will be presented to all the drivers' information channels (visual, aural, haptic), in real and augmented reality and in a variety of positions (in the instrument cluster, head up displays, rear view mirror and other displays shared with the front passenger).

The integration of sensor technology, cloud computing, vehicle to vehicle, and vehicle to grid provide opportunities to present information to the driver and other interested stakeholders. From an ergonomics perspective, issues of information presentation and information overload are critical. Is the right information being presented in the right way for the drivers to understand and act upon? Are they receiving too much or too little information? Both of these can make the driving task more difficult and stressful. Nissan has referred to the transition from automobiles to infomobiles. However, insufficient research is being conducted on the implications of these changes for vehicle safety, usability and more generally the driving process (Green, 2000, 2008).

2. THE STUDY

In order to understand the requirements of eco-drivers and make recommendations regarding the design of HMI, a qualitative study was conducted of the experiences of early adopters of electric vehicles using material from news reports, blogs and interviews (see Table1).

Table 1 Data sources

Data source	Details
News items	1. 106 news items relating to LCVs issues in November 2010
	2. Curran's (2010) reports made when travelling 4500 in Europe in the Think City electric car
	3. Lanning's (2008) experiences of the Smart Fortwo Electric Drive.
Blogs	1. Personal blogs of participants in the US Mini E trial (June 2009-June 2010)
	2. The Boxwell's blog (2006-2011). A UK family who tested the Reva G-Wiz and Mitsubishi iMiEV electric cars
	3. Kodama's (2011) blog. A Californian electric car enthusiast who has owned a GM EV1, Toyota RAV4, and who took part in the BMW Mini E trials.
Review of field trials	1. 12 month CABLED project in West Midlands, UK, reporting findings from 22 Mitsubishi iMiEV drivers. (Aston University, 2010a, 2010b).

2. Mini E trials in USA, Germany and UK from 2009. Changes in the functioning of the ecosystem which benefit and impact the well-being of humans

3. Davis University (California) PHEV trails with 34 participating households using adapted Prius vehicles (Kurani et al, 2009)

4. CENEX Smart Move Trial in North East England. 6 month study considering the feasibility of integrating Smart Fortwo Electric Drive cars in to fleets.

The aim was to understand the extent to which current HMI supports the needs of drivers of electric vehicles. Several writers, including Ljung et al. (2007) have written about the importance of testing HMI in a realistic setting in order to understand the problems faced by users. Carsten and Nilsson (2001) wrote about the importance of testing the safety aspects of HMI with trials in real world settings.

Akenhurst (2009) advocated the use of blogs and suggested that they may reflect more genuine attitudes of consumers than other forms of data collection. However, Ram and Jung (1994) suggested that early adopters may have different characteristics to people who acquire a product at a later stage of development. For example, early adopters are more willing to tolerate the inconveniences of new products than other users. Therefore the bias of the data sources used here should be noted; the participants were drawn from higher socio –economic bands, they had positive attitudes towards low carbon vehicles and were willing to change their behaviour in order to use the vehicles effectively.

The data sources outlined in Table 1 were thematically analysed for issues relating to user-experience, and human-machine interaction. Similar trends emerged from each data source; therefore the results have been presented in terms of these themes.

3.RESULTS

3.1 Range Anxiety

Range anxiety is the term used to describe drivers' anxiety about their car's ability to cover the distance required before it needs to be recharged. It occurs almost exclusively in drivers of fully electric vehicles because the limited charging infrastructure means they cannot easily 'refill'. It is also more prevalent amongst those who are less experienced in driving LCVs.

In the Mini E trials conducted in the Berlin metropolitan area, Cocron (2010) found that participants initially had concerns about the limited range of the car (95-125 miles). After three months ownership, almost all found the range suitable for their daily needs. Similarly, Hoffman (2010) in the Smart Electric Drive UK trial found that participants' anxiety disappeared with extended use

In the Cenex trial participants reported anxiety throughout the length of the trial (Carroll, 2010). Drivers tended to be overly-cautious about the distances they were willing to travel, and only 7% of journeys were undertaken when the battery was showing less than 50% of charge. The result may reflect the short time participants spent with the cars. Over the six month trial 264 different people drove the four available Smart Fortwo EVs and it is likely that they did not have time to develop confidence that the range would be sufficient for their needs.

Out of the 20 drivers interviewed as part of the Cabled trial, 55% had felt concerned about the range of their vehicle but the majority of these thought that their anxiety would reduce with experience. Not all vehicles in the Cabled trial provided drivers with information about the range and the majority of drivers thought that this would be useful. Providing drivers with a reliable indication of the range, preferably in an analogue form, was suggested as a way of reducing their anxiety.

After they had experience of driving their vehicles, 41of the Cabled drivers were asked how much they agreed with the statement, "I have trust in the vehicle's range". 61% of the participants trusted the feedback; others gave examples of times when it had not corresponded with the distance that had been covered. For low carbon vehicles, the range is affected by factors such as outside temperature and road conditions, not merely the distance travelled. This may make the feedback about range appear to be unreliable.

Even before commencing the trial, 54% of participants cited outside temperature as a factor that they thought would reduce range as being a factor that would reduce the range. Becoming an effective driver of a low carbon vehicle involves understanding influences on the range and being able to take account of them when planning a journey e.g. realising that the range will be reduced when the weather is cold. It would help drivers to have an estimated range figure that took account of factors like temperature and speed of travel to provide them with a realistic estimate of how far they could travel before needing to recharge.

3.2 The charging process

The charging process was novel for many LCV drivers, and resulted in a number of problems, the most serious of which was vehicles failing to charge. The experiences of early adopters show that feedback is needed about the charging progress and warnings needed if charging are interrupted. Some participants suggested a mobile phone app that would provide them with information about charging when they were away from the vehicle. It was also suggested that drivers could be provided with information on the location of nearby charging points.

Problems with charging were often exacerbated by the location of the charging socket. Charging could be awkward when the socket was on one side of the vehicle and the cable needed to be stretched to reach the charging point. Standardisation which takes into account the context of use would be useful (perhaps with a central charging socket).

3.3 User feedback

Reliable information about range and charging was needed. Some of the drivers have suggested an early warning of low charge. Participants in the Cabled trial suggested that that this should be given before the 10% state of charge was reached. Those who had used public charging points suggested that providing in-car information about the location and availability these would help them to better plan their journeys.

Most drivers were happy with the lack of engine noise but others pointed out the danger to pedestrians and wildlife. Some drivers reported being unsure whether their vehicle was `on` or `off` due to the lack of feedback when they switched on. Some drivers wanted feedback about how their driving style affected the range of the vehicle and it was suggested that this would help them to drive more effectively.

3.4 User adaptation to Eco-driving

The drivers of electric vehicles adopted eco-driving techniques. For example, Mini E drivers selected their route in order to optimise the range (Maloughney (2009), avoided highway driving, reduced the use of secondary controls and controlled their vehicle by using the throttle instead of the brake. Curran (2010) reduced his speed to conserve battery power, although that meant driving more slowly. In the US Mini E trial, 98% of Mini E drivers liked the way regenerative braking could be actively used to increase their range (Lentz, 2010). Additionally, drivers sometimes chose not to operate additional features, such as the radio, heating or electric windscreen wipers in order to save power.

4. DISCUSSION

The next 5 years is going to see a substantial increase in the prevalence of telematics and associated HMI in LCVs. Dr. Bob Schumacher, General Director of Advanced Engineering & Business Development, Delphi Electronics & Safety, stated the need to engage practices in user experience design, active safety technologies, manage information flow from the infotainment to the user, and monitor the driver's attention while driving is their priority (Delphi 2011).

The review of the experiences of early adopters of electric vehicles has shown that regardless of the model driven, there are a range of factors which need to be addressed to promote eco-driving, reduce anxiety and support the additional tasks that an alternatively fuelled vehicle creates. These have arisen because of the stage of development (with immature fuelling infrastructures), the failure to understand the context of use (how, when and where charging will occur) and the need for new types of information to support eco-driving.

The drivers consulted were enthusiastic and dedicated early adopters, many of whom share their experiences on social media. The use of such material may lead to a source of bias. In this study we have tried to balance these self generated comments with the experiences of users in more controlled trials. The results from all groups agree. However, the self selection and entry requirements to the studies have meant a significant bias in the respondents – who are drawn from a group who are interested in sustainability, automobiles and new technology, and from high socio economic groups. Regardless of this bias, the studies have the high level of ecological validity associated with field trials, and have led to the production of recommendations which will lead into the next generation of HMI.

One of the issues faced by all users was the need to understand the different kinds of equipment and information shared about the vehicle's status. Features like battery state of charge (SOC), charging states, power flow diagrams, starting/stopping charge, vehicle-feedback on driving styles, range available, charging points available etc. Users will need to adapt and become acquainted with such information to reduce chances of anxiety, worries and increase safety. User training may be needed to enable a "first time" user to understand and interpret the displayed information. The need for the user to identify any related issues with the help of vehicle HMI/telematics will be necessary. The users' perspective needs to be considered to understand the driver-vehicle interaction/experience to increase usability, reliability, safety and enjoyment.

5. RECOMMENDATIONS

The design flexibility of configurable instrument clusters will allow contextual information to be shown relevant to the current task. This will allow greater amounts of information to be displayed, whilst maintaining simplicity if managed carefully. However, there is no consensus on how new information (such as that relating to the state of charge) should be displayed, or how this can be displayed to encourage novice and expert drivers in more efficient and sustainable driving. Clearly, future automobiles and other modes of transport will be more connected and integrated into our lives, with technology transfer and integration with pervasive and mobile computing.

Design guidelines for eco-feedback interfaces include portraying information in an abstract or metaphorical manner to allow drivers to understand how their actions relate to the goal of greater efficiency; data should be presented simply, unobtrusively and displayed when required; positive reinforcement should be used to encourage behaviour change; information about user's past experiences as it relate to goals should be accessible and information should be presented in a grounded context so that drivers can quickly understand the relative impact of their behaviour (Kurani et al, 2009).

The automotive interfaces that have most successfully applied these principles are the Ford SmartGauge as found in the Fusion hybrid, and the LCD screens in the Chevrolet Volt. User feedback regarding the SmartGauge has been largely positive with drivers finding it rewarding, and useful in helping maximise fuel economy.

Specifically, we would recommend that the HMI of low carbon vehicles includes:

- an estimated range figure that takes into account factors such as temperature and vehicle speed. This should be in analogue form
- information showing the state of charge and a warning when the 10% state of charge is reached
- information about the charging progress while this is taking place and a warning if it is interrupted

- information about range and charging available remotely e.g. via a mobile phone app
- drivers with information about the location and availability of nearby charging stations
- a noise when the engine is on as a safety measure
- a clear indication of when the vehicle is `on`
- drivers with information about how their driving behaviour affects the range and state of charge.

ACKNOWLEDGMENTS

The Low Carbon Vehicle Technology Project (LCVTP) is a collaborative research project funded by Advantage West Midlands (AWM) and the European Regional Development Fund (ERDF, between Jaguar Land Rover, Tata Motors European Technical Centre, Ricardo, MIRA LTD., Zytek, WMG and Coventry University.

REFERENCES

Akenhurst, G., 2009. *User generated content: the use of blogs for tourism organisations and tourism consumers* [online] Service Business Springer Accessed 3 August 2011, http://akehurstonline.co.uk/Service%20Business%20paper%202009.pdf

Aston University, 2010a. *Data Analysis Report of ultra-low carbon vehicles from the CABLED trial - Q1*. Aston University, UK, March 2010.

Aston University, 2010b. Data Analysis Report of ultra-low carbon vehicles from the CABLED trial - Q2. Aston University, UK, June 2010

Boxwell, M., 2011. Owning an electric car [online] Accessed 13 July 2011. http://www.owningelectriccar.com/electric-car-blog.html

Carroll, S., 2010. The Smart Move trial: description and initial results. Loughborough, UK: Centre of excellence for low carbon and fuel cell technologies, Smartmove-10-010.

Carsten, O. M. J., and Nilsson, L. 2001. Safety assessment of driver assistance systems *European Journal of Transport and Infrastructure Research* 1 (3) 225-243

Cocron, P., 2010. Expectances and experiences of drivers using an EV: Findings from a German field study. *In*: Mrowinski, V., Kyrios, M. & Voudouris, N. (Eds.), *Abstracts of the 27th International Congress of Applied Psychology*, 11-16 July 2010 Melbourne, Australia. 250-251.

Committee on Climate Change (2009), *Meeting carbon budgets, the need for a step change*, Accessed 22nd February 2011, from http://www.theccc.org.uk/reports/1st-progress-report

Curran, P., 2010. An electric car odyssey: Around Europe by battery power in the transport of the future Mail Online [online] 26 July. Accessed 3 November 2010,: http://www.dailymail.co.uk/travel/article-1297510/Electric-car-adventure-Around-Europe-transport-future.html

Deloitte, 2010. Gaining Traction: A customer view of electric vehicle mass adoption in the U.S. automotive market. Deloitte Consulting LLP.

Delphi, 2010, Mitigating driver distraction, Accessed 22nd February 2011, http://delphi.com/news/featureStories/fs_2010_11_12_001/

Frost and Sullivan, 2009. *Telematics for electric vehicles: the key to reducing range anxiety and enhancing the overall EV ownership experience.* Frost & Sullivan Automotive Market Research

Green, P. 2000, Crashes induced by driver information systems and what can be done to reduce them. SAE, 2000, http://www.umich.edu/~driving/publications/SAE2000-01-C008.pdf

Green, P. 2008, Driver interface/HMI standards to minimize driver distraction/overload, UMTRI 2008-21-2002 http://deepblue.lib.umich.edu/bitstream/2027.42/65018/1/102437.pdf

Hoffman, J., 2010. Does the use of battery electric vehicles change attitudes and behavior? *In*: Mrowinski, V., Kyrios, M. & Voudouris, N. (Eds.), *Abstracts of the 27th International Congress of Applied Psychology*, 11-16 July 2010 Melbourne, Australia. 252.

just-auto, 2010. Intelligence set: Global market review of vehicle instrumentation and cockpits – forecasts to 2016. 83615.

Kodama, D., 2011. The EV Chronicles. [online] Accessed 13 July 2011, http://www.eanet.com/kodama/ev-chronicles/

Kurani, K. S., Axsen, J., Caperello, N., Davies, J., and Stillwater, T., 2009. Research Report UCD-ITS-RR-09-21: Learning from consumers: Plug-in hybrid electric vehicle (PHEV) demonstration and consumer education, outreach and market research program. [online] Institute of Transportation Studies, University of California. Accessed 15 November 2010, http://pubs.its.ucdavis.edu/publication_detail.php?id=1310

Lanning, P., 2008. First Drive: Smart Electric [online]. Accessed 15 November 2010, http://www.thesun.co.uk/sol/homepage/motors/phil_lanning/article1920106.ece

Lentz, A., 2010. MINI E – Regenerative Braking [online]. Institute of Transportation Studies, UC Davis University of California, Accessed 10 December 2010, http://phev.ucdavis.edu/news/mini-e-2013-regenerative-braking/?searchterm=mini%20e.

Ljung, M., Bloomer, M., Curry, R., Artz, B., Greenberg, J., Kochhar, D., Tijerina, L., Fagerstrom, M.,and Jakobsson, L. 2007. The Influence of study design on results in HMI testing for active safety [online] Ford Motor Company and Volvo Cars Accessed 30 August 2011, http://www-nrd.nhtsa.dot.gov/pdf/nrd-01/esv/esv20/07-0383-O.pdf

Maloughney, T., 2009a. My first real taste of range anxiety. minie250.blogspot: blog [online], Accessed 2nd December 2010, http://minie250.blogspot.com/2009_08_01_archive.html RAE, 2010. Electric Vehicles: charged with potential. London: Royal Academy of Engineering, ISBN 1-903496-56-X.

Stanton, N.A., Young, M.S., Walker, G.H., Turner, H., and Randle, S. 2001, Automating the driver's control tasks, International Journal of Cognitive Ergonomics, 2001, 5(3), 221–236

Treatise 2005, Ecodriving: the smart driving style, Produced by SenterNovem, Utrecht for the EC TREATISE project, September 2005. http://www.thepep.org/ClearingHouse/docfiles/ecodriving.pdf

Vaughn, A. 2011, Electric car capital race hots up as London adds charging points, The Guardian, 26th May , 2011

CHAPTER 4

The Error Prevention Effects and Mechanisms of Pointing

Takayuki Masuda[1], Masayoshi Shigemori[1], Ayanori Sato[1],

Gaku Naito[2], Genki Chiba[2], Shigeru Haga[2]

Railway Technical Research Institute[1]
Tokyo, JAPAN[1]
Rikkyo University[2]
Saitama, JAPAN[2]
masuda@rtri.or.jp

ABSTRACT

The error prevention effects of "point and call check" are known and it is used in several industries. It is thought that point and call check have several error prevention mechanisms. It is likely that the eye focusing effect of "pointing" is one of them. We investigated if "pointing" had the error prevention effect and the effect is due to the eye focusing. The task was to count the dots on the display. Participants counted the dots with or without pointing to the dots. The density of the dots was controlled (high or low). If error prevention effect of pointing is due to the effect of eye focusing, the error prevention effect was revealed more clearly in high density condition than low density condition that focusing the only target dots is more difficult. As a result, error prevention effect of pointing was seen in only low density condition. These results indicate that error prevention effect of pointing is not due to the effect of eye focusing, but may be due to other error prevention mechanism. One interpretation of these results is pointing has the memory promoting effect. Because the distance between dots is longer in low density condition than in high density condition, remembering the position of each dot is more difficult in low density condition. In consequence, error prevention effect of pointing was seen in only low density condition.

Keywords: pointing, human error, error prevention

1   INTRODUCTION

The error prevention effects of "point and call check" are known and it is used in several industries. In the railway field, various types of workers, not to mention train drivers and conductors, use "point and call check" in Japan.

It is thought that point and call check has five error prevention mechanisms: (1) eye fixation with pointing, which makes people closer to the object and deliver clear visual images to the retina (Iiyama, 1980), (2) memory enhancement with calling, which makes people be able to focus their attention on the object and memorize it when we call it to rehearse the object (Iiyama, 1980), (3) error awareness with calling, which makes people's accuracy of cognition increased due to both the visual and auditory senses (Iiyama, 1980), (4) arousal with pointing and calling, which stimulates the activity level of cerebrum by muscular movements of chin, hands arms (Iiyama, 1980), (5) delay of responses with pointing, which inserts time rag between perception and response to inhibit people's premature response (Haga, Akatsuka and Shiroto, 1996).

It has been reported that point and call check has error prevention effects (Shigemori, Saito, Tatebayashi, Mizutani, Masuda and Haga, 2009; Naito, Shinohara, Matsui and Hikono, 2011; Sato, Shigemori, Masuda, Hatakeyama and Nakamura, 2011), but the sufficient evidences of these five error prevention mechanisms have not been provided. Therefore we need to confirm each mechanism.

In this paper we focus on the mechanism of eye fixation with pointing. We experimentally investigate if "pointing" has the error prevention effect and the effect is due to the eye focusing.

We compare the error rate in the task counting the dots on the PC display in high and low density conditions. In high density condition, the distance between dots was shorter than in low density condition and it is more difficult to fix eyes only on the target dot in high density condition than in low density condition because eye was attached to non-target dots. If error prevention effect of pointing is due to the effect of eye focusing, the error prevention effect will be revealed more clearly in high density condition than in low density condition since focusing only on the target dot is more difficult. On the contrary, if the error prevention effect was revealed more clearly in high density condition than low density condition, pointing may have other error prevention mechanism.

For the purpose of confirming the eye focusing effect of pointing, eye movements are recorded. If error prevent on effect of pointing is due to the effect of eye focusing, the difference of the number of eye fixations on each dot between with pointing and without pointing will be more in high density condition than in low condition.

Procedure

2   PROCEDURE

2.1   Participants

Thirty four people (fifteen male, nineteen female) participated in the study; they had the mean age of 20.94 years. All participants were aware of their right to withdraw from the study at any time and had a full debriefing about the aims of the study.

2.2   Equipment

We collected data using the experiment software (developed with Microsoft Visual Basic 2008). Experiment software was installed to the PC (VAIO VPCB11AGJ). Output was shown on the display (DELL   E198FP) at 1024 × 768 pix. The responses of participants were recorded with a keyboard connected to the PC. The display was positioned 60 cm from the participants.

Eye movements were recorded using the eye movement tracking device (NAC EMR-8). The eye movement was tracked at 60Hz. The tracking data was recorded on digital HD videocassette recorder (SONY GV-HD700).

2.3   Task

The task was to count the dots on the display. This experiment consisted of two sessions. The first session was a trial session. Participants performed the task under different experimental conditions: (1) with pointing, (2) without pointing. The density of the dots was controlled (high or low: Figure 1). In the trial session, participants experienced one high density trial and one low density trial without pointing, and the order was randomly selected. One session consisted of twelve trials: six high density trials and six low density trials in each experimental condition. Each trial was selected randomly. The order of with or without pointing was counter balanced. The density of dots was controlled not to be the same in the six consecutive trials.

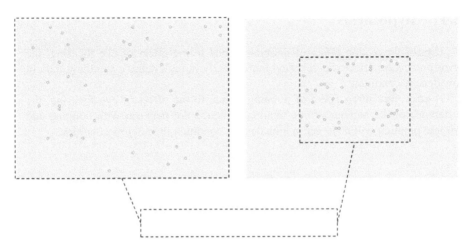

Figure 1 High and low density conditions and area dots were presented in each condition.

In the trial session, the number of dots on the screen in one trial was forty. In one session, the number of the dots on the screen was as shown in Table 1 and the average number of the dots was forty.

Table 1 Number of dots in trial sessions

experimental conditions		number of dots
without pointing	high density	40
	low density	40

Table 2 Number of dots in each experimental condition

experimental conditions		number of dots
with pointing	high density	37, 38, 39, 41, 42, 43
	low density	37, 38, 39, 41, 42, 43
without pointing	high density	37, 38, 39, 41, 42, 43
	low density	37, 38, 39, 41, 42, 43

The screen was divided into the 70 x 50 arrays of cells. In high density trials, the locations of the dots presented on the screen were selected from all cells. On the other hand, in low density trials, the locations of the dots presented on the screen were selected from limited cells (from rows eighteen to fifty three and from columns thirteen to thirty eight).

The dots were presented for 400ms per each dot. When forty dots were presented, dots were presented for 16000ms.

2.4   Hypothesis

If error prevention effect of pointing is due to the effect of eye focusing, the error prevention effect was revealed more clearly in high density condition than in low density condition.

If error prevention effect of pointing is due to the effect of eye focusing, the difference of the number of eye fixations on each dot between with pointing and without pointing will be larger in high density condition than in low condition.

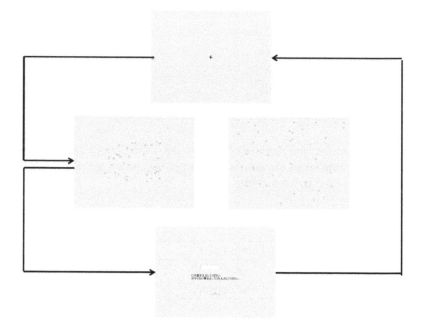

Figure 2 Flow of experiment

3   RESULTS

3.1   Control of the density of dots

In order to confirm the control of the density of dots, analysis of variance was conducted.

The results indicated that the main effect of the density and the average distance between dots in high density condition was fewer than that in low density condition ($F(1,33) = 20516.14$, $p < 0.01$). There was no main effect of with/without pointing and interaction between with/without pointing and density of dots.

These result validated the control of the density of dots.

3.2 Error rate

Analysis of variance (ANOVA) was conducted to examine the effects of with/without pointing and density of dots. The results indicated that the main effect of with/without pointing and the number of wrong count in with pointing condition was significantly fewer than that in without pointing condition ($F(1,33) = 9.81$, p <0.01). There was a marginally significant interaction between with/without pointing and density of dots ($F(1,33) = 3.47$, p <0.1). The simple main effect test indicated that in with pointing condition the number of wrong count in low density condition was fewer than that in high density condition ($F(1,33) = 6.25$, p <0.05 The simple main effect test indicated that in low density condition the number of wrong count in with pointing condition was fewer than that in without pointing condition ($F(1,33) = 10.88$, p <0.01).

Table 3 Error rate in each experimental condition

Experimental conditions	High density		Low density	
	Non-point check	Point check	Non-point check	Point check
Mean	0.61	0.55	0.62	0.45
SD	0.30	0.27	0.27	0.26

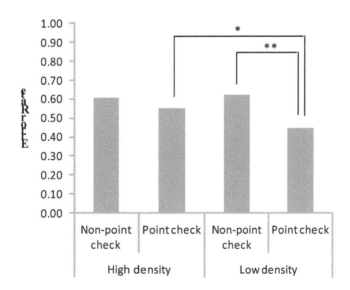

Figure 3 Error rate in each experimental condition (*: p<.05, **: p<.01).

3.3 Eye fixation

Because of the deterioration in the accuracy of the eye tracker calibration, the data was not analyzed.

36

4   DISCUSSION

As shown above, error prevention effect of pointing was seen only in low density condition. One interpretation of this result is that difficulty of the experimental task affected the result. Therefore, because of low difficulty, the error prevention effect of eye fixation with pointing was seen in low density condition, but that was not seen in high density condition because of the high difficulty. Because there was no difference in error rate in without pointing between high density condition and low density condition, it can't be assumed that each condition has different difficulty.

Another interpretation of this result is that error prevention effect of pointing is not only due to the effect of eye focusing, but may be due to other error prevention mechanism. One possibility is that pointing has the memory promoting effect. Because the distance between dots is longer in low density condition than in high density condition, remembering the position of each dot is more difficult in low density condition. Pointing accompanying physical movement may prompt spatial memory. In consequence, error prevention effect of pointing was seen only in low density condition.

In order to count the dots accurately, remembering what dots have already been counted is important. Therefore, it can hardly be assumed that eye fixation have no relation with error prevention effect. Because of the deterioration in the accuracy of the eye tracker calibration, we couldn't test the difference in eye fixation between each experimental condition. As a result, we can't test to what extent eye fixation has relation with prompting memory and error prevention. We need to perform further experiments under various density conditions and to research using eye camera.

REFERENCES

Haga, S., Akatsuka, H. & Shiroto, H. 1996. Laboratory experiments for verifying the effectiveness of "finger-pointing and call" as a practical tool of human error prevention. *Japanese Association of Industrial / Organizational Psychology Journal*, 9: 107-114.

Iiyama, Y. 1980. Utility and application of the point and call method: Its scientific background. *Safety*, 31: 28-33.

Naito, H., Shinohara, K., Matsui, Y. & Hikono. M. 2011. Effects of pointing action on visuo-spatial learning. *Journal of Institute of Nuclear Safety System*, 18: 21-27

Sato, A., Shigemori, M., Masuda, T., Hatakeyama, N. & Nakamura, R. "Memory accelerated effect of the point and call method." The 75th Conference of the Japanese Psychological Association. Tokyo, Japan, 2011.

Shigemori, M., Saito, M., Tatebayashi, M., Mizutani, A., Masuda, T. & Haga, S. "Human error prevention effect of the point and call method." The 6th Conference of Japanese Society for Cognitive Psychgology. Niiza, Japan, 2009.

CHAPTER 5

Implementation and Evaluation of Lane Departure Warning and Assistance Systems

Emma Johansson[1], Erik Karlsson[1], Christian Larsson[1] and Lars Eriksson[2]

[1]Volvo Technology Corporation
Gothenburg, Sweden
emma.johansson@volvo.com

[2]VTI, Swedish National Road and Transport Research Institute
Linköping, Sweden

ABSTRACT

Two driving simulator experiments were conducted to assess different alternatives of lane departure warning and assistance systems. In the first experiment, two types of Lane Departure Warning system (LDW) were compared with one Lane Keeping Assist system (LKA). In the second experiment, the benefit of an LKA was assessed as compared with the condition of no LKA. In addition, similar steering wheel support as for lane departures was implemented as a Lane Change Decision Aid system (LCDA) in the second experiment, giving feedback in so called blind spot situations. The purpose was to investigate the possibility to have the same type of support for different lateral threats in order to harmonize and simplify the overall support given to the driver. Results indicate some deviating driving behaviour for novice drivers and generally a positive objective and subjective results supporting LKA consisting of a combination of guiding steering force and vibrations of the steering wheel.

Keywords: Lane Departure Warning (LDW), Lane Keeping Assist (LKA), Lane Change Decision Aid (LCDA), Advanced Driver Assistance Systems (ADAS)

1   INTRODUCTION

In Sweden, around one hundred persons are killed every year in accidents involving heavy goods vehicles, which equals approximately 20% of all persons killed in road accidents in Sweden (Strandroth, 2009). The Large-Truck Crash Causation Study (LTCCS), conducted by the National Highway Traffic Safety Administration (NHTSA) and the Federal Motor Carrier Safety Administration (FMCSA) in the United States is based on US data collected 2001-2003. According to LTCCS, accidents involving heavy trucks that can be related to unintended lane departures are estimated to stand for a large amount of the overall number of accidents (Starnes, 2006). Thirty-four per cent of all single-vehicle accidents involving trucks are so-called Right Roadside Departure accidents and for Left Roadside Departure the figure is 27%. When looking at both single-vehicle and multi-vehicle crashes the figures are lower. Rear-end collisions stand for most of the accidents (23%) and Right and Left roadside departures combined for 12 % (ibid.).

Needless to say it is not easy to couple accident statistics to how well a technical system would resolve the situation. The coding of accidents is sometimes different between databases. Sometimes both the conflict (e.g. lane departure) and the outcome (e.g. head-on collision) are coded in the accident database (as is done for example in police reported accidents in UK, DfT, 2004) while in other databases only the latter is presented. A lane departure resulting in a head-on collision could be coded as a lane departure accident or a head-on accident or in some instances be coded as both. Thus, the effectiveness estimation of a system very much depends on the classification made in the accident database used. A lane departure situation resulting in a head-on collision could perhaps be avoided with a Lane Departure Warning system (LDW) or a Lane Keeping Assist system (LKA), perhaps also mitigated with an emergency braking system. With an LDW only a warning is provided to the driver and with an LKA additional steering wheel torque is given to guide the vehicle back in lane. The potential of different measures to increase safety was investigated by The Swedish Transport Administration (Strandroth, 2009), and it was suggested that truck systems detecting vulnerable road users would have the highest estimated effect with LDW coming second.

According to ISO 17361 an LDW should detect the lateral position of the subject vehicle relative to the lane boundary and, if the warning condition is fulfilled, warn the driver by either a visual, auditory or haptic warning. A standard for LKAs is currently not available but under development and will cover systems that provide additional steering torque. Several vehicle manufacturers today offer LDW to their customers (e.g. Citroën, Mercedes, Volvo and BMW). Additional lane keeping assistance is often offered as well (e.g. Honda, Audi, Ford, Lexus, Toyota and VW). In the truck segment lane departure warning systems are offered by Volvo, Renault, Mercedes, MAN, Iveco, Scania to name a few. At the time for this paper, no truck brands offer systems using applied steering wheel torque.

One reason for introducing other types of countermeasures than warning sounds in order to avoid unintended lane departures is that more active intervention in vehicle control now are possible to accomplish with current technology. Drivers would be able to get a more full support in a dangerous situation rather than a warning solely. Additionally, sounds sometimes are seen as disturbing by some drivers which was observed in a study by Braitman et al. (2010). Navarro et al. (2010) compared more traditional LDWs with motor priming, in which motor priming was considered to be a mix of LDW and LKA since it provided minimal direct actions on the steering control (i.e. an intervention at motor level but without intruding into vehicle control). This was done by the application of small asymmetric oscillations of the steering wheel. Comparison between the motor priming and other haptic and auditory systems was made to investigate a way to support the driver beyond just improvement of the situation diagnosis (i.e. warning only). The results showed improvements in recovery manoeuvres with motor priming, but that drivers preferred the auditory warning.

The present study includes two driving simulator experiments with professional truck drivers conducted to evaluate LDWs and LKAs.

2   TWO EXPERIMENTS

The first experiment was carried out in the moving base truck simulator at the Swedish National Road and Transport Research Institute (Figure 1). Two types of LDW were compared with one LKA. Both the effectiveness of the systems as well as the drivers' impressions and preferences were examined. The second experiment took place in the fixed base truck simulator at Volvo Technology (see Figure 1), where the benefit of an LKA was assessed. Similar steering wheel support as for lane departures was implemented as a Lane Change Decision Aid system (LCDA) which gave feedback in so called blind spot situations. The idea was to investigate the possibility to have the same type of support for different lateral threats in order to harmonize and simplify the support for the driver.

Figure 1 The simulator used in experiment 1 to the left. and the simulator used in experiment 2 to the right. Photos: VTI/Hejdlösa bilder (left) and Volvo (right).

2.1 Experiment 1

Three driver aids were used and compared: (i) the lane departure sound currently in the AB Volvo trucks, (ii) vibrations in the steering wheel, and (iii) guiding force in the steering wheel. The LDW sound was presented from the left or right side speaker depending on whether the left or right lane markings were crossed. The sound, which resembles the sound of driving on rumble strips milled into the road, contains 70ms long pulses, with a silence of 33ms in between the pulses. The fundamental frequency of the sound was around 133 Hz and the spectrum contained harmonics up to about 4kHz. The vibrations were presented in the steering wheel as a symmetric oscillation, with a period of 100 ms and an amplitude of +/-1 Nm. The outcome of the guiding force applied to the steering wheel was a torque ramped from 0 to 7 Nm, where the guiding direction was opposite the direction of lane crossing. The ramp up time depended on how the truck approached the lane marker (e.g. heading angle and speed). The guiding torque was active until the vehicle stopped drifting and was back in lane. The presentation order of the three driver aids was equally varied between subjects.

Twenty-four subjects (23 male, 1 female), novice, medium and very experienced professional drivers, participated. Their average age was 37 years old (SD=14.6, range: 18-61). All 24 subjects held C driving licences (i.e. allowed to drive a heavy truck with a total weight exceeding 3,500 kg and a light trailer in tow). Nineteen subjects held the additional CE driving licenses and had done so for an average of 17.5 years (i.e. they were allowed to drive heavy truck with a total weight exceeding 3,500 kg with one or more trailers of undesignated weight). Drivers were distributed evenly with regards to their driving experience and to which warning or assistance system they were exposed to first, second and thirdly. Only four subjects had previously driven with ADAS such as LDWs and Adaptive Cruise Control.

A secondary task was introduced in order to allow for lane departures. The subjects were prompted to perform a task on a radio/CD application presented on a 7'' touch screen mounted on the right side of the driver. The task consisted of changing tune each time a CD-player disturbance was activated (i.e. when the sound from the CD-player was replaced by noise). In order to further increase the likelihood of the vehicles to pass the lane markings an additional yaw motion from the simulator was given. The yaw motion was triggered during the secondary task.

At the end of each experimental drive subjects were asked to intentionally cross the left and right lane markings in order to provoke the just used driver aid and get a clearer idea of its functionality.

The following dependent variables were computed for the occasions where subjects performed the radio task and experienced an active driver aid: (i) **Duration of lateral excursion [s]:** Time from when vehicle's right/left front wheel crosses the right/left lane marker in the direction from the centre of the driving lane, until vehicle's right/left front wheel crosses the right/left lane marker in the direction towards the centre of the proper driving lane, (ii) **Overshoot [m]:** Movement past centre of the right lane after a lateral excursion defined as the distance between the centre of the right lane and the maximum lateral position opposite to the lane departure, (iii) **Peak acceleration of steering wheel motion [degrees/s^2]:** Once the subject's recovery steering manoeuvre has started, the sharpness of the steering is given by the maximum acceleration of the steering wheel motion.

A range of questions were presented to the subjects after each drive. Spontaneous comments were collected as well as opinions about whether the system felt complicated, would be easy to understand and learn, and would have any effect on perceived safety.

A repeated measures analysis of variance (ANOVA) on overshoot with experience groups as categorical variable (i.e. mixed design) showed a significant main effect of repetition, $F(2, 42) = 4.13$, $p< .05$, and significant interaction effects of group by repetition, $F(4, 42) = 3.43$, $p< .025$, and group by aid by repetition, F(8, 84) = 2.61, p< .025. The three-way interaction effect is illustrated in Figure **2**Figure 2. It is the group of low experience level that exhibits a higher mean overshoot only in the first trial with the vibration aid, as compared with the other repetitions and aids (except the first trial with the sound aid). The driver groups of medium and high experience show stable mean overshoot over aids and repetitions. ANOVAs on duration of lateral excursion and peak acceleration of steering wheel motion showed no significant effects.

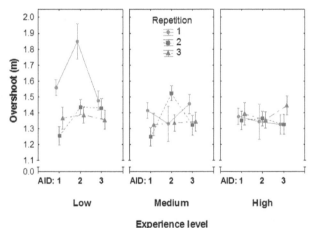

Figure 2 Novice drivers with significantly higher values of Overshoot with the LDW with vibrations. AID1= Sound, AID2=Vibrations, AID3: Guiding force.

Overall, the subjects were somewhat more positive towards the LDW consisting of vibrations and a bit more hesitant to the guiding force (see Figure 3).

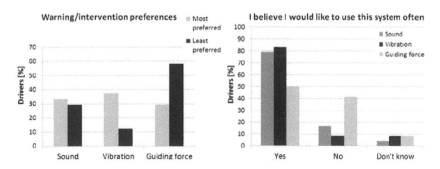

Figure 3 Drivers' preference ratings of the three alternatives (left) and their thoughts about using the systems if installed in their vehicles (right)

With activation of the guiding force, many drivers were sceptical about leaving over control to the vehicle and there were thoughts about whether drivers would automatically countersteer in real traffic. Positive comments brought up some aspects with regards to accidents and incidents where drivers felt the system would serve its purpose as a countermeasure. Additionally, there were many "what if" questions concerning a guiding force on slippery/icy road surface and/or with a heavy loaded trailer. Many subjects compared the vibration in the steering wheel to rumble strips milled into the road surface and thought this was good and made the warnings easy to understand. Subjects reflected on that the sound would probably create annoyance, after some time.

2.2 Experiment 2

A combination of the vibration and guiding force was evaluated in a second experiment in order to combine a more actively supporting system with the vibrations which were well perceived by subjects in the first experiment. 22 subjects were exposed to one version of an LKA. An additional drive was included where an LCDA was assessed as well. This was done in order to see if the same type of feedback in lane departure situations could be used also for situations where the subject had vehicles in blind spots. 21 subjects drove the experimental runs with the systems deactivated. This between group design was chosen due to the added LCDA comparison; level of expectancy of the blind spot scenario was thought to be difficult to keep low in a within group design.

The LKA was presented when the driver had crossed 50 cm over the right lane markings onto the curb (total curb width was 2.73 meters) and consisted of a combination of vibrations in the steering wheel and a guiding force. For the mid and left hand side lane markings LKA was presented already after 15 cm. The LCDA system used a similar combination of vibrations and force as well as an additional blind spot display located by the right hand A-pillar.

The vibrations were presented in the steering wheel as an asymmetric oscillation, with a period of 85 ms and an amplitude changing between -1, 0 and 1 Nm. The time of the torque was longer in the direction of the lane centre and shorter in the direction of lane departure in order to give a guiding vibration feeling (Figure 4).

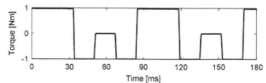

Figure 4 The vibrations part of the LKA in Experiment 2

The outcome of the guiding torque applied to the steering wheel was a torque ramped from 0 to 4 Nm for the LKA and 0 to 5 Nm for the LCDA, where the guiding direction was opposite the direction of lane crossing or blind spot object. The ramp up time depended on how the truck approached the lane marker (e.g.

heading angle and speed). The guiding force was active until the vehicle stopped drifting and was back in its lane.

Forty-four subjects (41 male, 2 female), all professional truck drivers, participated. One male subject had to be removed due to simulator sickness and there was some data loss due to occasional equipment or computer failures. The average age of the 43 remaining subjects was 40 years (SD=12, range 20-65). All subjects held valid C licenses and all except for one subject also held additional CE driving licences (M=17.8 years). Four drivers drove professionally less than 30.000 km/year and six drivers more than 100.000 km/year. Eleven subjects had some previous experience of ADAS.

The secondary task and the additional yaw motion used in Experiment 1 were also used in Experiment 2.

Each subject drove four drives. An initial test run in order to get accustomed to the simulator was followed by a drive with two sharp LCDA scenarios. In between the LCDA scenarios other critical scenarios were added in order to further lower the expectancy level of the LCDA scenarios. In the LCDA drive the subject was instructed by an experimental leader to follow a vehicle. The two sharp LCDA scenarios consisted of that both vehicles overtook a vehicle standing still in the right lane. When the two vehicles entered back into the right lane a vehicle suddenly appeared from behind into the subject's right side blind spot (see Figure 5). In the third experimental drive the LKA was assessed, in which the subject did not have a vehicle in front to follow and instead was prompted to perform the secondary task seven times. After the LKA drive the subject was instructed to intentionally cross the left and right markings in order to provoke the LKA. The subject was also instructed to deliberately change lane despite vehicles being present in the right lane in order to provoke the LCDA system. After each drive the subject were asked a range of questions by the experimental leader.

Figure 5 The LCDA scenario with host vehicle (truck) entering back into right lane, and hidden vehicle suddenly appears from behind into subject's right side blind spot.

Same measures were calculated for LKA in Experiments 1 and 2 (i.e. Duration of lateral excursion, Overshoot and Peak acceleration of steering wheel motion). Analysis of driving behaviour in the LCDA scenarios is not presented in the current paper, just drivers' opinions.

For the baseline group the same measures were calculated for the situations where subjects would have experienced the driver aids if they would have been activated.

One-way ANOVAs show significant effects of group on first event of Duration of lateral excursion, $F(1, 30) = 6.10$, $p < .025$, and first event of Peak acceleration of steering wheel motion, $F(1, 30) = 4.35$, $p < .05$. With the LKA (treatment) drivers are back in lane quicker than without (Baseline). See Figure 6. The treatment group also has significantly higher peak acceleration of steering wheel motion. (Non-parametric tests with Mann-Whitney U test show the same results as the ANOVAs.)

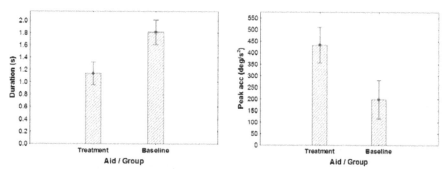

Figure 6 The significant effect on first event of Duration of lateral excursion (left) and Peak acceleration of steering wheel motion (right)

Seventy-two per cent of the subjects in the group with the LKA activated answered that they indeed felt the system in the lane departure events. Subjects who felt a system commented on that they felt it when they crossed the lane markers and that it warned them by vibration and force in the steering wheel. A majority of the drivers in both groups said they would like to have the LKA in their vehicle.

For the LCDA scenario a higher percentage of subjects in the treatment group stated that they saw one or both of the two blind spot vehicles on their right (see Figure 7). When asked if they wanted to have this LCDA system in their vehicle most subjects answered that they would (Figure 7). This question was asked after the last test run, and the baseline group could respond here as well.

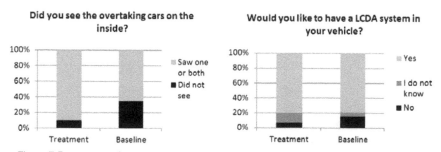

Figure 7 Percentage of subjects of each group detecting the blind spot vehicle (left) and that would like to have the LCDA system in their vehicle (right), respectively.

After the fourth and last drive when the subjects were asked if they perceived the LKA and LCDA they had tested as one system or several, 64% answered that they saw it as several, 26% saw it as one and 10% was unsure. A common sectioning were steering and vibration into one system and the additional blind spot display as another one but also warnings for staying in lane as one and warnings for vehicles in blind spot as one.

A vast majority (93%) of the subjects answered, after they had driven the fourth drive that they believed that the system/systems that they had tested would be of benefit for a driver in those situations. The percentage of subjects who wanted to have a system in their vehicle also increased to 95%. Some reservations were however made about the guiding force in the steering wheel, which was also the case in the first experiment. Still, when the subjects were asked if they would trust a system like this 71% answered that they would.

3 DISCUSSION

In the first experiment, the novice drivers showed more pronounced overshoot in the first repetition with the vibration-LDW. One explanation could be that vibrations were actually easier to distinguish compared to the guiding force when looking at the two options of presenting input in the steering wheel. This could have caused a surprise effect in the first repetition causing the overshoots. Experienced truck drivers are also more used to compensate for irregularities coming from the front axle, which could explain their stable overshoot behaviour over the three support systems. In the second experiment, the drivers with the LKA activated were faster back in lane and had a higher peak acceleration of the steering wheel movement. No significant effects were found for Overshoots, which would be one of the safety-critical metrics to observe when studying unintended effects. Even though no effects for Overshoots were found, it is important to include a combination of metrics that could capture both intended and unintended effects. In the analyses of the second experiment, however, only the first event, or repetition, of each performance measure was used.

When looking at the subjective data it is clear that drivers are rating the three support systems (Experiment 1) quite similarly but questions the realisation of the guiding force in terms of implications it might have during different driving conditions. The comments about the sound alternative supports the results found by Braitman (2010). The results of the second experiment show a high acceptance of the guiding force both for lane departure situations and blind spot events. However, the comments made in the first experiments were also made in the second with regards to possible negative effects the system could create in, for example, reduced friction situations or when driving with full load.

Some methodological lessons were made when running the two experiments. One critical aspect with regards to the steering wheel specification (e.g. type of steering wheel, cabin, links and lag) is that small differences in the simulator set up most likely create differences in the outcome. The forced yaw motion was considered to be a necessary methodological step in order to get subjects' out of their lane, especially since the subjects were professional truck drivers. However, in both experiments subjects clearly indicated that they often felt this force, often associated to be wind gusts or similar. In a follow up experiment it would be good to trigger the yaw motion when subjects have their eyes off the road or try and perhaps as combined with a different secondary task. A system-paced task would most likely force the driver to look down more from the road scene that in turn would create more lane departures. A down side would be that subjects would be unsure on how to prioritise between the primary and secondary tasks.

ACKNOWLEDGMENTS

The authors would like to thank Per Nordqvist, Peter Nugent, Stefan Bergquist, Mathilda Fulgentiusson at Volvo as well as Anders Andersson, Anne Bolling and Håkan Sehammar at VTI for their contribution. The authors would also like to acknowledge the projects funding the two experiments. Experiment 1 was carried out in the national project 'Driver and system controlled heavy vehicle steering' within the Swedish competence centre ViP (http://www.vipsimulation.se/). Experiment 2 was conducted within the EU project InteractIVe (http://www.interactive-ip.eu).

REFERENCES

Braitman, K., A. McCartt, D. Zuby and J. Singer. 2010. Volvo and Infinity drivers' experience with select crash avoidance technologies. *Traffic Injury Prevention, 11:3, 270-278.*

DfT, Department for Transport, UK. 2004. STATS20 - Instructions for the Completion of Road Accident Reports. Accessed September 2, 2011, http://www.dft.gov.uk/collisionreporting/Stats/stats20.pdf.

ISO 17361 Intelligent transport systems - Lane departure warning systems - Performance requirements and test procedures

Navarro, J., F. Mars, J. Forzy, M. El-Jaafari and J. Hoc. 2010. Objective and subjective evaluation of motor priming and warning systems applied to lateral control assistance. *Accident Analysis and Prevention 42 (2010) 904-912.*

Starnes, M. 2006. Large Truck Crash Causation Study: An Initial Overview, DOT HS 810 646, August 2006.

Strandroth, J. 2009. In-depth analysis of accidents with heavy goods vehicles – Effects of measures promoting safe heavy goods traffic. Swedish Road Administration. ISSN: 1401- 9612.

CHAPTER 6

On Learning Characteristics of Automotive Integrated Switch System Using Face Direction

Takehito Hayami*1 Atsuo Murata*1, Youichi Uragami*1, Makoto Moriwaka*1, Shinsuke Ueda*2, and Akio Takahashi*2

*1 Graduate School of Natural Science and Technology, Okayama University
Okayama, Japan
murata@iims.sys.okayama-u.ac.jp

*2 Department 2, Technology Research Division 8, Honda R&D Co., Ltd., Automobile R&D Center
Tochigi, Japan
Akio_Takahashi@n.t.rd.honda.co.jp

ABSTRACT

The development of switch system which leads not only to a faster response and less frequent visual off road but also to quick learning would be useful for the purpose of enhancing safety. The usability is, however, not evaluated on the basis of learning characteristics of such measures as operation time and frequency of visual off road. The evaluation of usability of switch systems by means of ease of learning is very useful for automotive designers of cockpit modules. An attempt was made to investigate the learning characteristics of the developed integrated switch system using face direction with the traditional touch-panel interface.

Keywords: automotive integrated switch, learning, face direction, usability

1   INTRODUCTION

Automotive cockpits are becoming more and more complicated (Murata et al., 2009a, Murata et al., 2009c). While in-vehicle information systems such as IHCC system and ITS surely support driving activities, the operation of such systems induces visual and physical workload to drivers. Therefore, it must be noted that in-vehicle information system has both positive and negative impacts on safety driving. With the development of by-wire technologies, automotive interfaces that control a display using computers have increased. Due to the widespread of such systems, a variety of switch systems can be installed to in-vehicle equipments.

The usability of switches are, in general, affected by many factors such as ease to operate, frequency of visual off road, and physical workload while operating, etc (Murata et al., 2007, Murata et al., 2008, Murata et al., 2009b, Murata et al., 2011). The improvement of such factors leads to the enhanced usability of switches, and eventually contributes to the safety driving. However, the usability of the developed switch system is rarely evaluated from the viewpoint of ease of learning (learning process). Therefore, the development of switch system which leads not only to a faster response and less frequent visual off road but also to quick learning would be useful for the purpose of enhancing safety. The usability is, however, not evaluated on the basis of learning characteristics of such measures as operation time and frequency of visual off road. The evaluation of usability of switch systems by means of ease of learning is very useful for automotive designers of cockpit modules.

Therefore, the development of switch system which leads not only to a faster response and less frequent visual off road but also to quick learning would be useful for the purpose of enhancing safety. The usability is, however, not evaluated on the basis of learning characteristics of such measures as operation time and frequency of visual off road. The evaluation of usability of switch systems by means of ease of learning is very useful for automotive designers of cockpit modules. An attempt was made to investigate the learning characteristics of the developed integrated switch system using face direction together with the traditional touch-panel interface.

2 METHOD

2.1 Participants

A total of 3 male participants licensed to drive took part in the experiment. All were young adults aged from 21 to 24 years. They took part in the same experiment consecutively for ten days. They hand no orthopaedic or neurological diseases.

2.2 Apparatus

Two switches were used in the experiment. One was a traditional touch-panel switch, and the other was a developed integrated switch which can realize many functions by making use of facial direction. This system enabled the participant to select one of many functions by changing the facial direction and pressing a "Confirmation" key placed around a steering wheel. Using a three-

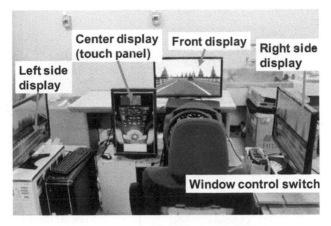

Figure 1 Outline of experimental setting.

dimensional magnetic-type location measurement system (POLHEMUS, 3-SPACE Fastrak), the switch system automatically recognized the facial direction (central, left, right, middle-left) and the combination of the recognition and the pressing of "Confirmation" key enables the participant to carry out a pre-determined switch operation such as the open and close of left-side window. In the simulator, according to Japanese and British standard, the participant location was on the right (The driver's seat was located on the right).

2.3   Experimental task

Viewing the road in Figure 1, the participants were required to carry out the simulated driving task. The straight road was used in the simulation (The curved road was not used). The participants were also required to carry out the switch operation task using either the touch-panel interface (See Figure 1) or the integrated switch using facial directions. The switch operation included the following tasks: (1) open/close of a window on the driver's side, (2) open/close of a front passenger's seat, (3) operation of a door mirror on the driver side, (4) operation of a door mirror on the front passenger side, (5) on-off of a hazard ramp, (6)adjustment of a seat, (7) adjustment of an air conditioner, (8) operation of an audio system, (9) operation of a car navigation system, (10) on-off of a door mirror light on the driver side, (11) on-off of a door mirror light on the front passenger side, and (12) information presentation to HUD (Head Up Display). It must be noted that the operation (12) was carried out only by the integrated switch operation using facial directions.

The operation principle of the integrated switch using facial directions is briefly explained below. The three-dimensional magnetic-type location measurement system used for recognizing facial direction is shown in Figure 2(a). Five areas shown in Figure 2(b) were discriminated using the measurement

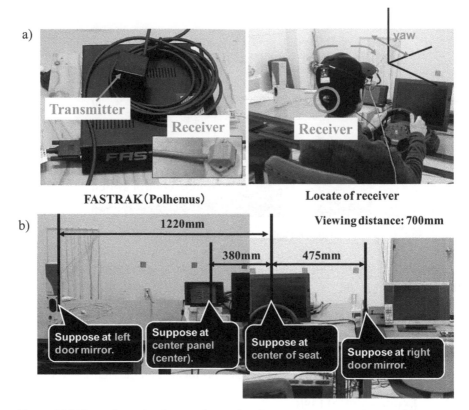

a)

Transmitter

Receiver

Receiver

FASTRAK(Polhemus)

Locate of receiver

b)

1220mm

Viewing distance: 700mm

380mm   475mm

Suppose at left door mirror.

Suppose at center panel (center).

Suppose at center of seat.

Suppose at right door mirror.

Figure 2 (a) Three-dimensional magnetic-type location measurement system used for recognizing facial direction, and (b) areas that are discriminated using the measurement system.

system. The discrimination was carried out on the basis of the yaw angle output from the three-dimensional magnetic-type location measurement system (Fastrak). When the participant moved his head toward the left door mirror and the system recognized this, he could carry out the switch operations (2), (4), and (11). When the head movement to center panel was recognized, the participant could carry out the switch operations (6), (7), and (8). When the system recognized that the head was directed to the center of a seat, the participant could conduct the operations (5) and (12). The system recognition of the head movement to the right door mirror enabled the participant to carry out the switch operations (1), (3), and (10). After selecting the function, the participant can move his head freely and continue and complete the switch operation using a steering-wheel mounted switch. Thus, a lot of functions can be realized by making use of facial directions. The contents of switch operations (1)-(12) were orally presented to the participant. In case of the touch-panel interface, the participant carried out switch operation using the touch-panel.

2.4 Design and procedure

Both switch type and experimental session were within-subject factors. The participants were required to carry out simultaneously a main tracking task and a secondary switch operation task such as the operation of CD and the putting on of hazard ramp. The duration of one experimental session was about 7 min. In one session, the participant conducted the switch operation 10 times. In one-day experiment, the participant repeated the session 30 times (a total of 300 switch operations). Between sessions, the participant was allowed to take a short break. The three participants conducted the experimental task above for ten days.

2.5 Evaluation measure

Making use of the output data from the three-dimensional magnetic-type location measurement system (Fastrak), the frequency of visual off road and the visual time off road were obtained. When the head movement was out of the center area in Figure 2(b), this was regarded as visual off road. The following evaluation measures were used to compare the usability between two types of switches:
(1)Task completion time in switch operation
(2)Percentage correct in switch operation
(3)Frequency of visual off road
(4)Visual time off road
(5)Workload evaluation by NASA-TLX (WWL: Weighted Workload Score)

3 RESULTS

In Figure 3, it is shown how the task completion time is shortened or reduced with the progress of learning (increase of repetition (day)). For both switches, the task completion time decreased on days 2, 3, 4, and 5 as compared with that on the first repetition (day 1). A two-way (switch type by day) ANOVA conducted on the task completion time revealed no significant main effects or an interaction. In Figure 4, the task completion time is compared among day1, days2-6, and days7-10. A two-way (switch type by block (day1, days2-6, and Days7-10)) ANOVA conducted on the task completion time revealed only a main effect of switch type ($F(1,2)=88.605$, $p<0.01$). The proposed integrated switch led to a faster operation. In Figure 5, the fitting of the task completion time to learning curve $y=ax^b$ is depicted. For both switches,

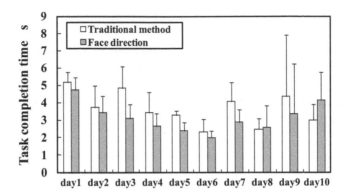

Figure 3 Task completion time as a function of switch type and day.

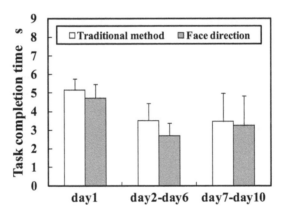

Figure 4 Task completion time compared between switch types and among day1, days2-6, and days 7-10.

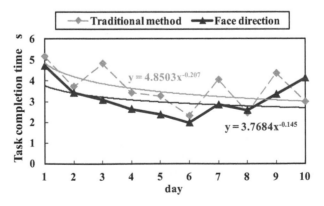

Figure 5 Modeling of learning process by $y=ax^b$.

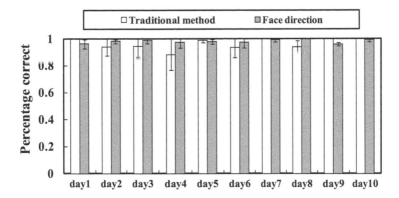

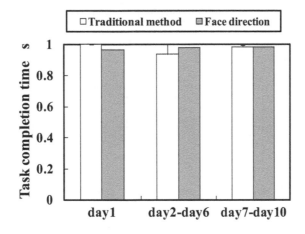

Figure 6 Percentage correct as a function of switch type and day.

Figure 7 Percentage correct compared between switch types and among day1, days2-6, and days 7-10.

the learning rate is very slow, indicating that the operation of both switches can be easily learned.

In Figure 6, the percentage correct is plotted as a function of switch type and day. A similar two-way ANOVA conducted on the percentage correct revealed no significant main effects or interaction. In Figure 7, is compared among day1, days2-6, and days7-10. A two-way (switch type by block (day1, days2-6, and Days7-10)) ANOVA conducted on the percentage correct revealed no significant main effects or interaction.

The learning curves of frequency of visual off road for both touch-panel switch and eye-gaze/facial direction switch are shown in Figure 8. The number of visual off road is remarkably higher in days 1 and 2 than in days 3, 4, and 5 for both types of switch. The learning might be helpful to reduce the visual off

54

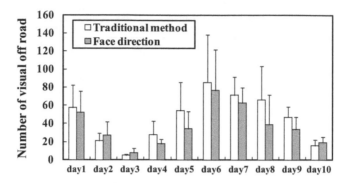

Figure 8 Frequency of visual off road as a function of switch type and day.

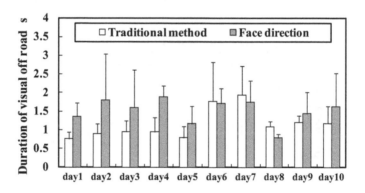

Figure 9 Visual time off road as a function of switch type and day.

road. In days 3, 4, and 5, the visual time off road decreased. Although the frequency of visual off road seems to increase on days 6-9, a similar two-way ANOVA revealed no significant main effects or interaction. In Figure 9, the visual time off road is plotted as a function of switch type and day. A similar two-way ANOVA revealed no significant main effects or interaction.

In Figure 10, the subjective rating on usability is shown as a function of switch type and day. In Figure 11, the NASA-TLX (WWL) score is plotted as a function of switch type and day. Kruskal-Wallis non-parametric statistical test revealed no significant main effects for both Figures 10 and 11.

4   DISCUSSION

The results from Figures 3, 4 and 5 indicate that the usability should be evaluated by taking the learning characteristics of task completion time into account. The switch systems used in this experiment should be compared and

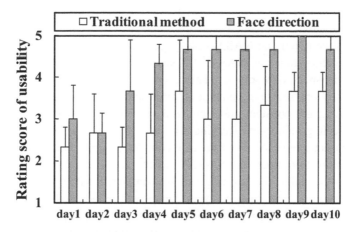

Figure 10 Rating score of usability as a function of switch type and day.

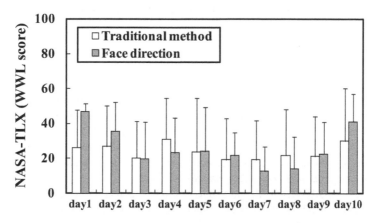

Figure 11 NASA-TLX score as a function of switch type and day.

evaluated on the basis of the data after day 2. Until day2, all participants experienced the switch operation task 600 times using each switch. Such trials (repetitions) are necessary to get accustomed to using the switch system. Although it might be very time consuming, abounding initial data (in this case, days 1 and 2) and using the stable data on days 2, 3, and 4 would be essential for the proper evaluation of switch systems.

From Figures 6 and 7, the learning effect is not observed in the aspect of percentage correct, which indicates that both touch-panel and proposed integrated switch can be operated easily (without errors) without learning periods. Figures 8 and 9 also indicate that the learning effects are less observed for both types of switches.

The 10-day experiment was imposed on the three participants. In Figure 3, the task completion time tended to increase on days 9 and 10. Similar tendency was observed for the NASA-TLX score (See Figure 11). NASA-TLX score got constant and stabilized after day3 in the same manner as the task completion time. It must be noted that NASA-TLX score increased on day10. The increase of task completion time and NASA-TLX score on days 9 and 10 must be due to the monotonousness, boredom, and habituation.

Different from the task completion time and the NASA-TLX score, the usability rating got constant and stabilized after day5. This indicates that the subjective feeling induced by learning does not necessarily correspond with the objective measure represented by the task completion time. Taking into account the task completion time, NASA-TLX score, and the subjective rating on usability synthetically, the learning can be completed within 5 days. For the sake of caution, the 10-day experiment was conducted. The learning characteristics of both types of switches seem to be grasped with a 5-day experiment. In future research, the number of participants should be increased. The learning characteristics of older adults should be clarified, which might reveal different characteristics from those of young adults, and help the universal design of switch systems.

REFERENCES

MURATA,A. and MORIWAKA,M. 2007. Applicability of Location Compatibility to the Arrangement of Display and Control in Human-Vehicle Systems -Comparison between Young and Older Adults-, *Ergonomics* 50(1): 99-111.

MURATA,A. and MORIWAKA,M. 2008. Evaluation of Control-Display System by means of Mental Workload, *Proceedings of 4th International Workshop on Computational Intelligence & Applications*, 83-88.

MURATA,A., MORIWAKA,M. and SHUGWANG,W. 2009a. Development of Thumb-Operated Dial-Type Integrated Switch for Automobile and its Effectiveness, *Proceedings of 5th International Workshop on Computational Intelligence & Applications*, 330-335.

MURATA,A., TANAKA,K. and MORIWAKA,M. 2011. Basic study on effectiveness of tactile interface for warning presentation in driving environment, *International Journal of Knowledge Engineering and Software Data Paradigm* 3(1): 112-120.

MURATA,A., UCHIDA,Y. and MORIWAKA,M. 2009b. Fundamental Study for Constructing A System to Assist The Left Visual Field of Older Drivers - Effectiveness of The Alternative of The Left Front Side-view Mirror by The Central Visual Field-, *Proceedings of 5th International Workshop on Computational Intelligence & Applications*, 320-325.

MURATA,A., YAMADA,K. and MORIWAKA,M. 2009c. Design Method of Cockpit Module in Consideration of Switch Type, Location of Switch and Display Information for Older Drivers, *Proceedings of 5th International Workshop on Computational Intelligence & Applications*, 258-263.

CHAPTER 7

Focusing on Drivers' Opinions and Road Safety Impact of Blind Spot Information System (BLIS)

Giulio Francesco Piccinini, Anabela Simões, Carlos Manuel Rodrigues

ISEC UNIVERSITAS
Lisbon, Portugal
g.f.piccinini@gmail.com

ABSTRACT

The crashes caused by the presence of a vehicle in the car's blind spot areas account for about 20% of the overall lane change crashes. Recently, in order to overcome the issue, car manufacturers introduced some Advanced Driver Assistance Systems (ADAS), detecting other vehicles in the blind spot areas and warning the drivers. Those systems, generally called Blind Spot Information System (BLIS) or Lane Change Warning (LCW) are supposed to be helpful for the drivers but, up to now, little information is available on drivers' opinions and the road safety impact of such systems. In order to fill this gap, focus groups interviews were conducted with the aim of collecting drivers' opinions about BLIS. Overall, the participants were satisfied about the help provided by the system during the lane change task. Furthermore, they admitted that the lane change behaviour is not modified by the system. However, it appeared that, in the long-term, behavioural adaptations might occur: drivers could rely on the system and carry out the lane change without checking the mirror, only based on the information provided by the system. Based on these results, further research is required on the topic.

Keywords: lane change, focus groups, road safety, behavioural adaptation.

1   INTRODUCTION

In a document recently prepared by the National Highway Traffic Safety Administration (NHTSA, 2009), 312000 crashes were reported during merging or changing lanes and, in those situations, 769 people died and 48000 were injured. Despite representing only 4 to 10% of the total number of crashes (Wang and Knipling, 1994), those figures could be further reduced if, during the lane change performance, drivers' attention could be enhanced by some means. In fact, as reported by several studies (Lavallière et al., 2011; Kiefer and Hankey, 2008; Tijerina et al., 2005; Olsen et al., 2005), drivers often fail to check the rear-view and the left side mirrors. Furthermore, according to a literature review on lane change crashes (Lee et al., 2004), in those situations, most drivers did not try an avoidance manoeuvre, suggesting that they were probably not aware of the presence of another vehicle on the side, when carrying out the lane change. Then, based on this outcome, a part of the conflicts might have been arisen with vehicles passing in the blind spot areas. Blind spot areas are the zones on the left and on the right of a vehicle (approximately 9.5 m long and 3 m width) which cannot be seen by the drivers through the rear-view and wing mirrors. In the USA, in the period 2002-2006, the crashes caused by the presence of a vehicle in the blind spot areas accounted for 26% of the overall lane change crashes (Farmer, 2008). A possible solution to the problem is to equip vehicles with some Advanced Driver Assistance Systems (ADAS) that, using cameras installed on the wing mirrors, detect the presence of another car/motorcycle moving (in the same direction) in the blind spot areas and alert the drivers. Those systems (generally called Lane Change Warning or Blind Spot Information Systems), developed by the Original Equipment Manufacturers (OEM), can warn the driver in several ways:
1. through a blinking warning light in the interior parts of the vehicle (close to the A pillars and nearby the mirrors);
2. through a light illuminating in the wing mirrors;
3. through a light illuminating on the wing mirrors and a vibration on the steering wheel when the driver uses the indicator to change lane.

In an on-line survey performed in the Netherlands (van Driel and van Arem, 2005), participants appeared to greatly favour the introduction of ADAS helping drivers in detecting vehicles in the blinds spots. Then, there are high expectations (from the users and from the institutions) that systems such as Blind Spot Information System (BLIS) and Lane Change Warning (LCW) could increase driving comfort and, therefore, have a beneficial impact on road safety. However, the introduction of a change in the driver-vehicle-environment (DVE) system might also bring some side effects, such as automation surprises (Sarter et al., 1997), overtrust in the system (Parasuraman et al., 1993), situation awareness impairment (Stanton and Young, 2000) and negative behavioural adaptations (OECD, 1990). In order to reduce (or avoid) the appearance of those issues, the Human-Centered automation approach has been introduced with the objective of establishing a cooperative collaboration between humans and machines (Inagaki, 2006).

2 THE HUMAN-CENTERED DESIGN

"The term Human-Centered (or user-centered) design (or automation) has been widely popularized as the proper approach to integrate humans and machines in a wide variety of systems" (Sheridan, 2002, pag. 155). The concept has been initially applied to aviation (Billings, 1991) and process automation (Hendrick, 1996) but it was later extended to the automotive field (Fancher et al., 2001; Tango and Montanari, 2006).

A clear objective of the Human-Centered design approach is to obtain a level of human involvement that allows the best performance for the human-machine system (Wickens et al., 2004, pag. 430). For example, during the Human-Centered design of an in-vehicle system, some aspects need to be taken into account (Tango and Montanari, 2006):

- understanding and coping with system limitations, since the beginning;
- clearly inform drivers about systems' limitations;
- find new way to support the driver, maintaining him/her in the control loop;
- carefully consider the interaction between users' needs and systems' needs.

The involvement of the users during the design process is the key element of the Human-Centered design approach. Eventually, once the system is released, it is important to keep on involving the user through an evaluation process which enables the designer to gain information about user satisfaction and any problems with the functionality of the system. Such after-market evaluation is, usually, performed through interviews and focus groups (Abras et al., 2004) and it is useful for the development of a future version of the system (Nielsen, 1993, pag. 71).

This article describes an effort to apply the Human-Centered approach for the evaluation of a Blind Spot Information System (BLIS). A qualitative study was carried out in Portugal to evaluate users' opinions and road safety impact related to the usage of the system.

3 METHOD

Focus groups interview technique was selected as research method since it was considered the most appropriate for the application of the Human-Centered approach. Indeed, such method enables the researchers to explore the widest range of ideas or feelings that the participants have about a specific system (Krueger and Casey 2009, pag. 19-20). Compared to single respondent interviews, this method allows generating a fruitful discussion between people who can agree or not with what expressed by another participant. Focus group interviews were already performed in the past to get a deeper understanding on various topics related to in-vehicle systems and, more in general, to road safety. Young and Reagan (2007) made use of focus groups to investigate the patterns of use of speed alerting and cruise control; Strand et al. (2010) investigated user experiences and road safety implications about the usage of Adaptive Cruise Control; Shams et al. (2011)

collected taxi drivers' views on risky driving behaviours in order to propose countermeasures for the improvement of road safety in Iran.

The present study was conducted in Braga, in the northern part of Portugal between September and November 2011. The participants were all occasional or regular users of Adaptive Cruise Control (ACC) and Blind Spot Information System (BLIS) and they were recruited through the help of a local dealer due to the difficulties in finding ADAS users in Portugal.

Overall, two focus groups sessions were performed. Each focus group interview lasted about two hours and was separated in 2 parts: the first one aimed at discussing the usage and effectiveness of Adaptive Cruise Control while the latter addressed the users' opinions and road safety impact of BLIS. In this paper, we exclusively focus on the analysis of the second part, being the findings about ACC already published (Piccinini et al., 2012).

The focus group interviews were performed by a research team made up of 3 people: a moderator, leading the discussion, an assistant moderator, helping the moderator for the administrative tasks (distribute consent forms and questionnaires, operating the video recorder, etc.) and a note taker, responsible for sketching participants' position and for noting down the most salient moments during the discussion. The participants were received by the research team as soon as they reached the designated location. In the room assigned to host the focus groups, a table with drinks and food was set up in order to favour the dialogue between the participants and the research team before the beginning of the discussion (during the waiting period previous to the arrival of the remaining participants). The research team took advantage of this time to introduce the study to the participants and to create a friendly atmosphere.

Finally, when everyone joined, the research team invited the participants to take a seat according to the assigned disposition. Then, the moderator introduced the members of the research team and described the purpose of the study and the modality of the session. Before beginning the focus group session, each participant signed a consent form and filled in a personal questionnaire requiring information such as age, gender, mileage driven since getting the driving license and car owned.

The part of the focus group session dedicated to BLIS started with the drivers filling in a questionnaire concerning the patterns of use for the system (usage of the BLIS in the different types of road, weather, luminosity and traffic conditions). Then, the discussion began and developed according to a questioning route prepared in advance by the research team and revolving around four topics: users' satisfaction of BLIS, critical situations occurred with the system, usage of BLIS and suggestions for the future implementation of the system (Table 1).

The focus groups sessions were video recorded and transcribed verbatim. The discussion was analysed qualitatively according to the thematic analysis approach. Thematic analysis has been defined as the "method for identifying, analysing, and reporting patterns (themes) within data" (Braun and Clarke, 2008). Based on this approach, the transcribed data were first coded and then, through an iterative procedure, themes were retrieved. Within the thematic approach, the analysis focused on articulated data. Articulated data are the information resulting from the discussion in direct response to the questions presented in the questioning route (Massey, 2011). The choice of directing the attention to articulated data is justified by the fact that the research team specifically prepared the study to get participants answering to the predefined questions.

Table 1: questioning route (questions put and relative topic)

Questions	Topic
1. How were your expectations about BLIS satisfied?	Satisfaction
2. Can you discuss with the other participants the critical situations you experienced while driving with BLIS?	Problems / critical situation
3. When you are driving, is the BLIS always activated or is there any specific situation in which you deactivate it?	Usage
4. When you want to change lane with BLIS activated, how do you behave with respect to the warning signal (light)?	Usage
5. Is there any situation in which you don't look at the warning signal (or you don't keep it into account)?	Usage
6. Are there any suggestions to improve the actual system or other functions that you would like to implement?	Suggestions

4   RESULTS

Overall 13 people took part in two focus groups discussions: 6 participants joined in the first session and 7 in the latter one. The sample predominantly included men due to the difficulties in finding women driving vehicles equipped with Adaptive Cruise Control (ACC) and Blind Spot Information System (BLIS): globally, 12 participants were males and only 1 was a female. The age of the participants ranged from 33 to 61 years old. They were all experienced and regular drivers, having driven more than 150000 km since they got their driving licence and using their vehicle daily. The participants were all users of BLIS: everyone drove more than 50 km with the system activated and the majority of them did it for more than 3000 km.

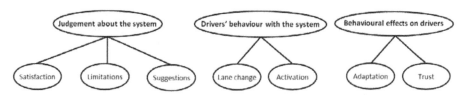

Figure 1. Themes emerged from the analysis of the focus groups interviews

From the qualitative analysis of the focus groups sessions, three main themes appeared: "Judgement about the system", "Drivers' behaviour with the system" and

62

"Behavioural effects on drivers". Each main theme was, then, subdivided in various subthemes as reported in Figure 1.

4.1   Judgement about the system

For the theme "Judgement about the system", three sub-themes were drawn: 'Satisfaction', 'Limitations' and 'Suggestions' (Figure 1).

Taking into account 'Satisfaction', drivers appeared to be very pleased about BLIS since the system went over drivers' expectations. One participant mentioned: "The system really exceeded my expectations. I was a bit reluctant, knowing other systems, but after using it, I think it is an added value, without any doubts". Overall, the system was judged by the participants as useful, comfortable and safe. A driver reported: "In all those years, I had many dangerous situations because I did not notice the vehicle in the blind spot. I had already several of those circumstances and I got scared. For me, BLIS is extremely beneficial".

Drivers were satisfied about the system even being aware of its limitations. During the discussion, some shortcomings were mentioned by the participants and they were collected in the sub-theme 'Limitations'. The issues more frequently reported were:
- the system does not work properly with hard rain;
- BLIS gets confused when driving close to a barrier between lanes in motorway;
- the system is late in detecting vehicles approaching fast on the side;
- the camera of the system detects a too small blind spot angle;
- BLIS has often false detections of vehicles.

In particular, the late detection of fast vehicles was considered a relevant limitation in terms of trust towards the system as reported by a participant: "My feeling is that, when I drive at 120 or 130 kph and a car overtakes me at 140 or 150 kph, the warning lights up when the car is already passing on my left [...]. I don't trust the system at 100%".

Despite the limitations, the participants seemed confident about the fact that the system won't create any critical situations (incidents, accidents, etc.). One driver stated: "I favour more those systems which inform the drivers and don't interfere in the functioning of the vehicle because they leave to the user the responsibility to intervene. [...] When I used the system, I did not notice any critical situation".

In the end of the discussion, participants proposed some solutions to further improve the system (included in the sub-theme 'Suggestions'):
- to increase the angle of the camera in order to detect a larger blind spot area;
- to enhance the efficiency of detection in order to reduce the false alarms;
- to arrange a solution to clean the camera when it gets dirty;
- to reduce the dimension of the camera;
- to adopt the system only on the left side of the vehicle;
- to introduce a warning sound associated to the warning light.

With respect to the last suggestion, there was not agreement in the group as the introduction of a warning sound might represent a bother for the driver. One participant stated: "Personally, I don't like it so much [...]. I get more stressed with the sound than with the light".

4.2 Drivers' behaviour with the system

The theme "Drivers' behaviour with the system" was split in two sub-themes: 'Lane change' and 'Activation'.

Regarding the sub-theme 'Lane change', it emerged that, in large majority, the sample considered the system as an assistant for taking the decision of changing lane. One participant summarized the lane change behaviour with the system as: "I think that the (lane change) behaviour must be separated in two situations: when the warning lights up and when it does not. When the warning light is on, my behaviour is to delay the overtaking, waiting for the light to turn off. On the other hand, when there is no warning light, I confirm in the mirror [...] and then, I start the overtaking manoeuvre". In describing the lane change behaviour with BLIS, the participants remarked a positive feature of the system: they don't need to move their head to get the system's warning signal. This aspect of the system is related to the location of the warning light which, overall, was considered as the proper one. A driver mentioned: "The warning of the system is located in the ideal position [...]. I think that it is not required to move your eyes from the road in order to perceive if the warning light is on or off".

With regards to the other sub-theme ('Activation'), the participants stated that they never switch off the system. Unlike other Advanced Driver Assistance System (such as the Lane Departure Warning), the BLIS is not bothering the driver and therefore, it is always kept activated. The drivers admitted to switch off the system only when the system gets confused. The utility of BLIS was considered especially relevant in reducing the risk of an accident when entering the motorway, in the acceleration lane. One participant reported: "There is a very critical situation in the acceleration lane in the motorways [...]. I never had an accident but when I have it, I already know how it will happen. And, it will be in a car without BLIS [...]. So, I think BLIS is very important".

4.3 Behavioural effects on drivers

The last theme emerging from the discussion was the "Behavioural effects on drivers" subsequently divided in two sub-themes: 'Adaptation' and 'Trust'.

The sub-theme 'Adaptation' refers, to a larger extent, to the definition of "behavioural adaptations" introduced by the Organization for Economic and Co-operation Development (OECD, 1990). This sub-theme gathered the parts of the discussion related to possible modifications of drivers' behaviour during the lane change task, as a consequence of the introduction of BLIS. In general, participants stated that the lane change task is not modified by the introduction of BLIS. One participant reported: "I think that it isn't (the lane change task does not change) because the lane change behaviour always passes from looking at the mirror". However, other participants mentioned about the possibility to incur behavioural adaptations in the long-term: "The tendency, at the beginning, is to confirm (with the mirror) [...]. But when you get used to the system, it is almost instinctive. [...] If the warning does not light up, there is a predisposition of trusting the system". Another participant referred about the possibility to lose attention for what is passing on ahead: "Having the light always blinking and having the fear that there is something which, in reality, is not there, draws the attention to the warning light and leads us to lose the attention for what there is in the front".

Concerning the second sub-theme ('Trust'), generally, people did not seem to completely trust the system when taking the decision to change lane. A participant stated: "I don't trust the system for what it concerns the decision of changing lane, based on the information that provides [...]. It was reported that, sometimes, there are situations in which the system detects vehicles that do not exist in reality. However, I think there is also the opposite risk that the system fails in detecting a vehicle that is coming". On the other hand, some participants admitted that, in some occasions, they started the lane change without looking at the mirror, based only on the information provided by the system: "The system always met my expectations up to the point that I have the encouragement to start changing lane before looking at the mirror. When BLIS is not detecting anything on the side, people have the tendency to begin turning the steering wheel and, only afterwards, looking at the mirror".

5   DISCUSSION AND FURTHER WORK

Recently, car manufacturers marketed vehicles equipped with warning systems detecting vehicles in the blind spot areas (BLIS or LCW), with the intent of reducing the possibility for a driver to be involved in a lane change crash. Despite being already used by drivers, little information is available about users' opinions and road safety impact of those devices. Up to now, few studies (e.g., Kiefer and Hankey, 2008) have been performed to analyse the alterations of human behaviour during lane change, following the introduction of such systems.

The objective of this study was to investigate users' opinions concerning Blind Spot Information System. Drivers expressed satisfaction about the help provided by the system during the lane change task, despite being aware of its limitations. According to the comments made by the participant, two situations need to be distinguished during the lane change with BLIS, depending on the state of the warning light. If the warning light is on, drivers wait for it to turn off. On the other hand, when the warning is off, drivers verify in the mirror and perform the lane change as they would regularly do in a traditional vehicle. Then, apparently, the introduction of BLIS does not modify drivers' lane change behaviour. However, as referred by some participants, in the long-term, some behavioural adaptations might occur: in fact, drivers could rely on the system and carry out the lane change without looking at the mirror when the warning doesn't light up. This behaviour would be extremely dangerous since the system is not completely dependable, especially in some situations which are clearly identified (hard rain, fast approaching vehicles, etc.). In order to study more in detail the topic, in the next months, a Field Operational Test (F.O.T.) with users of BLIS is planned to be carried out.

ACKNOWLEDGMENTS

This research received funding from the European Commission Seventh Framework Programme (FP7/2007-2013) under grant agreement no. 238833 (Marie Curie Initial Training Network ADAPTATION: 'Drivers' behavioural adaptation over the time in response to ADAS use').

The authors would like to thank the Volvo dealer "Auto Sueco Minho" for the help provided during the study and the colleagues Susana Rôla and Ana Ferreira who moderated the focus groups and greatly helped during the transcription.

REFERENCES

Abras, C., Maloney-Krichmar, D. and Preece, J. 2004. User-Centered Design. In. *Encyclopedia of Human-Computer Interaction.* Bainbridge, W., eds. Thousand Oaks: Sage Publications.

Billings, C. E. 1991. Toward a Human-Centered Aircraft Automation Philosophy. *The International Journal of Aviation Psychology,* 1(4), 261-270.

Braun, V. and Clarke, V. 2008. Using thematic analysis in psychology. *Qualitative Research in Psychology,* 3(2), 77-101.

Fancher, P., Bareket, Z. and Ervin, R. 2001. Human-Centered Design of an Acc-With Braking and Forward-Crash-Warning System. *International Journal of Vehicle Mechanics and Mobility*, 36:2-3, 203-223.

Farmer, C. M. 2008. Crash Avoidance Potential of Five Vehicle Technologies. Insurance Institute for Highway Safety.

Hendrick, H. W. 1996. The Ergonomics of Economics is the Economics of Ergonomics. *Proceedings of the Human Factors and Ergonomics Society Annual Meeting October 1996 vol. 40 no. 1 1-10.*

Kiefer, R. J. and Hankey, J. M. 2008. Lane Change Behavior with a Side Blind Zone Alert System. *Accident Analysis and Prevention*, 40 (2008), 683–690.

Krueger, R. A. and Casey, M. A. 2009. *Focus groups: A Practical Guide for Applied Research, 4th edition.* Sage publications, Inc.

Inagaki, T. 2006. Design of human–machine interactions in light of domain-dependence of human-centered automation. *Cogn Tech Work,* (2006) 8: 161–167

Lavallière, M., Laurendeau, D., Simoneau, M. and Teasdale, N. 2011. Changing Lanes in a Simulator: Effects of Aging on the Control of the Vehicle and Visual Inspection of Mirrors and Blind Spot. *Traffic Injury Prevention*, 12:2, 191-200

Lee, S. E., Olsen, E. C. B. and Wierwille, W. W. 2004. A comprehensive examination of naturalistic lane changes. Technical Report DOT-HS-809-702, National Highway Traffic Safety Administration, U.S. Department of Transportation, 2004.

Massey, O. T. 2011. A proposed model for the analysis and interpretation of focus groups in evaluation research. *Evaluation and Program Planning 34 (2011), 21-28.*

66

NHTSA 2009. Traffic Safety Facts: A Compilation of Motor Vehicle Crash Data from the Fatality Analysis Reporting System and the General Estimates System. Washington, DC: National Highway Traffic Safety Administration.

Nielsen, J. 1993. *Usability Engineering*. Academic Press, London, UK.

OECD 1990. *Behavioural adaptations to changes in the road transport systems*. Organization for Economic and Co-operation Development publications, 1990.

Olsen E. C. B., Lee S. and Wierwille W. "Eye glance behavior during lane changes and straight-ahead driving". Paper presented at: Transportation Research Board Meeting; January 9–13, 2005, Washington, DC.

Parasuraman, R., Molloy, R. and Singh, I. 1993. Performance Consequences of Automation Induced 'Complacency'. The International Journal of Aviation Psychology, 3, 1-23.

Piccinini, G. F., Simões, A. and Rodrigues, C. M. "Usage and effectiveness of Adaptive Cruise Control (ACC): a focus group study". Paper presented at the First International Symposium on Occupational Safety and Hygiene; February 9-10, Guimarães, Portugal.

Sarter, N. B., Woods, D. D. and Billing, C. E. 1997. Automation surprises. In. *Handbook of Human Factors & Ergonomics*, second edition, G. Salvendy (Ed.), Wiley, 1997.

Shams, M., Shojaeizadeh, D., Majdzadeh, R., Rashidian, A. and Montazeri, A. 2011. Taxi drivers' views on risky driving behaviour in Tehran: A qualitative study using a social marketing approach. *Accident Analysis and Prevention 43 (2011), 646-651.*

Sheridan, T. B. 2002. *Humans and Automation: System Design and Research Issues*. John Wiley & Sons, Inc., Publication.

Stanton N. A. and Young M. S. 2000. A proposed psychological model of driving automation. *Theor. Issues in Ergonomics Science*, 2000 Vol. (1) No, 4, 315-331.

Strand, N., Nilsson, J., Karlsson I. C. M. and Nilsson L. 2010. Exploring end-user experiences: self-perceived notions on use of adaptive cruise control systems. *IET Intelligent Transport Systems*, Vol 5, Iss. 2, 134-140.

Tango, F. and Montanari, R. 2006. Shaping the drivers' interaction: how the new vehicles systems match the technological requirements and the human needs. *Cogn Tech Work* (2006) 8: 215–226.

Tijerina L., Garrott W. R., Stoltzfus D. and Parmer E. 2005. Eye glance behavior of van and passenger car drivers during lane change decision phase. *Transp Res Rec.* 2005 1937: 37–43.

Van Driel, C. and Van Arem, B. 2005. Investigation of user needs for driver assistance: results of an Internet questionnaire. *European Journal of Transport and Infrastructure Research*, 5(4), 297-316.

Wang, J. and Knipling R. R. 1994. Lane change/merge crashes: Problem size assessment and statistical description, Final Report, DOT HS 808075, Washington, DC: National Highway Transportation Safety Administration, 1994.

Wickens, C. D., Lee, J., Liu, Y. and Becker, S. G. 2004. *An Introduction to Human Factors Engineering, Second Edition*. Pearson Education, Inc.

Young, K. L. and Regan, M. A. 2007. Use of manual speed alerting and cruise control devices by car drivers. *Safety Science*, 45 (2007), 473-485.

CHAPTER 8

Speed vs. Acceleration Advice for Advisory Cooperative Driving

Qonita Shahab, Jacques Terken

Eindhoven University of Technology
Eindhoven, The Netherlands
{q.m.shahab, j.m.b.terken}@tue.nl

ABSTRACT

Cooperation among vehicles is aimed to create a smooth traffic flow and minimize shockwaves and traffic jams. Cooperative adaptive cruise control (C-ACC) systems calculate acceleration values and exchange them between vehicles to maintain appropriate speed and headway. Before C-ACC technology gets mature, cooperative driving may already be made possible by advisory systems, keeping the drivers in the loop. While C-ACC systems are based on acceleration values, in conventional vehicles one of the main sources of information to the driver for maintaining appropriate speed is the speedometer. In this paper we present a study addressing the question of whether advisory systems should employ acceleration or speed values to advise the driver. Subjective results showed that preferences were approximately equally split between both systems. Objective results showed that acceleration advice caused more uniform speed in heavy traffic and more stable distance keeping, that speed advice led to more efficient accelerator pedal changes, and that letting drivers use their preferred advice resulted in a shorter time headway leading to a more effective traffic flow.

Keywords: cooperative driving, advisory system, driver behavior

1   INTRODUCTION

Cooperative Adaptive Cruise Control (C-ACC) systems extend existing cruise control systems by enabling wireless communication between vehicles, and between vehicles and the infrastructure, for the purpose of cooperative driving. The aim is to create a smooth traffic flow and minimize shockwaves and traffic jams. Algorithms for C-ACC calculate the acceleration/deceleration needed to optimize speed and distance. Since the technology for autonomous C-ACC is not yet mature, advisory systems have been explored i.e. keeping drivers in the loop, thus enabling faster market penetration. A field test demonstrated the promising effect of using an advisory C-ACC system in order to achieve better traffic flow (van den Broek, Netten, Hoedemaeker, & Ploeg, 2010). This field test adapted a C-ACC algorithm (van den Broek, Ploeg, & Netten, 2011) for advising drivers about the desired acceleration/deceleration in order to adjust their vehicles to the traffic.

In conventional vehicles, acceleration and deceleration are only the means by which the driver maintains appropriate speed and distance, and one of the main sources of information about speed is provided by the speedometer. In order to match the mental model of the drivers, we designed an advisory system for cooperative driving that provides drivers with speed target information, called Cooperative Speed Assistance (CSA). This raised the question whether advisory C-ACC systems should inform the driver about acceleration (as generated by C-ACC algorithms) or about desired speed.

We conducted an experiment to investigate whether speed information or acceleration information is preferred by drivers and which type of advice information is more effective for traffic flow. In this paper, we discuss the experiment setup in Section 2; the experiment results in Section 3; and discuss the results and state the conclusion in Section 4.

2　　EXPERIMENT SETUP

We developed two different interfaces (one for Acceleration advice and the other for Speed advice) for the purpose of the experiment. The experiment was conducted in a medium-fidelity driving simulator.

2.1　　System Design

Although the CSA interface gives either Acceleration or Speed advice, it takes into account the time headway (distance in seconds) of the driver's vehicle to the preceding vehicle. The interface only gives an advice whenever the driver is less than 6.5s distance from the preceding vehicle, i.e. 6.5s distance is when the interface tells the driver that there is a platoon ahead to follow.

In the Acceleration interface, a simple predictive feed-forward control algorithm is used. It takes into account acceleration, speed, and time headway values in order to create an acceleration target of -1.0 (full brake) to 1.0 (full throttle). The acceleration advice thus guides drivers to achieve the advised time headway of the platoon, which is 1.2s based on the average time headway value obtained from a previous field trial (van den Broek et al., 2010). The following set of formulas describes the algorithm, where $v1$ = driver's vehicle speed, $v2$ = preceding vehicle speed, a1 = driver's vehicle acceleration, a2 = preceding vehicle acceleration, D = current distance between driver's vehicle and preceding vehicle in meters.

(1) Doptimal = platoonHeadway * v1
(2) predictedDoptimal = platoonHeadway * (v1+a1*dt)
(3) predictedD = D + (v2-v1)*dt + 0.5*(a2-a1)*dt^2
(4) dD = 0.1 * Doptimal, where 0.1 is a hysteresis value
(5) aTarget = (predictedD − predictedDoptimal) / (predictedDoptimal)
(6) if (D < Doptimal − dD), then [Too Fast, shows aTarget]
(7) else if (D > Doptimal + dD, then [Too Slow, shows aTarget]

The algorithm finds Doptimal first as the advised distance in meters based on platoon headway and the speed of the driver's vehicle. Then it calculates predictedDoptimal, which is the Doptimal in the next time frame (dt = time slice). PredictedD is then calculated by taking into account speed and acceleration of both driver's vehicle and preceding vehicle to obtain the D in the next time frame (dt = time slice). dD is the difference between predictedD and D, which is calculated with 0.1 hysteresis value so the driver is allowed 10% error in achieving the predictedD. The aTarget (target acceleration) is calculated based on predictedD and the predictedDoptimal, and then comparing with the D the driver receives information that s/he is driving Too Fast or Too Slow, and an advice how much to accelerate or decelerate.

In the Speed interface, the system compares the driver's vehicle speed and preceding vehicle speed to provide the driver with information that s/he is driving Too Slow or Too Fast and an advice about the desired target speed. The advised time headway is taken into account, i.e. Too Slow condition is only informed when the driver maintains more than 1.2s distance, and Too Fast condition is only informed when the driver maintains less than 1.2s distance.

2.2   Interface Design

Two interfaces were created, providing information about the desired acceleration and speed, respectively (see Figures 1 and 2). Both interfaces employ the same background color scheme, creating a glanceable visual display (Matthews, 2006). A black background indicates the 'Appropriate' condition, i.e. the driver is not driving too fast or too slow, or there is no platoon detected ahead (when the time headway is larger than 6.5s). Red indicates the 'Too Fast' condition, i.e. the driver has to slow down. White indicates the 'Too Slow' condition, i.e. the driver has to speed up. The pie-like visualization shows slices to the right if the driver drives too fast, and slices to the left if the driver drives too slow. The number of slices indicates the size of the difference between the current speed/acceleration and the target speed/acceleration. The visual design of the Acceleration interface is illustrated in Figure 1, and the visual design of the Speed interface is illustrated in Figure 2.

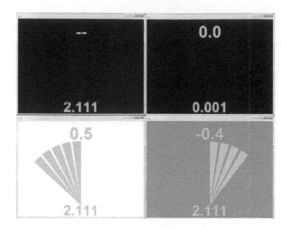

Figure 1 The Acceleration Interface. The number of pie slices shows the amount of acceleration i.e. 10% of full deceleration/acceleration per slice. The number above the pie ranges from -0.1 to 1.0, where -1.0 is a deceleration advice of 100% strength, and 1.0 is a full acceleration advice of 100% strength. The bottom number shows the current acceleration in m/s^2. The top left panel illustrates the condition of no preceding vehicle detected with current acceleration 2.111 m/s^2, the top right panel illustrates the Appropriate condition with current acceleration 0.001 m/s^2. The bottom left panel illustrates the Too Slow condition with a target acceleration of 0.5 (50% of full scale acceleration). The bottom right panel illustrates the Too Fast condition with a target acceleration of -0.4 (40% of full scale deceleration).

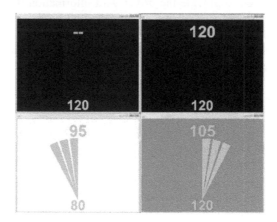

Figure 2 The Speed Interface. The number of pie slices shows the difference between the current speed and the target speed, i.e. 5 km/h per slice. The number above the pie shows the speed advice i.e. the speed target. The bottom number shows the current speed in km/h. The top left panel illustrates the condition of no preceding vehicle detected with current speed 120 km/h. The top right panel illustrates the Appropriate condition with current speed 120 km/h. The bottom left panel illustrates the Too Slow condition with a target speed of 95 km/h (15 km/h to increase). The bottom right panel illustrates the Too Fast condition with a target speed of 105 km/h (15 km/h to decrease).

The CSA interface also provides an auditory distance warning, which beeps whenever the driver is too close to the preceding vehicle. It consists of a burst of two pulses of 1600Hz each. The first pulse lasts 50ms, the second one 125ms, separated by an inter-pulse interval of 25ms. This warning sound was designed following a guideline on sound design (Edworthy, Loxley, & Dennis, 1991). The burst is displayed when the driver's vehicle is too close to the preceding vehicle, i.e. less than 0.5s time headway. It is displayed again after two seconds if the driver does not slow down.

2.3   Driving Simulator Test

The CSA interface was developed using Java language showing an application of 640x480 pixel size displayed on a 7-inch screen. The application was connected to the driving simulator in order to exchange real-time network messages every 50ms. The network message consists of speed, acceleration, time headway, preceding vehicle's speed and acceleration, brake and throttle pedal values, etc.

For the simulations we developed two highway scenarios: one with an Easy platoon (small fluctuations of the platoon's speed) and one with a more demanding (Hard) platoon (large fluctuations of the platoon's speed). The Easy platoon had a random fluctuation between 100 and 110km/h, and the Hard platoon had a predefined fluctuation of 120 km/h, 60 km/h, 105 km/h, 60 km/h, 105 km/h, which was adapted from the speed profile used in the field trial mentioned previously (van den Broek et al., 2010).

Twenty nine drivers (7 female, 22 male, age 20-38) with minimum 1.5 years driving experience participated in the experiments. The participants were required to drive with one interface first, and another interface later. Order of interfaces was balanced across participants, i.e. half of the participants drove with the Acceleration interface first, and the other half with the Speed interface first. For each interface, they first drove in the Easy platoon and then in the Hard platoon. They were asked to follow a platoon with the assistance of the interface. In total, each participant drove four 7-9 minute periods.

After each period of driving, each participant rated their mental effort based on the Rating Scale Mental Effort (RSME) (Zijlstra & van Doorn, 1985) of 0 (absolutely no effort) to 150 (more than extreme effort). After each usage of the interface, each participant rated the interface using Van Der Laan Acceptance scale (Van Der Laan, Heino, & De Waard, 1997), which consists of nine 5-point Likert-type scales: I find the system Useful-Useless, Pleasant-Unpleasant, Good-Bad, Nice-Annoying, Effective-Superfluous, Likeable-Irritating, Assisting-Worthless, Desirable-Undesirable, Raising alertness-Sleep inducing. At the end of the experiment, we interviewed participants to obtain a preference of interface and discuss the reasons and the difference between the interfaces. Participants received a small fee based on 40 minutes participation.

3   RESULTS

3.1   Subjective Results

The interview results showed that 18 participants preferred the Speed interface and 11 participants preferred the Acceleration interface. Participants preferring the Speed interface indicated that they considered it to be calmer, create less panic, and offer more freedom to control the vehicle, and that the rate of change of the information was lower than in the Acceleration interface. Participants preferring the Acceleration interface indicated that they considered the information to be more precise and the Speed interface to be less safe.

Participants also commented about the colors, pie-like visualization and the numbers on the interface. They mostly liked the color changes for the noticeability. They mostly agreed that the current acceleration information was meaningless compared to the current speed information. The target speed number was considered useful by those who preferred the Speed interface, and the graphical acceleration information was considered useful by those who preferred the Acceleration interface. Even though the preciseness (10% per pie slice) of the pie-like visualization was considered quite helpful, participants could not estimate the exact amount of acceleration required by the system, and they mentioned that some practice would be needed to get used to it.

Based on the RSME results, the Acceleration (Ac) interface was rated as requiring more mental effort than the Speed (Sp) interface. Multivariate tests showed an effect of Interface ($F_{1,28} = 5.591$, p= .025) and no Platoon effect or interaction. Both in the Easy and Hard platoons, Ac was more demanding than Sp (mean for Ac = 48.64, for Sp = 39.47 out of 150).

Based on the Van Der Laan scale ratings, generally all participants regarded both interfaces as somewhat unlikely to induce sleep (mean Ac = 2.00, SD = 0.93 and mean Sp = 2.21, SD = 0.90 out of 5.0 scale). Multivariate tests of each Van Der Laan item x Preference showed an interaction effect, except for Good-Bad and Likeable-Irritating. The means are listed in Table 2 below.

Table 2 Acceptance scores (5 points) as a function of preference

Item	Ac preference		Sp preference		Interaction effect
	Ac	Sp	Ac	Sp	
Useful	4.36	3.55	3.72	4.06	$F=5.131$, $df_1=1$, $df_2=27$, $p=.032$
Effective	4.18	3.45	3.78	4.17	$F=7.54$, $df_1=1$, $df_2=27$, $p=.011$
Assisting	4.27	3.73	3.89	4.28	$F=5.959$, $df_1=1$, $df_2=27$, $p=.021$
Desirable	3.45	3.36	3.11	3.94	$F=4.713$, $df_1=1$, $df_2=27$, $p=.039$
Unpleasant	2.09	2.73	3.11	2.11	$F=10.168$, $df_1=1$, $df_2=27$, $p=.004$
Annoying	2.18	2.73	3.00	2.17	$F=6.841$, $df_1=1$, $df_2=27$, $p=.014$
Good	4.09	3.91	3.56	3.94	$F=2.007$, $df_1=1$, $df_2=27$, $p=.168$
Irritating	2.64	2.73	2.72	2.22	$F=2.75$, $df_1=1$, $df_2=27$, $p=.109$

3.2   Objective Results

Speed. Analysis of Variance indicated that average speed was smaller in the Easy platoons than in the Hard platoons ($F_{1,28}$=2367.722, p=.000), as it was intended by the speed profiles. There was an Interface x Platoon interaction ($F_{1,28}$=8.732, p=.006), showing that in the Hard platoons average speed was higher using the Speed interface (mean=83.05 km/h) compared to using the Acceleration interface (mean=81.34 km/h). The average speed in the Easy platoons was 101.43 km/h. The higher average speed in the Hard platoon for the Speed interface may be due to more overshoot (less precision) compared to using the Acceleration interface. In order to remove the precision adaptation, the higher frequency data (high fluctuation of speed) were removed using frequency domain analysis (data of one participant had to be removed due to insufficient data for the computation). The difference between interface conditions in the Hard platoons still applied (t=-9.377, df=27, p=.000), i.e. higher average speed using the Speed interface (mean=82.96 km/h) compared to using the Acceleration interface (mean=81.05 km/h).

Multivariate analysis of variance indicated that there was also an interaction effect between Interface and Preference ($F_{1,27}$=7.048, p=.013). Participants who preferred the Acceleration interface did not drive differently using Acceleration and Speed interfaces, but participants who preferred the Speed interface drove faster (t=-2.887, df=27, p=.008) while using the Speed interface (mean=92.73 km/h) than while using the Acceleration interface (mean=91.17 km/h).

The standard deviation of the speed was also different between platoons, because the platoons were different as intended, i.e. there were more fluctuations of speed in the Hard platoon than in the Easy platoon ($F_{1,28}$=576.747, p=.000). There was an interaction effect between Interface and Platoon ($F_{1,28}$=10.463, p=.003), showing that in the Hard platoons average standard deviation was higher when using the Speed interface (mean=16.15 km/h) than when using the Acceleration interface (mean=14.91 km/h). The average standard deviation was lower in Easy platoons (mean=5.71 km/h), with no differences between interfaces.

Time Headway. Apart from variation of speed, distance to preceding vehicle also provides information about stability of a platoon. Time headway is a preferred measure for distance to preceding vehicle, because distance in meters varies depending on the vehicle speed, thus time headway provides more consistent information. Although the advised time headway was 1.2s, the average time headway maintained by participants throughout the experiment was larger than 1.2s. Participants maintained 1.41s average time headway with no effect of platoons or interfaces. However, Multivariate analysis of variance showed that there was an interaction effect between interfaces and participant's preference for Ac or Sp interface ($F_{1,27}$=4.894, p=.036). Participants who preferred the Acceleration interface maintained a shorter time headway while driving using the Acceleration interface (mean=1.44s) compared to driving with the Speed interface (mean=1.62s). Participants who preferred the Speed interface maintained a shorter time headway while driving using the Speed interface (mean=1.28s) compared to driving with the Acceleration interface (mean=1.39s). This shows that they maintained a shorter time headway with their preferred interface, as seen in Figure 3.

with the Acceleration interface (mean=1.39s). This shows that they maintained a shorter time headway with their preferred interface, as seen in Figure 3.

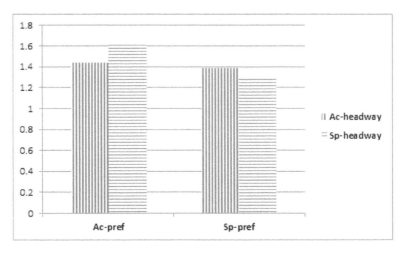

Figure 3 The interaction effect between preference and average time headway. Those with Ac preference maintained shorter time headway while driving with Ac, and those with Sp preference maintained shorter time headway while driving with Sp.

Standard deviation of time headway was measured in order to find out the precision in distance keeping. Both in Easy and Hard platoons, average standard deviation of time headway was smaller ($F_{1,28}$=12.43, p=.001) while driving using the Acceleration interface (mean deviation = 0.65s) compared to driving with the Speed interface (mean deviation = 0.77s).

Accelerator Pedal Analysis. Measurement of accelerator pedal movements provides information about the efficiency of throttle changes i.e. deeper and more frequent changes are considered less efficient. In the experiment, the recorded throttle data consisted of values ranging from 0 (no pressure on the acceleration pedal) to 1.0 (full pressure on the acceleration pedal). In order to analyze the frequency of the throttle changes, a Fast Fourier Transform (FFT) was applied for frequency domain analysis. The output was a plot of frequencies against amplitude (components of throttle depth) of each frequency.

Multivariate tests were done on the range of 0.0-0.1 Hz, because those frequency data have visible peaks on the FFT plot, as seen in Figure 4 below. The frequency of 0.1 Hz means a change of throttle value every 10 seconds. Based on Multivariate tests, using the Acceleration interface resulted in larger throttle changes compared to using the Speed interface, both in easy and hard platoons ($F_{1,25}$=9.637, p=.005). There was an interaction effect between Interface and Platoon ($F_{1,25}$=5.255, p=.031), indicating that the difference in the throttle changes was larger in Hard platoons (t=3.034, df=25, p=.006) than in Easy platoons (t=2.128, df=25, p=.043).

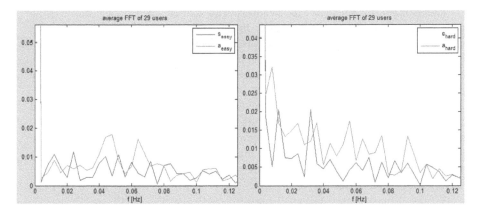

Figure 4 The 5000-point FFT computation of throttle data of each user averaged and plotted. X-axis = the frequencies (0 to 0.12 Hz), Y-axis = the amplitude (components of throttle depth). The left image shows the data from the Easy platoon, and the right image shows the data from the Hard platoon. The red line represents the Acceleration interface, and the blue line represents the Speed interface. It is visible that the red line has higher peaks than the blue line.

4   DISCUSSION AND CONCLUSION

We conducted an experiment for the purpose of cooperative driving to investigate subjective judgments and performance effects of acceleration and speed advice in Easy and Hard traffic conditions. We found that acceleration advice leads to more mental effort, as rated by the participants. This is supported by the average speed data, showing that participants maintained lower speed with the acceleration advice while driving in Hard platoons.

The acceleration advice resulted in a more uniform speed among drivers in the case of Hard platoons, shown by the lower standard deviation in speed compared to driving with the speed advice. Moreover, the standard deviation in time headway was also lower compared to driving with the speed advice. These findings show that acceleration advice may lead to fewer shockwaves due to less deviation in speed and time headway. However, driving with the acceleration advice required larger changes of the throttle pedal, both in Easy and Hard platoons. This indicates that drivers have problems in adjusting their speed precisely according to the acceleration advice. This means that acceleration advice may also lead to less efficient fuel consumption.

There was no clear preference for one type of advice, so we conclude that there are different types of drivers: those who prefer acceleration advice, allowing more precise control, and those who like speed advice, allowing more freedom in the implementation of the advice. From the objective data, the smaller standard deviation of time headway using the Acceleration interface indicates precision, and the higher standard deviation of time headway using the Speed interface indicates freedom. This is further supported by the fact that the participants who preferred the Speed interface drove faster in both types of platoons while using the Speed interface.

Interestingly, the average time headway was shorter when people drove using their preferred interfaces. Since shorter time headway is useful for better traffic throughput (van Arem, van Driel, & Visser, 2006), we argue that the usage of preferred interface may result in better traffic throughput. Practically this may be made possible by allowing drivers to choose between acceleration and speed advice in such an assistance system.

In conclusion, both types of advice have their own advantages and disadvantages. While the speed advice takes less mental effort and more efficient acceleration pedal changes, the acceleration advice is more useful for reducing the likelihood of traffic shockwaves. Looking at the differences among people, we argue that people can adapt to their preferred advice and have more confidence in driving with a shorter headway.

ACKNOWLEDGMENTS

The authors would like to acknowledge Thijs van den Broek of TNO Automotive for sharing the speed profile used in the A270 Field Trial (van den Broek et al., 2010). We also thank Jan Rouvroye of Eindhoven University of Technology for providing the algorithm of the Acceleration interface. This research was made possible through a grant from The Dutch Ministry of Economic Affairs to the HTAS project Connect & Drive.

REFERENCES

Edworthy, J., Loxley, S., & Dennis, I. (1991). Improving Auditory Warning Design: Relationship between Warning Sound Parameters and Perceived Urgency. *Human Factors: The Journal of the Human Factors and Ergonomics Society*, *33*, 205-231.

Matthews, T. (2006). Designing and evaluating glanceable peripheral displays. *Proceedings of the 6th conference on Designing Interactive Systems* (pp. 343–345). ACM.

Van Der Laan, J. D., Heino, A., & De Waard, D. (1997). A simple procedure for the assessment of acceptance of advanced transport telematics. *Transportation Research Part C: Emerging Technologies*, *5*(1), 1–10. Elsevier.

Zijlstra, F., & van Doorn, L. (1985). The construction of a scale to measure subjective effort. *Technical Report*. Delft University of Technology.

van Arem, B., van Driel, C. J. G., & Visser, R. (2006). The Impact of Cooperative Adaptive Cruise Control on Traffic-Flow Characteristics. *IEEE Transactions on Intelligent Transportation Systems*, *7*(4), 429-436.

van den Broek, T. H. A., Netten, B. D., Hoedemaeker, M., & Ploeg, J. (2010). The experimental setup of a large field operational test for cooperative driving vehicles at the A270. *13th International IEEE Conference on Intelligent Transportation Systems (ITSC)*.

van den Broek, T. H. A., Ploeg, J., & Netten, B. D. (2011). Advisory and autonomous cooperative driving systems. *Consumer Electronics (ICCE), 2011 IEEE International Conference on* (pp. 279–280). IEEE.

CHAPTER 9

LDW or Rumble Strips in Unintentional Lane Departures: Driver Acceptance and Performance

Lars Eriksson[1], Anne Bolling[1], Torbjörn Alm[2], Anders Andersson[1], Christer Ahlström[1], Björn Blissing[1] and Göran Nilsson[3]

[1]VTI, Swedish National Road and Transport Research Institute
Linköping, Sweden
lars.eriksson@vti.se
[2]HiQ
Linköping, Sweden
[3]Swedish Road Marking Association
Sweden

ABSTRACT

Using induced unintentional lane departure (ULD) in an advanced driving simulator, we investigated car drivers' acceptance of rumble strips and lane departure warning system (LDW), respectively, and driving performance with each of these warning types. Twenty-four participants drove with simulated rumble strips in one trial and with a simulated LDW in another trial. A forced yaw motion of the vehicle induced ULDs while the driver attended jammed music coming from a CD-player. Each drive took about 25 min to complete and was set up to include 13 events of ULD coupled with jammed music in which the driver had to immediately change point of gaze from the road to take care of the CD-player. The results show the drivers were more satisfied with the LDW, trusted the rumble strips more, and overall preferred the warning types about equally. All drivers considered it valuable to have assistance in ULD, either by the rumble strips or the LDW, and several (i.e. 25%) chose to have both types of warning in parallel. Response completion was faster with the rumble strips warning, but no difference was found in time to back in lane and lane exceedence area, respectively. Thus, although differences in driver acceptance and performance were found between using the LDW and the rumble strips there were no major overall differences. The clear preference for having a warning function further strengthens the positive opinion on the need for assistance systems in ULDs.

Keywords: lane departure warning, rumble strips, acceptance, performance

1   WARNING OF UNINTENTIONAL LANE DEPARTURE

Driving on rumble strips milled into the road at road markings generates vehicle vibration and sound that can warn for lane departure. For highways, it has been estimated that rumble strips at centerline have reduced accidents by about 15%, and that rumble strips on shoulder have effected an approximate 40-50% accident reduction (e.g. Anund et al., 2008; Mahoney et al., 2003; Persaud et al., 2004). Overall, it seems a sleepy or distracted driver can be significantly supported simply by the presentation of real-time warning for the current unintentional lane departure (ULD). Rumble strips and/or an in-vehicle lane departure warning system (LDW) can deliver such a warning. The vehicle can host a system that registers lane departure by using readings of road marking positions relative to the car and indicates lane departure by delivering, for example, a sound in the car cabin.

A likely LDW advantage is that the warning of an LDW can be suppressed by activation of the direction indicator prior to the exiting of the driving lane, as opposed to rumble strips that warn even in intentional lane departures. An LDW also normally includes other criteria for suppression of warning, such as speeds lower than a predefined value or high lateral and longitudinal accelerations that indicate active driver inputs. A possible rumble strips advantage may be that they can be perceived as generating a more aggressive or annoying warning that more promptly alert a sleepy or distracted driver. Important is that although an LDW warning can be adjusted to be similarly aggressive or annoying, the rumble strips are not dependent on vehicle technology in the same manner as an LDW.

In the present study, we investigated whether drivers more readily accept either rumble strips or an LDW from driving with induced ULDs in a driving simulator. We also investigated whether there are any noticeable differences in driving performance between using the two warning types.

2   METHODS

2.1   Participants

Eleven women and 13 men participated as drivers. The recruitment criteria included holding a driving license for at least 5 years, not to be a professional driver, and having a driving mileage of at least 5000 km per year. The participants had a mean age of 39 years, a mean time of 20 years of holding the driving license, and drove an estimated mean of 14350 km per year.

2.2   Apparatus

The VTI driving simulator III was used, which has a moving base with a high performance motion-base system (see Figure 1). It is mainly used for passenger vehicles and is the first-hand choice when, for example, realistic road sensations are prioritized. The forces generated by the motion-base system involve linear motion, tilt motion and vibration. The linear motion was used for the car's lateral motions, and the tilt motion was used for the roll and pitch motions to simulate sustained accelerations such as in curve driving. A vibration table provides additional capabilities to generate road roughness for higher frequencies.

Figure 1 The VTI driving simulator III with the moving base platform and a car cabin mounted on it.

A Saab 9-3 car cabin was used in all driving sessions. Three DLP projectors provide a 120° forward field of view with a spatial resolution of 1280 × 1024 pixels, and edge blending and geometrical correction are provided by a dedicated graphics card. Three LCD displays are used for the visual presentations in the rear view mirrors. Two loudspeakers are placed close to the windshield in the dashboard, one in each front door, and one in the rear of the cabin. A ButtKicker low-frequency transducer beneath the cabin was used to generate vibrations when driving over rumble strips. A touch screen with a spatial resolution of 800 × 600 presented the radio/CD-player control panel.

2.3   Overall Simulator Scenario

The overall scenario was a car drive performed in twilight on a road of varying straight stretches and curves with an LDW-equipped car or with rumble strips at the road's edge lines and centerline. The driver tasks were to drive normally at 90 km/h, keep both hands on the steering wheel and stay in lane while listening to music coming from the CD-player. At several occasions during the drive, the CD-player started to function incorrectly by jamming of the played music. At these moments, the driver's primary concern was immediately to take care of this by pushing the CD-player button for next tune. The intention was to make the driver change point of gaze from the road to the CD-player. To induce ULDs in each drive, forced yaw motions of the car were induced from outside the driver-vehicle control loop. This yaw motion started 0.2 s after the jamming of the music, and it could not be sensed through the steering wheel and only marginally or not at all through the rest of the vehicle. If the driver kept point of gaze on the road, the induced motion shown in the visual presentation was easily nulled by naturally quick and short steering movements. Each drive was setup to include several occasions of jammed music, of which most but not all were coupled with forced yaw motions together with warnings according to the LDW or the rumble strips. Each drive also included events of overtaking a car parked to the right slightly off the driving lane. No other vehicles or road-users were included in the scenario.

2.4   Design and Stimuli

The experiment had a within-subjects design with the two treatment levels of driving with the simulated rumble strips and the simulated LDW system. Half of the participants started driving with the LDW and the other half with the rumble strips. The within-subjects design with the two treatment levels was analyzed as a mixed design in some of the statistical analyses to test for any possible order (or group) effects. Thus, in the mixed design analyses the grouping factor was which type of warning the participants started with.

Twilight conditions were simulated and a simplified night driving mode was used that included only the own car headlights. The normal image as it would look during daylight conditions (i.e. the color buffer) is shown to the left in Figure 2, and on the right-hand side the final twilight scene is shown. (The presented virtual world in the simulator is of course more realistic and compelling than Figure 2 indicates.)

Figure 2   To the left is the scene showing the road in daylight conditions and to the right is the final scene showing the road in twilight conditions.

The simulated road was 8 m wide with each driving lane and shoulder 3 m and 1 m wide, respectively. To simulate rumble strips a sound was played when each wheel edge encountered the rumble strips. The sound played was recorded from driving a car at 90 km/h over the "Målilla" rumble strip wide (see further specification of the rumble strips in Anund et al., 2008). Measured sound level from the rumble strips at the driving position was approximately 77 dB(A). The rumble strips sound signal was also sent to the ButtKicker to initiate vibrations. Each rumble strip at the simulated road was 0.35 m. The middle rumble strips were centered on the road and the shoulder rumble strips were placed at a distance of 2.95 m from the center of the road.

A VTI LDW algorithm simulated the LDW functionality of an operational Volvo system. The algorithm outputs three separate alarm signals for each side, which are visual, auditory, and haptic alarms. Only the auditory alarm was used. The alarm is set to true if the system is in the active state and one or both of the following criterions are fulfilled: (i) The distance between the side of the vehicle and the road marking is smaller than a tuneable threshold value, and/or (ii) the time-to-line crossing is smaller than a tuneable threshold value. The LDW system is disabled when, for example, the turn signal indicator light is switched on or the vehicle speed is below a predefined threshold. After an alarm, the system will enter a post-alarm state and remain there for a predefined time. In this state, no further alarms can be issued. The LDW sound level was 74 dB(A).

Combined with the CD-player jamming a forced yaw motion was induced 0.2 s after jamming onset. The yaw was added according to:

$$\Psi(t) = \Psi_{crit} * \sin^2\left(\frac{t}{T_{distract}}\frac{\pi}{2}\right), where\ 0 \leq t \leq T_{distract}$$

in which $\Psi(t)$ is the yaw added at time t, Ψ_{crit} is the final angle and $T_{distract}$ is the time it takes for the yaw to finish. This motion was an added yaw to the vehicle and the extra motion is not noted in the steering wheel but will be noted in the motion system if the lateral acceleration motion becomes too large. The parameters used were Ψ_{crit} = 3.3° and $T_{distract}$ = 3.0 s. This was considered to induce a large enough yaw motion while normally not noticeable. For example, the induced lateral motion with 3 s yaw motion at constant speed of 80 and 90 km/h is 1.92 m and 2.16 m, respectively.

In each drive, there were 13 induced yaw motions of which one was coupled with no warning. This missing warning in each drive was in the LDW condition intended to reflect the LDW's inability to read road markings 100% of the time. In the rumble strips condition, the missing warning was intended to reflect that rumble strips are not allowed in Sweden when housing area is closer than 150 m from the road. The 12 induced yaw motions coupled with warnings of each drive were distributed with four yaw motions for each of three types of road segment comprising straight road, curve to the left, and curve to the right. For each type of curve, two yaw motions were to the left and two to the right. For the straight road, three yaw motions were to the left and one to the right.

The CD-player was a program running on a small computer in the car cabin of the simulator. A 7'' touch screen presented the CD-player and the sound was played in the touch-screen speakers. The placement of the radio was at a low right position of the middle console and approximately 10 cm above the seat edge. A total of 19 events of jammed CD-player occurred in each drive in which there were six jammed events without the car's lane positioning being perturbed by the induced yaw motion.

In summary, each drive was set up to include:

- 19 events of jammed music of which 6 without induced yaw motion
- 13 events of induced yaw motion coupled with jammed music
- 12 of the 13 events of induced yaw motion with warning of lane departure
  - 4 at straight road: 3 left and 1 right yaw motions
  - 4 at curve to the left: 2 left and 2 right yaw motions
  - 4 at curve to the right: 2 left and 2 right yaw motions
- 6 events of parked car overtake

2.5   Questionnaires

Questionnaires were administered before and after each drive. One was first introduced concerning background information about year of acquiring driving license, opinions on the two types of warning for lane departure etcetera. One was also administered after each of the two driving sessions, with the final questionnaire after the second drive including comparisons of the warning types.

2.6   Performance Indicators

The data used for analyses included the performance indicators response completion time, time to back in lane, and lane exceedence area.

Response completion time was the time from warning to the completion of response defined as a steering wheel reversal with a local maxima or minima in the steering wheel angle signal larger than an absolute value of 6°. A fifth order Butterworth filter with a cut-off frequency of 0.75 Hz was applied to remove small corrective steering wheel maneuvers.

Time to back in lane was measured from time of warning of lane departure to the moment of having all four wheels of the vehicle back within the driving lane.

Lane exceedence area was determined as the integral over the lateral position when the vehicle was outside the lane. The interval was chosen from the time when the vehicle left the lane and 10 s onwards.

2.7   Procedure

The participant first read a written description of the driving scenario and the tasks to be performed. After having signed the informed consent form, the questionnaire about background information was completed. The participant was then seated in the car cabin of the driving simulator and verbally instructed about the driving simulator, overall scenario conditions and the driver tasks. A short driving session was completed with the purpose of making the participant accustomed to the driving simulator and familiarized with the driving and the CD-player handling tasks. The experiment started and the participant drove the two road distances of which one was made with the rumble strips and the other with the

LDW. At the very beginning of each road distance, the participant made several tests of the warning type in use by exceeding the driving lane both to the left across centerline and to the right across shoulder road marking. If the participant waited on two consecutive occasions with redirecting his or her point of gaze to the CD-player when the music was jammed the experimental leader intervened the driving by commenting on this. That is, if disregarding the task to take care of the dysfunctional CD-player immediately on two occasions in a row, which in effect made it easy to null the induced yaw motion of the vehicle, the instruction for the CD-player handling task was repeated and emphasized. When lane departure occurred, the task was to directly correct for this and then continue to drive normally at 90 km/h, keep both hands on the steering wheel and stay in the proper driving lane. Each road distance took about 25 minutes to complete, after which the simulator was stopped and parked. After driving the first road distance, the participant was accompanied out of the driving simulator for a pause outside and for answering the questionnaire about the drive and the type of warning just used. After driving the second road distance, and having completed the questionnaire about the just used warning type, questions about comparisons of LDW and rumble strips were answered. Total time for each experimental session was about 1.5 hours.

3   RESULTS

3.1   Driver Assessments

Trust Trust in the functioning of rumble strips and LDW, respectively, pertains to the posed questions: *How much trust do you have in the functioning of such rumble strips? How much trust do you have in the functioning of such warning system (LDW)?* The responses were made on a 7-point scale of trust ranging from 1 = very little trust to 7 = very much trust. An analysis of variance (ANOVA) of trust ratings showed significant main effects of warning exposure, $F(1, 23) = 12.98$, $p< .01$, and warning type, $F(1, 23) = 19.08$, $p< .001$, with no significant interaction effect. The participants expressed more trust overall in warning functioning after being exposed to warning events in the driving sessions, 5.17 (SE = 0.26), than before, 5.81 (0.24). They also considered the functioning of the rumble strips overall more trustworthy, 5.77 (0.25), than that of the LDW, 5.21 (0.22).

The van der Laan acceptance scale for assessing the acceptance of technology is based on nine 5-point rating-scale items, of which the items load on two factors or scales (van der Laan, Heino, & de Waard, 1997). One scale indicates the usefulness of the technology and the other scale indicates the satisfaction rendered by the used technology. From the ratings of items of the 5-point rating scale scores are computed so that they range from 2 = very positive to -2 = very negative in terms of usefulness and satisfaction, respectively.

Usefulness An ANOVA showed no significant effects revealed by the usefulness scores, $F(1, 23) = 3.93$, $p= .06$. The mean rating of LDW before and after driving sessions was 1.23 (0.13) and 1.44 (0.09), respectively, and the mean rating of rumble strips before and after driving sessions was 1.30 (0.13) and 1.53 (0.10), respectively.

Satisfaction An ANOVA of the satisfaction scores showed significant main effects of warning exposure, $F(1, 23) = 7.54$, $p < .025$, and warning type, $F(1, 23) = 63.42$, $p < .0001$, with no significant interaction. The main effect of warning exposure shows the participants were overall more satisfied with warning of lane departure after being exposed to the warning events of the driving sessions, 0.82 (0.19), than before, 0.57 (0.16). The main effect of warning shows the participants were overall more satisfied with the LDW, 1.06 (0.16), than with the rumble strips, 0.33 (0.19).

Overall choice From the question "Which do you prefer?" the response frequencies of the possible responses "LDW", "rumble strips", "none", and "both" were 9, 8, 0, and 6, respectively. (One participant did not answer the question.)

3.2   Driver Performances

Data Because each drive included 12 planned potentially induced unintentional lane departures with warnings there were 24 planned possible responses to warnings for each driver. With 24 drivers, this amounts to 576 possible responses for each performance measure, corresponding to 288 possible responses for each type of warning. However, drivers nulled the forced yaw motion in many occasions and the total amount of responses to induced ULDs was 146. Response completion times shorter than 300 ms were considered too short to be complete responses to the delivered warnings and were excluded from the analysis. The data of the times to back in lane and lane exceedence area coupled to these too short response times were accordingly removed from the analyses of time to back in lane and lane exeedence area. After this, there were 140 responses for each measure. One missing value and one outlier value were replaced with the condition means. Thus, 141 data points were used for the analyses of performance. These 141 data points were distributed with 76 for the LDW drive and 65 for the rumble strips drive. Because of too few data values over the conditions of curves and directions of forced yaw motions, and the similar pattern of responses over these conditions with each warning type, the different conditions of ULD were collapsed into one condition.

Response completion time An ANOVA showed significant main effects of group, $F(1, 22) = 7.05$, $p < .025$, and warning type, $F(1, 22) = 9.71$, $p < .01$, with no significant interaction effect. The group starting with the LDW drive showed an overall longer mean response completion time. The group starting with the LDW drive had a mean response completion time of 0.739 s (0.037) and the other group a mean of 0.598 s (0.037). Irrespective of which warning type the participants started with they showed a longer mean response completion time driving with the LDW. The mean response completion time was 0.600 s (0.037) with the rumble strips and 0.737 s (0.032) with the LDW.

Time to back in lane An ANOVA showed no significant effects of the group and warning variables on time to back in lane.

Lane exceedence area An ANOVA showed a significant main effect of group, $F(1, 22) = 4.51$, $p < .05$, on lane exceedence area, with no other significant effects. The group driving first with the LDW exhibited an overall larger mean lane exceedence area, 0.58 m (0.06), than the other group, 0.41 m (0.06).

4   CONCLUSIONS AND DISCUSSION

While the drivers showed more satisfaction with the LDW, they also showed more trust in the rumble strips. The preferences of warning type revealed no clear overall choice in favor of the LDW or the rumble strips. There was, however, a clear preference for having the function of warning for ULD. Although response completion was faster with the rumble strips warning, no difference between the warning types was found in time to back in lane and lane exceedence area, respectively. Thus, as revealed by the drivers' assessments and performances from driving with a scenario including ULDs, the main conclusion of the present study is that no major overall differences between LDW and rumble strips were found.

Comments made by the participants are quite informative about what the motivations were for the preferences. Comments like "more pleasant" and "the sound not annoying" may indicate the preference for the LDW in terms of pleasantness/comfort. Comments like "it felt more serious" and "you awaken, get a faster reaction" may indicate the preference for the rumble strips in terms of efficiency in alerting. There may therefore be a reliance on two competing aspects of warning reflected in the preferences: (i) the aspect of pleasantness/comfort, and (ii) the aspect of alert efficiency. If efficiency in alerting is prioritized, the preference is in favor of the rumble strips, and if pleasantness/comfort is prioritized, the preference is in favor of the LDW.

Irrespective of with which warning type the participants started driving, the LDW induced overall longer mean response completion time. This thus indicates a quicker response completion from having rumble strips warning of ULD. As indicated above, this is also supported by the driver comments. Furthermore, there was an overall higher trust in the rumble strips. That there is no difference between the warning types in time to back in lane can be explained by the fact that the LDW warning generally is delivered somewhat earlier in a lane departure than the rumble strips warning. That is, with quicker response completion using the rumble strips, the times to back in lane are not different for the two warning types because of slightly earlier warning of the LDW. Although there was a significant group effect on lane exceedence area, more important is that there was no overall effect of warning type. Perhaps the drivers were more satisfied with the LDW because the LDW warning was considered more pleasant, inducing a somewhat calmer response, in conjunction with that no greater area was spent outside the driving lane compared with the rumble strips warning. (In addition, the times to back in lane were not different for the warning types.)

Our general conclusion is that the acceptance and performance levels were high for both warning types. All drivers considered it valuable to have assistance in cases of ULDs, whereas the kind of technical solution seems to be of less importance. Several participants (i.e. 6 out of 24) also thought it was good to have both solutions in parallel. This may reflect the driving scenario emphasized ULDs when looking away from the road, as opposed to when keeping eye and mind on the road (and getting nuisance alarms).

Still, rumble strips have the advantage of not being dependent on vehicle technology in the same manner as an LDW. On the other hand, an LDW has the potential advantage of using warnings tailored to driver and/or traffic states (e.g. Dong et al., 2011; Donmez, Ng Boyle, & Lee, 2007; Lee et al., 2011), with the added potential opportunity to use displays based on vibrotactile, multisensory, or motor-priming presentation (e.g. Navarro et al., 2010; Spence & Ho, 2008).

ACKNOWLEDGMENTS

We gratefully acknowledge the funding and otherwise support by the Swedish competence center Virtual Prototyping and Assessment by Simulation (ViP), and especially the ViP Director Lena Nilsson (VTI, Swedish National Road and Transport Research Institute) and the ViP Board for their efforts and support. For important project work, we especially acknowledge Jonas Ekmark (Volvo Car Corporation), Lars-Eric Svensson (Swedish Road Marking Association), Hans Holmén (Swedish Transport Administration), Johan Sjöstrand and Tobias Östlund (HiQ), and Mats Lidström, Laban Källgren, Kristina Kindgren, and Magnus Hjälmdahl (VTI).

REFERENCES

Anund, A., G. Kecklund, and A. Vadeby, et al. 2008. The alerting effect of hitting a rumble strip – A simulator study with sleepy drivers. *Accident Analysis and Prevention*, 40: 1970-1976.

Dong, Y., Z. Hu, and K. Uchimura, et al. 2011. Driver inattention monitoring system for intelligent vehicles: A review. *IEEE Transactions on Intelligent Transportation Systems*, 12(2): 596-614.

Donmez, B., L. Ng Boyle, and J. D. Lee. 2007. Safety implications of providing real-time feedback to distracted drivers. *Accident Analysis and Prevention,* 39: 581-590.

Lee, S. J., J. Jo, H. G. Jung, K. R. Park, and J. Kim. 2011. Real-time gaze estimator based on driver's head orientation for forward collision warning system. *IEEE Transactions on Intelligent Transportation Systems*, 12(1): 254-267.

Mahoney, K. M., R. J. Porter, and E. T. Donnell, E. T., et al. 2003. *Evaluation of centerline rumble strips on lateral vehicle placement and speed on two-lane rural highways* (PTI Report No. FHWA-PA-2002-034-97-04(111)). Harrisburg, USA: Pennsylvania Department of Transportation.

Navarro, J., F. Mars, J.-F. Forzy, M. El-Jaafari, and J.-M. Hoc. 2010. Objective and subjective evaluation of motor priming and warning systems applied to lateral control assistance. *Accident Analysis and Prevention*, 42: 904-912.

Persaud, B. N., R. A. Retting, and C. A. Lyon. 2004. Crash reduction following installation of centerline rumble strips on rural two-lane roads. *Accident Analysis and Prevention*, 36: 1073–1079.

Spence, C., and C. Ho. 2008. Tactile and multisensory spatial warning signals for drivers. *IEEE Transactions on Haptics*, 1(2): 121-129.

Van der Laan, J. D., A. Heino, and D. De Waard. 1997. A simple procedure for the assessment of acceptance of advanced transport telematics. *Transportation Research - Part C: Emerging Technologies*, 5: 1-10.

CHAPTER 10

Human Factors in a Compact Mobile Workspace: A case study of Patrol Vehicles

Martin Sepoori, John Hill

Michigan Technological University
Houghton, USA
mksepoor@mtu.edu

ABSTRACT

Human safety is of high priority in workspace design. The proximity of potentially hazardous objects is a threat in a compact workspace. It is even more challenging in a mobile workspace, owing to the inertial forces involved. A police car is one such example. Space restrictions due to airbag deployment, human trajectories in case of a collision and visibility issues and distractions limit the available space for in-car accessories such as laptop, radio, camera and radar controls. While safety may require that the equipment be moved farther away from the driver in case of a crash, more driver attention will be demanded for secondary tasks, leading to distraction. Hence it becomes necessary to find a balance between safety and ergonomics. In this present case study, far-side crashes to police vehicles are considered, which cause the driver to fall towards the in-car equipment. Sled tests were conducted on 50th and 95th percentile male dummies (Anthropomorphic Test Devices) seated in an SUV and a sedan, at 19.9mph (32kph), for 40° and 70° angles of impact. These trajectories are compared against the mounting locations of the equipment, and airbag deployment zones. 12 local police officers were surveyed regarding the human factor issues in the cockpit, using subject-based ratings. An analysis of this comparison and potential injury outcomes are discussed as well.

Keywords: police car crashes, sled tests, mobile workspace, automotive crash safety, ergonomics, far-side impact, driver injury

1   HUMAN FACTORS ISSUES IN POLICE VEHICLES

Police vehicles need a greater attention from design engineers as it involves both vehicle ergonomics and workspace design. Interviews with the local police officers revealed several safety concerns, complaints and suggestions. Questions were raised regarding the safety of the in-car equipment at the time of a crash. Moreover, a few complaints were put forth concerning the attention required to perform secondary tasks. Since secondary tasks such as conversing with the dispatcher, operating the siren and microphone controls are necessary for police officers, these concerns cannot be eliminated, but only minimized.

1.1   Ergonomic Concerns

The patrol vehicles currently in service are regular passenger vehicles with police equipment installed. Although there is not enough evidence to support or reject the argument that the police officers are bigger than civilians in North America, the partition behind the front seats does cause driver discomfort, limiting the longitudinal movement of the seat. Officers taller than 6 feet expressed the discomfort caused thereby, especially in sedans. This may be one of the reasons why they preferred SUVs to sedans.

Cars fitted with laptop and laptop mount have little or no room for free body movement. There often is a compromise in terms of location and orientation of the installed equipment such as visual displays and controls, requiring the driver to take eyes off the road for longer periods. Besides these, the discomfort caused by seat belt fastened over the utility belt needs to be addressed as well.

Cowles (2010) conducted a comparative study of overall ergonomics of the most commonly used police vehicles based on subjective ratings from 9 police officers, on a scale of 1 to 10; 1 for 'totally unacceptable', 5 for 'average' and 10 for 'superior'.

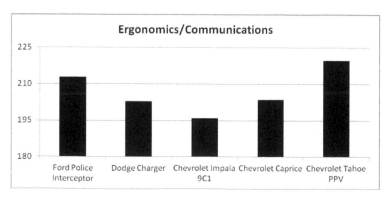

Figure 1 Comparison of overall vehicle ergonomics (Data extracted from Cowles, S., "2011 Model Year Police Vehicle Evaluation Program.")

The graph in figure 1 shows the combined overall ratings for various aspects such as ergonomics of front and rear seats, instrumentation, vehicle controls, visibility and communications. It can be noticed from this report that Chevrolet Tahoe was rated the most comfortable car overall. Impalas were rated the least welcoming.

1.2   Visual and Cognitive Loads

Position and orientation of display elements in the police car are not always ideal. From figure 2, it can be noticed that the objects that require visual attention while driving are not located in the field of view.

Figure 2 Typical arrangement of visual displays in a police car (Image: Ford Crown Victoria 2009, Michigan State Police Post # 90, Calumet)

While radar displays can be moved to any desired location, video display screens and camera controls in most cases do not entertain this freedom of customization. It can be concluded from Engström et. al., (2005) and Liang et. al., (2010) from their studies on the effects of visual and cognitive distractions on driving performance, that visual demands cause decreased speed, poor lane-keeping, steering over-correction and decreased hazard perception, while cognitive loads cause gaze concentrations at the center of the road, steering under-correction and better lane-keeping. It has been concluded that visual loads cause more distraction than cognitive loads. Therefore, when in pursuit or any other stressful driving conditions, the displays may cause a significant issue.

Cades et. al., (2011) states that driver distraction is not merely caused by off-road glances, but the driver's attention is greatly handicapped by cognitive loads as well, although the eyes are on the road. Interestingly, Brookhuis' (2010) study on

90

monitoring drivers' mental workload shows that driver mental workload should not be too high or too low. Harbluk et. al., (2007) and Recarte et. al., (2000) have concluded that drivers under cognitive workloads have an impaired ability of visual scanning of the environment, resulting in poor perception of road signs and other traffic cues. In addition to these distractions, another important element is communicating with the dispatcher, which may be likened to conversing on a cellular phone. Several studies conducted on the effects of cell phone conversations on driver abilities showed that the drivers' braking abilities and visual perception abilities were negatively affected (Strayer 2003), drivers failed to respond to some critical traffic events (McKnight 1993) and made poor judgment of gaps while driving (Brown 1969).

1.3   Safety Concerns

Several safety concerns were raised due to the presence of laptop and its mounting structure. For a far-side crash of delta-V 28 kmph, the intrusion of the body can be as high as 600mm for sedans (Digges 2005). From our survey of the police vehicles, this intrusion would not interfere with the laptop mounting zones seen in the police cars, although the intrusion is deep enough for a head injury (Figure 7). In addition to the laptop and the mount, the safety concerns regarding the center console (shown in the following images) will be discussed in results section.

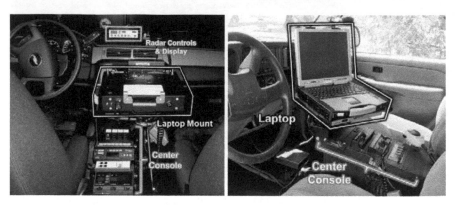

Figure 3 Laptop mounts and center consoles in Chevrolet Tahoe (left) and Chevrolet Impala (right) Images: Chevrolet Tahoe 2010, Michigan State Police Post # 86, Iron Mountain (left), Chevrolet Impala 2007, Phoenix PD (right)

Objects such as the microphones, remote controls for radar and radar display units may become high-speed projectiles during a crash. Adding to this, any equipment in the airbag deployment zones can cause even more severe damage. To eliminate this hazard, the automotive manufacturers have released upfitting documentation for their vehicles, indicating airbag deployment zones and where and how to install the equipment.

2   METHOD

The current study focuses on the hazards to human safety posed by far-side crashes in light of the human factors and ergonomic limitations in a compact mobile workspace such as a police car. The following sections describe the three important phases of the study:

2.1   Vehicle Survey

In the vehicle survey, the most commonly used police vehicles (Chevy Tahoe, Impala, Ford Crown Victoria, Dodge Charger and Ford Expedition) were studied. Several vehicles from across the Upper Peninsula of Michigan were considered for the study. The most commonly seen equipment were the siren controls, radar, radar controls, camera and its control panel, microphones, center console, spotlight and laptop. Measurements of the interiors and the equipment were taken to reconstruct detailed 3-dimensional models for further analysis. The layouts discussed in this paper are limited to those seen in Tahoes and Impalas only. Therefore, our study may be limited to these vehicles, and those with similar dimensions such as GMC Yukon, Chevrolet Suburban, Silverado, and Chevrolet Malibu for sedans.

2.2   Human Factor Survey

Twelve local police officers from police stations in Houghton and Calumet, Michigan, participated in subject-based rating survey on four aspects of each component in the car. Among them, 9 had at least 5 years of experience driving the patrol vehicles, and the other 3 had an experience of 2-5 years. The following four aspects were rated on a Semantic Differential rating scale:
  (1) Ease in reaching (1 = easy to reach; 5 = remote)
  (2) Frequency of use (5 = very frequent; 1 = almost never)
  (3) Importance of the component (5 = very important; 1 = not at all important)
  (4) User-friendly design (5 = simple/easy to use; 1 = complicated/complex)
Average ratings and further analysis are discussed in results.

2.3   Sled Tests

In collaboration with Kettering University Crash Safety Center, fourteen sled tests were conducted on 95th and 50th percentile male dummies, for a Chevrolet Impala and a Chevrolet Tahoe, at 19.9 mph (32 kmph), for 40° and 70° angles of impact (PDOF). Motion-tracking cameras were used to track the dummy trajectories for 200ms from the initiation of the impact. The geometric data was obtained from the sled tests, and compared with the interior dimensions obtained from vehicle surveys and airbag deployment zones. However, the data was obtained from 13 tests only, as the fourteenth test was conducted without a seatbelt, and the cameras could not collect the trajectory data.

92

3   RESULTS

The averages of the subject-based ratings are tabulated below (some of the components were not present in all the cars). Each reading ranges from 1 to 5.

Table 1 Average user ratings for in-car components

Component	Proximity lower=closer	Frequency lower=rare	Importance lower=unimportant	Complexity lower=easier
Mic/radio controls	1.58	4.67	5	1.33
Siren controls	2	2.67	4.83	1.5
Radar controls	1.45	4.18	4.09	1.45
Camera controls	1.82	4	4.27	2.27
Laptop	2	2.5	2	3.5
Spotlights	1	4.25	4.75	1.75
Seatbelt	5	5	5	1
Emergency lights	3	5	5	1
Cell phone	1	3	3	2.5
Cup holder	1	5	5	1

Components that interfere with neither airbag zones nor dummy trajectories were excluded from the following images, to improve legibility and clarity. The Tahoe in our survey did not have a camera, and hence not included in the analysis. Cell phones and seat belts were not included in the analysis.

The following plots were obtained using a 3D plotting software *Calc3D Pro*. Laptop is assumed to be in a closed position, and circular in shape, since it can be rotated about the mount pivot. But the mounting structure itself has a variable arm length, which varies between 60 mm and 180 mm. For the sake of this study, the laptop is considered fixed, with its center predefined, located as seen in the patrol vehicles during the survey.

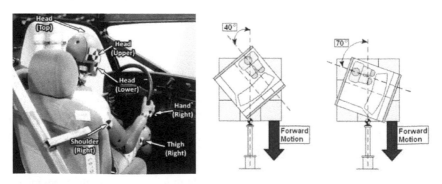

Figure 4 Tracking points on the dummy (left); Sled test setup - Angles of impact (right)

Figure 4a (right) shows the points on the dummy that are tracked. Figure 4b (right) shows the diagram of the setup for sled tests, and how the angles were measured.

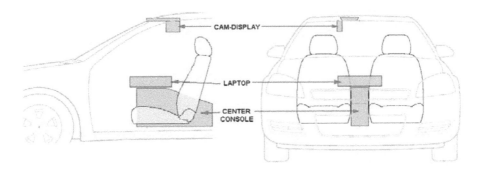

Figure 5 Common in-car equipment in an Impala (Original image source: www.the-blueprints.com; image modified based on geometric data from the survey)

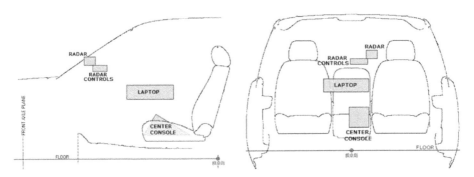

Figure 6 Common in-car equipment in a Tahoe (Original blueprint source: www.gmupfitter.com; modified based on geometric data from the survey)

Overall, the images may contain an error of ±15mm for Tahoe layouts and ±30mm for Impala's, due to limited availability and accuracy of blueprint images. The sled tests were conducted without any in-car equipment installed, therefore the dummy trajectories appear to pass through the equipment in the graphs. In the study, we look at head injuries only, since these are the major contributors to fatal incidents (Yoganandan, et. al. 2010).

HeadTOP refers to the top-most part of the head, *headUPR* approximately refers to the location of right ear and *headLWR* to right jaw (figure 4a). The graphs above include both 95th and 50th percentile dummies.

94

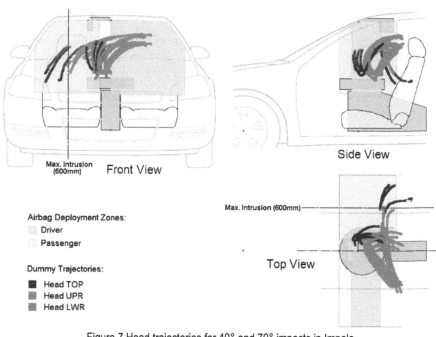

Airbag Deployment Zones:
- Driver
- Passenger

Dummy Trajectories:
- Head TOP
- Head UPR
- Head LWR

Figure 7 Head trajectories for 40° and 70° impacts in Impala

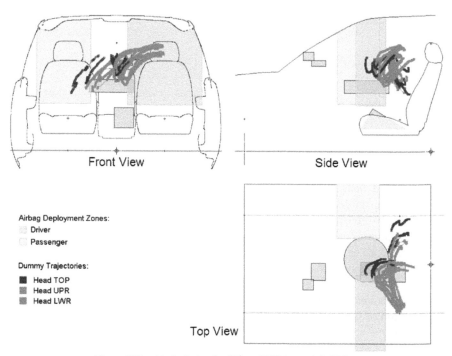

Airbag Deployment Zones:
- Driver
- Passenger

Dummy Trajectories:
- Head TOP
- Head UPR
- Head LWR

Figure 8 Head trajectories for 40° and 70° impacts in Tahoe

4   DISCUSSION & CONCLUSIONS

From the subjective ratings, it can be seen that almost all components are necessary for the duties of the job, except the laptop and the cell phone. Since not all vehicles had the same components, only the most commonly seen ones were shown in the layout drawings. The following key points can be concluded from the analysis:

- Laptops and their mount structures are a major threat for head injuries in far-side crashes of police vehicles. For oblique impacts, however, the head may be expected to hit the airbag before hitting the laptop. Even so, this raises a question whether the interference of the airbag with the laptop causes any unforeseen damage or injury.
- The airbag deployment zones are in conflict with the spatial location of some equipment, especially the laptop and its mount. Disabling the passenger airbag is a good solution in case of absence of the passenger. But this still does not address the driver airbag.
- Seat belts do not protect the driver enough in far-side crashes, which can be observed from the above trajectories. The dummy swung back due to the tension in the seat belt in the 40° tests (oblique front-right impacts), whereas in the 70° tests, the dummy flew almost to the passenger seat's H-point. This means that 3-point seat belts do not provide lateral support for far-side crashes. Similar results may be found in Kumaresan et. al., (2006) and Feist et. al., (2007). This would expose the head to body intrusion, resulting in head injuries (Digges 2005, Bostrom 2003).
- Redesign suggestions can be made to install all the visual display equipment in the field of view. The control panels, however, are already within the reach. Driver anthropometry needs to be considered to determine this workspace as well.
- For smaller displays, best available position is on the dashboard, behind the airbags, in a way not to obstruct the driver view.

Further research needs to be done to quantify the workspace ergonomics with regards to the restrictions on available workspace and potential driver distraction and unfamiliarity issues caused by repositioning of the equipment.

ACKNOWLEDGMENTS

The authors would like to acknowledge the support and collaboration of Dr. Janet Fornari and Justin Schnabelrauch of Kettering University Crash Safety Center. We would also like to acknowledge Lt. Derek Dixon of Iron Mountain Sheriff's Office, Capt. Blake Rieboldt of Marquette City Police Dept., Sheriff Brian McLean of Houghton County, Chief John Donnelly from Houghton City PD, Sgt. Jason Wickstrom of MSP Calumet, Tpr. Mary Groeneveld from MSP Iron Mountain, Brian Cadwell from Michigan Tech Police Services and other police officers for their valuable time cooperation with our survey. This study has been funded by the US National Institute of Justice.

REFERENCES

Bostrom, O., Fildes, B., Morris, A., Sparke, L., Smith, S., Judd, R., 2003. "A cost effective far side crash simulation." *International Journal of Crashworthiness* 8(3): 307-313.

Brookhuis, K., de Waard, D., 2010. "Monitoring drivers' mental workload in driving simulators using physiological measures." *Accident analysis and prevention* 42 (3): 898-903.

Brown, I. D., Tickner, A. H., Simonds, D. C., 1969. "Interference between concurrent tasks of driving and telephoning." *Journal of Applied Psychology.* 53(5), 419-424. Washingtom, DC: American Psychological Association.

Cades, D., Arndt, S., 2011. "Driver Distraction Is More than Just Taking eyes off the Road." *Institute of Transportation Engineers,* ITE J 81 no7 J1 2011 p. 26-33.

Cowles, S., 2010. "2011 Model Year Police Vehicle Evaluation Program", *Ergonomics and Communications Evaluation,* Department of State Police and Department of Management and Budget, State of Michigan.

Digges, K., Gabler, H., Mohan, P., Alonso, B., 2005. "Characteristics of the injury environment in far-side crashes." *49th Annual Proceeding, Association for the Advancement of Automotive Medicine.*

Engström, J., Johansson, E., Östlund, J., 2005. "Effects of visual and cognitive load in real and simulated motorwar driving." *Transportation Research Part F,* Vol. 8, 2, 97-120.

Feist, F., Gugler, J., Edwards, M. J. (2007). "Methodology to address non struck side injuries." *APROSYS Deliverable D1115A.* Rep. AP-SP11-0144, public., 2007.

Harbluk, J. L., Y. I. Noy, P. L. Trbovich, M. Eizenman, 2007."An on-road assessment of cognitive distraction: Impacts on drivers' visual behavior and braking performance." *Accident Analysis and Prevention,* 39(2), 372-379. Amsterdam, The Netherlands: Elsevier.

Kumaresan, S., Sances, A., Carlin, F., Frieder, R., Friedman, K., Renfroe, D., 2006. "Biomechanics of Side Impact Injuries: Evaluation of Seat Belt Restraint System, Occupant Kinematics and Injury Potential." *Engineering in Medicine and Biology Society, 2006. EMBS '06.* 28th Annual International Conference of the IEEE.

Liang, Y. L., Lee. J. D., 2010. "Combining cognitive and visual distraction: Less than the sum of its parts." *Accident analysis and prevention* 42 (3):881-890.

McKnight, A. J., McKnight, A. S., 1993. "The Effect of Cellular Phone Use Upon Driver Attention." *Accident Analysis and Prevention,* 25(3), 259-265. Amsterdam, The Netherlands: Elsevier.

Recarte, M. A., Nunes, L. M., 2000. "Effects of verbal and spatial-imagery tasks on eye fixations while driving." *Journal of Experimental Psychology: Applied,* 6(1) 31-43. Washington, DC: American Psychological Association.

Strayer, D. L., Drews, F. A., Johnston, W. A., 2003. "Cell phone-induced failures of visual attention during simulated driving." *Journal of Experimental Psychology: Applied,* 9(1), 23-32. doi:10.1037/1076-898X.9.1.23. Washington, DC: American Psychological Association.

Yoganandan, N., Baisden, J., Maiman, D., Gennarelli, T., Guan, Y., Laud, P., Ridella, S., 2010. "Severe-to-fatal head injuries in motor vehicle impacts." *Accident analysis and prevention,* 42(4):1370-1378.

CHAPTER 11

Modular Dashboard for Flexible in Car HMI Testing

Frank Sulzmann [1], Vivien Melcher [2], Frederik Diederichs[1], Rafael Sayar [2]

[1] Institute for Human Factors and Technology Management IAT,
University of Stuttgart, Germany
frank.sulzmann@iat.uni-stuttgart.de; frederik.diederichs@iat.uni-stuttgart.de

[2] Fraunhofer Institute for Industrial Engineering IAO,
Stuttgart, Germany
vivien.melcher@iao.fraunhofer.de; rafi@rafisautoshop.de

ABSTRACT

Driving simulators are important tools for the evaluation of in car infotainment and driver assistance systems with respect to their ergonomics, design, usability and user experience. To increase efficiency by integrating different devices and HMI prototypes into the driving simulator in a flexible way, a new type of modular testing environment has been conceptualized. It is based on a modular dashboard which was developed and integrated into the existing driving simulation car, a Renault Scenic. This approach eases the adaptation of the dashboard to any new HMI requirement on the one hand and still allows for evaluation in the realistic in-vehicle environment. The paper includes examples of use during different research projects which demonstrate the function and advantages of the new dashboard.

Keywords: modular dashboard, HMI prototypes, evaluation studies, driving simulator

1   INTRODUCTION

In-vehicle audiovisual information and entertainment systems have become an important factor to distinguish a brand in the automotive industry. Manufacturers compete to integrate as many functions as possible in the smartest way. Onboard diagnostics, navigation, smartphones, media-players, internet access and Apps provide an unlimited amount of warnings, information and entertainment for the driver. Consequently the demand for testing and evaluating these systems, their hardware positioning and their HMI is increasing. Studies in driving simulators are commonly used to evaluate the ergonomics, usability, user experience and distraction of information and entertainment systems in different kinds of driving conditions in the context of User Centred Design.

The driving simulator at the Fraunhofer Institute for Industrial Engineering (IAO) in Stuttgart, Germany is based on a real vehicle Renault Scenic (figure 1) which is equipped with a motion system. The simulator is used for evaluating concepts of ADAS, IVIS, entertainment, general HMI, design elements and the ergonomic positioning of devices. For each test the integration of the objects of investigation in the simulation car is time-consuming and requires a huge amount of effort. Often suppliers of in car HMI or design elements need a complete dashboard or interior for every single prototype to be tested. After the tests, the dashboard remains many times partly destroyed or punched and in generally bad conditions for following experiments. To increase efficiency when integrating devices and HMI prototypes into the driving simulator in a flexible way a new type of a modular testing environment had to be conceptualized and designed. In order to fulfill the new requirements for in car HMI testing the dashboard and the vehicle interior of the Fraunhofer IAO driving simulator was reconstructed and a new modular dashboard was developed and smoothly integrated into the existing car.

Figure 1 The Fraunhofer IAO driving simulator based on a Renault Scenic.

This paper describes the conception and development of the modular dashboard and interior and discusses its use case scenarios and advantages compared with standard dashboard solutions.

2   MODULAR DASHBOARD

The requirements engineering, conceptualization, construction and integration of the modular dashboard were carried out in the following steps.

2.1.   Requirements

The driving simulator is used for evaluating different devices. The devices' integration has to be done with a minimum amount of time and effort. This includes for instance cameras, tracking systems, displays, input devices, smartphones, speakers, warning lights and ambient light. The flexible placement of devices and HMI prototypes requires a dashboard with a modular design approach. Components have to be easily installed and replaced. Furthermore a new level of flexibility in the reconfiguration of the interior is needed to fulfill the requirements of future projects and studies. Components like the jets of the integrated air-conditioning and infotainment systems should be rearranged in the whole dashboard without limits.

The modular dashboard has to provide a flexible and realistic testground for new concepts of in–vehicle entertainment and driver information. Parameters of driver distraction, attention and design aspects have to be determined taking different arrangements into account.

In addition to the integration of test items a flexible mounting of tracking devices is needed. This means that various cameras for video capturing and eye-tracking have to be placed freely all over the dashboard wherever applicable. Cable management also has to be taken into account to connect the corresponding devices with the driving simulator.

To avoid dashboard reflections on the front windshield, the dashboard has to be designed in a neutral appearance and must not use any reflective materials.

2.2.   Concept

In order to enable easy reconfiguration of the dashboard a modular approach was chosen. The dashboard was separated in structural components, casing elements and the items to be integrated. A modular building kit system based on aluminum profiles and special connections was used to build up the dashboard's supporting structure. The skeleton structure was made of aluminum profiles. Replaceable panel elements serve as covering. Devices and other elements can be easily attached and integrated on the panel elements or in the slots of the aluminum profiles.

A flat computer monitor was used to display virtual speedometers and revolution counters. The center console is based on a huge multi-touch-monitor to display virtual prototypes of information systems. Their position can be changed using parameters in software. This approach is more flexible than using a smaller, moveable monitor with a rail system. The large display area can be used to virtualize the center stacks of different cars (figure 2).

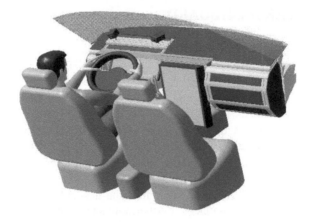

Figure 2: Concept of the modular dashboard.

In evaluation studies the subjects are monitored using video cameras and eye-tracking systems. In front of the driver a rail system was integrated into the dashboard to get a minimal installation height of the camera's stand (figure 3). Furthermore a flexible mounting of video cameras is possible using the nuts of the aluminum profiles all over the dashboard.

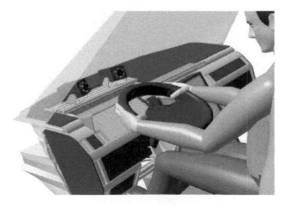

Figure 3: Concept of the camera rack above the head unit.

To ensure the flexible installation of the air-jets, the air-conditioning-system was built up using flexible tubes. The rearrangement of the air-jets can be easily done by replacing panel elements.

2.3.   Construction & Integration

First of all the car interior had to be measured exactly. To achieve this hand made laser scanning technique was used. A supporting frame was constructed to serve as a stand for a laser distance meter. In the first step the original dashboard was manually scanned. After the demounting of the dashboard the interior was scanned again to measure the available space for the new dashboard. The outcome of these measurements were used to reconstruct the shapes of the original dashboard and the interior of the car.

The new dashboard was designed in CATIA. The cross section of the original dashboard was kept to avoid adaptions in doors and other issues.

Prior to the construction process an appropriate building kit system needed to be determined. The "MB Building Kit for Mechanical Engineering" offered by "item Industrietechnik, Germany" was chosen, because a huge amount of different fastening elements is provided. First of all the supporting frame was designed in form of a skeleton structure built up with aluminum profiles.

After the design the profiles were assembled to a skeleton structure. To avoid dashboard reflections on the front windshield, the skeleton structure was lacquered matt black (figure 4).

Figure 4: Skeleton structure assembled and lacquered.

The panels were attached to the skeleton structure and covered with a black, beamless material. The complete dashboard was then integrated into the Renault Scenic (figure 5).

Figure 5: New dashboard completed and integrated.

The rail system for the camera stands was integrated directly in the dashboard taking advantage of the aluminum profile's characteristics. For instance the slots were used to build up an integrated rail system (figure 6).

Figure 6: Fixation for the cameras on the camera rack.

The ambient light system is based on flexible light tubes installed in grooves. It is possible to rearrange the ambient light because the grooves were designed as a building set and the tubes were assembled by clamping (figure7).

Figure 7: Installation of ambient light.

3   USE CASES

The modular dashboard has been used in several experimental studies which requested installation of different hardware. This chapter provides three examples. The testing of different ambient light in the drivers cabin, the testing of different eye-tracking camera systems and the integration of smartphones in different positions.

3.1. Ambient Lighting

Within the research field of Fraunhofer IAO the modular dashboard is used to analyze the influence of dashboard display light and ambient lightning on the vision of the driver while driving at night or twilight. Glare due to in car lighting can lead to a reduced ability of the driver to perceive contrast within the environment. An adverse position of the LED light and wrong light color or light intensity are critical for the road safety. Optimal light conditions within the car are important to assure that the driver perceives all relevant object of the road traffic and reacts on them within the appropriate time. The modular dashboard provides the best preconditions for flexible testing of light placements and thus in vehicle light conditions (figure 8).

Figure 8: Ambient light installed in different positions inside the driver cabin.

On this account a set of LED light tubes were integrated into the modular dashboard at different positions. The position of the light tubes can be rearranged to fit the requirements of the test setting. Light intensity and color can also be adjusted. The intensity and color of the light tubes are controlled via an interface, which allows a setting change while driving (figure 9).

Figure 9: Ambient in car lighting and display lighting of the Fraunhofer IAO driving simulator at different light conditions.

Beside the light tubes also light settings of the two implemented displays are adjustable. The modular dashboard comes with a large screen behind the steering wheel to simulate all kinds of head unit designs, including minimalized screens for night driving or environmental adaptive brightness. In the center stack a large monitor allows for simulating different center stack designs and various positions and sizes for simulated center screens.

3.2.   Eye-tracking Cameras

The modular dashboard supports the integration and placement of additional dashboard parts, such as cameras for driver monitoring or eye-tracking. In an experiment different eye-tracking systems have been compared which requested the installation of different cameras in various positions on the dashboard (figure 10).

Figure 10: Possible position for cameras on the dashboard.

Especially the distance to the driver's head had to be changed stepwise in order to find the best position for the cameras. The modular dashboard supported this experiment with its flexible surface and options for adding an additional rack to hold the cameras. The special rack was made and integrated above the head unit. The rack itself can be moved forward and backwards on a track within a range of 15cm and can be fixed in every position. It also allows changing the distance between the cameras and their orientation.

3.3. Smartphone Bracket

The use of aftermarket navigation devices and smartphones inside the car is heavily increasing. However the positioning inside the car is a crucial question still. Besides the limited possibilities on regular dashboards researches also have a strong interest in knowing more about the best position in terms of reachability, usability and distraction. The modular dashboard was used to install a smartphone in different positions within the arm-rang of the driver and perform tests for distraction while operating the phone. Figure 11 shows exemplary two possible positions that were used for testing.

Figure 11: Different positions for the smartphone.

CONCLUSIONS

This modular dashboard approach eases the adaptation of the dashboard to any new HMI requirement on the one hand and still allows for evaluation in the realistic in car environment.

Devices integration can now be done with a minimum amount of time and effort. This has been proven for cameras, tracking systems, smartphones and ambient light. The modular assembly of the dashboard allows a flexible placement of devices and HMI prototypes for studies on ergonomics, usability and distraction as well as for integrating driver monitoring systems. The common problem of destroyed or punched original dashboards within driving simulators has completely been eliminated. HMI prototypes of different developers can be easily installed and replaced without leaving unfixable damage to the dashboard. Furthermore a new level of flexibility in the reconfiguration of the interior has been reached. For instance the air jets of the integrated air-conditioning can be rearranged in the whole dashboard without limits. These features provide a flexible and realistic testground for new ergonomic concepts of in car entertainment and driver information. Parameters of driver distraction, attention and design aspects can be determined taking different arrangements into account.

REFERENCES

All picture material is property of Fraunhofer IAO. The studies mentioned in this paper were all carried out by Fraunhofer researches and have not been published yet.

CHAPTER 12

Impact of Feedback Torque Level on Perceived Comfort and Control in Steer-by-Wire Systems

Swethan Anand[1], Jacques Terken[1], Jeroen Hogema[2], Jean-Bernard Martens[1]

[1]Eindhoven University of Technology
Eindhoven, The Netherlands
s.anand@tue.nl

[2]TNO Mobility
Soesterberg, The Netherlands

ABSTRACT

Steer-by-Wire systems enable designers to offer completely personalized steering feel to drivers, unlike existing steering systems that offer limited or no personalization. In this paper we focus on feedback torque level, a significant factor for steering feel. Earlier studies indicate that the preferred feedback torque level may be related to the perceived comfort and control that different torque levels offer. However, there is limited understanding on how drivers perceive comfort and control in regard to feedback torque level and the relationship between comfort and control. An exploratory study was conducted on a driving simulator to understand the impact of feedback torque variations on driver perception of comfort and control. The study shows that comfort and control are perceived together and that their optima are not defined by physical effort but may instead be dependent on factors such as personal experience and mental effort.

Keywords: steer-by-wire, steering control, steering comfort

1   INTRODUCTION

The steering wheel is the primary human machine interface (HMI) with which the driver interacts to control lateral motion of a vehicle (Newberry et al 2007).

Conventional steering systems have been equipped with different mechanisms for power assist to lower the steering effort and increase the comfort (Toffin et al 2007) for the driver while exerting lateral control. Conventional steering systems however are limited in terms of the ability to apply the amount of power assistance which is preferred by drivers (Verschuren and Duringhoff 2006). Steer-by-wire systems, which are fully electric steering systems, allow us to adjust various parameters including feedback torque that a driver experiences on the steering wheel (Kimura et al 2005). With the design flexibility that by-wire technology promises, in theory it will be possible to design an algorithm with steering parameters for a steering feel that offers optima of comfort and control to different people as preferences for steering settings are known to vary across individuals (Bathenheier 2004). There is however limited understanding of what the terms comfort and control mean and the relationship between them. Hence, to gain an understanding and help create better informed designs for steer-by-wire systems, an exploratory study was conducted.

The study also wanted to test the hypothesis that comfort and control are distinct attributes and that there would be a clear distinction between the two with force feedback variations. The hypothesis is illustrated in Figure. 1. Considering the omnipresence of power assist systems, it is assumed that comfort decreases more or less linearly with increasing force, except for a secondary effect at very low to zero force levels, where comfort is also very low. On the other hand, considering that extremely low and high feedback torque impact steering input (Chai 2004), we assume that control increases more or less linearly with increasing force, except for a secondary effect at very high levels, where the force is so high that it is difficult to execute the intended steering maneuver. The hypothesized relation implies that optimal force levels for comfort and control are different. By exploring the relationship between comfort and control and in the process of testing the hypothesis, the study provides designers with insights on driver perception of comfort and control in designing the HMI for steer-by-wire systems. The study also is intended to provide designers with a methodology and questionnaire to effectively study the impact of feedback torque variations on perceived comfort and control in steer-by-wire systems.

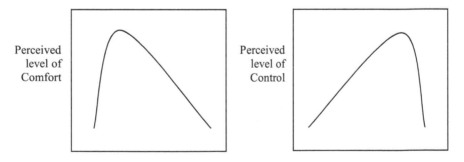

Figure. 1. Illustration of assumed relation between force level and perceived level of comfort and control.

2 METHODS

2.1 Experimental Design

The experiment followed a mixed methods design. Six levels of forces ranging from 0Nm to 7.2Nm were applied to a steering wheel in a driving simulator in order to allow participants to experience a wide range of forces that included feedback forces normally encountered in a vehicle as well as extremely low and high forces. The experimental design focused on quantitatively recording participants' judgments of comfort and control across the six different force levels.

Comparing all six different levels of forces after driving with each level of force produces methodological problems; as such judgments are likely to be heavily influenced by the order in which the forces were offered. As the time required to experience a particular force level is substantial (in the order of minutes), this would imply a large time interval between experiencing the first force level and making the judgment. It is known that such reliance on memory may favor more recent (Mayo and Crockett 1964) and more extreme experiences. Pairwise comparison is known to be a more stable experimental method and has therefore been adopted instead. This implies that participants perform the driving task with two different force levels and subsequently make a comparison between the two on a questionnaire. With six different force levels, a comparison matrix yields 30 differently ordered pairs. Each participant was offered only three pairs, selected at random, in order to prevent participant fatigue. 10 participants are therefore required to make all 30 (10 x 3) pairwise comparisons.

2.2 Participants

Thirty (N = 30) subjects who were regular drivers of a passenger car with a minimum of 1.5 years of driving experience were recruited to take part in the study. Participants recruited consisted mostly of students affiliated with Eindhoven University of Technology. Of the 30 subjects, 27 were male and 3 were female. Since an earlier study conducted on the simulator revealed that the effect of gender on preferred steering force was insignificant (Anand et al 2011), participant recruitment did not require gender equality. Of the 30 participants, 10 had driving experience ranging between 1.5 and 3 yrs, 7 between 3 and 5 years, 10 between 5 and 10 years and 3 over 10 years of experience.

2.3 Tools

2.3.1 Questionnaire. A post-task questionnaire to make the pairwise comparisons was developed for the study. Items included in the questionnaire were outcomes of a pilot study conducted earlier. In the pilot study, six doctoral students from Eindhoven University of Technology were recruited using convenience sampling to perform the same driving task as in the current experiment but were asked to compare forces in triads instead of pairs. Comparisons were expressed verbally via interview. The interviewing style followed the repertory grid technique and used laddering up and down. Interviews focused on gaining an understanding of how participants make comparisons with regard to steering feedback torque and also specifically into what comfort and control meant to subjects and the description of these terms in their own words. Content analysis of the interviews provided common terms that participants use in characterizing the experienced forces. Items that provided a general characterization of force and items that were relevant to comfort and control were first included in the questionnaire, to a total of twenty items. Seven items were included later on based on feedback from the first ten participants. Of the seven additional items, four provide subjective assessment of comfort and control while driving in straights and corners while the remaining three focus on attention, mental effort and physical effort required in performing the driving task.

For each item, the post-task questionnaire contained a binary response question and a 5-point Likert-type scale. The binary response question was to elicit whether the perceived effect of the attribute was "More" or "Less" when comparing the second force level in a pair with the first. The 5 point Likert-type scale below the binary response question allowed participants to indicate *how much* more or less the difference was between the two force levels for a particular attribute.

2.3.2 Driving Simulator. The driving tasks were performed in a fixed-base driving simulator manufactured by Green Dino Technologies Ltd in the Netherlands. The driving simulator offered participants a panoramic view of the driving environment as seen in Figure 2.

Figure2. Green Dino driving simulator.   Figure3.Driving task circuit.

A brushless DC motor with 8Nm rated torque fitted to the simulator was controlled to produce six levels of feedback torque. The feedback torque was varied from 0% - 90% of the rated motor torque. The feedback torque on the steering wheel was constant and was not varied in response to speed and steering angle in order to present participants with a single force sensation. The feedback torque applied in each of the six levels is as shown below in Table1.

Table1. Six feedback torque levels offered to participants.

Feedback Torque Level	Percentage of Rated Torque(8Nm) in %	Actual Torque Produced(Nm)
Level1	0	0
Level2	10	0.8
Level3	30	2.4
Level4	50	4
Level5	70	5.6
Level6	90	7.2

2.4 Driving Task

The experimental task required participants to navigate a 9km circuit designed with straight stretches, gradual curves and sharp curves. The circuit is as shown in Figure 3. The circuit had low density traffic in the opposite lane to simulate a realistic countryside driving environment. Participants were instructed to keep to the center of the lane while driving through the circuit.

2.5 Procedure

Upon obtaining informed consent, information such as gender and driving experience were collected from participants. They were then subjected to familiarization trials to get familiar with the experimental set-up. Familiarization involved participants driving on the driving task circuit for approximately five minutes. Following familiarization, participants were instructed to perform the driving task twice with two different force levels. On completion of the experimental tasks, participants were instructed to complete the post task questionnaire and make the pairwise comparisons. Upon completing the questionnaire, each participant performed the driving task with two other pairs of force levels and made comparisons on the post-task questionnaire as previously instructed. In total, each participant received 3 pairs of forces and expressed their pairwise judgments for the different items offered in the questionnaire.

3 RESULTS

Cluster analysis was performed across all participants on the pairwise comparison data which comprised of average scores for 27 attributes from 20 participants and 20 attributes from all 30 participants. The analysis was therefore performed with missing data for the 7 attributes that were not included in the post-task questionnaire offered to the first 10 participants in the study. Clusters that could not be described as one-dimensional (established by the size of the second singular value) were iteratively split into smaller clusters, but clusters that would only be supported by a single attribute were avoided. Whenever a cluster was split, all atributes were iteratively assigned to the cluster with which they had the highest correlation.The analysis revealed that the attributes could be grouped into four clusters. The clusters formed can be seen in Table 2. The table also presents the values of variance explained from data for each item within the cluster.

Table2. Four clusters formed from preference scores on clustering analysis with explained variance values for each item.

Cluster1	R^2	Cluster2	R^2
Trustworthy	0.34	Tiring	0.83
Secure	0.53	Physical Effort	0.87
Safe	0.53	Heavy	0.93
Control	0.61		
Exaggerated From Normal	0.70	**Cluster3**	R^2
Operable	0.71	Ergonomic	0.55
Normal	0.72	Straight Control	0.60
Stress	0.73	Natural	0.65
Mental Effort	0.75	Straight Comfort	0.79
Doable	0.76		
Requires Attention	0.81	**Cluster4**	R^2
Frustration	0.86	Sturdy	0.04
Pleasurable	0.87	Turn Control	0.42
Comfort	0.88	Turn Comfort	0.79

On observation of the individual attributes within the clusters, the clusters were assigned the following semantic terms: Cluster1 = Overall Comfort & Control, Cluster2 = Physical Effort, Cluster 3 = Straight Comfort & Control and Cluster4 = Turn Comfort & Control.

As can be seen from the *Overall Comfort & Control* cluster, comfort and control are not perceived distinctly by subjects. The high correlations between *Comfort* and *Control* ($r = 0.9$, $p < .001$) rating scores across participants reinforce this impression. The results further reveal other items which are related to *Comfort* and *Control*. The results therefore reject the initial hypothesis that comfort and control are perceived distinctly.

In order to understand the relationship between the clusters and the force levels, the preference scores of individual attributes were analyzed using multi-dimensional scaling (MDS) software. Since a two-dimensional solution did not explain more variance, we used a one-dimensional solution. The analyses treated participants as a homogenous group, although there were some outliers in terms of their judgment of forces. Results from the analysis produced the patterns shown in Figures 4a-4d as a function of force level for the different clusters. With exception of Cluster 2, *Physical Effort*, the relationships are non-linear as a function of force level.

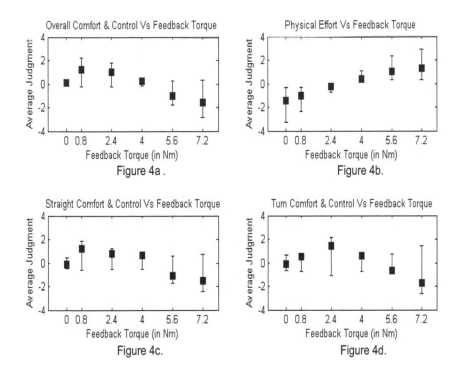

Figure 4a.

Figure 4b.

Figure 4c.

Figure 4d.

4   DISCUSSIONS

The non-linear relationships for Clusters 1, 3 and 4 produce a similar pattern as observed in Figure1, where it was assumed that a secondary effect may not produce a linear relation between force level and the perceived psychological construct. The patterns confirm the existence of secondary effects which influence perceived comfort and control, indicating that too high and too low force result in less perceived comfort and control.

The pattern in Figure 4a illustrates that maximum comfort and control are perceived at 0.8Nm and 2.4Nm followed by 0Nm and 4Nm. The least comfort and control appear to be perceived at 5.6Nm and 7.2Nm which indicates that extremely high forces may be detrimental to perceive comfort and control.

The pattern in Figure 4c for the *Straight Comfort & Control* cluster reveals that participant perception of ergonomics and naturalness of feedback torque increases from 0Nm to a peak at 0.8Nm, gradually decreasing at 2.4Nm and 4Nm and then significantly dropping at 5.6Nm and 7.2Nm. The cluster involves items such as *Straight Control* and *Straight Comfort* which indicated participants' ability to perceive comfort and control with changes in feedback torque while driving on straight sections of the circuit. Presence of such items and their overall relationship with comfort and control reiterate the point that comfort and control are perceived together.

The pattern in Figure 4d for the *Turn Comfort & Control* cluster illustrates that participant perception of increased sturdiness as well as comfort and control in the curve segments of the circuit (referred by items *Turn Comfort and Turn Control*, respectively) increases from 0Nm and peak at 2.4Nm from which point it decreases gradually to 7.2Nm. The relationship can be viewed as an inverted U-pattern. Since the cluster includes items relating to perceived comfort and control, it further reiterates that comfort and control are not distinct. The pattern in comparison with the *Straight Comfort & Control* pattern in Figure 4c shows that 2.4Nm is perceived to offer more comfort and control during cornering than in the case of driving in straights, where 0.8, 2.4 and 4Nm are perceived to offer similar amount of comfort and control. This suggests that drivers are more sensitive in their perception of comfort and control when significant steering actions need to be performed (in order to take turns) as opposed to driving in a straight line where only minimal corrections of the steering wheel are required. Therefore conventional steering systems that already offer increased torque during cornering when compared with on-center driving (by making feedback torque a function of steering angle) are perceived to provide optimal comfort and control. The findings further suggest that if such a conventional steering system had been used in the experiment, there might not have been two distinct clusters for perceived comfort in on-center driving and turning, but we would instead have just two clusters – *Overall Comfort & Control* and *Physical Effort*. Hence items specific to steering conditions, such as on-center driving and turn taking, can be excluded from questionnaires that assess perceived comfort and control if the feedback torque on the steering wheel is varied as a function of steering angle.

On observation of Figures 4a, 4c and 4d, it can be concluded that comfort and control are perceived to be the highest in the 0.8-2.4Nm region with a sharp drop on either side at 0 Nm and 4Nm followed by 5.6Nm and 7.2Nm, which are perceived to offer the least comfort and control.

The relationship between the *Physical Effort* cluster and feedback force as seen in Figure 4b shows that physical effort monotonously increases with increases in force. This shows that participants were able to clearly sense the different force levels offered to them. The absence of items relevant to comfort and control in this cluster as seen in Table2 suggest that participants' perception of comfort and control involve aspects apart from physical effort.

Looking at the *Comfort & Control* cluster in Table 2, it becomes evident that driver perception of comfort and control may not be significantly different from each other. Items in the cluster further suggest that perception of comfort and comfort may be influenced by how operable, normal, doable and pleasurable a feedback feels to the driver. The presence of *Normal* in the *Comfort & Control* cluster suggests that driver perception of comfort and control may significantly rely on the feedback that drivers have been familiar with. During interviews conducted in the pilot study, 'Normal' was used by participants to describe forces that they were most familiar with and used to in driving their cars on normal roads. This goes on to suggest that feedback settings that differ considerably from one's own perception of normality may impact their judgment of comfort and control that a setting may offer.

Presence of items such as *Mental Effort, Frustration and Stress* in the same cluster as *Overall Comfort & Control* suggests that perception of comfort and control may involve aspects relating to the cognitive processing abilities. And since they negatively correlated with items such as *Operable, Normal* and *Doable* in the same cluster, it suggests that increased comfort and control is perceived when the feedback force does not require significant cognitive resources to perform the steering maneuver.

Based on discussions emerging from the cluster analysis, explained variance from data and semantic meaning of the items, a questionnaire to understand the amount of comfort and control that is perceived with different steering settings must include the following items: *Comfort, Control, Normal, Operable, Pleasurable, Mental Effort* and *Stress* from the *Overall Comfort & Control* cluster and *Heavy* and *Physical Effort* from the *Physical Effort* cluster. And as discussed earlier, items relating to specific steering conditions in the *Straight Comfort & Control* and *Turn Comfort & Control* clusters can be excluded if feedback torque applied on steering wheel mimics conventional steering systems. It is important to note that the analysis performed contained missing data from 10 participants for 7 of the 27 attributes that were used, and this may have induced correlation bias in cluster analysis and influenced variance explained in the data. Therefore variance explained for items such as *Physical Effort, Mental Effort* and *Pleasurable*, which are among seven attributes with missing data, may vary from values seen in Table2 if data from the missing participants were available. However, high levels of variance explained in data from 20 participants for these items ($R^2 \geq 0.75$ as seen in Table2) reflect the need for their inclusion in questionnaires for experiments that study perceived comfort and control with different steering settings.

5   CONCLUSIONS

The study was able to successfully explore the relationship between perceived comfort and control with variations in feedback torque. The study found that subjectively assessed comfort and control are mutually dependent on each other and that their relationship with force feedback is non-linear. The study also finds that driver perception of comfort and control are dependent on what drivers perceive as normal based on their existing driving experience. This therefore shows that drivers rely on a personal baseline to judge comfort and control and this may significantly impact assessment of settings tested and developed for steer-by-wire systems. The reliance on a personal baseline indicates that participant familiarization with new settings is essential prior to evaluation. Further, the study contributes in developing a questionnaire to assess perceived comfort and control of steering systems. The study also provides designers with insights on what regular drivers convey when they state to have perceived a certain level of comfort and control.

However it may be noted that the study was conducted in a fixed base simulator and therefore was limited in its ability to produce lateral acceleration forces which may impact drivers' ability to sense speed and therefore their ability to sense differences in comfort and control. Further studies conducted on a test-car can help validate the findings as a test-car allows participants to experience the lateral forces and speed unlike in the simulator. And since steering in a test-car has bearing on personal safety, unlike the simulator; assessment of comfort, control and also preference for feedback torque settings may differ. In the study conducted, feedback torque was the only parameter varied. It may be interesting to study the effects of variation of other parameters such as damping, power assistance boost-curves and steering ratio to understand their effects on comfort and control under different driving conditions. The findings may help in generating a personalized algorithm with optimal steering feel for individual drivers.

ACKNOWLEDGMENTS

This research was made possible through a grant from The Dutch Ministry of Economic Affairs, Agriculture and Innovation to the HTAS project VERIFIED.

REFERENCES

Anand, S. Terken, J. and Hogema, J. (2011) *Individual Differences in Preferred Steering Effort for Steer-by-Wire Systems*. In. M. Tcheligi et al. (eds). Proceedings of the 3rd International Conference on Automotive User Interfaces and Interactive Vehicular Applications, Salzburg, 2011, pp. 55-62

Chai, Y.: *A Study of Effect of Steering Gain And Steering Torque on Driver's Feeling For SBW Vehicle.* In: Proceedings of Fisita world automotive congress, Barcelona (2004)

Barthenheier, T. and Winner, H. *Das persönliche Lenkgefühl*, Vortrag auf der Fahrwerk.tech 2003, Fachtagung des TÜV-Süddeutschland, München, 2003

Kimura, S., Segawa. M., Kada. T., Nakano, S., (2005). "Research on Steering Wheel Control Strategy as a Man-Machine Interface for Steer-by-Wire System[J]", Koyo Engineering Journal English Edition No. 166E(2005)29-33

Mayo, C. W., & Crockett, W. H. (1964). Cognitive complexity and primacy-recency effects in impression formation. *The Journal of Abnormal and Social Psychology, 68*(3), 335-338. doi:10.1037/h0041716

Newberry ,A.C., Griffin, M.J., Dowson M., Driver perception of steering feel, *Proceedings of the Institution of Mechanical Engineers, Part D, Journal of Automobile Engineering, 221, 405-415, 2007*

Toffin, D.; Reymond, G.; Kemeny, A. & Droulez, J. (2007). Role of steering wheel feedback on driver performance: driving simulator and modeling analysis. *Vehicle System Dynamics*, 45:4, 375 – 388

Verschuren, R., and Duringhof, H.-M. (2006) "Design of a Steer-by-Wire Prototype." *SAE Technical Paper No 2006-01-1497*

CHAPTER 13

Automotive Speech Interfaces: Concepts for a Better User Experience

Omer Tsimhoni, Ute Winter

General Motors Advanced Technical Center - Israel
Herzliya, Israel
Omer.tsimhoni@gm.com, Ute.Winter@gm.com

Timothy Grost

General Motors Global Vehicle Engineering
Warren, MI
Timothy.grost@gm.com

ABSTRACT

Automotive speech interfaces are an integral part of a driver's multimodal interface. The interface poses a unique technological and design challenge because of the automotive context. In this chapter, we describe an integrative approach to its design and productization. While our approach is based on User Centered Design, it combines with it technological innovation and theoretical research in areas related to the user, the machine, and the user interaction. Current development trends in speech applications, such as an increase in the number of features and tasks desired by users and the availability of new technologies lead to our vision: Mixed-initiative natural language speech interfaces as part of a multimodal interface allowing the user to combine or switch among the manual-visual and auditory modalities seamlessly. We suggest four areas in need of research: user and context modeling, global markets and cultural variation, dialog management, and multimodal speech interfaces. We introduce six research projects, five of which were performed in collaboration with General Motors and one by Nuance and IBM.

Keywords: Automotive Speech Technologies, adaptive technologies, user centered design, multimodal human-machine interfaces

1   INTRODUCTION

While mobile speech applications have been available for quite some time, only recently have new speech applications been reintroduced into the smart device environment (Apple and Android) for a growing variety of tasks. Although the automotive domain might appear to be a subset of the mobile environment, there is need for a dedicated set of solutions that are specific to the in-vehicle environment and in some cases are substantially different: Requirements for multimodal interaction differ substantially as a driver cannot touch or look at the screen for long; the car acoustic environment is far more challenging than that of a mobile phone; the user interface needs to serve both driver and passengers; mobility needs and connectivity are different than the mobile environment; unique automotive applications are required, e.g., safety, security, and vehicle assistance; and finally automotive regulation and voluntary guidelines must be adhered to.

In General Motors, we have a long-term goal of introducing robust and natural multimodal in-vehicle interfaces offering speech as an interaction modality to the user to provide a pleasing and valuable user experience. Our vision is to create an eco-system that builds upon and enhances existing speech applications, optimizes them for users' needs and desires in the car environment, and adds value in terms of context-understanding and user-modeling, personalization and adaptation, cross-cultural differences, and multimodal interaction between user and interface.

By integrating theoretic constructs and technology research with collected data and empirical testing, we believe we can address the challenge. In this paper, we first describe our holistic view of interface design and development. We then present an applied approach for the creation of automotive multimodal interfaces with focus on spoken interaction and speech technologies. We introduce six research projects, five of which were performed in collaboration with General Motors and one by Nuance and IBM. While our approach is applied to the automotive domain, it can also be applied to speech applications in other domains.

2   IN-VEHICLE USER INTERFACE DESIGN AND DEVELOPMENT

2.1   User Centered Design Approach

The design of an in-vehicle multimodal interface typically follows the paradigm of User-Centered Design (Nielsen, 1993). User-Centered Design (UCD) is a multidisciplinary design approach based on the active involvement of users to improve the understanding of user and task requirements, and the iteration of design and evaluation. Naturalistic observation, contextual inquiry, and other techniques (e.g., Dray and Siegel, 2009) are used to gain a better understanding of the user in the relevant context. It is widely considered the key to product usefulness and usability - an effective approach to overcoming the limitations of traditional system-centered design (Mao et al. 2005). Figure 1 illustrates a simplified but typical research and development cycle using the traditional UCD approach.

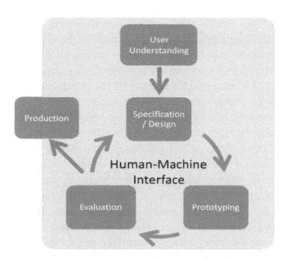

Figure 1 Design and development cycle using the traditional User-Centered Design approach

While Figure 1 may illustrate a reasonable simplification of the UCD process, one could argue that it is missing a clear process for innovation. Furthermore, it lends itself to incremental innovation in which user needs are heavily biased by the technologies that are available to them at the time and by their own expectations for future technologies. An alternative approach would combine user centered design with technology innovation so as to circumvent this difficulty.

The classic UCD approach typically assumes an existing set of technologies that has to be adapted and configured to a set of users. In the automotive industry due to the long development cycle for a next generation interface, the underlying technologies usually evolve in parallel over a period of years. Consequently, automotive interfaces while in a design phase will have to serve future user needs and fulfill prospective user expectations. Our approach therefore extends beyond the classic UCD approach in that the technologies have to be created or further developed in parallel to match the user research.

2.2 Application to Speech Interfaces

Suppliers of speech interfaces focus on technological challenges such as reaching acceptable accuracies and robust performance for Natural Language Understanding, making constant progress on the technology side. At the same time, our goal is to integrate the developed speech technology along with other input devices and displays into an in-vehicle multimodal interface as demonstrated in Figure 1.

To this end, several limiting factors must be considered: time to product, vehicle context of the user interface, and globalization (Table 1). First, on the time-to-product dimension, the design approach needs to take into account the relatively long time from initially developing the user interface specifications until the vehicle is produced. Because of the relatively long development cycle of vehicles, currently measured in years—not months—new in-vehicle interfaces need to be designed with a prediction for future needs and with a built-in flexibility to accommodate future trends and changes. Furthermore, vehicles remain on the road for many years after they are produced and user interface technologies need to remain relevant. This leads to a second dimension, the context of the interface. Speech applications are typically used for information and entertainment services. The automotive context poses a unique set of challenges. A driver may use the system while alone or when accompanied by other passengers. Those passengers may interfere or want to take part in the interaction, or in some cases need not be part of it because of privacy considerations. Furthermore, the driving situation varies substantially as a result of variations in road conditions, traffic, weather, and user goals. These challenges not only influence the design of the speech application but also pose a huge challenge on the integration of speech technologies into in-vehicle multimodal interfaces. A third dimension is the global distribution and huge variety of customers that use our vehicles. The user interface should be perceived as intuitive and pleasing in all markets. Speech interfaces are unique in this respect because they must be localized to accommodate language preferences and communication habits.

Table 1 Limiting factors for designing automotive speech and multimodal interfaces

Dimension	Implication for Development
Time to product	Estimate future user needs and expectations and explore upcoming technology and enabler progress
Vehicle context	Special user needs relevant to the automotive environment
Globalization	Considerations for localization to global global markets
Other constraints	Limitations in technology, integration, budget, safety, and legal

To accommodate these limiting factors and to provide the user with an intuitive and effortless experience answering his needs, basic UCD approach needs to be enhanced in several ways described in the following paragraph.

2.3   Integrative Approach to Speech Interfaces

Figure 2 illustrates an expansion of the process to allow concurrent analysis and innovation which do not only originate from the user's perspective but also from innovation in the technological world and from matching of technologies to user's future needs by way of a user interface analysis. We combine work on all three perspectives of the human-machine interface – the user understanding, the underlying technology development and the multimodal interface design. To achieve an optimal user experience, one component cannot be explored and specified without consideration of the other two components. Considering this dependency and adding the dimension of the long time line of a production cycle, GM is taking a holistic approach to successful multimodal interfaces: Our research and engineering labs constantly learn about the automotive context using contextual inquiry techniques (Holtzblatt, 2003; Holtzblatt and Wendell, 2004). Following the GM user experience design approach (Gellatly et al, 2010), we explore user preferences and goals, expected use cases, and scenarios. We develop personas for our design, as well as test scenarios and measures for evaluation of user satisfaction using usability testing (Nielsen, 1993) and user centered design (Norman and Draper, 1986; Vredenburg and Butler, 1996). In dependency and in parallel, we research innovative new technologies and their enablers, test their performance and match them with the forecast of user expectations in years to come.

This process is built upon a holistic view of the set of tasks a user will want to perform in the automotive context, and how the interface needs to be specified and designed to fulfill user wants and needs, and their sense of aesthetics, and what features the technology enablers will have to provide for the desired experience.

At a certain point, knowledge is taken into a specification and design phase integrating multiple subsystems from various suppliers and multiple disciplines into one. The phase is followed by an implementation and evaluation phase. Cyclic and iterative specification, implementation and evaluation facilitate the development from a preliminary interface prototype to a mature interface for production, thus addressing immediate market needs that could appear after the initial specification and design phase.

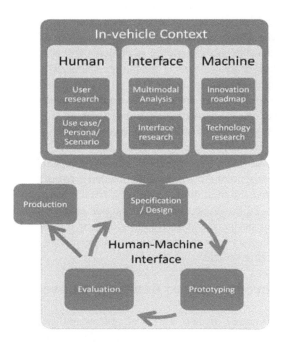

Figure 2 Integration of user, technology and interface research is at the core of our approach

3   CONCEPTS FOR IN-VEHICLE SPEECH INTERFACES

3.1   Speech Applications Development Trends

Because innovation in the area of speech interfaces has primarily originated in a technology perspective, a strong emphasis has traditionally been put on the top right box in Figure 2, ("machine-" based innovation) rather than the top left. Spoken dialog design often focuses on the selection of reasonably intuitive command words and templates, which offer the speech recognition engine diverse phonetic content for improved disambiguation, thus better recognition accuracy for a more satisfying user experience.

Our enhanced approach from Figure 2 leads to the path for future speech interface design. With a rapid increase in the number of features, domains, and tasks, traditional speech user interfaces require the user to remember an ever increasing list of commands. To simplify the interaction and reduce the user's responsibility, voice user interface designers have been adopting a more natural and flexible set of possibilities for interaction with in-vehicle speech systems. Some properties have been identified to be essential to achieve the desired flexibility (Jurafsky & Martin, 2008), such as the use of natural language utterances. The user will be able to provide information to the speech system in his preferred way from a single information piece to one-shot utterances containing all of the information necessary to complete a task. These properties will be combined with mixed initiative dialog strategies rather than system initiative applications. Additionally, speech interfaces can be personalized for a few possible drivers of each vehicle. Such flexibility enables any user to be a system expert by simply having a basic understanding of the system model and the features supported by speech.

Our vision is that the user experience in vehicles will be enhanced by allowing the user to combine or switch among modalities seamlessly. We view a spoken dialog with in-vehicle speech applications as embedded in a multimodal interface. For example, the user may have an option to seamlessly interweave within a speech session disambiguation via a touch screen, by touching the intended recognition result on a display showing the N-best list of results and then returning to the speech session or even concurrently attending to both.

Trends to include such a large variety of enhanced capabilities to model natural interaction doubtlessly point the way to some needed areas of research. We need to understand, build and test multimodal interfaces that are based on the user's natural and preferred ways of interaction between a human and a machine. Thus speech interfaces have to be designed and developed according to principles for multimodal interfaces that are based on spoken communication between a user and a machine in the context of a driving vehicle.

3.2   Key Concepts for Next Generation Speech Interfaces

A natural consequence of these trends of speech application development is a need for research in areas that have not received adequate attention in the past, or have not yet been integrated sufficiently into existing products. Several key concepts need to be explored and developed to advance towards the goal of a next generation speech interface. While the list below is by no means comprehensive, it represents four general research directions on which we have identified to be essential for the holistic speech user experience:

- *User and context modeling*
- *Global Markets and Cultural Variation*
- *Dialog management*
- *Multimodal speech interfaces*

User and context modeling

While communication partners engaged in a dialog try to make verbal progress to achieve their individual goals, whether mutually the same or not, both partners constantly adapt one to the other and to the dialog situation to optimize their chances for fulfilling the goals. Users of speech applications carry out a similar adaptation process intuitively based on their linguistic and social competencies. It is therefore plausible that speech technologies may benefit from adaptive technologies. While the user adapts to the machine and to the car environment, the machine in turn adapts to the user and context as well, potentially leading to a more accurate and satisfying interface and an improvement to the user experience.

Adaptive speech technologies are not new (Jurafsky & Martin, 2008). While speech adaptation typically refers to areas such as acoustic adaptation or frequency of user request, there is potential for adaptation of other elements within speech systems as part of an in-vehicle multimodal interface. Adaptive opportunities can be found in the areas of user preferences, user characteristics, expertise level, preferred dialog interaction style, and the user's mental model of the interface, among others. Any user behavior is additionally dependent on the environmental conditions in general, on the driving workload, and on the time and location of an interaction. The interface can thus adapt to various aspects of the interaction: typical and recurring user errors, recovery from errors, preferred dialog and multimodal interaction paths, presentation style, and initiative strategies are some examples. This is obviously done in the context of the environmental situation and the user's mental concepts during the time of the interaction.

Sun et al. (2012) explore user adaptation in automotive environments in several levels for a natural language spoken dialog system with the goal of making interaction more natural and services available in cars easier to use. They demonstrate these adaptive capabilities using a Point of Interest (POI) application.

Hecht et al (2012) explore the potential of adapting the Language Models (LM), which are at the core of speech recognition systems, for text messaging. They show that both driving workload and input method (speaking versus typing) affect the message language characteristics. Their findings suggest that dictation accuracy can be enhanced, for example, by adapting language models based on input method and driving workload.

Global Markets and Cultural Variation

If the auditory-vocal interaction with in-vehicle speech applications has to be understood as spoken communication between a human and a machine, there is need to first understand the design dimensions and range of parameters that may be affected by deploying an existing speech application into global markets. We need a method for discovery of cultural variability in interaction, which – according to our premise – has to be derived from cultural spoken communication theories in correlation with human-machine interface design dimensions for speech interfaces. Typical dimensions include cooperative principles for communication, control handling, dialog patterns and sequences, turn taking and grounding conventions, information distribution, choice of words and phrases in dependency of typical automotive tasks and domains, and conflict resolution. Carbaugh et al. (2012) discuss a framework for investigating cultural variation in communication, drawing on ethnographic and naturalistic data, and describe how this framework could be the basis of a model for analyzing the role of culture in in-car spoken interaction.

To complement this theoretic approach, there is need for empiric evaluation of existing and future designs in global markets. In most cases it could serve to calibrate existing systems so that they better conform with the expectations of users in global markets. As an example, Wang et al. (2012) describe an examination and analysis of turn-by-turn guidance for complex interactions. Although the study was conducted in China and pertained to Mandarin text-to-speech, it addresses the issue of complex road structures in urban settings, which is not so much cultural as it just signifies the different environments that are associated with global markets.

Dialog management

When users are allowed to request a task according to their preferred interaction style and in natural language, they may distribute information over one or more utterances and in many different ways and wording. The variety of possible dialog sequences thus increases. Today a smart device interface is mostly based on a single utterance pronunciation followed by interaction with the touch screen. This display-centric approach is less likely to work well in the car because recognition is more error prone, and because the driver can only make limited use of non-speech modalities. What is needed in the automotive environment is a robust multimodal but speech-centric dialog manager that does not distract the driver and handles speech recognition errors well, while supporting very flexible dialog pattern. One approach to this problem is the use of statistical techniques to model the dialog. Tsiakoulis et al. (2012) explore and optimize partially observable Markov decision processes (POMDPs), which have recently been proposed as a statistical framework for dialog managers, for an in-vehicle POI application. The dialog managers have explicit models of uncertainty, which allow alternative recognition hypotheses to be exploited, and dialog management policies that can be optimised automatically using reinforcement learning.

Multimodal speech interfaces

The unique automotive environment poses a challenge to the designers of speech user interfaces as part of a multimodal HMI. Voice User Interfaces can be supported or may cooperate with other modalities, such as a touch screen in the center stack. Although multimodal interfaces offer a variety of modes, the user experience at present is still limited due to a lack of flexibility among the modes. There are opportunities to improve and enhance the user experience, for example by allowing the user to seamlessly switch between modes or by the modalities mutually supporting each other. The latter can be achieved for instance by displaying pictorial or linguistic visual representations of the spoken prompt. Such multimodal interfaces can learn and understand the user, while at the same time they shape the mental model of the user. Labsky et al. (2012) discuss such cooperation between modalities for SMS dictation and editing in the car environment while spoken communication is the central mode.

4   CONCLUSION

We presented an approach that integrates user, speech interface and technology research. This approach builds upon a combination of theoretical research and empirical evaluation and testing and utilizes an iterative approach to the improvement of prototypes and designs. Applying this approach to automotive speech applications leads the path to concepts and research fields such as user modeling, dialog management and multimodality.

In many industrial settings there is an increased focus on empirical testing and repeated iterations, which tend to occur late in the process. While we fully understand the importance of user testing and evaluations that are empiric in nature, we also value the contribution of early user research based on the user centered design approach and theory-based research to help analyze the user interface issues, enhance technology, and guide the way to better user interfaces.

The goal of generating an effective, efficient, and pleasing user interface can only be attainable if the designers of the interface have a good holistic understanding of their user needs, the capabilities of existing and future technologies, the optimal interaction and modalities, and the integration of these factors.

REFERENCES

Carbaugh, D., Molina-Markham, E., van Over, B., and Winter U. 2012. Using Communication Research to address Cultural Variability in HMI Design, *in Proceedings of the 4th International Conference on Applied Human Factors and Ergonomics,* San Francisco, CA, July 21-25. Boca Raton, FL: Taylor & Francis Group.

Dray, S. and Siegel, D. 2009 Understanding Users in Context: An In-Depth Introduction to Fieldwork for User Centered Design. Lecture Notes in Computer Science (5727) 950-951, Human-Computer Interaction – INTERACT

Gellatly, A., Hansen, C., Highstrom, M., and Weiss J. 2010. Journey: General Motors' Move to Incorporate Contextual Design Into Its Next Generation of Automotive HMI Designs. Proceedings of the Second International Conference on Automotive User Interfaces and Interactive Vehicular Applications, November 11-12, 2010, Pittsburgh, Pennsylvania, USA

Hecht, R., Tzirkel, E., and Tsimhoni, O. 2012. Adjusting Language Models for Text Messaging based on Driving Workload, *in Proceedings of the 4th International Conference on Applied Human Factors and Ergonomics,* San Francisco, CA, July 21-25. Boca Raton, FL: Taylor & Francis Group.

Holtzblatt, K. 2003 Contextual Design in *The Human-Computer Interaction Handbook: Fundamentals, Evolving Technologies and Emerging Applications* Jacko, J. and Sears, A., Eds. Mahwah, NJ: Lawrence Erlbaum Associates, Inc.

Holztblatt, K., Wendell, J.B., and Wood, S. 2004. *Rapid Contextual Design: A How-to Guide to Key Techniques for User-Centered Design.* San Francisco, CA: Morgan Kaufmann.

Jurafsky & Martin, 2008 Jurafsky, D., Martin, J.H. 2008. *Speech and language processing: an introduction to natural language processing, computational linguistics, and speech recognition.* 2nd ed. Pearson Education, Upper Saddle River, NJ

Mao, J., Vredenburg, K., Smith, P., and Carey, T. 2005. The state of user-centered design practice. *Communications of the ACM 48(3) 105-109*

Labsky, M., Curin, J., Macek, T., Kleindienst, J., Young, H., Thyme-Gobbel, A., König, L., Quast H., and Couvreur, C., 2012, Dictating and editing short texts while driving *in Proceedings of the 4th International Conference on Applied Human Factors and Ergonomics,* San Francisco, CA, July 21-25. Boca Raton, FL: Taylor & Francis Group.

Nielsen, J., 1993 *Usability Engineering.* San Francisco, CA: Morgan Kaufmann Publishers Inc.

Norman, D., Draper, S., 1986 *User Centered System Design; New Perspectives on Human-Computer Interaction.* Mahwah, NJ: Lawrence Erlbaum Associates, Inc.

Sun, M., Rudnicky, A., Levin, L., and Winter U. 2012. User Adaptation in Automotive Environments, *in Proceedings of the 4th International Conference on Applied Human Factors and Ergonomics,* San Francisco, CA, July 21-25. Boca Raton, FL: Taylor & Francis Group.

Tsiakoulis, P., Gasic, M., Henderson, M., Prombonas, J., Thomson, B., Yu, K., Young, S., and Tzirkel, E., 2012, Statistical methods for building robust spoken dialogue systems in an automobile *in Proceedings of the 4th International Conference on Applied Human Factors and Ergonomics,* San Francisco, CA, July 21-25. Boca Raton, FL: Taylor & Francis Group.

Vredenburg, K., and Butler, M., 1996. Current Practice and Future Directions in User-Centered Design. *In Proceedings of Usability Professionals' Association Fifth Annual Conference,* Copper Mountain, CO.

Wang, P., and Tsimhoni, O. 2012. Turn-by-Turn Navigation Guidance for Complex Interactions, *in Proceedings of the 4th International Conference on Applied Human Factors and Ergonomics,* San Francisco, CA, July 21-25. Boca Raton, FL: Taylor & Francis Group.

CHAPTER 14

Remote Target Detection Using an Unmanned Vehicle: Contribution of Telepresence Features

Chris Jansen (2), Linda R. Elliott (1), Leo van Breda (2), Elizabeth S. Redden (1), Michael Barnes (1)

(1) US Army Research Laboratory
Human Engineering and Research Directorate
Fort Benning, GA, USA
(2) TNO Human Factors
Soesterberg, The Netherlands

ABSTRACT

Both audio and visual perceptions are critical for soldiers to gain situation awareness (SA) of their surroundings. However, when unmanned ground vehicles are used for reconnaissance, it has not been established whether high fidelity (i.e., stereoscopic vision and three dimensional audio) perceptions will enhance awareness and effectiveness relative to lower fidelity capabilities that are more common to existing systems. This report describes the effectiveness of robotic telepresence features that included enhanced visual and three-dimensional audio sensors, along with naturalistic head-mounted control of camera movement. Soldiers performed equivalent search and identify tasks with each controller interface. Measures included indices of performance (e.g., time, accuracy), workload (NASA-TLX), SA, and user experience. Results indicated that the integrated multisensory perception and naturalistic control provided by telepresence features contributed to better task performance and lower workload. This experiment was conducted as part of a research collaboration between the US Army Research Lab field element at Fort Benning, GA, and TNO Netherlands.

Keywords: Telepresence; Robot controller; 3-D audio; Naturalistic control

1   INTRODUCTION

Small ground robots, such as the PackBot and TALON, have been widely used by military personnel to hunt for terrorists and perform all types of reconnaissance duties (Axe, 2008). Their rugged small size and video capabilities make them very effective for non-line-of-sight reconnaissance tasks (e.g., search and assessment) in darkness or dangerous context, and their widespread use reduces the human risk in combat reconnaissance missions. While their contributions to soldier performance and well-being have been established, many vital robot tasks have been identified as high workload tasks given Army operational context (Mitchell 2005; Mitchell and Brennan, 2009). In addition, many tasks have been identified that can lead to failure (e.g., damage to operator, robot, and/or rescue victim) (Scholtz, Young, Drury, and Yanco, 2006). Exploratory missions in which the robot operator must teleoperate the robot while also attending to the environment in order to gain intelligence, has been associated with especially high operator workload (Chen et al., 2008). Such missions require many abilities, such as driving, sensing, and information evaluation, in order to perform successfully while maintaining SA

2   TELEPRESENCE AND 3-DIMENSIONAL AUDIO

Several approaches are expected to mitigate the high workload associated with robot control, to include higher levels of robot autonomy, intelligent decision support systems, and improved displays and controls (Chen and Barnes, in press; Chen, Barnes, & Harper-Sciarini, 2011). Displays and controls having telepresence features are expected to enhance perception while reducing cognitive workload (Van Erp et al., 2006). Telepresence features include an array of characteristics that enable the operator to feel "present" in robot teleoperation tasks, and that enable naturalistic control of devices such as cameras and arms for grasping and manipulating objects.

In this study, telepresence features were systematically varied to investigate effects on performance when compared to a baseline controller. Telepresence capabilities included three dimensional (3-D) vision, 3-D audio, and head-driven camera movement controls. The operators wore a head-mounted display (HMD) and head tracker to allow them to move the camera through head movements as if they were seeing through the robot's "eyes." This capability has also been demonstrated in a portable binocular format (Jansen, 2006). In additional, three-dimensional audio capability has been developed and demonstrated for robot control tasks (Keyrouz and Diepold, 2007) and in this investigation, was expected to contribute to object search and identification tasks when audio localization cues are provided.

3   EXPERIMENT DESIGN

Experiment tasks were performed indoors, within an environment with several partitions that the robot would have to navigate in order to visually locate targets of interest. The reconnaissance environment consisted of a large room (about 60m^2) subdivided in several sections, and a smaller adjacent room (about 8 m^2). Eleven possible target objects varying in size were positioned at different height levels in the reconnaissance environment.

Three human-robot interface setups of the control station were used in this experiment:

- Mono-Joystick: Mono audio and video on Head Mounted Display, with joystick control for robot movements and heading of sensor system. Participants were asked (and reminded when needed) not to move their heads.

- Mono-Headtracking: Mono audio and video on Head Mounted Display, with joystick control for robot movements and headtracking for directing the sensor system.

- Telepresence: Stereo audio and video on Head Mounted Display, with joystick control for robot movements and headtracking for directing the sensor system.

Figure 1 shows the general layout, with positioning of target items for the audio search task. It provides a map of the target environment depicting the robot's starting location, the decoy and practice targets (gray boxes), and real targets (F and I were used in practice trials). Targets K and J are in a room adjacent to the larger room with the other targets. The gray T-shapes subdivided the larger room.

The unmanned ground vehicle (UGV) used was TNO's robot called 'Generaal' (see figure 2). This UGV is a fully manually controlled UGV, with a fast and powerful pan-tilt-roll sensor system that can accurately mimic human head movements enabling remote perception of the UGV environment. On top are two cameras for providing stereo vision at the control station, and two microphone arrays that can be positioned at either side for spatial 3D audio, or next to each other in front thereby functioning as a mono-audio condition. The sensor unit is presented enlarged in the upper right panel, with the microphone array placed in their 3D audio position, at either side of the stereo cameras. The lower right panel shows how the two microphone arrays were placed in the center position right above the stereo cameras, for receiving mono sound. Each participant performed the sound localization task 18 times. Each of the six targets was used for each of three conditions. After each trial, the participant switched to one of the other two experimental conditions.

Participants. Twenty-two Soldiers were recruited from the OCS to participate in the study. Solders included those with prior service as well as those who entered OCS directly from college. Eighteen Soldiers were able to participate fully; other sessions had to be cancelled, two due to time constraints and two due to discomfort (e.g., nausea).

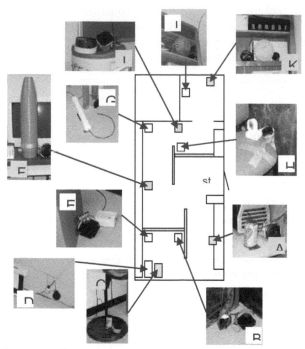

Figure 1. Experiment test layout

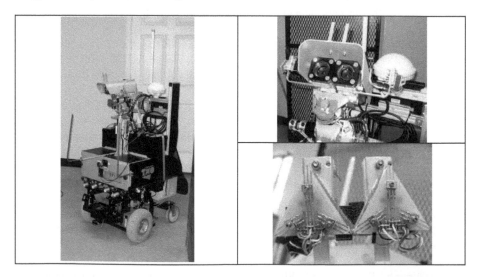

Figure 2: TNO's Unmanned Vehicle 'Generaal'. Left panel shows the vehicle with sensor unit on a pan-tilt-roll motion platform with 3D audio and stereo visual sensors.

4   RESULTS

Figure 3 provides the mean times and standard deviations for time to detect and identify target items for each controller condition. Time was shortest for full telepresence and longest for the mono joystick condition. Two-way repeated measures ANOVA of both display type and target type showed significant differences among the means for display ($F(2,14) = 12.42$, $p = 0.00$, $\eta^2_p = 0.64$) and for targets ($F(5,11) = 15.14$, $p = 00$, $\eta^2_p = 0.873$).

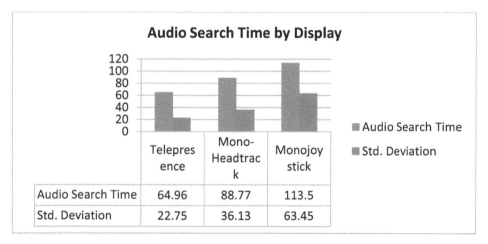

	Telepresence	Mono-Headtrack	Monojoystick
Audio Search Time	64.96	88.77	113.5
Std. Deviation	22.75	36.13	63.45

Figure 3. Mean time and standard deviations for time to approach for each ontroller condition.

Percent of correct identifications. Figure 4 shows the percentage of targets correctly identified by display and target. Two-way ANOVA (see table 2) shows that the differences due to display conditions are significant ($F(2,14) = 9.51$, $p = 0.002$, $\eta^2_p = 0.58$). In addition, differences due to target are significant ($F(5,11) = 4.104$, $p = 0.024$, $\eta^2_p = 0.651$), and while the interaction term was not significant, the effect size was very high ($F(10,6) = 3.30$, $p = 0.08$, $\eta^2_p = 0.846$). Post-hoc Holm's Bonferroni indicate that the telepresence was significantly higher in percentage of correct identifications. Figure 5 illustrates the interaction. Telepresence was generally associated with higher percentages but was particularly helpful for some targets.

132

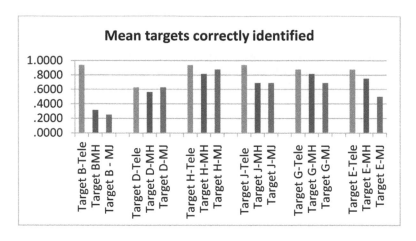

Figure 4. Percentage of targets correctly identified by display and target.

Mean Driving Errors A repeated measures ANOVA of display and target showed a significant difference due to display ($F(2,13) = 5.14$, $p = 0.02$, $\eta^2_p = 0.44$) but not for targets ($F(5,10) = 2.36$, $p = 0.12$, $\eta^2_p = 0.54$) or their interaction ($F(10.5) = 1.67$, $p = 0.29$, $\eta^2_p = 77$). However, effect sizes are large (see figure 5)

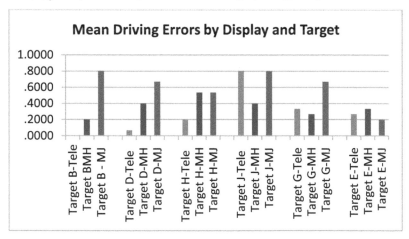

Figure 5. Mean driving errors by display and target.

Spatial Ability Assessment. Spatial ability was assessed through the Cube Comparisons Test [*] (Ekstrom et al., 1976). This test measures the ability to mentally

---

[*]Copyright 1962, 1976 by Educational Testing Service.

rotate a line drawing of a 3-D cube. Soldiers were allotted 3 min to mentally rotate and respond to 21 test items. The dependent measure is the correct identification of the mental rotation of each test item from a series of forced-choice line drawings. Soldiers read the instructions and performed sample items prior to the test. They were encouraged not to guess, as the final score is calculated by subtracting the number wrong from the number correct. Scores ranged from 0 to 19, with a mean of 7.22 (sd 4.79).

Spatial ability correlated significantly with audio search measures. For telepresence audio search times, the spatial scores correlated $-.52$ ($p = 0.02$); for mono joystick, spa correlated $-.65$ ($p = 0.00$), for mono head tracking, spa correlated $-.65$ ($p = .00$). There was a significant interaction between display and spa ($F = 4.67, p = .025$), as reflected in the different correlation values between spa and audio search times for the different display conditions. The correlation with performance was lowest for telepresence, with the implication that telepresence allowed participants with lower spa to perform somewhat better than the other conditions. This can be seen in figure 6, where Soldiers with low spa performed much faster in the telepresence condition as opposed to the other conditions.

NASA TLX. Soldiers provided direct ratings of the NASA-TLX workload scales. Ratings were significantly lower for the telepresence condition, for mental workload (workload ($F(2,15) = 20.98, p = 0.00, \eta^2_p = 0.74$); effort ($F(2,15) = 9.44, p = 0.00, \eta^2_p = 0.56$), and frustration ($F(2,15) = 7.82, p = 0.01, \eta^2_p = 0.51$) (see figure 7).

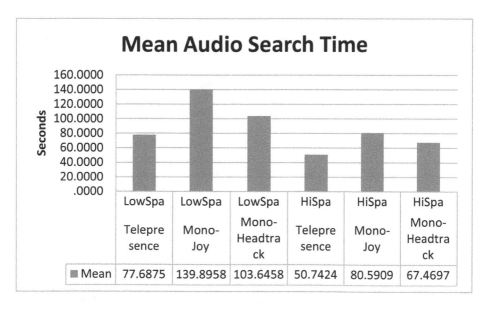

Figure 6. Mean audio search times by spa and display.

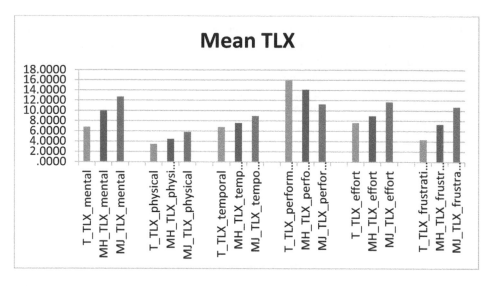

Figure 7. Mean NASA-TLX workload ratings by display.

5   DISCUSSION

Results supported expectations that telepresence features would aid performance and lower workload. Operators found targets more quickly and identified them more accurately with the telepresence system. In addition, self reports of workload using the NASA TLX showed lower ratings of workload, frustration, and effort associated with the telepresence condition. In addition, the telepresence condition enabled operators with lower scores on the spatial ability test to perform more effectively, more so than operators with higher scores. This is another indicator that the telepresence system lowers cognitive effort while enhancing ease of use.

Further details can be stated with regard to experiment conditions:

- Does headtracking control lead to improved performance as compared to joystick control? The results show no difference between the Mono-Headtracking condition and the Mono-Joystick condition in correctness of target identification. However, with joystick control, more time is needed for target identifcation: about 26% more time is needed when using a joystick for sensor control (here Mono-Joystick with 111.2 seconds on average) as compared to headtracking (here Mono-Headtracking with 88.2 seconds).

- Does a 3D audio system lead to improved performance as compared with a directional microphone? When comparing the Telepresence condition (having 3d audio) with the Mono-Headtracking condition (having a directional microphone), we see that with Telepresence the percentage of correctly identified targets is about 23% higher. In addition, target identification takes about 35% more time without having the 3D audio functionality available (here 88.2 and 65.0 seconds for Mono-Headtracking and Telepresence respectively).

- What would be the maximum performance benefit of telepresence functionality as compared with the currently mostly used control systems with joystick control and mono sensor information, provided it exists? Based on the results in this study, the use of a Telepresence human-robot interface results in identification/localization times for audio that are about 42% shorter than with current commonly used interfaces (65.0 sec and 111.2 sec for Telepresence and Mono-Joystick respectively). In addition, the target identification performance increases by about 26% when using the Telepresence human-robot interface.

These promising results encourage more elaborate testing in operational settings, following our initial field trials with telepresence UGV.

REFERENCES

Axe, D. Warbots: U.S. military robots are transforming war in Iraq, Afghanistan, and the future. Nimble Books LLC: Ann Arbor, MI, 2008.

Chen, Y. C.; Durlach, P. J.; Sloan, J. A.; Bowens, L. D. Human–Robot Interaction in the Context of Simulated Route Reconnaissance Missions. Military Psychology 2008, 20, 135–149.

Chen, J.Y.C., & Barnes, M.J. (in press). Supervisory control of multiple robots: Effects of imperfect automation and individual differences. Human Factors.

Chen, J.Y.C., Barnes, M.J., & Harper-Sciarini, M. (2011). Supervisory control of multiple robots: Human performance issues and user interface design. IEEE Transactions on Systems, Man, and Cybernetics, Part C: Applications and Reviews, 41(4), 435-454.

Ekstrom, R.; French, J.; Harman, H.; Derman, D. Kit of Factor-Referenced Cognitive Tests; Educational Testing Service: Princeton, NJ, 1976.

Jansen, C. Telepresence Binoculars (TBI): A Technology Demonstrator for a Telepresence Control Unit of Unmanned Vehicles. In Proceedings of the Human Factors of Uninhabited Military Vehicles as Force Multipliers; TNO Human Factors Research Institute: Soesterberg, The Netherlands, 2006.

Jansen, C., van Breda, L., & Elliott, L. (in press) Remote auditory tearget detection using an unmanned vehicle – Comparison between a telepresence headtracking 3D audio setup and a joystick-controlled system with a directional microphone. NATO OTAN Research and Technology Organization. Report number not yet assigned. , Neuilly-sur-Seine, France.

Jansen, C.; Van Erp, J. B. P. Telepresence Control of Unmanned Vehicles. In Human-Robot Interactions in Future Military Operations; Barnes, M. J., Jentsch, F. G., Eds.; Ashgate: UK, 2010.

Keyrouz, F.; Diepold, K. Binaural Source Localization and Spatial Audio Reproduction for Telepresence Applications. Presence-Teleoperators and Virtual Environments 2007, 16 (5), 509–522.

Kolasinski, E. Simulator Sickness in Virtual Environments; Tech. Rep. No. 1027; U.S. Army Research Institute for the Behavioral and Social Sciences: Alexandria, VA, 1995.

Mitchell, D. K. Soldier Workload Analysis of the Mounted Combat System (Mcs) Platoon's Use of Unmanned Assets; ARL-TR-3476; U.S. Army Research Laboratory: Aberdeen Proving Ground, MD, 2005.

Mitchell, D. K.; Brennan, G. Infantry Squad Using the Common Controller to Control An ARV-A(L) Soldier Workload Analysis; ARL-TR-5029; U.S. Army Research Laboratory: Aberdeen Proving Ground, MD, 2009.

Neerincx, M. A.; Mioch, T.; Jansen, C.; Van Diggelen, J.; Larochelle, B.; Kruijff, G.; Elliott, L. Multi-Modal Human-Robot Interaction for Tailored Situation Awareness. IEEE Transactions on Systems, Man and Cybernetics - Part C, Special Issue on Multimodal Human - Robot Interfaces, submitted for publication.

Pettitt, R.; Redden, E.; Pacis, E.; Carstens, C. Scalability of Robotic Controllers: Effects of Progressive Levels of Autonomy on Robotic Reconnaissance Tasks; ARL-TR-5258; U.S. Army Research Laboratory: Aberdeen Proving Ground, MD, 2010.

Scholtz, J., Young, J., Drury, J., & Young, H. Evaluation of human-robot interaction awareness in search and rescue. DARPA MARS report. MITRE Corporation: Bedford, MA. http://www.cs.uml.edu/~holly/papers/scholtz-young-drury-yanco-icra04.pdf (8 march 2012).

Van Breda, L.; Van Erp, J. Supervising UMVs: Improving Operator Performance Through Anticipatory Interface Concepts. In Proceedings of the Human Factors of Uninhabited Military Vehicles as Force Multipliers, Neuilly-sur-Seine, France, 2006; pp 22-1–22–12.

Van Breda, L. & Draper, M.H. (2012). Supervisory Control of Multiple Uninhabited Systems -Methodologies and Enabling Human-Robot Interface Technologies. NATO RTO technical report RTO-TR-HFM-170. Neuilly-sur-Seine, F: NATO Research and Technology Agency.

Van Erp, J.; Duistermaat, M.; Jansen, C.; Groen, E.; Hoedemaeker, M. Telepresence: Bringing the Operator Back in the Loop. In Virtual Media for Military Appliations Meeting Proceedings, Neuilly-sur-Seine, France, 2006; pp 9-1–9-18).

CHAPTER 15

Optimizing the Design of Driver Support: Applying Human Cognition as a Design Feature

Boris M. van Waterschoot, Mascha C. van der Voort

University of Twente
The Netherlands
b.m.vanwaterschoot@utwente.nl

ABSTRACT

The design of advanced driver assistance systems faces the challenge of providing automated support behavior that complements the human driver safe and efficient. In order to explore and evaluate the coordination between man and machine, human emulation as a simulation alternative is already acknowledged (cf. Wizard of Oz approach). However, validation for such an approach within the context of designing driver support, is currently missing. This paper reports a validation study concerning the use of human emulation during the design of advanced driver assistance. For this, an automated and an emulated version of a lateral support system were compared in a fixed-base driving simulator setup. Participants received a directional precue on the steering wheel and used this information to choose the safe direction in an upcoming time-critical situation in order to avoid a rear end collision. Differences in precue onset were observed for the automated and emulated version, but the distinct characteristics in terms of timing did not reveal an effect on driver reaction times. Present results not only reveal a relatively high tolerance for precueing onset, they also suggest human emulation as a valid simulation alternative during the design process of driver support.

Keywords: Advanced driver assistance systems (ADAS), driving simulation, human emulation, human centered design, cooperative driving, precueing

1   INTRODUCTION

The nature of the conventional driving task is changing at high pace due to the emerging trends in vehicle automation. One of these trends is the implementation of advanced driver assistance systems (ADAS), which are vehicle control systems using sensors, being able to recognize and react to traffic situations. Examples of ADAS applications currently available for commercial vehicles are longitudinal control and warning systems (e.g. Adaptive Cruise Control) and lateral support (e.g. Lane Departure Warning). In general, implementing these in-vehicle driver support systems are attempts to reduce the cognitive efforts placed on the driver. And while these support systems are aimed at providing relevant information and to execute driving (sub)tasks in order to make driving more comfortable and safe, they have the potential to contribute to traffic efficiency as well, e.g. in terms of increased traffic capacity and achieving string stability when using cruise control assistance (Piao and McDonald, 2008). However, as stated by the same auteurs, the full potential of such a system can only be revealed with a penetration level of 100 percent and new safety issues might arise because of drivers' reduced workload, increased response time and the sudden need for driver intervention when confronted with a vehicle changing lanes. The possibility of driver support eliciting unwanted effects on driver behavior can be shown by negative behavioral adaptation, which may occur when increased safety margins and drivers' personalities result in driving behavior that was not intended by the systems' designers. This shows that the full potential of driver support and all its possible effects on both driver and traffic are difficult to anticipate. When designing driver support systems, one should therefore be cautious concerning their predicted safety benefits (Rudin-Brown and Noy, 2002). Furthermore, because of a fundamental asymmetry in competencies for coordinated activity between human and automation (e.g. Woods et al., 2004), challenges lie in finding the answer how proper coordination between man and machine can be established and how this knowledge can be used as an aid for designing driver support.

Metaphorically speaking, contemporary driving can be viewed as man and machine coordinating their behavior in order to complement individual actions. Not surprisingly, a general shift in human factors research is being observed where adapting the automation to the human needs and shortcomings is replaced by an approach that tries to integrate man and computer within a single system. In this view man and machine cooperate or collaborate in order to accomplish a shared or joined task. One way of approaching the design of such a man-machine system can be explained as aiming at getting the automation being a team player (Christoffersen and Woods, 2002; Klein et al., 2004; Davidsson and Alm, 2009). Within this point of view, the supporting automated team players or agents are confronted with human abilities and cognition, different from their own, but they have to be interpreted in order to complement the human in an efficient manner. Ideally, the automated system should therefore be augmented with human abilities that enable a cognitive interaction between human and automated agents. Such systems, emulating human capabilities, would allow automation and humans to interact in a peer-to-peer fashion.

Several methodological approaches are available in order to anticipate the potential safety and usability issues associated with driving automation (for a review of automation issues, see Parasuraman and Riley, 1997). Among these approaches, the use of Virtual Reality (VR) in driving simulator studies enables the controlled presentation of different driving scenarios and support behavior. Moreover, driving simulators potentially serve as both design and research environment, combining the design and evaluation of driver support.

The notion of humans and automation being collaborating agents, combined with the availability of VR, is of special interest for the current study because it provides the possibility to establish a setting for human-agent teamwork within the context of anticipating and evaluating their collaboration.

Given such a setting, the current research proposes the use of human emulation as a simulation alternative in order to support the design of ADAS. It is claimed that during the development process of driver support, a human co-driver can simulate support behavior in order to explore and assess design alternatives. Moreover, by using a human co-driver, the simulation environment has access to maximized (i.e. human) cognitive abilities that potentially serve as a model for support behavior. In addition, such an approach by-passes automation limitations that would otherwise constrain the potential behavioral repertoire of the envisioned driver support.

While the use of human emulation (cf. Wizard of Oz studies, for a review of this approach, see Dahlbäck et al. 1993) is well covered in literature and its application is already acknowledged in the context of designing and evaluating driver support (Schieben et al., 2009), validation of such an approach is currently missing. The main objective of this study, therefore, was to investigate whether emulating driver assistance is a valid simulation alternative during the design process of ADAS.

Two qualities that - at least intuitively - distinguish humans from pre-programmed algorithms (i.e. an automated version) are timing and accuracy. On both variables humans are known for their inconsistency and as a result this might constitute one of the potential drawbacks of the current approach because it potentially influences the support's behavioral characteristics. However, while such inconsistency is inherent to human behavior, the variation of this characteristic might be of such a (small) degree that it complies with our claim of appropriately simulating automation by means of human emulators. An important prerequisite for using human co-driver behavior as a simulation alternative for driver support is that it should elicit driving behavior similar to that of an automated version. Because, when the assumed inconsistencies of the co-driver have a different effect on the driver's behavior this would prove the inability of humans to emulate specific driver support.

In order to address whether a human co-driver and an automated version have a different effect on drivers' behavior, we performed a driving simulator experiment in which both versions of a single support system were compared. While drivers received lateral support by means of a directional precue on the steering wheel, fifty percent of the trials were given by an automated version and fifty percent of the haptic feedback drivers received was provided by a human co-driver.

Given the assumed differences between an emulated and automated version, we expected the onset of the precue to be different for both conditions. Since this could result in different time courses, and therefore different support behavior for both versions, the drivers' responses on the imperative stimulus were expected to be different for both conditions as well.

2 METHOD

2.1 Participants

Twenty eight participants (23 male and 5 female, aged between 18 and 44) attended an experimental session of 45 minutes. Participants were divided into Drivers and Emulators. Three participants attended the experiment as Driver and Emulator successively in different sessions. All participants had normal or corrected-to-normal vision and were naïve about the purpose of the study.

2.2 Driving task and apparatus

Participants drove, with a short headway, behind an ambulance in the center of a three lane highway in ACC mode. This means that they drove with a fixed speed (approximately 84 km/h), not using accelerator, brakes and clutch. Because of the short headway, participants were deprived of upcoming traffic and this forced them to make a swerve manoeuver when the ambulance would press brakes due to upcoming stationary vehicles. During each run of approximately 1.7 km, drivers received driver support by means of a directional precue on the steering wheel, which indicated the safe direction in case of an inevitable lane change. As soon as the ambulance pressed brakes, drivers acted according to the earlier received cue and reaction times for initiating a lane change to the right or left were measured. The imperative stimulus consisted of the ambulance' brake lights turning red.

Driver support was either generated by a predefined automatic version or by a human co-driver that emulated the support behavior. Emulators were seated behind a curtain and Drivers were unaware of their presence and task. Emulators controlled a secondary steering wheel which was connected to the Drivers' steering wheel. An additional monitor showed an animated representation of the traffic situation and indicated the appropriate direction by means of a green arrow. Emulators' were asked to respond as soon as a visually presented cue (purple vehicle) appeared on the screen by turning the steering wheel in the precued direction.

The setup consisted of a fixed-base, medium-fidelity driving simulator and the car mockup-up was placed in front of a visual screen with 180 degrees field of view. The virtual driving environment was generated using Lumo Drive version 1.4 developed by Re-lion. Driving data was recorded with a frequency of 30 Hz and contained trial number, time, vehicle position, steering wheel angle, and codes for each event executed by Drivers and support system. Figure 1 shows the co-driver's interface and an animated impression of the current setup, respectively.

Figure 1 Left: the co-driver's interface and right: an animated impression of the current setup

2.3 Procedure

Participants were divided into Drivers (N = 25) and Emulators (N = 7) and were received separately in order to keep Drivers unaware of the Emulators' presence. After being informed about the general procedure of the experiment and after being familiarized with the driving task, Drivers performed 18 runs with a short break in between. During the experimental trials (67%) Drivers received a directional precue on the steering wheel, which indicated the safe direction for a future swerve manoeuver. After receiving the haptic cue, Drivers had to respond accordingly as soon as the ambulance would hit brakes due to upcoming stationary vehicles. Precues were induced either by the Emulator or by an automated version in a 50/50 ratio and were presented randomly. Since Emulators provided input during all experimental trials, neither they, nor the experimenter knew which version induced the driver support (double blind). During the remaining trails (33%) Drivers received no directional precue and they performed a two-choice reaction time task after the imperative stimulus appeared.

Since Emulators were assumed to show variable responses, the time courses for both versions were expected to be different. After cue onset (i.e. the Emulator's imperative stimulus) a fixed interval of 3.3 seconds followed before the Driver's imperative stimulus was presented. However, while the Driver's precue (target) was set within a fixed interval in the pre-programmed version, the onset of the Driver's precue depended on the (early or late) responses of the Emulator. This means that the time course for the automated condition was the same for all trials, while the time course for the emulated condition might differ for trials. In Figure 2, the time course for the emulated condition is given.

142

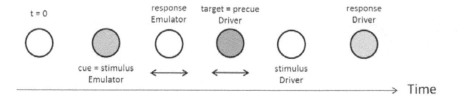

Figure 2 Time course for experimental trials in emulated condition. While the time interval between cue and target is fixed in the automated support version, here the onset of the target depends on the speed of the Emulator's response

2.4   Experimental design and data analysis

This study used a 2 x 3 repeated measures design. The first within-subject factor was Support Type (emulated vs. automated version), the second factor Support determined whether support was given and in which direction (no support and left vs. right). Dependent variable was reaction time (RT) and was recorded for both Drivers and Emulators. RT for both groups was defined as the time from their respective imperative stimulus onset to the moment in time at which the steering wheel angle was 10 degrees. In order to determine whether Drivers received the directional precue at the same time for each Support Type, RT Support compared the timing of emulated support and automated support. Trials in which participants responded before or at stimulus onset and trials with RT > 2 seconds were discarded from data analysis. The number of trials submitted to analysis was 388 (86%) and the probability level for statistical significance was $p < .05$.

3   RESULTS

3.1   RT Support

A 2 Version (emulated vs. automated support) x 2 Direction (left vs. right) repeated measures ANOVA revealed a main effect for Version, $F(1,24) = 401.7, p < .001$. This means that the emulated support ($M = 1.01$ sec, $SD = .23$ sec) was given faster than the automated version ($M = 1,78$ sec, $SD = .07$ sec). This confirms our hypothesis that the onset of the precue was different for both conditions and therefore resulted in different time courses for both versions. In addition, the results show a larger RT variability for emulated support.

3.2   RT Drivers

A 2 Version (emulated vs. automated) x 3 Support (no support and left vs. right) repeated measures ANOVA revealed a main effect for Support, $F(2,23) = 32.4, p < .001$. This implies faster responses for precued trails to left ($M = 1.18$ sec, $SD = .12$

sec) and right (*M* = 1.20 sec, *SD* = .12 sec) as opposed to driver responses that were not preceded by directional support (*M* = 1.38 sec, *SD* = .15 sec). The absence of an effect for Version shows that emulated and automated support behavior elicited similar driver responses. Summarized results are given in figure 3.

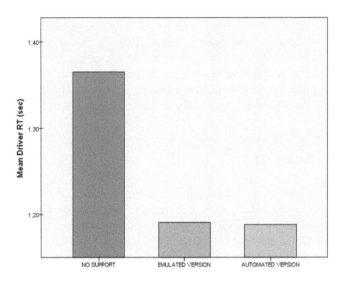

Figure 3 Mean reaction time (RT) for driver responses as a function of Support (no support: M = 1.38 sec, emulated version and automated version: M = 1.19)

4   DISCUSSION

The aim of this study was to investigate whether emulating driver assistance is a valid simulation alternative during the design process of ADAS. For this we employed a driving task in which drivers were supported by means of a directional precue on the steering wheel, which indicated the safe direction in case of a required lane change. In our experiment we compared two different versions of support. Because the emulated support version was induced by a human co-driver, the onset of the support in this condition depended on the speed of the co-driver's response to a visual cue. The automated version, on the other hand, was pre-defined and therefore not prone to variable timings of the support behavior.

The results show that, although both versions differed in terms of timing of the support behavior, this was not reflected by different driver responses. This not only implies a relatively high tolerance for directional precueing when used as driver support, but might also serve as a first claim in favor of human emulation as a simulation alternative. Furthermore, by applying the precueing paradigm in order to elicit driver responses, we were able to confirm the suggestion that drivers might benefit from response preparation as reflected by decreased reaction times (Hofman et al., 2010).

5   CONCLUSIONS

This study suggests that human emulation might serve as a valid simulation alternative during the design process of driver support. However, employing human characteristics by means of using a human co-driver during the design process of ADAS offers additional and comprehensive potential. While current ADAS can be characterized as rigid and reactive, future ADAS might be proactive and flexible. Not entirely a coincidence, the latter is one of the main characteristics of human behavior. When human qualities become available within a context of driver and support cooperating during driving, this would enable studying the requirements needed for an optimized coordination on a peer-to-peer basis. By observing human co-drivers, their behavior may eventually serve as a design feature or model for future ADAS.

ACKNOWLEDGMENTS

This study is part of the knowledge centre Applications of Integrated Driver Assistance (AIDA), a collaboration between the Netherlands Organisation for Applied Scientific Research (TNO) and the University of Twente. Special thanks to Gertjan Tillema for his valuable contribution to this work.

REFERENCES

Christoffersen, K. and D. D. Woods. 2002. How to make automated systems team players. In: E. Salas, ed. *Advances in Human Performance and Cognitive Engineering Research*, Volume 2. Oxford: Elsevier Science.

Dahlbäck, N., A. Jönsson and L. Ahrenberg. 1993. Wizard of Oz studies – Why and how. *Knowledge-Based Systems* 6: 258–266.

Davidsson, S. and H. Alm. 2009. Applying the "Team Player" approach on car design. In D. Harris, ed. *Engineering Psychology and Cognitive Ergonomics*. Berlin: Springer.

Fastrez, P. and J.-B. Haué. 2008. Designing and evaluating driver support systems with the user in mind. *International Journal of Human-Computer Studies* 66: 125-131.

Hofman, P., G. Rinkenauer and D. Gude. 2010. Preparing lane changes while driving in a fixed-base simulator: Effects of advance information about direction and amplitude on reaction time and steering kinematics. *Transportation Research Part F: Traffic Psychology and Behaviour* 13: 255-268.

Klein, G., D. D. Woods, J. M. Bradshaw, R. R. Hoffman and P. J. Feltovich. 2004. Ten challenges for marking automatio a 'team player' joint human-agent activity. *IEEE Intelligent Systems 19*: 91–95.

Parasuraman, R., and V. Riley. 1997. Humans and automation: Use, misuse, disuse, abuse. *Human Factors* 39: 230–253.

Piao, J. and M. McDonald. 2008. Advanced Driver Assistance Systems from Autonomous to Cooperative Approach. *Transport Reviews* 28: 659-684.

Rudin-Brown, C. M. and Y. I. Noy. 2002. Investigation of behavioural adaptation to lane departure warnings. *Transportation Research Record* 1803: 30-37.

Schieben, A., M. Heesen , J. Schindler , J. Kelsch and F. Flemisch. 2009. The theater-system technique: agile designing and testing of system behavior and interaction, applied to highly automated vehicles. Proceedings of the 1st International Conference on Automotive User Interfaces and Interactive Vehicular Applications, September 21-22, 2009, Essen, Germany.

Woods, D. D., J. Tittle, M. Feil and A. Roesler. 2004. Envisioning Human-Robot Coordination in Future Operations. *IEEE Transactions on Systems, Man, and Cybernetics - Part C* 34: 210-218.

CHAPTER 16

Dictating and Editing Short Texts while Driving

Jan Cuřín, Jan Kleindienst, Martin Labský, Tomáš Macek

IBM Prague Research Lab
V Parku 2294/4
14800 Praha 4, Czech Republic
Tomas_Macek@cz.ibm.com

Hoi Young, Ann Thyme-Gobbel, Lars König, Holger Quast, Christophe Couvreur

Nuance Communications
1198 East Arques Avenue
Sunnyvale, CA 94085, USA
Hoi.Young@nuance.com

ABSTRACT

Although several existing in car systems support dictation, there is none, which would systematically address dictation and error correction for automotive environment. Dictation and correction systems available for desktop and mobile are not suitable for the car environment where safety is the crucial aspect. This paper presents a multi-modal automotive dictation editor (codenamed *ECOR*), designed as a test-bed for evaluation of numerous error correction techniques. Results are presented both for a standard use of the application as well as for the case when a particular type of correction is enforced. Reported results are obtained on native US-English speakers using the system while driving a standard lane-change-test (LCT) low fidelity car simulator. The dictation editor was tested in several modes including operations without any display, with a display showing the full edited text, and with limited view of just the "active" part of the dictated text. The measured results are compared to SMS dictation using a cell phone and to destination entry using a GPS unit. The results indicate that the eyes-free version keeps the distraction level acceptable while achieving good task completion rate. Both multi-modal versions caused more distraction than the eyes-free version and were comparable to the GPS entry task. By far, the cell phone texting task was the most distracting one. Text composition speed using dictation was faster than cell phone typing.

Keywords: voice, speech recognition, multi-modal, automotive systems, user interface, error correction

1.   INTRODUCTION

The primary concern when developing automotive UIs is to keep driver's distraction minimal. The secondary aims are to minimize task completion time, to maximize task completion quality and achieve high user acceptance.

In this paper we evaluate a prototype text dictation UI codenamed ECOR. The ECOR system has been developed with the aim to act as a test-bed for evaluation of multiple error correction techniques. It implements several variants of multimodal user interface. We report here results of usability tests conducted with these different UIs. A mix of objective and subjective statistics was collected in attempt to capture UI performance and optimize the UI design.

The first part of the text refers to preexisting work, description of the system and of the testing setup. Then we present a summary of results obtained both for natural usage and for testing using selected correction mechanisms.

2.   RELATED WORK

The effect of using cell phones in a car was studied for example in (Barón and Green, 2006). Users tend to use phones in spite of law restrictions. The study reports significant negative impacts of cell phone use on driving performance.

Comparisons of driving performance degradation due to using conventional and speech-enabled UIs were addressed in several works; e.g. (Medenica and Kun, 2007); a good summary can be found in (Barón and Green, 2006). The general conclusion is that while speech UIs still impact driving quality, they do so significantly less than conventional UIs. Most distraction caused by conventional systems seems to be due to drivers looking away from the road, which can be measured e.g. by the number and duration of eye gazes. In addition, using speech was observed to be faster for most evaluated tasks.

A number of approaches were described to perform dictation in hands-busy environments (Oviatt et al., 2000). Previous work was also done on email messaging in a car (Jamson, Westerman et al., 2004). In particular, hands-free text navigation and error correction were addressed by (Suhm, Myers et al., 2001). The impact of the most prevalent correction method, re-speaking, was evaluated by (Vertanen, 2006).

We presented more detailed results regarding natural use of our system in (Curin, Labsky et al., 2011).

3.   APPLICATION INTERFACE

The ECOR dictation editor allows for entering short texts primarily using open-domain dictation. Alternate input modalities include spelling by voice and handwriting (e.g. to input out-of-vocabulary words).

Figure 1. Full message view   Figure 2. Strip view with text position indicator

The prototype can be used with or without a display (multi-modal and eyes-free modes). The user initiates dictation by pressing the speech button. Recording ends automatically after the user has stopped speaking or after the speech button has been pressed again. After dictating a phrase, the recognized text is echoed back using text-to-speech (TTS). The driver may navigate the text using previous/next buttons or a rotary knob while TTS plays back the active word(s). Text can be navigated by whole recognized phrases (called chunks), by individual words, and by letters (chunk, word and letter modes).

The active text item is always the one last spoken by the TTS. It is subject to contextual editing operations, which include deletion, replacement by the next or previous n-best alternate, and several voice commands. Context-free editing operations include undo and redo, and corrective voice commands (see Table 1).

Besides the eyes-free setup, there are two kinds of GUI available: the Full view (Figure 1), always showing the full dictated text, and the Strip view (Figure 2), only showing the active word(s) and, optionally, near context.

Besides dictation input, several voice commands listed in Table 1 are recognized by the system in order to carry out actions not mapped to physical controls such as buttons or knobs.

Table 1: Types of voice commands

Always available	After some text has been entered
Help	Send it
Chunk / word / spell mode	Read the whole message
Undo	Delete the whole message
Redo	Capitalize / To uppercase / To lowercase
Spelling A B C D ...	Replace <wrong> by <correct>
	I said <correct>

4. EXPERIMENTAL SETUP

A low fidelity LCT car simulator similar to (Mattes, 2003) was used to simulate driving in an office environment. The simulator was shown on a 40" screen and the ECOR screen (except for the eyes-free setup) showed on a separate 8", 800x480 touch-screen, positioned on the right hand side of the simulator screen. A Logitech steering wheel and pedals were used to control the simulator and 5 buttons (incl. push-to-talk) on the steering wheel controlled the prototype.

One LCT trip consisted of a 3 km straight 3-lane road with 18 irregularly distributed lane change signs. Drivers kept a fixed speed of 40 km/h during the whole trip. One LCT trip took approximately 5 minutes.

5. EVALUATION PROCEDURE

All user studies described in this paper were conducted with native US-English speakers. We conducted two types of tests. In natural usage tests the users interacted freely with the UI to enter messages of prescribed semantic content. During feature tests, users were interactively asked by the test conductor to modify or correct parts of dictated text using a specific correction method, such as voice commands.

A group of 28 novice test subjects (14 female, 14 male, age between 18 and 55) was used to measure objective statistics as well as to record subjective feedback. They used their cell phones regularly to send at least 10 text messages or emails per day. For the natural usage tests, these users had no previous experience in using voice-controlled dictation systems. The feature tests were carried out several months later with 12 users randomly selected from the initial group.

The Eyes-free, Full view and Strip view versions of ECOR were evaluated as well as cell phone typing and GPS address entry, and compared to undistracted driving. Each test subject was evaluated on the undistracted driving task, one of the three ECOR tasks, and one of the two reference tasks (cell phone texting or GPS). The order of the distracted driving tasks was counter-balanced.

First, each subject was allowed to train driving until they mastered the LCT and their driving performance did not further improve. Then, two undistracted LCT trips were collected. The second one was used to compute an adapted model of the driver's ideal path and the first was used to compute driving performance statistics using the adapted ideal path. The ideal path was modeled using a linear poly-line estimated using the LCT Analyzer tool (Mattes, 2003).

After the undistracted LCT trips, the users were first introduced to the selected task and then they were evaluated while performing the task while driving. Users were given enough time (up to 15 minutes) to practice the task while parked. The whole procedure had to fit into 1 hour for each participant.

The semantics of the text to be entered was prescribed without enforcing a particular wording (e.g. "instruct your partner to buy oranges, wine and chocolate"). For the cell phone texting task, the subjects were instructed to enter a sequence of text messages with the same semantic content as for the ECOR tasks. The subjects were using their own cell phones, so they were familiar with the phone UI. Use of predictive typing was left up to the choice of the user.

For the destination entry task, a single navigation unit (*TomTom XXL*) was used by all subjects to enter a set of prescribed addresses. The device used auto-complete, therefore the subjects did not typically need to enter a complete address.

Driving quality was evaluated using the LCT Analyzer measuring the following objective statistics. The Mean Deviation from the adapted ideal track (MDev) measured how much off the ideal track the car typically was (in meters). The Standard Deviation of Lateral car Position (SDLP) was computed as the standard deviation of the absolute deviation values, and measured how much the car weaved within its lane. Both statistics were computed in 2 blends: for the overall trip and just for the segments corresponding to lane keeping. We also computed reaction times as the delay between the moment when the lane change sign became visible and the moment when the driver started responding by an observable turn (1.5°) of the steering wheel in the correct direction.

6.   NATURAL USAGE TESTS

The results for natural usage tests are presented here just briefly. We refer the reader to (Curin, Labsky et al., 2011) for a complete set of results. Detailed graphs for overall and lane-keeping driving statistics, respectively, are shown in Figure 3, including 95% confidence intervals.

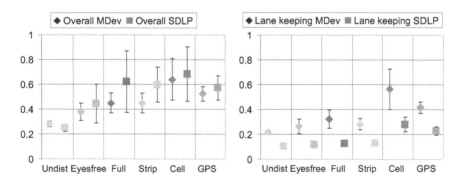

Figure 3: Averaged MDev and SDLP values for undistracted driving, for 3 types of ECOR UI and for 2 reference tasks.

During the evaluation, no lane change signs were missed for the Undistracted (28 subjects) and ECOR Eyes-free tasks (8 subjects). One sign was missed for the ECOR Full View (9 subjects), and also for the GPS task (14 subjects). For the Strip view, we recorded 3 missed signs in total (11 subjects) and for the Cell phone task, there were 4 missed signs (14 subjects). Each LCT trip contained 18 signs. There were no out-of-the-road excursions throughout all evaluated drives.

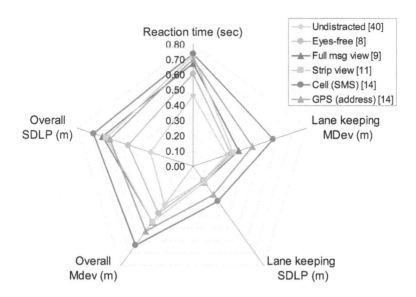

Figure 4: Comparison of various objective driving performance indicators for undistracted driving, for 3 types of ECOR UI and for 2 reference tasks. Least distracting tasks are described by curves located closer to the graph center.

Statistical significance in this paper was determined by the two-sample unequal variance Student's t-test for one-tailed distribution. Two sample means were considered significantly different for p < 0.05. Summary of the results:

- All ECOR setups were significantly less distracting than cell phone usage regarding most of the statistics.
- The Eyes-free ECOR setup was significantly less distracting than both cell phone and GPS usage.
- All ECOR setups with GUI were insignificantly less distracting than the GPS task in most of the evaluated parameters.
- All secondary tasks led to significantly higher MDev than during undistracted driving.

The radar plot in Figure 4 shows data in an easily comparable way. The best driving performance (closest to the center of the graph) was observed for the Undistracted task (light blue), followed by the Eyes-free ECOR task (light green), in most cases followed by the two ECOR tasks with display (in other two green colors). The GPS task (orange) is overlapping with both ECOR tasks with display for the Overall SDLP and for reaction time. The Cell phone task (in red) was associated with the worst driving performance for all statistics.

In addition to driving performance, we also evaluated the subjects' message composition performance. For distracted driving trips except GPS, we collected texts of the composed messages and scored them manually with "quality" based on semantic overlap between the message and its prescription.

Table 2: Text entry speed and quality, numbers of editing operations and dictation turns needed to compose a message, and average word error rate (WER) per task. Numbers in brackets indicate number of subjects per task

Task [subjects]	Avg. msgs sent	Operations per message	Dictations per message	Msg. quality	Avg. WER
Eyes-free [8]	5.5	3.2	1.8	96%	14.1%
Full msg view [9]	5.0	4.2	2.8	92%	14.7%
Strip View [11]	4.1	4.0	2.6	97%	16.2%
Cell (SMS) [11]	4.4	-	-	92%	-

Table 2 shows the average number of messages sent out during one LCT trip. We can see that using voice to dictate messages was on average faster than typing using cell phone. At the same time, the quality of messages did not significantly differ across all tested setups.

Subjects were able to send slightly more messages without the display, which we attribute to the subjects not spending extra time checking the display (perhaps both the text and visual UI elements).

The number of sent messages did not statistically differ for the three ECOR setups (eyes-free, full message, and strip view) and there were no significant differences in semantic quality of text across all ECOR setups and cell phone for native speakers. The average word error rate (14.1-16.2%) for individual ECOR setups did not differ with statistical significance and even though we did not ask the subjects explicitly they seemed to be satisfied or even positively surprised with the ASR performance.

After performing each measured distracted task during an LCT trip, the subjects were asked to fill-in the System Usability Scale (SUS) questionnaire (Brooke, 1996), the NASA Task Load Index (NASA-TLX) questionnaire (Hart and Staveland, 1988), and to answer four additional questions regarding usability and perceived accuracy of the UI. A more detailed summary of the subjective results was published in (Curin, Labsky et al., 2011).

Cell phone texting was the only significantly different outlier in the SUS rating. Using an A through F usability scale described in (Barón and Green, 2006), the ECOR systems and GPS scored A- or B+ levels whereas the Cell phone texting task fell close to the worst F level ranking. In terms of statistical significance, only the cell phone received significantly worse rating than all other UIs ($0.000 < p \leq 0.002$).

The NASA Task load Index (NASA-TLX) rating corresponded well with the objective parameter values: all ECOR tasks were slightly better than GPS and significantly better than Cell phone texting.

Note that among all ECOR tasks, the Eyes-free version received the highest (worst) rating for Mental Demand. This means that not seeing what is dictated was perceived as mentally demanding, but it seems to "pay off" as measured by most of the other objective and subjective indicators reported in this paper.

9.   FEATURE TESTING

Most novice subjects in natural usage tests tended to only use the simplest functions. We did not get enough feedback about driving performance when the subjects had to correct an error using a specific technique. About half of the subjects preferred word-level editing and the other half used re-dictation of a whole dictated chunk. To discover user acceptance of some of the more advanced editing techniques, we used an ECOR Full View setup similar to the natural usage tests but interactively asked the subjects to modify or correct parts of previously dictated text using a specific correction technique.

Besides dictation, the ECOR UI supports several input and correction methods: input by spelling, input by handwriting, correction using voice commands and browsing through n-best candidates. In the evaluation, we focused on three techniques:

- Re-dictation by deleting and then dictating again, or dictating over the active part of text (baseline).
- Correction using voice commands; e.g. "Replace pyjamas by bananas", or just "I said bananas".
- Input by voice spelling, e.g. "spelling A C D C..."

During feature testing we reused 12 random participants from the previous natural usage tests. After undistracted driving, each subject performed three distracted LCT trips, each using a different correction technique. The order was randomized and counter-balanced. Subjects were asked by the test conductor to first dictate a message and then to modify a chosen word. For example, after dictating "Buy 5 bananas and 3 oranges", the subject was asked to change "bananas" to "onions". The test conductor kept the driver busy dictating and correcting throughout the LCT trip.

During evaluation, we compared the three correction techniques mainly in terms of their impact on driving performance. Figure 5 shows that the difference between correction techniques was very small while lane keeping. The results for the whole trip however confirm our initial expectation that the 3 correction methods listed above are ordered from the easiest to the hardest. Still, the driving performance degraded less for the harder tasks (esp. for spelling) than we originally expected.

Overall, the subjects participating in this study drove very well both when undistracted and when under distraction. This we attribute to the fact that all participants were previously exposed to both ECOR and the driving simulator and they could "sleep over" their experience.

Regarding parameters measured for the whole trip (both lane keeping and lane changes), there was a statistically significant difference between MDev measured for undistracted driving and the group of all distracted driving tasks ($0.001 < p \leq 0.049$), and similarly for SDLP ($0.001 < p \leq 0.025$). For lane keeping, there was a statistically significant difference between SDLP measured for undistracted driving and the group of all distracted driving tasks ($0.010 < p \leq 0.045$), but the observed difference in MDev did not reach statistical significance ($p > 0.055$).

154

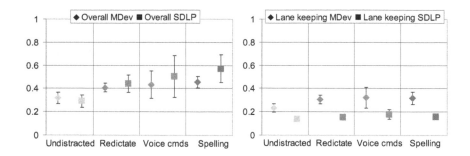

Figure 5. MDev and SDLP values for 3 different correction strategies. Values for the whole trip are shown on the left, lane-keeping values on the right.

As expected, the lane-keeping MDev and SDLP show less significance for the different secondary tasks, while the overall parameters seem to serve as the best indicator. In other words, most of the driving degradation related to the more difficult correction techniques seems to get projected into the quality and latency of lane change maneuvers.

The difficulty of each correction technique also got projected to the participants' subjective feedback; specifically the SUS score showed that the subjects significantly preferred re-dictation and voice commands over spelling as shown in Figure 6. This difference in SUS reached statistical significance ($0.028 < p \leq 0.043$).

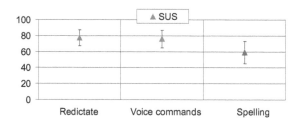

Figure 6. Average SUS scores reported by 12 participants for 3 different correction strategies.

10. CONCLUSIONS

We presented a system for dictation and error correction of short messages intended for in-car use. We evaluated both natural usage by novice users, where participants were free to dictate and correct using techniques of their choice, and specific correction techniques, where returning participants were asked to modify and correct previously dictated text using a particular correction method.

Three different ECOR UIs were evaluated and compared to sending instant messages by phone and to entering destination manually into an off-the-shelf navigation unit. All ECOR setups were significantly less distracting than cell phone

155

usage; the Eyes-free ECOR was significantly less distracting than both cell phone and GPS usage and all ECOR setups with GUI were (insignificantly) less distracting than the GPS task.

With respect to user acceptance (subjective feedback based on the NASA TLX, SUS score and custom questionnaires), all ECOR setups scored significantly better than cell phone usage. Although participants in their subjective ratings significantly preferred re-dictation and voice commands over spelling by voice, the difference between the corresponding objective driving performance statistics did not reach statistical significance.

11. REFERENCES

Barón, A. and Green, P. 2006, Safety and Usability of Speech Interfaces for In-Vehicle Tasks while Driving: A Brief Literature Review. *Tech. Report UMTRI-2006-5. University of Michigan Transportation Research Institute.*

Brooke, J., 1996, SUS: a "quick and dirty" usability scale. In P. W. Jordan, B. Thomas, B. A. Weerdmeester, & A. L. McClelland. *Usability Evaluation in Industry.* London: Taylor and Francis.

Curin, J., Labsky, M., Macek, T., Kleindienst, J., Koenig, L., Young, H., Thyme-Gobbel, A., Quast, H. 2011, Dictating and Editing Short Texts while Driving: Distraction and Task Completion. In: Proceedings of AUI'2011, Salzburg, Austria, November 2011.

Hart, S. and Staveland, L.1988, Development of NASA-TLX (Task Load Index): Results of empirical and theoretical research. In P. Hancock & N. Meshkati (Eds.), *Human mental workload*, pp. 139-183. Amsterdam: North Holland.

Jamson, A.H. and Westerman, S.J., et al., 2004, Speech-based E-mail and driver behavior: effects of an in-vehicle message system interface. *Human Factors.*; 46(4): 625-39.

Labsky, M. and Macek, T. et al. 2011, In-car Dictation and Driver's Distraction: a Case Study. In: J. Jacko (Ed.) *Human-Computer Interaction, Lecture Notes in Computer Science*, Volume 6763/2011, pp. 418-425, Springer Berlin / Heidelberg, 2011.

Mattes, S., 2003, The Lane-Change-Task as a Tool for Driver Distraction Evaluation. *Proc. Annual Spring Conference of the GFA/ISOES*

Medenica, Z. and Kun, A. 2007, Comparing the Influence of Two User Interfaces for Mobile Radios on Driving Performance, *Proceedings of the 4th International Driving Symposium on Human Factors in Driver Assessment, Training, and Vehicle Design*, Stevenson, Washington, July 9-12,

Sauro, J., 2011, Measuring Usability with the System Usability Scale, http://www.measuringusability.com/sus.php

Suhm, B. and Myers, B. et al. 2001, Multimodal error correction for speech user interfaces. *Proc. ACM Transactions on Computer-Human Interaction*, 2001.

Oviatt, S. et al. 2000, Designing the user interface for multimodal speech and pen-based gesture applications: State-of-the-art systems and future research directions. *HCI*, Vol. 15 (4), p.263-322

Vertanen, K. 2006, Speech and Speech Recognition during Dictation Corrections. *Proc. Interspeech.*

Wang, Y. and Chen, Ch. et al. 2008, Measurement of Degradation Effects of Mobile Phone Use on Driving Performance with Driving Simulation, *Proceedings of the Eighth International Conference of Chinese Logistics and Transportation Professionals.*

CHAPTER 17

User Adaptation in Automotive Environments

Ming Sun, Alexander Rudnicky

Carnegie Mellon University
School of Computer Science
Pittsburgh, USA
{mings, air}@cs.cmu.edu

Ute Winter

General Motors Advanced Technical Center - Israel
Herzliya, Israel
ute.winter@gm.com

ABSTRACT

Speech interaction provides a natural form of communication that is intuitively familiar to people. Such interaction can provide drivers with easy access to a variety of services. A problem with such systems is that they are designed to work with a diverse population of users. However, an individual's understanding of services and the language they might use to express their goals will vary. Given that a privately owned car is driven regularly by a small number of individuals, it makes sense to have speech interfaces adapt to the needs and characteristics of specific user(s). We describe user adaptation that may be possible at several levels of a spoken dialog system (such as language and use history). We also describe a prototype based on these principles and an initial user study.

Keywords: *spoken dialog system, speech recognizer, parser, dialog manager, natural language generator*

1   INTRODUCTION

Contemporary car drivers and their passengers (collectively, users) have at their disposal a large variety of services, including access to a car's environmental and information resources and, more recently, remote information sources through wireless internet access. As such services proliferate and acquire additional functionality; traditional forms of interaction (e.g. mechanical) become less adequate. Moreover, their operation becomes less intuitive and may require users to both consult manuals as well as remember particular operations in detail. The unfortunate consequence of this may be that people might actually be less likely to invoke services, even as they appear to grow more useful. Spoken language provides useful counterbalance to this complexity, by allowing users to express their needs verbally, in natural (but goal-relevant) language.

As mentioned in abstract, many spoken dialog systems target at serving a large amount of users in general. This implies that the models within those dialog systems are very generic. For example, the speech recognizer requires a collection of speech data from a variety of speakers or recording environments so that the recognition model can handle speech input of different styles. Since we are dealing with the automotive environment, it would be preferable to use models that are trained specifically for very few users, e.g., the car owners. A well-adapted dialog system can better understand a specific user's commands, making the dialog interaction more efficient and natural. Ultimately, the user experience would be improved.

Other researchers have examined adaptive dialog systems. Some research group discovered that adaptation of system prompts according to user expertise improved the usability of an in-car spoken dialog system (Hassel and Hagen, 2005). Other researchers minimized the frequency of questions about the user by storing user information in a user model (Ishimaru, et al., 1995). In addition, a user model is also adopted to represent whether a particular user has speech recognition problems (Litman and Pan, 1999). It is also pointed out that a dialog system can adjust speaking rate and voice message content to suit the skills of each individual user (O'Sullivan, 2009).

In our work, we explored several approaches towards user adaptation at different levels in a spoken dialog system. In this paper, we describe an integrated approach to adaptation in aspects such as understanding speech, triggering appropriate dialog actions, generating natural language and making use of history.

The rest of this paper is organized as follows. In section 2 we describe the dialog system prototype that we implemented. In section 3 we discuss component-level adaptation. The experimental setup and analysis is presented in Section 4, followed by a Conclusion.

2 CMU POI SYSTEM

We have developed a POI (point of interest) dialog system that allows users to acquire information about travel and locations in the Pittsburgh area (defined for our purposes as about a 10 km radius of Carnegie Mellon University). POIs include gas stations, hotels, restaurants and other likely destinations for a car driver. An internal database and online resources such as the Google Maps API are used to manage information such as distance and driving route. The POI system is built within the

Olympus/Ravenclaw framework (Bohus and Rudnicky, 2008). A multi-modal interface was developed that incorporates both speech interaction and a visual display. The latter is a browser window that displays information retrieved from the web (such as a map).

The Ravenclaw/Olympus framework provides a platform with the necessary components for building a complete dialog system (see Figure 1).

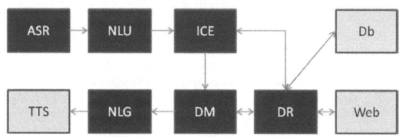

Figure 1 Diagram for Ravenclaw/Olympus framework

The Automatic Speech Recognition (ASR) component decodes speech signal to text, which is passed to Natural Language Understanding (NLU) to extract the semantics; Input Confidence Estimation (ICE) assigns belief values to input concepts. Based on user's input and the current context, the Dialog Management (DM) component decides the next action; this may involve communicating with the Domain Reasoner (DR) that interfaces to information resources. Finally the system generates an output for the user by speaking information or asking a question. Appropriate visual output may also be generated. Spoken output involves Natural Language Generation (NLG) to construct a sentence and Text-to-Speech synthesis to speak an utterance.

Dialog systems can be designed to provide adaptation at different levels of dialog processing:

- Recognition (ASR) makes use of three models: acoustic, lexical and language. The first can be adapted to take into account how individuals produce speech sounds. For example, the acoustics of a male voice are different from those of a female voice, which in turn are different from those of a child's. Lexical models code a talker's pronunciation of different words, and language models represent the statistics of word sequences.
- Language understanding (NLU) parses input using a pre-specified grammar. However the parser may be adapted by tracking the relative frequency of different locutions and words to capture speaking style (information that is coordinated with the lexical and language models in ASR).
- Input confidence (ICE) reflects a judgment as to whether a user input should be believed. For example a poorly-recognized word might be accepted if it's commonly chosen by the user.
- The Dialog Manager (DM) can adapt the dialog plan to reflect common

interaction patterns, for example by trying to skip questions with predictable answers.

- Language Generation (NLG) can be adapted by considering factors such as length of usage (output can become terser with experience) and difficulty (information that seems difficult to assimilate can be presented more verbosely). Similarly spoken material can be slowed down (or sped up) in the synthesis (TTS) stage.

In this paper we will focus primarily on adaptation for the modules in Figure 1 that are in dark boxes.

Interaction Structure

We first consider the structure of an interaction that is generic and not adapted to the needs of a particular individual. This might be the form which is engineered to be as broadly accessible as possible, and would be appropriate for the initial behavior of an adaptable system.

The POI system introduces its capabilities and expects the user to request something. In Ravenclaw, as in many contemporary systems, tasks are represented as forms, with empty slots, that need to be filled. Once a form is filled the system can execute the corresponding request. A typical user input may contain information about the form identity ("how far is...") and perhaps a value for one of the slots ("the river"). The initial input may not be sufficient to carry out the question; that is, other slots remain empty. The dialog manager then comes back with a question related to a missing slot value ("where are you right now?"). Once the form is filled it can be submitted to the Domain Reasoner which in turn can query its data sources. If successful, the answer is returned to the DM, which then communicates it to the user; if unsuccessful, the DM is tasked with clarifying the request ("which river do you mean? There are three..."). Given this, a generic POI interaction might unfold this way:

S: Hello, welcome to the my POI system.
S: I can help you find neighborhoods and businesses in Pittsburgh, and provide you with directions.
S: What can I do for you?
U: How far is downtown?
S: OK. You want downtown. Where are you right now?
U: I am at CMU.
S: OK. You are at CMU and you want downtown.
S: I found that the distance is 3.6 miles.

From the example above, we can envision several potential improvements that would make the conversation more natural and concise. For example, the first two system utterances (the introduction) might not be necessary if the user is already experienced with this system. Moreover, instead of asking "where are you right now?" on each occasion, it could ask "are you at CMU?" if it has learned that this particular user always starts out from CMU. As a result the interaction above might unfold as follows.

S: Welcome back. What can I do for you?
U: How far is downtown?
S: OK. You want downtown. Are you now at CMU?
U: Yep.
S: OK. The distance is 3.6 miles.

The second "expert" interaction is clearly more efficient for the user: less time is spent accessing information. Note that such adaptations need to be dynamic. For example, if the user doesn't interact with the system for a while (say a month or more), it makes sense to restore some of the orienting features of the dialog (such as reminding the user of the system's capabilities). Likewise, if the user appears to have problems with certain outputs (say by always asking for a repetition), the system might regress adaptation in certain contexts, or even switch to a more structured and verbose dialog.

3 USER ADAPTATION STRATEGIES

As noted in the previous section, adaptation can be introduced at several different levels in a spoken dialog system. In this section we describe adaptation strategies that can lead to better understanding spoken language, more efficiently managing dialog states, and the generation of more natural responses.

3.1 Understanding spoken language

In a dialog system, speech recognition translates speech signal into text. From our experience with spoken dialog systems, the inevitability of speech recognition errors lowers the user's passion to continue interacting with the systems. In this work, we aim to improve the recognition accuracy for a specific user by adapting speech recognition models to that user.

But for a computer, it is difficult to interpret meaning from raw text. Therefore, there is another important component, parser, which takes the text and transforms it into structured information with the help of grammar. In this section, we address the adaptation strategies on speech recognizer and parser.

3.1.1 Adaptation in the Speech Recognizer

A speech recognizer requires three models to decode human speech, namely lexicon model (LX), acoustic model (AM) and language model (LM). The lexicon specifies the words that can be recognized together with their pronunciations. The AM provides statistical information about speech sounds. Finally, the LM provides the likelihood of seeing given sequences of words.

In this project we do not include Am adaptation. However the Olympus framework does include AM adaptation functionality.

From the LM perspective, to convey the same meaning, different people construct different sequences of words. However, for a particular user, the construction of word sequences may be predictable. For example, a user might prefer saying "how far is downtown from CMU" to "could you tell me the distance between downtown and CMU". With an adapted LM, ASR can predict how likely this specific user is going to construct this sentence. In this work, we built an adapted LM based on analyzing the specific user's input history. The model will give more weight to the words "how", "far" and "from", as well as their sequence if the user often produces queries such as "how far is A from B". The language model in the POI system is derived automatically from the grammar which itself is adapted. Direct LM adaptation at the ASR level is also possible through the accumulation of user speech. The process of such adaptation is well-understood and is not studied in the current project (Xu and Rudnicky, 2000).

3.1.2 Adaptation in the Parser

The Phoenix parser [Ward1994] uses a semantic grammar to extract concepts from the recognition string produced by the decoder. The Phoenix grammar specification allows the developer to specify concepts and their expression in words. A fragment of the POI grammar is shown below:

```
[QueryHowFar]
    (HOW_FAR [Destination]) %%0.40%%
    (HOW_FAR [Destination] from [Origin]) %%0.50%%
    (HOW_FAR from [Origin]) is [Destination]) %%0.10%%

HOW_FAR
    (how far is)
    (how far away is)
    (how far)
;
```

Macros such as HOW_FAR can be expanded to real words such as "how far is".

In this work, we keep track of the frequency with which a particular speaker uses the different forms in their language. In other words, we compute how often a user is going to say "how far is A from B" as opposed to "how far is A". Based on the statistics, we can assign a probability to that instance of grammar.

In the Olympus framework, grammar is related with LM in speech recognizer. LM needs training data of how people construct a sentence to convey certain semantics. However, large corpora of real data are not always available. In Olympus framework, the system samples the grammar file to generate a training corpus. The probabilities next to the grammar (e.g., 0.40 and 0.60 in the example above) will influence the frequency of those sentences following that particular grammar. The LM generated from the example grammar shown above expects that a user says "how far is A from B" more often than "how far is A". Therefore, in our project, when calculating the frequency of a specific user firing each of the grammars, we are able to generate a user-adapted LM training corpus. In the future, we can use N-best lists as adaptation material and confidence measures to exclude unreliably recognized sentences (Souvignier and Kellner, 1998).

3.3 Dialog State and Response Generation

A conventional dialog system will normally interact according to a pre-specified flow. Directed dialog systems use a fixed flow and restrict changes of focus. A system such as Ravenclaw supports a flexible flow that allows the user to change topic (i.e. task) of to vary the order or form in which information is provided to the system. Nevertheless such a system will be configured by the developers to follow some particular order of exchanges, usually reflecting the "natural" order, at least as perceived by the developers. The pre-specified order serves the very important function of guiding a user who is not completely familiar with the task thorough the necessary steps. However it is not necessarily the case that only a single flow is appropriate to the task, and that different users may have different mental models of the activity, leading them to naturally want to complete the task following a different order of steps.

Accordingly, a useful form of adaptation might be to take into account the order in which a user prefers to (say) enter information. A mixed-initiative, non-directive dialog system would allow the user to try different approaches, presumably closer to their internal model of the task. We believe that preferred order can be inferred from interaction history, and that the system's prompting behavior can be modified to be more compatible with the user's understanding of the task.

In addition to the order of steps in a dialog, the prompts generated by the system (in the NLG module) can be modified as a function of history. We consider two sources of information that would feed into adaptation: the length of time that the system has been used by a particular user and evidence about their success at different stages of the task.

A long-time user of the system will have become familiar with the requirements of a particular task (e.g., what information needs to be provided). Accordingly, the system can begin to progressively shorten its prompts so that they serve as quick orientation to task progress rather than an explanation of what's requires. For example, the system can start with "Please tell me the neighborhood or business that you are travelling to", progress to "Where do you want to go?" and eventually to "Where to?" or "Destination?" The system needs to update history and track use patterns, adjusting response accordingly. For example, if the user has not accessed a particular task for a long time, the longer prompts can be reintroduced. Reintroduction does not mean starting over but might simply involve a faster progression through the prompt sequence. The system can use other cues for moving in the prompt sequence. For example, if the user appears to have trouble at a particular step (for example asking for the prompt to be repeated), perhaps just that step's prompt might be regressed (then progressed again). Responsiveness to the (inferred) state of the user may be perceived positively by that user.

3.4 History of Use

Based on history, the system can try to make tasks more efficient. For example, if the system finds out that for 70% of time the user starts navigation from CMU and 25% from Squirrel Hill, The DM can pre-load hypotheses (e.g., CMU and Squirrel Hill) for particular concepts (e.g., Origin) with confidence scores derived from this history with some discounting. The benefit is realized in two ways: having a prior probability for an input can boost its confidence in cases when recognition confidence is low. This in turn can allow the system to bypass an explicit confirmation of the input ("did you say you were leaving from CMU?") and simply combine the relevant information with the next prompt ("Leaving from CMU, where are you going?") Depending on the history (which in addition to frequency could take into account time of day or even GPS data) the opening prompt might be further reduced to "Where are you going to from CMU?" It is not essential that the system correctly guess this information at all times, just the proportion of occasions that allows the expected time cost of the transaction to be lowered.

4 EXPERIMENTS

We performed a preliminary study to determine the effectiveness of the user adaptation strategies described above. Six users were recruited for this experiment. One participant finished three-forth of the experiment and left. For the rest of them, they interacted with the POI system and carried out a total of 32 tasks, equally divided among the following goals: namely finding the distance between A and B, looking up the address of A, searching for the route from A to B, and locating services in the vicinity of a location. The tasks were described in pictorial form and depicted the desired information as it might appear in a browser. The goal was to minimize the use of language so as not to prime the participants to use particular language forms. One of the experiment tasks is shown in Figure 2 (this one asks the participant to obtain the route between A and B):

A: CMU
B: Homestead

Figure 2 one example of task

Although we are investigating system adaptation, it is the case that users will themselves adapt to the system as they become more familiar with the tasks they are carrying out. To differentiate these two types of adaptation, the experiment includes two conditions: adaptation and control.

Participants were brought into a quiet room and the tasks were explained to them in the same way for both groups. They were given a sheaf of questions (one to a page) and asked to proceed. After 16 tasks, both groups were asked to take a

164

break. During this time the experimenter sat in front of the computer and did something; in the adaptation condition the system processed the data obtained in the first part and was restarted; in the control condition, the system was simply restarted. Then the participants finish the remaining 16 tasks.

System performance can be assessed using several different metrics. In the present study we focus on two: The time taken to complete the tasks and users' sense of the change of the dialog system. Additional metrics, such as success rate provide additional information. At the end of the session we asked participants to fill out a questionnaire to assess their attitude towards the application and to determine their awareness of any changes in system behavior. Figure 3 shows the change of average time that is required to finish each part of the experiment. Figure 4 shows on average how many tasks are successful within each part. For both evaluation metrics, we asked the participants to move on to the next task if they could not complete the current one within 10 dialog turns. To show the dynamics of the change in more detail, we separate 32 tasks into four parts (for adaptation group, adaptation happens in part 3 and part 4).

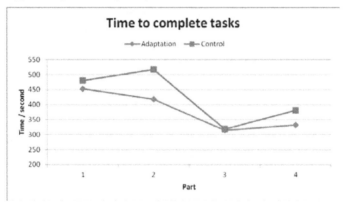

Figure 3 Time used to complete each part

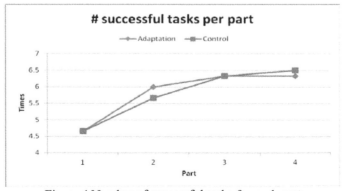

Figure 4 Number of successful tasks for each part

From Figure 3 and Figure 4, we can see that the completion rate is comparable for both adaptation group and control group. However, adaptation makes the interactions faster. More research work is needed to examine the effectiveness of each of the adaptation strategies as described in previous sessions.

We also report the participants' answers to the questionnaire: for all the three participants in adaptation group, they did not notice the change of the system although the time they spent on the next 16 tasks was reduced. In the control group, two participants think that the system worked better (one participant left the experiment after finishing 24 tasks without filling the questionnaire).

5 CONCLUSIONS

We find that the techniques that we evaluated may promote efficiency in a user's interaction with a geographic information system, as might be used in an automobile. Adaptation can provide a benefit beyond just familiarity with a system and the adaptive system may produce improvement in efficiency than the non-adaptive system (whose performance reflects only user experience). Our preliminary evaluation does not show a clear effect; however this is an initial effort on our part and several improvements are possible, for example the design of the initial language may have a significant effect. We also suspect that the length of the adaptation period may have been too short. We are continuing our work on developing adaptation techniques and on developing a better understanding of the process.

REFERENCES

Bohus, D. and Rudnicky, A., 2009. The RavenClaw dialog management framework: architecture and systems. *Computer Speech and Language*, 23(3), 332-361.

Hassel, L. and Hagen, E., 2005. Adaptation of an automotive dialogue system to users' expertise. *Proceedings of SIGdial 6*, pp. 222-226.

Ishimaru K. and Furukawa H., et al., 1995. User model for intelligent cooperative dialog system. *Proceedings of 1995 IEEE-International Conference on Fuzzy Systems*, Vol. 2, pp. 831-836.

Litman D. J. and Pan S., 2002. Designing and evaluating an adaptation spoken dialogue system. *User Modeling and User-Adapted Interaction*. Vol. 12, pp. 111-137

O'Sullivan D., 2009. Using an adaptive voice user interface to gain efficiencies in automated calls. *White Paper, Interactive Digital.*

Souvignier B. and Kellner A., 1998. Online adaptation for language models in spoken dialogue systems. *Proceedings of ICSLP '98.*

Ward, W. and Issar, S., 1994. Recent Improvements in the CMU Spoken Language Understanding System. *Proceedings of the ARPA Human Language Technology Workshop*, pp. 213-216.

Xu, W. and Rudnicky, A., 2000. Language modeling for dialog system. *Proceedings of ICSLP '00*. Paper B1-06.

CHAPTER 18

Traffic Light Assistant – Evaluation of Information Presentation

Michael Krause, Klaus Bengler

Technische Universität München - Institute of Ergonomics
Boltzmannstraße 15, 85747 Garching, Germany
krause@tum.de

ABSTRACT

To reduce stops at traffic lights and related fuel consumption, a car driver would need information about the state of upcoming signals. Previous research has shown the potential for this (Thoma et al. 2008, Popiv et al. 2010). One aspect of the German pilot project (*KOLIBRI* cooperative optimization of traffic signal control) is to deliver this information to the car via already installed mobile phone networks. The information needs to be displayed to the driver while driving. So, on one hand, special care must be taken for suitability while driving; on the other hand, the interface needs a pleasant design for acceptance by the driver. In a static driving simulator, five human-machine interface designs for a smartphone were tested according to objective measurements (glance duration) and subjective ratings (SUS, AttrakDiff).

Keywords: nomadic device, in-vehicle information system, IVIS

1   KOLIBRI Project

In KOLIBRI (Kooperative Lichtsignaloptimierung – Bayerisches Pilotprojekt; English: cooperative optimization of traffic signal control – Bavarian pilot project), a bidirectional transmission based upon already installed second- and third-generation mobile communication networks (GSM, UMTS) between a car and a communication center will be set up. The center gets information from traffic lights (also via mobile networks) about upcoming signal states, and forwards an aggregated and convenient data file to a test carrier and a smartphone. The information received needs to be properly conditioned for on-board display, and on a smartphone for use while driving.

Speed advice and other tools for traffic lights are not new. For a short review, see Krause et al. (2012). For on-board integration, former studies proposed and successfully evaluated a combination with the speedometer, so the actual focus is on the smartphone visualization.

2.1 Experiment Setup

The study was conducted in the fixed-base driving simulator of the Institute of Ergonomics Technische Universität München. The simulator environment consists of six beamers and six sound channels. Three Projectiondesign F1 beamers at a resolution of 1400 x 1050 offer a nearly 180° front view on three 3,4 m x 2,6 m screens and three additional beamers (different manufactures) render the projection of appropriate mirror images on screens behind the BMW E64 mockup. The driving simulation software is SILAB V3.0 (WIVW GmbH, Würzburg), in conjunction with CarSim V7.11 (Mechanical Simulation, Ann Arbor), and an active steering wheel with software from Simotion, München for feedback. For gaze tracking, the simulator is equipped with a head-mounted Dikablis system from Ergoneers, Manching. The processing performance is provided by 15 rack-mounted computers.

In the experiment, a Samsung Galaxy Ace S5830 smartphone (display: 3.5 inch, 320 x 480, Android 2.3.3) was used. The smartphone was installed with clamp mounting in the air vents right beside the steering wheel (see Fig. 1). The human machine interfaces (HMIs) were implemented with Adobe Flash Professional CS5 and packed for the phone with Adobes AIR-SDK 2.7.1. The Adobe AIR environment was not available for this phone via market, so an emulator version from inside the SDK was installed via USB.

Figure 1: (left) static driving simulator with BMW mockup, (right) mobile phone mounted beside the steering wheel

For the simulation, a section of the Federal Road B13 in the north of Munich was utilized as two simulator courses (north-south; south-north) for a length of approximately seven kilometers with seven traffic light-controlled intersections.

The same section from *48°17'56 N 11°34'36 E* to *48°14'48 N 11°36'7E* is also one of the real test tracks for experiments. For the traffic lights, the standard signaling scheme 'slow' of SILAB was used (green: 20s; yellow: 1.8s; red 30s; red/yellow: 2.2s). Due to restrictions in the simulation, the traffic light cycle of intersection N was started when the test person reached a waypoint in the section before intersection N-1. The signals were not coordinated or optimized. The road has two lanes per direction and constructional separation via guard railing.

Speed is limited to 100 km/h most of the time; in front of intersections, it is reduced to 70 km/h. The simulator generated some traffic in the opposite direction. In the driving direction, the three cars behind (50 m, 110 m, 140 m) and the one in front (200 m) of the driver were locked at fixed distances.

The speed recommendation for the driver in this experiment is based on the simple equation *distance to the stop line of next traffic ligh*t divided by *remaining time*. For *remaining time,* there are two important times: *start of green* and *end of green* (the states yellow and red/yellow were treated as red). When the car passes the stop line at intersection N, in the next ten constant traffic light cycles at intersection (N+1), a search loop tries to find a pair comprised of

maximum speed value (v_max; based on *start of green)* and
minimum speed value (v_min; based on *end of green)*
which meets the conditions:
(v_min < actual_speed_limit) and
(v_min > 0) and
(v_max < 0.7* actual_speed_limit)

The 70% limit is introduced to show only speed recommendations, if it is possible to drive not too slowly and thus, to reduce the relative speed of cars on the road for acceptance by the driver and for safety. If a matching pair is found, v_max is limited to the actual_speed_limit. The calculation is continuously carried out while driving at a rate of about 50 – 100 times per second.

If no matching pair is found, i.e. one had to drive too slow or too fast, *"Ankunft bei Rot"* (*Arrival at red*, see Fig. 2) is displayed. The driver normally doesn't know if the *Arrival at red* display is due to the lower or upper limit; this is to intentionally prevent speeding.

In this early version, only the actual valid speed limit is considered in the calculation; upcoming speed limits are neglected. If appropriate v_min and v_max are found, they are displayed with the HMIs, as described in the following section. If the driver arrives at a red traffic light and has a speed below 5 km/h, a countdown display shows the residual red light time.

For gaze analysis, D-LAB (Ergoneers, Manching) was used and data were exported to MATLAB for further processing. Three areas of interest were defined: the windscreen (i.e. traffic scene), the speedometer and the smartphone. The smartphone reported its actual display state to the eye tracking system. Only the times when a speed recommendation HMI was shown are considered in this paper.

2.2 Human Machine Interfaces

HMI1 "fisheye": The recommended range of speed is highlighted in green. The rest of the numbers are white. The actual speed is marked by two brackets "> <" and has a bigger font size.

HMI2 "speedo" uses a familiar speedometer. The needle shows the actual speed and the speed recommendation is highlighted by a green sector.

HMI3 "carpet" uses in one lateral dimension the same concept as the speedometer in the radial angle dimension. The position of the car is its actual speed, and the green area is the speed recommendation.

HMI4 "recom" shows a single number for the speed recommendation. The advice is the average of v_min and v_max rounded to steps of five. The intention is to guide the driver to the middle of the green phase. The design is similar to a German/European traffic sign for speed advice without speed limitation (a white font on a blue ground).

HMI5 "roll", the actual speed rounded to 5, is the big number on the front of a roll. The recommended speed range is highlighted in green. If the green speed range is barely visible, an up or down arrow offers some help.

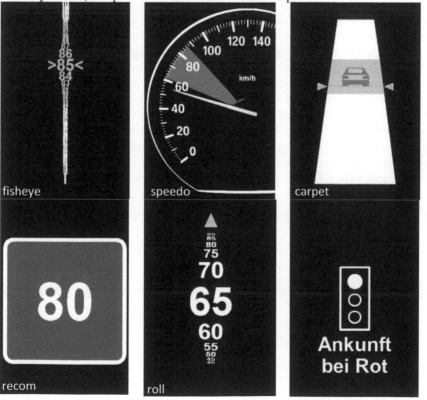

Figure 2: The five used HMIs at different speeds and recommendations: HMI1 "fisheye", HMI2 "speedo", HMI3 "carpet", HMI4 "recom, HMI5 "roll"". Last image: "Ankunft bei Rot" (Arrival at red)

Font sizes: The number in HMI4 is 13.9 mm tall. The actual speed in the middle of HMI5 is 10.8 mm. The marked actual speed in HMI1 has a height of 4.6 mm. The numbers in the speedometer have a size of only 3.1 mm, which falls below the recommended level of DIN EN ISO 15008:2009 at a viewing distance of 535 mm, and below the acceptable limit at 660 mm.

2.3   Experiment

After being welcomed, the test subjects got an explanation on the driving simulator. Afterwards, they signed a letter of consent and filled out a questionnaire on demographic information. Seated in the car, the Dikablis eye tracking system was adjusted. After an acclimatization round on one of the simulator courses, they were introduced to the HMIs by a moving demonstration on the phone screen and drove each HMI for about a minute. In the experimental runs, each person drove each of the five HMIs and a baseline run in random order. The six runs were randomly assigned to the two simulator courses (north-south; south-north), with the constraint that each course is used three times. After each run, except for the baseline run, the test subjects filled in a system usability scale (SUS) and an AttrakDiff2 questionnaire (Hassenzahl et al., 2003); additionally, they were verbally asked for their opinion. The SUS form was a bilingually combined, careful translation of the original (Brooke, 1996) with an instruction from Bangor et al. (2009)

The 20 participants that finished the experiment were from 19 to 55 years in age (M=25.8; SD=7.6; 85% male; 15% female). One person quit due to simulator sickness. All were licensed drivers and received their driver's license mainly at the age of 18. The annual mileage ranged from 800 km to 30,000 km (M=11,575 km; SD=7,703 km). One person had a color deficit. Forty-five percent had a corrected-to-normal visual acuity. Fifty-five percent had some previous experience with driving simulators, and 40%, especially with the driving simulator at the Institute of Ergonomics. Ninety-five percent had driven an automatic car, like the one used in the experiment, before. Drivers needed mainly 90 – 120 minutes per session.

3.1   RESULTS (OBJECTIVE MEASUREMENTS)

Figure 3 shows the combined times of the areas "speedometer" and "smartphone" if not intermitted by "windscreen" when the smartphone shows a speed recommendation HMI, labeled as "Eyes off the Road".

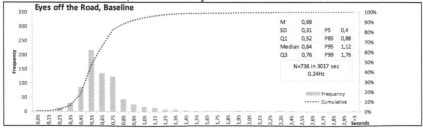

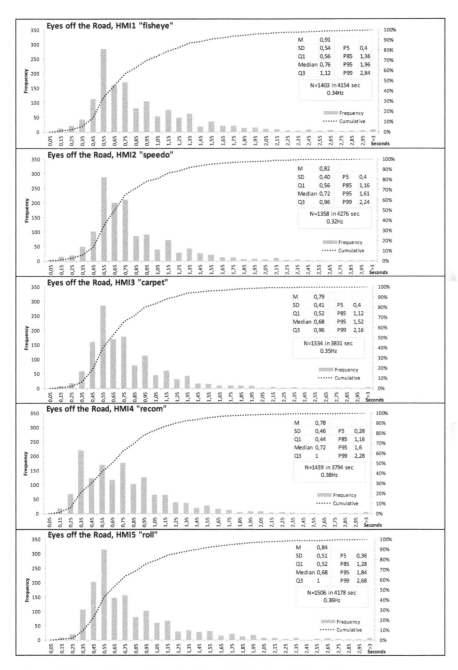

Figure 3: Eyes-off-the-road times for different HMIs from all 20 participants

Inspired by Horrey & Wickens (2007) "[…] vehicle crash do not reside at the mean […], but rather in the tails of the distribution", the whole histograms with

172

some typical key data are reported (Figure 4). Two anchors are marked: the 2 seconds from AAM (2006) and 1.5 seconds as mentioned in DIN EN ISO 15005:2003, which are referenced several times within the ESoP (2006).

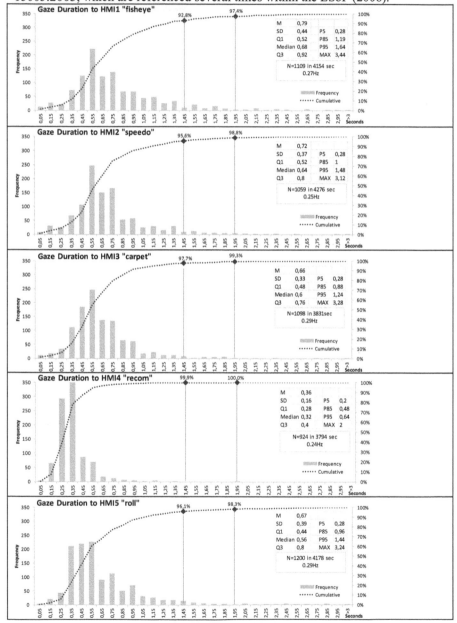

Figure 4: Gaze duration for different HMIs from all 20 participants

3.2 RESULTS (SUBJECTIVE RATINGS)

The subjective ratings mainly show the same ranking (Figure 5 and 6): (1) speedo, (2) carpet, (3) roll, (4) fisheye and (5) recom.
According to Figure 6 "carpet" has the highest hedonic quality.

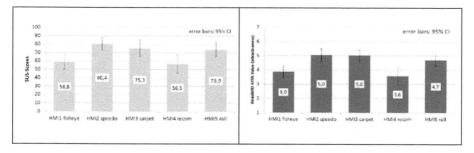

Figure 5: (left) SUS-Scores of the HMIs; (right) AttrakDiff2 ATTR Value (attractiveness)

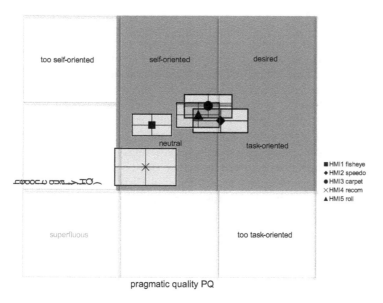

Figure 6: AttrakDiff2 HQ-PQ-portfolio diagram; with rectangles for 95% CI

4 DISCUSSION

The glance durations for HMI4 "recom" are remarkable low (Figure 4). The values of the others are comparable to speedometer checks, as reported in a literature review by Green (1999). The widely used 85th percentile level for 2 seconds and 1.5 seconds is excelled by all variants with values greater than 90 percent.

Except for HMI1 "fisheye", the 85th percentile time values even meet the 1-second limit as recommended by Stevens et al. (2002) and mentioned by the ESoP (2006) for well-structured graphics (only example status).

It's imaginable that the variants have different "eyes-off-the-road" times due to speedometer checks and gaze strategies. The hidden link between the histograms of Figure 3 and 4 are the gaze transition probabilities. For comparison, the "eyes-off-the-road" times for the baseline run are also reported. In the baseline run, the smart phone has a blank screen, but still calculates the speed recommendation algorithm in the background and sends appropriate triggers to the eye tracking system; what the driver would see, if the screen were not blank. So, this should give a comparable set of road sections. Due to the blank screen, the driver could not follow any advice, and hence, the total time for these sections is shortened.

If one assumes that the 85th percentile time value for "eyes-off-the-road" is a good measure for the tail of distribution and multiplies it by the mean frequency (Hz-value), this measure of severity would be:

(0.37) speedo, (0.39) carpet, (0.44) recom, (0.46) roll and fisheye

5 CONCLUSION

Solely based on the gaze duration measurements (Figure 4), the HMI4 "recom" would be preferable. On the other hand, HMI4 gets the worst subjective ratings.

All systems meet the requirements of accepted glance duration limits, so there should be no safety issue and room for a combination or balance, as mentioned in the well-known triple of ISO 9241-11 (effectiveness, efficiency and satisfaction).

So, we propose the HMI2 "speedo" for onboard integration in the speedometer of the test car and the HMI3 "carpet" for implementation on a smartphone. Both HMIs have acceptable glance durations; the subjective ratings of both are good.

If HMI2 were to be implemented on a smartphone, there would be two different-looking speedometers in one car. Another drawback of HMI2 on a smartphone is the small numbers, or, to put in other words: an appealing advantage of HMI3 is that it uses no digits and thus supports portability among different screen sizes.

6 OUTLOOK

In this paper, the subjective ratings and the glance durations that enabled to proceed in the project are discussed. Further publications will take a closer look at gaze transition probabilities, compliance with the shown speed recommendations, behavior when the display shows "Arrival at red", fuel consumption, speed violations, display dynamics and other interesting issues with the first experiment.

In a second simulator experiment, subjects customized the favored HMI in the first experiment to their preferences. Further, the tactile detection task was used to get benchmark values for comparison with later trials on the real test tracks. With the results of both experiments and adequate countermeasures for the drawbacks mentioned by the test subjects, the interface could be improved further.

In parallel, the use of mobile phones in relation to cars and traffic is actually revealed with a paper-based and online survey.

ACKNOWLEDGMENTS

The project is founded by *Bayerische Forschungsstiftung* (Bavarian Research Foundation). The authors would like to thank the project partners TRANSVER, BMW Group, *Board of Building and Public Works in the Bavarian Ministry of the Interior*, Thomas Moll (data acquisition), Patrick Gontar (gaze recording), Sebastian Obermeier (simulator track) and the institute's graphic artist Julia Fridgen.

REFERENCES

AAM 2006. Driver Focus-Telematics Working Group 2006, Statement of Principles, Criteria and Verification Procedures on Driver Interactions with Advanced In-Vehicle Information and Communication Systems.

Bangor, A., T. Staff, P. Kortum, and J. Miller. 2009. Determining What Individual SUS Scores Mean: Adding an Adjective Rating Scale. Journal of Usability studies, 4(3), pp. 114 – 123.

Brooke, J. 1996. SUS: A quick and dirty usability scale. In P. W. Jordan, B. Weerdmeester, A. Thomas, & I. L. Mclelland (Eds.), Usability evaluation in industry, pp. 189 – 194 London: Taylor and Francis.

ESoP 2006. Commission recommendation of 22 December 2006 on safe and efficient in-vehicle information and communication systems: update of the European Statement of Principles on human machine interface (2007/78/EC).

Green P. 1999. Visual and Task Demands of Driver Information Systems. UMTRI 98-16.

Hassenzahl, M., M. Burmester, and F. Koller. 2003. AttrakDiff: Ein Fragebogen zur Messung wahrgenommener hedonischer und pragmatischer Qualität. In: J. Ziegler & G. Szwillus (Hrsg.), Mensch & Computer 2003. Interaktion in Bewegung, 187 – 196, Stuttgart, Leipzig: B.G. Teubner.

Horrey, W., C. Wickens. 2007 In-Vehicle Glance Duration: Distributions, Tails, and Model of Crash Risk. Transportation Research Record Vol. 2018, pp. 22 – 28.

Krause, M., C. Rommerskirchen, and K. Bengler. 2012. Ampelassistent – Entwurf und Evaluation der Informationspräsentation. In: 58. Frühjahrskongress der Gesellschaft für Arbeitswissenschaft (pp. 627 – 630). Dortmund, Germany: GfA Press, 2012.

Popiv, D., C. Rommerskirchen, M. Rakic, M. Duschl, and K. Bengler. 2010. Effects of assistance of anticipatory driving on driver's behaviour during deceleration situations. In: 2nd European Conference on Human Centred Design of Intelligent Transport Systems (HUMANIST), Berlin, Germany, April 2010.

Stevens, A., A. Quimby, A. Board, T. Kersloot, and P. Burns. 2002. Design guidelines for safety of in-vehicle information systems. Transport Research Laboratory, Project Report PA3721/01.

Thoma, S., T. Lindberg, and G. Klinker. 2008. Speed Recommendation During Traffic Light Approach: A Comparison of Different Display Concepts, in D. de Waard, F.O. Flemisch, B. Lorenz, H. Oberheid, and K.A. Brookhuis (Eds.) (2008), Human Factors for assistance and automation (pp. 63 – 73). Maastricht, the Netherlands: Shaker Publishing, 2008.

CHAPTER 19

Using Communication Research for Cultural Variability in HMI Design

Donal Carbaugh, Elizabeth Molina-Markham, Brion van Over

University of Massachusetts Amherst
Amherst, USA
carbaugh@comm.umass.edu

Ute Winter

General Motors Advanced Technical Center - Israel
Herzliya, Israel
ute.winter@gm.com

ABSTRACT

The "speech application" is one part of the larger human machine interface (HMI) which allows for verbal interaction with a vehicle in the accomplishment of particular tasks. In order for the "speech application" to be usable in a satisfying fashion, it must be able to recognize the speech of the occupants, which will vary culturally given local flow of dialogue, regional accents, cultural varieties in intonation, prosody, and pacing. Research on the influence of culture on design and usability has frequently relied on theories that consider culture in terms of macro-level universal dimensions. In this work, we discuss a framework for investigating cultural variation in communication drawing on ethnographic and naturalistic data. This framework, based in the Ethnography of Communication and Cultural Discourse Analysis, examines the details of everyday communication in order to discover the cultural patterns, norms, and preferences of culturally based interaction that is active in a specific community. In this paper, we describe how this framework could be the basis of a model for analyzing the role of culture in in-car communication.

Keywords: HMI, culture, communication, cars

1   INTRODUCTION

For some time now those who are concerned about human-computer interaction have struggled with conceptualizing and investigating cultural variation in the ultimate hope of creating user interfaces that cross cultural boundaries in highly usable ways. Attempts to research the role of culture in design and usability have often involved the use of macro-level cultural dimensions provided by Hofstede (1991), Hall (1989), or Ting-Toomey (1998). These dimensions are attractive because of their universality, and the ease with which one can begin to design for a given culture once one knows if they are a "high" or "low" context, their long term orientation, and a variety of other such attributes. However, such approaches have recently been criticized for their tendency to essentialize diverse cultural groups along national boundaries, and fail to effectively guide the development of culturally satisfying interfaces (Winschiers & Fendler, 2007; Sun, 2009). It is also unclear how these broad cultural dimensions can consistently inform the design of user interfaces that require the production of particular actions and sequences with which real people will interact.

In what follows we present a framework based in the Ethnography of Communication (Hymes, 1962, 1972; Philipsen, 1992), and one of its developments, Cultural Discourse Analysis (CuDA) (Carbaugh, 2007), for researching issues of culture, usability, and design. The framework is a general theory yet yields, when used, a bottom-up understanding of usability rather than a top-down view based in the cultural dimension models. We believe the utility of such an approach lies in its ability to identify cultural models of design with these being based upon local conceptions of personhood, relations, emotion, place, and action, as well as routine sequential actions, norms of interaction, and cultural preferences for channels and instruments, all of which are required for the understanding and development of truly localized design.

In addition to outlining how such a theory and method can effectively be used to address issues of design and usability in HCI, in what follows we present a research procedure for the investigation of human-computer interactions in an automobile. In a 2009 article, GM investigators discovered and discussed how cultural dimensions of activities were not only possible but perhaps even pervasive throughout various automotive speech applications (Tsinhomi, Winter, and Grost, 2009). Cultural features were suggested as part of the variety of possible utterances people produce in the car, the vocabulary they used, the variety of sounds they made as well as the variation in the phonemes produced. The importance of these features became apparent in several phenomena including the flow of conversation, response times, how prompts were formulated, and how errors were handled. The role of cultural variability in automotive communication then, is pervasive and powerful, yet understudied if at all. Such work, in our assessment, is necessary.

The remainder of this paper is organized to first discuss the field research cycle as we conceptualize it, followed by a set of considerations for establishing a field site, including both the means of collecting data and the procedures for its analysis.

2 FIELD RESEARCH CYCLE

The framework for field-based, cultural research into communication that we propose here involves a sequential design. To begin, we summarize this briefly in four phases, from activities prior to entering the field, to activities completed after leaving the field. This overview will be the foundation for a discussion of the specific application of the framework to research on in-car communication and the various means of collecting and analyzing naturalistic data in this context.

2. 1 Pre-fieldwork Activity

This phase of the project involves several activities, which are preliminary to doing the fieldwork itself. Before entering, it is crucial to be as knowledgeable about a field site as possible. If there is a literature available about the site, researchers should consult it in order to become more knowledgeable about the history of the area, its people, their occupations, customs, and so on.

A second set of activities involves planning about the fieldwork itself. If there is a special focus envisioned, then a preliminary conceptual map of that focus should be formulated. For example, if one were most interested in users' tactile manipulation of a navigation system, then what that is and how it works should be carefully thought through. Focusing on specific concerns in this way, establishes a theoretical position from which to observe communication in a particular context. This equips one for study and reflection while in the field, enables a systematic approach to one's observations, as well as provides ways of designing interviews.

2.2 Fieldwork Activity

This phase involves periods of observation of communication in a cultural context. Observations such as these inevitably lead to questions which researchers can ask participants during interview sessions. The accumulation of a descriptive record about observational and interview events creates a corpus of data which then needs to be analyzed. The analyses involve the distinct modes of investigation that we will discuss in the next sections.

2.3 Post-fieldwork Activity (Phase One): Descriptive and Interpretive Analyses

The activities conducted after leaving the field involve continuing phases of analysis, which focus on how particular communicative activities are done in a specific cultural context and what the meanings of those activities are in this context for the people who engage in them. These analyses lead, often, to additional questions about dynamics one observed, or heard about while in the field. This phase of the research can lead, when possible, back to the field for more detailed observations and analyses.

2.4   Post-fieldwork Activity (Phase Two): Comparative Analyses and Critical Assessments

These final phases of a field project are crucial as a sharper view of the cultural dimensions of communication get better understood through comparative analysis. Also, through critical study with participants, researchers can contribute ideas about better design of social practices, products, policies, or other human creations (Carbaugh, 2008).

Note, and we emphasize, that the phases we present here are sequential in their design but cyclical in their possibilities. A phase of fieldwork can result in revising one's earlier conceptualizations; post-fieldwork activities can lead one back to the field focused on other observations, with different questions, and so on.

3.   SOME CONSIDERATIONS OF THE FIELD SITE

The framework described above has been used in a variety of contexts to study culturally-informed communication (e.g., Carbaugh, 1988, 1996, 2005; Saito, 2009; Boromisza-Habashi, 2011). Here we describe a model for applying this framework to an examination of communication in automobiles. This description is the basis for future research that could be undertaken at various locations around the world.

3.1   The Field Sites: Field sites in this case would be selected in order to enable comparison of driving practices in different cultures. These should include rural and urban driving in various metropolitan areas. As will be discussed later in this report, comparison is an essential phase of investigation in Cultural Discourse Analysis and the selection of a variety of field sites would facilitate this process.

3.2   Selection of Participants: Several criteria would guide the selection of participants for each field site of the study. A pre-questionnaire could solicit basic demographic information about each potential participant, about their willingness to have their data downloaded into the automobile computer system, and other information concerning their driving experiences and habits. This preliminary information would be used to ascertain the suitability of a participant for the study. Additionally, and most importantly, each participant of the study should be a native speaker of the language in which the speech application has been designed. This is important as a control on the speech recognition and dialog flow capabilities of the system being tested. Additional criteria to be met could include some balance between male and female drivers, tech savvy and not, young and old, urban and rural, and if possible, including early adopters of technology.

3.3   General Schedule at each Field Site: In this section we provide a detailed discussion of the steps of the research cycle described above as they could be undertaken in studying communication in automobiles.

We envision a potential pre-fieldwork period of 10-14 days which would involve the research team in contacting possible participants and making local arrangements. The fieldwork would be designed to occur over a 10 day period which could be set as follows: Day 1: Arrival, settle, plan and meet with local contacts, establish possible routines, conduct initial observations of settings; Day 2: Conduct a preliminary pilot study of 1-2 participants using the *a priori* framework; do some very tentative descriptive and interpretive analyses, make any revisions that may be needed in the research design; Day 3: Observational work in the car with 2 participants (recorded); interview afterward; conduct preliminary analyses of initial data; review and revise the framework as needed; Days 4-8: Conduct driving, interview, and debriefing sessions with 2 participants each day including observations in the car; preliminary analyses of data; eventually integrating data and analyses; create an initial descriptive account; revise framework as needed; Day 9: Final data gathering (as needed); final in-field analyses (translations, interpretations); Day 10: Depart.

A post fieldwork period of 10-20 days would involve the development of the descriptive report about the car including analyses of specific sequences and norms. This would also include the development of a very preliminary interpretive report of the cultural terms, meanings, and aesthetics of the car as a communication situation; assessments of theory, methodology, and findings.

4.   MEANS OF DATA COLLECTION

Drawing on the framework described above, multiple schematic ways of collecting data become possible including an observational scheme, an interview guide, and technical logs.

4.1   Gathering Data from Study Participants: Upon first meeting
the research team, selected participants would be introduced to the study, and given a pre-questionnaire to assess their suitability for the study. If meeting the inclusion criteria, the participant would be asked to sign a consent form, which makes explicit conditions for participating in the project. They would then be introduced to the capabilities of the car's speech recognition and dialog flow, would engage in an off-road (i.e., parking lot) test drive, would engage in a driving session, and finally would conclude with an interview session. We anticipate that the overall session with each participant would last around 120-150 minutes.

4.2   Field Researchers: In the context of this particular application, the
primary task of researchers would be to monitor participants' driving and to formulate questions about the driving to be asked in the subsequent interview session. During the "drive-along," researchers would be advised not to interrupt the driver while driving, but they could respond to questions the driver asks. The researchers also would be involved in sharing observations about the drive-along and subsequent interview during a debriefing session among researchers that could be held after each driving-interview session. Driving sessions may involve one or two researchers in the automobile with the driver.

4.3   Observational Scheme: The observational scheme equips the researcher with a specific, structured way of watching and listening while in the car with study participants. The scheme would involve selective attentiveness to the following components:

1. **Setting**: In what physical environment is the communication taking place?
2. **Participants**: Who is involved in the communication practice?
3. **Ends**: This component would have two parts: What are the participant's goals of the practice (e.g., to send an email)? What are the outcomes of the practice (e.g., the email was sent, or the effort to do so was unsuccessful, or the user got irritated at the car)?
4. **Act/Sequence**: What specific communication acts got done, and in what sequence?
5. **Key**: What was the emotional pitch, or tone of the communication (e.g., perfunctory, serious, frustrated)?
6. **Instrumentalities**: What multiple mode(s) or cues were used in this communication (e.g., voice, gesture, pressing a button, words)?
7. **Norms** (see below): What were the norms - stated and/or implied - for this interaction?
8. **Genre**: Is there a generic form to this communication practice which participants use, and if so, what is it?

These eight components would provide a basic investigative tool for analysts to systematically describe communication in and about the car. A sub-set of the concepts could be more useful in some cases than in others.

A recommended way of using the observational scheme would be to create a chart on a piece of paper (or recording device), to be filled in by the researcher. This scheme could be used by the researcher with each component across the top of the paper and a time-line down the left margin, so the researcher could record observations about particular components at specific times. For example, during the 21st minute (recorded on the left side) the driver attempted to change radio stations (the A, I-through touch, K-frustrated) but failed (the End). These observations could then be used to formulate questions during the later interview session.

4.4   The Interview Session: After the driving session, the researchers will ask the driver a series of questions. An interview guide will be provided that draws upon the framework described above. The wording of the questions will be shortened and modified to fit the particular field participants and sites of concern.

The first section addresses the driving session itself, and anything which had been noticed during it and warranted probing. A second set focuses on the participant's (the driver's) background using dialog systems. In addition to giving information about the participant, these questions establish a wider context for the particular communication situation of the driving session. Next, the questions focus on the sequencing of the driving session –for example preferences for how a dialog should be initiated, the degree and form of feedback that the system should provide while a particular exchange is taking place, the repairing of communication errors, and the ending of the interaction. The final questions are designed to explore various modes of cueing that the driver may engage in with the system; these modes may or may not have come up during the discussion of sequencing and these questions encourage further exploration of those that may not have been discussed. The questions in each section are designed to prompt the participant to produce talk about dialog systems that could be analyzed for cultural propositions and premises that the participant uses to interpret and produce their interactions with the dialog system. Questions that more explicitly explore cultural premises the participant may hold about the ideal driver will be posed, including relations with the interface as well as others in the car, feelings about the car/interface, and in-car communication generally. The interview session closes with a catch-all question that would allow the participant to share anything about their experience that was not explicitly requested in prior questions.

4.5   The Debriefing: The researchers will meet together to discuss the overall session. The purposes of the debriefing session are to reflect upon the specifics of the driving session, the interview session, particular observations made during each, to identify useful focal concerns for further attention, perhaps to modify the methodology in subsequent sessions, and so on.

4.6   Technology: For any particular study, the car's human-machine interface could be designed to explore a system in general, or specific features of any one system. It is desirable to add logging capabilities to enrich the corpus of data for analysis. For exploring spoken communication with the described methodology, the HMI needs to contain a speech application as part of a multimodal interface, which supports the use of natural language utterances and a pre-defined variety of typical in-vehicle tasks. This allows examining cultural variation of any verbal, as well as non-verbal acts as part of the communication.

5.   PROCEDURES FOR THE ANALYSIS OF DATA

After collecting the data above, there are specific procedures that are employed for their analysis. These are summarized here in four phases.

5.1   Phase One: Descriptive Analysis

A descriptive phase of study would respond to the question, how is that done here? What are exact instances of the phenomenon of concern to participants in real time and place? Care would be taken to record multiple instances of the phenomenon of interest such as selecting music, making a call, errors and their (attempted) correction, or direction giving. Detailed descriptions would be created of each instance. A group of instances would comprise a focused corpus for study. Note that the analytic tasks here would involve the following: 1. noticing an instance of a phenomenon; 2. making a collection of multiple instances of that phenomenon; 3. transcribing the instances so that a descriptive record of the pattern is created.

After creating a descriptive record, researchers would be able to search across multiple instances for linguistic and non-linguistic qualities which recur in a patterned way. The descriptive analysis would lead to claims such as: This is how direction giving, or attempted error correction, or volume control, or implementation, or speech prompting is actually done in this corpus of data.

A key part of the descriptive analysis would be the recording of observational data onto electronic devices (such as video and audio recorders) and the subsequent transcription of those data. The latter are done using specific transcription conventions and orthographic techniques. These include the ways nonverbal positions and movements are active during the activity being studied. Also synchronized with these would be the verbal data logged into the system at the exact time of the instance. The careful documentation of these data and their inscription would provide the toe-hold of the field study in actual communication events.

5.2   Phase Two: Interpretive Analysis

An interpretive phase of study would respond to the question, what does this specific communication practice (or finding from the descriptive analyses above) mean to participants? What cultural significance and value (or lack thereof) does this have for them? While a descriptive analysis gives a field investigator evidence of patterns of practices people create together; an interpretive analysis tells us the meanings of those practices. An interpretive analysis would follow exacting procedures which take different trajectories, but in all cases, this analysis would begin:

1. take one pattern of practice that has been documented and analyzed through descriptive study;
2. examine the corpus of instances of that practice;
3. if available select participants' terms which are active in the corpus and use them to formulate a statement about that practice (the participants' terms are called cultural terms and should be placed in quotation marks; the statement formulated through *cultural terms* is called a *cultural proposition*).
4. if salient to participants, trace the meanings of this proposition about the action getting done, the feeling of it, identity issues, social relations;
5. interpretation of norms: What is being presumed about good or proper practice in this phenomenon? This involves a four-part analysis, through which specific communication norms can be explicated - and later compared - in a prototypical formula of these four parts:
    A. Context: When in (setting or context of the car, with participants P, or participant relations R),
    B. Condition of Identity (related to conduct): if one wants to (e.g., be a good driver, get directions to a place),
    C. Force: one (must, preferably should, permissibly could, must not),
    D. Conduct: do action X.

184

Communication norms are analyzed further by positing explicit or implicit imperatives, along dimensions of intensity and crystallization.
6. additional trajectories of interpretive analyses build on the above, and can be elaborated by examining cultural metaphors and formulating semantic dimensions or cultural premises.

These analytic procedures would provide bases for claims about the meaningfulness of practices to participants.

5.3 Phase Three: Comparative Analysis

A comparative phase of study would respond to the questions, to what degree is this practice the same and to what degree is it different from others? If one were to study direction-giving in Atlanta and in Shanghai, or speech prompting of entertainment, we would expect some similarities and some differences in this practice in the two locations. Through such study we could get a better idea about what indeed is culturally distinctive in one set of practices, as opposed to those in another; we also would get a better sense of what is similar across such practices. Through comparative study, and based upon descriptive and interpretive analyses, one would be well-placed for such assessments.

5.4 Phase Four: Critical Analysis

A critical assessment would begin by asking what works well and what does not? After carefully investigating phenomena through the above modes, researchers could better reflect upon what is working well and what is not. Critical assessment and subsequent considerations could lead to better designed capacities, and, in this case, more usable speech applications.

6. CONCLUSION

The framework presented here provides a model for ethnographic investigation of issues of design and usability in human-computer interaction in automobiles. Applying this framework to a particular context, we have outlined cyclical phases of fieldwork activity, schematic means for collecting data, and procedures for the analyses of these data. The success of a speech application depends greatly on its situated use in a particular cultural context. Our framework seeks to uncover the distinctive cultural models of communication that are active in various communities to ascertain means of adapting human-machine interfaces to local contexts of use.

The framework for naturalistic field studies is designed to be seamlessly integrable into a user-centered design approach (Dray & Siegel, 2009; Norman & Draper, 1986). It should clearly be part of the research phase before any design, implementation and evaluation cycle. Thus designers have an overview on cultural norms and preferences for all necessary interface design dimensions, such as cooperative principles for communication, control handling, dialog patterns and sequences, turn taking and grounding conventions, information distribution, choice of words and phrases in dependency of typical automotive tasks and domains, and conflict resolution.

REFERENCES

Boromisza-Habashi, D. 2011. Dismantling the antiracist "hate speech" agenda in Hungary: An ethno-rhetorical analysis. *Text &Talk*, 31: 1-19.

Carbaugh, D. 1988. *Talking American: Cultural Discourses on DONAHUE*. Norwood, NJ: Ablex Publishing Corporation.

Carbaugh, D. 1996. *Situating Selves*. Albany: State University of New York Press.

Carbaugh, D. 2005. *Cultures in conversation*. Mahwah, NJ: Lawrence Erlbaum Associates.

Carbaugh, D. 2007. Cultural discourse analysis: Communication practices and intercultural encounters. *Journal of Intercultural Communication Research* 36: 167-182.

Carbaugh, D. 2008. Putting policy in its place through cultural discourse analysis. In E. Peterson (ed.), Communication and Public Policy Proceedings of the 2008 International Colloquium of Communication (pp. 55-64). Digital Library and Archives, University Libraries, Virginia Tech. Retrieved from http://scholar.lib.vt.edu/ejournals/ICC/2008/ICC2008.pdf

Carbaugh, D., Nuciforo, E. V., Molina-Markham, E., & van Over, B. 2011. Discursive reflexivity in the ethnography of communication: Cultural Discourse Analysis. *Cultural Studies <-> Critical Methodologies*, 11(2): 153-164.

Dray, S. and Siegel, D. 2009. Understanding Users in Context: An In-Depth Introduction to Fieldwork for User Centered Design. Lecture Notes in Computer Science (5727) 950-951, Human-Computer Interaction – INTERACT

Hall, E. T. 1989. *Beyond culture*. New York: Doubleday.

Hofstede, G. 1991. *Cultures and organizations: Software of the mind*. London: McGraw-Hill.

Hymes, D. 1962. The ethnography of speaking. In. *Anthropology and human behavior*. eds. T. Gladwin, and W. Sturtevant. Washington, DC: Anthropological Society of Washington.

Hymes, D. 1972. Models of the interaction of language and social life. In. *Directions in sociolinguistics: The ethnography of communication*, eds. J. Gumperz, and D. Hymes. New York: Holt, Rinehart and Winston.

Norman, D. and Draper, S., 1986. *User Centered System Design; New Perspectives on Human-Computer Interaction*. Mahwah, NJ: Lawrence Erlbaum Associates, Inc.

Philipsen, G. 1992. *Speaking Culturally*. Albany, NY: State University of New York Press.

Sun, H. 2009. Designing for a dialogic view of interpretation in cross-cultural IT design. *Lecture Notes in Computer Science*, 5623: 108-116.

Saito, M. 2009. *Silencing Identity through Communication: Situated Enactments of Sexual Identity and Emotion in Japan*. Germany: VDM Publishers.

Ting-Toomey, S. 1989. Intercultural conflict styles. A face-negotiation theory. In. *Theories in intercultural communication*, eds. Y. Y. Kim, and W. B. Gudykunst. Newbury Park: Sage.

Tsimhoni, O., Winter, U., & Grost, T. 2009. Cultural Considerations for the Design of Automotive Speech Applications.

Winschiers, H. and Fendler, J. 2007. Assumptions considered harmful. *Human Computer Interaction*, 10: 452-461.

CHAPTER 20

Language Models for Text Messaging based on Driving Workload

Ron M. Hecht, Eli Tzirkel, Omer Tsimhoni

General Motors Advanced Technical Center - Israel
Herzliya, Israel
ron.hecht@gm.com, eli.tzirkel@gm.com, omer.tsimhoni@gm.com

ABSTRACT

One way in which drivers can send messages (e.g., text messages and emails) is using voice dictation. The technology that enables dictation in a vehicle is challenging given the noisy environment and the need for high accuracy. In this paper, we investigate the impact of input method and driving workload on the structure and complexity of dictated messages. We describe findings from a driving simulator study (Green et al., 2011) in which dictated messages were collected and analyzed. We explore the hypothesis that driving workload or input method affects the message language characteristics. We show that both driving workload and the input method significantly impact the message vocabulary and its perplexity. This may be exploited for enhancing dictation, for example by creating language models that are specific to driving and by adapting them according to workload.

Keywords: texting, SMS, language models

1   INTRODUCTION

Text messaging is one of the most popular methods of communication world-wide. It is estimated that in 2010 alone a total of about six trillion messages were sent (ITU, 2010). This number reflects a large growth in the use in Short Messaging Service (SMS), which has tripled from 2007 to 2010. Given the magnitude of usage per subscriber and the diversity of subscribers; the nature of the SMS messages and alternative messaging systems has to be understood in the context of the user environment. We specifically focus on some conditions of SMS dictation in a car environment under different driving workload circumstances.

Dictation and dialog in a car environment was studied previously (Lindstrom et al. 2008; Villing 2009a; Villing 2009b). Lindstrom et al., for example, conducted a user study in order to understand the effect of driving workload on speech. In that work, however, the grammatical structure of messages was not explored. In this work we focus on grammatical differences in text messages under different workload conditions. We evaluate grammatical differences by comparing statistical language models estimated from message data. The statistical language model used here, known as N-gram (Ney et al. 1992; Jurafsky, and Martin, 2000), represents the probability of a sequence of words to occur, under a Markovian assumption.

We begin the study by reconfirming that language models corresponding to text messages and speech messages differ. We continue by studying the differences of language models associated with two different workload conditions.

2 CORPUS

A corpus of text messages was collected in a driving simulator environment, transformed into text, and normalized. The normalized data were then used as input to an experimental language model system.

2.1 Corpus Collection

An experiment was conducted in a driving simulator study in order to generate a database of dictated text messages with and without simulated driving. The experiment as designed and executed by Green et al (2011). (See Green, Lin, Kang, and Best, 2011 for a detailed description of the experimental procedure and findings.) Overall, 24 pairs of participants took part in the experiment for two hours each. Participants were pairs of friends in their late teens or early twenties. In the study, they were instructed to send messages to each other, while one of the participants was seated in the driving simulator (parked or driving) and the other was seated outside. During the study, they communicated via text messaging while under four conditions. In the first condition, the driver typed text messages while parked. In the other three conditions the driver communicated via a speech system in three difficulty levels (parked, driving in low driving workload, and driving in moderate driving workload). In these speech-based conditions, the driver's messages were converted in real-time to text by a Wizard of Oz system. Wizard of Oz was used in order to avoid additional variability in the data due to speech recognition errors. In all of the conditions the participant sitting outside typed their text messages. All conditions were divided into sections. Each section started by a predetermined message sent by the driver, which was then followed by a free form SMS "conversation" between the pair of participants.

2.2   Corpus Normalization

The corpus of text messages sent by the driver was normalized. Each text message was manually reviewed and converted to a canonical form in order to achieve a common ground for comparison from a vocabulary perspective. The first normalization act was to remove all initial predetermined messages that drivers were asked to send. In addition, punctuation marks were removed in order to make the text and the speech messages more easily compared.

As for the rest of the messages, additional variations were detected by matching all the words against a dictionary of 50,000 common words in English. This matching yielded a set of about 1,000 instances of words that were different from canonical form and were handled manually. A summary of the number of messages in each condition is shown in Table 1.

Table 1 Number of messages in each condition

Message type	Number of messages
All Messages	5,196
Sent by the driver	2,629
Sent by driver excluding predetermined messages	2,157
Text (parked)	469
Speech (parked)	573
Speech (low workload)	535
Speech (moderate workload)	580

3   EXPERIMENTAL LANGUAGE MODEL SYSTEM

An experimental language model system, consisting of a training phase and a test phase, was used in the experiment, as shown in Figure 1. Because of the relatively small number of messages in the corpus, a jackknife v-fold approach (Duda, Hart and Stork 2000) for dealing with lack of data was used. After normalization of the text and omission of the predetermined text messages, each of the four experimental conditions was divided into six sets, creating a total of 24 sets. For each condition, six training sets and six testing sets were created. Each of the test sets was one sixth of the condition's data and together the test sets were mutually exclusive. The conjugate training set for a test set was the rest of the condition's data.

The LM evaluated in this paper was a Markov chain of order N known as *N-Gram model*, where N is the number of consecutive words considered. According to this model the probability of a word depends approximately on the $N - 1$ previous words (Eq.1).

$$P(w_n, \dots, w_1; \lambda) = \prod_{t=1}^{n} P(w_t | w_{t-1}, \dots, w_{t-N+1}; \lambda) \qquad (1)$$

where λ is the language model and n is the sequence length.

As speech has a relatively broad vocabulary (usually tens of thousands of words), for practical reasons, the order of the model was kept low. In our case, we chose the order to be one or three; called a Unigram model and a Trigram model, respectively.

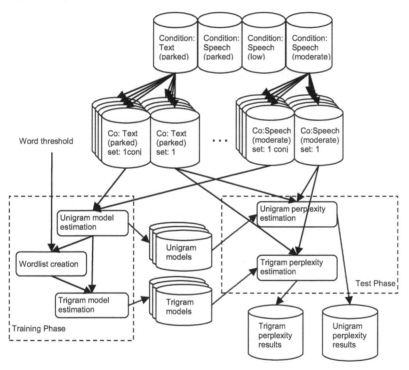

Figure 1 Experimental System Description – model estimation is depicted on the left side and the testing is described on the right side.

3.1 System Training Phase

The system training phase involved language model estimation and is shown on the left hand side of Figure 1. Language model estimation is the process of estimating the likelihood of a word sequence to appear given a corpus of data. We are interested in the conditional likelihood $P(w_t | w_{t-1}, \dots, w_{t-N+1}; \lambda)$. Let's denote the number of occurrences of a sequence of words $w_t, \dots, w_{t-N+1}$ in a corpus as $C(w_t, \dots, w_{t-N+1})$. Over a large database a good estimation of the likelihood is shown in Eq. 2.

$$P(w_t | w_{t-1}, \dots, w_{t-N+1}; \lambda) = \frac{C(w_t, \dots, w_{t-N+1})}{C(w_{t-1}, \dots, w_{t-N+1})} \tag{2}$$

In practice, however, such an estimate would assign a zero probability to words not seen in the training data, and inaccurate probabilities where occurrence is low. A common method to overcome this challenge is using discounting. In this paper the likelihoods were Kneser-Ney (KN) (Kneser and Ney 1995) discounted. The goal of KN is to discount based on reducing the probability of seen sequence of words $(c(w_t, \dots, w_{t-N+1}) > 0)$ by a fixed value D and redistributing this portion among the unseen sequence of words $(c(w_t, \dots, w_{t-N+1}) = 0)$. For seen sequence, the probability is defined as:

$$P_{\substack{KN \\ s.t. \\ c(w_t, \dots, w_{t-N}) > 0}} (w_t | w_{t-1}, \dots, w_{t-N}; \lambda) = \frac{C(w_t, \dots, w_{t-N}) - D}{C(w_{t-1}, \dots, w_{t-N})} \tag{3}$$

The total probability mass collected from the observed sequences is then spread among the unseen sequences.

The training phase was composed of three stages. In the first stage, a Unigram model was estimated. The likelihood of appearance was estimated for each word in the training data (about 1,000 distinct words out of 5,000 instances). In the second stage, a list of the most common words was created based on the likelihoods derived in stage one. Only words that were more likely to appear than a certain threshold entered the list. (About 200 words appeared more than five times in the data, with a likelihood greater than 0.001 each). Finally, in the third stage, a Trigram model was estimated. This model was KN discounted, as discussed above, based on the list of common words from stage two. Overall, the output of the training phase was a set of 48 language models ($[\text{Unigram}, \text{Trigram}] \times [\text{condition}] \times [\text{set}] = 2 \times 4 \times 6 = 48$). The Unigram and Trigram models are shown in the center of Figure 1 as the output of the training phase and input to the test phase.

3.2   System Test Phase

The effectiveness of the LMs was estimated by conducting a set of comparisons between test sets and LMs. The test phase was composed of two parallel stages. Each of these stages is a similarity estimation stage. overall, each of the 48 models was tested against four different conditions. Test sets were omitted in order to prevent overlap of the training and the testing data and to regularize the test.

4   RESULTS

LM data from the four conditions were compared using three methods: Perplexity, log likelihood ratio, and detection hypothesis testing. Although our goal was to check whether the likelihood under two models differ i.e. $P(w_t | w_{t-1}, \dots, w_{t-N+1}; \lambda_1)? = P(w_t | w_{t-1}, \dots, w_{t-N+1}; \lambda_2)$; in this case we had to conduct indirect testing i.e. $\hat{P}(\lambda_1 | w_{t-1}, \dots, w_{t-N+1})? = \hat{P}(\lambda_2 | w_{t-1}, \dots, w_{t-N+1})$ since the accuracy of the models could not be verified. $\hat{P}$ denotes the estimated model. This notation is used here in order to emphasize the difference between true and estimated distributions.

4.1 Perplexity Estimation

The effectiveness of the LMs was estimated using the perplexity measure (Jurafsky, and Martin, 2000). Perplexity is defined per language model and a sequence of words:

$$PP(W; \lambda) = PP(w_1, \dots, w_n; \lambda) = \frac{1}{\sqrt[n]{P(w_1, \dots, w_n; \lambda)}} \tag{4}$$

where $W = w_1, \dots, w_n$

Perplexity, in the context of a Unigram model, can be viewed intuitively as a weighted branching factor. It can be considered as a measure of the question: "How many possible words are considered to be said next?" A low perplexity value signifies a low branching factor which indicates a constrained vocabulary. For example, in a Unigram model low perplexity indicates that a small set of words is very likely to appear, while the rest of the words are less likely to appear. Low perplexity denotes 'simple' language.

As described in 3.2, we compared between LMs and relevant test sets. Each perplexity result was assigned to its relevant condition and averaged, yielding results for each condition. Experiment results for Unigrams are shown in Table 2. The highest perplexity was obtained for text messages. More importantly, perplexity decreased when workload increased, indicating a correlation between driving workload and the simplicity of the language used. Drivers used the 'most complex' language when they texted while parked and the 'simplest' language when they used speech in the moderate workload condition.

Table 2 Perplexity for Unigram models within condition. The perplexity was averaged among all the six pairs of test data and conjugate model.

Condition	Perplexity within condition
Text (parked)	221.0
Speech (parked)	209.2
Speech (low workload)	207.0
Speech (moderate workload)	199.8

4.2 Log Likelihood Ratio Hypothesis Testing

A more effective Unigram representation of perplexity can be achieved since this model is an independent identical distribution (i.i.d) model. In addition, by applying a logarithmic scale to the perplexity measure an intuitive representation can be achieved (Cover and Thomas 1991).

$$\log PP(W;\lambda) = -\frac{1}{n}\log P(w_1,\dots,w_n;\lambda) = -\frac{1}{n}\sum_{i=1}^{n} logP(w_i|\lambda)$$

$$= -\sum_{w\in \mathrm{w}} \frac{C(w|w_1,\dots,w_N)}{n} logP(w|\lambda) \qquad (5)$$

$$= -\sum_{w\in \mathrm{w}} P(w|w_1,\dots,w_N) logP(w|\lambda)$$

where w is the set of all words in the vocabulary.

Let's assume that two LMs λ_1, λ_2 were trained. Given a test set composed of a set of messages, we would like to check whether the messages were generated from the first or second LM. The first challenge is to define which events are i.i.d. Unigram models are i.i.d by definition. Generally speaking, however, speech is not i.i.d. Intuitively, assuming that a person's LM is similar in two points in time, might suggest that this person didn't absorb any new information between those two occasions. Given that people absorb information all the time the i.i.d assumption does not hold and therefore an approximation has to be made. In this work we assume that messages are generated in an i.i.d manner; however, per the definition of the N-gram model the words are not i.i.d.

According to Neyman-Pearson lemma (Cover and Thomas 1991) the optimal decision criterion test for a message M and two i.i.d hypotheses is of the form:

$$T < \frac{P(M|\lambda_1)}{P(M|\lambda_2)} = \frac{P\left(w_{M_1},\dots,w_{M_n}|\lambda_1\right)}{P\left(w_{M_1},\dots,w_{M_n}|\lambda_2\right)} \qquad (6)$$

where T is a threshold, and the words of the message M are $w_{M_1},\dots,w_{M_n}$.

This criterion is known as the likelihood ratio between the two models. Ideally and intuitively, the point in which T equals one, is the equilibrium point between the two models ($P(M|\lambda_1) = P(M|\lambda_2)$).

This criterion is equivalent to:

$$\frac{1}{M_n}logT <$$
$$D_{kl}\left(P\left(M|w_{M_1},\dots,w_{M_n}\right)\|P(M|\lambda_2)\right) - D_{kl}\left(P\left(M|w_{M_1},\dots,w_{M_n}\right)\|P(M|\lambda_1)\right) \qquad (7)$$

where D_{kl} is the Kullback Leibler divergence (Cover and Thomas 1991).

Given that $D_{kl}\left(P\left(M|w_{M_1},\dots,w_{M_n}\right)\|P(M|\lambda_i)\right)$ can be rewritten for Unigram as:

$$D_{kl}\left(P\left(M|w_{M_1},\dots,w_{M_n}\right)\|P(M|\lambda_i)\right) =$$

$$= \sum_{w\in \mathrm{w}} P(w|w_{M_1},\dots,w_{M_n})log\frac{P(w|w_{M_1},\dots,w_{M_n})}{P(w|\lambda_i)} =$$

$$= \sum_{w\in \mathrm{w}} P(w|w_{M_1},\dots,w_{M_n})logP(w|w_{M_1},\dots,w_{M_n}) \qquad (8)$$

$$+ \log PP(M;\lambda_i)$$

Given that $\sum_{w\in \mathrm{w}} P(w|w_{M_1},\dots,w_{M_n})logP(w|w_{M_1},\dots,w_{M_n})$ is not dependent on the model and appears in both divergences in Eq (7), it can be omitted.

$$\frac{1}{M_n}logT < \log PP(W;\lambda_2) - \log PP(W;\lambda_1) \qquad (9)$$

where $logT = 0$ is the equilibrium point.

We tested the hypothesis that a pair of language models shared the same distribution $(P(w_1, ..., w_N | \lambda_1) \approx P(w_1, ..., w_N | \lambda_2))$, i.e., the null hypothesis was the equilibrium point. For each condition, and all corresponding test messages, the log likelihood ratio of that model and an alternative model were estimated. The distribution of these scores for the different messages was tested against the null hypothesis. The 95% confidence intervals of $logT$ are shown in the Table 3. All the confidence intervals for log ratios did not include zero, indicating that the differences were significant. Trigram models were not found to be different.

Table 3 95% confidence intervals for comparison among Unigram models (Log likelihood ratio where 0=same)

LM1 condition	Text (parked)	Speech (parked)	Speech (low workload)	Speech (moderate workload)
Text (parked)		[0.58,1.27]	[0.47,1.16]	[0.54,1.23]
Speech (parked)	[0.77,1.39]		[0.15,0.77]	[0.42,1.04]
Speech (low workload)	[0.71,1.36]	[0.01,0.65]		[0.51,1.15]
Speech (moderate workload)	[0.70,1.31]	[0.12,0.73]	[0.37,0.98]	

4.3   Condition Detection

We looked at workload detection in order to further quantify the significance of the difference between language models obtained in different conditions. This is the indirect testing that was described in section 4 above. Condition detection was applied to a set of messages by comparing their log likelihood ratios. Significance of detection results was calculated as described in (Gillick and Cox 1989). Results for Unigram models are presented in Figure 2. The Figure shows the detection errors tradeoff for pairs of conditions, which is the probability of extracting the messages of one condition from a pool containing messages of the two conditions together versus the extraction of the other condition from the same pool. A good classifier would be close to the axes origin. The black solid line represents random detection, i.e. detection by tossing a biased coin. This is the null hypothesis. The black dotted line represents 95% confidence interval for the null hypothesis. The other three lines represent a comparison between pairs of conditions. It can be seen that the detection is significantly better than random for the majority of the detection rates, although the absolute detection values are quite modest. As expected, the detection between speech (parked) and text (parked) is much more significant than among the three speech conditions (parked vs. low workload and parked vs. moderate workload) but all are significant.

194

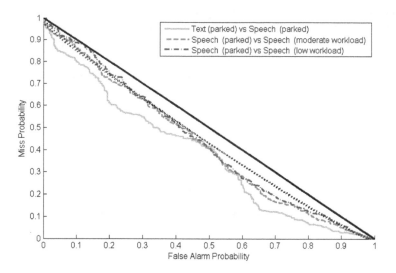

Figure 2: Significance of detection in for three model-pair conditions. A random classifier is represented by the solid black line, and model-pair classifiers in green red and blue.

5   CONCLUSIONS

We showed that the grammar and choice of words in spoken messages were correlated to the input method and driving workload. Language models derived from spoken messages under low workload and moderate workload were different from models derived from spoken messages while parked. In addition, we verified that the grammar of spoken messages and text messages were different.

Although we saw significant differences in LMs for Unigrams, we could not show significant differences in our results for Trigram language models. This result can be explained by the small dataset used in this experiment. Accurate estimation of Trigram typically requires a much larger number of pairs and triplets of words. Such repetitions did not occur in this corpus of several thousand of messages.

A possible implication of this work is that text-message dictation using speech recognition can be improved by taking into account the driving workload and perhaps other similar factors affecting the driver's choice of words. In this study even though the number of messages was quite limited, we found significant differences in language grammar for various conditions of driving workload. It is expected that larger numbers of messages could result in a greater understanding of the relation among consecutive words under various such driving conditions.

A methodological implication for the study of speech applications is that people change the way they speak as a function of various factors, such as driving workload. Therefore, it is important to vary these factors when testing automotive systems so as to span a large spectrum of possibilities.

REFERENCES

Cover, T. and Thomas, J. A. 1991. *Elements of information theory*, pp. 304-311. New York, NY: John Wiley & Sons, Inc.

Duda, R.O, Hart, P. E., and Stork, D. G. 2000. *Pattern Classification*, pp. 472-474. New York, NY: John Wiley & Sons, Inc.

Gillick, L., and Cox, S. J. 1989. *Some statistical issues in the comparison of speech recognition algorithms*. Proceedings of International Conference on Acoustics, Speech, and Signal Processing (ICASSP).

Green, P. A., Lin, B., Kang, T., and Best, A., 2011. *Manual and speech entry of text messages while driving*. Technical Report UMTRI-2011-47. Ann Arbor, MI: University of Michigan Transportation Research Institute.

International Telecommunication Union (ITU) 2010. *The World in 2010: ICT facts and figures*. International Telecommunications Union. Accessed Feb 26, 2012. http://www.itu.int/ITU-D/ict/facts/2011/material/ICTFactsFigures2010.pdf

Jurafsky, D. and Matrin, J. H. 2000. *Speech and language processing*, pp.191-234. Upper Saddle River, NJ: Prentice-Hall, Inc.

Kneser, R., and Ney, H. 1995. *Improved backing-off for N-gram language modeling*. Proceedings of International Conference on Acoustics, Speech, and Signal Processing (ICASSP).

Lindstrom, A., Villing, J., Larsson, S., Seward, A., Aberg, N., and Holtelius,C. 2008. *The effect of cognitive load on disfluencies during in-vehicle spoken dialogue*. Proceedings of Interspeech.

Ney, H., Haeb-Umbach, R., Tran, B. H., Oerder, M., 1992. *Improvements in beam search for 10000-word continuous speech recognition*. International Conference on Acoustics, Speech, and Signal Processing (ICASSP).

Villing, J. 2009. *In-vehicle dialogue management - Towards distinguishing between different types of workload*. Proceedings of SiMPE.

Villing, J. 2009. *Dialogue behaviour under high cognitive load*. Proceedings of SIGDIAL.

CHAPTER 21

Turn-by-Turn Navigation Guidance for Complex Interactions

Peggy Wang

General Motors China Science Lab
Pudong, Shanghai, China
peggy.wang@gm.com

Omer Tsimhoni

General Motors Advanced Technical Center - Israel
Herzliya, Israel
Omer.tsimhoni@gm.com

ABSTRACT

Most in-vehicle navigation systems can provide drivers with turn-by-turn speech instructions. Users not only expect their navigation system to guide them to unfamiliar destinations but also to provide real-time guidance based on traffic congestion. In order to avoid time delays, stress, and frustration, navigation systems need to provide accurate, effective, and efficient information. Various studies have investigated user interface considerations for better speech guidance. Only a few studies, however, considered the effect of road characteristics on speech guidance and fewer yet focused on complex interactions that carry a high risk for navigation errors. In this paper we describe and analyze speech guidance for a unique set of complex interactions, namely, elevated roads. We focus on the components of speech instructions provided in current systems (maneuver and wording) and other potential information content. We report a field study in which we examined two navigation systems to record turn-by-turn speech instructions on elevated roads in Shanghai: GM OnStar, a server-based built-in navigation system, and Eroda, a portable GPS device. This study represents the first step towards enhancing the comprehensibility of a turn-by-turn navigation guidance for complex interactions.

Keywords: in-vehicle navigation, turn by turn speech instructions

1   INTRODUCTION

Some argue that the usability of a navigation system is the most important aspect of its design (Burnett, 2000). Most in-vehicle navigation systems, including built-in systems and portable GPS devices, can provide drivers with turn-by-turn speech instructions. Users not only expect their navigation system to guide them to unfamiliar destinations but also to provide real-time guidance based on traffic congestion. In congested traffic, even a single guidance error may result in long delays. In order to avoid time delays, stress, and frustration for the driver, navigation systems need to provide accurate, effective, and efficient information. A usability evaluation of turn-by-turn navigation systems (Burnett and Joyner, 1997) has concluded that compared with getting instruction from an "informed" passenger, drivers using turn-by-turn navigation system made more navigational errors, took longer time to complete the route and had higher mental workload. Recent customer surveys of turn-by-turn navigation in China have revealed several areas in which speech instructions can be improved for better system performance and user satisfaction.

Various studies have described user interface considerations for better speech delivery and specifically, better turn-by-turn speech instructions. Some studies identified common usability problems of the speech instructions by both heuristic evaluation and formal user testing, for example, using incorrect terms to describe the road geometry (Nowakowski, Green and Tsimhoni, 2003). Some studies analyzed all potential navigation information for use in the speech instruction given by the in-vehicle navigation systems. The range of different information types included landmarks, road sign information and road geometry (Burnett, 1998). A large number of research activities have addressed the use of natural language generation for route description (e.g., Dale, Geldof and Prost, 2005) and destination description (e.g., Tomko, 2007; Tomko and Winter, 2009). Several principles are used to generate speech instructions, providing adequate information and decreased ambiguity while minimizing drivers' mental workload.

Describing locations by landmarks (visual, cognitive, or structural) has been shown to result in route descriptions that are much easier to follow than route described by distance (Burnett, 2000; Sorrows and Hirtle, 1999). In addition, the maneuvers followed the concept of grouping hierarchical organization as it has been shown to contribute to comprehensibility (Taylor and Tversky, 1992). Segmentation is a way to give hierarchical structure of the route and reduce user's cognitive load. There are two segmentation strategies: landmark-based and path-based. The landmark based segmentation determines optimal break points in the sequence of paths. The path-based segmentation aggregated several paths into a higher level entity based on the characteristics of the element road. The road status hierarchy, path length and turn typology are three features that can be used for path-based segmentation (Dale, Geldof and Prost, 2005). Segmentation is used to help a user get a clear route view by a higher level structure. Agrawala and Stolte (2001) also suggested not giving an equal status to all parts of the route description. Salience and relevance in content selection and linguistic knowledge are also considered when generating spoken instructions. All the landmark, path length and turning information are retrieved from real GIS data.

However, it is unclear whether landmark or segmentation information is needed in the speech instruction for all kinds of roads. While for roads with simple structure, speech instructions of current navigation systems work relatively well, for roads with complex structure, such as an entrance into an elevated road, the road geometry is non-standard. These may consist of mixed road types including both elevated roads and surface roads at a given decision point. For these cases, the current speech instruction is not sufficiently clear. Users tend to have difficulty following the instructions, which may lead to navigation errors. In this paper we consider road characteristics in order to improve speech instructions given by navigation systems. The focus of this study is on the components of speech instructions in complex interactions, including maneuver, wording and information content. More specifically, it addresses the needs associated with elevated roads in China.

2   FIELD STUDY

2.1   Method

A field study was conducted to explore turn-by-turn speech instructions on elevated roads and to collect information about relevant scenarios. Two navigation systems were compared, the OnStar Gen 9 model (Onstar, 2012) and the portable Eroda EV5 (Eroda, 2012). The OnStar is a telematics service developed and deployed by General Motors. OnStar provides subscription-based communications, hands free calling, turn-by-turn navigation and remote diagnostics throughout the United States, Canada and China. OnStar is a speech system and works on voice command to enable the driver to communicate without using their hands. In China, the most widely used OnStar feature is navigation, from which 80% of the user demand comes. Once users need navigation, they call an advisor at the call center and ask for a destination to which they want to go. The advisor processes the user's demand and uploads the navigation instructions to the user's car. After that, the user follows the speech turn-by-turn instructions. The Eroda EV5 is a portable navigation device running Windows CE 6.0. It has a 7 inch screen displaying a birds-eye virtual street view but it also has a text-to-speech module that prompts the speech turn-by-turn instructions.

In Shanghai there are about eleven elevated road. Two elevated roads were selected for this study. The first route was from Zhoujiazui Road to Dongjiao Hotel. It went through the Yangpu Bridge, which was part of the Inner-ring elevated road. Along this route, three places were considered in which the speech instructions could be ambiguous and difficult to understand. The second route of the field study was from No. 5 Jinian Road to Sinan Road. Video recordings were collected on two sunny afternoons from 1 to 5 pm.

2.2 Results

The first location was a three-way fork, shown in Figure 1. The guidance route is indicated by a red arrow. The OnStar speech instruction associated with this maneuver was "200米后走中间岔路，并入内环高架路". [Translation] "In 200 meters, drive on the middle road, merge into the Inner-ring elevated road." (The literal instruction is actually "200 meters later on," but simplified for English readers in this and all following instructions.) The OnStar system displayed a left turn, as the variety of possible turn indicators was limited. The portable Eroda displayed the amplified street view at this spot to indicate the direction. The screens of OnStar and Eroda are shown in Figure 2.

Figure 1. Road structure of the first location   Figure 2. Screens of OnStar and Eroda

For the second location, OnStar speech instruction is "500米后靠右，沿内环高架路行驶." [Translation] "In 500 meters, drive on the right side and along the Inner-ring elevated road." The turn indicator on the OnStar screen is a right oblique, while the Eroda indicated go straight. The road structure is shown in Figure 3. The screens of OnStar and Eroda are shown in Figure 4.

Figure 3. Road structure of the second location   Figure 4. Screens of OnStar and Eroda

The road structure for the third location is shown in Figure 5. OnStar speech instruction is "700米后靠左驶入龙东大道后，靠右驶入龙东大道." [Translation] "In 700 meters, drive on the left side to Longdong Avenue. Then drive on the right side to Longdong Avenue". The indicator on OnStar screen is turning left. The screens of OnStar and Eroda are shown in Figure 6.

Figure 5. Road structure of the third location   *Figure 6. Screens of OnStar and Eroda*

The second route of the field study was from No. 5 Jinian Road to Sinan Road. This route was highly complex as two elevated roads were included. Furthermore, along part of the route, GPS signal on the surface road was blocked by the above elevated road. OnStar lost GPS signal and induced an error message. This failure took several seconds. Figure 7 shows the turning point of the second route with irregular structure. At this location, the OnStar speech instruction was "5.2公里后，向右斜前方转弯，驶入南北高架路。" [Translation] "In 5.2 kilometers, turn to the right oblique to the South-north elevated road." Actually, the road was on the lefthand side. As indicated by the orange rectangular in Figure 7, there was a traffic sign at the intersection, showing destinations from each road.

Figure 7. Turning point of the second route with irregular structure

3   ANALYSIS OF ISSUES

In each of the locations described above, a driver may have experienced difficulties in understanding the speech instructions. In this section, we analyze the speech instructions of both OnStar and Eroda from a wording and maneuver perspectives. Table 1 summarizes these findings.

First, Table 1 demonstrates that the speech instructions of both devices use the distance-to-turn format with ego-centered directions and absolute distance, such as "In 500 meters, drive straight onto" and "In 200 meters, drive onto the middle road." However, research has indicated that this is not the optimum solution. As described earlier, the use of notable landmarks can increase the usability of vehicle navigation (Burnett, 1998). Second, all the roads in the above locations have irregular structures, either intersection of surface and elevated roads or with a non-standard geometry. Advanced instructions of these irregular structures are difficult to comprehend out of context. For example, it is much more difficult to comprehend two consecutive spoken instructions like "In 700 meters, drive on the left side to Longdong Avenue. Then, drive on the right side to Longdong Avenue." Third, effective speech instructions of navigations system must match the user's mental model of the world and provide informative pieces of data for decision making. Obviously, both the words and the maneuvers in the two tested systems, even in most existing navigation systems, could not meet this requirement.

Table 1. Summary of the speech instructions

OnStar	Eroda
"200米后走中间岔路，并入内环高架路" "In 200 meters, <u>drive on the middle road,</u> merge into the Inner-ring elevated road"	"前方岔路，请走中线上高架" " In the front road, <u>drive on the middle</u> onto the elevated road"
"500米后靠右，沿内环高架路行驶" "In 500 meters, <u>drive on the right side and along</u> the Inner-ring elevated road."	"前方500米，请直行上高架" "In 500 meters, <u>drive straight onto</u> the elevated road"
"700米后靠左驶入龙东大道后，靠右驶入龙东大道" "In 700 meters, <u>drive on the left side</u> to Longdong Avenue. Then <u>drive on the right side</u> to Longdong Avenue".	"前方一公里，请向右前方行驶" "In 1000 meters, <u>drive on the right front side</u>. *After almost reaching the turning point:* "前方路口，请向右前方行驶" "In the front road, <u>drive on the right front side</u>"
"5.2公里后，向右斜前方转弯，驶入南北高架路" "In 5.2 kilometers, turn to <u>the right oblique</u> to the South-north elevated road"	*There was no prompt before turning. After turning:* "前方200米，请直行上高架" "In 200 meters, drive <u>straight onto</u> the elevated road"

Typically, the speech instructions consisted of three parts: (i) Positioning of the maneuver (e.g., in 500 meters); (ii) description of the maneuver, which is the most important piece of information in the instruction, telling the driver what to do at the required position; and (iii) continuing direction, providing information about where to go next. The maneuver in each instruction, shown in Table 1, is underlined.

For the first instruction, the maneuver "drive on the middle", is identical for both OnStar and Eroda. Figure 1 Shows that the middle road is the ramp to the elevated road, while the other two lanes are surface roads. As a result, the driver may need some effort to ascertain the lane. As to the wording, the OnStar's speech instruction " 并入-merge into " is not an entirely accurate word describing the connection of the ramp and the *upcoming* Inner-ring elevated road.

For the second instruction, the maneuvers are "drive on the right side" (OnStar) and "drive straight" (Eroda), which are quite different. Considering the road shown in Figure 3; the instruction "drive on the right side" is confusing since there is another lane on the right side, which leads to the exit of the Inner-ring elevated road. In this situation, when there is no direction change at the fork, the instruction of "keep driving along" should be the critical information for the driver. As this location also has an irregular road geometry, other potential information such as "to the direction of the Inner-ring elevated road" maybe needed for confirmation.

For the third instruction, the maneuver in OnStar is "drive on the left and then drive on the right", which is very confusing without knowing the road structure. The wording of "Drive on the left side of the upcoming fork, then drive on the right of the next fork" is much clearer and easier to understand. The Eroda only prompted the second half of the instruction, and repeated it when almost reaching the turning point. However, the maneuver of "drive on the right front side" is not enough. Adding the instruction "to Longdong Avenue" would yield a better match with the user mental modal of the road.

For the last instruction, the maneuver in OnStar is "turn to the right oblique", which assumes that the driver has turned left at the intersection. Immediately after turning left, the driver was supposed to turn to the right oblique side and drive onto the North-south elevated road. The maneuver in Eroda "drive straight onto elevated road" will likely lead to an error.

Further inspection of the above four turning instructions (both speech and display-based) in the two routes revealed additional usability issues, some of which are mentioned below:

(1) Using only right and left for direction can induce confusion in the scenarios when there are more than two lanes at the intersection. Both OnStar and the Eroda used right and left to indicate the direction at the turning point.

(2) The spoken instructions for the irregular road structures could be mentally difficult to understand, especially in conjunction with elevated road.

(3) The OnStar turn by turn display icons provided unclear illustrations, which sometimes even contradicted the spoken directions. In the above three decision locations, the OnStar icons did not facilitate driver's understanding of the speech instructions. Even the best-matched icons could not illustrate the complex road structure, rather they confused or even misled the driver.

(4) Another issue which came up in this session was address ambiguity. When the driver asked the OnStar advisor for instructions to the Shanghai TV University, the advisor checked with the driver for environmental information, such as surrounding roads, to confirm the address. At least two buildings in the area had the same name: Shanghai TV University. In fact, OnStar navigated to the unintended one. The destination was also set on Eroda, which got the right place by entering and selecting the address from the map. Conventionally, destination in China is

conveyed by the name of the building or the entity (for example, the university), rather than the street number.

(5) The algorithm of selecting a route is not intelligent. First, on the elevated road OnStar instructed the driver to drive along the right or left lane along the road and not change the lane. This was unnecessary. Second, the turn-by-turn navigation instructed the driver to exit the main road and then drive back to the same road within about 1000 meters.

(6) The command "Repeat last maneuver" was frequently used. During the testing, it was the most frequently delivered command.

4   DISCUSSION

Our analysis clearly points out that speech instructions for navigation systems should be improved, specifically when describing complex road structures. The field study has revealed that turn-by-turn guidance on elevated highways poses a potential usability problem. Further, a sample of two navigation systems revealed similar issues, as well as some differences and inconsistencies. This research clears the way for usability testing of real participants and it remains to be seen what kind of confusion will occur when navigating via the above intersections and decision points, and what additional points will be revealed. Based on the analysis of the findings, we propose several ideas worth exploring. First, we propose to use potential information of landmarks, road properties and segmentations to improve the speech instructions of turn by turn navigation, specifically for elevated roads in Shanghai. Second, for complex interactions the maneuvers should have more flexibility to more accurately describe the actual layout and required maneuver. Specifically, the choice of turning or not, entering or not, and exiting or not should be explicit and unambiguous. Third, for some consecutive speech instructions, the use of the phrases: "the upcoming fork, and "the next fork" would help the driver to better understand the layout of the route. What is more, separating the speech instructions into more pieces, if possible, will make it more easy to understand and easy to follow, but the tradeoff between accuracy and overloading the driver's memory should be taken into account. Fourth, more information should be added for further confirmation. Consider integrating other potential information into the speech instructions of navigation systems, we hypothesize that using the road property and the road name (ramp of Inner-ring elevated road) in the speech instruction, will improve drivers comprehension. As an example, the proposed modification for the first speech instruction would be: "In 200 meters, drive onto the ramp to the Inner-ring elevated road", which translates to: "200 米后，走匝道上内环高架路" in Chinese. With this new set of suggestions, we expect the usability study to not only to provide additional insights for the problems with existing guidance in complex systems, but also to confirm and fine-tune the advantages afforded by such modification to the speech interface.

REFERENCE

Agrawala, M. and Stolte, C. 2001. Rendering effective route maps: improving usability through generalization. In *Proceedings of Siggraph*. Los Angeles, CA.

Burnett, G.E. 1998. *"Turn right at the King's Head": Drivers' requirements for route guidance information*. PhD dissertation, Loughborough University, UK

Burnett, G. 2000. Turn right at the traffic lights: The requirements for landmarks in vehicle navigation systems. *The Journal of Navigation 53* 499-510.

Burnett, G.E., Joyner, S.M..1997. An assessment of moving map and symbol-based route guidance systems. In Y. Ian Noy (Ed.), *Ergonomics and safety of intelligent driver interfaces* (pp. 115-136). Mahwah, NJ: Lawrence Erlbaum Associates.

Dale, R., Geldof, S., Prost, J.P. 2005. Using natural language generation in automatic route description. *Journal of Research and Practice in Information Technology 37* 89–105

"Eroda" Accessed February 26, 2012 http://www.eroda.asia

Nowakowski, C., Green, P., and Tsimhoni, O. 2003. Common Automotive Navigation System Usability Problems and a Standard Test Protocol to Identify Them, in *ITS-America 2003 Annual Meeting. Intelligent Transportation Society of America*: Washington, D.C.

"OnStar" Accessed February 26, 2012 http://www.onstar.com

Ross, T. and Burnett, G. 2001. Evaluating the human-machine interface to vehicle navigation systems as an example of ubiquitous computing. *International Journal of Human-Computer Studies 55* 661-674.

Sorrows, M.E., Hirtle, S.C. 1999. The nature of landmarks for real and electronic spaces. In Freksa C. and Mark, D. M. (eds), *Spatial Information Theory*, number 1661 in Lecture Notes in Computer Science, Springer.

Taylor A. H., Tversky B. 1992. Description and depiction of environment. *Memory and Cognition 20* 483-496.

Tomko, M.: 2007. *Destination descriptions in urban environment*. Ph.D. thesis, Department of Geomatics, University of Melbourne.

Tomko, M., Winter, S. 2009. Pragmatic Construction of Destination Descriptions for Urban Environments *Spatial Cognition & Computation 9* 1-29.

CHAPTER 22

Identifying Customer-oriented Key Aspects of Perception with Focus on Longitudinal Vehicle Dynamics

Thomas Müller, Christian Gold, Armin Eichinger, Klaus Bengler

Technische Universität München - Institute of Ergonomics
Boltzmannstraße 15, 85747 Garching, Germany
tmueller@lfe.mw.tum.de

ABSTRACT

Performance of the vehicle's longitudinal dynamics essentially determines the character and the perception of a vehicle by the customer. To place a vehicle at a specific niche in the market or the product portfolio, it is necessary to know the relations between the variable parameters of the power train and their perception by the user. Therefore, as a necessary basis, this study deals with the identification of customer-oriented key aspects of perception with focus on longitudinal vehicle dynamics.

For this purpose, a group of subjects took part in a study, where they drove selected cars implementing a variety of different vehicle concepts. Being interviewed using a repertory grid technique, subjects were assisted in structuring and verbalizing their subjective perception. Subjective data for different aspects of the vehicle's longitudinal dynamics were collected by standardized questionnaires. With the help of recorded driving dynamics data different driver types were identified. The influence of driver type on subjective perception was investigated. These aspects of perception could be described using a two dimensional structure identified by principal components analysis.

Keywords: perception, longitudinal vehicle dynamics, repertory grid technique, driver type, principal components analysis

1   INTRODUCTION

The power train of an automobile represents the central component of the vehicle and dominates its classification in a particular segment or the assignment to a specific vehicle type, based on both, objective and subjective criteria.   In the automotive industry standards, such as the 10 point scale and standardized maneuvers, are used to assess driving and the driving behavior of a vehicle. Some of these standardized maneuvers are start-up behavior, straight ahead driving or driving comfort (Heißing, 2002). Furthermore, in recent years the user and its interaction with the product are more and more pushed in the focus of the development process of technological products (Garrett, 2011). Therefore, an appropriate consideration of specific customer needs and customer perception is also becoming gradually more important for the development of modern cars.

The individual perception of a vehicle as a sports car or luxury car is indeed strongly influenced by various aspects of vehicle design, product and brand environment, personal background, experience and much more (Haffke, 2010). However, successful positioning of a vehicle at a specific area in the product portfolio stands or falls with an appropriate design of technical vehicle parameters and their adequate perception by the customer. Therefore, the present study uses a particularly important factor out of this spectrum and deals in more depth with the influence of the power train and the resulting characteristics of the longitudinal dynamic behavior on the individual vehicle-perception of everyday drivers.

How do customers perceive technological parameter of longitudinal dynamics? To answer this question, the presented method will connect the technical parameters with the customer-oriented key aspects. Therefore, a perception-oriented experimental design was developed. A group of subjects compares different types of power trains in different vehicles, in order to categorize and evaluate them from their individual point of view.

2   METHOD

2.1   Framework conditions of the study

The study was conducted with 20 participants, who were selected according to the following criteria: Age from 30 to 60 years; driver license since at least 5 years; mileage minimum 5.000 km/year; experience with automatic transmissions; driver of a premium brand (BMW, Audi, Mercedes Benz or similar) vehicle built in 2005 or later.

Three of them were female, 17 male. The average age of the subject group was 45 ($SD = 10.23$), ranging from 30 to 60. There were five different vehicles available to carry out the experiment. The vehicles were intentionally very different in regard to the vehicle class (small car – luxury car), as well as the motor type (standard/powerful gasoline engines – standard/powerful diesel). To blank out the subject's individual gear-shifting behavior, all vehicles were equipped with automatic transmissions. Figure 1 shows the vehicles used in this experiment.

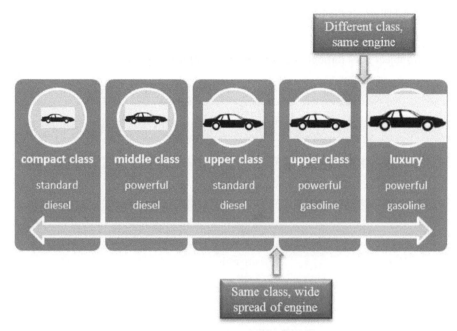

Figure 1 Spread of used test vehicles

2.2   Procedure

The structuring and verbalization of subjective perceptions is quite difficult for the majority of subjects. Therefore, the Repertory Grid Technique is used in this study as a supportive element from the field of knowledge management. This method is a combination of qualitative and quantitative data collection and is used to determine subjective perceptions and individual thinking (Kühl, 2009). In addition, subjects were faced with specific questions, regarding relevant aspects of the dynamic driving performance of the different cars during two intensive test drives. The characteristics and subjective quality of the vehicles were evaluated by a 10 point scale, always combined with the possibility of individual free text answers.

After welcoming, the subjects were first asked to fill out a questionnaire, regarding their personal background. It also contained some questions for a driver type assessment. Two questions are shown exemplary in Figure 2:

In situations with high traffic density I usually behave...	more aggressive	□□□□□□	more defensive
I would describe my driving style as:	sportive/dynamic	□□□□□□	calm/balanced

Figure 2 Questions for driver type assessment

208

The evaluation of the driver type is based on both, the subject's self-assessment and the recorded driving dynamics data and will be handled more detailed in the chapter results.

Each subject goes through the experiment as follows in Figure 3:

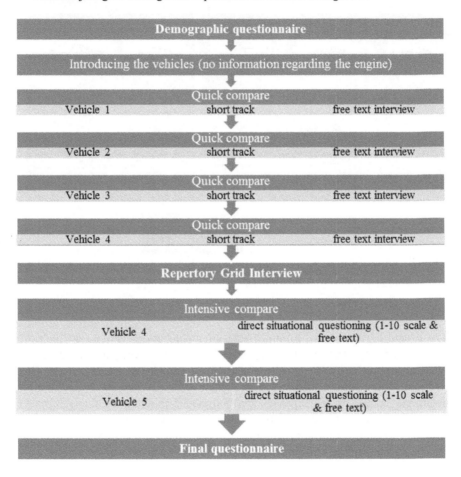

Figure 3 Experimental procedure

After giving instructions to the subjects regarding the test vehicles (no specific information about the engine), each subject was evaluating four out of five vehicles on a short test track (Figure 4). The order of vehicles was permuted for each subject. The route was chosen this way, that the subjects automatically had to drive through various typical situations (e.g. traffic light start, acceleration out of the roundabout, highway acceleration lane, as shown in Figure 4 (red circles)), where the engine and its characteristics become evident. While these short track drives, the investigator on the co-driver's seat wasn't giving any instructions to the subject except to describe the characteristic of the engine/vehicle in his own words. Immediately after these four 10-minutes test drives, the subject was asked to verbalize and organize his experienced perceptions with the help of the Repertory Grid Interview. The aim and the approach of this method will be explained more detailed in the next chapter. After the interview subjects were driving two cars on a 60 km drive for a more intensive test (Figure 5). Here the driving dynamics data were recorded for the driver type evaluation. For further investigations there were also a couple of specific driving maneuvers around the track, where the subjects had to answer appropriate questions regarding their perception in these situations by using a 10-point scale.

Figure 4 (left) short track, 10km, country road and highway, each red circle represents an appropriate point to evaluate the power train (Google Maps, 2012)

Figure 5 (right) long track, 60km, each number represents a driving maneuver with appropriate questioning immediately after (Google Maps, 2012)

2.3 Repertory Grid Technique (short track)

The Repertory Grid Technique (RGT) is based on A. Kelly's Personal Construct Theory from 1955 and, in general, it is intended to capture how people experience things, what the experiences mean for them and why they give meaning to specific things or attributes (Hassenzahl, 2000). According to Personal Construct Theory every individual is construing all elements of everyday life by relating them to former experiences and grouping them with the aid of personal constructs.

So the main parts of the RGT are elements and constructs. Elements are the objects of investigation (here: different vehicles and engines). The intention of research is to get to know the attributes, which people use to distinguish different vehicle concepts from each other. Using RGT helps the participants of the study to verbalize their individual constructs (Fromm, 1995). Figure 7 gives an overview of the procedure of a Repertory Grid Interview:

210

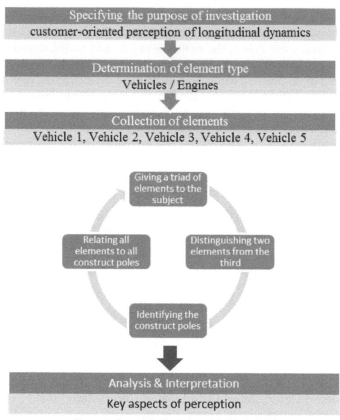

Figure 6 Overview Repertory Grid Technique

To develop these constructs the participant is given a triad (three randomly selected elements (in this case: vehicles) of usually about 10, (Fromm, 1995)) by the experimenter. The participant has to specify how two of the elements are similar but different from the third (Harrison, 2007). In case of vehicles such personal constructs related to experience may be powerful – weak, sportive – comfortable, etc. The attribute of the two similar elements is called the *implicit construct-pole*. Its antipode is called *emergent construct-pole*. These two construct poles are building a construct in the sense of a semantic concept (Lyons, 1968). After finishing this triad of elements, the participant has to repeat the process of comparing similarities and differences with another triad until no more constructs can be identified (Fallman, 2005). In this way the participant can name all the attributes which are the most important for him, to distinguish the different elements.

This fact makes an advantage of this research method evident: compared to pre-structured approaches (e.g. questionnaires) it is possible to investigate individual paradigms and the participant's individual focuses of perception (Fromm, 1995). The RGT helps to verbalize them and to make them accessible to further investigations.

Through relating all elements to all developed constructs by using a 5-scale rating, the Repertory Grid finally can be generated as exemplary seen in Figure 7:

emergent pole (similarity)						implicit pole (antipode)
hard	1	3	4	3	2	soft
aggressive	2	5	2	1	1	defensive
agile	5	1	3	1	4	lazy
dynamic	2	4	5	1	1	slow
sportive	2	2	1	4	2	comfortable
safe	1	4	1	3	3	unsafe
inspiring	5	2	3	1	5	boring
	5	4	4	3	1	

(left label: Constructs; columns: Vehicle A, Vehicle B, Vehicle C, Vehicle D, Vehicle E)

Elements

Figure 7 Exemplary result of a Repertory Grid Interview

2.4   Driver Type Assessment

For the comparative analysis of the results with regard to different types of drivers, key performance indicators are essential to enable a reliable determination of the driver type. These indicators refer to both, the subject's self-assessment and its matching with the measured driving dynamics data.

Depending on the driver type, drivers use different amounts of their available adhesion potential for handling their driving task (shown in Figure 8). A normal or more defensive driver is using lower accelerations (lateral and longitudinal), a very sportive driver can drive close to the physical limit. The investigation of the different level of utilization of the adhesion potential thus seems to be suitable for the driver type assessment.

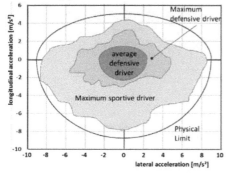

While driving, 14 acceleration values per second were recorded. It is not useful to interpret the longitudinal or the lateral acceleration separately, but only the combination makes sense for drawing conclusions regarding the driver type. The utilization of the adhesion potential (a_{pot}) is calculated by

$$a_{pot} = \sqrt{a_{long}{}^2 + a_{lat}{}^2}$$

Figure 8 Utilization of adhesion potential for defensive and sportive drivers (according to Lienkamp, 2011)

3   RESULTS

3.1   Driver Type Categories

Since low acceleration values are irrelevant for the driver type evaluation, all values were sorted by size and only the 1.500 highest values were used for further investigations (shown in Figure 9). Already by a qualitative visual check of the values it is possible to distinguish a defensive (Figure 9, left) and a sportive driver (Figure 9, right).

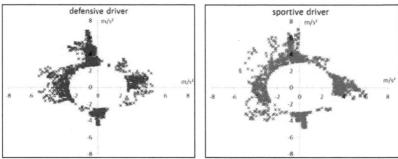

Figure 9   same vehicle, same track, different driver types: the 1500 highest accelaration values of a "defensive" driver (left) and a "sport" driver (right)

For a more reliable and quantitative classification of drivers, it is necessary to compare the area under the curve for each subject, exemplary shown again for the defensive and the sportive driver in Figure 10.

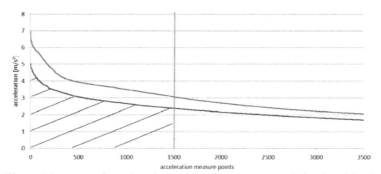

Figure 10   curve of accelaration values (lower curve = defensive driver)

In this way all subjects were categorized in a certain driver type „sport", „normal" or „defensive". Now it was possible to investigate if there is any connection between the driver type and their key aspects of perception.

3.2   Results of Repertory Grid Interview

The results of the Repertory Grid Interviews were analyzed and the constructs were grouped thematically. Regarding the perception of sportive and defensive drivers, there are some mentionable results, which suggest making a difference in this point is reasonable.

Only the group of sportive drivers mentioned the relation between engine sound and engine performance respectively sportiness of the car. In general, through all the interviews and comments it was obvious, that the group of more defensive drivers assesses the engine sound critically and would prefer the sound to be as discreet as possible. Furthermore this group didn't pick out the engine sound and engine performance as a central theme at all, whereas all issues regarding controllability and maneuverability came up more often compared to sportive drivers. So, in this case their subjective feeling of safety seems to be more influenced by the engine's "character".

As another result different driver types seem to be able to distinguish some aspects of their individual perception much more detailed than other groups do. For example sportive drivers were able to evaluate the automatic transmission systems more nuanced.

4   SUMMARY AND DISCUSSIONS

This study identified key aspects of the perception of vehicle dynamics. Furthermore, a method was developed to classify driver type based on, subjective and objective data. A connection between these two study aspects could be established by analyzing the influence of driver type on the perception of a vehicle's or, more precisely, an engine's character.

This study described the diversity of consumer-oriented perception while driving a car by qualitative and quantitative subjective ratings. With the help of factor- and cluster analysis the relevant attributes were combined into meaningful groups in order to derive the relevant dimensions, regarding the perception of a vehicle power train. Figure 11 shows the interpretation of the investigated key aspects of perception and the vehicles used in this study.

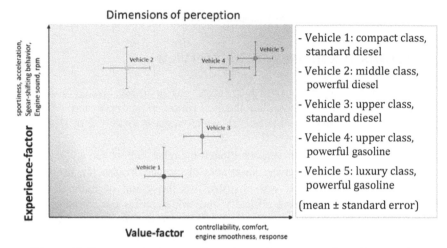

Figure 11 Factor analysis: two dimensions of customer-oriented perception

The identified correlational structure was explained by an experience- and a value-oriented factor. The experience factor combines all aspects of sportiness, acceleration, gear-shifting behavior and engine sound. The value/quality-factor combines aspects like controllability, comfort, engine smoothness and response.

These results lay the foundation for further studies, dealing with the relation between the specification of objective vehicle parameters and the resulting subjective human perception.

ACKNOWLEDGMENTS

The presented research and development work is carried out in the project "Energiemanagement III" within the framework of the cooperation CAR@TUM ("Munich Centre of Automotive Research") between the BMW Group and the Technische Universität München. This particular research was supported by BMW Group Research and Technology. The responsibility for this publication is held by authors only.

REFERENCES

Fallman, D. and Waterworth, J. 2005. Dealing with User Experience and Affective Evaluation in HCI Design: A Repertory Grid Approach, Umeå Institute of Design.

Fromm, M. 1995. Repertory Grid Methodik – Ein Lehrbuch, Deutscher Studien Verlag, Weinheim.

Garrett, J. 2011. The Elements of User Experience: User-Centered Design for the Web and Beyond, Second Edition, New Riders

Google Maps. 2012. Grafiken © 2012 AeroWest, COWI A/S, DDO, Digital Globe, GeoBasis-DE/BKG, GeoContent, GeoEye, Kartendaten ©2012 GeoBasis-DE/BKG

Haffke, A. 2010. Marktabgrenzung in der Automobilindustrie – Identifikation von Produktkategorien aus Kundensicht, Diplomica Verlag

Harrison, M. 2007. An empirical assessment of user perceptions of feature versus application level usage, Communications of the Association for Information Systems (Volume 20, 2007) 300-321.

Hassenzahl, M. 2000. Capturing Design Space From a User Perspective: The Repertory Grid Technique Revisited, International Journal of Human-Computer Interaction, 12(3&4), 441-459.

Heißing, B. and Brandl, H.J. 2002. Subjektive Beurteilung des Fahrverhaltens, Vogel Fachbuch.

Kühl, S. and Strodtholz, A. 2009. Handbuch Methoden der Organisationsforschung, VS-Verlag.

Lienkamp, M. 2011. Dynamik der Straßenfahrzeuge, Vorlesungsunterlagen, Lehrstuhl für Fahrzeugtechnik, Technische Universität München.

Lyons, J. 1968. Introduction to Theoretical Linguistics. Cambridge University Press.

Section II

Distraction of the Driver

The Role of Situation Awareness in Classifying Driver Cognitive Distraction States

Yu Zhang[1], David Kaber[2]

[1]General Motors Technical Center
Warren, MI, USA
yu.1.zhang@gm.com
[2]North Carolina State University
Raleigh, NC, USA
dbkaber@ncsu.edu

ABSTRACT

Cognitive distractions affect driver central cognitive processes. Visual distractions influence perceptual sensors and, to some extent, motor responses. Consequently, driver overt behaviors do not provide a sensitive or reliable basis for detecting driver cognitive distraction. This study demonstrated the value of indices of internal cognitive processes, specifically situation awareness measures, for cognitive distraction classification using a support vector machine (SVM) technique. In addition, the study establishes the importance of domain knowledge on the effects of distraction for promoting state classification accuracy. It also establishes the need to develop classifiers for different driving control modes.

Keywords: situation awareness, cognitive distraction, support vector machines, operational driving; tactical driving; distraction mitigation

1   INTRODUCTION

Two principle forms of distraction are generally associated with in-vehicle information systems, including visual and cognitive (Lee, Regan, & Young, 2009).

Knowing when drivers are visually or cognitively distracted is essential to develop effective distraction mitigation technologies. Prior studies suggest visual distraction can be effectively detected based on the dependency of visual attention on eye behavior (Donmez, Boyle, & Lee, 2007). In contrast, cognitive distraction detection based on overt behavior measures has revealed far from satisfactory results (Liang, Reyes, & Lee, 2007; Y. Zhang, Owechko, & Zhang, 2004). One potential reason for this is that cognitive distraction interferes with central cognitive processes, which depend on more than a single sensor (e.g., the eyes). Another reason is that cognitive distraction may not lead to uniform behavior changes, as drivers may adapt their control behaviors to account for distraction (Horrey & Simons, 2007; Liang & Lee, 2010). Therefore, detecting cognitive distraction may require knowledge of drivers' internal cognitive processes.

Beyond this, driving is a multi-level control process posing different cognitive demands on drivers according to required actions and decisions. Strategic control (e.g. planning a trip) and tactical control (e.g. passing a lead vehicle) have been found to be more cognitively demanding and to require greater situation awareness (SA) than simple operational control (e.g., steering or braking; Jin & Kaber, 2009; Matthews et al., 2001; Rogers et al., 2011). However, the majority of prior research has only attempted to identify cognitive distraction states when drivers performed operational maneuvers (e.g., lead-car following; Liang et al., 2007). In addition, previous efforts in estimating driver cognitive distraction level either adopted a data-driven perspective (i.e., using limited domain knowledge; Liang et al., 2007; Y. Zhang et al., 2004) or relied on linear combinations of features, limiting the capability to integrate predictive data from multiple sources (Angell et al., 2002).

To address the above issues, the present study: 1) examined the utility of internal process measures in driver distraction classification; 2) considered a broader range of driving tasks, including operational and tactical control in the classification problem; and 3) assessed the value of domain knowledge in distraction classification processes.

1.1 Internal cognitive process measures

There are few direct measures of driver internal cognitive processes, particularly for SA assessment (Endsley, 1995) and perceived workload analysis (e.g., the NASA Task Load Index (TLX); Hart & Staveland, 1988). Existing workload measures indicate the degree of cognitive resource competition between secondary and primary tasks (Angell et al., 2006). Indices of SA reflect changes in driver processes at three stages, including perception of environmental elements (Level 1 SA), comprehension of the driving situation (Level 2 SA) and projection of future states (Level 3 SA; Endsley, 1995). Unlike workload measures, SA indices provide a unique structured understanding of driver internal cognitive processes and may not be inferred by external performance or physiological metrics.

There are two major techniques available to assess driver SA in real-time, including the Situation-Present Assessment Method (SPAM: Durso & Dattel, 2004) and the use of real-time SA probes based on the Situation Awareness Global

Assessment Technique (SAGAT; Jones & Endsley, 2004; Jones & Kaber, 2004). In both procedures, participants answer queries about concurrent driving tasks while task displays remain in full view. Drivers' answers to queries are compared with true system state information and subject-matter expert (SME) responses to the same queries posed to the drivers (Endsley, 1995). Outcome measures usually include response latency to queries and response accuracy.

1.2 Support Vector Machines

Machine learning approaches, such as support vector machine (SVM), are commonly used to solve classification problems involving data from multiple sources and do not require data inputs to have Gaussian distributions (Y. Zhang et al., 2004). In a typical classification problem, each observation in a data set contains a "target value" (y_i, a class label) and several associated "attributes" ($\mathbf{x}_i$, predictor variables related to the class label). The complete data set is separated into training and test sets. Machine learning methods produce models to describe the relation between the "attributes" and the "target label" based on the training set and then predict the class labels for the test set. Model performance is usually measured based on classification accuracy for the test set. The present study applied one such technique, SVM, to an existing data set developed by Rogers et al. (2001) in order to identify factors that might influence the detection of driver cognitive distraction.

SVMs map an attribute vector $\mathbf{x}_i$ for a data point i into a high-dimensional feature space $\mathbf{Z}$ through a mapping function Φ. The mapping function is defined by a kernel function $\kappa(x_i, x_j) = \Phi(x_i)\Box\Phi(x_j)$ (i and j denote two different entries). Data belonging to different classes are separated by a hyper-plane $\mathbf{P}$, which is developed through the learning processes (Cortes & Vapnik, 1995). In order to avoid attributes in a greater numeric range dominating those in smaller numeric ranges, the input vectors need to be scaled to a common range (Hsu et al., 2003). Proper kernel functions must also be selected according to properties of the attributes. Prior studies suggest that radial basis function (RBF) kernels generally produce satisfactory classification accuracy and require less computational resources than other kernels. "Grid search" processes are typically performed to select kernel parameter(s), as well as a penalty coefficient for misclassification, which is required for building the SVM classifier. In such processes, pairs of kernel parameter(s) and penalty coefficients are used to fit a predetermined training data set. The pair that generates the best results is selected to construct the final classifier. A cross-validation (CV) process is used to make fair comparisons (minimum bias in error rate estimation) among the pairs of parameters. In a k-fold CV, the training data set is divided into k mutually exclusive subsets of equal size. Data from k-1 subsets are used to train the classifier and testing occurs with the remaining subset. This process returns k estimates from k repetitions and estimates the CV accuracy as the average of the k estimates (Japkowicz & Shah, 2011). A k=10 CV has been suggested to produce reasonably good performance with a relatively large sample-size (Kohavi, 1995). Related to this, previous studies also suggested that 10 repetitions of a 10-

fold CV (i.e., 10×10-fold) usually can improve replicability of the classification results (Japkowicz & Shah, 2011).

2 METHODOLOGY

2.1 Existing Data Set

Ten male and ten female drivers participated in Rogers et al.'s experiment (2011). All drivers were between 16 and 21 years of age (M=18.8 yrs, SD=1.4 yrs) with driving experience between 0.5 to 4 years (M= 2.45 yrs, SD= 1.62 yrs). All participants had a valid driver's license and 20/20 vision without correction. An interactive STISIM Drive™ M400 driving simulator was used for the study and provided drivers with a 135-degree field of view of a simulated roadway through three HDTVs. A 12-inch HP tablet computer was used to present a visual distraction task (described below), and was positioned approximately 15 degrees below the natural line of sight and 30 degrees to the right of participants. An auditory cognitive distraction task (also described below) was presented using a Dell desktop computer with hi-fi speakers. An ASL EYE-TRAC®6 Series head-mounted eye tracker with a head motion tracker was used to record driver gaze behavior at 60Hz.

The experiment followed a 2x2x2 within-subject design with two levels of visual distraction (with and without), two levels of cognitive distraction (with and without), and two primary driving tasks (following and passing). Each participant drove in eight 8-minute test trials on a simulated four-lane interstate highway while following roadway regulations. Speed limits varied between 55mph and 65mph at six virtual locations in each trial. One of eight unique combinations of the three experimental factors described below was presented in each trial.

Primary driving tasks. The following task, representing operational control, required participants to follow a lead vehicle at a safe distance and change lanes when the lead vehicle changed lanes. The passing task, demanding both tactical and operational control, required participants to first remain in a right lane behind a lead vehicle, pass the vehicle when they detected the lead vehicle was 10 mph slower than the speed limit, and return to the right lane to follow a new lead vehicle. For analysis purpose, both tasks were divided into two phases, including: 1) monitoring, when drivers were observing lead car actions, and 2) maneuvering, when drivers were making lane changes or passes.

Visual distraction. The visual task simulated the visual demands of using a navigation device enroute. The interface presented three arrows pointing in different directions and the graphics display was refreshed every 10 seconds. When all arrows were gray in color, no action was required; when two of the arrows were yellow, participants were required to identify an upward arrow in yellow.

Cognitive distraction. The cognitive task simulated cognitive demands of listening to auditory instructions from a navigation system. The audio messages described the path of a car traveling on an octagon highway loop with an exit occurring at each segment of the loop, facing south, southeast, and so on. Each message was 5 seconds long, specifying the starting exit, the orientation (clockwise or counterclockwise) and the number of exits passed by the virtual car. Participants were asked to verbally identify the finishing exit of the car within 15 seconds.

Response measures included the driver performance metrics (steering entropy, headway distance) and visual behavior measures captured with the eye tracker (off-road glance frequency, glance duration, etc.) with respect to the monitoring and maneuvering phases. The NASA-TLX was used to assess cognitive workload and real-time probes (based on SAGAT) were used to measure SA in each trial.

2.2 Support Vector Machine Modeling

The SVM modeling process involved two stages. First, statistical analyses, specifically ANOVAs, were used to identify response measures that were sensitive to cognitive distraction. Second, those measures significantly affected by cognitive distraction, including external behavior and internal cognitive process measures, were used to construct the SVM models. To demonstrate the value of internal cognitive process indices and domain knowledge on the distraction effects in the classification process, three other types of SVM classifiers were developed, including: 1) classifiers including all response measures as inputs, 2) classifiers including only external behavior measures as inputs, and 3) classifiers including only external behavior measures that were sensitive to cognitive distraction.

The research hypotheses were based on Rogers et al. (2011) findings on the influences of cognitive distraction on driver SA: Hypothesis (H) 1 - The inclusion of measures of driver internal behaviors in SVM models was expected to be significant. H 2 - It was also expected that models with a reduced set of response measures might produce comparable performance to full models. H 3 - As the influence of cognitive distraction on driver behavior may depend on the type of primary driving task, it was expected that SVM classifiers developed separately for the following and passing tasks would show better performance than classifiers based on measures across the two types of driving tasks.

In Rogers et al.'s study, the visual and cognitive tasks were constant sources of distraction for drivers in the test trials. Therefore, this study used the average values for the various response measures for each trial, as a single observation for SVM development. All measures were scaled to a common range [-1, 1]. For some measures with extreme ranges, the 98th percentile of those measures was used for scaling purposes instead of the maximum value. (For example, the range for time-to-collision (TTC) was between 2.3 and 778.5 seconds; however, the 98th percentile of TTC was only 161.4 seconds, which was significantly smaller than the maximum value.) All SVM classifiers of cognitive distraction were developed using the 10 × 10-fold CV process with RBF kernels regardless of the presence of visual distraction. Ten estimations of model accuracy and prediction variance were collected on 10 replications of the process for a single type of SVM model. All classifiers were developed by the "e1071" SVM package (Karatzoglou et al., 2006).

The evaluation criteria for the SVMs included the overall accuracy of the classifier, which was calculated as the percentage of correctly classified observations in a data set. Measures of Signal detection theory (SDT; Green & Swet, 1988) were calculated as the percentages of entries in a data set belonging to the four potential SDT outcomes, including: 1) a hit, when the presence of cognitive distraction was correctly identified; 2) a miss, when the presence of cognitive distraction was not identified; 3) a false alarm, when cognitive distraction was absent but was identified as present; and 4) a correct rejection, when cognitive distraction was absent and was not identified as present. In addition, Cohen's κ (1960) statistics were also calculated to account for the possibility that correct classification was a result of mere coincidental concordance between the classifier output and "true states" of driver distraction.

Pair-wise comparisons of the performance measures mentioned above were conducted to evaluate the differences between the two types of SVM models. The total number of observations on model prediction accuracy was 20 (i.e., 2 types of models * 10 observations per model). Hence, the error degrees of freedom for each pair-wise comparison were 20 - 1 (for bias) = 19. Beyond this, sensitivity analyses were applied to the classification process to assess how uncertainty in model inputs might influence model outputs. The results of the sensitivity analysis were used to compare the influence of each input feature in predicting driver distraction states. The present study used a feature-based sensitivity of posterior probabilities (FSPP) technique (Shen et al., 2007), which calculates the absolute difference of the probabilistic output of a SVM with and without the feature. A larger difference in the probabilistic output, due to exclusion of a single feature, indicates a greater influence of that feature in predicting distraction states (Platt, 1999).

3 RESULTS

The classification results are summarized in Table 1. Fitting SVM models according to different primary tasks showed advantages over fitting models across driving task types. There was a significant improvement in the classification accuracy for the following task model with all response measures vs. a model in all measures applied across task types (F(1,19)=3.05, p=0.007). The SVM model for the passing task, including as inputs only those response measures that were significantly influenced by cognitive distraction, also showed superior classification accuracy as compared to models with all response measures applied across task types (F(1,19)=2.11, p=0.049). However, such improvement was not observed for the SVM models of the passing tasks including all response measures.

In general, SVMs achieved high accuracy (greater than 90%) in classifying cognitive distraction with all response measures and with measures that were significantly influenced by cognitive distraction. In contrast, with only external measures, the prediction accuracy of models was low (69.68% for following and 57.70% for passing). Cohen's κ statistics also suggested weak agreement between the true distraction state and classifier predictions when external behavior measures were the only inputs (0.39 for following and 0.15 for passing). SVMs outputs also revealed significant improvements in classification accuracy by including internal process measures under all three analysis conditions, i.e., the following task only, the passing task only, and across the two primary tasks. The improvement in prediction accuracy was most substantial for the passing task at approximately 35%. Interestingly, reduced SVM models for passing, including as inputs only those measures that were significantly influenced by cognitive distraction, demonstrated equivalent or even slightly better prediction accuracy compared to SVMs with all measures (95.05% vs. 92.18%; $F(1,19)=1.44$; $p=0.17$). In contrast, retaining only those measures that were significantly influenced by distractions in SVMs showed significantly degraded accuracy, as compared to models with all response measures, for the following task (87.89% vs. 96.06% ; $F(1,19)=3.94$; $p=0.001$).

Table 1. Prediction outputs of SVM models for classifying cognitive distraction

		Following	Passing	Overall
All Measures	Accuracy	96.06%±3.49%	92.18%±4.79%	91.76%±2.79%
	Hit	94.84%±4.72%	89.85%±7.42%	91.40%±4.00%
	Correct rejection	93.65%±4.92%	90.78%±6.36%	90.44%±4.30%
	False alarm	6.35%±4.92%	9.22%±6.36%	9.56%±4.30%
	Miss	5.16%±4.72%	10.15%±7.42%	8.60%±4.00%
	Cohen's κ	0.92	0.84	0.84
External Measures	Accuracy	69.68%±6.99%	57.70%±3.24%	59.47%±4.51%
	Hit	68.79%±10.47%	58.15%±12.57%	57.34%±7.58%
	Correct rejection	68.17%±9.65%	58.30%±11.53%	60.59%±8.43%
	False alarm	31.83%±9.65%	41.70%±11.53%	39.41%±8.43%
	Miss	31.21%±10.47%	41.85%±12.57%	42.66%±7.58%
	Cohen's κ	0.39	0.15	0.19
Measures were significantly influenced by cognitive distraction	Accuracy	87.89%±5.56%	95.05%±4.07%	
	Hit	85.43%±7.83%	93.51%±6.27%	
	Correct rejection	87.15%±7.26%	92.95%±5.72%	
	False alarm	12.85%±7.26%	7.05%±5.72%	
	Miss	14.57%±7.83%	6.49%±6.27%	
	Cohen's κ	0.88	0.90	
External measures were significantly influenced by cognitive distraction	Accuracy	48.40%±7.23%	61.91%±7.48%	
	Hit	63.18%±18.11%	54.50%±11.30%	
	Correct rejection	34.04%±14.30%	67.09%±10.84%	
	False alarm	65.96%±14.30%	32.91%±10.84%	
	Miss	36.82%±18.11%	45.50%±11.30%	
	Cohen's κ	0	0.24	

Note: Response measures that were significantly influenced by cognitive distraction in following tasks included time-to-collision, steering entropy during monitoring and maneuvering, accuracy of Level 1 SA, latency of Level 2 and 3 SA, and TLX workload scores. Response measures that were significantly influenced by cognitive distraction in passing tasks included speed variance, steering entropy during monitoring and maneuvering, number of off-road glances per minute, off-road glance percentage, accuracy and latency of Level 2 and Level 3 SA, and TLX workload scores.

To assess the influence of each response measure on the uncertainty of model outputs, FSPP sensitivity analyses were applied to SVM models with only measures that were significantly influenced by cognitive distraction (see Table 2). For both tasks, internal measures, SA measures and TLX scores, all showed considerable influence in output uncertainty with a minimum of 4.41%. In following tasks, the exclusion of any of the measures that were included in the SVM model led to more than 10% change in estimated posterior probabilities. Similarly, for the passing task, the absence of the accuracy of Level 3 SA and TLX scores caused the estimated posterior probabilities of the model to change by more than 10%.

Table 1. Sensitivity analysis of SVM models with only measures that were significantly influenced by cognitive distraction

Following			Passing		
Measure	FSPP	Rank	Measure	FSPP	Rank
TLX	21.85%±3.17%	1	TLX	10.18%±2.15%	3
Accuracy of Level (L) 1 SA	14.28%±3.05%	4	Latency of L2 SA	5.58%±1.98%	7
Latency of L2 SA	15.65%±2.79%	2	Accuracy of L2 SA	10.83%±3.26%	2
Latency of L3 SA	15.24%±2.46%	3	Latency of L3 SA	7.49%±1.83%	4
			Accuracy of L3 SA	12.01%±2.82%	1
Time-to-collision	10.64%±2.79%	7	Speed variance	6.53%±1.89%	5
Steering entropy			**Steering entropy**		
(In monitoring)	11.94%±3.20%	6	(In monitoring)	5.20%±1.70%	8
(In maneuvering)	12.43%±1.94%	5	(In maneuvering)	6.51%±2.16%	6
			# of off-road glance per minute		
			(In monitoring)	5.03%±2.93%	9
			(In maneuvering)	4.97%±2.51%	10
			Off-road glance percentage		
			(In monitoring)	4.71%±3.54%	11
			(In maneuvering)	4.41%±2.20%	12

Interestingly, as TLX scores accounted for over 20% of the distraction state estimation uncertainty in the following task, it is possible that higher prediction accuracy may be due to driver perception of increased workload under distraction vs. actual changes in safety behavior or SA. In contrast, SA measures became the leading elements for detecting cognitive distraction in passing, especially the accuracy of Level 2 and 3 SA. This supports the notion of increasing driver reliance on explicit awareness as the level of complexity of driving control increased.

1   DISCUSSION AND CONCLUSION

In summary, the SVM classifiers achieved satisfactory performance in identifying cognitive distraction when drivers were performing both operational and tactical driving tasks. As expected, applying different classifiers for the different driving tasks revealed equivalent or superior performance, as compared to applying classifiers across driving tasks (supporting H3). In addition, when driver internal process measures (SA and workload) were included as model inputs, classifier performance significantly improved for both operational and tactical modes (supporting H1). Sensitivity analyses further confirmed the value of the SA measures as inputs. Especially for passing tasks, which were more cognitively demanding than simple following tasks, high levels of SA (Level 2 & 3) were the leading features in cognitive distraction classification. Last but not least, classifiers developed based on domain knowledge of the effects of distraction (the reduced SVM models) revealed superior performance than classifiers including all measures for driving tasks that were cognitively demanding (i.e., passing; supporting H 2). Such benefit may not occur for simple operational following tasks. This may be due to a lack of driver adaptation of driving behaviors when tactical control is required in the presence of distraction (Horrey & Simons, 2007) or the behavior changes due to cognitive distractions may not be effectively countered by adaptation.

As suggested by this study, driver (cognitive) distraction alerting systems may need to integrate internal cognitive processes measures for satisfactory performance. Additionally, in Rogers et al.'s experiment, drivers were only exposed to one level of cognitive distraction. Future research may use machine learning techniques to distinguish among levels of cognitive distraction exposure. In this way, the distraction effects posed by in-vehicle devices may be compared to the distraction effects of standard cognitive loading tasks. On this basis, new safety standards might be developed to protect drivers from excessive cognitive distraction.

ACKNOWLEDGMENTS

The research was supported by the Ergonomics Laboratory at North Carolina State University and was completed while the first author worked as a Research Assistant in the Edward P. Fitts Department of Industrial & Systems Engineering. The authors would like to acknowledge Meghan Rogers and Yulan Liang for assistance in the data collection process.

REFERENCES

Angell, L. S., et al. (2006). *Driver workload metrics project pask 2 (final report).* Crash Avoidance Metrics Partnership (CAMP) .Washington, D.C.

Angell, L. S., et al. (2002). An evaluation of alternative methods for assessing driver workload in the early development of in-vehicle information systems. *Proceedings of SAE Government/Industry Meeting* (pp. 1-18). Washington, D.C.

Cohen, J. (1960). A coefficient of agreement for nominal scales. *Educational and Psychological Measurements*, (20), 37-46.

Cortes, C., & Vapnik, V. (1995). Support-vector networks. Machine Learning, (20), 273-297.

Donmez, B., Boyle, L. N., & Lee, J. D. (2007). Safety implications of providing real-time feedback to distracted drivers. *Accident; analysis and prevention*, 39(3), 581-90.

Durso, F. T., & Dattel, A. R. (2004). SPAM: The real-time assessment of SA. In S. Banbury & S. Tremblay (Eds.), *A cognitive approach to situation awareness: Theory and application* (pp. 137–154). Hampshire, UK: Ashgate Publishing.

228

Endsley, M. R. (1995). Direct Measurement of SA in Simulation of Dynamic Systems: validity and use of SAGAT. *Procedding of the international conference on experimenal analysis and measurement of situation awareness*. Daytona Beach, FL.

Green, D., & Swet, J. (1988). Signal detection theory and psychophysics. New York: Wiley.

Hart, S. G., & Staveland, L. E. (1988). Development of NASA-TLX (Task load Index): Reault of emplirical and theoretical research. *Human Mental Workload* (pp. 139-183). Amsterdam: North-Holland.

Horrey, W. J., & Simons, D. J. (2007). Examining cognitive interference and adaptive safety behaviours in tactical vehicle control. *Ergonomics, 50*(8), 1340-50.

Hsu, C.-wei, Chang, C.-chung, & Lin, C.-jen. (2003). *A practical puide to support vector classification (Technical Report)*. Department of Computer Science, National Taiwan University. Retrieved Sep 10, 2009 from http://www.csie.ntu.edu.tw/~cjlin/papers

Japkowicz, N., & Shah, M. (2011). Error estimation. *Evaluating learning algrithms: A classification perspective* (pp. 161-205). New York, NY: Cambridge University Press.

Jin, S., & Kaber, D. B. (2009). The role of driver cognitive abilities and distractions in situation awareness and performance under hazard conditions. *The proceedings of the IEA 2009 17th World Congress on Ergonomics*. Beijing, China.

Jones, D. G., & Endsley, M. R. (2004). Use of real-time probes for measuring situation awareness. *International Jornal of Aviation Psychology*, 14(4), 343-367.

Jones, D. G., & Kaber, D. B. (2004). Situation awareness measurement and the situation awareness global assessment technique. In N. A. Stanton et al. (Eds.), *Handbook of Human Factors and Ergonomics Methods* (pp. 42-1~42-8). Boca Raton: CRC Press.

Karatzoglou, A., Meyer, D., & Hornik, K. (2006). Support vector machine in R. *Journal Of Statistical Software*, 15(9).

Kohavi, R. (1995). A study of cross-validation and bootstrap for sccuracy estimation and model selection. *Proceeding of International Joint Conference on Artifical Intelligence (IJCAI)* (pp. 1137-1145). Lawrence Erlbaum Associates.

Lee, J. D., Regan, M. A., & Young, K. L. (2009). What drives distraction? distraction as a breakdown of multilevel control. In M. A. Regan et al. (Eds.), *Driver distraction: Theory, effects, and mitigation* (pp. 42-56). Boca Raton: CRC Press.

Liang, Y., & Lee, J. D. (2010). Combining cognitive and visual distraction: Less than the sum of its parts. Accident; analysis and prevention, 42(3), 881-90. Elsevier Ltd.

Liang, Y., Reyes, M. L., & Lee, J. D. (2007). Real-time detection of driver cognitive distraction using support vector machines. *IEEE Transactions on Intelligent Transportation Systems*, 8(2), 340-350.

Matthews, G. M., et al. (2001). Model for situation awareness and driving: Application to analysis and research for intelligent transportation systems. Transportation research record, (1779), 26-32.

Platt, J. C. (1999). Fast training of support vector machines using sequential minimal optimization. In B. Schölkopf, C. J. C. Burges, & A. J. Smola (Eds.), Advances in Kernel Methods- Support Vector Learning (pp. 185-208). Cambridge, MA: MIT Press.

Rogers, M., et al. (2011). The effects of visual and cognitive distraction on driver situation awareness. In D. Harris (Ed.), *Engineering psychology and cognitive ergonomics, HCII 2011* (pp. 186-195). Berlin Heidelberg: Springer-Verlag.

Shen, K.-Q., et al. (2007). Feature selection via sensitivity analysis of SVM probabilistic outputs. *Machine Learning*, 70(1), 1-20.

Zhang, Y, Owechko, Y., & Zhang, J. (2004). Driver cognitive workload estimation A data - driven perspective. *IEEE Intelligent Transportation Systems Conference* (pp. 642-647).

CHAPTER 24

Couples Arguing When Driving, Findings from Local and Remote Conversations

A.N. Stephens & T.C. Lansdown[1]

University College Cork, Cork, Republic of Ireland
[1]Heriot-Watt University, Edinburgh, United Kingdom

ABSTRACT

It is well established in the literature that inappropriate use of in-vehicle information systems, aftermarket nomadic devices or cellular telephones can result in reductions to driving performance. Further, convincing data exist that driver inattention is a contributory factor in nearly four-fifths of traffic incidents. However, less is known about the potential distractions for the driver from social factors.

This paper reports findings from a study seeking to explore the socio-technical potential for driver distraction. Twenty couples who were romantically involved participated in the experiment. The study was conducted using a driving simulator and the participants were required to engage in emotionally difficult conversations. One partner was driving while the other was either i) talking while sat next to the driver, or ii) conversing with the driver from a remote location using a telephone. The driver used a handsfree telephone during the remote conversations. A revealed differences protocol was employed to identify contentious conversation topics. This required identification and subsequent discussion of sources of on-going disagreement in their relationships.

Results indicate driver performance was adversely effected for both longitudinal and lateral vehicle control. Further, the revealed differences tasks were subjectively viewed as emotionally more difficult relative to a control. Performance was worst with the partner present during the contentious conversations. It seems during difficult conversations, drivers may be better able to regulate task demands with their partner not in the vehicle. Findings from this study suggest that difficult discussions with a romantic partner have a measureable detrimental effect on the ability to drive safely. The data encourages further work to be undertaken to gain an understanding of the potential for socio-technical emotional distractions.

Keywords: Couples, argue, driving, revealed differences, phone

1   INTRODUCTION

Mobile phone use is now ubiquitous throughout the world. World Health Organisation recently reported that 67% of the global population have cellular telephone subscriptions, with many users having multiple devices (WHO, 2011). Indeed, in a survey of British drivers one in four were prepared to read text messages while driving, and one in seven would write text messages in a typical week (Lansdown, 2012). Data have shown that driver inattention is responsible for approximately 79% of all traffic collisions and 65% of near misses (Neale, Dingus, Klauer, Sudweeks, & Goodman, 2005). Driving performance has been found to be reliably degraded by the use of cellular telephones (Fairclough, Ashby, Ross, & Parkes, 1991; Haigney & Westerman, 2001; Lamble, Kauranen, Laakso, & Summala, 1999). Both the cognitive task of engaging in conversation (Drews, Pasupathi, & Strayer, 2008) and the control display task of interacting with a mobile telephone (Brookhuis, de Vries, & de Waard, 1991) can have debilitating effects on vehicle control. Burns, Parks and Lansdown (2003) looked at the relative implications of hands-free telephone use versus conversation with a passenger. Results indicated reaction times and subjective mental workloads were significantly compromised with respect to a control drive.

The cellular telephone is just one example, although a widely prevalent one, of non-driving critical secondary tasks which may be undertaken while driving. Most research has considered the impact of technological systems. For example, route guidance systems, congestion warning devices, or parking aids. There has been relatively little consideration in the literature of the emotional and socio-technical demands on the driver.

Emotion can have a powerful effect on the ability to drive the vehicle safety. Angry drivers have been shown to be more distracted by emotionally-charged advertising and also to drive at excessive speed (Megìas, Maldonado, Catena, Di Stasi, Serrano, & Cándido, 2011). Further, they are slower to recognise, and subsequently react to driving hazards (Stephens & Groeger, 2011) and have higher risk of involvement in traffic accidents and incidents (Underwood, Chapman, Wright, & Crundall, 1999). Arguments with a passenger can also be a significant source of distraction, leading to accidents. Indeed, young drivers have been shown to be at particular risk from interactions with passengers or telephones (Neyens & Boyle, 2008). Interactions between parents and children have also been found to result in increased levels of driver distraction (Koppel, Charlton, Kopinathan, & Taranto, 2011).

Studies suggest that an in-vehicle conversation is less distracting than one undertaken using a telephone (for example, Drews, et al., 2008; Haigney & Westerman, 2001; Lamble, et al., 1999). It may be that the remote conversation is more difficult for the driver as it typically has a specific purpose, whereas, an in-vehicle conversation may be less directed and more general in nature (Crundall, Bains, Chapman, & Underwood, 2005). Further, the passenger conversing with the driver has the opportunity to observe prevailing traffic conditions and therefore moderate their behaviour to limit discussion during challenging periods. Crundall et al. (2005) found that during challenging driving circumstances conversation was suppressed when a passenger was present, but not during a remote conversation.

The emotional state of the driver has been considered in several studies. Some of those that have focused on conversational tasks specifically are reviewed the below. Driver anxiety was manipulated by Briggs, Hole, & Land (2011). They used either arachnophobic or non-arachnophobic drivers in a telephone conversational task in a driving simulator. Results showed increased errors and elevated heart rates for the arachnophobes when discussing spider-related topics. Further, reduced visual scanning was also observed for the spider-fearful drivers. Reduced visual scanning has been previously shown to be associated with decreases in driver performance (Underwood, Chapman, Brocklehurst, Underwood, & Crundall, 2003).

Conversational ease while using a mobile telephone during simulated driving was investigated by Rakauskas, Gugerty, & Ward (2004). The conversations were controlled by asking questions of different complexity. For example, "what did you study at university?" would be an easy example. An example of a hard question might be "Explain why you think world will be a better place than 100 years?" Any conversational task was found to compromise driving relative to a control by imposing significantly greater workload and reducing vehicle speed. Dula, Martin, Fox, & Leonard (2011) also used a driving simulator to consider the impact of conversation on driving. The study had three conditions, a control, a mundane conversation and an emotionally challenging one. The emotional conversation resulted in increased speeding, collisions and a reduction in lateral control.

Producing a meaningful, yet empirically reliably conversational task is an experimental challenge and numerous different approaches have been taken. For example, the tight control of the speech syntax and required responses (Burns, Parkes, and Lansdown, 2003) or selection of specifically sensitive individual differences (Briggs, et al., 2011). Studies have largely employed somewhat artificial conversational tasks. For example, Rakauskas, et al. (2004) and Dula, et al. (2011), both used rather contrived question and answer dialogues. It would be preferable to utilise an experimental design that facilitates natural emotional interactions between a driver and passengers. Such an approach may offer a more robust mechanism to investigate genuine social interactions for drivers.

Strodtbeck developed a protocol in 1951 to "reveal differences" between partners in order to resolve sources of on-going disagreement. Specifically, it assumes the differences are derived from conflicts in their relationship. In this approach, each partner is required to produce a prioritised list of contentious issues. These lists are then integrated by the experimenter to develop a script to assist in the resolution of the couple's concerns. The published literature reveals little research pertaining to romantically-attached couples experiencing conflict while driving. This study sought to characterise the cognitive and performance effects of emotionally-charged conversation while driving. Further, the modality of the conversation was manipulated, either with one participant situated next to the driver, like a passenger; or remotely using a hands-free cellular telephone. It was hypothesised that contentious conversations would be more disruptive of driving performance. Moreover, remote conversations would impose the highest workload on the driver, resulting in the worst driving performance and the greatest effect on emotional state, relative to the other conditions.

2   METHOD

All participants provided informed consent for the study, which had been previously ethically approved by the School of life sciences ethics committee, Heriot-Watt University. According to the revealed differences protocol (Strodtbeck, 1951), in the laboratory each member of the couple was independently asked to rank order five 'sources of on-going disagreement' in their relationship. This approach was adopted to identify issues of particular contention specific to the couple. The lists were then interleaved by the experimenter into a single five item list prioritising the driver first. Due to the sensitive nature of the discussions contrived the experimenters were not present in the laboratory during the conversations. However, for experimental validation the conversations were recorded so that the participants could not be identified but could be heard, to determine that there was an on-going dialogue for experimental control. Content was not explicitly reviewed.

All drivers undertook a familiarisation drive, in which they were instructed to drive as they would using your vehicle, complying with the instructions to the best of their ability. The primary driving task was to follow a white sports utility vehicle at a safe distance, if the simulated vehicle lagged too far behind the lead vehicle, an audible message was played asking the driver to reduce the gap. If this gap exceeded 200ft collection of following data was suspended.

2.1   Participants

Twenty romantically involved couples were recruited for the study from Edinburgh Universities. The most frequent driver was selected to be the driver in the simulator. Two of the drivers were female. Average driver age was twenty-five years (SD = 3.91). Average mileage was 6597 miles (7,064), and the drivers had been licensed for 6.47 years on average (SD = 4.22). Only one of the non-drivers did not hold a driving license. Average age of the non-drivers was twenty-three years (SD = 3.72), and of those who were licensed, the average length for holding a license was 4.7 years (SD = 3.81). The participants who were in the non-driving role, had an average mileage of 3,922 miles per year (SD = 5,743). Each participant received £10 for their contribution to the study.

2.2   Equipment

An STI SIM Model 100 simulator was used for the experiment. This is a fixed base driving simulator which presents a 60° forward view using 1280 ▫ 800 pixels. A 7.1 channel Dolby surround system is used for auditory feedback. Drivers interact with the system using a Logitech G25 steering wheel with a manual gearbox. The typical driving scenario for each condition was a dual carriageway with centralisation reservation. Drivers were asked to accelerate to 60 miles an hour and follow a white sports utility at a safe distance of approximately 2 seconds at headway; In the left lane (driving on the left side of the road). The white SUV lead vehicle's speed varied by a predictable sign historical pattern increasing and decreasing by up to 10 miles an hour. Participants drove for 9.5 miles per condition, however, the initial 0.5 miles of data was not recorded. The mildly changing bends in the road and variation in the vehicle speed results in a driving task with both lateral and longitudinal variability for the driver to content with. Other vehicles in the simulation were seen to overtake the driven vehicle periodically and could be seen travelling in the opposite direction on the other carriageway. Vehicles in the simulation behave in a semi-intelligent manner, for example, slowing and accelerating to avoid hazards or collisions.

2.3   Experimental Design

A repeated measures design was employed with three levels, i) local conversation, ii) contentious remote conversation, and iii) a control. In the Control condition, the participants were required to conduct a conversation on a topic of interest, but not a contentious one, throughout the data collection period. During the Local conversation, the partners were sat next to each other, like driver and passenger, while discussing throughout the drive, the 'Sources of On-going Disagreement' identified previously. The Remote conversational tasks were the same as the Local condition, except the passenger was located in another room, communicating with the driver via telephone. The driver used a handsfree telephone in this condition. Across both contentious conditions, the conversation topics were introduced by the partner who worked off the compiled list of identified topics.

The following post-hoc measures were taken after each condition. NASA R-TLX subjective mental workload (Byers et al., 1989), ratings of anger, conversational ease, and distraction. The drivers were also required to undertake an embedded divided attention task during the conditions. This consists of red diamonds appearing in the upper left and right corners of the projected display. These change to equilateral triangles periodically, at which time, the driver is required to respond by pressing a corresponding button on the left or right of the steering wheel as quickly as possible. The non-drivers were also required to provide ratings for anger levels and their ease of participation in the conversation, and their rating of the driver's distraction level.

The experiment specifically sought to encourage the couples to address challenging and potentially sensitive issues in their relationship. Therefore, after the driving conditions had been completed, and as part of the 'revealed differences' protocol, it was important that time was set aside for the partners to explicitly appreciate the attributes which they value in the other person. Ten minutes at the end of the study was allocated for the participants to explicitly state to each other the things they enjoy and love.

3   RESULTS

3.1   Driver Performance

The driver's ability to maintain vehicle control was found to be significantly disrupted when a passenger was present. Standard deviation of both Lane position (F (1.07, 19.17) = 5.62, p <.05) and Speed (F (2,38) = 7.56, p <.005) were found to be significantly different, see Figures 1a and 1b respectively. Post-hoc comparisons revealed that the contentious Remote Drive had best performance, followed by the Control, and the Local Contentious Drive had worst performance.

The embedded divided attention tasks undertaken by the drivers were not found to produce any significant differences between conditions.

3.2   Subjective Ratings

Ease of conversation was rated after each condition. Drivers rated the non-contentious conversation (control) as easier to participate in than either of the other conditions (F(2,38) = 5.65, p <.01), see Figure 2. Further, no significant differences were found in the rated ease of participation in conversation between Local or Remote discourse.

Both drivers and non-drivers rated their current levels of anger after each condition. The drivers were significantly more angry after the contentious conversations when compared to the non-contentious conversation (F (1,19) = 6.95, p <.01), see Figure 3. No significant differences were found for driver anger between the two contentious conditions. Non-drivers also reported significantly higher levels of anger (F(1.19,22.62) = 7.99, p <.01) in the contentious tasks. Further, post-hoc testing revealed that they were most angry during the remote contentious drive, followed by the local drive, with least expressed anger in the control condition. No significant differences were found between the control drive and the local contentious drive for the non-drivers. The non-drivers were most angry during conversations when they could not observe the driver. Moreover, the non-drivers rated their emotional state to be significantly more insecure, discouraged, irritated or stressed in the contentious conversations than the control (F (2,38) = 7.03, p <.01). This was particularly true for the remote contentious conversation.

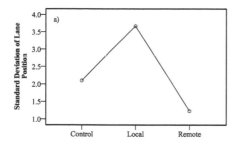

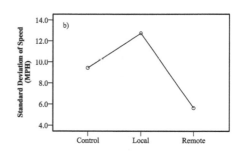

Figure 1a Standard deviation of lane position Figure 1b Standard deviation of speed

During the Control condition, a significant correlation was found between passenger's ratings of the ease of participation in the conversation, for themselves and the driver (r (18) = 0.61, p <.005). Under low emotional demand, the participants appeared to agree on the difficulty of engaging in discussion. In both the Local and Remote conversations, significant correlations were no longer found for ratings of conversational ease between the driver and passenger, suggesting as the conversational task became more taxing, differences may have began to emerge in the effort required to converse.

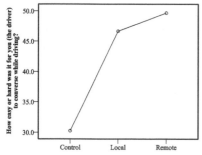

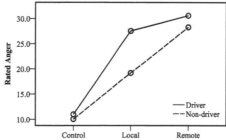

Figure 2 Conversational Ease   Figure 3 Rated Anger

The reported effect of emotional state on driver performance differed across the three conditions (F (2,38) = 7.61, p <.01), see Figure 4. Drivers reported their emotional state influenced driving behaviour significantly more so after contentious conversational tasks than the control condition, in post-hoc testing. When asked to rate their level of distraction, both the contentious conversations were found to be significantly more distracting (F (2,38) = 12.96, p <.0001). In post-hoc testing, the remote contentious conversation was rated most distracting and significantly more so than the local contentious conversation.

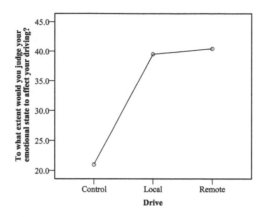

Figure 4 The influence of driver emotional state on performance

The subjective mental workload data reveals the same trends as the other subjective reports. The contentious conversations imposed significantly greater mental workload than the control (F (2,38) = 10.87, p <.0001), see Figure 5. Post-hoc testing revealed again that the contentious conversational tasks significantly imposed more mental workload than the control condition. No other significant differences were found in post-hoc testing. Analysis of the main effects for sub-scales, Mental Demand, Physical Demand, Temporal Demand, Effort, Performance and Frustration; all revealed the same pattern of significant differences as the overall mean mental workload scores. Performance and Physical Demand showed the same main effect, however the conservative nature of post-hoc testing failed to reveal significant differences between conditions.

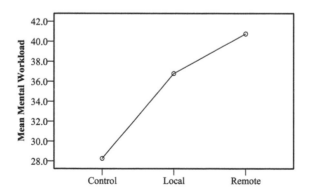

Figure 5 Subjective mental workload

4   DISCUSSION

This study aimed to explore the influence of emotionally difficult conversations on the driver's performance. It examined the impact of romantically-attached couples discussing contentious topics either while sitting next to each other or via a handsfree telephone conversation. The study had two hypotheses, i) that the Remote conversation would be most taxing of the driver's resources, and ii) that the combined Remote and Contentious conversations would be more taxing that the Control. The results revealed that the contentious tasks were indeed the most disruptive for the driver. However, no support was found for the Remote conversational task as a more difficult one than the Local contentious one.

The Local contentious conversation was found to be most disruptive of driver performance. Both longitudinal and lateral control were found to have significantly higher variance during the Local contentious drive when compared to either the Remote or Control drives. The most consistent driving performance was found in the Remote contentious condition. It would appear that the presence of a romantically-attached passenger was sufficient to disrupt the driver's control of the simulated vehicle, over and above, the distracting potential of the nature of the conversational topic. It may have been that the special bond between the individuals was sufficient to compromise the driver's motivation to control the vehicle. Their particular 'consideration' of their partner's wishes, may have influenced task scheduling. The data further suggest that when the conversational topic was difficult, this effect was more pronounced. It would seem that 'in-vehicle', the presence of a passenger, was sufficient to act as a visual distractor. It would be interesting to compare the specific findings from romantically-attached parties with the disruption potential from other individuals. The finding is interesting in that other studies have shown a relative safety benefit from having a passenger in the vehicle (Drews, et al., 2008; Haigney & Westerman, 2001; Strayer & Johnston, 2001). When the conversation is emotionally-challenging, the driver may be able to control the demands from a remote conversation more effectively than a local one.

Of further interest, is that when the conversation becomes contentious, the passenger appears to underestimate the demand it places on the driver. Both passenger's and driver's ratings of conversational participation were significantly correlated for the Control condition. However, when the conversations became more demanding, ratings of ease of participation were no longer significant. Passengers rated the difficulty of participating in the conversation as relatively lower for themselves than for the driver. The drivers rated conversational participation as relatively more taxing for themselves than their passengers, with the exception of the Control, which they rated at a similar level. Suggesting perhaps, that the drivers were accurately reporting the increasing demands of the Contentious driving conditions, whereas the passengers were not aware of the level of differential difficulty the drivers were experiencing. It would be interesting to know whether the partner would have regulated the conversation had they been more aware of the difficulty placed on the driver

Either of the contentious conversations were found to result in significantly higher anger levels than the Control. In contrast, the non-drivers were less angry during the local contentious conversations than the driver, but found that the Remote contentious conversation provoked significantly more anger than the control. Relatively similar levels of provoked anger were found for both drivers (M = 30.60, SD = 31.39) and non-drivers (M = 28.30, SD = 28.78). The non-drivers were better able to manage their levels of anger in the Local contentious condition. It may have been that their ability to watch the driver enabled them to moderate their anger relative to the demands on their partner. The non-drivers could have been less concerned by the demands of the driving task during the Remote condition were they could not observe it, as has been suggested previously (Maciej, Nitsch, & Vollrath, in press). Further, the driver may have attended more to the conversation during the Local condition. In contrast, finding it easier to protect the primary driving activity in the Remote condition when the emotional distractions were more distant. Mental workload and vehicle control data would support these suggestions.

Subjective mental workload findings revealed that either of the Contentious conversational tasks imposed significantly greater workload than the Control. The Remote condition imposed a higher mean score than the Local condition, but not significantly so. NASA R-TLX sub-scales presented the same trend as the overall workload score, with the exception of Performance and Physical Demand that did not reach post-hoc significance as a result of the conservative nature of testing. It would appear clearly that a contentious conversation makes life more difficult for the driver, but we cannot assert that such tasks are exacerbated by a telephone conversation. Indeed, the vehicle performance data suggest that in such circumstances, the driver is able to 'tune out' the conversation in order to actively focus on vehicle control. This finding was unexpected and published data suggests that the Remote conversation should have been the most taxing on the driver (Crundall, et al., 2005). In the Remote condition, the passenger is less able to mediate the conversation to consider the driver's demands, and it was assumed the degraded conversational auditory fidelity, from the cellular telephone, would further enhance this effect. Perhaps during the Control condition the passenger presented an additional visual distraction, absent from the remote conversation. It is suggested that perhaps the conversation was the most demanding component of the driving scenario. Consequently, in the Remote environment the driver was able to prioritise the driving task more than when experiencing a difficult conversation with a loved one. It would be interesting to repeat the study with a driver only control condition.

Subjective findings from this study support the revealed differences protocol as an effective way to introduce a personal and emotionally-taxing conversational task for the participants. The contentious tasks were consistently rated by the driver as more difficult to take part in than the neutral Control drive. Thus, the 'revealed differences' protocol appeared successful in its application in a driving context. The driver's anger ratings revealed the same significant trend as the participation difficulty ratings. The contentious tasks were rated to be significantly harder to participate in.

The scope of this study was limited in that it only enabled consideration of the member in the relationship who drove most frequently. In all cases, except two, this resulted in males being the driver. Of the two female drivers, one's partner did not drive and the other was from a same-gender relationship. Further research will seek to address this limitation in the investigation of emotionally-derived distractions.

Results support the notion that the social component of a relationship can result in emotionally-derived driver distractions. These distractions may, in turn, lead to measureable changes in state and task performance. Romantically-attached adults were considered in this study but the approach could be applied in other settings. For example, to investigate how the acquisition of skill in younger drivers may be effected by emotional distractors. It would be interesting to investigate task maturity for both driving and relationship components, particularly so, for recognised at risk groups, e.g., young male drivers.

5   CONCLUSIONS

The study reported in this paper investigated difficult conversations between romantically-attached couples, during simulated driving. 'Revealed differences' was found to be an effective protocol to meaningfully and emotionally stimulate the conversations. In contrast to previous published work, we found passenger presence was more disruptive of driving. The passengers were shown to find the contentious conversations most difficult when conducted over a telephone. Evidence for the functionally-disruptive influence of couples arguing while driving is presented. Further research should investigate this socio-technical demand and its impact on others, e.g., the young driver.

ACKNOWLEDGEMENT

The authors would like to give special thanks to Dr Bjarne M. Holmes for his advice regarding the application of the revealed differences protocol.

REFERENCES

Briggs, G. F., Hole, G. J., & Land, M. F. (2011). Emotionally involving telephone conversations lead to driver error and visual tunnelling. *Transportation Research Part F: Traffic Psychology and Behaviour, 14*(4), 313-323.

Brookhuis, K. A., de Vries, G., & de Waard, D. (1991). The effects of mobile telephoning on driving performance. *Accid. Anal. & Prev. Accid. Anal. & Prev., 23*(4), 309 - 316.

Burns, P. C., Parkes, A. M., & Lansdown, T. C. (2003, August 24 -29). *Conversations in cars: the relative hazards of mobile phones.* Paper presented at the XVth Triennial Congress of the International Ergonomics Assocation and the 7th Joint Conference of Ergonomics Society of Korea / Japan Ergonomics Society, Seoul, Korea, August 24 - 29.

Byers, J. C., Bittner, A. C. J., & Hill, S. G. (1989). Traditional and raw task load index (TLX) correlations: are paired comparisons necessary? A. Mital (Ed.), In: *Advances in industrial ergonomics and safety 1* (pp. 481 - 485: Taylor & Francis.

240

Crundall, D., Bains, M., Chapman, P., & Underwood, G. (2005). Regulating conversation during driving: a problem for mobile telephones? [doi: DOI: 10.1016/j.trf.2005.01.003]. *Transportation Research Part F: Traffic Psychology and Behaviour, 8*(3), 197-211.

Drews, F. A., Pasupathi, M., & Strayer, D. L. (2008). Passenger and Cell Phone Conversations in Simulated Driving. *Journal of Experimental Psychology: Applied, 14*(4), 392-400.

Dula, C. S., Martin, B. A., Fox, R. T., & Leonard, R. L. (2011). Differing types of cellular phone conversations and dangerous driving. *Accident Analysis & Prevention, 43*(1), 187-193.

Fairclough, S. H., Ashby, M. C., Ross, T., & Parkes, A. M. (1991). *Effects of handsfree telephone use on driving behaviour.* Paper presented at the Proceedings of the ISATA Conference (International Symopium on Automotive Technology and Automation), Florence, Italy.

Haigney, D., & Westerman, S. J. (2001). Mobile (cellular) phone use and driving: a critical review of research methodology. *Ergonomics, 44*(2), 132.

Koppel, S., Charlton, J., Kopinathan, C., & Taranto, D. (2011). Are child occupants a significant source of driving distraction? *Accident Analysis & Prevention, 43*(3), 1236-1244.

Lamble, D., Kauranen, T., Laakso, M., & Summala, H. (1999). Cognitive load and detection thresholds in car following situations: safety implications for using mobile (cellular) telephones while driving. *Accident Analysis & Prevention, 31*(6), 617-623.

Lansdown, T. C. (2012). Individual Differences & Propensity to Engage with In-Vehicle Distractions - a Self-Report Survey. *Transportation Research Part F: Traffic Psychology and Behaviour, 15*, 1-8.

Maciej, J., Nitsch, M., & Vollrath, M. (in press). Conversing when driving: The importance of visual information for conversation modulation. *Transportation Research Part F.*

Matthews, G., Jones, D. M., & Chamberlain, A. G. (1990). Refining the measurement of mood: The UWIST Mood Adjective Checklist. *British Journal of Psychology, 81*, 17-42.

Megìas, A., Maldonado, A., Catena, A., Di Stasi, L. L., Serrano, J., & Cándido, A. (2011). Modulation of attention and urgent decisions by affect-laden roadside advertisement in risky driving scenarios. *Safety Science, 49*(10), 1388-1393.

Neale, V. L., Dingus, T. A., Klauer, S. G., Sudweeks, J., & Goodman, M. J. (2005). *An overview of the 100-car naturalistic study and findings* (Research Report No. 05-0400). Washington DC, USA: NHTSA.

Neyens, D. M., & Boyle, L. N. (2008). The influence of driver distraction on the severity of injuries sustained by teenage drivers and their passengers. *Accident Analysis & Prevention, 40*(1), 254-259.

Rakauskas, M. E., Gugerty, L. J., & Ward, N. J. (2004). Effects of naturalistic cell phone conversations on driving performance. *Journal of Safety Research, 35*(4), 453-464.

Stephens, A. N., & Groeger, J. A. (2011). Anger-congruent behaviour transfers across driving situations. *Cogn Emot*, 1-16.

Strayer, D. L., & Johnston, W. A. (2001). Driven to Distraction: Dual-Task Studies of Simulated Driving and Conversing on a Cellular Telephone. *Psychological Science (Wiley-Blackwell), 12*(6), 462.

Strodtbeck, F. L. (1951). Husband-wife interaction over revealed differences. *American Sociological Review, 16*(4), 468-473.

Underwood, G., Chapman, P., Brocklehurst, N., Underwood, J., & Crundall, D. (2003). Visual attention while driving: sequences of eye fixations made by experienced and novice drivers. *Ergonomics, 46*(6), 629-646.

Underwood, G., Chapman, P., Wright, S., & Crundall, D. (1999). Anger while driving. *Transportation Research Part F: Traffic Psychology and Behaviour, 2*(1), 55-68.

WHO (2011). *Mobile phone use: a growing problem of driver distraction* (ISBN: 978 92 4 150089 0). Geneva, Switzerland: World Health Organization.

CHAPTER 25

An On-Road Examination of the Errors Made by Distracted and Undistracted Drivers

Kristie L. Young, Paul M. Salmon, & Miranda Cornelissen

Monash University Accident Research Centre, Monash Injury Research Institute
Monash University, Victoria, Australia. Email: kristie.young@monash.edu

ABSTRACT

This study explored the nature of errors made by drivers when distracted versus not distracted. Participants drove an instrumented vehicle around an urban test route both while distracted (i.e. performing a visual detection task) and while not distracted. Two in-vehicle observers recorded the driving errors made, and a range of other data were collected, including driver verbal protocols, and video and vehicle data (speed, braking, steering wheel angle, etc). Classification of the errors revealed that drivers were significantly more likely to make errors when distracted; although driving errors were prevalent even when not distracted. Interestingly, the nature of the errors made by distracted drivers did not differ substantially from those made by non-distracted drivers. This suggests that, rather than make different types of errors, distracted drivers simply make a greater number of the same error types they make when not distracted. Avenues for broadening our understanding of the relationship between distraction and driving errors are discussed along with the advantages of using a multi-method framework for studying driver behavior.

1. INTRODUCTION

The concept of driver distraction has been the focus of intense research over the past two decades. A large body of research has confirmed that distraction can affect a range of driving performance measures (Young, Regan, & Lee, 2009). These degradations translate into an increased safety risk, with estimates indicating that secondary task distraction is a contributing factor in up to 23 percent of crashes and near-crashes (Klauer et al., 2006).

Distraction is a complex, multifaceted phenomenon and, despite the immense research effort, there is still much to understand about its mechanisms and its relationship with other aspects of human cognition and behavior (Regan, Hallett, & Gordon, 2011). One aspect of distraction for which there is limited knowledge is its relationship with driving errors (Young & Salmon, 2012). Although both are popular areas of road safety research and are both seemingly related, there has been little exploration of *how* they are related and how they may interact. That is, there has been little systematic exploration of the causal role of distraction in driving errors that are prevalent in road crashes. Although various detailed driving error taxonomies exist (e.g., Stanton & Salmon, 2009) role that driver distraction plays in the range of driving errors documented is unclear.

From a practical perspective, understanding the role of distraction in driving errors can inform the development of better countermeasures to mitigate both distraction induced error through training and technology (e.g., Advanced Driver Assistance Systems) and wider road system design. From a theoretical perspective, understanding the relationship between driver distraction and driving errors and the mechanisms by which distraction contributes to different errors is important for improving knowledge of the consequences of distracted driving and the mechanisms through which it leads to crashes.

The present study aimed to explore, using a suite of human factors methods in an on-road context, the nature of errors made by distracted drivers and how these may differ, in both number and kind, to those made when drivers are not distracted. Although largely exploratory in nature, based on existing distraction literature and theory a number of predictions can be made about the nature of the errors that may be made by distracted drivers. First, it was predicted that drivers would make a greater number of driving errors when distracted. Also, given the largely cognitive nature of the distracter task used and previous evidence suggesting that cognitive distraction impairs drivers' visual scanning behavior and hazard perception (e.g., Harbluk et al., 2007), it was predicted that drivers would make a greater number of perceptual and observation-based errors when distracted.

2. METHOD

2.1 Participants

Twenty-three drivers (10 males, 13 females) aged 19-51 years (mean = 28.9, SD = 8.6) participated in the study. Of the participants, 17 held a valid Full driver's license while the remaining 6 held a valid Probationary (P2) license. Participants had an average of 10.1 years (SD = 8.9) driving experience and drove an average of 11.9 hours (SD = 15.6) a week. Participants were recruited through a Monash

University newsletter and were compensated for their time and travel expenses. The study was approved by the Monash University Human Research Ethics Committee.

2.2 Materials

2.2.1 On-Road Test Vehicle

The On-Road Test Vehicle (ORTeV) is an instrumented 2004 Holden Calais equipped to collect two types of data: vehicle-related data (e.g. speed, GPS location, accelerator and brake position, steering wheel angle) and eye tracking data. Driver eye movements can be tracked and overlaid on a driver's-eye camera view using the FaceLab eye tracking system. ORTeV is also equipped with seven unobtrusive cameras recording forward and peripheral views spanning 90° each respectively as well as three interior cameras and a rearward-looking camera. Analysis and reporting of the vehicle and eye-tracking data collected is beyond the scope of this paper and will not be reported.

2.2.2 Visual Detection Task

The Visual Detection Task (VDT; (Engström & Mårdh, 2007) was used as the distracter task. This task was selected as it is a measure of cognitive load, but does not require a verbal response from drivers. Drivers were unable to give a verbal response as that they were providing continuous verbal protocols (see 2.2.3). The VDT is an extension of the Peripheral Detection task (PDT) where the visual stimulus is presented at increased intensity and in the driver's central field of view. The VDT was piloted prior to testing to ensure that it provided a sufficient level of cognitive demand , but did not place drivers at undue risk.

The set-up for the VDT followed the general specifications described in Engström and Mårdh (2007). A single red LED light was positioned so that the stimulus was presented in the driver's central field of view via reflection in the windshield (Figure 1). Each visual stimulus was presented for a maximum of two seconds and turned off as soon as the response button was pressed. Stimuli were presented with a random temporal variation of 3 to 5 seconds. Participants responded to each stimulus by pressing a button attached to their left index finger. The presentation of the visual stimuli was controlled by an HP laptop which also recorded mean response time to the stimuli and hit rate.

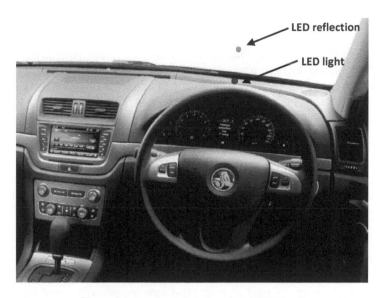

Figure 1 Approximate location of the VDT stimuli

2.2.3 Driver Verbal Protocols

Verbal Protocol Analysis (VPA) was used to elicit data regarding the cognitive processes undertaken by drivers while en-route. VPA is commonly used to investigate the cognitive processes associated with complex task performance and has been used to explore a range of concepts (e.g., situation awareness, decision making) in various domains, including road transport (Walker, Stanton, & Young, 2007). In the present study, participants provided verbal protocols continuously as they drove the instrumented vehicle around the test route. The verbal protocols were recorded using a digital Dictaphone and transcribed verbatim post-trial. For this paper, the VPA data were used as an additional source of information to the observers' error records for the classification of errors.

2.2.4 Driving Route

The driving route comprised an 8.7 km urban route in the suburbs surrounding the Monash University Clayton Campus in Melbourne. The test route contained a total of 11 intersections with a combination of fully, partially and un-signalized traffic controls and 50, 60, 70 and 80 km/h speed zones. The route took approximately 15-20 minutes to complete. All drives were completed on weekdays between the hours of 9.30am and 2.30pm to avoid peak traffic and school pick-up and drop-off times.

2.3 Procedure

A demographic questionnaire was completed by participants prior to the study. After a VPA training session, participants completed a static baseline run of the VDT. Participants then completed a 20 minute practice drive around the Clayton campus to familiarize themselves with the ORTeV, the VPA method and the VDT. At the end of the practice route, participants were informed that the test had begun and that data collection had commenced. Participants then completed the driving route twice, once while performing the VDT and once with no secondary task. The order in which the distracted and undistracted drives were completed was counterbalanced across participants using a standard Latin square design. Participants provided verbal protocols during each test drive. Driving errors made during each test drive were recorded by two in-vehicle observers. Participants were told to concentrate on driving safely, but not to ignore the VDT.

2.4 Error Recording and Classification

The in-vehicle observers used an error pro-forma to manually record the errors made during the test drives. Prior to testing a set of criteria were developed to assist with the identification of errors. These criteria were informed by the Victorian Road Rules and previous driving error literature (Gstalter & Fastenmeier, 2010; Risser & Brandstätter, 1985). On-route, the observer located in the front passenger seat provided directions. Both observers recorded error details, including the error type, where on the route it occurred (e.g., intersection, mid-block), the context in which it occurred (e.g., braking for corner, turning onto freeway), and the error recovery strategy used (e.g., reduced speed, braked). Upon completion of the drive, the two observers checked agreement on the errors recorded. Both observers received extensive training in error definition, observation and recording and both had prior experience with on-road error recording.

The errors recorded during the on-road study were classified post-hoc using a driving error taxonomy (Stanton & Salmon, 2009). The taxonomy specifies 24 error types across five error categories. Two researchers with significant experience in error classification independently classified the errors into external error types using the taxonomy. Based on a comparison of both analysts' classifications, a very high level of agreement was found (Cohen's Kappa = .889), with any disagreement being resolved through further discussion and reclassification if necessary.

2.5 Data Analysis

The nature of errors was analyzed using Generalized Estimating Equations (GEE). Three models were fitted: the first examined differences in the total number of errors made across distracted and undistracted drives, and the second and third differences across the distracted and undistracted drives in the types of specific and external error types made, respectively. Drive order was included as a covariate. All GEE models were specified with negative binomial error function given signs of over dispersion a large number of zeros (Hibe, 2008). Given the data was count data, a log link function was used to keep cell estimates non-negative. The inter-correlation between the repeated measures was specified as unstructured.

3. RESULTS

Drivers made a total of 268 errors when distracted and 182 errors when undistracted. The first GEE model indicated that drivers were 48% more likely to make an error when distracted compared to when not distracted ($Exp(B) = 1.478$, $p<.001$). The driving errors observed were classified into 18 specific error types. A breakdown of these different error types is presented in Table 1, which includes the frequency with which each error was made under each condition. The specific errors committed were similar regardless of whether drivers were distracted or undistracted, as evidenced by the different error types accounting for similar proportions of the total errors made. The most common error was exceeding the speed limit, with 90 instances of participants exceeding the speed limit when distracted and 67 when undistracted. This was followed by lane excursions (where any part of the vehicle moved outside of the marked lane) (42 distracted and 34 undistracted), activating the indicators too early before a turn (31 distracted and 22 undistracted) and travelling too fast for turn (20 distracted and 13 undistracted).

Table 1 Top 10 specific error types (frequency and proportion of all errors) made by drivers when distracted and not distracted.

Error	Distracted		Not Distracted	
	No.	%	No.	%
Exceeding speed limit	90	33.46	67	36.81
Lane excursion	42	15.61	34	18.68
Activated indicator too early	31	11.52	22	12.09
Travelling too fast for turn	20	7.43	13	7.14
Accelerated too fast	16	5.95	8	4.40
Lane excursion into adjacent turn lane (when double turn lanes)	11	4.09	2	1.10
Failed to notice indicator had turned off before turn	10	3.72	7	3.85
Braked late and hard	4	1.49	4	2.20
Encroached into intersection when stopping (over stop line)	7	2.60	4	2.20
Activated indicator too late	4	1.49	1	0.55
Other	33	12.3	20	10.9
TOTAL	268	100	182	100
Mean no. errors per driver	11.6 (range:6-17)		7.9 (range: 5-7)	

A GEE model was fitted to examine differences across the distracted and undistracted drives in the five most common specific error types made. The results of the model are contained in the left of Table 2. As displayed, drivers were significantly more likely to exceed the speed limit, activate their indicator too early and accelerate too fast when distracted than when not distracted. The odds of making a lane excursion and travelling too fast for a turn did not differ significantly across the distracted and undistracted conditions. For the 'too fast for turn' category, the high parameter estimate value suggests that the non significant result may be due to low power given the relatively small number of this error type.

Table 2 Results of GEE models for the specific error types and external error modes

Specific Error Type	Exp(B)	p	External Error mode	Exp(B)	p
Exceed speed limit	**1.337**	**.001**	**Violation**	**1.396**	**.000**
Lane excursion	1.240	.291	**Misjudgment**	**1.379**	**.045**
Activated indicator too early	**1.627**	**.038**	**Action too much**	**1.621**	**.014**
Travelling too fast for turn	1.663	.113	Action mistimed	1.339	.233
Accelerated too fast	**1.847**	**.020**	Perceptual failure	1.326	.546

Significant results (p < .05) in bold

Based on information collected by the in-vehicle observers and provided through the drivers' verbal protocols, the specific error types made during the on-road study were independently classified into external error modes (and associated underlying psychological mechanisms) by two researchers using Stanton and Salmon's (2009) driving error taxonomy (Table 3). This further set of classification allowed for the identification of the possible mechanisms underlying the specific errors types (e.g., whether a 'braked hard' error was due to a perceptual failure because the driver did not see the red signal or due to driver misjudgment regarding required stopping distance).

A third GEE model was fitted to examine differences across the distracted and undistracted drives in the five most common external error modes made. The results of the model are contained in the right side of Table 2. When distracted, drivers were significantly more likely to violate the road rules and make misjudgment and action too much errors compared to when not distracted. The odds of making errors associated with mistimed actions or perceptual failures did not differ significantly across the distracted and undistracted conditions; however, the relatively high parameter estimates again suggests that a lack of statistical power may be underlying these results.

Table 3 External error modes (frequency and proportion of all errors) across the distracted and not distracted conditions

Underlying mechanism	External error mode	Distracted		Not Distracted	
		No.	%	No.	%
Violation	Violation	107	42.9	78	32.9
Cognitive and decision-making	Misjudgment	59	22.0	42	23.1
	Perceptual failure	11	4.1	8	4.4
	Inattention	2	0.7	2	1.1
	Wrong assumption	3	1.1	1	0.5
Action errors	Action too much	36	13.4	22	12.1
	Action mistimed	35	13.1	24	13.2
	Fail to act	9	3.4	4	2.2
	Right action on wrong object	2	0.7	0	0
	Wrong action	1	0.4	0	0
	Inappropriate action	2	0.7	1	0.5
Observation errors	Failed to observe	1	0.4	0	0
	TOTAL	268	100	182	100

4. DISCUSSION

This study examined the nature of errors made by distracted drivers and how these differ to those made when drivers are not distracted. As expected, drivers made a significantly greater number of driving errors when they were distracted by the visual detection task. Interestingly, however, driving errors were prevalent even when not distracted. When not distracted, drivers made almost one error for every kilometer travelled. Taken together, these findings indicate that although driver distraction does have a significant influence on error causation, it is just one of many factors that contribute to driving error. This is consistent with the systems approach to error causation (e.g., Reason, 2000), which implicates system wide factors in error causation as opposed to merely aberrant behavior. Indeed, the role of road system design in error prevention was also indicated with a number of drivers reporting in their verbal protocols that factors such as unclear road markings and limited or inappropriately placed speed and traffic signage contributed to a number of the errors made.

Although drivers made a slightly wider range of error types when distracted, in general the profile of errors made by drivers when distracted and undistracted was similar. Drivers made the same types of errors when distracted and not distracted and, for the more common error types at least, each error type accounted for a similar proportion of the overall errors made under each condition. Therefore, rather than making different types of errors when distracted, distracted drivers made a greater number of the same error types they made when not distracted. In other words, distraction-induced errors differed in degree, but not in type. This result is counter to what was predicted and also the literature on driver distraction which suggests that some error types may be more prevalent when drivers are distracted. In particular, previous research has found that cognitive distraction affects visual scanning behavior, particularly to the periphery (Harbluk et al., 2007; Lee et al., 2001; Strayer & Johnston, 2001), which suggests that drivers would make a greater number of perceptual and observation-based or late braking errors when distracted. This was not the case, with the occurrence of mistimed actions or perceptual failures no more prevalent when distracted.

In terms of the common error types made by drivers, exceeding the speed limit was the most prevalent error committed, accounting for over one third of all errors made when distracted and not distracted. Other common errors included a range of misjudgment and action-based errors such as lane excursions, activating indicators too early and travelling too fast when turning. The error types commonly observed in the current study are in line with those found in previous on-road (Salmon et al., 2010; Staubach, 2009) and crash analysis (Sandin, 2009; Staubach, 2009) studies, where speeding, misjudgment and action-based errors were prevalent.

Given its exploratory nature, this study contained a number of limitations that may have attenuated the effects of distraction on the nature of the errors observed. The study utilized a small sample size, only explored one type and level of distraction and, importantly, the drivers were exposed to only a limited range of driving situations and traffic scenarios under light traffic conditions. When exposed to a greater range of traffic conditions and scenarios, differences in the nature of distraction-induced errors may indeed become apparent. Likewise, the use of different types and levels of distracting tasks may also reveal differences in the nature of errors made distracted.

The current study has provided unique insight into the nature of errors made by distracted drivers under real-world driving conditions. It was found that driving errors are common even under undistracted conditions, but are significantly more pronounced when drivers are distracted. It was also revealed that the profile of errors made by distracted and undistracted drivers was very similar, suggesting that, at least for cognitively distracted drivers under light traffic conditions, the errors made differ in degree, but not in type to the errors made when not distracted.

ACKNOWLEDGEMENTS

This study was funded by the Monash University Researcher Accelerator Program. Dr Salmon's contribution was part funded by his Australian National Health and Medical Research Council Post Doctoral Fellowship. Thanks to Johan

Engström for advice on the VDT, Nebojsa Tomasevic and Ashley Verdoorn for assistance with ORTeV, and Stuart Newstead for statistical assistance.

REFERENCES

Gstalter, H., & Fastenmeier, W. (2010). Reliability of drivers in urban intersections. *Accident Analysis & Prevention, 42*(1), 225-234.

Harbluk, J.L., Noy, Y.I., Trbovich, P.L., & Eizenman, M. (2007). An on-road assessment of cognitive distraction: Impacts on drivers' visual behavior and braking performance. *Accident Analysis & Prevention, 39*(2), 372-379.

Hibe, J.M. (2008). *Negative binomial regression*. Cambridge, UK: Cambridge University Press.

Klauer, S.G., Dingus, T.A., Neale, V.L., Sudweeks, J.D., & Ramsey, D.J. (2006). The impact of driver inattention on near-crash/crash risk: an analysis using the 100-Car Naturalistic Driving Study data. Blacksburg, Virginia: Virginia Tech Transportation Institute.

Lee, J.D., Caven, B., Haake, S., & Brown, T.L. (2001). Speech-based interaction with in-vehicle computers: The effect of speech-based e-mail on drivers' attention to the roadway. *Human Factors, 43*, 631-640.

Reason, J. (2000). Human error: models and management. *BMJ, 320*(7237), 768-770. doi: 10.1136/bmj.320.7237.768

Regan, M.A., Hallett, C., & Gordon, C.P. (2011). Driver distraction and driver inattention: Definition, relationship and taxonomy. [doi: 10.1016/j.aap.2011.04.008]. *Accident Analysis & Prevention, 43*(5), 1771-1781.

Risser, R., & Brandstätter, C. (1985). *Die Wiener Fahrprobe*: Freie Beobachtung, Kleine Fachbuchreihe des Kuratoriums für Verkehrssicherheit, Band 21.

Salmon, P.M., Young, K.L., Lenné, M.G., Williamson, A., Tomesevic, N., & Rudin-Brown, C.M. (2010). *To err (on the road) is human? An on-road study of driver errors*. Paper presented at the Australasian Road Safety Research, Policing and Education Conference, Canberra, ACT.

Sandin, J. (2009). An analysis of common patterns in aggregated causation charts from intersection crashes. *Accident Analysis & Prevention, 41*(3), 624-632.

Stanton, N.A., & Salmon, P.M. (2009). Human error taxonomies applied to driving: A generic driver error taxonomy and its implications for intelligent transport systems. *Safety Science, 47*(2), 227-237.

Staubach, M. (2009). Factors correlated with traffic accidents as a basis for evaluating Advanced Driver Assistance Systems. *Accident Analysis & Prevention, 41*(5), 1025-1033.

Strayer, D.L., & Johnston, W.A. (2001). Driven to distraction: Dual-task studies of simulated driving and conversing on a cellular telephone. *Psychological Science, 12*(6), 462-466.

Walker, G.H., Stanton, N.A., & Young, M.A. (2007). Easy rider meets knight rider: an on-road explanatory study of situation awareness in car drivers and motorcyclists. *International Journal of Vehicle Design, 45*(3), 307-322.

Young, K.L., Regan, M.A., & Lee, J.D. (2009). Measuring effects of driver distraction: Direct driving performance methods and measures. In M. A. Regan, J. D. Lee & K. L. Young (Eds.), *Driver distraction: theory, effects and mitigation*. Boca Raton, FL: CRC Press.

Young, K.L., & Salmon, P.M. (2012). Examining the relationship between driver distraction and driving errors: A discussion of theory, studies and methods. *Safety Science, 50*, 165-174.

CHAPTER 26

Driver's Distraction and Inattention Profile in Typical Urban High Speed Arterials

Nikolaos Eliou, Eleni Misokefalou

University of Thessaly
Volos, Greece
emisokef@uth.gr

ABSTRACT

Over the last years, distracted driving constitutes a considerably increasing road safety problem with disastrous results and it possesses a leading position among the accidents causes. The present study deals with driver's distraction due to out of the vehicle factors. Considering exterior factors as the most significant, we can group them in four categories: built roadway, situational entities, the natural environment, and the built environment. Regarding the fourth category, it is related to the wide variety of civil infrastructure, the commercial land use combined with the high vehicle speeds. All these contribute to the setup of a very dangerous environment by increasing driver's distraction and inattention. This research is based on a medium scale experimental procedure. The distraction of the driver's attention is evaluated via a continuous recording of his gaze. The main objective of this paper is to assess the side effects of roadside advertising and overloaded informational signs to driver's distraction and inattention. The results of this type of research procedures are very useful as a tool to prevent the forthcoming pressure for more and more billboards and trademarks on the roads as well as to encourage the adaptation of more precise regulations with regard to the road infrastructure, the placement of roadside elements, etc.

Keywords: Driving, distraction, advertising, billboards, naturalistic, research

1   INTRODUCTION

The distraction of driver's attention during the implementation of the driving task is not simply a theory. It is a procedure which is activated and developed depending on many factors. It is detected in all drivers, with varying extent and frequency of appearance, but, in every case, the results of this distraction are intense for the driving task, the driver's safety and, finally, for the rest of road users. Distraction at all forms, has become object of research recently, with distraction from a secondary task concentrating most of the research on the subject, particularly after the widespread use of mobile phones and the integration of driver assistance systems in modern vehicles. Naturally, priority is given to drivers of passenger cars without overlooking the other road users' categories such as truck drivers, motorcyclists, bicyclists etc (Misokefalou et al., 2010).

1.1   Definition and characteristics of driver distraction

The first step to a proper approach is to understand the basic characteristics of distraction as it appears in general. Distraction may be visual, cognitive, biomechanical and auditory (Ranney et al., 2001). In the first International Conference on Distracted Driving (2005) the scientific community agreed on a definition for distracted driving: "Distraction involves a diversion of attention from driving because the driver is temporarily focusing on an object, person, task, or event not related to driving, which reduces the driver's awareness, decision-making, and/or performance, leading to an increased risk of corrective actions, near-crashes, or crashes" (Hedlund et al., 2006).

The main causes of distraction are classified into two categories: Those coming from the interior of the vehicle and those from the external environment. In the second category, one finds some very important potential sources of driver distraction. In the case of causes related to advertising, it should be particularly emphasized that the purpose of their presence at some point at the roadside, or even in a moving vehicle in the road, is to capture the driver's gaze in order for him/her to devote the required time so as to assimilate the information obtained. Roadside advertising billboards are designed by their very nature to attract attention. Crucially, though, the related potential threat to road safety is generally not acknowledged by the industry (Crundall et al., 2006).

1.2   Frequency of driver distraction

The importance of this issue emerges from data which shows distraction from a secondary task as a cause of serious accidents as well as crashes. A characteristic research was carried out by the Virginia Tech Transportation Institute (VTTI) for NHTSA, the "100- Car Naturalistic Driving Study" (Klauer et al., 2006). During the 100-Car Naturalistic Driving Study, driver involvement in secondary tasks contributed to over 22 percent of all crashes and near-crashes recorded during the study period. These secondary tasks, which can distract the driver from the primary task of driving (steering, accelerating, braking, speed choice, lane choice, manoeuvring in traffic, navigation to destination, and scanning for hazards), are manifold and include such things as eating/drinking, grooming, reading billboards, using and adjusting in-vehicle entertainment devices, conversation with passenger(s), viewing the scenery, tending to children and pets, smoking, cell phone use and related conversation, use of other wireless communication devices, and note-taking, to name a few (Hedlund et al.,2006). Not all distracters involve secondary tasks initiated by driver –they can be events, objects, activities or people both inside/outside the vehicle (Tasca, 2005). At this point it should be noted that, as near crash is defined the subjective judgment of any circumstance that requires, but is not limited to, a rapid, evasive maneuver by the subject vehicle, or any other vehicle, pedestrian, cyclist, or animal to avoid a crash (Klauer et al., 2006). The statistics are confirmed by the data from accidents in many countries (e.g. accident data from United States in 2008 (NHTSA, 2009) and Greek Police for 2009 and 2010 (Greek Police, 2010)).

Particularly for billboards, in the Crundall et al. study (2006) is supported that though it is acknowledged that research into advertisement distraction has been extremely limited (Beijer et al., 2004), the few studies that have been conducted have demonstrated that drivers do look at and process roadside advertisements (Hughes and Cole, 1986), and that fixations upon advertisements can be made at short headways or in other unsafe circumstances (Smiley et al., 2004). Previous studies of accident statistics have also identified external distractors, including advertisements, as a significant self-reported cause of traffic accidents (Stutts et al., 2001). Particularly, for roadside distractors, evidence is mounting that roadside distractions (and advertising in particular) present a 'small but significant' risk to driving safety (Lay, 2004). Conservative estimates collated from a review of several accident databases put external distractors responsible for up to 10% of all accidents (Wallace, 2003). This is confirmed also by a recent simulator study (Young et al., 2009) in which there is a tentative suggestion that more crashes occur when billboards are present.

1.3   Methods of evaluating driver distraction

The only certain way for the researcher to detect driver's distraction is via the results that distraction produces. The use of standardized methods gives the researchers the possibility to exchange data, conclusions and best practices (Eliou & Misokefalou, 2009). Therefore, it is important to detect the most suitable method of data collection (Young & Regan, 2007). This aim can be achieved via a comparative study between the allocated methods, examining the advantages and disadvantages of every method separately as well as the usefulness and necessity of the results that every one of them produces. An analysis of this kind was made in the study of Eliou and Misokefalou (2009). The most popular among the available methods are based on elements of accidents, on experiments, on observation and surveys. Furthermore, there are some kinds of methods that are not included in any of the previous categories like Peripheral Detection Task and Visual Occlusion.

2   METHOD

2.1   Selection of the appropriate method

The method considered the most appropriate is an observational-naturalistic study, which takes place in the field, using specially equipped vehicles with regard to record the driver's eye movements in order to measure the frequency and the duration of the glances at every object considered potential source of visual distraction. The available equipment (Facelab machine) is capable of making continuous data recording. The main advantage is that with this method, in contrast with all the others categories, driving comes as close to the real thing as possible which is important for the research when we study human reactions. Naturally, there are some limitations both in designing and carrying out the experiment. The most important of these is the limited number of participants in comparison with other methods like questionnaires study, the unfamiliar vehicle which causes stress to the driver, the anxiety because of the sense of being monitored as the vehicle is equipped with cameras and, finally, the subjective discretion of the analyst-observer at the data processing.

Captiv software, which is compatible with FaceLab L2100, was used for the analysis of the results. This software gives the opportunity to analyze the data in detail by recording the total time that the billboard captured driver's gaze during driving. At this point it should be noted that as distraction, in this study, is considered the continuous or intermittent but repeated capture of the gaze from a theme for longer than a total of two seconds as glances that last more than this time are related to driving errors (Rockwell, 1998).

2.2   Participants

Using volunteer drivers, who are required to drive a car on the Thessaloniki's Ring Road, under the supervision of the researcher, who was always in the passenger seat checking the proper function of the system, the obtained results are characterized by a high degree of reliability and validity. Ten drivers(mean age =28.3 years, range = 25 to 30 years) participated in the survey (7 males, 3 females). The drivers were selected with criterion their age. All drivers were familiar with the road, as they use it in a daily basis, but the subject of the study was completely unknown to them. Each one of them, in order to become familiar the vehicle, drove the selected route 2 times before the third run which was the one that we focused our attention at during the analysis process.

2.3   Experimental site

The research took place from January 2010 to April 2010, in Thessaloniki's Ring Road which is a suburban road. The route under observation has a total length of 12.5 km and 12 intersections. For the purpose of the research, drivers drove both directions of the total of 12,5km. The flow of vehicles is continual without being interrupted by traffic lights. The speed limit of the road is 90km/h. The most significant problem of the road is the speed of the passing vehicles in relation to the road geometry as well as the absence of an emergency lane (Atzemi, 2007).

2.4 Material - Data collection

The equipment used in the survey was very carefully chosen in order to produce the optimal quality, completeness and integrity of results. It includes a passenger vehicle and a monitoring and recording system, which detects and records every single movement of the driver's gaze and the driver's head. It is composed of two cameras for the recording of the above and an external camera for the recording of the road scene. All measurements for the experiment took place during the day, under normal traffic conditions as well as normal weather and lighting conditions.

3 RESULTS

In this study, the information isolated and analyzed in depth, is related to the external impulses that cause driver distraction and concentrates interest mainly on billboards near the road and the role of their position in driver's distraction of attention. For this purpose, all billboards along the road were identified and mapped for both directions of the route. Additionally, we noticed a section at a specific junction of the Ring Road, where a large number of illegal posters are placed in disorder which, at first view, leads to a sharp visual disturbance (marked as advertisement billboard number 8 and 20). The analysis included an examination of driver behavior, as far as concern the reactions of drivers' pupils of the eyes while driving under the existence of these potentially evocative distraction elements of the road environment (Chrisostomou, 2010).

The following Fig. 1 shows the percentages from the analysis of the gaze direction to the advertisement billboards of the route. Each driver drove the selected route 3 times but we decided to focus our attention at the third one because of the familiarization of the driver with the vehicle which we analyzed at the method section. As it shows, distraction from advertisement billboards possesses a high percentage of the driving time which ranges from 6 to 8.85% with an average of 7.84%.

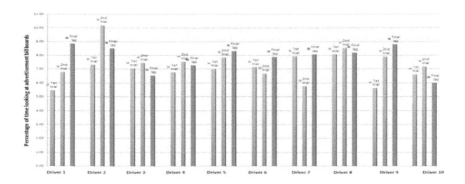

Figure 1 Percentage of driving time looking at advertisement billboards

The detailed analysis of the data came from the eye gaze, in terms of glance duration and frequency, led us to the following conclusions:

- All roadside billboards of the route distract the majority of the drivers (gaze captured for more than 2 seconds).
- At the points of the route where many billboards are placed in a short distance (e.g. advertisement billboard number 9 and 21), the majority of drivers are distracted as their gaze is captured by more than one billboard. At these points, it is observed intermittent but repeated capture of the drivers' gaze.
- Billboards attracts women's gaze more than men's. The average of percentages of the total time that women look at advertisement billboards is 8.7%, while for men is 7.5%.
- The billboards found in the center or near the central field of vision are more likely to attract the driver's gaze.
- At the section which contains the high gather of posters placed in disorder, the large number of illegal posters attracts multiple glances from drivers so that the visual disturbance leads to confusion.
- During the third route, 50 percent of drivers' gaze is captured by more advertisement billboards than during the first route.
- At Fig. 2 is obvious that there are certain advertisement billboards that capture drivers' gaze during all three rides. The percentage of billboards that capture drivers' gaze mostly at the second and the third route is limited.

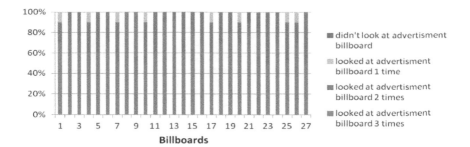

Figure 2 Percentages of drivers distracted from billboards at all three, at two, at one and at none of the routes

From the survey results arises that the presence of distraction, and more specifically the kind caused due to billboards, is common in drivers and depends largely on the characteristics of the billboards and their position in the field of vision.

4   DISCUSSION

Distraction of driver's attention during driving is a major road safety problem, which threatens not only the driver's safety but also the safety of other drivers and road users. The focus of the research on drivers of passenger vehicles is due to the fact that those drivers consist the largest category of road users with growing involvement in accidents, which are caused by the distraction of driver's attention. The goal of the research is to identify and clarify the causes, the frequency of appearance and the way that certain factors influence the distraction of attention of each driver, focusing on the role played by roadside advertising in Greece as a parameter of the distraction of the driver's attention.

The methods commonly used in a study of driver distraction aren't all feasible or effective to the same extent. The chosen method allows the continuous data recording with its main advantage being the fact that driving is as close to the real thing as possible. Thus, the results are characterized by a high degree of reliability and validity. It, also, gives the opportunity to the participant to have an adjustment period with the vehicle in order to obtain a normal driving behavior. The small possibility of the researcher to control the situations and create desirable driving scenarios is among the disadvantages of this method. The environmental conditions, also, cannot be controlled. Another disadvantage is the increased cost of the method due to the eye tracker. Finally, as disadvantage of the eye tracker we could mention the difficulty of the installation in the car as well as its sensitivity to changes (e.g. lightness conditions).

This research concluded that all roadside billboards of the route distract the majority of the with signs in the raw causing the greater distraction. Also, the more centrally positioned in the field of vision the signs are placed, the more eye-catching they are. There is a need to relate the drivers' distraction to specific aspects of advertising signs (size, message content, position by the road).

Much of the data analysis requires collaboration with experts such as psychologists and doctors in order to provide an integrated approach. Furthermore, a comprehensive policy to reduce the visual pollution near the roads, such as billboards, can help not only to improve the road aesthetic but also to significantly improve road safety by eliminating driver's visual distraction of attention.

To sum up, it is a fact that driver distraction is a major cause of accidents; therefore, the responsibility over the issue translates into efforts to reduce the number of injured and dead drivers. This research will be extended, in the future, to Urban Freeways in the major Greek cities (Athens and Thessaloniki. Already in progress is the same experiment in Athens, with a sample of 20 participants.

REFERENCES

Atzemi, M. 2007. Quality assessment of the internal and eastern Thessaloniki's Ring Road. Diploma thesis in the program of postgraduate courses engineering project management, Department of Civil Engineers, Aristotle University of Thessaloniki, Greece.

Beijer, D.D., Smiley, A., Eizenman, M. 2004. Observed driver glance behavior at roadside advertising signs. *Transport Research Record*, 1899, 96–103.

Chrisostomou, K. 2010. Investigation of driver distraction of attention. Diploma thesis in the program of postgraduate courses design, organization and management of systems of means of transport, Department of Civil Engineers, Aristotle University of Thessaloniki, Thessaloniki.

Crundall, D., Van Loon, E., & Underwood, G. 2006. Attraction and distraction of attention with roadside advertisements. *Accident Analysis & Prevention*, 38, 671–677.

Eliou, N., Misokefalou, E. 2009. Comparative analysis of drivers' distraction assessment methods. *22nd ICTCT Workshop, Towards and Beyond the 2010 Road Safety Targets-Identifying the Stubborn Issues and their Solutions*, Leeds, UK.

Greek Police. 2011. Accident Statistics for the Year 2011. Ministry of Citizen Protection, Greece.

Hedlund, J., Simpson, H.M. & Mayhew, D.R. 2006. International conference on distracted driving-summary of proceedings and recommendations. *International Conference on Distracted Driving*, Toronto (2–5October 2005).

Hughes, P.K. & Cole, B.L.: What attracts attention when driving?, Ergonomics, 29, 311–391, 1986.

Klauer, S.G., Dingus, T.A., Neale, V.L., Sudweeks, J.D., Ramsey, D.J. 2006. The impact of driver inattention on near-crash/crash risk: an analysis using the 100-car naturalistic driving study data. Report No. DOT HS 810 594, National Highway Traffic Safety Administration, Washington, D.C

Lay, M.G. 2004. Design of traffic signs. In C. Castro & T. Horberry (Eds.), *The human factors of transport signs*, 25–48. Boca Raton, FL: CRC Press.

Misokefalou, E., Eliou, N., Galanis, A. 2010. Driving distraction of attention audit of motorcycle and bicycle users in urban areas. Case study in the city of Volos, Greece. *5th International Congress on Transportation Research*, Volos, Greece.

Ranney, T.A., Garrott, W.R. & Goodman, M.J. 2001. National Highway Traffic Safety Administration driver distraction research: past, present and future. *17th International Technical Conference on Enhanced Safety of Vehicles*, Amsterdam.

Rockwell, T.H. 1998. Spare visual capacity in driving- revisited. In A.G. Gale, H.M. Freeman, C.M. Haslegrave, P. Smith & S.P. Taylor (Eds.) *Vision in Vehicles II*, 317-324. Amsterdam: Elsevier.

Smiley, A., Smahel, T., Eizenman, M. 2004. Impact of video advertising on driver fixation patterns. *Transport Research Record*, 1899, 76–83.

Stutts, J.C., Reinfurt, D.W., Staplin, L.W., Rodgman, E.A. 2001. The Role of Driver Distraction in Traffic Crashes. *AAA Foundation for Traffic Safety*, Washington, DC

Tasca, L. 2005. Driver distraction: towards a working definition. *International Conference on Distracted Driving*, Toronto (2–5October 2005).

Young, K. & Regan, M. 2007. Driver distraction: a review of the literature. In: Faulks, I.J., Regan, M., Stevenson, M., Brown, J., Porter, A. & Irwin, J.D. (Eds). *Distracted driving*. Sydney, NSW: Australasian College of Road Safety, 379-405.

Young, K., Mahfoud, J.M., Stanton, N. A., Salmon, P.M., Jenkins, D.P., Walker, G.H. 2009. Conflicts of interest: The implications of roadside advertising for driver attention. while driving: Skill and awareness during inspection of the scene. *Transportation Research Part F* 12, 381–388.

Wallace, B. 2003. Driver distraction by advertising: Genuine risk or urban myth? *Municipal Engineer*, 156, 185–190.

Section III

Environmental Impact of Vehicles

Evaluation of Interaction Concepts for the Longitudinal Dynamics of Electric Vehicles — Results of Study Focused on Driving Experience

T. Eberl*, R. Sharma, R. Stroph, J. Schumann, A. Pruckner

BMW Group Research and Technology
80788 Munich, Germany

*corresponding author:
Thomas.TE.Eberl@bmw.de

ABSTRACT

Electric vehicles offer a new degree of freedom in the design of driving torque in coast load conditions. Therefore, new possibilities for the vehicle longitudinal control are obtained. Different driver control strategies are derived and driving experience aspects of the longitudinal dynamics are presented in this contribution. The driver experiences can be classified by the dimensions safety, discomfort, energy efficiency and further specific aspects of the longitudinal control. The characteristic of these driving experience dimensions is analyzed using the results of a subject study (27 subjects) in a realistic city traffic environment. The results of the study show that besides of the well known driving experience dimensions discomfort and safety the newly identified dimensions energy efficiency and the experience of the longitudinal control are responsible for the differing driving experience in the concepts. Furthermore, the results of the study show an adaption of the driver to the tested interaction concepts, which acts as a stimulation (e.g. One-Pedal-Challenge) and in addition can be used to increase the vehicle efficiency.

Keywords: Electric vehicles, Regeneration, Longitudinal Dynamics, Driving Experience, Human-Machine-Interaction (HMI)

1   MOTIVATION AND OUTLINE

According to a report of the United Nations in 2030 more than 60 percent of the world population will live in cities (United Nations, 2007). To enable individual mobility in this scenario, electric vehicles are one possibility to fulfill the emission targets. Additionally, the efficiency of electric vehicles can be increased further by regenerative braking which allows kinetic energy to be converted and stored in the battery during deceleration events.

Regenerative braking can be controlled by different interaction concepts for longitudinal dynamics, which necessitates different driver inputs in specific driving tasks. An appropriate interaction concept for the longitudinal control can offer a positive driving experience (cf. BMW brand slogan: "Sheer driving pleasure") and thus can be one factor for the success of a vehicle concept. This article presents a set of driving experience dimensions that characterize the driving experience of longitudinal dynamics and shows their characteristic in different interaction concepts for the longitudinal dynamics.

1.1   INTERACTION CONCEPTS FOR THE LONGITUDINAL DYNAMICS OF ELECTRIC VEHICLES

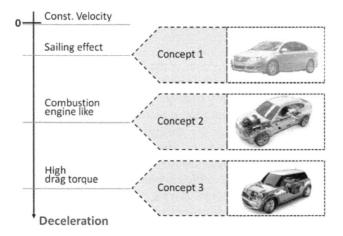

Figure 1 Possible interaction concepts for the longitudinal dynamics of electric vehicles

Figure 1 shows different interaction concepts which characterize the deceleration of a vehicle in a driving condition in which the driver neither actuates the gas nor the brake pedal. Concept 1 results in a low deceleration of the vehicle caused only by the drag resistances and is hereafter labeled as "Sailing effect". This interaction concept can also be achieved by a driver of a combustion engine powered vehicle by setting the gear to the neutral position in a driving situation and achieving therewith a smooth freewheeling.

When lifting the gas pedal in interaction concept 2 the vehicle decelerates similar to a combustion engine vehicle. This interaction concept is common in current hybrid and electric vehicles and is labeled here as "combustion engine like".

The interaction concept 3 was introduced by BMW during the MINI-E consumer study in 2008 (Turrentine, 2011). This concept represents a high deceleration level that allows the driver to achieve the majority of necessary decelerations only by lifting the gas pedal. When the driver requires rapid deceleration, like emergency braking situations or close to standstill on a steep downhill, the driver needs to actuate the brake pedal. This concept is labeled here as "High drag torque".

In order to achieve high vehicle efficiency the regeneration needs to be utilized during the maximum number of vehicle deceleration events, independently of their activation. Therefore, using concepts 1 or 2 the vehicle should be designed to regenerate when the driver actuates the brake pedal for the desired deceleration. Concept 3 regenerates the same amount of energy when decelerating, here triggered by the gas pedal. Hence, the different interaction concepts presented here can be classified as equal in terms of energy flow.

2 DRIVING EXPERIENCE OF THE LONGITUDINAL DYNAMICS

In order to decide how to create a positive driving experience of the longitudinal dynamics one main task was to understand which experience dimensions drivers observe while driving. The experience dimensions result mainly from the interaction for longitudinal control, i.e. accelerating and braking. To characterize this interaction some interaction process models have been applied. Especially the interaction model according to Krueger et al. (2000) focuses on the interaction between driver input and vehicle reaction. Krueger et al. differentiate between two different interaction loops, one interoceptive and one exteroceptive interaction loop. The exteroceptive interaction loop contains the anticipation of a target location and whether the current vehicle path differs from that compensational driver input. The interoceptive interaction loop focuses on the interaction between driver and vehicle. Concerning the vehicle longitudinal control the interoceptive interaction loop illustrates that different interaction concepts result in diverse driver inputs required for a given driving situation. The goal of the present study is to demonstrate the impact of the different interaction concepts on the experience of the interaction.

2.1 Experience dimensions of the vehicle longitudinal dynamics

Using previous work, containing an explorative subject study (Eberl et. al., 2011), the experience dimensions in Figure 2 can be derived. During the subject study the driving situation was studied using a course with low speed and high traffic density as well as a second course with higher speed and low traffic. To analyze the impact of the interaction concepts in the mentioned study two interaction concepts were evaluated that had a significant difference in vehicle reaction (Concept 1: Sailing Effect, Concept 3: High drag torque).

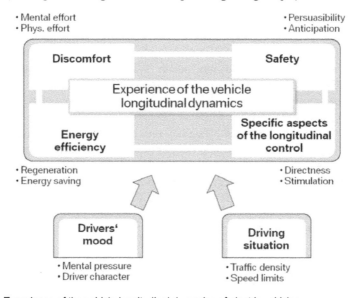

Figure 2 Experience of the vehicle longitudinal dynamics of electric vehicles

The experience dimensions can be classified into four categories (see Figure 2), which are dimensions concerning discomfort, safety, energy efficiency and specific experience of the longitudinal control. Besides these experience dimensions the driving situations and the drivers' mood have a huge impact on the driving experience of the vehicle longitudinal dynamics.

The characterization of the dimension discomfort focuses on the effort to control the longitudinal dynamics. The drivers experienced an effort both mentally and physically. The perceived safety however, can be substantiated by the level of longitudinal control which can be characterized according to Haider (1977) by the persuasibility and the anticipation. Another category of experience dimensions results through the usage of the control elements, here the pedals of the vehicle. Two specific aspects can be distinguished: The stimulation through the control elements, allowing drivers to expand their driving expertise and the directness, therefore the capability to control a situation quickly. Considering electric vehicles, the energy efficiency might be an additional and very valuable experience dimension. This dimension can be subdivided in the perception of saving energy by the regeneration process and by an efficient driving behavior.

2.2   Study focus

Results of the previously mentioned subject study (chapter 2.1) show that the above mentioned driving experience dimensions vary significantly for different longitudinal control interaction concepts and driving situations. To create a positive driving experience one main task is to identify the value of the experience dimensions in different interaction concepts and driving situations. Using this evaluation, strategies for the vehicle longitudinal dynamics can be derived resulting in a positive driving experience. To structure the research approach one first analysis should be done in an urban environment, which might be the main usage area of electric vehicles in near future. .

Thus the goal of the following study is to determine the value of the introduced driving experience dimensions in different longitudinal control interaction concepts and driving situations in an urban environment.

3   EXPERIMENTAL DESIGN

The longitudinal control interaction concepts (see Figure 1) were used as the independent variables. Dependent variables were the experience dimensions and the driving behavior, which characterizes additionally the subjective driving experience.

Quantitative as well as qualitative methods are used to characterize the dependent variables. The subjects received a questionnaire before driving to characterize the driving style and additionally a questionnaire to characterize the driving experience dimensions. Mainly self developed rating scales were used for this questionnaire. Furthermore, the AttrakDiff mini (Hassenzahl, 2010) and the SAM-scales (Bradley, 1994) were used. As qualitative measures, the method thinking-aloud was used, as well as a structured interview. During the study objective driving data was captured by a data logger, which measured signals from the vehicle data bus, and additionally installed measurement equipment.

The study included 27 subjects (25 men, 2 women) between 18 and 61 years of age. The subjects offered a variable level of driving expertise with electric vehicles, especially the MINI E vehicle (mean 650 km).

To characterize the driving experience in the city environment a within-design of the study was used. The subjects drove a specific selected city course representative of urban driving situations (5.5 km) in each of the four examined longitudinal control interaction concepts following a short apprenticeship of all examined interaction concepts. To avoid learning effects concerning the order in which the subjects experienced the different interaction concepts, the sequence was altered as well as the driving direction of the course. The technical characteristics of the different interaction concepts are shown in Table 1. In addition to the concepts of Figure 1, two diverse characteristics of a drag torque are evaluated during the study, one noticeable level and one strong level.

Table 1 Characterization of analyzed concepts for longitudinal control

Level-Nr.	Description	Deceleration
1	Sailing effect	Deceleration only due to driving resistances, i.e. at v_veh = 80 km/h a_veh = - 0.2 m/s²
2	Combustion engine like	Constant deceleration of approximately a_veh = -0.8 m/s² over vehicle speed
3	Noticeable drag torque	Constant deceleration of approximately a_veh = -1.5 m/s² over vehicle speed
4	Strong drag torque	Constant deceleration of approximately a_veh = -2.25 m/s² over vehicle speed

4   RESULTS

The results are classified in the four experience dimensions (see Figure 2), a holistic experience evaluation and an outline concerning the driving behavior. To present the results mainly box plot graphs are used, in which a grey box represents 50% of the middle values allocated around the median, which is shown with the black diamond. The small triangles represent the highest and lowest shape that was not identified as an outlier. An 'X' between two box plots represents the level of significance (X $p<=0.05$; XX $p<= 0.01$; XXX $p<=0.001$). In the following figures the analyzed interaction concepts for longitudinal dynamics are defined by the numbers which represent the deceleration levels according to Table 2.

4.1   Discomfort

To characterize the dimension discomfort the results of the physical effort executing a pedal change are presented. In Figure 3 only the results of drivers familiar with electric vehicles are shown, as they are used to handle regenerative decelerations and can therefore evaluate the physical effort of pedal changes. The results show decreasing annoyance by an increasing drag torque. Clearly, when driving with a higher drag torque less pedal changes need to be carried out for longitudinal control. On the other hand some

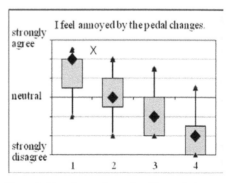

Figure 3 Experience dimension Discomfort (Results of the drivers familiar with the regeneration)

subjects reported that they did not feel annoyed by any pedal changes and experienced the pedal changes positively as a sort of sports activity.

4.2   Safety

As shown in section 2.1 the dimension safety can be characterized by the level of longitudinal control. This aspect is experienced at a high level for all analyzed concepts. The level of longitudinal control can furthermore be concretized by the persuasibility of the driving situation (see Figure 4). The persuasibility of the longitudinal dynamics is rated higher for a higher drag torque. As shown in Figure 4 the subjects experienced the absence of pedal changes as an increase in safety. Lifting the gas pedal in concepts 3 and 4 results in a noticeable vehicle deceleration, which left the impression of a reduction of their stopping distance for emergency events since the response time is considerable abbreviated.

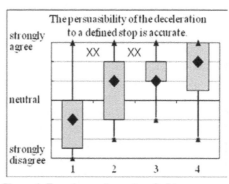

Figure 4 Experience dimension: Safety

4.3   Specific aspects of the longitudinal control

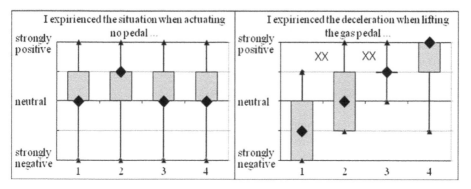

Figure 5 Experience dimension: Specific aspects of the longitudinal control - The figure on the left shows the experience during a Feet-Off-Situation whereas the experience when controlling the deceleration with the gas pedal is shown on the right

The experience of the longitudinal dynamics is highly influenced by the interaction of driver and vehicle by the pedals. To specify this interaction, the experience during a feet-off situation was analyzed. A feet-off-situation is characterized by a driving state in which the driver neither actuates the gas nor the brake pedal. The subjects experience a feet-off situation in the different concepts on average as neutral (see Figure 5). Solely the deceleration level 2, "combustion

270

engine like", was rated more positively, since the drivers were familiar with a deceleration similar to a combustion engine powered vehicle.

Regarding longitudinal control, another important interaction facility is the gas pedal. The results show that this interaction is experienced more positively with a higher deceleration. One main experience aspect of this interaction is the stimulation. While driving with level 3 and 4 the subjects experience the deceleration when lifting the gas pedal as a sort of challenge and like to adapt their driving behavior to perform almost all decelerations only by the gas pedal (One-Pedal-Challenge). This adaption already takes place after a really short time (Krems, 2011). Therefore this positive experience can both be found in the reports from the electric vehicle beginner and experienced drivers.

Considering strategies for longitudinal control, a specific aspect is the directness. The subjects experience an increasing directness when enhancing the deceleration level (Figure 6). According to the subjects, this directness is based on two facets. Having an increased deceleration level, less pedal changes need to be done by the driver and therewith the driver needs to spend less time on the pedal change. Another experience facet for the direct interaction is the ability to accelerate and decelerate by interacting

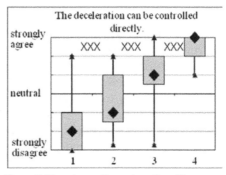

Figure 6 Experience dimension: Specific aspects of the longitudinal control - Directness

only with the gas pedal which also results in high situational awareness. The subjects characterize this usage of the gas pedal for accelerating and decelerating as a symmetric interaction they feel very comfortable with and which helps them to estimate the current pedal.

4.4   Experience of energy efficiency

Considering electric vehicles, the regeneration ability is one important technical function to enlarge the efficiency. Therefore, it is of major interest to examine the experience of this process. As plotted in Figure 7, the results show an increasing regeneration experience by a raising deceleration level. Although, the subjects were told before the study that all four concepts offer an equal deceleration level by regeneration independent of the interaction concepts, the subject

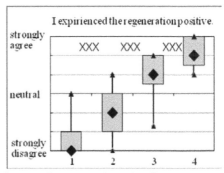

Figure 7 Experience dimension: Energy efficiency

experienced the regeneration at highest deceleration most positive. The subjects argue when driving in concept 1 they decelerate by actuating the brake pedal and cannot distinguish between regeneration and conventional friction braking.

4.5 Holistic experience quantificaton

The results of the above outlined experience dimensions can be substantiated by the results of the AttrakDiff mini as shown in Figure 8. The dimensions pragmatic and hedonic quality can take values between 1 and 7 for a positive experienced characteristic in both dimensions. The results show an increasing pragmatic quality beginning at level 1 with a maximum at level 3 and a similar pragmatic quality of level 4. According to the subjects, this increasing pragmatic quality results from the ability to perform more

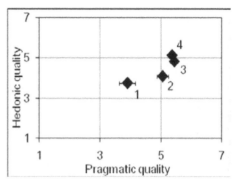

Figure 8 Holistic experience quantification: AttrakDiff mini

decelerations with the gas pedal and to avoid a pedal change. The pragmatic quality of concept 2 with a value of 5.05 is already prominent, as the subjects are used to this deceleration level based on the experience of their combustion engine powered vehicles that afford a similar level. The hedonic quality also rises with an increasing deceleration level and has a maximum for concept 4. According to statements from the electric vehicle beginner drivers level 3 offers the highest hedonic quality. These subjects complain when evaluating level 4 on the amount of deceleration and feel slightly unsafe when controlling such a high deceleration only by lifting the gas pedal. Opposite to this, the drivers familiar with the MINI-E vehicle rate the highest hedonic quality for concept 4. It is substantiated that this level gives them a high degree of freedom of controlling the deceleration. The experts argue these decelerations are rarely carried out but are available for safety reasons.

4.6 Characterization of the driving behavior

To characterize the driving behavior sensor data were logged during the study and were analyzed. Preliminary results show, that the subjects fulfilled higher vehicle decelerations when driving with concepts 1 and 2. When driving with level 3 and 4 the subjects concentrated to do almost all decelerations only by the gas pedal and adapted their driving behavior accordingly. By this adaption, especially in level 3 with a clearly noticeable drag torque of 1.5 m/s², the drivers longitudinal control interaction was more efficient as they required lower decelerations and drove more smoothly.

5 CONCLUSION

Regeneration offers new possibilities for the vehicle longitudinal dynamics which can be controlled by different interaction concepts. These longitudinal control interaction concepts can be characterized by different experience dimensions. The dimension values of these interaction concepts were examined by a subject study in a realistic city traffic environment (27 subjects). The experience dimensions could be grouped into well known and often cited dimensions concerning discomfort and safety but additionally into new dimensions characterizing the energy efficiency and specific aspects of the longitudinal control interaction. The results of the study show that mainly these new experience dimensions are responsible for the differing experience of the longitudinal control concepts. This can especially be observed for the aspect directness of the interaction which relates to the deceleration levels. The holistic driving experience is furthermore characterized by the results of the AttrakDiff mini that show an increasing pragmatic and hedonic quality by a raising deceleration level. This positive hedonic quality results from a stimulatory challenge offered to the driver to adapt his own driving behavior to do almost all decelerations only with the gas pedal (One-Pedal-Challenge). This adaptation results in reduced pedal change actions and objectively higher vehicle efficiency. This positive effect is supplemented by the results of an analysis of objective data logged during the study. A detailed analysis of the objective driving data will be carried out in the future.

REFERENCES

Bradley M. M. and P. J. Lang, 1994. Measuring emotion: the Self-Assessment Manikin and the Semantic Differential. In. *Journal of Behavior Therapy and Experimental Psychiatry*. Vol. 25 (1) pp. 49-59. Elsevier Science Ltd. Great Britain.

Eberl, T., M. Jung, R. Stroph and A. Pruckner, 2011. Interaction concepts for the longitudinal dynamics of electric vehicles. In: *Fahrer im 21.Jahrhundert*. VDI-Berichte 2134, VDI-Verlag GmbH, Düsseldorf

Haider, E., 1977. Beurteilung von Belastung und zeitvarianter Beanspruchung des Menschen bei kompensatorischer Regeltätigkeit. In. Simulation, Fallstudien, Modelle. VDI-Verlag Düsseldorf.

Hassenzahl M. and A. Monk, 2010. The inference of perceived usability from beauty. *Human-Computer Interaction*, 25 (3), 235-260

Krems, J. F., 2011. MINIEVatt Berlin – Freude am umweltgerechten Fahren, Verbundprojekt Klimaentlastung durch den Einsatz erneuerbarer Energien im Zusammenwirken mit emissionsfreien Elektrofahrzeugen – MINI E 1.0 Berlin, *16EM0003*, Chemnitz,

Krüger, H.-P., A. Neukum and J. Schuller, 2000. Bewertung von Mensch-Maschine-Systemen. In: *3. Berliner Werkstatt Mensch-Maschine-Systeme*, ZMMS Spektrum Band 11. Düssldorf: VDI-Verlag GmbH, Berlin

Turrentine, T., G. Dahlia, A. Lentz and J. Woodjack, 2011. The UC Davis MINI E Consumer Study. In: *Research Report Institute of Transportation Studies*

United Nations Human Settlements Programme (UN-HABITAT) 2008. State of the World's Cities 2010/2011 - Briding the Urban Divide. London : Earthscan

CHAPTER 28

Innovative Environmental Design in Means and Systems of Transport with Particular Emphasis on the Human Factor

Grabarek Iwona, Choromanski Wlodzimierz

Warsaw University of Technology
Warsaw, Poland
igr@it.pw.edu.pl, wch@it.pw.edu.pl

ABSTRACT

The Article contains an analysis of the human factor and ergonomic principles in designing new means and systems of transport. The particular emphasis was put on the PRT (Personal Rapid Transit) system vehicle as well as on the innovative electric car that can be used by persons with different level of physical disability (as a driver or passenger).

Keywords: innovative vehicles and transport systems, human factor, disabled persons, ergonomics

1   INTRODUCTION

The XXI century is an era of new challenges for designers of means and systems of transport. This is particularly true in case of urban agglomerations. Many research papers are aimed at proposing solutions that ensure energy efficiency, investment and operational cost-savings, economic profitability, lowest emissions, but above all, user-friendliness and ability to meet the user's needs. These requirements can be difficult to meet in current transport systems, as the modernization efforts are not always successful. It is simply easier to fulfill the requirements in newly designed systems and means of transport where there is no need to adjust to the existing design restrictions. New requirements generate the need for new solutions or adaptation of old concepts that could have not been used before due to lack of proper technology. The following two transport systems were described in the present paper:

1. PRT (Personal Rapid Transit) /the abbreviation APM transport system means Automated People Movers/,
2. Urban, environmentally-friendly, electric vehicle (Eco Car).

The key feature of both systems is their versatility understood as conformity with requirements of disabled users. The percentage of population with disabilities, including elderly people in European countries, is increasing to very significant levels. Thus, improving mobility of these people is becoming important from the sociological and economic perspective and can be seen as a moral obligation of designers and constructors. While the currently existing modes of transport require individual adaptation to the needs of disabled people, the idea behind the proposed solution presented in the paper is that they are widely accepted and allow unassisted operations to all people, including the disabled. The article discusses the issues related to the so-called human factor in the designing of transportation systems. Noteworthy, the proposed solutions aim at limiting the fleet of individual means of transport used within urban agglomerations. The Eco car is intended for transport in city centers that can be, for instance, rented in park-and-ride parking lots. The article does not elaborate on issues such as the propulsion system or modern control systems, which can also be seen as challenges for transport system design. The research work is carried out within the Eco-Mobility Project that is being co-financed by the European Regional Development Fund/ Operational Program for Innovative Economy.

2   URBAN ELECTRIC CAR – ECO Car

As mentioned before, the urban electric car has a number of unique features also in terms of ergonomic design. The basic requirements for the car designed are as follows:

1. The targeted group of users has a significant share of elderly people which is typical for populations of most developed countries. The disability can result not only from diseases and injuries (unrelated to age), but also from a wide spectrum of indispositions that are a consequence of ageing.
2. The car is to be environment-friendly, therefore it should be a zero-CO_2-emmission and easily recyclable device.

The article elaborates on the first requirement, yet the environmental friendliness of an electric car is by no means obvious. Although an electric car emits no CO_2, its batteries require loading with electricity that can be produced with either clean or pollutant technologies. Moreover, lithium-ion batteries or their derivatives are very difficult to recycle. The following assumptions were made regarding the first requirement mentioned above (Choromanski, Grabarek and et al, 2011):

a) The car can carry up to four people, including one in a manual or electric wheelchair

b) The disabled person can drive the car from his or her wheelchair provided it is a manual wheelchair with disability restricted to lower limbs. According to the research conducted in Poland, disabled people in electric wheelchairs tend to have higher level of physical disability (upper limbs disability) and thus, cannot drive a car.

c) The passenger/driver in a manual wheelchair is to enter the car from the rear (by using a special ramp) or from sides. The automobile platform can be lowered with the pneumatic suspension system. The resulting threshold of 12 centimeters (4 and ¾ inch) is easy to cross. In case of an electric wheelchair, the entrance is located in the rear of the car.

d) The car has been equipped with special fixing systems that lock and immobilize the wheel-chaired passenger and/or driver.

e) The driver's seat is located in a car's axis

f) Drive-by-wire system has been used, with no mechanical connection between the steering system and the steering arms of the wheels. This solution enables various types of steering system solutions (steering wheel of various shapes, joystick).

The overall functional structure of the vehicle is shown in figure. 1

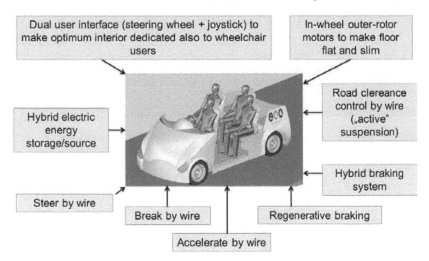

Figure. 1 The eco-car structure

Functionalities shown in figure 2 can be achieved by a system of movable and foldable seats. This design, along with the microprocessor-control system, is considered a significant innovation by the authors. The steer-by-wire technology applied in the design of the Eco-car controls wheels' turning angles, enhances driving comfort and allows the vehicle control system to intervene independently from the driver. The brake-by-wire technology controls the braking system and

276

enables interventions independent from the driver, in order to adjust the braking force of each wheel to road conditions. The pneumatic suspension allows the lowering of the vehicle platform, shock-absorption and maintaining a constant ground-clearance.

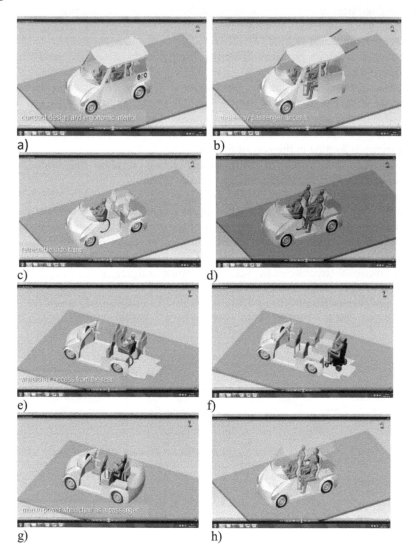

Figure 2 Visualization of Eco-car functionality: a) vehicle structure, b) access to the vehicle's interior, c), d) location of the disabled user (driver) and locations of four users (including one disabled person) e),f) access via the rear ramp for a manual and electric wheelchair g),h) location of one electric-wheel-chaired passenger and three non-disabled users

After a thorough analysis of the requirements for the designed vehicle, the decision has been to place the driving motors in the wheels and install a hybrid battery of electrochemical cells to store electric energy. In order to cut the vehicle's costs and enhance the range, the number of lithium-ion batteries has been reduced, but a super condenser (electric double-layer capacitor or EDLC) has been added.

2.1 Selected issues of the ergonomic design of the eco-car cabin

One of the key challenges in the ergonomic design of the Eco-car interior was to adapt it both to non-disabled and disabled users in wheelchairs. A special system of a movable and foldable front seat was developed to enable the user in a wheelchair to drive the Eco-car. This solution meant to provide sufficient space in the front part of the vehicle to maneuver and immobilize the wheelchair as shown in figure 2c. One of the assumptions was that a disabled driver should be able to enter the car, lock and immobilize the wheelchair and start driving without any external aid. Access to the cabin through the rear door by a disabled person requires the folding of the mid-rear seat and a system of wheelchair locking and immobilizing. One distinction from standard cars is the height of Eco-car's seats. The vehicle's interior was designed for a wide group from C5 females operating from the driver's seat to C95 males operating from the largest wheelchairs. According to current norms, wheelchairs height is between 40 and 52 centimeters. The driver's seat height has been set at 40 centimeters which implies that the maximum difference between the driver's seat and wheelchair level is 12 centimeters. As seen in the design, the driver's seat has neither longitudinal nor vertical adjustment system. It is the same space in the cabin that is occupied by a driver in a wheelchair and by a non-disabled one sitting in a driver's seat. This implies a specific dashboard design with a longitudinal and vertical adjustment system to adapt to drivers of various body size. Figure 3. presents the driver's scope of view. The determined area takes into account the following factors: utmost vertical positions / the highest – male C95 sitting in a wheelchair and the lowest – female C5, sitting in the driver's seat / line of sight and the maximum turn of head to the left and to the right / in sagittal plane / with eyeball movement for a female C5. The size of this area determines the location and area of the windshield.

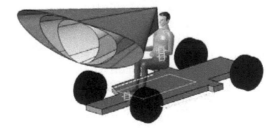

Figure 3 Maximum scope of view determined for a variety of users with various size

278

The active wheelchair design has a characteristic low backrest, whereas driving a car requires a constant back- and head-support. In the Eco-car design this support is to be provided by the backrest of a folded driver's seat. A disabled driver is also seat belted to the backrest. Those functionalities required a specific design of the backrest that would allow a close contact between the wheelchair and the backrest despite the protruding elements on the back of the wheelchair (supporting shelf or pushing handles).

After the driver is seated in a wheelchair or a driver's seat, the adjustment of a dashboard with its controls can take place. This process is activated by a remote control and ends when the steering wheel along with the entire dashboard reaches the driver who is sitting in a wheelchair. The design of a 'steering wheel' is one of the key issues, as it needs to enable the driving and controlling of the car by using only upper limbs. Moreover, it should be intuitive to use for all users. The following three options of ' a steering wheel' have been proposed: multi-function steering wheel, scooter-like handle bars and joystick with additional buttons placed on the dashboard. All three solutions are being tested in a test station during driving simulation. Figure 4 presents one of the variants of the multi-function steering wheel.

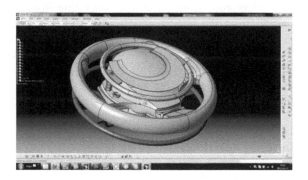

Figure 4. Multi-function steering wheel

An inner and an outer rims control the car's acceleration and breaking. Buttons on the wheel activating: indicators, windshield wipers and washers, low and high beam lights and horn. The dashboard is also equipped with a touch screen – a driver's interface that enables communication with the vehicle, being very useful before and during driving. The information delivered by a visual display and voice instructions help a disabled user enter the car, lock and immobilize the wheelchair and locate the control board. The execution and confirmation of all instructions is necessary before the car can be started and driven. At each stage of the preparation and driving, the touchscreen displays adequate information. The touchscreen location on the dashboard is variable and dependent on the version of a steering device. Figure 5 presents an example of touch screen position in two of the proposed options.

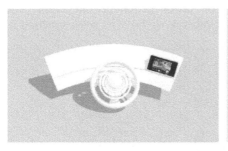

Figure 5 Touchscreen location on the dashboard in two proposed versions of a steering device.

3   PERSONAL RAPID TRANSIT

Personal Rapid Transit (PRT) is a public transportation mode featuring small automated vehicles operating on a network of specially-built guide ways. PRT is a type of automated guide way transit (AGT), a class of system which also includes larger vehicles all the way to small driverless subway systems. In PRT designs, vehicles are sized for individual or small group travel, typically carrying no more than 3 to 4 passengers per vehicle. The guide ways are arranged in a network topology, with all the stations located on sidings, and with frequent merge/diverge points. This approach allows for nonstop, point-to-point travel, bypassing all intermediate stations. The point-to-point service has been compared to a taxi (transportation idea, which means picking up passengers from the place where they stay at the beginning of the journey and taking them directly to the destination point).

The basic assumptions of the system built at the Warsaw University of Technology are as follows (Choromanski and Grabarek, 2009; Choromanski and Kowara, 2011; Choromanski and Kowara 2001):

- Maximum speed of the vehicles and minimum separation between the vehicles in motion is considered in the research and testing stage at 50 km per hour and 10 meters respectively.
- Four-passenger, electrically driven vehicles moving under the overhead guideway.
- Travel direction-switches are critical for PRT systems. A special mechanism designed to change travel direction will be installed in the vehicle.
- Rail network for the PRT vehicle is two-layered and consists of layer I (main grid with constant speeds of 50 km per hour) with 10 meter separation gap between vehicles and layer II (with guide-ways suitable for speeds from 0 to 50 km per hour that allow for point-to-point travel, bypassing intermediate stations).
- The overall dimension of the PRT vehicle should comply with the ergonomic requirements and its design must allow for accommodation of wheelchair users with a guide (foldable seats).

The ergonomic design of the PRT cabin refers not only to its size, taking into consideration users' anthropometric measurements, but also specific solutions enabling people with limited efficiency of lower extremities to 'dock' the wheelchair while the vehicle is in motion (Grabarek, 2009). The seats structure in the PRT cabin is shown in the figure 6.

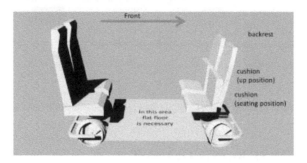

Figure 6 Seats structure in the PRT cabin

The safety reasons of support the solution of 'docking' the wheelchair facing backwards, which is shown on the figure 7, although it is possible to fix the wheelchair facing forwards but only in case of an individual ride. An crucial issue was to equip the cabin with an interface allowing for, among others, choosing the route and its modification. The interface system (implemented to a great extent by a touch monitor) is a part of a more general PRT computer system presented in the figure 8. Because passengers might be of a different level of physical efficiency, the interface's management has to be intuitive, simple and taking into consideration different kinds of passengers' limitations. Ensuring the legibility of the content shown is necessary to detect and understanding of the possible information. For some people, especially the elderly ones, any information presented in a form of icons or pictures is completely incomprehensible, unlike the text they got used to during the major part of their lives.

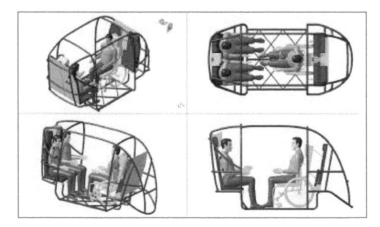

Figure 7 Interior design of PRT cabin

Passenger services:	Vehicle control:
• requesting and assigning vehicle • choosing destination • passenger information services • safety supervision • etc.	• choosing route • real-time control of velocity and direction • secure following preceding vehicle and • synchronizing on guideway joints

Management:
• planning of overhauls • predictive sending free vehicles to zones of expected heightened demand • collecting technical, statistical and business data • providing interface for supervising staff

Figure 8 PRT computer system

In order to make the use of the monitor comfortable and intuitive for the majority of the aging society, the graphic elements (icons, symbols) aimed at communications with the user must come with a proper comment. The users must be given the option of choosing the destination point in different ways.

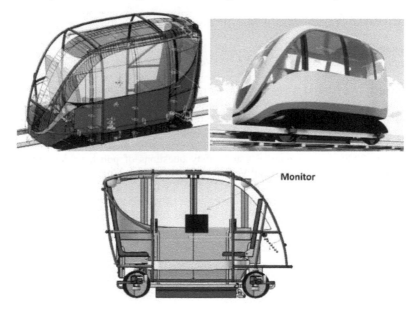

Monitor

Figure 9 Proposal of the PRT cabin design

The figure 9 shows the shape of the PRT cabin. The limited scope of this work does not allow for the analysis of the innovative vehicle's propulsion based on the electric linear motor and wireless energy transfer. Proper shaping of the track (track transition curve), limiting the impact of accelerations on human body, also constitutes an important element of the project.

CONCLUSIONS

Constructions presented in this article constitute an attempt of introducing of new solutions to transport modes and systems that take into account often formulated postulates, especially those regarding urban transport. The adjustment to users with different degrees of disability, environment friendliness, low energy consumption and its economy, are important features of innovative transport. This text focuses only on the problems related to the human factor, however the project includes all the above-mentioned factors. Within the frames of the Eco-mobility project, an experimental track and vehicles' prototypes are being constructed at the moment, which will allow for verification of the adopted assumptions.

ACKNOWLEDGMENTS

The authors would like to acknowledge, that the research has been carried out as part of the project ECO-mobility, co-funded by the European Regional Development Fund under the Operational Program Innovative Economy (UND-POIG.01.03.01-14-154) – Project Co-ordinator Prof. W. Choromanski

REFERENCES

Choromanski, W., and J. Kowara, 2011. PRT–Modeling and dynamic simulation of the track and the vehicle. *Proceedings of the 13 International Conference of Automated People Movers* , 23-26 May 2011, Paris, ASCE – American Society of Civil Engineers.
Choromanski W., and J. Kowara, 2011. Modeling and simulation of prt vehicle with polyurethane wheels. *Proceedings of the 22nd Symposium on Dynamics of Vehicles on Roads and Tracks,* ISBN 978-1-905476-59-614-19 August, Manchester.
Choromanski W., and I. Grabarek ,2009, Ergonomics considerations in the design of the personal rapid transit., *Proceedings of the International Society of Biomechanics , XXII*
Grabarek I., 2009. Some ergonomics and safety questions of polish vehicle prt (personal rapid transit) construction . Rail Vehicles., Poznan, Issue 2, pp. 6- 9,
Choromanski W., and I. Grabarek, et al. 2011, New Concept of an Electrical City Car and its Infrastructure. Paper presented at the Polish – Chinese Seminar: "ECO-Mobility - the innovative technologies" on the 16 – 17 of June 2011

CHAPTER 29

Multifactor Combinations and Associations with Performance in an Air Traffic Control Simulation

Tamsyn Edwards[1], Sarah Sharples[1], John R. Wilson[1] & Barry Kirwan[2]

1. Human Factors Research Group
Faculty of Engineering
University of Nottingham
University Park, Nottingham
NG7 2RD, UK

2. Eurocontrol Experimental Centre,
Bretigny/Orge,
France

ABSTRACT

In air traffic management (ATM) knowledge of the impact of human factors on performance is critical to address safety incidents. Due to developments in ATM systems the performance-related incidents that occur in an air traffic control (ATC) environment are generally complex and multi-causal in nature. This research uses an en-route air traffic control simulation to investigate the relationship between factors known to influence controller performance, and the association between multifactor combinations and performance. Results indicate that several factors known to affect controller performance do covary. In addition, results suggest that factors may interact to produce a cumulative impact on performance. Further research is needed to investigate the nature of multiple factor co-occurrences and the association with performance, both in air traffic control and other safety-critical domains.

Keywords: air traffic control, human performance, multifactor combinations

1 INTRODUCTION

Air traffic control is a safety critical system. To ensure flight safety, air traffic controllers (ATCOs) are required to maintain a consistently high standard of performance. Human factors have been repeatedly evidenced to affect human performance (Chang & Yeh, 2010). Therefore, knowledge of the impact of human factors on human performance and error is critical in addressing safety incidents in air traffic control.

Due to developments in ATM systems, controllers are now relatively well-defended from single human factor influences, such as fatigue or workload, on performance. The performance-related incidents that do still occur are generally complex and multi-causal in nature, or are seen as having no direct causes but many contributors, as highlighted by so-called 'Swiss Cheese' and Resilience Engineering models (e.g. Reason, 1990). Therefore, the residual threats for incidents often result from the interaction of multiple human factors and the resulting cumulative impact on performance.

However, relationships between factors have received scant attention in the literature (Glaser, Tatum, Nebeker, Sorenson, & Aiello, 1999).This lack of research may have therefore restricted an ecologically valid understanding of the possible interactions between factors, and the subsequent combined impact on performance (Cox-Fuenzalida, 2007). Therefore, several calls have been made for research on the interrelations between factors, and the subsequent effects on human performance and error (Chang & Yeh, 2010; Murray, Baber, & South, 1996). The present research was conducted as an initial step to addressing this research gap.

The aims of the current study were therefore two-fold. First, the study aimed to investigate relationships between factor dyads within the context of an air traffic control domain, within controlled laboratory conditions. Secondly, the study aimed to investigate the association of co-occurring factor dyads with performance.

2 METHOD

An en-route air traffic control (ATC) simulation exercise was utilised to investigate the relationship between factors and the subsequent association with performance. Participants were students who had been trained in ATC principles during a dedicated training session and who had demonstrated their knowledge in a purpose-designed competency test. The study used a within measures design and so all participants completed the same en-route ATC simulation exercise. The session consisted of 5 periods of alternating taskload, beginning with a low taskload. Each taskload period was 20 minutes long (Galster, Duley, Masalonis, & Parasuraman, 2001) interspersed with a 4 minute transition period between each taskload. The taskload level was created by changing the number of aircraft in the controlled sector (Tenney & Spector, 2001) and the complexity of the task by the number of aircraft requiring vertical movements and the number of aircraft pairs set on a conflicting flight path (Brookings, Wilson, & Swain 1996). Participants were required to complete all control actions, including assuming aircraft, coordinating aircraft, and transferring aircraft. To ensure participants were motivated to perform to the best of their ability, all participants were instructed to try to be as safe and efficient as possible and a cash voucher prize was made available to the top 3 performing participants. Several covariate factors, which had previously been identified as critical factors that frequently negatively influenced controller performance (Edwards, Sharples, Wilson & Kirwan, 2012) were measured.

2.1   Measures

Covariate factors were measured using subjective, self-report scales. Mental workload was measured using the uni-dimensional Instantaneous Self Assessment scale (ISA) (Tattersall & Foord, 1996), a reference tool in the industry. Fatigue was measured through the use of a visual analogue scale (Lee, Hicks, Nino-Murcia 1990), and stress and arousal were measured using the Stress Arousal Checklist (Mackay, Cox, Burrows, & Lazzerini, 1978). Finally, situation awareness (SA) was measured using the Situation Present Assessment Method (SPAM) (Durso et al., 1995). Every 4 minutes, participants were asked to verbally rate workload and asked to respond to a SA-related question (Durso et al., 1995). Every 11 and 13 minutes alternately (5 minutes in, and 5 minutes before the end of each taskload period), participants were asked to pause the simulation and complete a fatigue (VAS) and stress/arousal (SACL) measure.

Performance was assessed by safety (number of alerts that aircraft minimum separation is about to be lost) and efficiency (time to assuming control of aircraft entering the subject's airspace 'sector', time to responding to aircraft needs, time to transfer aircraft, and frequency of routing aircraft on a more direct route to save fuel). Measures were recorded continuously in the simulation software.

2.2   Participants

A total of 29 male participants took part in the simulation. Age ranged from 18 years – 27 years with an average of 21 years (SD=2.6). All participants had normal or corrected vision and did not have any known colour-blindness or hearing impairments. All participants were native English speakers.

3.   RESULTS

Due to the quantity of analyses and results, only results for the strongest and most relevant data trends will be presented in this article.

3.1   Taskload and subjectively experienced factors

Descriptive statistics support that the manipulated taskload levels created variation in the averaged experience of subjectively measured factors. The averages varied in the direction expected for all factors apart from SA (Table 1).

Table 1. Mean and standard deviation for single factors averaged across taskload sections

	Low taskload 1 (0-20 minutes)		High taskload 1 (25-44 minutes)		Low taskload 2 (49-68 minutes)		High taskload 2 (73-92 minutes)		Low taskload 3 (97-116 minutes)	
	M	SD	M	SD	M	SD	M	SD	M	SD
Workload	2.7	0.6	4.3	0.5	2.5	0.7	4.4	0.5	2.0	0.6
Fatigue	19.0	15.4	28.1	18.1	22.4	13.9	30.7	19.9	32.7	21.0
Stress	37.0	6.4	46.8	8.7	34.7	6.4	44.9	10.7	32.6	7.3
Arousal	35.9	5.9	38.1	5.8	35.7	4.4	36.4	5.6	30.7	5.3
SA (Secs)	2.7	1.1	2.2	1.0	1.7	0.9	1.6	0.8	1.5	0.6

3.2   Covariance between factors

The covariance between factors was examined. Workload and stress appear to have the strongest association (Figure 1). Spearman's rank correlation coefficient revealed a highly significant relationship, $r_s=0.6$, df=287, p<0.001.

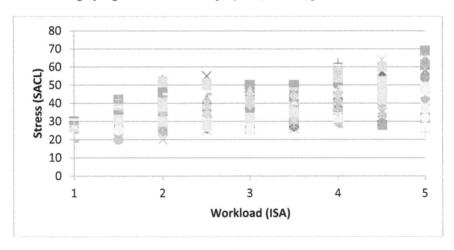

Figure 1. Scatterplot of stress against workload (ISA scale is five-point, i.e. 1-5, 5 being very high)

The relationship between workload and fatigue ($r_s=0.2$, df= 287, p<0.01) and workload and arousal ($r_s=0.4$, df=287, p<0.001) although significant, had a low shared variance between factors, suggesting a weaker relationship. The relationship between workload and SA was significant but weak ($r_s=-0.1$, df=236, p<0.05).

Fatigue correlated significantly with both stress (r_s=0.3, df=287, p<0.001) and arousal (r_s=-0.4, df=287, p<0.001). In addition, SA was found to be significantly related to arousal; as arousal rose, time to respond to SA-related questions increased (r_s=0.2, p<0.005). This is believed to be an artifact of the experimental set up (see discussion).

3.3   Factor interactions and associations with performance

The authors extended the analysis to investigate interactions between factor dyads, and associations of factor dyads with efficiency and safety-related performance. To facilitate this analysis, a median split approach was utilized (Denollet et al. 1996; Miles and Shevlin, 2001) to transform the continuous data into discrete factor groups. For the sake of brevity, results characterising significant data trends are reported below.

3.3.1   Safety-related performance

First, the co-occurrences of workload and SA, and the association with safety performance, are reported. Measures indicating a high level of SA appear to have been associated with less conflict alerts, and therefore safer performance (Figure 2). It is interesting to note that in this study, lower perceived workload was associated with more safety alerts. The association of combined factors on performance is interesting. The number of alerts is significantly more when low workload and poor SA combine, M=2.2, SD=0.8 (two factors which appear to be independently negatively associated with performance) compared to the number of alerts when participants report both high workload and good SA (M=1.3, SD=0.6), two factors which appear to be independently positively associated with performance, W=33, p<0.05, r=-0.4.

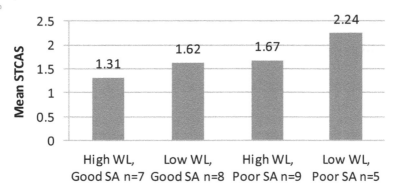

Figure 2. Median split application showing the average performance (frequency of STCAs) under different levels of workload and SA in high taskload phase 1.

3.3.2   Efficiency-related performance

The same trend was observed in efficiency-related performance. One example of this was the association of stress and SA on time to assume aircraft. Again, a median split transformation was conducted on the data. Performance (time to assume aircraft) was averaged across each factor dyad group. Figure 3 reveals that the combination of high stress and poor SA, which in this study are negatively associated with performance, is associated with a longer period of time to assume aircraft (M=108.6, SD=53.7). A Mann-Whitney U analysis confirmed that time to assume aircraft was significantly faster in group 1 (high stress with good SA) compared to group 4 (high stress and poor SA), W=52.5, p<0.05, r= -0.5. Factor groups 1-3 formed a homogenous subset.

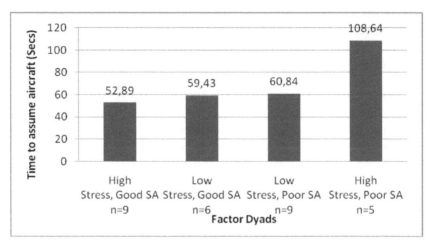

Figure 3. Median split application showing the average performance (time taken to assume aircraft) under different levels of stress and SA in high taskload 1.

4   DISCUSSION

4.1   Taskload and subjectively experienced factors

The taskload manipulation created variation in participants' self-reported experience. All factors varied in the direction expected apart from the time to respond to SA-related questions. Here, response time decreased (suggesting an improvement in SA) on average throughout the simulation regardless of taskload phase.

These results are likely to be an artifact of the experimental design. Approximately 30% of participants in post-trial debriefs admitted that especially during phases of high taskload, and towards the end of the simulation when feeling fatigued, they simply responded without much care in an attempt to eliminate the distraction as quickly as possible. Alternative measures of SA, such as the Situation Awareness Global Assessment Technique (SAGAT) (Endsley, 1988) would have avoided this issue. However, in the present study, the application of SAGAT would have created a disruption to the simulation task, and also to an accurate measurement of workload.

4.2   Covariance between factors

Results suggest that multiple factor dyads covaried. These findings support previously-identified relationships in the literature between workload and stress (Glaser et al., 1999), workload and fatigue (Corrandini & Cacciari, 2002), workload and SA (Ma & Kaber, 2005) fatigue and stress (Park, Ha, Yi & Kim, 2006) and arousal and SA (Taylor & Selcon, 1991). Many of these results were expected, and confirm the suggestion in the literature that many human factors are related.

The strongest identified relationship existed between the factors of stress and workload. Causation cannot be inferred, although from previous research (e.g. Bellorini, 1996) it may be suggested that these factors may influence each other in a bi-directional relationship. This finding suggests that workload and stress may frequently co-occur, and human operators managing a high workload should be monitored for such influences on performance especially within safety-critical systems.

Results suggested that the relationship between workload and SA was relatively weak, which was unexpected, as was the direction of the relationship between arousal and SA. As discussed in section 4.1, the authors believe this to be an artifact of the simulation set-up.

4.3   Factor interactions and associations with performance: a cumulative impact?

A key finding of interest is that specific factor dyad groupings appear to be associated with a cumulative impact on both safety related and efficiency related performance. It is only when negative performance-influencing factors combine (i.e. high stress in combination with poor situation awareness, Figure 3) that performance significantly declines. Performance appears to remain stable when associated with several other factor combinations.

This result may be interpreted in the context of 'Limited Resources' (Wickens, 1992). An accumulation of negatively-influencing factors may lead to resources which are rapidly depleted. Of course, controllers are not passive in their environment, and when a task is becoming too cognitively demanding, strategies or priorities will be utilised to support performance (Sperandio, 1971). However, if factor co-occurrences further continue to negatively influence performance, and/or resources become substantially depleted, the remaining cognitive resources may not be sufficient to meet the demands of the task, resulting in reduced performance. Therefore, it may be hypothesised that under some circumstances, it is not necessarily the extent of a single factors' influence on performance (unless in an extreme form, such as overload), but the number of interacting factors which when combined have a negative influence on performance, that determines the overall association with performance.

This finding is important for the control room. Supervisors may need to monitor the occurrence of multiple factors, and implement supportive strategies when factors which are negatively associated with performance co-occur. In addition, the finding that performance can remain stable under several multifactor influences suggests that if this period of performance could be identified prior to a steep performance decline, supportive strategies may prevent performance from declining further. Future research should therefore be conducted on the 'markers' that might indicate when performance is about to decline, potentially preventing a performance-related incident.

It is acknowledged that these results are provisional. Results should be interpreted with caution, due to the small, homogenous sample, and lack of a high fidelity simulation with trained ATCOs potentially reducing the valid generalisation of results. Future research should replicate these results using a full-scale simulation with trained ATCOs as participants.

5. CONCLUSION

The relationship between factor dyads, and the association of co-occurring dyads on performance was investigated within the context of an air traffic control task. Initial findings suggest that relationships do occur between factors, and interactions between factors can occur, potentially creating a cumulative impact on performance. Previous research has infrequently considered multiple factor co-occurrences and performance-effects. The authors suggest that the current findings, although provisional, support a move towards examining multi-factor co-occurrences and their subsequent impact on performance within safety-critical environments. Future research is needed to further inform the current results (for example, specifying which are most critical to controller performance). Further research is also necessary to extend the current results by investigating the association of 'triadic' and 'quad-factor' co-occurrences with performance, and exploring multifactor co-occurrences in other safety-critical domains. Last, these results need to be translated into the identification of practical real-time markers of the limits of human performance under multi-factor conditions, in order to assure safety in air traffic control operations.

ACKNOWLEDGMENTS

The authors would like to thank EUROCONTROL for funding and supporting this research.

DISCLAIMER

The contents of this article are however the opinions of the authors and do not necessarily represent those of parent or affiliate organizations.

REFERENCES

Bellorini, A. (1996). Task-related factors of stress: the analysis of conflict resolution in air traffic control. *1st International Conference on Applied Ergonomics*. Istanbul: Turkey.

Brookings, J. B., Wilson, G. F., & Swain, C. R. (1996). Psychophysiological responses to changes in workload during simulated air traffic control. *Biological Psychology, 42*, 361-377.

Chang, Y. & Yeh, C. (2010). Human performance interfaces in air traffic control. *Applied Ergonomics, 41*, 123-129.

Corrandini, P. & Cacciari, C. (2002). The effect of workload and workshift on air traffic control: A taxonomy of communicative problems. *Cognition, Technology & Work, 4*, 229-239.

Cox-Fuenzalida, L. E. (2007). Effect of workload history on task performance. *Human Factors, 49(2)*, 277-291.

Denollet, J., Rombouts, H., Gillebert, T. C., Brutsaert, D. L., Sys, S. U., Brutsaert, D. L., & Stroobant, N. (1996). Personality as independent predictor of long-term mortality in patients with coronary heart disease. *The Lancet, 347(8999)*, 417-421.

Durso, F. T., Truitt, T. R., Hackworth, C., Crutchfield, J, Nikolic, D., Moertl, P., et al. (1995). Expertise and chess: A pilot study comparing situation awareness methodologies. In D. J. Garland & M. R. Endsley (Eds.), *Experimental Analysis and Measurement of Situation Awareness* (295–304). Daytona Beach, FL: Embry-Riddle Aeronautical University Press.

Edwards, T., Sharples, S., Wilson, J. R., & Kirwan, B. (2012). Factor interaction influences on human performance in air traffic control: The need for a multifactorial model. *Work: A Journal of Prevention, Assessment and Rehabilitation, 41(1)*, 159-166.

Endsley. M. R. (1988). Situation awareness global assessment technique (SAGAT). *Proceedings of the Aerospace and Electronics Conference*, Dayton, OH, 23-27 May.

Galster, S. M., Duley, J. A., Masalonis, A. J., & Parasuraman, R. (2001). Air traffic controller performance and workload under mature free flight: Conflict detection and resolution of aircraft self-separation. *International Journal of Aviation Psychology, 11(1)*, 71-93.

Glaser, D. N., Tatum, B. C., Nebeker, D. M., Sorenson, R. C., & Aiello, J. R. (1999). Workload and social support: Effects on performance and stress. *Human Performance, 12(2)*, 155-176.

Lee, K. A., Hicks, G., & Nino-Murcia, G. (1990). Validity and reliability of a scale to assess fatigue. *Psychiatry Research, 36*, 291-298.

Ma, R. & Kaber, D. B. (2005). Situation awareness and workload in driving while using adaptive cruise control and a cell phone. *International Journal of Industrial Ergonomics, 35*, 939-953.

Mackay, C., Cox, T., Burrows, G., & Lazzerini, T. (1978). An inventory for the measurement of self-reported stress and arousal. *British Journal of Social & Clinical Psychology, 17(3)*, 283-284.

Miles, J. & Shevlin, M. (2001). *Applying Regression & Correlation: A Guide for Students and Researchers*. London, UK: Sage Publications.

Murray, I. R., Baber, C., & South, A. (1996). Towards a definition and working model of stress and its effects on speech. *Speech Communication, 20*, 3-12.

Park, F., Ha, M., Yi, Y., & Kim, Y. (2006). Subjective fatigue and stress hormone levels in urine according to duration of shiftwork. *Journal of Occupational Health, 48*, 446-450.

Reason J. (1990). *Human Error*. New York: Cambridge University Press

Sperandio, J. C. (1971). Variation of operator's strategies and regulating effects on workload. *Ergonomics, 14(5)*, 571-577.

Tattersall, A. J. & Foord, P. S. (1996). An experimental evaluation of instantaneous self-assessment as a measure of workload. *Ergonomics, 39(5)*, 740-748.

Taylor, R. M., & Selcon, S. J. (1991). Subjective measurement of situational awareness. In Y. Queinnec & F. Daniellou (Eds.), Designing for everyone. *Proceedings of the 11th Congress of the International Ergonomics Association* (pp. 789-791). London: Taylor & Francis.

Tenney, Y.J. & Spector, S. L. (2001). Comparisons of HBR models with human-in-the-loop performance in a simplified air traffic control simulation with and without HLA protocols: Task simulation, human data and results. *Proceedings of the 10th Conference on Computer Generated Forces & Behaviour Representation*, Norfolk, VA, 15-17 May.

Wickens, C.D. (1992) *Engineering Psychology and Human Performance*. NY: Harper Collins.

CHAPTER 30

Driving Faster or Slower? Biased Judgments of Fuel Consumption at Changing Speeds

Gabriella Eriksson[1,2], Ola Svenson[3,4]

[1]Department of Behavior and Learning, Linköping University, Sweden
[2]Swedish National Road and Transport Institute, Linköping, Sweden
[3]Decision Research, Eugene, Oregon 97 401 USA
[4] Risk Analysis, Social and Decision Research Unit, Department of Psychology
Stockholm University, Sweden

ABSTRACT

Reduced fuel consumption by lower speeds on roads would reduce emissions. Do drivers, who choose vehicle speed, realize the gain in fuel of a reduced speed? Judgments of fuel consumption at increasing and decreasing speeds were made by professional truck drivers and student groups when they were not driving. For decreases in speed, truck drivers underestimated fuel saved significantly. Engineering and psychology university students' judgments also tended to underestimate fuel saved but not statistically different from the correct values. For increases in speed, the truck drivers judged the fuel they would waste close to correctly. The psychology student group overestimated the fuel wasted following an increase in driving speed as did the engineering students but only for speed increases greater than 30 km/h. The results indicate that eco driving systems need to support drivers' judgments of fuel saved or lost if they change their driving speed.

Keywords: Fuel consumption, speed, truck drivers

1   INTRODUCTION

The present contribution aimed to explore how drivers' judgments of fuel consumption at varying speeds differ from empirical data. We also wanted to investigate these judgments in three groups with different experience of driving and fuel consumption. The three groups were professional truck drivers, engineering students and psychology students. Professional truck drivers are very experienced in driving and fuel consumed is often discussed in their line of work so they are assumed to be aware of it to a larger extent than the other groups. Engineering students were assumed to have a better knowledge of the relationship between fuel consumed and speed than non-engineering students since mathematics and physics is a large part of their education. We hypothesized that the professional truck drivers would make more accurate judgments than the other groups since they were far more experienced in driving. Our second hypothesis was that psychology students would perform worse than the other two groups because their lack of experience of driving and knowledge about the physical relationship between speed and fuel consumption.

2   THE STUDY

2.1   Method

2.1.1   Participants

Twenty-four professional truck drivers, 50 engineering students and 48 psychology students participated in this questionnaire study.

2.1.2   Materials and procedure

The participants were asked to estimate the amount of gasoline consumed in liters of a passenger car. There was one speed change in each problem. The distance was always 100 kilometers. Participants were told that the conditions were the same between problems. Hence, the only change between problems was the speed. The participants were told not to make any formal mathematical calculations and that their answers should be based on intuitive judgments. All of the participants were given a reference speed with its corresponding gasoline consumption. Half of the participants in each of the three groups were given the reference speed 120 km/h (8.8 liters per 100 km). They were asked to judge the fuel consumed for decreasing speeds of 110, 100, 90, 80, 70 and 60 km/h. Similarly, the other half of the participants were given the fuel consumption for 60 km/h (5.6 liter per 100 km) and made estimates at increased speeds 70, 80, 90, 100, 110 and 120 km/h. Judgments were compared to actual measured values.

2.2 Results

In the following, the results will be analyzed separately for the two reference speeds. First, the fuel saving condition where participants judged the fuel consumption at decreasing speeds (reference speed 120 km/h) will be presented. Last, we will report results of the fuel lost condition with participants judging fuel consumption at increasing speeds (reference speed 60 km/h).

2.2.1 Fuel Saved by Decreasing Speed

Table 1 shows the average judgments of fuel consumed at decreasing speed for the three groups separately. Correct fuel consumption at each speed according to Carlsson, Hammarström and Karlsson (2008) is also shown. The table shows which judgments differ significantly from correct values. In all problems, the three groups' average judgments are greater than correct fuel consumption. This means that the amount of fuel saved when slowing down a car is underestimated. Professional truck drivers were more biased in their judgments than both student groups.

Table 1 Average judged fuel consumption at different speeds for truck drivers, engineering and psychology students respectively when car speed is slowed down and correct fuel consumption according to Carlsson et al (2008)

Speed (kph)	Truck drivers' estimates	n	Engineering students' estimates	n	Psychology students' estimates	n	correct Carlsson et al
110	7.74**	10	7.41	20	7.52	20	7.5
100	7.51***	10	7.07	20	7.08	21	6.9
90	7.27***	10	6.74	22	6.55	21	6.4
80	7.00***	10	6.35*	21	6.14	21	5.9
70	6.91***	10	5.97	21	5.63	20	5.5
60	6.83***	10	5.75	21	5.22	20	5.2

Note: Estimates are mean values of the groups. Fuel consumption is expressed in liters/100km. Significant deviation of mean from correct speed *p=0.05, **p=0.01 and ***p=0.001.

2.2.2 Fuel Lost by Increasing Speed

The average judgments of fuel consumption of all three groups judging increasing speeds is shown in table 2. Corresponding statistical test results are also given in the table. The average judgments of both student groups are greater than the correct fuel consumption. They overestimate the amount of fuel lost when driving faster. Truck drivers made more accurate judgments of fuel lost than the students.

Table 2 Judgments of fuel consumption at increasing car speeds given by truck drivers, engineering and psychology students and the correct consumption

Speed (kph)	Truck drivers' estimates	n	Engineering students' estimates	n	Psychology students' estimates	n	correct Carlsson et al
70	5.78	11	7.21	21	6.34**	21	5.5
80	6.06	11	8.19	21	7.31*	21	5.9
90	6.59	11	9.19	21	8.32*	21	6.4
100	7.04	11	10.11*	21	9.34*	21	6.9
110	7.58	11	10.93*	21	10.35*	21	7.5
120	8.33	11	11.74*	21	12.04*	21	8.1

Note: Estimates are mean values of the groups. Fuel consumption is expressed in liters/100km. Significant deviation of mean from correct speed $*p=0.05$, $**p=0.01$ and $***p=0.001$.

Summing up, the professional truck drivers' judgments were significantly lower than correct values. Their judgments of increased fuel demands at increasing speeds were less biased and close to correct values.

The engineering and psychology student groups made accurate judgments of fuel consumed when speed was reduced. At increasing speeds, the psychology groups' judgments were significantly greater than correct values. The judgments of the engineering groups were also significantly greater when the speed was 40 km/h or more above the reference speed.

To sum up, the results show that truck drivers underestimated the savings of fuel consumed at decreasing speeds. However, when judging the amount of fuel lost by increasing speed their judgments were more accurate.

3   DISCUSSION AND CONCLUSIONS

The present study shows that professional truck drivers are biased in their judgments of fuel saved at decreasing speeds. They underestimate the fuel saving effect of a decreased speed. At increasing speeds, their judgments of the amount of fuel lost are more accurate. The two student groups were similar in their judgments. Their estimates differed from the truck drivers' in that they made more accurate judgments of the fuel consumption when speed was reduced. Hence, they were better at estimating the amount of fuel saved by decreasing speed.

It is important that drivers realize the magnitude of the fuel saving effect of lowering speeds. This study shows that the judgments of fuel saved at decreasing speeds are biased. It is important that eco driving system aids support drivers so that they realize the gain in fuel and the decrease in driving costs of keeping a lower speed.

ACKNOWLEDGMENTS

This study was supported by funds from the FFI (Strategic Vehicle Research and Innovation)-project EFESOS- Environmental Friendly efficient Enjoyable and Safety Optimized Systems and by funding from the Swedish Research Council to Ola Svenson. The authors want to thank Lars Eriksson, Magnus Hjälmdahl, Linda Renner, Jan Andersson, Martina Stål, Per Henriksson and Ulf Hammarström for advice and support.

REFERENCES

Carlsson, A., Hammarström, U., Karlsson, B., 2008. Fordonskostnader för vägplanering, VTI PM 2008-10-24.

CHAPTER 31

Merging Navigation and Anticipation Assistance for Fuel-saving

Christoph Rommerskirchen, Klaus Bengler

Technische Universität München - Institute of Ergonomics
Boltzmannstraße 15, 85747 Garching, Germany
rommerskirchen@tum.de

ABSTRACT

This work, which was performed as a part of the EC funded project eCoMove, describes the investigation of an advanced driver assistance system (ADAS) which helps to improve fuel consumption of automobiles by using a combination of anticipation and navigation assistance.

The anticipatory assistance system is based on the work of Popiv (2012). The aim of the system is to pursue the driver to extensively exploit the motor torque during deceleration phases instead of conventionally applying pressure on the brakes (Popiv, 2012). To this existing system two different types of navigation advices are included. This raises the complexity of the ADAS which can lead to new problems. The goal is to reduce fuel consumption in addition with high user acceptance by also using navigation instructions. To test the new system a driving simulator study was performed.

Keywords: driver assistance, fuel efficiency, driving simulator, anticipation, navigation, deceleration

1   INTRODUCTION

The main goal of the EU funded project eCoMove is to reduce fuel consumption by 20% in traffic e.g. by helping the driver to drive more efficient using cooperative systems. One important part of the project is to develop a so called ecoHMI for driver assistance. Due to the successful studies of among others Popiv (2012) it was decided to use the anticipation assistance system based on her work. The fuel reduction which is already shown with this system is the result of an extension of coasting phases in deceleration scenarios. This system is extended by a set of navigation instructions. This paper deals with the investigation of the effects of two different possibilities to integrate a navigation system in an anticipatory assistance system. The goal is to evaluate the possible fuel reduction and the influence of the system on the driver to see if it is possible to use such an assistance system for fuel reduction. Based on the results of a driving simulator study the problems and the benefits of this ADAS are being discussed.

2   OBJECTIVE

The objective of this study is to investigate if it is possible to combine the anticipatory ADAS in the instrument cluster described e.g. by Popiv (2012) with navigation instructions. Therefore two different variants of an ADAS were developed and tested in the driving simulator.

2.1   The assistance system

Popiv et al (2010) showed that "Not in every deceleration situation the driver's natural anticipation horizon suffices for the most efficient, comfortable, and safe action." And according to the work of Reichart et al (1998) between 15% and 20% fuel saving can be reached theoretically throughout a journey, when coasting. Therefore an a ADAS was developed to extent the driver's natural anticipation horizon in classical deceleration scenarios like e.g. upcoming speed limits. What has not been researched before is the aspect of the influence of two advices like deceleration and navigation at the same time.

The system is based on the information of the cooperative sensor and systems like Car2Car, Car2X and ecoMap information developed and improved by eCoMove. This leads to an electronic horizon which can be used for an anticipatory assistance HMI concept (Loewenau et al. 2010).

The anticipatory assistance system which only gives visual information on the instrument cluster consists of the following elements: The ego vehicle on the occupied lane is constantly displayed in the instrument cluster between speedometer and tachometer. The deceleration situation is imposed on the virtual road at the point of assistance activation, e.g. the speed limit. The color of the ego car gives the information when the driver should start to coast (green car) or when he needs to brake (orange). (Figure 1). (Popiv, 2012)

Figure 1 Anticipatory assistance system during an approach of a speed limit advice with coasting advice (green ego car)

Newly added are two different versions of navigation advices in this system. One version the so called integrated navigation (Figure 2 left in combination with speed limit) shows the advice in the same manner as the anticipatory advices. This has the advantage of one common design for navigation as well as anticipatory advices. Here the navigation instructions are used as an integrated part of the anticipatory ADAS. It should prolong the coasting phases in navigation advices. This combination increases the complexity of the system because of having two advices on the virtual road.

The other more conservative version shows the navigation advice separately below the anticipatory advice (Figure 2 right in combination with speed limit). Additionally there the distance to the intersection has to be shown.

Figure 2 Situation with speed limit approach and navigation instruction; Left: Integrated navigation instruction; Right: Separated navigation instruction

2.2 Main objectives

In the following the three main objectives of the study are described:

Fuel consumption: The main question is how much fuel consumption can be reduced by using the different systems and how it is reduced.

Subjective acceptance: How is the subjective acceptance, measured through a questionnaire, of the test persons?

Integrated or separated navigation assistance system: Which system should be preferred due to acceptance and fuel consumption?

3 EXPERIMENTAL DESIGN

In the following the description of study in the driving simulator, including software, hardware, test course and participants is provided.

3.1 Subjects

30 participants aged between 19 and 64 years (mean: 34 years, standard deviation: 12 years) took part in the experiment. The 18 male and 12 female drivers are all holding a valid category B European driver's license. Nine participants drive less than 5,000 km per year, 14 between 5,000 and 20,000 km and seven more than 20,000 km per year. The participants were recruited through notices posted in the local stores. For most of the participants it was the first study in a driving simulator. They received a comparison of 30€.

3.2 Driving simulator

The experiment was held at the fixed based driving simulator, located at the Institute of Ergonomics, Technische Universität München. The 6-channel fixed-based driving simulator consists of three projection screens for the front view with a 180° field of view. For the rear-view mirrors projection three additional projection screens are implemented.

302

Figure 3 Static driving simulator of the institute of Ergonomics (www.lfe.mw.tum.de)

The Mock-Up used at the Institute is a BMW 6 Series (E64) with automatic gear shift and full access to the CAN-Bus. Instead of the standard instrumental cluster the car is equipped with a free programmable TFT-display.

The simulation software used in the driving simulator is SILAB (www.wivw.de) which allows precise and flexible creation of driving situations as well as full control over simulated traffic. The software runs together with the driving dynamics model CarSim from Mechanical Simulation. Part of the driving dynamics is a simulation of the fuel consumption validated by Rommerskirchen et al (2011) for this driving simulator to calculate the fuel saving with the assistance system.

3.3   Test course

The test course the subjects had to drive in the simulator is about 15.5 km long and includes typical situations on rural roads (8.5 km), German highway (5.5 km) and urban scenarios (1.5 km). Ten different situations for the assistance system were tested (Table 1).

Table 1 List of tested situations

Road type	Situation
Rural road	Approaching a construction site
Rural road	Approaching a speed limit of 70 km/h and 50 km/h
Rural road	Approaching a speed limit and navigation advice to turn left or right
Rural road	Navigation advice to turn left

Rural road	Navigation advice at Roundabout
Rural road	Town entry
Highway	Approaching a speed limit of 120 km/h
Highway	Navigation advice to exit the highway
Urban	Navigation advice at traffic light intersection
Urban	Approaching a red traffic light

The course was designed after the German guidelines of road building. Particular attention was paid to have a course that gives a realistic feeling to the subjects. This includes a varying simulation environment as well as a realistic traffic simulation.

After an introduction and a practical drive in the simulator of about 20 min every subject had to do three runs, which were Baseline, a drive with separated navigation and integrated navigation system. The order of the runs for the different test persons was permutated. Additionally to the visual navigation system there were speech instructions for all the test runs, including Baseline is given. After the test the subjects had to fill in a questionnaire asking for their acceptance, their likes and dislikes and open questions about what should be improved. They also had to state which version of the system they prefer to use and why they decided for one system or another. The duration of the test including all drives and questionnaires was for each participant about 2h.

4 RESULTS

In the following the results of the three main objectives are described. The descriptive analysis of the data which was recorded by the driving simulator is done with the help of Matlab® and Excel®. Statistical analysis is done via the one-way repeated-measures analysis of variance (ANOVA) with post hoc comparisons using Bonferroni corrections.

4.1 Fuel consumption

It can be shown that in this test the average fuel consumption could be significantly reduced ($F(2,58) = 6.6$, $p < .05$) by up to 8%. The post hoc analysis showed that the integrated navigation was significantly better than the separated and the Baseline. But no significant difference could be detected between Baseline and separated navigation. Also no significant extension of the duration of the test drive could be observed ($F(2,58) = 1.6$, $p > .05$).

304

Table 2 Fuel consumption and duration of the whole test drive

	Baseline	Separated Navigation	Integrated Navigation
Average fuel consumption from all drivers	100% (SD: 23%)	96% (SD: 23%)	92% (SD: 17%)
Duration	10.3 min	10.5 min	10.5 min

As an example of how the fuel consumption is reduced the results of one typical situation is given. It consists of a combination of speed limit advice and navigation advice. The driver approaches a speed limit of 70 km/h 100 m before an intersection on a rural road, where the driver has to turn right (see Figure 2).

Figure 4 Simulated deceleration situation – Approaching speed limit and intersection

The coasting instruction in this situation is given 850m before the intersection (advices are shown in Figure 2). It was followed by a long coasting phase, starting 4 to 4.5s after the first appearance of the visual advice, whereas in the Baseline the subjects kept speed or even accelerated as shown in Figure 5. This led to a mean fuel reduction in this situation compared to the Baseline by 21% for the separated navigation system and to 29% for the integrated one. This is comparable to nearly all the other situations where a high fuel saving gain through coasting could be observed.

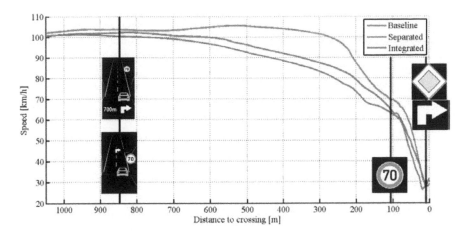

Figure 5 Speed and distance to the crossing of Baseline, with integrated and separated navigation advice

A difference in fuel consumption between the two versions of the system can be observed in situations where only navigation without any other anticipatory instruction is shown. There the integrated system has a significantly higher fuel reduction (Table 3). The subjects reacted earlier and therefore had a longer coasting phase. There are two possible explanations. It could be that it is easier to estimate the distance by showing the situation on the virtual road than by showing the distance in meters to the site. The other possible explanation is that in the integrated version the driver feels more secure to follow the system because of the integration it seems to be more helpful in fuel-saving.

Table 3 Fuel consumption in scenarios with navigation advice

Situation	Baseline	Separated Navigation	Integrated Navigation
Approaching a speed limit and navigation advice to turn left or right	100% (SD: 28.9%)	79% (SD:29.2%)	71% (SD:14.6%)
Navigation advice to turn left	100% (SD: 30.2%)	94% (SD: 17.7%)	90% (SD: 14.8%)
Navigation advice at Roundabout	100% (SD: 29.0%)	98% (SD: 45.6%)	77% (SD: 25.2%)
Navigation advice to exit the highway	100% (SD: 40.5%)	110% (SD: 56.3%)	74% (SD: 34.3%)
Navigation advice at traffic light intersection	100% (SD: 43.3%)	103% (SD: 24.6%)	95% (SD: 20.1%)

306

4.2   User Acceptance

The subjective user acceptance was measured through a questionnaire. On a scale from one to five most subjects stated that both systems are helpful or very helpful. Only one person said that the integrated system is slightly annoying. Also 60% of the subjects declared that they want to have this system in their car. Some people had problems to understand the color-coding of the egocar for the coasting advice, but even so they followed the advice given. Nobody seems to have a problem with more than one instruction shown at the same time on the display like it is done for example in Figure 2.

4.3   Preferred system

The subjects were asked in the questionnaire after the test about their preferred system. The result is very heterogeneous. Half of the subjects preferred the integrated system, the other half the separated one. The main reasons for the decision made by the subjects is shown on Table 4. The most discussed point is how to show the distance to the deceleration scenario. Some people really preferred to know the meters (separated navigation) and some said it is easier to estimate the distance by showing the situation on the virtual road (integrated navigation).

Table 4 Main reasons for the decision which system is preferred

Separated system preferred	Integrated system preferred
Meters are shown	the counting meters are too distracting
it seems to be more accurate	easier to estimate the distance
more familiar (it looks like existing navigation systems on the market)	looks nicer
bigger and clearer	more lively
	less exhaustive
	less precise

5   CONCLUSION

The results of this study show that it is possible to combine two advices like navigation and anticipation in the ADAS. The subjects do like and use the systems shown. Although the drivers have no preference for one particular system the integrated system should be used because of the better results of fuel saving. It could also be demonstrated that the fuel reduction of about 8% with the system, calculated in the validated driving simulator, is significant. It could not observed that the higher complexity of the combination of navigation and anticipatory advises lead to any problems for the subjects. But although the system itself is of a higher complexity the situations where the system was tested were less complex. The next step is to look whether the results still are the same in more complex scenarios like in an urban environments with much more traffic and pedestrians.

ACKNOWLEDGEMENTS

This study is part of the eCoMove project (Cooperative Mobility Systems and Services for Energy Efficiency) funded by the 7th framework program – information society technologies for clean and efficient mobility of the European Commission.

REFERENCES

Loewenau, J. et al. 2010. *eCoMove – EfficientDynamics Approach to Sustainable CO2 Reduction*. Paper presented at the 17th ITS World Congress, Busan, Korea, 2010

Popiv. D., Rakic, M., Bengler, K. 2010. *Driver's Natural Anticipation Horizon in Deceleration Situations,* Paper presented at AHFE International Conference, Miami, 2010

Popiv, D., 2012. *Enhancement of Driver Anticipation and Its Implications on Efficiency and Safety*, Dissertation, Technische Universität München, Munich, Germany, Institute of Ergonomics

Reichart, G. et al. 1998. *Potentials of BMW Driver Assistance to Improve Fuel Economy*. Proceedings of International Federation of Automotive Engineering Societies World Congress, Paris.

Rommerskirchen, C., Müller, T., Greier, M., Bengler, K., 2011. Validierung des Kraftstoffverbrauchs eines Fahrsimulators und Unterschiede im Fahrverhalten bezüglich des Kraftstoffverbrauchs. In: *Ergonomie im interdisziplinären Gestaltungsprozess*, eds. Grandt, M. & Schmerwits, S. 53. Fachausschusssitzung Anthropotechnik der Deutschen Gesellschaft für Luft- und Raumfahrttechnik Lilienthal-Oberth e.V. Neu-Isenburg, Germany: 187-198

Section IV

Integrated Transport Systems

Impact of an Integrated Transport System: An Ethnographic Case Study of the Experiences of Metrocable Users in Medellín, Colombia

Adriana Roa-Atkinson, Peter Atkinson, Jane Osmond, Andrée Woodcock,

Coventry University, UK
P.Atkinson@coventry.ac.uk

ABSTRACT

For those with impaired mobility, inclusive, accessible transport systems can represent a real change, transforming lives and livelihoods. Such systems not only meet accessibility requirements but also a code of conduct, which 'allows people with disabilities to access the world of equality, and promotes a behaviour based on principles of respect and solidarity among users' (World Bank 2009). During the last decade, governments in developing countries have been implementing integrated transport systems in their main cities, and in order to generate real social and environmental impact, these systems have included elements of transport, urban regeneration and social programmes. One such system – the Metrocable - can be found in Medellín, Colombia. This ethnographic study aims to contribute towards the understanding of the impact of transport design on inclusivity, through the experiences of the Metrocable's users and other local inhabitants.

Keywords: Transport Design, Inclusivity, Ergonomics, Metrocable, Sociology of Technology

1   INTRODUCTION

The pioneering Medellín Metrocable system is seen as a good example of sustainable integrated transport in Latin America (Green City Index 2010). First replicated in Caracas and then Rio de Janeiro, the success of this system as a viable transport alternative in low-income areas and hilly geographical conditions has been recognised nationally and internationally, not least for its suitability for integration with other transport options (ITDP 2012).

The integration of mass transport projects with other municipal policies through coordination programs such as PUI (Integral Urban Project) has gained public recognition as a means of alleviating social problems (Reitman, 2011; Fukuyama, 2011). Thus the Metrocable has acquired its reputation as a driver for social change. EDU (2011) defines a PUI as:

An intervention model which takes into account social and physical aspects, as well as, the interinstitutional coordination seeking the integration of resources from different sources, in order to perform an urban intervention of profound impact and which can transform a territory or a zone with specific problems.

Canon-Rubiano (2010) raised concern about the lack of primary research and the need to evaluate further the social impact of the Metrocable project on the local population combining elements of social inclusion, mobility and transport. Looking at these elements, Blanco and Kobayashi (2009) pointed out the important role that the approach of the PUI has played within the areas where the first metro cable system was developed. The 'holistic' approach of PUI, coordinating investment in transport infrastructure with investment in housing, environment and public spaces was based on community oriented planning with the active involvement of diverse institutions. Thus it started 'a radical transformation on internal and external perception within the area' (IBID). Intangible impacts and lack of information from informal sectors create challenges to the measurement of impact. Ongoing research by Brand and Davila (2011) looking at the PUI 'social urbanism' approach suggests that the social impact of specific projects is still unclear, especially on age, gender equality and disability access.

This ethnographic study contributes to the understanding of the impact of transport design on inclusivity, through the experiences of the Metrocable's users and other local inhabitants.

2   METHODOLOGY

The fieldwork took place in Medellín in August 2011. Neighbours and users of the system were interviewed in the Park Libraries: Spain (Santo Domingo) and San Javier. Additionally, Metrocable Line L (ARVI Station) was included in order to capture both the socioeconomic impact of tourism on the local population and perceptions from the surrounding sectors.

Specifically, interviews and/or focus groups took place with communities in San Javier and Santo Domingo, neighbourhood women and men, tourists, a transport operator line manager, customer service staff (including those with disabilities) a transport security member, disabled users, organised groups and social workers.

In addition, taxi drivers were interviewed as were Line L market stall owners, and observation was undertaken in the streets and substations of the metro system, with a particular focus on mobility problems.

3   FINDINGS

The findings from the qualitative data gathering process have been grouped into the following themes: Change of Behaviour (Metroculture); Socioeconomic impact; Passenger Experience and Transport Integration and Inclusivity.

3.1   Change of Behaviour and Metroculture

The Metro transport system was planned and designed between 1979 and 1984, and began operation in 1995. The subsequent development in 2004 of the Metrocable, conceived as a method of connecting Medellín's least developed suburbs to the wider metro system, was launched with Line K operating from 2004, and Line J from 2008. A long campaign ran (from 1988) prior to the operation of the first Metro trains, to gain 'buy-in' and acceptance by the local population. Since then the institution has focused on building infrastructure and social networks to develop projects in more isolated areas and to connect them with the Metro system in conjunction with planned policies of the Municipality (Metro, 2011).

The campaign aimed to educate and train citizens in the use of a mass transport system and behavior appropriate within it, and was implemented by the metro operator with the aim of building a social, educational and cultural model. Using this 'public good' ethos, now known as 'Metroculture' (Cultura Metro), the information was 'cascade-disseminated' to groups and organisations from different sectors and beliefs (Metro 2010a; Metro 2011). The campaign included straplines such as: 'Love our big project and love the project from now on' (1988-1992); 'We have seen it being born, growing up and taking the first steps' (1994-1995); 'Let us take you' (2004); 'Bring metro culture to our city' (2006) and 'Leaving others to go out makes for an easier entrance' (2009-2010) (Metro 2010b). The results of this campaign are now embedded into the community:

> *Although it has been operating all these years, the metroculture has persisted. All the users look after it. There has always been a sensitisation process directed at children as the market of future users. Metro 'small friends' (sic) was a programme which taught the children what the Metro did and how to behave through puppets and a variety of games. (Neighbour and worker, Santo Domingo)*

3.2   Socioeconomic Impact

After implementing Line K, Medellín created the PUI in order to improve quality of life in the poorest neighbourhoods, focusing on housing improvements, public equipment and facilities, and public space and mobility (Blanco and Kobayashi 2009; Medellín's Mayor's Office, 2006; Metro 2011). After the positive impact of the PUI in Santo Domingo, it is now being both considered and introduced in other areas of the city.

> *The metrocables are very important, for the development of the area, for the environment, for the security, for culture – confronting the perception that people have about the areas where these developments have reached. The metrocable arrived in comuna 1 and 2 and then the PUI. Now there are improvements to the public spaces, the streets. Then other facilities came, such as the libraries. You will see that in the areas with metrocables these days the children play with computers... before, they played with pistols. (Focus Group, Disabled - FGW)*

Jobs and economic opportunities

For some community members, past experience was one of limited access to jobs and education. However, the attendant recreation and social programs included in the Metrocable development from the planning and operational stages have succeeded in involving the population in the project. Therefore, one of the first direct impacts of the Metrocable was access to employment that included benefits and rights. These job opportunities triggered the active participation of community members when family members began to work in the construction sector on infrastructure projects, and then other members of the community began to access more highly-skilled jobs.

Opportunities and quality spaces provided by the metro stations have also enabled the development of small businesses, particularly for some female interviewees. For example, the support and training offered by CEDEZO (local development business centre), and city council programs to encourage small business development have generated regular income for families.

> *Thanks to metrocable... I opened a small shop to sell candies and refreshments to the locals and tourists. Now the products are delivered to the shop and I don't need to go to the city to buy the products. (Focus Group Women – Santo Domingo)*

Housing and improvements

Despite some concerns about the loss of close neighbours in the Line K surroundings, the Metrocable is seen as a bonus in increasing the value of local housing stock due to home improvement loans offered to enhance the general 'look' of the surrounding area. For those who moved due to need for construction space, there were new improved living spaces in public housing programs on Line J.

> *Nowadays we have a lot of housing projects which the Medellín Council and other organisations ('vivienda popular') created... a lot of people were re-located, for example those from Vallejuelos should be happy because they left their shacks because it was a zone at high risk. (Neighbour, Juan XXIII, San Javier line)*

Schools and new infrastructure projects

After the implementation of Line K, complementary infrastructures were developed with community involvement to improve neighborhood education: the

Park Library Spain was one of the most popular, as was the bridge linking the Metro station with Park Library San Javier, also new schools were mentioned on Line J. Respondents confirmed that they saw an increase of scholar attendance, motivation for learning and a change of behavior towards public services in general.

We people of 'comuna 13' feel proud of living here, they are improving the look of the sector, delivering more schools, parks, open spaces and we have participated in saying what we want, presenting our needs to the council and then checking that things were done to our requirements. (Neighbour, San Javier line)

The perception of library workers is that the library services are more popular with children and youngsters. Both libraries work as part of a network with other libraries in the city, running programs for the elderly, women, extended families and promoting the city's cultural programs.

Before the Metrocable and the library the lead used to fly. The children (now) spend their time in the library...now all the world comes and takes away a very good image of this area...before it was a huge war, now children have a lot of recreation and can play outside until late. (Neighbour, Santo Domingo)

Since the opening of the libraries the strategy has been to work with the Metro operator to extend the Metroculture to expand the culture for reading through programmes such as "Bibliometro" and "rolling words" in the transport system.

3.3   Passenger experience

In general, local users are now satisfied with public transport. In the Line K sector, respondents commented that before the Metrocable, reliance on the relatively few public transport services could be problematic. Overcrowding often resulted in 'being squashed' or left behind. In terms of comfort, the Metrocable is seen as an 'elegant travel experience' because the gondolas are comfortable and made 'you feel all the time as if you are a tourist'.

Metrocable is a good way to start the day. The waiting time in the peak hours doesn't matter to me because the journey is very enjoyable and the waiting time is manageable with the option offered by the system – books, metro-journals. (Neighbour, Sto. Domingo)

Due to the family environment ethos, the gondolas are kept spotlessly clean because 'all of us - the users - look after these spaces'. Some concerns were mentioned, particularly the advertisements on the gondolas, because they make respondents 'feel dizzy' or disturbed enjoyment of the view. Respondents would also like to see more peak-hour gondolas to alleviate time-consuming queuing.

Although the Metrocable was initially conceived to improve public transport for the local population, J and K lines have also created new tourism potential. Tourists interviewed mentioned the views offered from the gondolas in particular and the ability to travel without extra cost using a single Metro ticket. The tourists also found that the attendants and locals, although initially shy, were friendly. Metrocable users commented that they like to see and travel with tourists for the opportunity to speak positively about the improvements that have taken place.

Additionally, the possibility to extend journeys without getting out at any of the

316

stations was mentioned. Improvements cited were more maps of the locality around the stations: this was echoed by some attendants who consider this as a way to improve customer information in general.

The implementation of maps specifically for each station, where the person arriving can get without charge and find their way quickly in this space...it is easier for people to ask rather than follow the signage available in the transport system. (Metro Attendant)

3.4   Transport integration and inclusivity

Metro

The Metro began operation in 1995 but at that time there were no inclusive access mechanisms for those with reduced mobility. Disabled users commented that they were often physically carried into the system by attendants. After a process of negotiation, access mechanisms were included and the first retro-fitted solution was introduced in 2003 in San Javier Station. With finance from the City Mayor's office and the Metro operator, nearly 40% of the metro stations were fitted with lifts, and in 2009, over 3000 passengers with reduced mobility and visual impairments made use of lifts, stair-lifts and other mechanisms (Metro, 2010c). Today, 96% of stations have accessibility options and, via a regional scheme to integrate disabled people into the workplace, ten people with disabilities are employed to help passengers and to raise awareness of facilities for disabled users and the correct use of the transport system by all (Nuestro Metro, October 2011).

There are still some accessibility issues: e.g. the stair-lifts do not provide the 100% comfort, security and speed principles touted by the system (Figure 1):

These improvements are of a modest nature but these are not comfortable and don't comply with the three principles that the Metro system proclaims: 'safety, speed and comfort'. These platforms are not fast, leave you feeling insecure and are not comfortable. When people get onto stair-lifts everyone looks at you as if you're different...it's exclusive because it is not used by everyone who needs it, because the person with the dodgy knee doesn't use it, the pregnant woman doesn't use it, the elderly are afraid to step onto it. It takes up to 3 or 4 minutes and in some instances up to 7 minutes. In that amount of time you should have got a lot further. (FGD)

Figure 1: wheelchair user on electromechanical platform (stair-lift) and using footbridge approach to Metro station Industriales (photograph by: Peter Atkinson)

Nuestro Metro magazine (January 2012) outlines company investment in modifications to improve disability access, including modified station entrances and tactile, high-visibility flooring surface to help guide people with visual impairments.

A recent addition is the 'isquiatic' support which allows users to adopt a semi-seated position. This was designed in collaboration with local universities using universal design principals. People with reduced mobility have priority, and users found that it helps when waiting for a train in a congested station, but that there is a lack of them in Metrocable stations.

Metrocable

According to passengers with mobility problems, the mobility and comfort offered by the Metrocable is good. Typically they do not need to queue and receive personalised attention when needed. The attendants and/or policemen help when they enter the station and, although they can travel alone, an attendant can also accompany them on request.

However, there are difficulties when entering and exiting Metrocable stations due to the immediate surroundings which feature streets and paths with sharp inclines and curbs, and steep flights of steps. In addition, despite the improvements some users with mobility problems are still not 100% comfortable when travelling.

When the gondolas are in operation one or two station attendants are present to control the number of passengers waiting to board – typically 8 to 10 (Figure 2). People with mobility problems and the elderly feel they need more time to board than is offered at present.

The gondola seating takes the form of two facing benches which can be folded up to accommodate a wheelchair: in this case only two passengers may board at the same time. When the seats are down, the space between seats is comfortable and features plenty of leg-room, but there is insufficient space for large luggage items.

Figure 2: Passengers boarding gondolas at San Javier Station. (photograph by Peter Atkinson)

The sub-stations do not present too many accessibility problems because the mechanisms were put in place from the beginning of the Metrocable project and were then fine-tuned based on the experiences of disabled users. Additionally, there is always either an attendant or a policeman to help those with mobility problems in and out of lifts, to exit the station, or to help the transfer into relevant metro stations.

Taxis

For taxi drivers, the perception is that despite the Metrocable having created competition, it has reduced congestion. There is still a big demand for short journeys between pick-up points to the closest metro, and although they found that some passengers still request long journeys, they advise them to use the Metro system to save time. In general, the drivers did not object to carrying disabled users, but some preferred accompanied passengers: the female drivers particularly mentioned a preference for a co-pilot because of the physical constraints of lifting wheelchairs into cars, for example.

Buses

Despite the presence of the System of Integrated Transport (SIT), which provides a feeder bus service to Metro stations, access for disabled users is not easy. Even the relatively new buses are boarded through a narrow entrance door via steep steps and therefore wheelchair access is problematic.

> *Here we say we have an integrated mass-transportation system, but a disabled person can arrive at any of the substations with 'integrated bus routes' (SIT) and that's where the integration stops! Because the buses are not accessible. (Focus Group, Disabled)*

Other local bus services provide a more precarious, if regular, service, but many do not integrate directly with the Metro system although they may stop nearby. They will sometimes pass by the disabled. Buses also tend to stop outside safe stopping points: this is particularly problematic for visually impaired users who cannot anticipate this:

> *This should be improved, drivers should be trained because they will stop anywhere and as someone without sight I cannot just get off in the middle of the road. (Library worker with severe visual impairment)*

More facilities for those with mobility problems have been designed into a new Metroplus (Bus Rapid Transit) system. Line 1 began operation in December 2011 (Nuestro Metro, December 2011). It integrates directly with the Metro network. Evaluation of its initial operation has led to some concern about the difficulties experienced by those with limited mobility and people travelling with children. (Nuestro Metro, January 2012). Despite this, there do appear to be benefits: Nuestro Metro reported on the experiences of a female wheelchair user who can now attend medical appointments near the Hospital Metro station thanks to Metroplus.

4   CONCLUSION

The interviews with some of the most disadvantaged communities in Medellín revealed the positive impacts of the Metrocable development and its integration with other municipal interventions such as PUI. This positive impact was enhanced through the extensive pre-launch campaigns aimed at the local population, through what is now known as 'Metroculture'. These campaigns were underpinned with an educational ethos about treating the system with respect, an ethos that is passed onto the young people of Medellín.

In terms of the socioeconomic impact, there is evidence that access to manual and skilled jobs has improved the economic status of the local population, as has access to housing loans to improve housing stock. In addition, the link to a fast and efficient transport system coupled with funding schemes for small businesses has enabled new development. Further, the building of an infrastructure around the Metrocable system – for example, schools and cultural spaces - has also improved the life choices of the local population.

On a more practical level, the passenger experience reveals that the Metrocable is comfortable, clean and offers a pleasant transport journey. Ongoing provisions have been made for those with mobility problems, with lifts, stair-lifts and seating spaces, underpinned with physical aid from attendants throughout the system.

However, despite recognition of the Metrocable as one of the most user-friendly transport modes, access to it can be limited for the elderly and those with mobility problems especially at transport gateways. There are insufficient lifts and escalators at station entrances so access requires negotiation up steep flights of stairs. Also, improvements are needed to the whole journey experience, as in some cases the main barrier of access to transport is the journey from the house to the transport gateway. There is still a high reliance on the local taxi services which can be expensive and do not always have space to deal with a wheelchair, for example. In addition the local bus services are not access friendly and tend to stop at random points, which cause difficulties for those with mobility and visual impairments.

This speaks to all aspects of inclusivity, not just in public transport, but also the accessibility of the streets and built environment surrounding the Metrocable: improvements need to be made at street level for those with reduced mobility.

Despite the problems outlined above, there is no doubt that the Metrocable project has provided a successful and pleasant transport solution for the populations of San Javier and Santo Domingo. Furthermore, the Metrocable system has received public acknowledgement as an articulator and facilitator of social projects, development and improvement to the infrastructure and environment of the region. This would not have been possible without the coordinated and planned policies of the Municipality, and strategic alliances with private and public institutions from different sectors (i.e. education, industry, health, recreation and culture).

From a transport ergonomics perspective this study clearly shows the wide scale, social and economic benefits that can be accrued by transport intervention, and the continued need to address the requirements of the whole journey experience of the excluded.

5   ACKNOWLEDGMENTS

Our thanks to members of the public, the staff of Comite de Rehabilitacion, the staff and users of the Park Libraries; to Empresa Metro and to the GRID research group from Universidad EAFIT. Project funded through Integrated Transport and Logistics Grand Challenge, Coventry University.

REFERENCES

Blanco, C. and Kobayashi, H. 2009. Urban Transformation in slum districts through public space generation and cable transformation at northeastern area: Medellin, Colombia. *The Journal of International Social Research*, V. 2/8, Summer, pp 75-90

Brand, P & Davila, J. 2011. Mobility innovation at the urban margins. *City: analysis of urban trends, culture, theory, policy, action*. V.15, Issue 6, December. pp 647-661.

Canon, Rubiano, L. 2010. *Transport and Social Exclusion in Medellín. Potential, Opportunities and Challenges*. Accessed 19 February 2012. http://www.bartlett.ucl.ac.uk/search/search.cgi?query=canon&collection=ucl-bartlett&school=dpu

EDU. 2011. Empresa de Desarrollo Urbano. *Municipio de Medellín*. Accessed November 2011. http://www.edu.gov.co

Fukuyana, F and Colby, S. 2011. Half a miracle: Medellín rebirth is nothing short of astonishing. But have the drug lords really been vanquished? *Foreign Policy*. http://www.foreignpolicy.com/articles/2011/04/25/half_a_miracle

Green City Index, 2010. *Assesing the environment performance of Latin America major cities. A research project conducted by the Economists Intelligence Unit.* Accessed 19 February 2012. http://www.siemens.com/press/pool/de/events/corporate/2010-11-lam/Study-Latin-American-Green-City-Index.pdf

ITDP 2012. *San Francisco and Medellín win 2012 Sustainable Transport Award*. Accessed 9 February 2012. http://www.sutp.org/index.php

Medellín Mayor's Office. 2005 "Proyecto Urbano Integral Nororiental". BID agreement number 4800000830, year 2005. Accessed 19 February 2012. http://edu.gov.co/pdf/sistematizacion_metodologia_PUI_web.pdf

Metro 2010a. *Cultura metro, nuevas experiencias en el trabajo comunitario*, Issue 2 http://www.metrodemedellin.gov.co/index.php?option=com_flippingbook&view=book&id=51&lang=es Accessed 19 February 2012.

Metro 2010b. *Cartilla Cívica: Mi patria desde el Metro*. Numero 2.

Metro 2010c. Revista Nuestro METRO 15 años. Accessed 19 February 2012. http://www.metrodemedellin.gov.co/index.php?option=com_flippingbook&view=book&id=40&lang=esNuestro

Metro. 2011. *Metrocable: An axes of development for the Valle de Aburrá. Magazine to promote awareness of mass transit systems*. Issue 3. Accessed 19 February 2012. http://www.metrodemedellin.gov.co/index.php?option=com_flippingbook&view=book&id=50&lang=es

Nuestro Metro, October 2011. Periódico Nuestro METRO, Issue 111. Accessed 3 Feb 2012. http://www.metrodemedellin.gov.co/index.php?option=com_content&view=article&id=97&lang=es

Nuestro Metro, December 2011. Periódico Nuestro METRO, Issue 114. Accessed 3 Feb2012.

Nuestro Metro, January 2012. Periódico Nuestro METRO, Issue 115. Accessed 3 Feb 2012.

Reitman, A. 2011. *Social Urbanism*. The Report Company. Accessed 13 Feb 2012. http://the-report.net/features-archive/26-g-colombia/184-development-g-social-urbanism

World Bank, 2009. *Colombian Transportation project on a fast track: Transport on a Human Scale*. Accessed February 19 2012. http://siteresources.worldbank.org. p. 59

CHAPTER 33

The Multi-faceted Public Transport Problems Revealed by a Small Scale Transport Study

Andree Woodcock, Slobodan Topolavic, Jane Osmond

Integrated Transport and Logistics
Coventry School of Art and Design, Coventry University, UK
A.Woodcock@coventry.ac.uk

ABSTRACT

An evaluation was made of a weeklong demonstration trial of an electric bus operating in a small, historic town in the UK, to examine the extent to which the bus fulfilled the needs of sustainable transport provision.

The town has a steady stream of tourists who come in to the city centre on diverse forms of transport to visit tourist attractions and attend the theatre. The interaction between the various vehicular and pedestrian traffic is chaotic. Associated problems such as high levels of congestion, pollution and inconvenience have led some people to stop visiting the centre altogether. A survey was undertaken to gather insights into different facets of the transport problem and to determine the acceptability and usefulness of an electric Park and Ride (PnR) bus. In terms of the ergonomics of the bus, although passengers liked the new design and rated it more highly in terms of ride quality. Interviews and observations of the bus drivers revealed design problems with the layout and operational issues. The 'near silence' of the bus was a continual worry to the driver, requiring greater vigilance.

The study concluded with a set of recommendations for the transport provider, the council and vehicle designer. The hexagon spindle model of ergonomics (Benedyk, Woodcock and Harder 2009) is used to explain the findings and take a holistic approach to evaluation.

Keywords: ecosystem, concept mapping, coral reef

1   INTRODUCTION

In this doctoral thesis on sustainable transport, Folkesson (2008, p29) argued that improvements should 'meet the demands of society in general and bus passengers in particular. The vehicles should provide high comfort with little vibrations, low noise, smooth drive and a minimum of exhaust pollutants. They should be fuelled with renewable fuels and still be as energy efficient as possible. Vehicles should also be and feel safe. And not to underestimate, the exterior and interior design of vehicles should be attractive and the interior layout should fit the travellers need.'

Sustainability and sustainable development are part of the ongoing debate about global warming and other environmental problems. Today's transport solutions are nonsustainable on a number of dimensions - CO_2, climate change and oil dependency, traffic congestion and road space, exhaust emissions, noise, as well as in terms of human factors. Litman and Burwell (2006), argue the need for total transport planning that takes in to account the perspectives and preferences of different stakeholders, such as: pedestrians, residents, aesthetics and environmental quality as well as the needs of commuters. Although the creation of sustainable urban transport systems should begin with urban planning and the embedding of a green infrastructure (Wolf, 2003), improvements can be made to existing public transport at a reasonable cost.

With this in mind, an evaluation was made of a weeklong trial of an electric bus operating in a small, historic town in the UK, which is a popular tourist destination. The trial was conducted at the request of the Park and Ride (PnR) service operator to inform their decision to use electric vehicles on the service. The bus ran as part of a scheduled 'Park and Ride service' transporting car occupants from an out of town car park to destinations within the city.

The study shared many attributes of similar studies which have used urban buses as high profile technology demonstrators, to raise awareness about sustainability, new forms of energy, gather attitudes of the public towards specific issues about public transport and people's willingness to adapt their travel behaviour. In particular the aims were to ascertain the opinions of different stakeholders towards traffic problems and evaluate the driver and passenger experience of the proposed electric bus.

A range of methods was used including
1. One to one semi - structured interviews with drivers to consider issues which may have arisen specifically because of the use of the e-bus
2. Semi structured interviews to ascertain attitudes towards the use of electric buses, with:
    a. Passengers on the park and ride buses
    b. Pedestrians/residents of Stratford
    c. Retailers
    d. Tourists

3. Observations of journeys on both electric and diesel Park and Ride buses to discover any usability problems or issues on the journey which might be related to vehicle design or fuel type.
4. Ergonomic audit of the vehicle cab

The city in which the study took place is one of the most popular tourist destinations in the UK. It is an area of historic and cultural interest, with small narrow streets. The study was conducted over 6 days to coincide with the weeklong trial of the electric bus. The electric bus was designed to make a statement, advertising its presence with low energy logos on a white livery. Sampling was opportunistic bearing in mind the need to include participants from a wide range of demographics.

2   RESULTS

This section presents an overview of the results from the surveys, followed by the observational study and the ergonomics audit. As these have been presented previously (e.g. Woodcock and Topolavic, 2010), only an overview is presented here, with the emphasis being placed on the usefulness of ergonomics models to explain the interactions an direct multi stakeholder investigations. In line with the Hexagon Spindle model (Benedyk, Woodcock and Harder, 2009) model, which in a similar way to most models put the user at the heart of the analysis, the results will be presented in terms of the effects of the transport system on different user groups
Results were based on over 300 interviews – 167 with passengers (134 on the electric bus), 108 pedestrians (visitors and residents) and 41 retailers – and 150 independent observations of approximately 70 bus journeys, 116 of which were on the electric Park and Ride (PnR) bus. Although the city has a conservative, senior profile, it attracts many younger visitors.

2.1   Pedestrians and tourists

The city attracts international visitors who are sophisticated and have high levels of expectations, as well as bus trips of school children. Pedestrian flows in the narrow streets are hindered by tourists stopping to take photographs, large vehicles mounting the pavements to make tight curves; long queues for public and tourist transport which clog up the streets; few areas have been pedestrianised which means that pedestrians have to be constantly aware of traffic at closely spaced intersections. The central bus station has been removed, thereby removing a central place for coaches to wait for their passengers.

2.2   City users

The lack of an integrated approach to transport was seen as a contributory factor for the demise of retail outlets. Pedestrians and vehicle owners expressed dissatisfaction with access to the centre; the train station was not in the centre of town and the train did not easily connect to other urban areas; the city centre and route to it were congested, not only by visitors to the town, but through cars which passed through, or which circulated round the town searching for parking spaces; parking provision was expensive and poor; metered parking restricted the length of time vehicles could remain in a parking bay (in some cases one hour) which meant that potential customers were continually anxious and could not spend time long periods of time in the city (e.g. for hairdressing and lunch engagements); public transport was not highly regarded, with many mentioning poor services and empty buses idling in the streets; the 'Park and Ride' scheme although praised by its users, was situated to one side of the town, necessitating a drive through the town to use it, and had restricted hours of operation limiting its use for those visiting the theatre, restaurants or working late. Many people therefore simply chose to visit nearby towns with better access.

2.3 Retailers

Prior to the study, retailers had been very vocal about the effects of poor transport planning on their businesses. They were the group most concerned about the transport problems in the city (followed by the PnR passengers). In the interviews many of the shop owners referred to the city centre as dying, and the reduced footfall which had led to a number of closures. They cited direct experience of cancelled bookings (because people were in traffic jams), loss of business attributed to traffic regulations, congestion in doorways caused by bus queues and vehicular pollution – fumes entering shops, high levels of noise and particles settling on outside dining areas.

2.4 Drivers of private vehicles

Vehicle drivers were frustrated by the transport situation. Many simply avoided driving into the centre at certain times, or went to other regional centres. They expressed frustration and anxiety directly related to their journeys; congestion, missed appointments, needing to allow extra time for journeys at certain times of day, vehicles cutting in front of the etc. Few (of any of the respondent groups) expressed an unprompted concern over pollution (less than 5%). For a slightly more elderly and in higher socio- economic demographics, public transport, regarded as a markedly inferior service, was not an option.

2.5 Park and ride (PnR) passengers

Only a small percentage of the population used the Park and Ride service. Those who did tended to be older (as they could travel to the centre free of charge) and had a higher level of concern over pollution. Passengers rated the electric bus more favourably on all measures: ride quality, noise, vibration, overall speed, acceleration/braking and comfort. On a score of 1 to 7 (with 7 being the highest), the average score for the diesel bus was 5.6 the electric bus 6.2. This was further confirmed by the observations on the bus. Some passengers, notably in the rear of the bus mentioned that they found the new whining sounds disturbing.

2.6   Electric bus driver's perspective

Interviews were held before and after the trial with the bus drivers, and observations made of the operation of the vehicle. The overall design of the instrument panel and the driver work station was better when compared to the older diesel bus. The electric bus had very little noise and no drive train vibrations or harshness, thereby improving the driver workstation.

The initial worries drivers had about the length of charge were realised. Problems with charging (relating to the size of generator and lack of feedback about charging success) meant that the bus could not be used for one day, and was taken out of service early on other. The instruments provided did not give enough information as to range left, effect of use of on-board electrical systems (e.g. lights, demister) and number of passengers on energy consumption. There was some help in the form of an economy gauge on longer driving sections, but this was not useful for urban driving.

Unforeseen operational issues emerged. Firstly, the electric bus 'cut out' a number of times. This seemed to be when the driver applied the brakes. The need to restart the bus held up traffic, alarmed passengers (by the beeping noises) and was a cause of considerable anxiety to the driver. As a consequence drivers were observed to adapt their driving style, by reducing the need to brake (e.g. through going through traffic lights). After repeated stoppages, one driver took the bus out of service. Secondly, the bus rolled back during hill starts, thereby putting pedestrians walking behind the bus in danger. This is not possible with diesel buses.

The major issue reported by all drivers was concern over the 'silent running' of the bus. Pedestrian awareness of traffic in general was a problem for all drivers; exacerbated by over spilling of pedestrians on to pavements, those unfamiliar with transport layout, English driving system and visitors stopping to take photographs. All drivers had to be extra vigilant driving the electric bus, as they believed that the pedestrians, used to relying on auditory channels would not be able to hear it. However, all drivers remained enthusiastic about the move to electric vehicles and were advocates of the use of sustainable fuels and were willing to promote this to their passengers.

2.7   Potential transport solutions

Each of the groups was asked to rank order proposed transport solutions, as shown in Figure 1. The most preferred solution for each group was given a score or 8, the least preferred method was scored 1. The figure clearly shows the differences between the stakeholder groups, for example visitors rated more bicycle paths more highly, the residents preferred public transport fare reduction, whereas the retailers and PnR passengers preferred an extended PnR service. Of the proposed measures traffic calming was the least preferred, as this would do nothing to restrict the amount of traffic coming into the centre. Car sharing was regarded as too difficult to operate; and cycling schemes may exacerbate the transport problems as there was a lack of room for cycle paths.

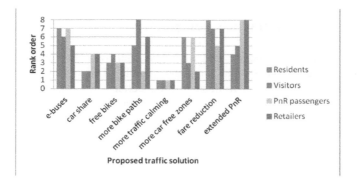

Figure 1 Preferences for different transport solutions

If the overall totals for each of the methods are considered, for all stakeholders, the proposed solutions were rated as follows (from most to least preferred): extended PnR, fare reduction, e-buses, more cycle paths, more car free zones, free bikes, car sharing, traffic calming. However, it may be assumed that the score for e-buses was inflated because of the focus of the study.

The poor service provided by public transport was seen as offering little incentive for potential travellers to switch modes of transport. Many of the respondents would have liked to have used public transport if it formed a viable alternative to private transport. It was criticised in terms of the frequency, cost, reliability and scheduling of the services, the failure to provide good road and rail connections to other cities, buses which ran empty yet congested the centre. The most frequent criticism concerned the lack of a bus/coach station. Without this, tourists did not have a focal or gathering point, buses and coaches congested the streets when they were waiting for passengers, or to commence their service, frequently with their engines running – thereby contributing to noise and pollution. Given that many of residents were used to a certain level of independence and luxury in travel – they were seen as being unlikely to want to switch to a service that was so markedly inferior to their preferred mode of transport. As such, parking restrictions were not seen as an effective deterrent. If people were concerned about the parking and congestion they avoided the town and took their business elsewhere.

3.   THE HEXAGON –SPINDLE MODEL APPLIED TO TRANSPORT RESEARCH

3.1   Description of the model

The hexagon spindle model of ergonomics (Benedyk et al, 2009) is an adaptation of the enhanced concentric rings model (Girling and Birnbaum, 1988). It sought to refresh outdated terminology, create differentiation and clarity between the sectors and to acknowledge temporality. Although developed in the context of learner centred educational ergonomics (Woodcock et al, 2009), it was intended to be applicable to other domains, in this case transport research. Its application to the transport research is shown in the following figures.

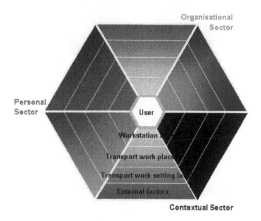

Figure 2 Hexagon applied to transport interactions

Models of ergonomics take as their starting point the user and represent the factors which influence the successful completion of the task in layers. For example, the main task of a student is to learn – this can be impeded of enhanced by many external factors, such as the design of the classroom, teaching materials and national agendas. Translating this to transport design, the user becomes either the driver or the transport service user, who may have a superordinate goal of reaching their destination (as safely, comfortably or conveniently as possible), and embark on a series of tasks to enable the completion of this goal. This can be influenced by factors such as the design of the vehicle, the transport infrastructure, behaviour of other passengers upto the level of investment placed in the transport system. The factors which need to be considered for design and evaluation of interactions are shown in Figure 3a. 6 factors are differentiated into 3 sectors – personal, organizational and contextual.

Figure 3b recognises the need to not just have one model to represent the whole journey or even a task. Task settings, products used to meet task needs and contexts change. Therefore the representation of any complex task requires the optimisation of a number of hexagons, placed for convenience along a central spindle. For example, in a transport environment, such tasks may be planning a journey using a transport information system, buying a ticket at a station or sitting in a vehicle. Although such tasks have the same individual in common, they may occur using different products, in different contexts and at different times. To represent this, hexagons arranged as plates along a time spindle are used. Each hexagon can represent a different stage of the journey.

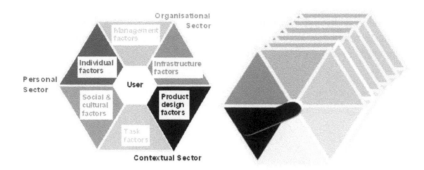

Figure 3a Sections in the model; figure 3b representation as a series of hexagons on a spindle

3.2   Aspects of the model applied to the case study

The aim of transport design should be optimize the user experience through understanding the factors that affect each stage of journey. The different sectors and factors help in the systematic identification of key issues. By using the Hexagon Spindle model, the quality of the user experience can be broken down into discrete stages and each factor examined. If the quality of the entire user experience (or the prediction of it by the user)is negative, it may outweigh the enjoyment of the visit itself, in which case it will be cancelled.

This is clearly demonstrated in the case study presented. If the journey is broken down in to basic stages – e.g. preparation, journey to destination, arrival, and return journey- in some cases the overall unpleasantness of the sum of the two journeys outweighs the benefits accrued at the destination. In such cases, journeys are made elsewhere. Looking at the sectors which contribute to the negative weightings for car users, most of the unfavourable comments are levied against the organizational /infrastructure sectors in terms of the work setting (e.g. road congestion and delays) and external factors (e.g. parking restrictions).

Clearly different user groups will be influenced by different factors. For the hexagon–spindle model, this at present requires the development of different models for each user group to represent the factors influencing the interaction with the transport system for drivers, pedestrians, users of public transport, tourists etc. This is certainly a complication, but not insurmountable. The differences between the groups, explains the different transport preferences and the way in which they rated the severity of the transport system. For example, tourists may have experienced much worse traffic problems in other cities and therefore scored the transport problems as being not so significant; they may also have felt that the experience at the destination outweighed the journey difficulties; for example, a tourist bus which waits directly outside a major department store for them, rather than having to walk to, and locate their coach at a central depot, was an ideal solution.

4. CONCLUSIONS AND RECOMMENDATIONS

4.1   From the case study

The evaluation of the PnR bus service was conducted from a multistakeholder perspective. This is required for the evaluation of all transport services. No transport service can be developed and evaluated in isolation. This makes real life trials so important. In terms of the evaluation, the starting point had been sustainability.

The study showed that although when prompted people were concerned about this, there were many more issues which needed to be resolved. Electric buses operating on the PnR, although welcomed, need to be introduced as part of a much wider scheme, which would require structural changes to the transport system and an awareness campaign that transport did contribute to the problem, and that local initiatives could make a difference.

Recommendations included the reinstatement of a bus station, reappraisal of all bus services (with subsidised service for residents), pedestrianisation of the city except for permit holders, penalties for those using the city as a 'shortcut', satellite car parks, introduction of train and bus services to regional cities, 'Land train' from the train station to the city centre.

The PnR service needed to be extended, with an additional terminus, better directions to it, extended hours of operation and city wide incentives, so it was seen as a viable means of transport for a larger number of people. The design of the bus was favourably regarded by the passengers and the drivers. However, as a demonstration vehicle, essential driver feedback on the state of the vehicle was missing and operational issues needed to be resolved. The recharging was more problematic than had been anticipated, with little feedback being provided as to whether the vehicle was charging successfully, or how much charge it had taken. Given that this was occurring overnight and at some distance from the main depot, personnel and extra buses had to be on stand-by at all times.

4.2 In terms of the Hexagon –Spindle model

The final section of the paper has demonstrated the ways in which the H-S model can be used to interpret the results of the evaluation. Mapping comments and results on to the different sectors can show where most effort is required in improving the transport system for different users. For example, the lack of awareness about the importance of local transport initiatives in reducing climate change is a social and cultural issue, which could be addressed by local awareness raising campaigns, with a view to encouraging uptake of more sustainable transport solutions and shifting attitudes towards public transport. Interaction with transport services in this model is seen as multifaceted, not just in terms of the direct interaction between the person and the machine

The model can be used to plan for the investigation of, and integrate diverse elements in a study such as this, from the design of the driver's instrument panel in the product /workstation sector, through to the effects of poorly designed road networks (infrastructure) on commuters and wider issues relating to government policy. While the focus of much transport research has been on the product sector of the model (the driver, machine and system), there is a pressing need for research which addresses issues in all levels and all sectors and which retains the overall vision of an improved transport experience for all

REFERENCES

Benedyk, R., Woodcock, A. and Harder, A. 2009, The hexagon spindle model for educational ergonomics, *Work,* 32, 3, 237-248

Folkesson, A. 2008, *Towards Sustainable Urban Transportation*, Unpublished Doctoral Thesis, KTH Royal Institute of Technology, Stockholm, Sweden.

Girling, G and Birnbaum, R. 1988, An ergonomic approach to training for prevention of musculoskeletal stress at work, *Physiotherapy,* 74, 9.

Litman, T., Burwell, D. 2006, Issues in sustainable transportation, *International Journal of Global Environmental Issues*, 6, 4, 331–347

Wolf, K. L. 2003, Ergonomics of the City: Green Infrastructure and Social Benefits. In C. Kollin (ed.), *Engineering Green: Proceedings of the 11th National Urban Forest Conference*. Washington D.C.: American Forests

Woodcock, A. and Topalovic, S. 2011, An investigation of the perception of transport problems and the role of electric buses, M.Anderson (ed), *Contemporary Ergonomics and Human Factors 2011*, p139-146

Woodcock, A., Woolner, A. and Benedyk, R. 2009, Applying the hexagon –spindle model for educational ergonomics to the design of school environments for children with autistic spectrum disorders, *Work,* 32, 3, 249-239.

CHAPTER 34

Identifying the Information Needs of Users in Public Transport

Stephan Hörold, Cindy Mayas and Heidi Krömker

Ilmenau University of Technology
Ilmenau, Germany
stephan.hoerold@tu-ilmenau.de

ABSTRACT

The development process of user-centered passenger information in public transport requires decisions in the areas addressees, location and information characteristics. To support these decisions, this paper describes a framework for identifying information needs of users of public transport systems based on the results of an information classification and an extensive task analysis.

Keywords: public transport, information needs, user-centered design, passenger information

1   INTRODUCTION

Transportation companies, as providers of transport services and passenger information, obtain new possibilities for the development of passenger information. Passenger information may not only be stationary, collective and static but mobile, individual and dynamic in order to fulfill the information needs of their users (Norbey, Krömker, Hörold and Mayas, 2012).

Identifying these information needs to develop user-centered passenger information systems, is difficult. Devadson and Lingam point out that users have different reasons for not answering questions about their information needs, precisely. Users may not know their real information needs in a special case or they don´t want to reveal their need for information, due to social reasons (Devadason and Pratap Lingam, 1997).

Nevertheless, identifying the information needs is critical for designing user-centered passenger information in the future, when technical development allows the communication of more, and more accurate information. The process of identifying these information needs in public transport requires four elements:

- The workflow of the user within a journey.
- The tasks that have to be performed in order to reach a destination.
- Available information which support the tasks.
- The passengers as users of public transport systems.

The developed framework is a first approach to overcome the mentioned difficulties and to bring the results of the four elements together. It combines the results of formal analysis and empiric methods within the four elements and allows experts to identify the information needs of passengers in public transport. Figure 1 shows the basic elements of the framework with a list of tasks sorted along the journey workflow and an information classification including available information in public transport systems. The knowledge of the passengers is required to allow experts to shift into the passengers' perspective to decide if an information is needed for all or some of the passengers.

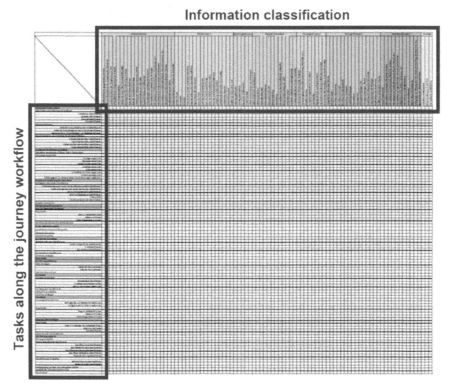

Figure 1 Framework elements for identifying information needs in public transport

2 DEVELOPING THE FRAMEWORK

The development of the framework requires analysis in the four elements, beginning with the journey workflow representing the journey from its starting point to its destination. Based on the results of the workflow analysis, the tasks and available information can be identified and structured. Passenger descriptions, as result of the passenger analysis, complete the framework.

Workflow of the passenger

Traveling with public transport is different from using individual means of transportation. Passengers have to adjust to the system and its specifications, e.g. stop points for entering or leaving vehicles. For the identification of tasks and information, it is necessary to understand the procedures and coherences within the journey workflow.

The travel chain (Verband Deutscher Verkehrsunternehmen, 2001) is a common tool to describe all steps of a passenger's journey workflow. As shown in table 1, the workflow is divided into three phases. The first phase covers the step 'journey planning' as preparation for the journey while the second phase describes all steps of the usage of the public transport system including the way to stop points and to the desired destination. The third phase influences all steps of the travel chain and covers all tasks, when dealing with disturbances, within these steps.

Table 1 Travel chain in public transport

Phase			Step	Location
Preparation	Dealing with disturbances	1	Planning the journey	Starting point: any location
Travel		2	Starting the journey	En route to stop point
		3	Waiting for the vehicle	At the stop point
		4	Entering the vehicle	Outside of the vehicle
		5	Travel with the vehicle	On board the vehicle
		6	Transfer to another vehicle	At the stop point
		7	Alighting from the vehicle	At the stop point
		8	Heading towards the destination	En route to destination

Tasks along the journey workflow

Each step within the travel chain consists of several tasks and subtasks which a passenger performs along the journey. Identifying these tasks and subtasks is crucial for the development of the framework and the identification of information needs. Tasks shape the need for information based on the personal knowledge of the passenger who needs the information.

The hierarchal task analysis (HTA) (Annet, 2005) is a structured approach for the analysis of a workflow. Each step along the journey workflow consists of several tasks that shape the need for information based on the personal knowledge. Subtasks, e.g. choosing the right travel times and connections, refine the identified tasks.

Finally 16 tasks, identified by experts performing a cognitive walkthrough (Smith-Jackson, 2005) for every step along the travel chain, and 78 subtasks as result of conducted focus groups with passengers, define the results of the hierarchal task analysis. Table 2 presents the results within the journey *preparation* phase and the *travel* phase, as well as the tasks that arise when *dealing with disturbances.*

Table 2 Hierarchal task analysis including examples for subtasks

Phase	Task	Subtasks
Preparation	Trip planning	Specify starting point
	Choosing from different routes	Specify destination
	Document trip planning	Specify departure time
	Purchase a ticket	...
Travel	Path finding to stops or station	Finding stop point
	Bridge waiting times	Check remaining time
	Entering the vehicle	...
	Travel inside the vehicle	Follow route information
	Alighting	Check remaining travel time
	Transfer	Identify next stop points
	Path finding to destination	...
Dealing with disturbances	Realizing a disturbance has occurred	Check available passenger information
	Check of the disturbance influences on own route	...
	Identify consequences	Identify delays
	Decide to act based on alternative travel options or routes	Identify altered transfers
		Identify changes within the route
	Complain about the disturbance afterwards	...

Information classification

Identifying the information needs of passengers in public transport requires an understanding of different kinds of information and knowledge of actual or in the near future available information.

For an information classification the actual information has to be gathered in an analyzing process first. In a case study within the German public transport we gathered available information and enriched the results with future passenger information developed in a case study within the IP-KOM-ÖV project (Mayas, Hörold and Krömker, 2012). Afterwards the resulting list of passenger information had to be structured and classes of information had to be found. For the German public transport system seven classes within the categories location, time and transportation system were defined. Table 3 shows these seven information classes. For international use parts of the classification have to be validated first as they depend on the public system in use.

Table 3 Information classification for German public transport with examples for included information

Location	*Time*	*Connection*
Actual position in geographic context	Departure time	Route information
Stop point information	Arrival time	Number of transfers
Directions to stop points	Real-time information	Means of transport
Ticket	*Vehicle*	*Network plan*
Ticket price	Accessibility	Number
Validity	Load factor	Name of stop point
Terms of use	Eco-friendliness	Direction
Disturbance		
Reason for disturbance	Impact	Duration

Passengers of public transport

Public transport systems address a broad range of different people. Passengers are a heterogeneous group that has different daily routines, various reasons for using the public transport and diverse knowledge of a place and the public transport system itself. Especially the diverse knowledge leads to different information needs to perform tasks along the travel chain. Therefore the different passengers have to be described so that experts using the framework can empathize themselves with the passengers.

One way for describing passengers, their expectations and knowledge, their feelings and personal daily routines is the persona technique by Allan Cooper (Cooper, Reimann and Cronin, 2007). It is essential to use a description technique that illustrates the passenger and covers all necessary information to understand the passenger's task performance process.

For the standardization and research project IP-KOM-ÖV seven personas for public transport were developed (Krömker, Mayas, Hörold, Wehrmann and Radermacher, 2011). The developed personas cover a range from high information needs and low knowledge to low information needs and high knowledge:

- Michael Baumann is a commuter from Stuttgart with a good knowledge of a place and of the public transport system. Michael has a tight schedule when traveling to work and cannot afford long delays.
- Carla Alvarez is a tourist from Barcelona who is new to the special public transport system of the city she is visiting.
- Casual user Hildegard is a 69 year old widow with mobility impairments from the north of Germany.
- Ad-Hoc user Bernd prefers his car over public means of transportation so that his knowledge of the transport system is limited.

3 THE FRAMEWORK

The results of the journey workflow analysis, the hierarchal task analysis and the information classification serve as basis for the identification of the information needs and shape the developed framework. The passenger analysis extends the framework and allows a passenger specific identification of information needs.

The framework combines 94 tasks and subtasks with 87 information types in seven information categories. For every task a passenger might perform experts evaluate with the aid of the information classification and information types, if a passenger needs an information or not. Performed for all or part of the tasks, the information needs can be extracted.

Not all users perform the same tasks with the same intensity and need the same information. Some passengers already know the next stops in the vicinity of their own position or the name of the stop they will alight from the vehicle. Some passengers might need information about accessibility while others need tourist information.
Including personas into the framework allows experts to refine the common information needs per task with individual aspects of different kinds of passengers.

Figure 2 shows an example for the identification of information needs within the definition of starting and destination points during the trip planning.

	location information					
X – Information needed for task 0 – Information not needed for task	actual position in geographical context	next stop points in the vicinity	Points of interest in the vicinity	stop points in the vicinity of the destination	description of the vicinity	Location of stops
trip planing	X	X	X	X	X	X
Define starting point and destination	X	X	X	X	X	X
Identifiy stop points	X	X	0	X	0	X
Identifiy adresses	X	0	0	0	X	0
Identifiy own position	X	0	0	0	X	0
Identify points of interest	X	0	X	0	X	0

Figure 2 Extraction from the framework for defining starting point and destination

The advantage of the described framework is the interrelation between the results of the hierarchal task analysis along the travel chain and the information classification. This interrelation allows the extraction of information needs and further information for the development of passenger information systems. The following questions may be answered based on the results of the framework:

(1) *Which tasks are performed when and where?*
Based on the results of the travel chain, tasks and subtasks can be mapped to every step of the travel chain.

(2) *Which tasks and subtasks require which information?*
The identification of the required information per tasks and subtasks is based on the interrelation between the hierarchal task analysis and the information classification.

(3) *Who needs which kind of information?*
If extended with passenger descriptions, the framework allows the identification of passenger dependent information needs.

4   RESULTS OF THE FRAMEWORK APPLICATION

For the development of passenger information systems the decision-making process of developers has to cover three characteristics:

- Identifying the addressees of the information.
    - *Range:*   Individual to collective
    - *Example:* For all passengers at a stop or on board a vehicle or individual via mobile devices.
- Placing the passenger information system at the right location.
    - *Range:*   Mobile to stationary
    - *Example:* At a stop point, inside a vehicle or with a mobile device.
- Deciding which characteristics the information has to fulfill.
    - *Range:*   Dynamic to static
    - *Example:* Public display with real-time information or network-plans at stops.

For example, when confronted with the integration of disturbance information into passenger information systems a developer has to decide who needs the information, where and when along the travel chain the information has to be displayed to reach the passengers quickly and how the information has to be shaped to fulfill the information needs.

The identification of the *addressees of the passenger information* and the decision about individual or collective information requires a passenger dependent extraction of information needs. With these results the presentation of the passenger information can be developed according to the information needs of different passenger types.

Figure 3 shows an example for the results of our work with the four personas Michael, the commuter from Stuttgart, Carla, the tourist from Barcelona, ad-hoc user Bernd and causal user Hildegard (Krömker, Mayas, Hörold, Wehrmann and Radermacher, 2011).

As seen in the graphic, commuter Michael has high information needs in case of disturbances. Due to his tight schedule at work he cannot afford long delays. As commuter he has a good knowledge of his daily route, the transport system as well as the ticket and location information. In contrast, ad-hoc user Bernd and tourist Carla need this information as they are unfamiliar with the places and the public transport system.

Casual user Hildegard has common knowledge about the system and the locations. Her mobility impairments result in an increased demand for special vehicle information.

Based on these results a developer may decide if an information is relevant for all or part of the personas.

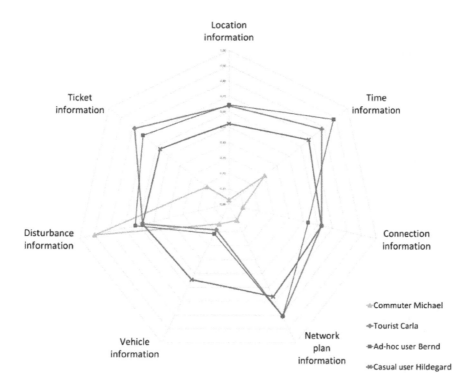

Figure 3 Example for the information needs of different personas

Placing the information at the right location depends on the passengers and the type of information. Disturbance information is needed in all phases of the journey workflow so that a mobile information system will serve this purpose. If the information should contain information about leaving a damaged vehicle, an information system inside the vehicle might be best.

Decisions on information characteristics depend mainly on the type of information. Real-time disturbance information needs to be dynamic to fulfill its purpose while a network-plan is a static information over a long period.

Furthermore, transportation companies and developers are able to compare their information system in general or at specified locations along the travel chain with the extracted information needs to identify information gaps within the journey workflow. This may help to develop a steady flow of information and reduce inconsistencies within the presented information.

340

5 CONCLUSION

International public transport systems differ from each other in aspects of network structure, ticketing, vehicles and organization. However the journey workflow and the included tasks as well as resulting information needs apply to most of them. Therefore the described framework can be applied in international context. In addition the results of the framework can be used to evaluate existing and develop new visions for future passenger information systems including human factors.

Another field that should be analyzed is the connection between information types in order to identify mandatory and optional information for the completion of a task.

ACKNOWLEDGMENTS

Part of this work was funded by the German Federal Ministry of Economy and Technology (BMWi) grant number 19P10003L within the IP-KOM-ÖV project. The project develops an interface standard for passenger information in German public transport with focus on the connection between personal mobile devices, vehicle systems and public transport background computer systems.

REFERENCES

Annet, J. 2005. Hierarchical Task Analysis (HTA). In. *Handbook of Human Factors and Ergonomics Method*, 33-1 – 33-7. Boca Raton, FL. CRC Press.

Cooper, A., R. Reimann and D. Cronin 2007. *About face 3: the essentials of interaction design*. Indianapolis. Ind.: Wiley.

Davadason, F. J. and P. Pratap Lingam 1997. A Methodology for the Identification of Information Needs of Users. In. *Conference Proceedings of the 62nd IFLA General Conference*, B 23: 41-51.

Krömker, H., C. Mayas, S. Hörold, A. Wehrmann and B. Radermacher 2011. In den Schuhen des Fahrgastes - Entwickler wechseln die Perspektive. In. *DER NAHVERKEHR*, 7-8/2011: 45-49.

Mayas, C., S. Hörold and H. Krömker (eds.) 2012. *Das Begleitheft für den Entwicklungsprozess - Personas, Szenarios und Anwendungsfälle aus AK2 und AK3 des Projektes IP-KOM-ÖV*. Available: urn:nbn:de:gbv:ilm1-2012200028.

Norbey, M., K. Krömker, S. Hörold and C. Mayas 2012. *2022: Reisezeit – Schöne Zeit!* Usability Day X – Fachhochschule Vorarlberg (accepted).

Smith-Jackson, T. L. 2005. Cognitive Walk-Through Method (CWM). In. *Handbook of Human Factors and Ergonomics Method*, 82-1 – 82-7. Boca Raton, FL. CRC Press

Verband Deutscher Verkehrsunternehmen 2001. *Telematics in Public Transport in Germany*. Düsseldorf. Alba Fachverlag.

CHAPTER 35

Accessibility in Public Transport – A Psychosocial Approach

Farnaz Nickpour, Patrick W. Jordan

Brunel University, London
Uxbridge, Middlesex, UB8 3PH
Farnaz.nickpour@brunel.ac.uk
patrick@patrickwjordan.com

ABSTRACT

Public transport is facing major challenges in the current economic and social climate. Public transport providers need to respond to increasing demand for service provision both in terms of volume and diversity of service users. They also need to increase the quality of the public transport service they provide. Issues of accessibility, reliability and quality of service are key indicators that are sometimes in conflict and need to be revisited. A research project was conducted in order to address accessibility issues associated with bus travel in London. The aim of the project was to assess the current situation and to make recommendations for improving accessibility of public buses through investigating barriers to a diverse range of people using (and not using) public buses. Findings suggest accessibility and inclusivity issues affecting public bus services fall into three broad categories: Physical, Psycho-social and Operational. Previously, the emphasis of accessibility research and improvements schemes in public transport had been on the physical elements of accessibility. While these are certainly vital, the outcomes of the research suggest that psychosocial issues are equally, perhaps even more, so. Findings highlight the need for a 'Mentality Shift' when addressing accessibility in public transport. Recommendations are made for addressing these in order to make bus travel inviting and enjoyable for mobility-challenged users.

Keywords: public transport, psychosocial, accessibility, inclusive

1 INTRODUCTION

1.1 Public Transport – the new climate

Public transport is facing major challenges in the current economic and social climate; a considerable rise in demand for public transport and an ageing population that is mainly dependant on public transport and is increasingly in need of specialised and door-to-door services. The above challenges double, considering the raised public awareness and the pressure from user organisations to improve the equality and quality of public transport for all.

Public transport providers need to respond to increasing demand for service provision, both in terms of volume and diversity of service users. Transport for London (TfL), a major public transport provider in UK, is currently facing over-subscribed door-to-door services and an increasing demand for accessible and usable public transport by conventionally marginalised groups such as older people and people with disabilities. Issues of accessibility, reliability and quality of service are key indicators that are sometimes in conflict and need to be revisited. There is need to keep the quality of service consistent and at the same time redefine and prioritise the areas of focus and improvement.

1.2 Public Bus Services

Buses will continue to be -probably for many years- the main and only form of public transport that can be accessible to almost all (London Travel Watch, 2010). There is also evidence that bus services are often more frequently used by disadvantaged or vulnerable sections of society, therefore poor performance is more likely to impact on these groups (London Travel Watch, 2009). Thus, the bus service proves to be the single most powerful transport tool in terms of inclusivity and equality potential and provision in a mega-city like London.

There have been great improvements in terms of making buses fully accessible. In London, all buses are now low-floor vehicles and have a space for one wheelchair (TfL Accessibility Guide, 2011). However, an 'accessible bus' does not necessarily guarantee an 'accessible bus service'. An accessible bus service requires not only an accessible bus and an accessible bus stop but also an empathic well-trained driver and a user-friendly environment. As well as improving inclusivity, making local bus services more accessible brings wider benefits including facilitating social inclusion in the local community, making bus travel easier and more pleasurable for every member of the local community and reducing the need for dedicated services (e.g. Dial a Ride) which are not cost-effective.

1.3 The Project

Commissioned by Transport for London and one local London borough, a research project was conducted in order to address issues associated with bus travel in London. The aim of the project was to produce recommendations for improving the accessibility of bus travel through investigating barriers to a diverse range of people using (and not using) public buses and what makes a journey either pleasant or unpleasant. A variety of approaches and techniques were used in order to understand the barriers to accessibility and inclusivity and how these could be overcome. The research project aimed to assess and improve the accessibility of public buses through a holistic and comprehensive service-oriented approach, focusing on an accessible bus service as a whole rather than focusing on segments of the whole service such as bus or bus-stop.

1.3.1 Bus service – key stakeholders

Broadly, with respect to bus service, three major stakeholders were defined:
- Service user - mobility challenged people
- Service provider - bus drivers
- Service operator - bus companies

Addressing accessibility and inclusivity issues, the project focused on mobility challenged people as the critical bus service users. For the purpose of this project, a mobility-challenged person was defined: "*A mobility challenged person is someone whose mobility has been challenged due to age, physical or mental impairment, or an external physical condition; each of the above could have substantial and long-term adverse effect on the person's ability to use public transport.* (Nickpour and Jordan, 2011)". This definition includes, but is not limited to, wheelchair users and those with other impairments that affect mobility. Other major groups with other mobility restrictions that may make it more difficult to use public transport are: older people, blind or visually-impaired people, deaf people or people with hearing difficulties, those with learning difficulties or social phobias, and guardians with buggies.

1.3.2 Bus service - stakeholder issues

Key issues concerning each stakeholder included:
- *For bus passengers:* Positive experience from start to finish – every stage of the journey should be efficient, enjoyable and smooth, and the user should be and feel safe at all times.
- *For bus drivers:* Pleasant working environment – drivers should be treated politely and respectfully by all passengers. They should be equipped with the skills needed to carry out all aspects of their duties competently and receive the full support of both bus users and their employers in doing so.
- *For bus operators:* Profitable business – operators should be encouraged and enabled to perform the service requirements against suitable performance targets in a manner which is commercially viable.

2   METHODOLOGY & METHODS

2.1   Methodology

The research project followed a combined primary and secondary research methodology, with a heavy focus on primary research conducted through a diverse range of field research methods. A major focus for the project was consultation with people who had a wide range of mobility challenges. Many other stakeholders were also included in the consultation process. This included bus drivers and representatives from bus operating companies, TfL, police and advocacy groups representing mobility-challenged people.

In addition to this consultation process, members of the project team gained first-hand experience of some of the issues faced by mobility-challenged people by taking bus trips while using wheelchairs. Information was also collected through observing mobility-challenged people travelling on buses and asking mobility-challenged residents of London Borough of Hillingdon – where the study was conducted – to take bus journeys and report their experiences.

2.2   Methods

A wide range of methods were used in order to collect first-hand information regarding the existing barriers and issues regarding accessibility and inclusion of bus services. All primary research was undertaken in the local London borough. In some cases, similar services were observed in other London boroughs as well. Due to space limitation, specifc details in terms of participants' process of selection, age, demographics, position, etc. is not included here. More detail on the above is provided in a Technical Report (Nickpour & Jordan, 2011).

2.2.1 Focus groups

Three focus group sessions with different focuses were run in order to provide a holistic understanding of the existing issues.   Each session focused on one stakeholder group. Firstly, a focus group session was held with nine representatives of service providers and a cross-section of other stakeholders aiming to look at organisational and big-picture issues. The participants included representatives from TfL, local Council, bus companies, Dial a Ride, Age UK, Metropolitan Police, Hillingdon Community Transport and Access and Mobility Forum. Then, one session was held with a diverse group of service users with a focus on mobility-challenged passengers. This included nine participants; one blind person, one person with learning difficulties, one wheelchair user and six older people. Finally, a session was held with service non-users including seven mobility-challenged members of the public who currently did not use public buses for a variety of reasons. These included previous negative experience with using public buses and lack of trust and confidece in the public bus service.

2.2.2 Access audits

Two sets of access audits were planned and carried out. The emphasis was on both immersion (Moore, 1985) and direct observation (Dray, 1997). The first series of audits included eight local bus journeys and were carried out by the project research team, role-playing by using a wheelchair, aiming to look at specific mobility issues. Each observation session was attended by two members of research team. The second series of access audits were carried out by a diverse group including five local participants with mobility impairments. Participants included one male older person aged 72, two wheelchair users, one with electric wheelchair and one with normal wheelchair. Also, one person with learning difficulty aged 21 and one blind person aged 42 carried out the access audits. All audit sessions were documented through various applicable audio, visual and textual formats.

2.2.3 Interviews & Meetings

A number of meetings and interviews were held with individuals from various organisations and groups in order to look into a number of issues in more detail. Altogether, five interview sessions were held, these included interviews with three bus drivers, meetings with Hillingdon Community Transport general manager, accessibility officer in Hillingdon Council, two officers from Disablement Association of Hillingdon and six members of the local Youth Council.

2.2.4 Observations

Two major observation sessions were held. One session focused on special services aimed at mobility-challenged passengers; The project team spent a day working with Dial-a-Ride service that provided door to door transport for mobility-challenged people. Another observation session took place at Bus Mentoring Day – a training day aimed at helping those who assist mobility challenged people with their travels.

2.2.5 Literature review

The literature review drew on a number of sources, reports and documents including reports by the Disabled Persons Transport Advisory Committee (DPTAC), Direct Gov, The Department of Transport and London Travel Watch.

The main source for the literature review was the new report by the Greater London Authority (GLA), titled "Accessibility of the Transport (GLA, 2010) which looked at the accessibility of all public transport within the capital including buses. The report drew on inputs from a wide variety of advocacy groups representing mobility challenged people as well as on a wide array of statistics quantifying accessibility of buses and other modes of transport.

3   FINDINGS

3.1   Physical Issues

From a physical accessibility point of view, users tended to find the most problematic part of the journey was getting from home to the bus stop and getting from the bus to their final destination. Examples of problems here included: narrow pavements, loose paving stones, steep roads and difficult crossings. There were also accessibility difficulties at some bus stops – for example, the positioning of litter bins and other street furniture sometimes made deploying and using the ramp somewhat inconvenient.

However, despite such difficulties, it was possible for mobility challenged people to board the bus at all of the stops examined in the audit. Improvements in the design of buses meant that, in general, once the user had reached the stop, the bus could be accessed OK and the on-board part of the journey completed.

3.2   Psychosocial Issues

Various observatory and immersive methods used also uncovered a number of other difficulties – mostly psychological and social – that users faced. These included:

3.3.1 Uncertainties

There were many aspects to this including uncertainties as to whether users would be able to get on and off the bus OK, whether they would have a long wait at the stop and whether their interactions with others would be positive.

3.3.2 Overcrowding

The start and end of the school day are times when the bus gets particularly crowded. This can sometimes mean that the bus is too crowded to let a wheelchair on. Even if it is possible to board, overcrowding can make it difficult for wheelchair users to get to the wheelchair bay and to move their chair into the proper position within it. Overcrowding is becoming an increasingly problematic issue as more and more people are using buses. This is due in part to the difficult economic conditions that we have had recently (bus travel tends to increase in times of financial hardship) and in part to the issuing of free bus passes to schoolchildren and older people.

3.3.3 Negative Experiences with Drivers

Many users had also mentioned that they had had problems with the drivers. This could be because of inconsiderate driving – for example pulling away too quickly – or because they were perceived as having an unfriendly or surely attitude towards the user. Indeed, during the access audits there were a number of incidents of drivers not stopping at bus stops when they saw a wheelchair user waiting to get on. Bus drivers mentioned that there were often problems with ramps failing to deploy and cited this as a reason why they could not always pick up wheelchair users.

3.3.4 Negative Behaviour of Other Passengers

A number of participants reported being annoyed or intimidated by the behaviour of other passengers. In particular they mentioned teenagers who they said could be very loud and often used foul language. A number of participants also mentioned that they also found it annoying when people had loud conversations on mobile phones or played music so loudly that it could be heard through their headphones.

The behaviour of other passengers when getting on and off the bus was also a source of annoyance and intimidation. In particular they mentioned pushing and shoving and people not waiting their turn in the queue. Other users had reported that they are wary of using buses in the evening or night because of the risk of encountering drunk or threatening people.

3.3.5 Off-putting stories

In some cases, participants were put off using the bus because of stories they heard about other people having bad experiences, in particular, stories of violent or frightening incidents. These stories may have been told to them by friends or they may have read or heard about them in the media.

3. 3   Operational Issues

An issue that may be contributory factor is the key performance indicators (KPIs) used to measure the performance of the bus operators. Currently, emphasis is mostly on reliability – that has to do with timeliness of bus service. There are no measures in place to monitor either the number of mobility challenged people using buses or the quality of their experience as one performance indicator.

It was observed that it can take some time for a mobility challenged person, such as a wheelchair user, to board the bus. This may lead to the bus running behind schedule with the consequence that it affects reliability. As reliability is the basis on which the bus companies are judged and the pressure for to run on time, drivers sometimes feel unenthusiastic about picking up mobility challenged passengers and hence may have a hostile attitude towards mobility challenged passengers or may try to avoid picking them up altogether.

4   DISCUSSION

4.1   Physical Versus Psychosocial Issues

Overall the research suggested that good progress had been made in terms of addressing the physical issues. There could be problems getting to and from the bus stop and sometimes there were problems with ramps and small wheelchair spaces. However, it was generally the case that it was physically possible to complete a journey without excessive difficulties.

Perhaps the most striking issue to emerge from the research was the role that psycho-social factors played in affecting mobility-challenged people's quality of experience of using public buses. In particular, the impact of the attitudes and behaviour of the driver and of other passengers.

Bad experiences of this nature were the most frequently cited reasons for not enjoying a bus journey or for not using the bus at all. Previously, the emphasis of accessibility research and improvements schemes has been on the physical elements of accessibility. While these are certainly extremely important, the outcomes of our research suggest that psychosocial issues are equally, perhaps even more, so. This observation mirrors those within the field of design generally where there has been increasing attention on psychosocial issues and their emotional consequences in recent years (Norman, 2005).

4.2 Special Service Versus Public Service

As part of this research we also looked at people's experiences with door to door transportation schemes for mobility challenged people within London. These included Dial-a-Ride, a minibus-based service which picks up passengers at their home and takes them to a pre-requested destination. This service was very popular with users. In particular they enjoyed the friendly atmosphere on the minibus and the friendly, attentive and considerate behaviour of the driver.

Mobility-challenged users praised the drivers for their empathy and understanding, for their cheerfulness and for making them feel valued and welcome whenever they used the service. They mentioned how much they looked forward to the social aspects of using the service and for the enjoyable conversations with other passengers. A challenge is to try and recreate some of these benefits on public buses and to put into place approaches and schemes that will help to foster a positive ambience.

4.3 Negative Interactions

It should be emphasised that the picture is not entirely negative; Field research supported the fact that many of the drivers have an excellent approach to interacting with mobility-challenged people. They are friendly, welcoming, informative and help make the journey a great experience. Similarly, many teenagers are polite, well-behaved and kind towards other passengers. However, this was mainly result of individual's intrinsic motivation and personal codes of conduct.

Nevertheless, it is also important to recognise that there are genuine problems with some bus drivers' and teenagers' attitudes and behaviours. Negative drivers' attitudes was observed and reported such as being rude and uncommunicative towards mobility challenged people. Also, in cases, some teenagers' behaviour appeared inconsiderate and liable to make people feel uncomfortable.

The effects of this negative behaviour tend to extend beyond the specific incidents that occur. When service users encounter a bad experience, they will remember this and will have a doubt in their minds about the quality of their experience next time.

This uncertainty can have a negative effect. The memory of the previous bad experience can create a sense of doubt – will this happen again? This doubt can make people question whether they want to use the bus again and leave them with some negative feeling for the duration of their travel. Moving forward, the challenge is to find effective ways of improving the ambience on board and tackling some of the psychosocial issues that have been identified.

5 CONCLUSION & RECOMMENDATIONS

5.1 Conclusions

There is need for a 'Mentality Shift' when addressing accessibility in public transport. This study suggests and highlights 'psycho-social' inclusion as the key area of focus. The findings suggest accessibility and inclusivity issues affecting public bus services fall into three broad categories: Physical, Psycho-social and Operational.

Physical issues are to do with the design of the bus and the built environment and are the 'typical' issues considered when looking at accessibility. Findings suggest the key identified physical barriers include *Getting to bus-stop, Space availability and priority on bus* and *Ramp technology & reliability.*

Psycho-Social issues are the 'soft' issues associated with the quality of people's travel experience. Findings suggest the key identified psycho-social barriers are *Ambience, Awareness and empathy* and *Communication.*

Operational issues concern the running of the service and cross-organisational strategies and regulations. The key identified operational barriers are Key performance indicators. Public bus service KPIs currently appear to focus only on efficiency rather than quality, inclusivity and pleasurability of service.

The results indicate that it is the psycho-social issues that seem to be proving the biggest barrier to using public buses; in particular for mobility-challenged people. Addressing these issues requires a focus on people. It involves making them aware of the effect that their behaviour is having, convincing them to change it and giving them the skills and insights needed to do so. It also involves creating a desirable ambience throughout the bus journey, making the public transport experience not only efficient but also pleasurable.

5.2 Recommendations

Overall – including both physical and psychosocial factors – the following nine recommendations are proposed as key principles for improving the quality of mobility challenged passengers' experience of public bus travel.

1. Create an inviting and friendly experience of the bus service
Perceptions about bus travel influence people's decisions about whether to take the bus and the emotions associated with anticipating using it. Mobility challenged people should be confident that their bus journey will be a positive experience.

2. Make bus stops reachable
Getting to and from the bus stop is, generally, the biggest physical barrier to bus travel for mobility challenged people. Making bus stops more reachable would significantly increase the numbers of people who could access public buses.

3. Make all bus stops fully accessible
Once at the stop, mobility challenged people should be accurately informed about when the bus will arrive. The design of the stop should also facilitate quick and easy ingress for mobility challenged people.

4. Promote and facilitate positive behaviour amongst passengers
Interactions with other passengers should be positive and friendly throughout the bus journey.
5. Ensure that key aspects of the bus are fully operational
The aspects of the bus that affect accessibility should be fully operational at all times. Mobility challenged people should be confident that their journey will run smoothly and efficiently.
6. Ensure that all users have a safe and comfortable space
All mobility challenged users should have a safe and comfortable space in which to complete their journey. They should be able to move into and out of this space easily.
7. Welcome mobility challenged people aboard
Drivers should warmly welcome mobility challenged people aboard the bus. They should communicate clearly and cheerfully with them throughout the journey.
8. Set off and drive smoothly
Ensure that mobility challenged people are settled before moving off. Make sure that this is done smoothly and that the drive is smooth and controlled throughout the journey.
9. Provide information clearly through multiple channels throughout the journey
Mobility challenged people should be clear about when the bus is approaching their stop and have plenty of time to prepare to exit.

ACKNOWLEDGMENTS

This research project was commisioned by London Bourough of Hillingdon and Transport for London. The authors would like to thank all local participants in the project and the user research team including Murtaza Abidi, Penelope Bamford, Thomas Wade and Jennifer McCormack.

REFERENCES

Dray S.M. (1997) Structured Observation: Practical methods for understanding users and their work in context, Proceedings of CHI'97, ACM, Atlanta, Georgia

Greater London Authority (GLA) (2010) Accessibility of the Transport Network (November 2010)

London Travel Watch (2009) TfL Performance Reprot (September 2009)

London Travel watch (2010) Bus Passengers' Priorities for Improvement in London (May 2010)

Moore P, Conn C. P. (1985) Disguised: A True Story, Word Books, 174 pages, ISBN 0849905168

Norman D (2005) Emotional Design: Why We love (or hate) Everyday Things, Basic Books, USA, 272 pages, ISBN: 978-0465051366

Transport for London (2011) TfL Accessiblity Guide, Available at: http://www.tfl.gov.uk/gettingaround/transportaccessibility/1171.aspx (accessed 13 August 2011)

CHAPTER 36

Human Behavior, Infrastructure and Transport Planning

Harald FREY [1], Christine CHALOUPKA-RISSER [2]

[1] Research Center of Transport Planning and Traffic Engineering
Institute of Transportation
Vienna University of Technology, Austria
harald.frey@tuwien.ac.at

[2] FACTUM Chaloupka & Risser OG
Vienna, Austria
christine.chaloupka@factum.at

ABSTRACT

Since traditional transport planning has adjusted it's planning principles to motorized traffic, it deals with the reprimand of road users. Thereby the consequences on the attractiveness of walking and cycling and hence resulting behavioral changes and needs of road users, esp. pedestrians and cyclists, were not considered in transport planning on a strategic level. Regarding actual planning standards focusing mainly on motorized traffic from a historical point of view, some more "human behavior-" and less car oriented tools for infrastructure planning are discussed. Biological as well as socio-psychological concepts will be reflected to change the understanding of human behavior and needs of different groups in traffic. The pedestrian resp. the unprotected road users as a new meter are carried into the focus of concern.

Keywords: Human behavior, Quality of life, speed, pedestrian, biology, needs, infrastructure

1 CHANGES IN THE INFRASTRUCTURE, CHANGES IN BEHAVIOR

Transport infrastructures have always been forming people's behavior and their (social, economic, cultural, etc.) structures (Knoflacher, 2007). Within towns and villages the streets were conceived as public space, not separated in sidewalks and lanes. Soon after the initiating mass motorization in the early 20th century, finding adequate space for the movement and parking of cars in the cities became a problem. Traditional roads which were organically linked to the settlements and the landscape formed a barrier for the effortless and fast car traffic. In the 19th century cars were subordinated to the general traffic. In England in 1865 a man with a red flag had to precede the car in public spaces (Red Flag Act) (Bailey, 1994). There was a lack of experience with the new traffic medium car, which allowed the drivers to experience individual speed levels in a dimension for which they are not evolutionary equipped (Knoflacher, 1987).

The amount of information in the manifold streets at that time went far beyond the possibilities of drivers in order to implement responsible actions at high speed levels. Transport and city planners therefore reduced the information density by removing any other road users out of the movement space of the cars. At the same time national traffic rules were passed and new control devices for the organization of the motorized individual traffic itself were required. These elements, like traffic signals, were also assigned on non motorized road users. Under the disguise of traffic safety, barriers for pedestrians, like protective gratings, pedestrian underpasses and crossovers were built (see below), in order to prevent pedestrians from crossing the car lanes in a disturbing manner. Detours and waiting times for pedestrians reduced the number of walking people rapidly and go with sometimes high social costs. As an example: In the city of Chapel Hill (USA, North Carolina) a pedestrian crossing over a major collector road was forbidden for pedestrians to be crossed. According to this children had to take busses to cover the last 400 meters distance to school which took more time and cost money (Methorst, 2010).

Figure 1 Pedestrian crossings in Ankara (Turkey) (Öncü, 2007) and Vodice (Croatia) (Wurz, 2009).

Pedestrian over- or underpasses are always affiliated with time- and energy losses - especially for the elderly; they work in very rare circumstances and only under constraints and sometimes make the other side unreachable for the disabled.

2   MOBILITY AND THE KEY ROLE OF SPEED

Mobility is defined by the freedom to choose to travel and sojourn in public space. The amount of distance that one can cover is less important than being able to make a trip (Methorst, 2010). Under this aspect the following discussion about speed has to be considered.

By determining speed as a normative command variable in standards and guidelines, the meaning and significance of the (straight) line and - what is less clear - of right and justice was set (compare French: "la droite" - line, straight, right; "le droit" – justice, droit) (Virillio, 2001). The straight line as a planning element and the influence of speed on planning parameters (e.g. the length of the straight line) changed human behavior regarding the choice of means of transportation (Knoflacher, 1996; Schafer, 2000; Schafer, 2005).

The right of those (motorists) who are moving along the roads with their individual speed becomes injustice - a barrier - for all other non motorized road users who want to cross the street. Social and economic relationships often extend transversely to the direction of lanes. Transport or city planning that ignores these principles promote a form of technological fundamentalism (Virillio, 1993).

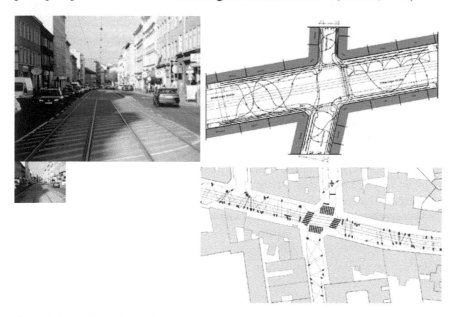

Figure 2 Comparison of actual walking lines of pedestrians within an hour (right) and the design of road space (left) shown on the examples of Ottakringer Street (above) and Josefstädter Street (below) in Vienna. (Buder et. al., 2008; Jokanovic et. al., 2011)

354

The requirements for the indicators of project planning (design speed, functional significance and, subsequently, all the technical parameters, such as cross slopes, etc.) were derived from the vehicle dynamics and result from the claims of a small portion of road users. The new measurement unit of traffic planning became the passenger car unit (PCU). All other road users were converted to PCUs.

The corresponding Austrian guideline uses the so-called operational speed as basis for the assessment and as an indicator for the level of service in the street section. The operational speed is the average speed of the car traffic at the significant traffic load.

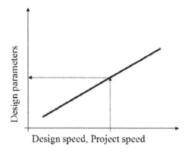 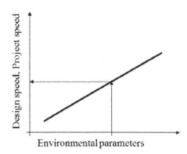

Figure 3 In the traditional method of road design the design parameters were derived from the design speed (defined by the road category). Environmental parameters (including social, economic, ecological indicators, etc.) were subordinated. Consistently these environmental parameters should determine the design speed and in series the design parameter of the road. This would have enormous effects on the typical standard cross-section of a road. (Schopf and Mailer, 2004)

The rapidly increasing number of pedestrian road accidents was not only a warning signal for the discrepancy between the expectations of the planners and the real behavior but also for the devastating effects of mechanical mobility for the social and economic structure.

The findings obtained so far show us, that much of what has been planned, has to look different, if the mechanisms of human behavior, based on needs and motives of very diverse traffic participants, would be taken into account. The planning in the past has expected a certain form of behavior, for instance, that pedestrian crossing are only necessary in certain places, which resulted in the penalization of pedestrians who do not use the provided crossing infrastructure. The prejudice of the planners was therefore confirmed by practical experience and reprimanded by laws.

355

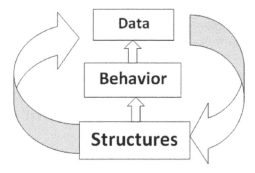

Figure 4 Relationship between structures, behavior and data. Data is a representation of the behavior which is influenced and determined by (physical, financial or organizational) structures (Knoflacher, 2007).

It would have been necessary to check in each occurring transport planning decision whether and how certain groups of road users were constrained, restricted or suppressed (see figure below regarding problems of elderly road users).

Pedestrians are a very heterogeneous group. With regard to walking and sojourning performance, in general young and healthy male adults have the least limitations. But they are a minority. Asmussen (Asmussen, 1996) showed that a remarkably large proportion of citizens (almost 40%) can be considered to belong to a vulnerable group. On top of this, even competent persons can be temporarily impaired by being influenced by alcohol or medical drugs, the use of a mobile phone or MP3 player, having fogged glasses, heavy bags, or simply distracted by their companions or interesting objects in shop windows (Fuller, 2005).

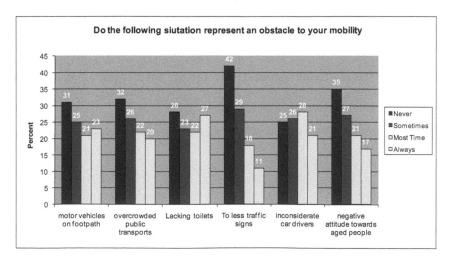

Figure 5 Problems of seniors in traffic, according to EU-project SIZE (Risser et al, 2010).

In the transport system human beings and their evolutionary predispositions as well as their needs and different inter- and intrapersonal motives need to be dealt with. Therefore, the knowledge of human behavior legalities provides the foundation for understanding cause-effect mechanisms in transportation systems.

3 TOOLS FOR INFRASTRUCTURE AND TRANSPORTATION PLANNING CONSIDERING THE NATURE OF HUMAN BEINGS

The tools, which are to be used, are based on the theory of evolution and epistemology and on the concept of Quality of Life (QoL) regarding mobility, including sojourning. The former have their roots in the 19th century Darwinian Theory and general system theory and were described in a popular way by Konrad Lorenz, Rupert Riedl and other biologists. The later are results of recent social psychological research.

3.1 Biological theories

During the last 200 years, the transportation system was developed as an attractor for mechanical transportation system users. The effects were increasing distances and hence a change from human related structures to inhuman, mechanic means related agglomerations, from free flow of pedestrians to congestions of car traffic (Knoflacher, 1997).

One of the deepest evolutionary levels, that affects all levels above is related to body energy (Knoflacher, 1995). It can be regarded as the main level. Energy saving was - and still is - the most successful strategy for survival in evolution. It is probably the deepest rooted driving force for behavior in general and human behavior in particular. A car driver, for example, requires only half of the body energy per unit of time compared to a pedestrian. But in the same time he moves ten to twenty times faster with ease. This acceleration creates an unimaginable and wonderful effect of strength and superiority which is much stronger than culture and ethics or anything else which can be derived from the later and weaker levels of consciousness in human evolution (Macoun et al, 2010).

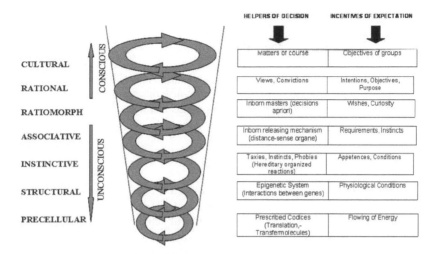

Figure 6 Evolutionary layers of expectation and experience (Macoun et al, 2010 adapted by Riedl, 1985).

3.2 QoL and implications on mobility on non- motorized traffic participants

People's behaviour emerges in interaction between the person and the environment. The decision to walk or not to walk is influenced by individual factors as well as by their perception of the physical and social environment, e.g. the perceived safety (Fyhri et al, 2010). One of many other essential findings within the COST 358 project "PQN" (Pedestrian quality needs) was the following:

"Still, for some groups, and in some situations, feelings of uncertainness might influence people's decision about whether to walk or not. There seem to be considerable national and regional differences in how perceived safety influences children's opportunities for walking in everyday life. In some inner city areas, especially in large cities, fear of crime or fear of accidents might give considerable limitations to children's independent mobility. In more rural areas and especially in Northern Europe, this seems to be a more marginal explanation."

3.3 QoL - Indicators for pedestrian handicaps

There exist some indicators, e.g. the OECD Indicators for Quality of Life of pedestrians which help to measure whether some activities can cause more or less problems in pedestrians´ performance in traffic. The items are:

- following a group discussion
- have a conversation
- read small letters
- recognise a person from a distance of 4 meters

- carry a 5 kg load for 10 meters
- bend to pick up an object from the floor
- walk 400 meters without stopping.

Hearing, eye sight and agility problems all aggravate the risk of accidents. Hearing and eye sight limitations particularly diminish discerning risks of collisions with other road users. Eye sight and agility problems increase the risk of falling. All three will feed increasing insecurity feelings.

3.4 Sojourning

The concept of sojourning is rather unknown, especially in the Anglo Saxon countries. Sojourning in public space is important because it is an indicator for quality of public space and it encourages all kinds of activities, which humans need for their well-being. There are many kinds of sojourning: professional activities, recreational activities, waiting, hanging out, but public space is also the home of the homeless and sometimes the scene of crime and violence. The average amount of time spent on sojourning is about 100 hours per person per year (Methorst, 2010).

4   IMPLICATIONS FOR PLANNING

Considering the previous chapters about tools for planning of infrastructure and the traffic system in general we could give the following advice.

In each system, the transport volume M of the traffic element i can be determined by the elementary transport equation $\boxed{M_i = D_i * v_i}$ (Knoflacher, 1997).

D_i is the number of road users in the system [road users / area], and v_i is its current average speed. The number (density) of road users is primarily determined by the attractiveness of the system. One indicator is the speed.

There are 2 options to increase the share of pedestrians (Knoflacher, 1997):

a) Increase speed (v), which means to remove all kind of barriers that constrain the movement of the pedestrians like curbstones, traffic signals, protective gratings, etc. and/or

b) Increase D through an attractive design of the environment, diversity and nearness of attractors for people.

Figure 7 Paradigm change with transport planning for people. Maximizing mobility in an efficient and sustainable way shown by the example of Eisenstadt (Austria). Before (left) and after (right) reorganization (Knoflacher, 1977; Knoflacher, 1993).

To increase the relative importance of pedestrian traffic, other means of transportation which disturb this mode, have to be knowledgeably reduced. Again, there is only the choice between D and v. The speed of vehicle traffic must be reduced significantly and the number of cars can be reduced (Knoflacher, 1996).

Based on the knowledge of our evolutionary features as well as the socio-psychological (QoL aspects) we can derive two important parameters for transport planning (Knoflacher, 1995):

- "fast" is everything above 3 - 4 km / h
- "far" is everything over 200-300 m

Therefore transport planning has to be based on the following hierarchy (Knoflacher, 1996):

1. Pedestrian
2. Cyclists
3. Public transport
4. Individual motorized traffic (cars, motor cycles)

This order will affect - among other things - the zoning of road space, steering of traffic lights, decisions for the expansion of car lanes due to so-called capacity problems, the arrangement of parking spaces in public space, the parking laws, etc., and is consistent with the road safety targets, aspects of sustainability at local, national and global level, the environment, the local economy, the social system, the efficiency of the system and the economical use of public tax funds. The responsibility of engineers as well as the knowledge of mechanisms of action in the transport system is of crucial importance in this process.

REFERENCES

Asmussen, E. 1996. De nieuwe normmens . Mens _ maat der dingen. Op weg naar integrale veiligheid en toegankelijkheid voor iedereen. POV Zuid-Holland, Den Haag.
Bailey, T., and Kennedy, D. 1994. The American Pageant. Lexington: D. C. Heath.

360

Buder, J. et al. 2008. Student work within the lecture „transport planning" at the Research Center of Transport Planning and Traffic Engineering, Institute of Transportation, Vienna University of Technology.

Fuller, R. 2005. Towards a general theory of driver behaviour. In: Accident Analysis and Prevention, vol. 37, nr. 3, p. 461-472

Fyhri, A., Hof, T., Simonova, Z., and Jong, M. De. 2010. The influence of perceived safety and security on walking. Draft WG2 paper for the Final Report of the COST358 Pedestrians' Quality Needs Project, Leeds.

Jokanovic, D. et al.2011. Student work within the lecture „transport planning" at the Research Center of Transport Planning and Traffic Engineering, Institute of Transportation, Vienna University of Technology.

Macoun, T. et al. 2010. A successive approach to a more objective and issue-related assessment method – in consideration of limited human perception and evaluation capabilities.,XVI PANAM, July 15-18, 2010 – Lisbon, Portugal

Methorst R. 2010. Evaluation of the pedestrians´ performance and satisfaction, COST 358, Final report, B.5.2, Rijkswaterstaat Centre for Transportation and Navigation, Delft, The Netherlands

Knoflacher, H. 1977. Generalverkehrsplan der Freistadt Eisenstadt. Wien 1977.

Knoflacher, H. 1987 Verkehrsplanung für den Menschen – Band 1: Grundstrukturen, Orac Verlag Wien.

Knoflacher, H., Linortner, G. 1993. Fußgängerzone Eisenstadt – Neue Maßstäbe in der Realisierung moderner städtischer Verkehrsplanung. In: Österreichische Gemeindezeitung, 5/1993. Jugend & Volk Verlag. Wien 1993.

Knoflacher, H. 1995. Fußgeher und Fahrradverkehr, Wien, Böhlau Verlag, 1995

Knoflacher, H. 1996. Zur Harmonie von Stadt und Verkehr, 2.Auflage, Wien, Böhlau Verlag.

Knoflacher, H. 1997. Landschaft ohne Autobahnen, Wien, Böhlau Verlag.

Knoflacher, H. 2007. Grundlagen der Verkehrs- und Siedlungsplanung: Verkehrsplanung, Verlag Böhlau.

Öncü, E. 2007. Making cars free in Turkish cities. Presentation at the conference "Towards Carfree Cities VII", Istanbul. (http://www.worldcarfree.net)

Riedl R. 1985. Die Spaltung des Weltbildes- Biologische Grundlagen des Erklärens und Verstehens. Verlag Paul Parey, Berlin und Hamburg.

Risser, R., Haindl, G. and Ståhl, A. 2010. Barriers to senior citizens' outdoor mobility in Europe: European Journal of Ageing 7, Berlin, Springer

Schafer, A. and Victor, D.G. 2000. The future mobility of the world population. In: Transport Research Part A 34 (2000), p.171-205.

Schafer, A. 2005. Transportation, Energy, and Technology in the 21st Century. GCEP Advanced Transportation Workshop. October 2005. Stanford University.

Schopf, J.M. and Mailer, M. 2004. Road design principles. Lecture of the seminar "Transport planning" at the Research Center of Transport Planning and Traffic Engineering, Institute of Transportation, Vienna University of Technology.

Wurz, O. 2009. Pictures from Vodice, Croatia.

Virilio, P. 2001. Fluchtgeschwindigkeit, Fischer Verlag, Frankfurt/M.

Virilio, P. 1993. Revolution der Geschwindigkeit, Merve Verlag, Berlin.

Section V

International Differences in Drivers

CHAPTER 37

The MINI E Field Study — Similarities and Differences in International Everyday EV Driving

Roman Vilimek, Andreas Keinath, Maximilian Schwalm

BMW Group
80788 Munich, Germany
roman.vilimek@bmw.de

ABSTRACT

Understanding how electric vehicles (EVs) are driven in the real world has taken an important step forward with the implementation of the MINI E field trials in the United States, Europe and Asia. An enormous amount of data was collected from extensive driver research carried out by various scientific partners, by data-loggers and from charging points both at home and within the public infrastructure. The results had direct influence on the development of the BMW i3 and helped to inform policy-making decisions as well as other EV market stakeholders.

The field trials discovered that everyday use of the MINI Es did not radically differ from typical driving patterns of conventional vehicles in the same segment. Requirements on EVs are internationally largely the same. Before conducting the trials, the limited range and charging durations were seen as major barriers. Results show that this is far less the case than could have been expected. In summary, an electric vehicle with slightly larger range and more space for passengers and cargo will be able to suit the mobility needs for urban use to a very large extent.

Keywords: Electric Vehicle, longitudinal survey, MINI E field trial

1   THE MINI E FIELD TRIALS

The MINI E is a conversion of the MINI Cooper. It was developed for field trials with customers and deployed in several test sites. The trials have been

364

carefully planned and executed by the BMW group in cooperation with public, private, and university affiliates. Field trials were held in the United States, Germany, the United Kingdom, France, Japan and China. Since 2009 over 600 vehicles were – and still are in certain study areas – on the road. As of January 2012, approximately 15 million miles have been logged, 15,095 people applied to be part of the MINI E trial, 430 private MINI E customers were surveyed, 14 fleet companies actively participated in the study and 15 research institutions gathered and analyzed data from 10 different cities in 6 countries on 3 continents. The compilation of data from the MINI E field trials has yielded arguably the most extensive results regarding everyday usage of electric vehicles worldwide.

The MINI E serves as a key learning project for the BMW i3, the first purpose-built EV from the BMW group. It is a two-seat development of the familiar MINI hatch with a 204 hp electric motor, a torque of 220 Nm and a 35 kWh Lithium-Ion battery containing 5,088 cells. The range in real terms is roughly 100 miles, depending on the driving style. Charging takes about 3.8 hours at 32 ampere and 10.1 hours at 12 ampere. The study was undertaken to address key questions on e-mobility. The results presented here focus on the following subset:

- What is the profile of users currently interested in EVs?
- What expectations do users have on the technology and suitability?
- How is the MINI E actually used on an everyday basis? What are the likes and dislikes on electric driving?
- What has to be changed in future in relation to charging and infrastructure?
- How du users perceive specific vehicle characteristics of the MINI E?
- How important is the ecological added value of an EV to MINI E users?

2 METHODS

The first field trials started in June 2009 in the United States and Germany. In the United States, 240 vehicles were used in fleets and 246 private customers drove the MINI E for at least one year in Los Angeles and New York / New Jersey. The University of California at Davis surveyed the household customers and in detail 54 of those. Of the remaining private users, 72 participated in a survey designed for international comparisons. In Germany, 80 users participated in the first Berlin trial, each for half a year between June 2009 and August 2010. The second Berlin trial started in April 2011 (30 users, 6 months). Additionally, 10 vehicles in the first trial and 38 vehicles in the second trial were used in private and public fleets. All Berlin trials were conducted by the Chemnitz University of Technology.

In terms of methodology, the Berlin projects are the blueprint of the MINI E trials. The Institute of Cognitive and Engineering Psychology at the Chemnitz University of Technology has a strong expertise in human-machine interaction and user research for in-vehicle systems. Scientists from Chemnitz conducted knowledge-sharing workshops in the early project planning phases with the Institute of Transportation Studies at the University of California at Davis, which has a long tradition in exploring alternative fuel vehicles and plug-in hybrid electric vehicles.

The research group in Chemnitz led by Josef Krems and researchers from BMW's development department then shaped a methods tool set that was used as far as possible in similar form in all following MINI E projects (cf. Cocron et al., 2011) and allowed to compare results between the countries involved.

The field trial in the UK – in the Oxford region and London – started in December 2009, involving 2 x 20 private users for half a year each and 20 vehicles in fleet usage. Research was conducted by the Oxford Brookes University. An additional field trial in Germany was conducted by Psyma Marketing Research in Munich with 26 MINI Es in households and 14 vehicles for fleet customers for 6 months, beginning in September 2010. In France, 2 x 25 (6 and 5 months, respectively) private users took part in the Paris field trial carried out by the French Institute of Science and Technology for Transport, Development and Networks (IFSTTAR) between December 2010 and December 2011. Additional 25 vehicles were placed in private fleets. The vehicle hand over for the Asian field trial took place in China starting in February 2011 and in Japan at the beginning of March 2011. Beijing and Shenzhen served as locations for the Chinese study with 2 x 25 private users in total (6 and 5 months, respectively) and 25 vehicles in fleets. Research cooperation partners included the Chinese Automotive Technology and Research Center and the marketing research company INS. In Japan's first phase 14 privately used MINI Es were driving through Tokyo for 5 months. In the second phase the same amount of participants were recruited in Osaka and Tokyo, again for 5 months. During the complete period 6 EVs were part of fleet car pools. Research was supported by cooperation projects with the Waseda University and the marketing research company IID, Inc. Following the explicit wish of the customers the study was kept up after the Great East Japan Earthquake in March 2011.

2.1 Application Process / Field Trial Participants

Persons interested in taking part in the MINI E field trial applied via an online application form. Applicants gave information about relevant aspects of socio-demographic and psychographic background and had to fulfill certain criteria to be selectable for the study (e.g., be willing to actually use the car on a regular basis, be willing to pay a monthly leasing fee of typically about $ 850). Details of the selection procedure are reported by Krems, Franke, Neumann and Cocron (2010).

The large number of applicants (on average 500-3500 per country per phase) allowed to deduce the profile of customers currently interested in buying an EV. This profile was internationally very similar. Typical applicants were male (approx. 80%), around 40 years old (except for China: mean age 32), well-educated with above-average income and an affinity for new technology. Their most important motivation was to experience a new clean and sustainable technology. Supporting environmental protection in general was in direct comparison a weaker, but still meaningful factor. Also relevant was, especially stated by US applications, to gain independence from mineral oil and to reduce local emissions. Details on these motivations and the "sustainability meets technology" factor are described in Turrentine, Garas, Lentz and Woodjack (2011). The user sample was formed to represent the characteristics of this early adopter profile. The MINI E field trial did not try to represent EV usage patterns of the population of the countries involved but intended to provide insights into preferences, attitudes and behavior of the target group of early buyers that will dominate the first years of EV usage.

2.2 Data Collection

In the following, the generic data collection procedure is described. A complete description and discussion on these methods is given by Krems et al. (2010), Bühler, Neumann, Cocron, Franke and Krems (2011) and Cocron et al. (2011).

After the application screening procedure, participants were interviewed by telephone for an assessment of motivation and attitudes. Further interviews were conducted before the participants received their MINI E, after a period of three months and at the end of the usage period. These interviews were typically done face-to-face or partially by telephone and supplemented with online questionnaires.

Travel and charging diaries were deployed to understand how the vehicle is used, which trips can or cannot be undertaken and when / how users charged their vehicle. The travel diary was administered three times for one week: before receiving the EV, in the middle and at the end of the usage period. The first travel diary served as a baseline. The charging diary was administered twice.

Onboard data-loggers delivered objective data by recording variables like trip length, speed, acceleration, frequency and duration of charging and battery status. The vehicle position was not tracked. This source of information compared to a control group of conventionally powered vehicles helped to get a complete picture in combination with subjective and diary data. Between 50-100% of vehicles were equipped with these loggers in Germany, France, UK and China. In Japan and the United States data-loggers were not put in place. Data on driving distances was read out from AC Propulsion chips. The combustion engine vehicle control group consisted of conventional vehicles from the same vehicle segment, altogether 40 privately owned MINI Cooper and BMW 116i.

3 RESULTS

Unlike in a laboratory experiment it is not possible to keep conditions constant in such a large scale field study. With the rapidly changing field of e-mobility over the last three years relevant questions changed and new topics arised, leading to an evolution of the surveys. Additionally it proved to be impossible to translate every question in an interculturally proper way into the Asian, European and American context. Therefore, although the study design as described above was always implemented as far as possible, it was not possible to pose the exact same set of questions in all locations. Moreover, major incidents like for instance the earthquake in Japan, the debate on nuclear power, CO_2 and renewable energy as well as ongoing changes in transport and environmental policy directly influenced attitudes, experiences and the lives of the field trial participants.

All care had been exercised to take those effects into account. However, a certain amount of imprecision is unavoidable. This should be kept in mind when reading the results below. Numerical values (% agreement) should be regarded as tendencies. Hence, only descriptive statistics are presented. Percentage values refer to answers on a Likert Scale from 1 (do not agree at all) to 6 (fully agree). Top three values are accumulated to "agreement", bottom three values to "disagreement".

The research reported here presents work in progress. France, China and Japan data relate to the first phase only. Second phase data analysis is currently ongoing.

3.1 Attitudes and Expectations

Before actually driving the MINI E, the majority of users expected to be constrained by the range and the missing cargo and back-seat passenger space. Only between 21-54% of users assume that they will be as flexible as with a conventional combustion engine car. However, on average about 90% of users in all test sites expected that they will be able to satisfy their mobility needs with the MINI E.

There were hardly any concerns regarding safety issues of the electric vehicle. Somewhat higher values are found in Japan on concerns about battery size and position (43%) and on battery chemistry (36%) before the trial. These numbers dropped during the trial considerably to 31% and 15%.

China is the only country involved were users reported considerable concerns on safety related aspects. Before the trial, 56% stated that they are concerned about battery size and position, 68% stated that they are concerned about safety in terms of battery chemistry and 60% reported that they see a danger in the high voltage systems in the vehicle. These values are 2-3 times higher than in Europe. Instead of declining like in all other countries, the values remained stable or even increased during the trial. Agreement rates on all three scales went up to 72%. The initial high values can be explained by the fact that individual mobility is a rather new phenomenon in China which is accompanied by a high level of information demand and the feeling of uncertainty. As there were no safety related incidents with the MINI E, the increased concerns about safety can only be explained by incidents in China caused by lower product safety standards of other products.

3.2 Everyday Usage and Range

Everyday use of the MINI Es does not differ considerably from the typical driving patterns of conventionally powered vehicles in the same segment. While the mean accumulated journey distance of a control group, a fleet of 40 BMW 116i and MINI Cooper, added up to 27.5 miles, MINI E customers in most locations even used their EV for larger distances. The mean daily driven distance in China was 28.6 miles, in France 29.3 miles, in the UK 29.7 miles, in the United States 31.6 miles (West Coast), and 29.0 miles (East Coast). The MINI E drivers in Germany used the car slightly less, between 22.8 to 24.0 miles in the two different trials. In the second Berlin trial 20 users were not equipped with a private wallbox and needed to rely on public charging infrastructure. When looking at the data of these customers in isolation a stronger reduction in daily driving to 20.0 miles becomes visible. MINI E drivers with private wallbox used the car for 28.5 miles on average.

Drivers stated that they were able to undertake about 80% of their trips with the MINI E (min.: 78%, Japan; max: 83%, Germany). This can be increased up to 90% if rear seats and an adequate boot would have been available. In round terms 10% of intended trips remain for which an EV is not suitable. Therefore, OEMs will need to find ways to offer EV customers solutions for those 10% use cases. However, considering the articulate skepticism of the users on limitations because of the range at the beginning of the trial, these results are amazing and encouraging.

Another line of results should be related to these initial expectations regarding the limited range as a possible source of discomfort. After having used the MINI E for some time the users' impression changes. The very first results on desired range were gathered by the Chemnitz University of Technology 2009 by interviewing participants after three months. The customers stated unanimous (94%) that a range between 87 miles and 100 miles is sufficient for everyday needs, especially for urban use (Neumann, Cocron, Franke & Krems, 2010). Looking at these early results in the light of internationally available comparisons after two years of research is quite astonishing as the pattern remains the same. Average ratings for acceptable ranges rank from 86 miles (China) and 87 miles (Germany) to 91 miles (Japan and France) and up to 115 miles (UK) and 117 miles (USA). When asked for optimal ranges for future EVs, customers typically request ranges between 125 miles and up to 155 miles (Neumann et al., 2010). Nevertheless, it is remarkable how well the MINI E users were able to cope with the available range and that a range at this level was well accepted after some months of experience.

Franke, Neumann, Bühler, Cocron and Krems (2011) add a very interesting perspective to this discussion: They studied how range is experienced in an EV and how this relates to other variables, notably stress. Franke et al. found that users were indeed able to adapt to limited range, but that they utilize the available range suboptimally. They were able to show that certain personality traits and coping skills moderated the experience of comfortable range and conclude that it may be possible to change this perception by information, training and suitable HMI design, allowing EV drivers to get able to use the available range to full extent.

One factor of discomfort in range was related to cold temperatures during winter. The MINI E has air cooled batteries and no battery heating system. A large number of users reported that severely low temperatures affected the distance that could be driven between charges. It was also much more difficult to get the vehicle fully charged. These problems will be addressed already with the BMW ActiveE, an electric vehicle based on a BMW 1 series coupé, which features a liquid battery cooling and heating system.

3.3 Charging

Charging a car was a completely new activity for almost all MINI E drivers. They mastered it very well from the beginning and stated that is easy to learn. There is a characteristic change in charging pattern as soon as drivers build up expertise on how much they can do with the available range. In the first one or two weeks, users are typically plugging in the car whenever possible, sometimes searching nervously for charging stations even when the remaining range is much higher than needed in the given situation. As the typical daily driven distances are about 30 miles, users quickly realize that they do not need to charge every night. Although there are also customers who make nightly charging a habit, data-logger results show clearly that users normally only charge once every two or even three days (average charging events per week: Germany 2.8, UK 2.9, China 2.5). This is not only a MINI E user phenomenon. The Ultra Low Carbon Vehicle Demonstrator Programme in the UK, involving 340 vehicles (thereof 40 MINI E), also found a charging frequency of less than once in every two days (Everett, Walsh, Smith, Burgess & Harris, 2010).

The standard method of charging in the field trial was using a 32 amps wallbox. In France however, charging at 32 amps would have been quite expensive for the customers. Thus, the MINI E users there charged at 12 amps, implying about 9-10 hours for a full charge. It is interesting to see how this affected charging behavior. For technical reasons, data logger results on this cannot be compared directly. Charging diaries and user reports point out that the extended charging duration did not allow for the strategy switch reported above. French users charged almost every night, on average 5.2 times per week (cf. Labeye, Hugot, Regan & Brusque, 2011).

Public charging infrastructure is something typically requested by EV customers as essential, however is comparatively seldom used. The highest proportion of public charging events was found in the UK (7%) and Germany (6%). In France, this was reduced to 3% and in Japan to 1%. Many reasons account for this. Japan has to be considered as being different because there were only few possibilities for the MINI E to charge in public. In the other countries, users stated that the private wallbox was sufficient, that charging stations were occupied, defect, in an unfavorable location or that the parking period would have taken too long. Especially the Berlin subsample of drivers without private wallbox pointed out how these difficulties hindered them from charging. More convenient public charging stations with shorter charging times and the option to reserve are highly demanded.

Turrentine et al. (2011) indicate that charging can even be a convenience factor compared to refueling a conventional vehicle. According to their results, MINI E drivers enjoyed the simplicity of home recharging. The drivers developed a charging routine that fits their specific lifestyles and were happy to be able to avoid the otherwise necessary trips to gas stations. Moreover, charging gave the drivers a feeling of control over the "fueling" behavior, the cost and the source of energy.

3.4 The MINI E Experience

The driving characteristics of the MINI E are seen to be outstanding by all of the users. That proved to be a key factor to acceptance as EVs have had a reputation for low performance in the past. The MINI E is very sporty compared to most EVs and also remarkable in terms of combustion engine standards. Clean vehicles can be fun.

Regenerative braking is a unique feature of electric vehicles. It refers to recapturing energy otherwise lost or unutilized during braking, coasting, or downhill driving. In the MINI E it was integrated in the accelerator pedal. This differs from most electric or hybrid vehicles and allows to drive the car with one pedal, using the brakes only for unplanned or emergency breakings. Feedback of all customers was unanimous that single-pedal driving was fascinating, almost game-like: It enables sporty driving and at the same time allows for a direct experience of efficient driving and energy saving. Turrentine et al. (2011) discuss factors that should be adjusted in regenerative braking to enhance this experience. Labeye et al. (2011) relate regenerative braking to models of driving strategies and depict in how far it may have advantages and disadvantages on the different levels of driving tasks.

Another key characteristic of EVs is that they do not emit engine noise. The silent interior conditions are an unquestioned unique advantage. Concerns about the low outside noise level are often expressed in public discussion. In an in-depth analysis on the silence of electric vehicles, Cocron, Bühler, Franke, Neumann and Krems (2011) argue, that drivers appreciate the low noise emission but are well aware of potential dangers and are able to adapt to it. Labeye et al. (2011) come to the same conclusion. Drivers in Europe rate the low noise level at the beginning of the field trial as a potentially problematic (Germany: 50%, France: 72%) but far less after several months (Germany: 16%, France: 50%). In Asia this pattern is quite different: Before getting acquainted with an EV, the rating of Japanese users is similar to Europe (57%) and the rating of Chinese users is even substantially lower (24%). After three months of experience in Asian traffic conditions, Japanese and Chinese customers estimate the potential danger of silent driving even higher than Europe customers at the beginning of the trial (77% and 64%, accordingly).

In general, a successful learning process is a crucial prerequisite for establishing a satisfactory environment in using EVs. During the first one to two weeks of usage, MINI E customers have a precipitous learning curve behind them. The effect becomes highly apparent in charging when users begin to switch from daily charging to charging only every other day or even only twice a week. Adapting to the range available, using regenerative braking and getting used to the lower noise level of the vehicle are key elements. This is not at all a process of constantly going through difficult situations. Users describe it as gaining experience in a new field. The relevant skills are easy to build up, making the drivers quickly feel competent. The learning process is reflected by MINI E drivers' requests for long test drives when it comes to actually buying an EV. The users were firmly convinced that typical test drive durations of up to half a day are not sufficient to learn enough about an EV. A large proportion demanded several days or even several separate test drive events, which is currently very uncommon in vehicle sales. While typically around 1/3 of the users stood behind this request in Europe (UK: 31%, France: 34%; Germany: 40%) and in Japan (38%), the Chinese customers again showed the highest need for information (72% demand extensive test drives). Bühler et al. (2011) discuss findings that show that in general experience has a high impact on acceptance. Based on these results they hypothesized and were able to demonstrate that the users' attitudes become more positive towards driving electric vehicles if the EV was available for their daily routine for a longer period of time.

To summarize, programs designed to interest a broader audience for e-mobility should above all provide ample opportunity for electric driving to enable potential customers to build up experiences.

3.5 Ecological Aspects

The absolute majority of drivers think that renewable energy should play an important role in recharging EVs (over all countries: 84-96% agreement). But the opinions differ whether an EV should be exclusively charged with green energy. While this is a prerequisite for German customers (74%) and surprisingly interesting to Chinese users (52%), all other countries do not see it as an essential part of e-mobility (France: 40%, UK: 39%, Japan: 31%, USA: 20%). A similar picture emerges when looking at the acceptance for nuclear energy as part of the energy mix for EVs: This source of energy is relatively popular in France (67% agreement), the US (63%) and the UK (54%), but it is not regarded as adequate in Germany (18%) and Japan (15%). Note, however, that before the Fukushima disaster, Japanese users had the highest acceptance for nuclear energy (80%).

Especially in the United States, the combination of electric vehicles and renewable energy is very unfamiliar (Kurani, Caperello, Bedir & Axsen, 2011). Kurani et al. (2011) argue that MINI E drivers with prior commitment to green electricity view EVs and renewable energy as overlapping in their motivation. Those without prior commitment wonder why they should do more if they are already driving an EV, which they see as an ecological contribution in itself (e.g. because it reduces local CO_2 emissions). The authors conclude that there is only little demand in linking EVs and green electricity in terms of monthly subscription services because of the wide variation among EV drivers in their motives.

2   CONCLUSIONS

The most intriguing finding in the international comparison of everyday EV use is the absence of major differences between the countries involved. Especially interesting is that urban EV use cases lead to very similar daily driven distances and a similar coverage of mobility needs. Two results need to be highlighted here: Differences in acoustics and ecological aspects. While the European customers saw silent driving as unproblematic after having adapted to it and valued this EV characteristic, MINI E drivers in Asia evaluated the low noise emission and the risk of not being noticed as potentially dangerous. This is most likely caused by the different traffic situations the drivers experienced. Solutions for emitting sounds for electric vehicles should reflect these differences in international perception. On ecological aspects, the study results clearly show that although renewable energies are a value in itself, the customers in different countries (even within Europe) do not all expect a link between green electricity and EVs. However, as soon as this link is regarded as relevant, like in Germany, the acceptance of EVs will strongly be related to an authentic connection between e-mobility and sustainability.

ACKNOWLEDGMENTS

The authors would like to acknowledge the project i team at BMW for making this study possible, the BMW data-logger team and the MINI E research team (Andreas Klein, Michaela Lühr, Pamela Ruppe, Felix Esch and Juliane Schäfer) for their valuable contributions to this project. We would like to thank our international research partners for bringing in their expertise and for conducting the research with extraordinary commitment and outstanding efforts. The research was funded in Germany by the Federal Ministry for the Environment, Nature Conservation and Nuclear Safety and by the Federal Ministry of Transport, Building and Urban Development. In the UK it was funded by the Technology Strategy Board.

REFERENCES

Bühler, F., I. Neumann, P. Cocron, T. Franke, and J. F. Krems. 2011. Usage patterns of electric vehicles: A reliable indicator of acceptance? Findings from a German field study. In: *Proceedings of the 90th Annual Meeting of the Transportation Research Board*. Paper retrieved from http://amonline.trb.org/12jj41/1

Cocron, P., F. Bühler, I. Neumann, T. Franke, J. F. Krems, M. Schwalm, and A. Keinath. 2011. Methods of evaluating electric vehicles from a user's perspective – the MINI E field trial in Berlin. *IET Intelligent Transport Systems 5*, 127-133.

Cocron, P., F. Bühler, T. Franke, I. Neumann, and J. F. Krems. 2010. The silence of electric vehicles – blessing or curse? In: *Proceedings of the 90th Annual Meeting of the Transportation Research Board*. Paper retrieved from http://amonline.trb.org/12jj41/1

Everett, A., C. Walsh, K. Smith, M. Burgess, and M. Harris. 2010. Ultra Low Carbon Vehicle Demonstrator Programme. In: *EVS 25*. Proceedings of the 25th World Battery, Hybrid and Fuel Cell Electric Vehicle Symposium & Exhibition. Shenzhen, China.

Franke, T., I. Neumann, F. Bühler, P. Cocron, and J. F. Krems. 2011. Experiencing range in an electric vehicle: understanding psychological barriers. *Applied Psychology doi: 10.1111/j.1464-0597.2011.00474.x.*

Krems, J. F., T. Franke, I. Neumann, and P. Cocron. 2010. Research methods to assess the acceptance of EVs – experiences from an EV user study. In: *Smart Systems Integration*, ed. T. Gessner. Proceedings of the 4th European Conference & Exhibition on Integration Issues of Miniaturized Systems. Como, Italy: VDE Verlag.

Kurani, K. S., N. Caperello, A. Bedir, J. Axsen, (in press) *Can markets for electric vehicles and green electricity accelerate each other? Initial conversations with consumers.* (Submitted to Transportation Research Records).

Labeye, E., M. Hugot, M. Regan, and C. Brusque. Electric vehicles: an eco-friendly mode of transport which induce changes in driving behaviour. Paper presented at the Human Factors and Ergonomics Society Europe Chapter – Annual Meeting, 2011.

Neumann, I., P. Cocron, T. Franke, and J. F. Krems. 2010. Electric vehicles as a solution for green driving in the future? A field study examining the user acceptance of electric vehicles. In *Proceedings of the European Conference on Human Interface Design for ITS*, eds. J. F. Krems, T. Petzold, and M. Henning. Berlin, Germany.

Turrentine, T., D. Garas, A. Lentz, and J. Woodjack. 2011. *The UC Davis MINI E Consumer Study*. Davis: University of California.

CHAPTER 38

Normative Influences across Cultures: Conceptual Differences and Potential Confounders among Drivers in Australia and China

Fleiter, JJ1, Watson, B1, Lennon, A1, King, M1, Shi, K2

1Centre for Accident Research and Road Safety-Queensland
Queensland University of Technology, Brisbane, Australia
2 Management School of Graduate University, Chinese Academy of Sciences,
Beijing, China

ABSTRACT

Normative influences on road user behaviour have been well documented and include such things as personal, group, subjective and moral norms. Commonly, normative factors are examined within one cultural context, although a few examples of exploring the issue across cultures exist. Such examples add to our understanding of differences in perceptions of the normative factors that may exert influence on road users and can assist in determining whether successful road safety interventions in one location may be successful in another. Notably, the literature is relatively silent on such influences in countries experiencing rapidly escalating rates of motorization. China is one such country where new drivers are taking to the roads in unprecedented numbers and authorities are grappling with the associated challenges. This paper presents results from qualitative and quantitative research on self-reported driving speeds of car drivers and related issues in Australia and China. Focus group interviews and questionnaires conducted in each country examined normative factors that might influence driving in each cultural context. Qualitative findings indicated perceptions of community acceptance of speeding were present in both countries but appeared more widespread in China, yet quantitative results did not support this difference. Similarly, with regard to negative social feedback from speeding, qualitative findings suggested no embarrassment associated with speeding among Chinese participants and mixed results among Australian participants, yet quantitative results indicated greater embarrassment for Chinese drivers. This issue was also examined from the perspective of self-identity and findings were generally similar across both samples and appear related to whether it is important to be perceived as a skilled/safe driver by others. An interesting and important finding emerged with regard to how Chinese drivers may respond to questions about road safety issues if the answers might influence foreigners' perceptions of China. In attempting to assess community norms associated with speeding, participants were asked to describe what they would tell a foreign visitor about the prevalence of speeding in China. Responses indicated that if asked by a foreigner, people may answer in a manner that portrayed China as a safe country (e.g., that drivers do not speed), irrespective of the actual situation. This 'faking good for foreigners' phenomenon highlights the importance of considering 'face' when conducting research in China – a concept absent from the road safety literature. An additional noteworthy finding that has been briefly described in the road safety literature is the importance and strength of the normative influence of social networks (*guanxi*) in China. The use of personal networks to assist in avoiding penalties for traffic violations was described by Chinese participants and is an area that could be addressed to strengthen the deterrent effect of traffic law enforcement. Overall, the findings suggest important considerations for developing and implementing road safety countermeasures in different cultural contexts.

Keywords: speeding, norms, cross-cultural research, traffic law enforcement, face

1   INTRODUCTION

Driving has been described as a socially-regulated behaviour (Stradling, 2007) and previous research on the influence of other people on driving and on other road user behaviours has explored a broad range of socially-based factors. Normative influences on road user behaviour are well documented and include personal, group, subjective and moral norms. Commonly, however, normative factors have been examined within one cultural context (and usually in highly motorised, developed countries) although limited examples have explored the issue across cultures and have added to our understanding of differences in perceptions of the various normative factors that may exert influence on road users. These can assist in determining whether successful safety interventions developed and employed in one location/cultural context may be successful in another.

Notably, the literature is relatively silent on normative influences in many of the countries that are experiencing rapidly escalating rates of motorisation. China is one such country where new drivers are taking to the roads in unprecedented numbers. China's recent economic growth has been accompanied by one of the highest annual motorisation growth rates in the world (Pendyala & Kitamura, 2007) and authorities are grappling with the associated challenges including appropriate traffic law enforcement, traffic congestion and extremely high levels of road trauma

(WHO,2009). Road crashes are reportedly the number one non-disease killer in China, ahead of other disasters, such as flood, fire and earthquake (Pendyala & Kitamura, 2007). It is clear that China must focus on curbing road crashes but it is equally important to note that while highly motorised countries like Australia have implemented countermeasures to reduce road trauma for decades with good success, there is still a need to reduce road-related deaths and injuries (Australian Transport Council, 2011). Therefore, efforts to better understand the impact of things such as normative influences on road user behaviour are warranted across cultural contexts to provide better information with which to develop new safety countermeasures.

1.1 The concept of culture

Applying the concept of culture to study road use has been approached in two ways. The first way (sociological approach) views culture in light of interactions between various groups of road users (e.g. according to age, ethnicity, education etc). It considers the differential impact of broader societal influences (e.g., traffic law) on specific groups, focusing on group interactions. However, this approach lacks consideration of personal factors such as driving history and personality traits. The second way that culture is examined is with cross-cultural studies where the same behaviour or group is studied in two or more culturally-bound contexts. Here, culture refers more to the characteristics that are specific to people because of their geographic location and common national history than to the way in which groups of people within one country interact with each other. This approach certainly encompasses broader societal influences (e.g. the normative influence of legislation) but also commonly examines the more individually-based characteristics such as risk perceptions, attitudes (Lund & Rundmo, 2008), driving style, and driving skills (Ozkan, Lajunen, Chliaoutakis, Parker, & Summala, 2006).

The current research fits in the latter grouping by examining speeding in two distinct cultural settings (Australia and China). Speeding is a recognised high-risk factor in road crashes globally (WHO, 2009) and Rothengatter and Manstead have highlighted cultural differences in intentions to violate traffic laws including speeding, citing social norms as a likely explanation. This highlights the need to better understand factors that influence speeding across cultural settings before recommending countermeasures developed in one cultural context for another.

1.2 Challenges in cross-cultural research

Methodological and theoretical challenges must be considered when conducting road safety research across contexts. From a methodological perspective, cultures may have fundamental differences that make comparisons difficult or inappropriate. For instance, King (2007) identified a range of factors (e.g., economic, institutional, social, and cultural) that can impact on the transfer of road safety knowledge across cultures. Theoretical challenges, however, relate more to issues of understanding behaviour across groups. The challenge is determining how best to operationalise theoretical components developed in one context so as to maintain the essence of their original meaning in another. The few published Chinese driving-related studies indicate that the addition of culturally-specific issues has helped understand driver behaviour across cultures. For example, Xie and Parker (2002) described factors that were relevant in China but not in Britain. Sense of social hierarchy, challenge to legitimate authority, and value of interpersonal networks were all found to important in how Chinese drivers deal with the road and enforcement environment.

In considering the generalisability of findings from one context to another, consideration must be given to the capacity of theories to predict behaviour across different cultural settings. The current research employed Akers' social learning theory (SLT) (Akers, 1977) to investigate speeding in Australia and China. SLT is a psychological-sociological hybrid of reinforcement/learning and differential association theories. It has been suggested that it "mandates the inclusion of cultural variables in the explanation of crime through its emphasis on 'definitional' learning" (Jensen and Akers, 2003, p.22). That is, the attitudes, rationalisations, and moral beliefs of a cultural group should orient the learning mechanisms described by the theory. As such, it appears useful to cross-cultural investigations.

We also included the concept of self-identity; the process of identifying as the type of person who performs a particular behaviour. When such identification takes place, the behaviour is said to become important to our self-identity, reflecting our values and motivations (Hogg, Terry and White 1995). Investigations of self-identity are limited in driver research although the literature suggests it can be useful in understanding the influence of one's own values. Further, the literature on the concept of self from a Chinese perspective suggests potentially large differences in the impact of this concept across cultures. It is suggested that in Chinese culture, relations with others strongly influence the concept of self; described as 'recognized, defined and completed by others' (Gao, 1998, p. 165). It appears that this concept is unexplored in relation to driver behaviour in China.

This research aimed to investigate normative influences and self-identity concepts on self-reported speeding among car drivers in Australia and China using qualitative and quantitative methods in order to explore similarities and differences between the two cultural contexts. This study was part of a larger research program investigating a range of factors influential on speeding, some findings of which are reported elsewhere (Fleiter et al, 2009; 2010; 2011).

2   METHOD

This research ulitised qualitative and quantitative investigations involving car drivers from Beijing (n=35 in focus groups and 299 questionnaire respondents) and Queensland (n=67 in focus groups and 833 questionnaire respondents). The method, sample, recruitment strategy and analysis plan is identical to that described in Fleiter et al (2009).

3   FINDINGS

The qualitative findings are presented first with participants identified according to gender, age and nationality (e.g., F30CN represents a 30-year-old Chinese woman and M19AU represents a 19-year-old Australian man).

Community norms: In both countries, responses indicated the perception of community support of exceeding speed limits. For instance, Australian participants commonly described perceptions that the majority of drivers speed:

> *"I just assumed that everybody speeds. Most on the road are keeping up with me and I'm sometimes trying to keep up with them." Male49AU*
>
> *"Speeding, well everybody does it." Male20AU*

Chinese participants also described the perception of speeding as commonplace (see Fleiter et al 2009) and were asked a specific question to assess community norms: '*Imagine that someone visits you from another country. They have never been here before. What would you tell them about driving speeds in Beijing?*' (based on Perkins and Wechsler 2006). An interesting outcome relevant to conducting research in China emerged; participants said they would tell visitors that there was no speeding in Beijing, despite their prior comments reporting it as commonplace.

> *Speaker 1:I would tell them that there is no speeding in Beijing. It's the same as when we tell others that our University is the best in China. When we go to foreign countries, we will tell others that China is the best country in the world. So when we talk about speeding with foreigners, you are not just yourself, you must...M30CN*
>
> *Speaker 2:...take the reputation of our country into consideration. M26CN*
>
> *Speaker 1:Yes, you are not only yourself. More importantly, you represent the honour of the whole country. In our country, we don't take speeding as seriously as foreigners [do]. In their eyes, speeding is a big deal so if they ask me about this issue, I will tell them speeding is not a common thing.*
>
> *Speaker :I agree. I would say that I have never noticed people speeding. We should show others the good side of our country and city. We should make them feel Beijing is a safe city to live in.*

Such comments highlight an important concept in Chinese culture. Face, or *mianzi* is an integral concept in Chinese conduct and refers to 'an individual's public or social image gained by performing one or more specific social roles that are well recognised by others' (Luo, 2007, p. 14). The current example relates more to face saving or reputation saving of the entire nation than of individuals. This issue is also related to the concept of self-identity.

Self-identity: When asked "*Is it important that you are known as someone who drives above the speed limit?*", Australian participants generally reflected the desire to project the image of oneself as a responsible or safe driver, even if they are not. Mixed views were reported regarding whether speeding violations were a source of embarrassment (see Fleiter et al 2010 for details). However, among Chinese participants, speeding engendered little negative social feedback. Overwhelmingly, participants indicated that they were not embarrassed to inform others about receiving speeding tickets. Rather, responses indicated a perception that it is just bad luck to be caught speeding, and further, that drivers are happy to tell others since it allows them to warn about enforcement locations.

> "It isn't embarrassing in our culture, unlike other countries. The [Chinese] public don't think speeding is a big issue." F27CN
> "It is not bad, not embarrassing. We should discuss with others [to warn them]. I was out of luck." Male29CN

The most common response in Chinese groups suggests that being known as someone who exceeds speed limits was considered of no real consequence and something that is not worthy of discussion because speeding is so commonplace:

> "It isn't important [to be known as someone who exceeds speed limits]. Speeding is your own business. Others won't care about whether you speed or not." F30CN
> "Speeding is a kind of illegal behaviour. The point is to get the public to accept this. If [they] have this awareness, they will discuss something about speeding tickets; but now the public doesn't have this awareness. They think speeding is okay.M38CN

Less commonly, responses indicated that to be known as someone who speeds was akin to being seen as an untrustworthy or unsafe driver; one with whom people may not wish to travel. However, contrasting sentiments were also described:

> "[To be known as someone who does not speed] is to admit you drive very slowly and your driving skill is not good""F28CN
> "I think driving at the speed limit will make others trust you. Friends and colleagues will have the impression that you are a trustworthy man." M33CN

The next section reports responses from quantitative items assessing self-identity and normative influences. Firstly, as a measure of identity and to provide an overall snapshot of how participants perceived their speeding, we asked 'Which of the following statements best applies to you as a driver?' according to 5 categories of compliance (1. I generally ignore speed limits and drive above them on the majority of occasions; 2.I don't pay much attention to speed limits and regularly drive above them; 3. I generally keep an eye on my driving speed but often go over the speed limit; 4. I keep a fairly close watch on my travel speed but occasionally go over the limit; 5. I always watch the speed I travel and never deliberately go over the speed limit) (modified from Elliott 2001). The two samples were compared using a chi-squared test for independence and post-hoc analyses were undertaken using standardised residuals to identify cells with observed frequencies significantly lower or higher than expected (i.e., outside the range of +1.96 to -1.96). Results revealed statistically significant differences between the Australian and Chinese samples, $\chi2$ (4) = 72.81, $p<.001$, $\varnothing_c$ = 0.26 (Cramer's V indicates a small effect size, Field, 2005). A significantly greater proportion of Chinese drivers (39.7%) described themselves as 'never deliberately driving above the speed limit' compared to only 19.8% of Australian drivers. Conversely, a significantly greater proportion of Australian drivers (24.8% compared to only 7% of Chinese drivers) described themselves as 'often driving above the speed limit'. See Figure 1.

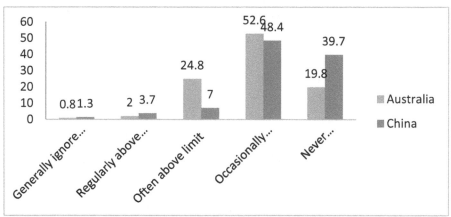

Figure 1. Proportion of Australian and Chinese drivers endorsing descriptions of speed limit compliance

Community norms towards speeding were assessed by two items scored on a 7-point Likert scale (1=*Strongly disagree* to 7=*Strongly agree*). For the first question '*Drivers here think it is okay to exceed the speed limit*', mean scores were significantly higher for Australian (*M*=3.1, *SD*=2) than Chinese drivers (*M*=2.2, *SD*=1.2); $t(820) = 8.08$, $p<.0001$, $\eta=.54$. This result suggests that Australian drivers perceive that it is more acceptable among the general community to drive above the speed limit although the mean scores indicate that for both samples, the majority of participants disagreed with the statement.

For the second question, '*Drivers here think you are unlucky if you get caught speeding*' (scored as above), mean scores were significantly higher for Australian (*M*=3.58, *SD*=2.1) than Chinese drivers (*M*=2.36, *SD*= 1.3); $t(827)=11.58$, $p<.0001$, $\eta=.69$. This result suggests Australian drivers reported greater agreement that being caught speeding is just bad luck. Both of these results on community norms were contrary to expectations based on the qualitative findings reported earlier.

In a related item, the level of embarrassment related to speeding violations was assessed by asking: '*I would be embarrassed to tell people if I got caught for speeding*' (scored as above). Responses revealed that Chinese drivers (*M*=4.57, *SD*=1.7) reported greater agreement than Australian drivers (*M*= 4.45, *SD*=2.1) although the difference was not significant; $t(653)= -.912$, $p=.362$. Thus, despite the non-significant outcome, and contrary to expectations based on the qualitative findings, Chinese drivers reported more likely embarrassment associated with receiving a speeding violation than did Australian drivers.

Previous research and findings from the Chinese focus groups (as discussed in Fleiter et al 2011) suggest the importance of social networks in avoiding traffic violation penalties. As such, for the Chinese sample only, two items assessed use of personal relationship by self and others to avoid punishment if caught speeding. The first question asked '*How often have you used your relationships to avoid the punishment after you are caught speeding?*' (scored 1=*Never* to 6=*Always*). While the majority of participants reported never using social relationships in this way (M=1.6, SD=1.2), over one quarter (27.4%) reported having done so (see Figure 2).

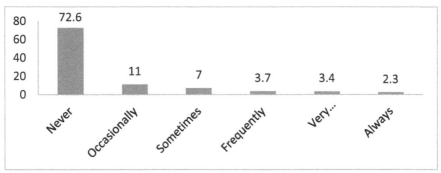

Figure 2. Self-reported frequency of using relationships to avoid punishment for speeding

The second question asked 'How many other people do you know who have used their relationships to avoid the punishment after being caught speeding?' (scored 1= None to 5 = All; M=2.63, SD=0.8). As can be seen in Figure 3, approximately one third of the sample reported knowing a few other people who use their networks to avoid penalties once caught and over half the sample (58.2%) reported knowing many others who do this (ie, 'some', 'most' or 'all' others).

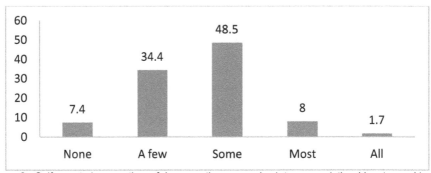

Figure 3. Self-reported proportion of known others perceived to use relationships to avoid punishment for speeding

4   DISCUSSION

This paper examines normative influences on speeding in Australia and China from qualitative and quantitative perspectives with the aim of examining similarities and differences across both cultural contexts. Importantly, on several key concepts, discrepancies existed between findings from focus group discussions and quantitative results. For instance, in focus groups, speeding was described as common in both countries but was readily noted as more widespread and as nothing unexceptional by Chinese participants. However, quantitative results for community norms indicated that Australian drivers reported greater agreement than Chinese drivers that in general, drivers believe that is it okay to speed and that they are unlucky if caught. This suggests that community norms are perceived as more favourable towards speeding in Australia than in China. Similarly, the qualitative explorations revealed mixed responses among Australian participants about whether it was embarrassing for others to know about one's speeding tickets while also revealing a total lack of embarrassment surrounding this issue for Chinese participants. However, the quantitative results were to the contrary with Chinese drivers reporting more embarrassment than Australian drivers.

Together these findings suggest that perceptions of community acceptance of speeding are greater in Australia and that Australian drivers are less compliant with speed limits, both of which are contrary to the qualitative findings. This may indeed be an accurate reflection of reality. It may also, however, reflect the situation described by Chinese participants where they would intentionally misrepresent a situation to foreigners to promote a positive national reputation. This 'faking good for foreigners' phenomenon is possibly related to the concept of face and the welfare of the collective having priority in countries such as China that are at the collectivist end of Triandis' individualism-collectivism dimension (2001). This phenomenon has not previously been described in the road safety literature yet is worthy of consideration for those outside China wishing to conduct research there and develop subsequent safety interventions. The Chinese phase of the current research was conducted by Chinese researchers but was sponsored by an Australian university whose name appeared on all forms viewed by participants. Therefore, it is possible that some responses may have been shaped by this phenomenon.

Another concept relevant to the findings relates to reports of the use of social relationships to avoid penalties once caught (as previously described by Xie and Parker 2002). One quarter of respondents reported having used social relationships in this way and reports of knowing others who also do this were prevalent. The concept of *guanxi* seems relevant here; the build up and transfer of social capital via a network of people which is central to every aspect of life in China (Yang, 1994).

A number of limitations are noted. The use of self-report data is acknowledged as potentially biased due to socially desirable responding. In addition, as discussed above, the 'faking good for foreigners' phenomenon may have biased results. However, a foreign researcher (first author) was present when this was discussed in China so it is difficult to assess how influential this may be. There are also potential limitations regarding the generalisability of results. This is particularly relevant in the Chinese context. The relatively small quantitative Chinese sample limits the relevance of findings to the broader driving population. However, this limitation is tempered by fact that this research was exploratory and that, encouragingly, results demonstrate some consistency with previously published investigations of Chinese driver behaviour, thereby adding support to the current findings. Overall, these results highlight the importance of considering culturally-specific societal influences in behavioural research (Triandis, 1997) and represent important considerations in traffic law enforcement and road safety countermeasure design.

5   REFERENCES

Akers, R. L. 1977. Deviant behaviour: A social learning approach (2nd ed.). Wadsworth.

Australian Transport Council. 2011. National Road Safety Strategy 2011-2020.

Elliott, B. 2001. The application of the Theorists' Workshop Model of Behaviour Change to motorists speeding behaviour in WA. Perth: Office of Road Safety.

Field, A. P. 2005. Discovering statistics using SPSS . London: SAGE

Fleiter, J. J. Lennon, A. and Watson, B. 2010. How do other people influence your driving speed? Exploring the 'who' and the 'how' of social influences on speeding from a qualitative perspective. Transportation Research Part F: 13:49-62.

Fleiter, J. J. Watson, B. Lennon, A. King, M. J. and Shi, K. 2009. "Speeding in Australia and China: A comparison of the influence of legal sanctions and enforcement practices on car drivers" Paper at Australasian Road Safety Research Policing Conference.

Fleiter, J. J. Watson, B, Lennon, A, King, M. J. and Shi, K. 2011. Social influences on drivers in China. Journal of the Australasian College of Road Safety 22:29-36.

Gao, G. 1998. "Don't take my word for it." - Understanding Chinese speaking practices. International Journal of Intercultural Relations22:163-186.

Hogg, M. A. Terry, D. J. and White, K. M. 1995. A tale of two theories: A critical comparison of identity theory with social identity theory. Social Psychology Quarterly 58:255-269.

Jensen, G. F. and Akers, R. L. 2003. "Taking social learning global": Micro-macro transitions in criminological theory. Social learning theory and the explanation of crime In R. L. Akers & G. F. Jensen eds. New Jersey: Transaction Publishers.

King, M. J. 2007. "A method for improving road safety transfer from highly motorised to less motorised countries. 14th Conference of Road Safety on Four Continents Bangkok.

Lund, I. O. and Rundmo, T. 2009. Cross-cultural comparisons of traffic safety, risk perception, attitudes and behaviour. Safety Science 47:547-553.

Luo, Y. 2007. Guanxi and business. New Jersey: World Scientific.

Ozkan, T. Lajunen, T. Chliaoutakis, J. E. Parker, D. and Summala, H. 2006. Cross-cultural differences in driving behaviours: A comparison of six countries. Transportation Research Part F 9:227-242.

Pendyala, R. M. and Kitamura, R. 2007. The rapid motorisation of Asia: Implications for the future. Transportation 34:275-279.

Perkins, H. W. and Wechsler, H. 1996. Variation in perceived college drinking norms and its impact on alcohol abuse: A nationwide study. Journal of Drug Issues 26: 961-974.

Rothengatter, T and Manstead, A. 1997. The role of subjective norm in intention to commit traffic violations.Traffic and Transport Psychology,p389-394 Amsterdam: Elsevier.

Shinar, D. 2007. Traffic safety and human behavior. Amsterdam: Elsevier.

Stradling, S. 2007. Car driver speed choice in Scotland. Ergonomics 50:1196-1208.

Triandis, H. 1997. A cross-cultural perspective on social psychology. In The message of social psychology: Perspectives on mind in society, Cambridge: Blackwell

Triandis, H. 2001. Individualism-Collectivism and Personality. Journal of Personality 69: 907-924.

World Health Organization. (2009). Global status report on road safety: Time for action.

Xie, C. and Parker, D. 2002. A social psychological approach to driving violations in two Chinese cities. Transportation Research Part F 5:293-308.

Section VI

Investigations into Safety and Accidents

A Critical Review of the STAMP, FRAM and Accimap Systemic Accident Analysis Models

Peter Underwood, Patrick Waterson

Loughborough Design School
Loughborough University
Loughborough, UK
p.j.underwood@lboro.ac.uk, p.waterson@lboro.ac.uk

ABSTRACT

The systems approach is arguably the predominant method used by researchers for analysing accidents, however, it is not being used in industry. An analysis of the STAMP, FRAM and Accimap systemic models was performed to understand which features of the techniques influence their selection. A combination of limited validation, usability, analyst bias and the implications of not finding an individual to blame for an accident is likely to inhibit the use of systemic analysis techniques.

Keywords: systems approach, accident analysis, STAMP, FRAM, Accimap

1   INTRODUCTION

The systems approach is arguably the dominant paradigm in accident analysis research (e.g. Salmon et al., 2010). It views accidents as the result of unexpected, uncontrolled relationships between a system's constituent parts. In the case of modern safety-critical systems these components include technical and social elements which interact with each other and the environment they exist in, be it physical, commercial or operational etc. The systems approach proposes that understanding how these interactions lead to accidents requires the study of systems as whole entities, rather than considering their parts in isolation.

Various disciplines, such as engineering, psychology and ecology, have used differing interpretations of this required holistic view, however, they all share a worldview focused on complex dynamic systems (Schwaninger, 2006). This worldview can be understood as the core components of the systems approach, as described in the systems theory literature (e.g. Skyttner, 2005), which focus on:

- System structure – the hierarchy of subsystems and the functions they perform
- System relationships – emergent behaviour, resulting from the interaction of system components, cannot be explained by studying these elements in isolation, i.e. the whole is greater than the sum of its parts
- System behaviour – inputs are converted into outputs, via regulated transformation processes, in order to achieve system goals. Dynamic system behaviour means that a goal can be achieved from a variety of initial starting conditions (equifinality). Alternatively, systems can produce a range of outputs from an initial starting point (multifinality). Open systems, e.g. socio-technical systems, are also influenced by their environment.

Systems theory has been advocated in accident analysis research since the 1980's (e.g. Leplat, 1984). However, its popularity did not significantly increase until the late 1990's, when researchers (e.g. Hollnagel, 2004; Leveson, 2004; Rasmussen, 1997) identified that the linear cause-effect basis of existing analysis models represented a theoretical limitation. Describing accidents in a sequential fashion is arguably inadequate as it is unable to sufficiently explain the non-linear complexity of modern-day socio-technical system accidents (Hollnagel, 2004, Lindberg et al. 2010). It can also guide analysts to search for the 'root cause' of an accident and stopping an investigation when a suitable culprit is found to blame may result in too superficial an explanation to correctly inform the development of safety recommendations (Leveson, 2004).

The systems approach was employed to resolve these limitations and has been used as the conceptual foundation for various accident analysis methods and models, e.g. STAMP.

1.1   Aims and Objectives

Despite the proposed advantages of the systems approach, there is evidence to suggest that methods and tools employing a systemic perspective are not being adopted in practice (e.g. Salmon et al., 2012).

The aim of this study is, therefore, to identify and understand the features of the systemic models which potentially hinder their adoption by practitioners. This will be achieved by meeting the following objectives:

- Review the scientific literature to identify the available systemic models
- Conduct a citation analysis to assess their relative popularity within the research community
- Evaluate the models to identify factors which may influence their usage

2   METHODS

2.1   Model Identification

A systematic electronic search for documents referencing a systemic model was conducted in 22 safety, systems engineering and ergonomics related journals. The results were combined with those gained from reference and citation tracking of key review and model evaluation articles (e.g. Sklet, 2004) and personal knowledge of the literature. A manual examination of the documents followed, in order to identify examples of systemic analysis tools.

A search of the Science Direct, PsychINFO, MEDLINE and Google Scholar databases was conducted to identify the number of citations received by each model. The most frequently cited techniques were shortlisted for further analysis, via the evaluation framework described in section 2.2. Other selection criteria, such as whether the tool was recently developed, have been used in previous studies (e.g. Sklet, 2004). However, citation count ranking was chosen as it provides a measure of a model's relative popularity and, therefore, the likelihood of its awareness within the practitioner community.

2.2   Model Evaluation

Previous studies have developed methods to evaluate various theoretical and practical aspects of accident analysis tools, some of which incorporate elements of the systemic analysis approach (e.g. Lehto and Salvendy, 1991). None, however, consider the systems approach in its entirety and an evaluation framework comprising three sections was designed to resolve this.

The development of each technique was considered, with regards to the general process of creating any system model (e.g. Bamber, 2003 p.240), to identify whether any stage had not been fulfilled and if this would affect its selection.

The ability to employ the systems approach is governed by the number of its concepts incorporated within a model. Therefore, each technique was analysed to identify how they address system structure, relationships and behaviour.

Additional theoretical and practical factors which affect model usage were also incorporated to highlight other potential limitations. These issues were selected by identifying relevant components within existing evaluation methods (e.g. Sklet, 2004) and including/excluding them based on their predominance within the systemic model literature. Figure 1 shows the framework structure.

The models were examined by performing a combined theoretical and inductive thematic analysis of the systemic technique literature, as described by Braun and Clarke (2006), with the different evaluation framework criteria forming the topics of interest. The analysis was conducted using NVivo 9.

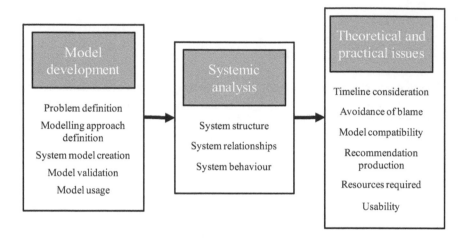

Figure 1- Evaluation framework structure

3   RESULTS

3.1   Model Identification

A total of 13 systemic models were identified within the 449 non-duplicated articles collected during the literature search. Performing a citation search for these techniques revealed a total of 476 documents (excluding duplicate articles and those unrelated to accident analysis), which were manually searched for explicit references to the models. The three most cited models, STAMP, FRAM and Accimaps, accounted for 52.0%, 19.9% and 17.9% of the 302 identified references respectively and were selected for additional evaluation. The remaining models were discounted from further analysis.

3.2   Model Evaluation

This section provides details of the STAMP, FRAM and Accimap evaluations.

STAMP

STAMP describes systems as a 'hierarchy of control based on adaptive feedback mechanisms' and provides an understanding of how a lack of system safety control, at both the design and operational stages, produces accidents (Leveson, 2004). The model was formally introduced by Leveson (2004) and its generic nature has seen it used in retrospective and prospective analyses within various domains, e.g. aerospace (Johnson and Holloway, 2003) and led outdoor activities (Salmon et al., 2012). Little work has been done to formally validate the model and an evaluation conducted by Hollnagel and Spezali (2008) led the authors to declare that STAMP must still be considered as in need of further development.

The system approach elements of (control structure) system hierarchy, component interactions and regulation are visually incorporated into the STAMP diagram. The majority of the remaining concepts are implicitly addressed.

Conflict exists within the literature regarding the training and expertise requirements and the usability of STAMP. For example, it is claimed that the model requires considerable effort to use and is only suitable for experienced users with extensive theoretical and domain knowledge (Hollnagel and Spezali, 2008; Johansson and Lindgren, 2008). However, Johnson and Holloway (2003) comment that STAMP analysis is simple, easy to follow and quick to learn. The lack of formal guidance also provides flexibility for the analyst and encourages them to consider interactions across the whole system, look beyond the proximal accident events, and consider the context of the actors involved (Hovden et al. 2011; Johnson and Holloway, 2003; Kontogiannis and Malakis, 2012).

FRAM

The FRAM model graphically describes systems as a network of interrelated sub-systems and functions which, although designed otherwise, will exhibit varying degrees of performance variation (Hollnagel and Goteman, 2004). The performance variability of any given system component can 'resonate' with that of the remaining elements, produce emergent variation that is too high to control, and result in an accident. FRAM was developed by Hollnagel (2004) to act as both an accident analysis and risk assessment tool. Consequently there are number of examples for both methods of application, e.g. aircraft collisions (Herrera and Woltjer, 2010) and rail network control (Belmonte et al., 2011). No formal evaluations of the validity and reliability of FRAM have been conducted to date and the developmental nature of the model has been highlighted in the literature (e.g. Herrera and Woltjer, 2010).

The vast majority of the systems theory concepts are explicitly represented by a FRAM diagram, with the main focus centred on the interaction of system components and the associated effects.

No formal usability assessment has been conducted on the model, however, various researchers have highlighted benefits and drawbacks to using it. The application of FRAM is structurally simple but, due to its theoretical grounding, requires an initial learning period coupled with extensive domain and human factors knowledge (Hollnagel and Spezali, 2008). While Herrera and Woltjer (2010) remark the user is guided towards explicitly identifying the systemic factors associated with the accident and why they occurred, a need for a more structured approach has been identified (Herrera and Woltjer, 2010; Stringfellow, 2010).

Accimap

The Accimap is a graphical representation of a particular accident scenario which shows the causal flow of events (acts and decisions) throughout the various

390

system hierarchical levels, as described in the Risk Management Framework devised by Rasmussen (1997) (Svedung and Rasmussen, 2002). It was created and first used by Rasmussen (1997) and has subsequently been employed to analyse accidents in various domains, e.g. aerospace (Johnson and de Almeida, 2008), and led outdoor activities (Salmon et al., 2010; Salmon et al., 2012). As with STAMP and FRAM, the model's generic nature means that it can be applied in any domain (Salmon et al., 2010). Although a formal validation process has been applied to the model, its validity and reliability remain in dispute (see Branford, 2007).

System hierarchy and the interactions, inputs and outputs, transformation processes and regulation of system components are explicitly represented in the Accimap diagram. The remaining systems theory concepts are either implicitly considered or not accounted for.

Use of Accimaps requires training and formal education and there is a current lack of usage guidance, which affects the model's accessibility and consistency across studies (Branford et al., 2009; Salmon et al. 2010; Sklet, 2004). It is argued, however, by Branford (2011) that it provides a clear and concise summary of the accident and propagation of events across the entire system structure can be visualised, which facilitates the creation of high-level safety interventions (Branford et al., 2009; Johnson and de Almeida, 2008).

4   DISCUSSION

The following sections present the identified themes which are believed to have the potential to significantly affect the selection of systemic models.

4.1 Model Validation

A lack of validation would appear most likely, in the first instance, to be the aspect of systemic model development which affects their selection by practitioners. Indeed, the other criteria have been met: the creators of the models have explained the objectives and analysis approaches of the techniques, provided a means of modelling any system and have seen their tools applied across multiple domains. Although all three models explicitly incorporate several systemic concepts and therefore provide a degree of face, content and construct validity, as described by Branford (2007, p.97-98), these forms of validity cannot be proven. Despite this, the research community is still advocating the use of systemic models, based on the assumption that they are conceptually valid

However, Reason (2008 p.95) comments that there is no single right view of accidents and finding the 'truth' is less important than practical utility. In this context, it is arguable that empirical validity is the dominant influence on a practitioner's model selection. Whilst empirical validation of systemic tools has occurred within research, via a number of accident analysis case studies (e.g. Salmon et al, 2012), it is far from extensive. As most practitioners in safety-oriented businesses tend to prefer well established methods and concepts, it is unlikely that they would use a relatively unproven systemic technique, unless a business case could be produced to justify otherwise (Johansson & Lindgren, 2008).

4.2 Usability and Analyst Bias

The usability of an analysis technique is affected not only by its features but also by the characteristics of the users, the tasks they are carrying out and the technical, organisational and physical environments in which it is used (Thomas and Bevan, 1996). This raises several issues, outlined below, which may affect model selection to a greater or lesser extent, depending on the analyst and their environment.

Usage guidelines

The perceived benefits and drawbacks of the limited model application guidance provided in the literature (see section 3.2) are indicative of the varying usability requirements of analysts. It is arguable that individuals who prefer the flexibility offered by a lack of usage guidance are more likely to adopt a systemic technique, as opposed to methods employing a more structured approach (e.g. HFACS; Wiegmann and Shappell, 2003). However, given that flexible non-systemic methods are also available, e.g. Why-Because Analysis (Ladkin, 2005), it seems other factors, described below, may also influence systemic model selection.

Cognitive style and previous experience

Although each model is based on aspects of the systems approach, they differ significantly in their theoretical underpinnings, means of application and graphical output (Salmon et al., 2012). There is little guidance on the relative benefits of the techniques and selection of one over another is likely to depend on the analyst's cognitive style and, therefore, their perception of which model is most usable. This helps explain their relative popularity but also suggests that the systems approach itself will not be aligned with the cognitive styles of some practitioners. This mismatch, therefore, may contribute to the general lack of systemic model use within industry.

In addition to cognitive style, an individual's previous experience will also affect their analysis approach and, arguably, their choice of model (Svenson et al., 1999). An analyst who is experienced in the use of sequential techniques may experience cognitive dissonance when presented with the systems approach and resist employing the systemic models.

Resource and regulatory constraints

Practitioners working in any industry will be faced with various resource constraints, e.g. time and financial budgets. Given that effective use of systemic tools requires a substantial amount of theoretical and multi-disciplinary knowledge, the time and cost required to train an individual (or a team) in the systems approach maybe unjustifiable. In addition, the use of systemic models is comparatively time-demanding in relation to other methods used in industry, which creates an extra barrier to their application (Johansson & Lindgren, 2008)

The level of regulation within an industry is also likely to play a significant part in model selection. Regulators of the nuclear industry, for example, have well established processes for probabilistically demonstrating the risks associated with operating a nuclear power plant. These processes must be adhered to by the power plant designers and operators and the introduction of systemic techniques, particularly given their qualitative nature, is unlikely.

4.3 Lack of Blame

The systems approach actively promotes the avoidance of blaming a single individual for causing an accident. However, searching for a human error makes it easier to find out who is responsible for the accident and who should be held accountable (Rieman & Rollenhagen, 2011). The financial and legal implications of apportioning blame can be vast and analysts may, therefore, be incentivised to use non-systemic techniques to ease the identification of culpable personnel.

4.4 Implications for The Systems Approach

Whilst one of the issues listed above may be sufficient for an individual to discard a systemic technique, it is more likely that all of these factors, to a greater or lesser extent, combine to inhibit the adoption of the systems approach. If it is to be adopted, this multi-faceted barrier must be reduced to the point where, from the viewpoint of practitioners, the benefits of using the systems approach outweigh costs. Given that that the numerous elements of the practitioner community (e.g. accident investigators, safety managers, regulators) have varying objectives, resource constraints and perspectives etc., solving this problem will be a significant challenge. A first step in identifying a solution should be improving the communication between the research and practitioner communities, with the aim of reaching a common understanding of the analysis model requirements of each party.

5   CONCLUSION

The systems approach is being promoted within the research literature as the conceptually preferred means of analysing socio-technical system accidents. However, the systemic analysis models developed to apply it are not being used within industry.

A systematic literature search was performed to identify examples of these systemic tools. A thematic analysis was subsequently performed on the literature related to the highest profile models (STAMP, FRAM and Accimap) to understand which features of the techniques are influential on their selection.

Model validation, usability, analyst bias and the implications of not apportioning blame for an accident were identified as the key issues which may influence the use of the systems approach within industry. It is likely that all of these factors, to a greater or lesser extent, combine to inhibit the use of systemic analysis techniques.

6   REFERENCES

Bamber, L. 2003. Risk management: Techniques and practices. In. *Safety at work.* 6th edn, eds. J.R. Ridley and J. Channing. London: Butterworth-Heinemann, pp. 227-262.

Belmonte, F., Schoen, W., and Heurley, L., et al. 2011. Interdisciplinary safety analysis of complex socio-technological systems based on the functional resonance accident model: An application to railway traffic supervision. *Reliability Engineering & System Safety* 96(2): 237-249.

Branford, K. 2011. Seeing the big picture of mishap: Applying the AcciMap approach to analyze system accidents. *Aviation Psychology and Applied Human Factors* 1(1): 31-37.

Branford, K. 2007. *An investigation into the validity and reliability of the AcciMap approach.* Australian National University.

Branford, K., Naikar, N., and Hopkins, A. 2009. Guidelines for accimap analysis. In. *Learning from high reliability organisations.* ed. A. Hopkins. Sydney, Australia: CCH Australia, pp. 193-212.

Braun, V. and Clarke, V. 2006. Using thematic analysis in psychology. *Qualitative research in psychology* 3(2): 77-101.

Herrera, I.A. and Woltjer, R. 2010. Comparing a multi-linear (STEP) and systemic (FRAM) method for accident analysis. *Reliability Engineering & System Safety* 95(12): 1269-1275.

Hollnagel, E. 2004. *Barriers and accident prevention.* Aldershot: Ashgate Publishing Limited.

Hollnagel, E. and Goteman, Ö. 2004. The functional resonance accident model. In. *Cognitive System Engineering in Process Control 2004.*

Hollnagel, E. and Speziali, J. 2008. *Study on developments in accident investigation methods: A survey of the'state-of-the-art.* SKI Report 2008:50. Sophia Antipolis, France: Ecole des Mines de Paris.

Hovden, J., Størseth, F., and Tinmannsvik, R.K. 2011. Multilevel learning from accidents – case studies in transport. *Safety Science* 49(1): 98-105.

Johansson, B. and Lindgren, M. 2008. A quick and dirty evaluation of resilience enhancing properties in safety critical systems. In: *Third Symposium on Resilience Engineering,* eds. E. Hollnagel, F. Pieri and E. Rigaud. Juan-les-Pins, France.

Johnson, C.W. and de Almeida, I.M. 2008. An investigation into the loss of the Brazilian space programme's launch vehicle VLS-1 V03. *Safety Science* 46(1): 38-53.

Johnson, C.W. and Holloway, C.M. 2003. The ESA/NASA SOHO mission interruption: Using the STAMP accident analysis technique for a software related 'mishap'. *Software: Practice and Experience* 33(12): 1177-1198.

Kontogiannis, T. and Malakis, S. 2012. A systemic analysis of patterns of organizational breakdowns in accidents: A case from helicopter emergency medical service (HEMS) operations. *Reliability Engineering & System Safety* 99: 193-208.

Ladkin, P.B. 2005. *Why-because analysis of the Glenbrook, NSW rail accident and comparison with Hopkins's accimap.* Research Report RVS-RR-05-05. Bielefeld, Germany: Bielefeld University.

Lehto, M. and Salvendy, G. 1991. Models of accident causation and their application: Review and reappraisal. *Journal of Engineering and Technology Management* 8: 173-205.

Leplat, J. 1984. Occupational accident research and systems approach. *Journal of Occupational Accidents* 6(1-3): 77-89.

Leveson, N. 2004. A new accident model for engineering safer systems. *Safety Science* 42(4): 237-270.

Lindberg, A., Hansson, S.O., and Rollenhagen, C. 2010. Learning from accidents – what more do we need to know? *Safety Science* 48(6): 714-721.

Rasmussen, J. 1997. Risk management in a dynamic society: A modelling problem. *Safety Science* 27(2-3): 183-213.

Reason, J. 2008. *The human contribution: Unsafe acts, accidents and heroic recoveries.* Farnham: Ashgate.

Reiman, T. and Rollenhagen, C. 2011. Human and organizational biases affecting the management of safety. *Reliability Engineering & System Safety* 96(10): 1263-1274.

Salmon, P., Williamson, A., and Lenné, M., et al. 2010. Systems-based accident analysis in the led outdoor activity domain: Application and evaluation of a risk management framework. *Ergonomics* 53(8): 927-939.

Salmon, P.M., Cornelissen, M., and Trotter, M.J. 2012. Systems-based accident analysis methods: A comparison of Accimap, HFACS, and STAMP. *Safety Science* 50(4): 1158-1170.

Santos-Reyes, J. and Beard, A.N. 2006. A systemic analysis of the Paddington railway accident. *Proceedings of the Institution of Mechanical Engineers, Part F: Journal of Rail and Rapid Transit* 220(2): 121-151.

Schwaninger, M. 2006. System dynamics and the evolution of the systems movement. *Systems Research and Behavioral Science* 23(5): 583-594.

Sklet, S. 2004. Comparison of some selected methods for accident investigation. *Journal of Hazardous Materials* 111: 29-37.

Skyttner, L. 2005. *General systems theory: Problems, perspectives, practice.* 2nd edn. London: World Scientific Publishing Ltd.

Stringfellow, M.V. 2010. *Accident analysis and hazard analysis for human and organizational factors.* Cambridge, USA: Massachusetts Institute of Technology.

Svedung, I. and Rasmussen, J. 2002. Graphic representation of accident scenarios: Mapping system structure and the causation of accidents. *Safety Science* 40(5): 397-417.

Svenson, O., Lekberg, A., and Johansson, A.E.L. 1999. On perspective, expertise and differences in accident analyses: Arguments for a multidisciplinary integrated approach. *Ergonomics* 42(11): 1561-1571.

Thomas, C. and Bevan, N. 1996. *Usability context analysis: A practical guide.* Version 4.04. Teddington: National Physical Laboratory.

Wiegmann, D.A. and Shappell, S.A. 2003. *A human error approach to aviation accident analysis: The human factors analysis and classification system.* Burlington, USA: Ashgate Publishing Ltd.

CHAPTER 40

Learning Safety via Experience-based Software (SIM-ERROR)

Masayoshi Shigemori, Ayanori Sato, Takayuki Masuda

Railway Technical Research Institute
Tokyo, JAPAN
gemo@h9.dion.ne.jp

ABSTRACT

The authors developed an experience-based learning software named SIM-ERROR, on which we can learn human error prevention by using the point and call method. Many workers in Japan use the point and call method primarily to recognize objects and verify their correct operation. The SIM-ERROR includes five tasks, in which workers can experience various error prevention functions of the point and call method, by comparing their performances with vs. without the point and call method. The authors validated the effectiveness of each task in the SIM-ERROR software using training for train operators. A total of 736 apprentice train operators participated in the training. They performed one of five tasks and were shown their error rate graph. Next, the instructor explained to them the methodology of the task. Then the participants answered a five point scale questionnaire about the five human error prevention effect of the point and call method. A comparison of the respective mean subjective ratings for five point and call method functions, before and after the training session, indicated that the participants were more convinced of the human error prevention effectiveness of the point and call method after the training than before in the three of five tasks. In the other two tasks, there were no differences between pre- and post ratings, but very high ratings were shown.

Keywords: safety, training, accident prevention

1   PREVENTION OF HUMAN ERROR

Human error causes most accidents in various industries; aviation, railway, and medical industry etc. (Shappell et al., 2006, Kohn et al., 2000, Baysari et al., 2008). Although several countermeasures have been used since the beginning of recorded time, human errors remain one of the most common causes of accidents. People have developed diverse safety technology; automated operation system, warning device, human-centered design. Nonetheless, we couldn't remove human error factors from worksite completely. Therefore, we should take measures not only deploying hardware but also software, and flesh-ware training for skill, knowledge, and attitude. In fact, the CRM training and non-technical skill training have gained recognition for a long time and have achieved noteworthy result (Flin et al., 2003, Goeters, 2002).

This study introduces one countermeasure to human error used widely in Japan and a software developed for learning this countermeasure named the point and call method. In addition, this study will highlight the usefulness of the training, which uses the SIM-ERROR in the experiment.

2   POINT AND CALL METHOD

The importance of recognizing objects and accurately operating something has become an essential skill set for life. Few if any, of our daily activities do not involve reading signs, telling the time, recognizing each manipulation in the procedure. Therefore, in order to accurately perceive, recognize and verify, we sometimes indicate target object by pointing it with our finger or operate something while saying the procedure out loud.

Train operators in Japan use a similar point and call method consciously in order to prevent their human error. For example, the train operators in Japan check the railway signal while pointing at it with their finger and calling its phase.

The human error prevention effect of the point and call method has been validated using choice reaction task experiments (Kiyomiya et al., 1965, Haga et al., 1996). Those experiments produced lower mean error rate during task execution with the point and call than without the point and call. Possible reasons for these finding include the following: (1) the point accelerates and maintains for prolonged periods eye-gaze, (2) the call strengthens the memory of the action, (3) the call makes a person more aware of human error, (4) muscle stimulation by the point and call enhance arousal level (Iiyama, 1980), and (5) delay associated with the point suppress impulsive behavior (Haga et al., 1996). Each human error prevention effects of the point and call method has been assessed (Shigemori et al., 2009, Sato et al., 2011, Shinohara et al., 2009). This method should play a useful role in preventing human error not only for train operators but also anyone who perceives objects or engages in complicated operations. Indeed, various workers, including power plant operators, workers of the factory, nurses, use the point and call method in Japan.

The managers of railway companies frequently complain that train operators use the point and call method poorly or that someone did not use it at all, although the managers and companies recommend or stipulate the point and call method as a rule. There might be a number of reasons why train operators don't do the point and call method in the approved manner, embarrassment or fatigue, for instance. A more plausible reason could be that it is difficult for train operators to recognize the effectiveness of the point and call method while at work. In fact, even in the experimental situation, error rates tend to be generally low. In the experiment by Haga et al. (1996), for example, the mean error rate without the point and call condition was 2.38%, compared to with the point and call condition score of 0.38%. Although the difference of error rates between conditions appears statistically large, the participants hardly recognize 2% differences.

Therefore, the authors developed software (SIM-ERROR) designed to experience the impact of the point and call method. Workers could better realize the human error prevention effect of the point and call method via the SIM-ERROR, thus urge workers to use the point and call method properly.

3   SIM-ERROR

The SIM-ERROR software allows workers experience the human error prevention effect of the point and call method. It includes five tasks corresponding to five human error prevention functions of the point and call method; (1) the dots counting task for optical acuity by pointing, (2) the n-Back task for memory effect by calling, (3) the go/no-go task for monitoring effect by calling, (4) the clock task for arousal maintenance effect by pointing and calling, and (5) the wait-paper-rock-scissors task for slow reaction effect by pointing. For all tasks, first the workers perform trials without pointing or calling and then second they do the similar trials with pointing, calling, or pointing and calling. After the task, they can recognize the effectiveness of the point and call as seen in graphs showing their error rates.

In the dots counting task, the instructor asks the workers to count dots randomly-scattered on the screen. Actual trials follow two practice trials. The actual trials consists five on the first half and five on the second half. The number of dots differs trial by trial from 26 to 34 (30 on average). The time interval for each trial depends on the number of dots, or 40 ms/ dot. In the second session, the workers are asked to count dots while pointing them.

In the n-Back task, the instructor asks the workers to select a color circle by clicking alternatives on the lower side of the screen for a target circle at the center as quickly as possible. It includes five color circles; red, blue yellow, black, white. After some choice reaction trials, the instructor asks the workers which color they responded in one and two trials before (n = 1, 2), by cricking the alternatives. One recognition judgment trial is included in each block. The numbers of the choice reaction trials differ from 12 to 16. In the first session, the workers are asked to select a color circle without saying a word, while in the second session, the workers are asked to do them with saying the color name aloud and they are asked to do them with saying aloud "the the the the...." in the last session. The last procedure is called the articulatory suppression. This task employs this method for bothering workers' internal speech, because some workers might do the choice tasks with saying color name in their head. If the workers employ the internal speech strategy in the first session, difference between first and second sessions will disappear. The instructor can explain the human error prevention effect of the call even in such a case by showing them different scores between the second and the third sessions.

In the go/no-go task, the instructor asks the workers to push a key as quickly as possible when a target stimulus (go stimulus) appears at the center of the screen, while they are asked to stay still when a distractor stimulus (no-go stimulus) appears. The distractor stimulus is a simple line drawing like Smiley Face, while the target stimulus is also a similar drawing but it has tight-lipped mouth with both ends turned down. Because the screen has a lot of target stimuli, the workers tend to push the key by mistake even when a distractor stimulus appears. The distractor stimulus appears after 15 target stimuli appear on average, in the 12 to 18 ranges. It consists seven distractor trials on the each half. In the second session, the workers are asked to push the key for the target stimuli while saying "Osu (it means "push" in Japanese)" and to stay still for the distractor stimuli while saying "Osanai (it means "don't push" or "stay still" in Japanese)".

In the clock task, the instructor asks the workers to monitor a sweep hand and to push a key if they detect an irregular two-steps jump of the hand, which appears once on Lap 2 and 4. The sweep hand needs two seconds for moving one step. It consists four laps on the each half. In the second session, the workers are asked to monitor the sweep hand while pointing it and saying "Yoshi (it means "checked" in Japanese)". The workers are also asked to answer two questionnaires after each half. One concerns fatigue and the other concerns anxiety about missing the irregular jump. The workers answer both questions by moving a sliding switch on a continuum scale on the screen.

In the wait-paper-rock-scissors task, the instructor asks the workers to choose their hand to lose to a dealer's hand, which appears at the center of screen, from among three choices on the lower side of the screen at the last word of command, "Jan, Ken, Pon, Pon". It consists nine trials on the each half. In the second half session, the workers are asked to choose their hand after pointing dealer's hand and the hand they intend to choose. They don't have to do it at the command word at this time.

4 EXPERIMENT

4.1 Purpose

In order to test the effect of the training using the SIM-ERROR, the authors conducted the experiment at a railway company. We developed the SIM-ERROR with the aim of appreciating the human error prevention effect of the point and call method with workers through the training. In addition, we'd like to test it in the practical situation. Therefore, we tested it with workers in railway and supervisors in the railway company also acted as trainers.

4.2   Method

Participant

A total of 736 apprentice train operators participated in the training. 118 participants did the dots counting task, 123 participants did the n-Back task, 182 participants did the go/no-go task, 155 participants did the clock task, and 158 participants did the waiting paper-rock-scissors task. One of five supervisors conducted the trainings.

Procedure

The experiment was conducted in a group setting. One session included about 35 participants. Each of them seated in front of a PC. Before the task, the participants answered a five point Likert-type scale (1. Not at all acceptable – 5. Very acceptable) questionnaire about the five human error prevention effect of the point and call method as follows; (1) the point accelerates and maintains for prolonged periods eye-gaze, (2) the call strengthens the memory of the action, (3) the call makes a person more aware of human error, (4) muscle stimulation by the point and call enhance arousal level, and (5) delay associated with the point suppress impulsive behavior. The questionnaire was printed on an A4 size paper.

The participants performed one of the five tasks of the SIM-ERROR at the direction of the instructor and were shown their error rates on each half in bar graph form after the task. The tasks were allocated in random order from group to group. Subsequently, the instructor explained more about the task, the general trend of the results, human error prevention effect of the point and call method in relation to the conducted task, relation between experience in the task and that of the actual job, and other human error prevention effect of the point and call method. At the end of the session, the participants answered the same questionnaire, which they did before the task. Each session, including performing the task, answering the questionnaires, and the explanation by the instructor, took about 30 minutes on average.

4.3   Result

We compared the respective mean subjective ratings for five point and call method functions between before and after the training session (Figure 1).

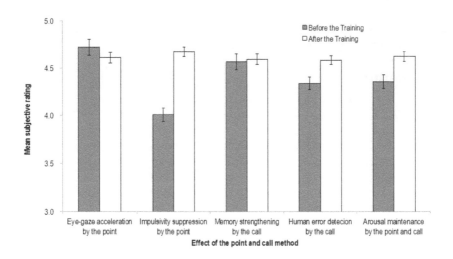

Figure 1 Mean subjective rating before and after the training on each task. Bars mean standard errors.

    The 2 x 5 mixed configuration ANOVA showed the main effect of the training (F (1, 731) = 15.88, MSe = 0.27, p < .01, η^2 = 0.02), the task (F (1, 731) = 6.49, MSe = 0.68, p < .01, η^2 = 0.02), and the interaction effect (F (4, 731) = 23.38, MSe = 0.27, p < .01, η^2 = 0.03). These results mean that the trainings improved participants' subjective estimate of the point and call method holistically but the effect depend on the tasks.

    In order to bring out interaction effect, we conducted the simple main effect test for each task. The result indicated that the mean subjective ratings after training increased in the wait- paper-rock-scissors task (F (1, 731) = 118.90, MSe = 0.27, p < .01, η^2 = 0.16), the go/no-go task (F (1, 731) = 15.73, MSe = 0.27, p < .01, η^2 = 0.02), and the clock task (F (1, 731) = 18.84, MSe = 0.27, p < .01, η^2 = 0.03). But there were no differences in the dots counting task (F (1, 731) = 3.27, MSe = 0.27, p > .05, η^2 = 0.004) and in the n-Back task (F (1, 731) = 0.16, MSe = 0.27, p > .05, η^2 = 0.0002). These results indicate that, at least, the training concerning impulsivity suppressive effect by the point using the wait-paper-rock-scissors task, the training concerning human error detection effect by the call, and the training concerning arousal maintenance effect by the point and call, have beneficial effect on awareness of the point and call method.

    Why did the trainings used in the other two tasks didn't have an effect on awareness? We conducted the simple main effect test to access contrast between before and after the trainings. The result showed that difference among tasks before the trainings (F (4, 731) = 19.09, MSe = 0.55, p < .01, η^2 = 0.10) while no difference after the training (F (4, 731) = 0.43, MSe = 0.55, p > .05, η^2 = 0.002). The paired comparison tests among the tasks before the training showed that both eye-gaze acceleration effect by the point and the memory strengthening effect by the call had the highest subjective ratings among other effects and that there was no difference ratings between the eye-gaze acceleration effect and the memory strengthening effect ($MSe = 0.55$, $p < .05$, $LSD = 0.17$). The results indicated that participants had high awareness concerning both effects before the trainings. Therefore, no differences of subjective ratings between before and after training might indicate the ceiling effect.

4.4 Discussion

In summary, the results indicated that the participants were more convinced of the human error prevention effectiveness of the point and call method after the training than before it. Of course, increments were not shown in the two tasks; the dots counting task for optical acuity by pointing and the n-Back task for memory effect by calling, but even in those cases, participants were sufficiently aware of the effect even before the training.

The reason that experience via the SIM-ERROR training enhanced participants' awareness of the point and call method must be due to empathy. Prior researches indicate that empathy needs similar experience (Batson et al., 1996, Eklund et al., 2009, Hodges et al., 2010). It is difficult for workers to experience human error prevention effect of the point and call method at work because human errors themselves rarely occur. The participants could experience human errors and their prevention effect via the SIM-ERROR. Those experiences must enhance awareness of the point and call method. Although we didn't describe scores of the tasks in this manuscript, mean error rates on the condition with pointing and/or calling were lower than those on the condition without pointing and/or calling on all tasks. These results indicated that the participants appeared human error prevention effect in the training.

5   GENERAL DISCUSSION

The authors developed the experience-based software for learning safety named SIM-ERROR. It included five tasks in order to experience human error prevention effect of the point and call method. The point and call method is one countermeasures for preventing human error, which is widely used by workers, in particular, train operators in Japan. Its effects on human error prevention include eye-gaze acceleration, impulsivity suppression, memory strengthening, human error detection, and arousal maintenance. These functions prevent human error not only in rail industry but also in other industries, including aviation, power-generating, and medical industry. In fact, the software has been already used by various industries; railway companies, electric power companies, and other factories.

The SIM-ERROR software currently functions at half capacity since it is incomplete. The authors intend to develop the software as the SIM-ERROR, in which workers can experience not only human error prevention effects of countermeasures but also human errors themselves. Fundamentally, the name of the software means simulation of human error. The authors are developing tasks on which workers can experience human errors and learn their mechanisms.

In addition of the tasks themselves, the training method is also important. In the experiment, the participants initially performed one of the tasks, secondly checked out their performance in graph form, and finally received an explanation and have discussion. In the discussion, the instructor made participants think about relation between experience on the software and their real work. Such a discussion had two significant roles. One was voluntary derivation of answers and the other was to find relation between experience on the software and the actual world. The experience on the SIM-ERROR must impact on safety behavior at work. For this, the experience on the SIM-ERROR must be generalized down to behavior at work. If workers regard the experience on the SIM-ERROR as temporary experience, the training becomes meaningless. The discussion about the relation between experience on the software and work will assist in bringing reality to the training.

REFERENES

Batson, C. D., Sympson, S. C., Hindman, J. L., Decruz, P. & Et Al. 1996. "I've been there, too": Effect on empathy of prior experience with a need. *Personality and Social Psychology Bulletin;Personality and Social Psychology Bulletin,* 22, 474-482.

Baysari, M. T., Mcintosh, A. S. & Wilson, J. R. 2008. Understanding the human factors contribution to railway accidents and incidents in Australia. *Accident Analysis & Prevention,* 40, 1750-1757.

Eklund, J., Andersson-Stråberg, T. & Hansen, E. M. 2009. "I've also experienced loss and fear": Effects of prior similar experience on empathy. *Scandinavian Journal of Psychology;Scandinavian Journal of Psychology,* 50, 65-69.

Flin, R., Martin, L., Goeters, K.-M., H☐Mann, H.-J. G., Amalberti, R., Valot, C. & Nijhuis, H. 2003. Development of the NOTECHS (non-technical skills) system for assessing pilots' CRM skills. *Human Factors and Aerospace Safety,* 3, 97-119.

Goeters, K.-M. 2002. Evaluation of the effects of CRM training by the assessment of non-technical skills under LOFT. *Human Factors and Aerospace Safety,* 2, 71-86.

Haga, S., Akatsuka, H. & Shiroto, H. 1996. "Shisa kosyo" no error boushi kouka no shitsunai jikken niyoru kensyo [Laboratory experiments for verifying the effectiveness of "finger-pointing and call" as a practical tool of human error prevention. *Japanese Association of Industrial/Organizational Psychology Journal,* 9, 107-114.

Hodges, S. D., Kiel, K. J., Kramer, A. D. I., Veach, D. & Villanueva, B. R. 2010. Giving birth to empathy: The effects of similar experience on empathic accuracy, empathic concern, and perceived empathy. *Personality and Social Psychology Bulletin;Personality and Social Psychology Bulletin,* 36, 398-409.

Iiyama, Y. 1980. Shisasyoko no kouyou to ouyou: Sono kagakuteki haikei [Utility and application of the point and call method: Its scientific background]. *Anzen [Safety]*, 31, 28-33.

Kiyomiya, E., Ikeda, T. & Tomita, Y. 1965. Fukuzatu sentaku hanno niokeru sagyohouhou to performance tono kankei ni tsuite: Sisakanko no kouka ni tsuiteno yobiteki kento [Relation between method and performance in a choice reaction task: Introductory investigation of the point and call method] *Tetsudo Rodo Kagaku [Railway Laver Science]*, 17, 289-295.

Kohn, L. T., Corrigan, J. M. & Donaldson, M. S. (eds.) 2000. *To err is human: Building a safer health system,* Washington, D.C.: National Academy Press.

Sato, A., Shigemori, M., Masuda, T., Hatakeyama, N. & Nakamura, R. 2011. Memory accelerated effect of the point and call method. *The 75th Conference of the Japanese Psychological Association.* Tokyo, Japan.

Shappell, S. A., Detwiler, C. A., Holcomb, K. A., Hackworth, C. A., Boquet, A. J. & Wiegmann, D. A. 2006. Human error and commercial aviation accidents: A comprehensive, fine-grained analysis using HFACS. Washington, D. C.: Federal Aviation Administration.

Shigemori, M., Saito, M., Tatebayashi, M., Mizutani, A., Masuda, T. & Haga, S. 2009. Human error prevention effect of the point and call method. *The 6th Conference of Japanese Society for Cognitive Psychgology.* Niiza, Japan.

Shinohara, K., Morimoto, K. & Kubota, T. 2009. The effect of "finger-pointing and call" on orientation of visual attention. *The Japanese Journal of Ergonomics,* 45, 54-57.

CHAPTER 41

Differential and Unique Roles of Off-road Assessments to On-road Safety

Aksan, N.[1], Dawson, JD[2], Anderson, SA[1], Uc, E.[1,3] & Rizzo, M[1,4,5]

[1]Department of Neurology, University of Iowa
[2]Department of Biostatistics, University of Iowa
[3]Veterans Affairs Medical Center
[4]Department of Mechanical & Industrial Engineering, University of Iowa
[5]Public Policy Center, University of Iowa
Iowa City, IA USA
nazan-aksan@uiowa.edu

ABSTRACT

Crash statistics from state records are often considered the gold-standard in judging the on-road safety of older drivers in the US. However, it is becoming increasingly evident state records have limitations and that on-road tests can better capture the safety risk potential of older drivers. We examined performance metrics derived from an 18-mile on-road test in relation to functioning off-road in a large sample of 345 adults with common neurodegenerative diseases (N=160) associated with aging (e.g. Alzheimer's, Parkinson's and Stroke) and controls without neurodegenerative diseases (N=185). The measures of driver safety included overall safety errors, serious errors that could have resulted in crashes or near-crashes had the circumstances been different at the time they occurred, in addition to performance in essential driving tasks such as navigation to a specified destination and identifying landmarks while driving. Off-road functioning measures were derived from visual, motor and cognitive domains. The findings showed that adults with neurodegenerative conditions committed greater numbers of overall safety errors, serious errors, and performed more poorly in navigation and landmark tasks compared to control adults. Findings in the multivariate framework showed that differentiable components of cognitive functioning were relevant to on-road driver safety risk over and above disease status and age. For example, speed of processing, visuospatial construction were particularly relevant to overall and serious errors, while memory and vision were particularly relevant to navigation related secondary driving tasks. Those findings inform design of road tests for elderly and suggest that clinical assessments of driver fitness should consider several domains of off-road functioning.

Keywords: older driver safety, neuropsychology, on-road tests

1   INTRODUCTION

Crash statistics from state records are often considered the gold-standard in judging the on-road safety of older drivers in the US. However, recent reviews and studies are beginning to articulate limitations of crash statistics more explicitly (Aksan et al., 2012; Anstey et al., 2005; Anstey & Woods, 2011; Rizzo et al., 2011). The first goal of this study was to quantify driver safety more comprehensively than crash statistics so as to reflect the complexity of the driving task better. We captured on-road safety of older drivers with a standard road test that included not only safety errors but also secondary tasks critical to driving safety such as ability to navigate through a route and ability to identify landmarks such as restaurants while driving. The second goal was to examine the differential predictive power of off-road assessments, including functioning in vision, motor, and cognitive domains to both safety errors and secondary task performance.

In discussing the limitations of crash statistics in capturing driver safety, Wood and colleagues (Wood, Anstey, Kerr, Lacherez & Lord, 2008) have noted that crash statistics fail to capture safety risks in which crashes were avoided because of the evasive maneuvers of other drivers. Ball and colleagues (Ball, Owsley, Stalvey, Roenker & Graves, 1998; Edwards, Myers, Ross, Roenker, Cissell, McLaughlin & Ball, 2009) have noted that crashes may fail to capture safety risks due to appropriate self-restrictions many older adults make to their risk exposure by avoiding bad weather, night driving, and heavy traffic hours/roads. Crashes may be manifestations of late emerging declines in driving performance, an outcome of long-term accumulated declines in motor, visual, and cognitive functioning (Aksan et al., 2012). Crash statistics also fail to capture performance in essential tasks secondary to driving that bear on road safety. In addition to the ability to control the vehicle while observing the rules of the road, we often need to engage in secondary tasks that are essential to why we drive at all. For example, the ability to navigate to a destination based on directions, ability to rely on dead reckoning to self-correct for navigation errors, identifying landmarks such as restaurants, grocery stores are critical secondary driving tasks that bear on driver's safety risk. On-road tests can be designed to capture a driver's performance in those essential secondary driving tasks in addition to safety errors that stem from inability to control the vehicle while observing rules of the road.

In this study, we used data from a standard 18-mile on-road test that included rural and urban routes, and captured safety errors of drivers. In addition, driver's ability to navigate through a route and identify restaurants were measured. We examined variability in all those performance metrics in unselected samples of older adults including those with and without neurodegenerative diseases. This is a reasonable sampling strategy on the premise that clinicians are often asked to make fitness to drive decisions in unselected populations (Anderson et al., in press) and that often knowledge of diagnosis is inadequate to judge driver safety on the road (Rizzo, 2011). In the current study, both aging adults free of neurodegenerative disease and those with stroke, Parkinson's, and Alzheimer's diseases were included.

We first examined mean differences in performance metrics as a function of disease status. Consistent with previous studies we would expect patients with neurodegenerative diseases to perform worse than those without neurodegenerative disease on all performance metrics (Dawson, Anderson, Dastrup, Uc & Rizzo., 2009; Uc, Rizzo, Johnson, Emerson, Liu, Mills, Anderson & Dawson, 2011). We then examined whether off-road assessments including vision, motor, and cognition tests differentially predicted safety risk outcomes. We would expect vision to be particularly important to performance in landmark identification, and cognition, and in particular memory, to be important to performance in route following. Finally, we examined whether aspects of cognitive functioning, speed of processing, memory, visuospatial construction differentially predicted safety risk outcomes.

2   METHOD

2.1   Sample

The participants were 345 (220 M, 125 F) active drivers between the ages of 50 and 89 (mean age = 71 years), including 185 with no neurologic disease, 40 with probable Alzheimer's disease (AD), 91 with Parkinson's disease (PD), and 29 with stroke. All participants held a valid state driver's license and were still driving. They were recruited from the general community by means of advertisements and from outpatient clinics. Exclusion criteria included alcohol or substance abuse, major psychiatric disease, use of sedating medication, and corrected visual acuity less than 20/50. All participants provided informed consent according to the policies of the Institutional Review Board at the University of Iowa.

2.2   Procedure

Participants took a battery of standardized neuropsychological tests lasting less than 2 hours. Scores from the battery of tests constituted the predictor variables in all analyses. All outcome measures were derived from an 18-mile on road drive test around Iowa City that lasted about 45-minutes in an instrumented vehicle. This drive test took place on a separate day, included both urban and rural routes, and was conducted on days when weather did not lead to poor visibility or road conditions. The test began after a brief acclimation period to the vehicle, and a trained experimenter sat in the front passenger seat to give instructions and operate the dual controls, if needed.

2.3   Off-road Functioning Measures

The battery of standardized neuropsychological tests was selected on the basis of their conceptual relevance to driving and demonstrated sensitivity to brain dysfunction (for test descriptions, see Lezak, Howieson & Loring, 2004; Strauss, Sherman & Spreen, 2006). The tests included: Trail Making Test Part A and Part B (TMT-A, TMT-B), Judgement of Line Orientation, Complex Figure Test-Copy (CFT-Copy), Complex Figure Test-30 Minute Delayed Recall (CFT-Recall), WAIS-III Block Design, Benton Visual Retention Test (BVRT), Controlled Oral Word Association (COWA), Rey Auditory Verbal Learning Test (AVLT), Grooved Pegboard, and Useful Field of View (UFOV – total from all four subtests) (Ball, Owsley, Sloane, Roenker & Bruni, 1993).

The raw scores on the individual tests were reversed when necessary so that high scores represented better functioning on each test, and subsequently transformed to z-scores. Guided by findings from a recent confirmatory factor analytic study (Anderson et al., in press), the scores were then formed into composite scores representing speed of processing (Grooved Peg Board, TMT-A, UFOV), visuospatial construction (CFT-Copy, Judgment of Line Orientation, Blocks), memory (BVRT-E, COWA, AVLT, CFT-Recall). Z-score of TMT-B time to completion score was used to capture attentional set shifting, an aspect of executive functioning. Those four scores were used in all bivariate and multivariate analyses.

Motor functioning was assessed with Functional Reach and Get up and Go tests (Alexander, 1994). Visual sensory functioning was assessed with near, far acuity (Ferris, Kassoff, Bresnick & Bailey, 1982) and contrast sensitivity (Pelli, Robson & Wilkins, 1988) tests. The tests in each domain were standardized and after appropriate reversals averaged into a composite score of motor and visual sensory functioning, in which high scores represented better functioning.

2.4   On-road Safety Outcome Measures

A certified driving instructor reviewed the videotapes of the drive to score safety errors according to the standards of Iowa Department of Transportation (September 7, 2005 version). The scoring generated information on frequency and types of safety errors the participants committed. The taxonomy of 76 errors types (e.g., incomplete stop, straddles lane line) are organized into 15 categories (e.g., stop signs, lane observance). 30 of these errors were classified as critical errors (e.g. entering an intersection on a red light), meaning under a different set of circumstances such errors would lead to crashes. The remaining errors were classified as non-critical errors. A single reviewer evaluated all drives in this study. To evaluate the reliability of this scoring system, a sample of 30 drives was re-reviewed by this instructor and was independently reviewed by a second driving instructor. For total number of errors per drive, the primary reviewer's intra-rater correlation was .95, and the inter-rater correlation was .73.

In addition to safety errors made during the drive, performance in two critical secondary driving tasks was tested during the drive: route following and restaurant identification. In the **route-following task**, the participants were asked to memorize a 4-turn route immediately before navigating through it. The number of incorrect turns along with whether the participant got lost to the point that they could not take self-corrective action to get back on the route was tallied. These two scores were pooled into a weighted composite score such that participants who could self-correct following incorrect turns received more credit for their performance in the secondary task. In the **landmark identification task**, the participants were asked to identify restaurants on either side of a 1.5 mile suburban commercial strip with speed limit ranging from 30 to 45. Some of the restaurants were determined to stand out more against their background, and were labeled as highly salient. The number of high and low salient restaurants identified by participants was tallied. These two scores were pooled into a weighted composite score such that participants who could pinpoint less salient restaurants received more credit for their performance in the secondary task. High scores represented better performance in both the route-following and landmark identification tasks.

3 RESULTS

The findings showed that adults with neurodegenerative conditions committed greater numbers of overall safety errors, $F(1,312) = 36.77$, $p < .001$, serious errors $F(1,312) = 27.36$, $p < .001$, and performed more poorly in navigation, $F(1,312) = 33.94$, $p < .001$, and landmark tasks, $F(1,312) = 37.97$, $p < .001$ compared to control adults. Figure 1 shows the standard score (z-score) in all four performance metrics for diseased and non-diseased groups. Table 2 shows that the bivariate correlations of those performance metrics with visual sensory function, motor function, and cognitive functioning were all significant.

Table 1 Bivariate Pearson correlations among safety risk outcomes and off-road assessments.

Predictors	Safety Risk Outcomes			
	Overall Errors	Serious Errors	Route-Navigation	Landmark Identification
Visual Sensory Function	-.31	-.26	.34	.31
Motor Function	-.22	-.18	.19	.15
Cognitive Function:				
Speed of Processing	-.45	-.38	.46	.35
Visuospatial Construction	-.37	-.34	.37	.25
Memory	-.33	-.32	.49	.32

Note. All correlations are significant at $p < .01$ or better.

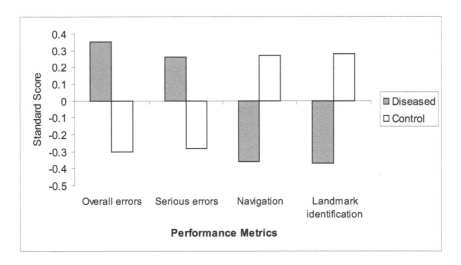

Figure 1 Standard Score on safety risk metrics as a function of disease status.

Table 2 shows the findings in the multivariate framework, where we examined which specific components of cognitive functioning (speed of processing, memory, visuospatial construction, and executive functioning) was the most relevant to each of the four performance metrics while controlling for disease status, age, visual sensory and motor functioning. The findings showed that speed of processing (UFOV-total loss, TMT-A, Grooved Pegboard) and visuospatial construction (Block Design, Judgment of Line Orientation, CFT-Copy) continued to add significantly to prediction of overall errors along with disease status. Visuospatial construction rather than speed of processing continued to significantly add to prediction of serious safety errors over and above disease status. Memory performance (CFT-Recall, COWA, AVLT, BVRT) predicted performance in the navigation and landmark tasks. And visual sensory functioning added uniquely to prediction of landmark identification.

Table 2. Standardized regression coefficients from the final step in multiple regressions predicting all four safety risk outcomes.

Predictors	Safety Risk Outcomes			
	Overall[a] Errors	Serious[a] Errors	Route-Navigation[b]	Landmark Identification[b]
Age	.09	.15*	-.02	-.06
Disease Status	.11+	.14*	.18**	-.18**
Visual Sensory Function	-.01	.01	.06	.14-
Motor Function	.03	.04	.07	-.05
Speed of Processing	-.32**	-.15	-.12	.11

410

Visuospatial Construction	-.16*	-.17*	.05	-.05
Memory	-.02	-.06	-.24**	.16*
Executive Function	-.11	-.07	.06	.02
Overall F	11.7**	8.5**	7.7**	8.0**

+ p < .10, * p < .05, ** p < .01 or better. [a] Overall F(8,331) for overall errors and serious errors. [b] Overall F(8,304) for route-navigation, and landmark identification tasks.

4   DISCUSSION

Elderly with common neurodegenerative conditions performed worse than those without neurodegenerative diseases, consistent with previous findings (e.g. Dawson et al., 2009; Uc et al., 2011). While all domains of functioning had significant associations with on road safety risk metrics, motor functioning had only fair while visual and cognitive functioning had moderate associations with the outcomes in the bivariate setting, consistent with recent reviews by Anstey et al., 2005. In the multivariate setting, the findings show that differentiable aspects of off-road functioning were relevant to the safety risk outcomes. Disease status in general predicted worse performance in navigation and landmark identification tasks, grater numbers of serious safety errors during the drive. Visual sensory functioning continued to add uniquely to performance in landmark identification over and above cognition. Memory was critical to performance in both secondary driving tasks. While both speed of processing and visuospatial construction were relevant to safety errors during the drive, it was only visuospatial construction that predicted serious errors during the drive over an above disease status.

Those findings suggest standard on-road tests which are typically designed to test driving skills of novice drivers, may need to be expanded to include additional safety relevant assessments that may be particularly challenging for older drivers. Finally, the findings suggest that clinicians who are asked to make fitness to drive judgments need to consider functioning in several domains of functioning.

ACKNOWLEDGMENTS

This study was supported by awards R01 AG 17717 and R01AG 15071 from the National Institute on Aging (NIA), by award R01 NS044930 from the National Institute of Neurological Disorders and Stroke (NINDS), and by award R01 HL 091917 from National Heart, Lung, Blood Institute (NHLB). The authors would like to thank the entire Neuroergonomics research team and all participants in the study. Please address all correspondence to Nazan Aksan, Nazan-aksan@uiowa.edu.

REFERENCES

Aksan, N., Anderson, S.W., Dawson, J.D., Johnson, A.J., Uc, E.Y. & Rizzo, M. (2012). Cognitive functioning predicts driver safety on road tests one and two years later. *Journal of the American Geriatric Society*, 60, 99-105.

Alexander NB (1994). Postural control in older adults. *Journal of the American Geriatric Society*, 42, 93–108.

Anderson, S.W., Aksan, N., Dawson, J.D., Uc, E.Y., Johnson, A.M. & Rizzo, M. (in press). Neuropsychological assessment of driving safety risk in older adults with and without neurologic disease. *Journal of Clinical and Experimental Neuropsychology*.

Anstey KJ, Wood J, Lord S, & Walker JG. (2005). Cognitive, sensory and physical factors enabling driver safety in older adults. *Clinical Psychology Review, 25*: 45-65.

Anstey KJ & Wood J, (2011) Chronological age and age-related cognitive deficits are associated with an increase in multiple types of driving errors in late life. *Neuropsychology, 25*, 613-621.

Ball K, Owsley C, Sloane ME, Roenker, D & Bruni, JR (1993). Visual attention problems as a predictor of vehicle crashes in older drivers. *Investigations in Ophthalmology, Visual Sciences, 34*, 3110– 3123.

Ball K, Owsley C, Stalvey B, Roenker DL, Graves M. (1998). Driving avoidance and functional impairment in older drivers. *Accident Analysis & Prevention, 30*, 312-322.

Dawson, J.D., Anderson, S.W., Uc, E.Y., Dastrup, E., and Rizzo, M. (2009). Predictors of driving safety in early Alzheimer's disease. *Neurology*, 72, 521-527.

Edwards, J. D., Ross, L. A., Wadley, V. G., Clay, O. J., Crowe, M., Roenker, D. L., & Ball, K. K. (2006). The useful field of view test: Normative data for older adults. *Archives of Clinical Neuropsychology, 21*, 275–286.

Edwards, J. D., Myers, C., Ross, L. A., Roenker, D. L., Cissell, G. M., McLaughlin, A. M., & Ball, K. K. (2009). The longitudinal impact of cognitive speed of processing training on driving mobility. *Gerontologist, 49,* 485–494.

Ferris FL III, Kassoff A, Bresnick GH & Bailey, I (1982). New visual acuity charts for clinical research. *American Journal of Ophthalmology, 94,* 91–96.

Lezak, MD, Howieson, DB., & Loring, DW. (2004). *Neuropsychological Assessment*, 4th Ed. New York: Oxford University Press.

Pelli DG, Robson JG, & Wilkins AJ. (1988). The design of a new letter chart for measuring contrast sensitivity. *Clinical Vision Sciences, 2*:187–199.

Rizzo M. (2011). Impaired driving from medical conditions: A 70-year-old man trying to decide if he should continue driving. *Journal of the American Medical Association, 305*, 1018-1026.

Strauss E, Sherman EMS, Spreen O. (2006). *A Compendium of Neuropsychological Tests: Administration, Norms, and Commentary*, 3rd Ed. New York: Oxford University Press.

Uc, E. Y., Rizzo, M., Johnson, A. M., Emerson, J. L., Liu, D., Mills, E. D., Anderson, S. W., & Dawson, J. D. (2011). Real-life driving outcomes in Parkinson Disease. *Neurology, 76*, 1894-1902.

Wood, J. M., Anstey, K. J., Kerr, G. K., Lacherez, P. F., & Lord, S. (2008). A multidomain approach for predicting older driver safety under intraffic road conditions. *Journal of the American Geriatrics Society, 56,* 986–993.

CHAPTER 42

Why Do We Move our Head during Curve Driving?

Daniel R. Mestre[1], Colas N. Authié[1,2],

[1] Institute of Movement Sciences, UMR 7287 CNRS / Aix-Marseille University,
Marseilles, France (daniel.mestre@univ-amu.fr)
[2] LPPA, UMR 7152 CNRS & Collège de France,
Paris, France (colas.authie@gmail.com)

ABSTRACT

During curve driving, drivers often tilt their head laterally towards the interior of the curve. Such head behavior may be elicited in response to centripetal forces, with drivers aligning their head axis to the resulting gravito-inertial force. However, previous work has suggested that this behavior might depend less on vestibular than on visual input. In the present study, we tested this hypothesis, using a fixed-base simulator in which, by definition, no centripetal forces are generated during driving. The experimental task consisted in driving along a winding road, constituted of a randomly ordered succession of curves of various radii separated by portions of straight lines. Head orientation was recorded, using an electromagnetic sensor, in synchrony with driving behavior. Results indicated that subjects exhibited systematic head roll tilts toward the interior of curves and that the amount of tilt angle increased with road curvature. However, head roll amplitude was inferior to that observed in real driving conditions. Head yaw orientation was clearly correlated with the tangent point location and represented about 56% of the tangent point eccentricity. Finally, the magnitude of roll and yaw movements of the head were significantly correlated. These results argue in favor of the hypothesis that visual determinants trigger head movements during curve driving, and suggest that head orientation may be part of a global gaze-head visual anticipatory strategy during curve driving.

Keywords: curve driving, simulation, optical flow, head movement

1   INTRODUCTION

During curve driving, it has been frequently observed that drivers tilt their head (roll movement) towards the inside of the curve. It is generally assumed that such head motion is elicited in response to centripetal forces, whereby drivers align their head axis to the resulting gravito-inertial force (vectorial sum of the earth gravitational force and the inertial force due to lateral acceleration during curvilinear motion; cf. MacDougall & Moore, 2005). However, previous work suggested that this behavior might depend less on vestibular than on visual input, being linked to the road's curvature (Zykovitz & Harris, 1999).

We further tested the hypothesis that head behavior is involved in an active visual information gathering process and that it is triggered by self-motion in a stable environment to direct gaze orientation toward a salient point in the global optical flow. Land & Lee (1994), by recording gaze behavior during car driving on a road clearly delineated by edge-lines, reported frequent gaze fixations toward the inner edge-line of the road, near a point they called the tangent point. This point is the geometrical intersection between the inner edge of the road and the tangent to it, passing through the subject's position. This result was subsequently confirmed by several other studies (Chattington et al., 2007; Kandil et al., 2009; Authié & Mestre, 2011) with more precise gaze recording systems. In a recent psychophysical experiment (Authié and Mestre, 2012), we demonstrated that, in a curve driving simulation, the tangent point of the curve can act as a singularity in the dynamic visual field, enabling optimal perception of the direction of self-motion. In the present experiment, we tested the hypothesis that head behavior during curve driving was linked to the tangent point location in the visual field. We tested this hypothesis, using a fixed-base simulator in which, by definition, no acceleration forces are generated during curve driving. Throughout the driving task, subjects were therefore exposed to a constant gravito-inertial force, equal to earth gravity.

2   METHODS

2.1   Participants

Nine participants (21-25 years old) participated in the experiment after completing an informed consent form. They were all experienced drivers with a driving license held since 4.3±1.7 years. Only active drivers, for the last two years, with normal or corrected to normal vision with contact lenses for myopia, could participate. No participant was aware of the experimental hypotheses. The experiment was conducted in accordance with the Declaration of Helsinki.

2.2   Apparatus and procedure

Participants drove on a SIM^2 driving simulator equipped with a seat, a steering wheel with force feedback, pedals and automatic drive (see Espié, Mohellebi, &

Kheddar, 2003 for detailed description). This device enables full control of driving scenarios, real time interactive driving, visual and auditory feedback, and on-line recording of simulated trajectories. The visual environment was generated at the rate of 60 Hz and projected onto a screen (see Figure 1).

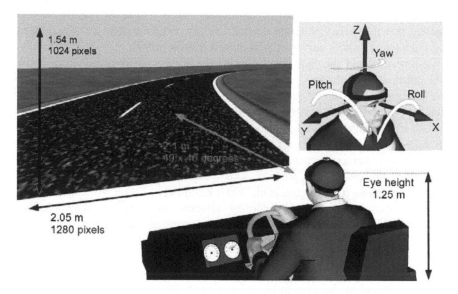

Figure 1. Experimental setup and visual scene used in the experiment. The up-right box indicates head axes of rotation – yaw, pitch and roll – as measured in the present experiment.

Participants were instructed to drive safely, but at the highest speed they could, without ever leaving the right lane (3.5 meters wide) of a two-lane road. The road edge-lines were continuous while the center line was discontinuous. The simulated track was composed of eight curves of various radii (50, 100, 200 and 500 meters), with each radius appearing in both right and left directions, and separated by portions of straight lines. The curves were set in a pseudo-random order of the succession of curve radii and directions, which was kept constant during the experiment. The visual scene consisted of a textured ground plane, a gray road and green roadside (Figure 1). Position and orientation of the head were measured at a rate of 120 Hz throughout the experiment with a Flock of Birds electromagnetic motion sensor (Ascension Technology, Burlington, USA). The Flock of Birds sensor was fixated on a headband on the top of the participants' head. The sensor position is reported as the (x, y, z) position with respect to the transmitter, and orientation as roll, yaw and pitch (Figure 2). This system has a precision of 0.003 cm for position and 0.005° for orientation, when the receiver stays within 40 cm of the transmitter (Pelz *et al.* 2001). At the beginning of each trial, participants were asked to align their head upright by fixating a target at the center of the screen. The head orientations reported in this paper are expressed from this initial position. Participants performed three sessions. They drove seven runs in each session, the first trial being for familiarization purposes.

2.3   Data analysis

Data were analyzed for each curve section of the experimental runway, for each participant. We separated the data into 8 curve sections (4 radii of curvature and two directions –left and right-), corresponding to the 8 curves of the runway. For each curve, we computed the average, minimum and maximum amplitudes of head orientation (yaw, pitch and roll, Figure 1) and its maximal speed. We also computed the average angle between the current direction of heading and the direction of the tangent point (tangent point angle).

3   RESULTS

3.1   Head orientation

Table 1 shows the relationships between the radius of curvature (RC) of the curves and the head yaw, pitch and roll angles. Negative radius of curvature values correspond to left-oriented curves. The average values of the averaged head orientations are displayed in the gray row. The peaks correspond to the maximum and minimum orientation values averaged across drivers. The speed rows indicate the maximum speed of the head averaged across drivers. The peaks and average orientation values are in degrees whereas the speed is given in degrees per second.

Table 1. Summary of head movements in the curve along the three axes of rotation.

Head rotation axis		Radius of curvature (m)							
		-50	-100	-200	-500	500	200	100	50
Yaw	Peaks (°)	-8.75 / -3.77	-6.74 / -2.88	-5.69 / -2.15	-3.24 / -0.83	3.02 / 0.27	5.01 / 1.53	7 / 2.84	9.21 / 4.18
	Average (°)	-7,63	-5,81	-4,63	-2,37	2,16	4,02	6,16	8,1
	Speed (°.s^{-1})	9,52	6,42	5,17	3,36	3,39	4,74	6,71	10,22
Pitch	Peaks (°)	-1.37 / 0.3	-1.45 / 0.1	-1.8 / 0.08	-2.08 / 0.02	-2.06 / 0.08	-1.64 / 0.26	-1.61 / 0.11	-1.87 / -0.21
	Average (°)	-0,59	-0,78	-0,87	-1,03	-1,02	-0,78	-0,92	-1,17
	Speed (°.s^{-1})	3,14	2,56	3,09	2,95	3,28	2,75	2,77	3,15
Roll	Peaks (°)	-4.25 / -1.18	-3.22 / -0.69	-3.26 / -0.67	-1.97 / -0.03	1.59 / -0.78	2.26 / -0.1	2.98 / 0.8	4.21 / 1.37
	Average (°)	-3,08	-2,09	-2,04	-0,93	0,58	1,16	2,09	3,01
	Speed (°.s^{-1})	5,04	3,77	3,32	2,6	2,99	2,95	3,41	5,18

Head yaw was always oriented toward the curve direction: the drivers always oriented their head in yaw to the left in left curves, and to the right in right curves.

Head pitch did not vary much. It remained rather stable across curve radii, with small tilt headway of ~1–2°. Head pitch will therefore be set aside in the following analyses. We observed significant head motion during curve driving, mainly for the yaw axis (from ~3.3°.s-1 to ~10.2°.s-1) and the roll axis (from 2.6°.s-1 to 5.2°.s-1).

Roll. Drivers tilted their head in roll with a maximal averaged peak of 4°, observed for the sharpest curves (50 meters radii). A three-way analysis of variance (ANOVA) with the absolute values of roll showed no statistical effect of (task) Repetition ($F(5,40)=1.70$, $p>.10$, $\eta2p=.17$), nor of curve Direction ($F(1,8)=0.04$, $p>.50$, $\eta2p=.01$). However, there was a significant effect of the curve Radius of Curvature –RC- ($F(3,24)=6.68$, $p<.05$, $\eta2p=.46$). Post-hoc analysis indicated that the head roll angle was significantly higher in the 50 meter RC curves than in the 500 and 200 meter curves. We also observed considerable between-subject variability. One subject tilted his head in roll with a magnitude of ~8° in average, whereas two other drivers were no-roll movers (roll tilt of ~1°).

Yaw. Regarding the magnitude of head yaw, analysis of the variance showed no significant effect of Repetition ($F(5,40)=1.18$, $p>.10$, $\eta2p=0.13$) or of the curve Direction ($F(1,8)=.61$, $p>.50$, $\eta2p=.07$). Here again, the analysis of variance showed a large effect of the curve Radius of Curvature ($F(3,24)=112.74$, $p<.05$, $\eta2p=.93$). The yaw amplitude systematically increased as a function of the sharpness of the curve (2.39°, 4.32°, 5.98° and 7.87° for the curves of 500, 200, 100 and 50 meter RC, respectively). Contrary to the observed head roll behavior, all drivers oriented their head in yaw with a comparable magnitude.

3.2.   Tangent point (TP) angle

The TP angle (θ) is defined as the absolute angle between the current direction of heading and the direction of the tangent point. TP angle depends on both the radius of curvature of the curve and the lateral position in the driving lane. It is therefore not surprising that this angle was affected by Repetition ($F(5,40)=22.34$, $p<.001$, $\eta2p=.76$), the curve Radius of Curvature ($F(3,24)=808.12$, $p<.01$, $\eta2p=.99$) and the curve Direction ($F(3,24)=808.11$, $p<.001$, $\eta2p=.99$). The TP angle systematically increased as the RC of the curve decreased (4.66°, 8.64°, 10.64° and 13.45° for curves of 500, 200, 100 and 50 meter RC, respectively) and was more eccentric for right curves (9.78°) than for left curves (8.91°). This last result was due to the decrease of the lateral distance to the curve's inside edge in left curves.

3.3.   Correlations between head and TP angles

We observed strong correlations, for all participants, between TP angle and head roll and yaw angles. Head roll was also significantly correlated to head yaw (figure 2).

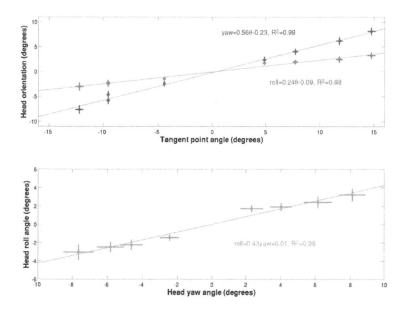

Figure 2. Averaged values, over trials and participants, of the head orientation (yaw and roll axes) and the tangent point (TP) angle, for the eight curve sections. Top subfigure shows the linear correlations between TP angle and head roll (red markers), as well as between TP angle and head yaw (blue markers). The bottom subfigure shows the linear correlation between yaw and head roll angles (orange markers). Vertical and horizontal bars indicate between-subjects standard error along the considered axis.

4. DISCUSSION

In this study, we analyzed participants' head movements while driving in simulated conditions. We used a fixed-base simulator, which implies that subjects only received visual and audio feedback from their driving performance. In particular, they had no driving-related vestibular and/or tactilo-proprioceptive information. Under these conditions, we focused our analysis on head motion during the task, consisting basically in driving safely along a winding road. Both head roll and yaw movements were analyzed, in conjunction with the tangent point angle (TP).

We found significant head roll movements during curve driving (Table 1), which were significantly correlated to the road geometry (i.e. to the radius of curvature of the curve). This result corroborates Zikovitz & Harris' (1999) results, obtained in actual driving conditions, suggesting that head roll movement is more dependent on visual than on vestibular inputs. We observed similar head roll in a fixed-base simulator, in which no gravito-inertial transformation was present during curve driving. However, head roll magnitude was rather small (4° maximum from peak to peak). In Zikovitz & Harris' (1999) study, although the conditions across studies cannot be fully compared, head roll angles up to 15-20° were observed. As such, the vestibular hypothesis of head tilt cannot be entirely rejected. Rogé (1996) hypothesized that drivers need to combine the dynamic information from the vehicle with their own corporal head roll while driving a simulator. She predicted that the lack of inertial forces would progressively suppress this behavior. However, we did not observe any reduction of the head roll magnitude as a function of the task repetition. As a consequence, we suggest that, although a vestibular contribution to head roll certainly exists, visual determinants are necessarily present.

Head yaw movement was also directed in the curve direction and affected by the curve radius. This result is less surprising than the head roll and was already observed by Land & Tatler (2001) with a racing driver. The head yaw (10° in sharp curves) reached a higher magnitude than head roll (4°). The TP angle and the head yaw orientation exhibited a comparable pattern (Figure 2). Head yaw orientation represented about 56% of the TP angle (Figure 2). From previous analyses of gaze data - recorded with the same driving simulator and track - we found that drivers gazed in the TP direction (Authié & Mestre, 2011). The present results suggest, here again, that subjects orient their gaze toward the TP while driving in the curve, and that head yaw orientation accounts for more than half of the gaze direction. It also suggests that the tangent point constitutes a valuable source of information in the dynamic visual field.

Finally, we propose that head roll motion is functionally linked to head yaw motion. The general idea is that multi-axis head orientation in space contributes to active information gathering, with the combined existence of biomechanical and physiological head-stabilizing mechanisms and anticipatory orienting mechanisms. This possibility is endorsed by the existence of a correlation between head yaw and roll during curve driving (Figure 2). Kim *et al.* (2010) found, in an eye-head saccade task, that a roll movement was elicited in conjunction with yaw movement in order to orient the gaze toward an eccentric target displayed horizontally. However, head roll movements observed in the present driving task were highly different from one subject to the other. This large inter-subject variability was also reported in other tasks (Kim *et al.* 2010; Stahl 1999; 2001) and its reason remains unknown (Stahl 1999; 2001). For the curve driving task, Rogé (1996) proposed that "head-mover" drivers might be more dependent on the visual reference frame. However, the small number of participants in the present experiment did not allow us to properly test this possibility.

5. CONCLUSION

We analyzed participants' head behavior while driving in simulation conditions, in which no change of inertial information was present during the curve. Results clearly indicate that drivers exhibited systematic head roll tilt toward the interior of curves, and that the amount of tilt angle increased with road curvature. However, head roll amplitude was inferior to that observed in a real driving situation (Zikovitz & Harris, 1999). Head roll orientation was correlated with head yaw orientation, this latter being correlated with the tangent point angle and representing about 56% of the tangent point eccentricity. This result confirms that drivers orient their gaze toward the tangent point while curve driving, and that the tangent point constitutes a valuable source of information in the visual field. These results also argue in favor of the hypothesis of visual determinants of head movements during curve driving. Moreover, roll head roll and yaw movement were correlated, suggesting that head movements during driving may be part of a global gaze-head visual anticipatory strategy during curve driving.

REFERENCES

Authié, C.N. & Mestre, D.R. (2011). Optokinetic nystagmus is elicited by curvilinear optic flow during curve driving. *Vision Research* , 51, 1791-1800.

Authie, C.N. & Mestre, D.R. (2012). Curvilinear heading discrimination: dependence on gaze direction and optical flow speed. PloS ONE, (accepted, doi:10.1371/journal.pone.0031479)

Chattington, M., Wilson, M., Ashford, D., & Marple-Horvat, D. E. (2007). Eye-steering coordination in natural driving, Experimental Brain Research, 180, 1–14.

Espié, S., Mohellebi, H., & Kheddar, A. (2003). A high performance/low-cost mini driving simulator alternative for human factor studies, Paper presented at the Driving Simulation Conference 2003, North America, Dearborn, Michigan, 1–10.

Kandil, F. I., Rotter A., & Lappe, M. (2009). Driving is smoother and more stable when using the tangent point, *Journal of Vision,* 9, 1–11.

Kim, K. H., Reed M. P., & Martin B. J. (2010). A model of head movement contribution for gaze transitions, *Ergonomics,* 53 (4), 447–457.

Land, M.F., Lee, D.N., 1994. Where we look when we steer. *Nature* 369, 742–744.

Land, M., & Tatler B. (2001). Steering with the head The visual strategy of a racing driver, *Current Biology* 11 (15), 1215–1220.

MacDougall, H.G., & Moore, S.T. (2005). Functional assessment of head-eye coordination during vehicle operation, *Optometry Vision Science,* 82, 706–15.

Pelz, J., Hayhoe, M., & Loeber, R. (2001). The coordination of eye, head, and hand movements in a natural task, *Experimental Brain Research* 139 (3), 266–277.

Rogé, J. (1996). Spatial reference frames and driver performance, *Ergonomics,* 39 (9), 1134–1145.

Stahl, J. S. (1999). Amplitude of human head movements associated with horizontal saccades, *Experimental Brain Research,* 126 (1), 41–54.

Stahl, J. S. (2001). Adaptive plasticity of head movement propensity, *Experimental Brain Research,*139, 201–208.

Zikovitz, D. C., & Harris, L. R. (1999). Head tilt during driving, *Ergonomics,* 42 (5), 740–746.

CHAPTER 43

Evaluating Novice Motorcyclists' Hazard Perception Skills at Junctions Using Naturalistic Riding Data

Mohd Khairul Alhapiz Ibrahim, Ahmad Azad Ab Rashid,
Muammar Quadaffi Mohd Ariffin

Malaysian Institute of Road Safety Research
Kajang, Malaysia
mkhairul@miros.gov.my

ABSTRACT

In Malaysia, young and inexperienced motorcyclists are greatly over-represented in the annual road accident statistics. In the existing rider training program, the skills of the novice motorcyclists were not entirely measured and examined prior to the license approval because the current curriculum does not include training and testing on the actual road. This study used an instrumented motorcycle to capture the riding performances of novice motorcyclists on actual road. Hazard perception skills of 103 novice motorcyclists during unprotected right turns (equivalent to the US left turns) at unsignalized T junctions were evaluated. Significant differences were found in participants' choice of speed and use of turn signals in responding to oncoming vehicles at junctions. However, participants did not significantly reduce their speed nor give longer turn signals when only crossing vehicles present at the junctions. It was found that novice motorcyclists in our sample did not respond sufficiently to the vehicles crossing the junctions and underestimated the potential risk of accident. The effectiveness of current rider training program in improving hazard perception skills of novice motorcyclists is discussed.

Keywords: Novice motorcyclists, hazard perception, naturalistic study, rider training

1   INTRODUCTION

Turning across the path of oncoming traffic demands time-critical decision making because any delays in deciding gap acceptance influence the chances of accident (Mitsopoulos-Rubens et al., 2002). Clarke et al. (1999) in explaining junction road accidents suggested that older drivers' failure to observe situation at junctions was the main reason they crashed while risk taking was the key factor in younger drivers' accidents involvement. Lack of skills was also found to have contributed to crashes at junctions. Clarke et al. (1998) analyzed 5 years overtaking accidents data in the UK and found that this type of accidents frequently involved overtaking a vehicle that was turning right. The researchers concluded that the cause of accident was mostly because of faulty right turn made by older drivers and poor overtaking decision by young drivers.

Spek et al. (2006) studied the effects of speed of oncoming vehicle at intersections to accident probability. They concluded that drivers tend to accept shorter time gaps as the speed of oncoming vehicle increases. They suggested that the probability of a junction accident will increase with the increase of speed of approaching vehicles. With older drivers tend to accept shorter gap more than younger drivers, it was concluded that older driver is prone to be involved in an accident with speeding vehicles at the junction. When gender is considered, female appeared to be at greater risk of involving in accidents at junctions. Alexander et al. (2002) conducting simulator studies on car drivers concluded that older or female drivers were less likely to accept shorter gaps compared to younger and male drivers. However, the probability of involving in an accident was higher for older or female drivers at larger gaps because they took longer time to cross compared to younger and male drivers.

The severity of the problem is also apparent in motorcycle safety because young and inexperienced riders are also over-represented in motorcycle accidents. Recent development shows increasing literatures focusing on motorcycle crashes at junctions (e.g.; Crundall et al., 2008; Pai and Saleh, 2008; Pai et al., 2009; Pai, 2009; Crundall et al., in press). Particular focuses of these studies were on issues of right of way violation, hazard and risk perception and injury severity of motorcyclists. In the developing countries, where most of the burden related to motorcycle casualties are borne (Cheng et al., 2011), motorcycle safety is a grave concern; yet there seems to be a paucity of literatures in many areas of motorcycle safety research including studies on cross-flow turns. In the low and middle income countries, motorcycle deaths are alarmingly high and steadily increasing. In Malaysia alone, accidents at junctions accounted for 30.1% of total motorcycle accidents with 4,161 fatalities and 25,128 injuries recorded in 2006-2010 period (RMP).

As part of the Malaysian rider training improvement plan beginning in 2008, series of naturalistic data collections using an instrumented motorcycle were conducted to evaluate the effectiveness of the current training program to pave the ways for possible modifications and upgrades. In general, it sought to investigate whether the current training curriculum and program had effectively improve the riding skills of Malaysian novice motorcyclists. Sets of driving performance measures and hours of video recording were compiled from these naturalistic data collections.

2   METHOD

2.1   Participants

Participants were candidates for Malaysian motorcycle license from the states of Malacca and Negeri Sembilan who had completed all the tutorial and practical training required for the license application. In the current study, a screening process was done to include participants with right turns maneuvering at unsignalized T junctions from the database. Riding performances of 64 males and 39 females novice motorcyclists ranging in age from 16 to 58 years old (M = 21.5, SD = 6.9) were analyzed to study their hazard perception skills. Seventeen participants who were above the age of 25 were classified as older motorcyclists.

2.1   Task and Apparatus

A 100 cc motorcycle was modified to include a data acquisition system capable of recording riding performance measures and surrounding video unobtrusively. The video data included front and rear view recording which permitted an analysis of participants' reaction to the oncoming and crossing vehicles at junctions. Outputs from sensors and electrical signals were obtained for measurement of the motorcycle's speed and period of turn signal activation.

Participants were asked to ride the instrumented motorcycle on an 8 km route of actual road which included turns at unsignalized junctions. The experiment route included riding in the city and residential area with mixed-traffic settings. For security reasons, participants were escorted by a safety motorcycle rode by an experienced rider who was instructed to keep a reasonable distance from the participants to avoid influencing their riding performances. Before each data collection session, health screenings were conducted to ensure that participants were free from medical problems and influence of alcohol or drugs. Participants read and signed informed consent forms before participating in the data collection. After the riding session, participants completed demographic survey forms and were given a safety helmet for their participation.

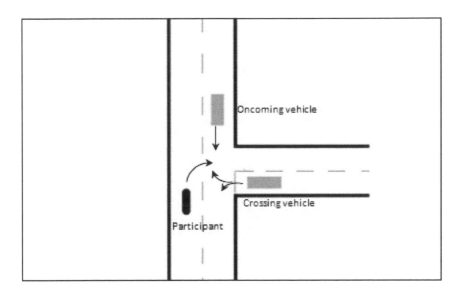

Figure 1 Illustration of the junction in the present study showing the directions of travel

3 RESULTS

Data were analyzed with one-way analysis of variance (ANOVA) with maneuvering speed and period of turn signal activation as within-subjects variables and type of hazards as between-subjects variables. Types of hazard were classified as presence of oncoming vehicle from the opposite riding direction, presence of vehicle crossing the junction and presence of both oncoming and crossing vehicle (see Figure 1). The values of means for each variable are summarized in Table 1, Table 2, Table 3 and Table 4. The plots of means are also presented in Figure 2 and Figure 3.

3.1 Maneuvering Speed

Significant differences were found in the speed chosen by participants in responding to the different hazards at junctions (F (3, 36) = 6.647, p = 0.001). A Tukey post-hoc test indicated that participants reduced their speed significantly in the presence of oncoming vehicle (M = 12.66 km/h, SD = 11.10, P = 0.02) and in the presence of both oncoming vehicle and crossing vehicle (M = 11.51 km/h, SD = 8.36, P = 0.02) compared to when there were no vehicle at the junction (M = 20.73 km/h, SD = 7.90). However, participants' maneuvering speed did not significantly differ (M = 19.59 km/h, SD = 9.69, P = 0.979) when only crossing vehicles were present at the junction. Significant effects of hazards on maneuvering speed were observed in the group of male (F (3, 60) = 5.959, p = 0.001) and younger (F (3, 82) = 4.958, p = 0.003) participants. Male and younger participants reduced their speed significantly in the presence of oncoming vehicle and both oncoming and crossing vehicle. No significant reductions of speed were found for female and older participants in the presence of all types of hazards.

Table 1 Mean maneuvering speed (km/h) by type of hazards and gender

	Presence of other vehicles at junction (hazards)							
	No traffic		Oncoming vehicle		Crossing vehicle		Presence of both	
	M	SE	M	SE	M	SE	M	SE
Male	21.9	1.550	11.9	2.332	20.6	2.560	12.2	2.769
Female	19.1	2.165	13.9	2.997	15.5	7.993	10.5	4.477
Overall	20.7	1.282	12.7	1.825	19.6	2.501	11.5	2.319

Table 2 Mean maneuvering speed (km/h) by type of hazards and age group

	Presence of other vehicles at junction (hazards)							
	No traffic		Oncoming vehicle		Crossing vehicle		Presence of both	
	M	SE	M	SE	M	SE	M	SE
Young	20.6	1.400	13.2	2.068	20.9	2.610	12.0	2.462
Old	22.1	2.460	10.8	4.094	14.2	7.148	8.8	8.800
Overall	20.7	1.282	12.7	1.825	19.6	2.501	11.5	2.319

3.2   Turn signal activation

Significant differences were also observed in the use of turn signals in responding to the different hazards at junctions ($F_{(3, 60)} = 2.942$, $p = 0.04$). The result of post hoc comparisons using the Tukey HSD test however revealed a significant interaction only during the presence of oncoming vehicles. The participants gave the turn signals significantly sooner ($p = 0.043$) before making the turn ($M = 15.4$ s, $SD = 5.44$) in the presence of oncoming vehicles compared to when there were no traffic present ($M = 11.7$ s, $SD = 4.12$). No significant differences were observed during the presence of crossing vehicles ($M = 11.3$ s, $SD = 4.06$, $p = 0.994$) and in the presence of both crossing and oncoming vehicles ($M = 13.7$ s, $SD = 5.19$, $p = 0.759$). Significant effects of hazards on the use of signals were observed only in the group of male participants ($F_{(3, 60)} = 3.992$, $p = 0.012$) with the interaction approached significance ($F_{(3, 82)} = 2.610$, $p = 0.057$) in the group of younger participants. Male participants gave longer turn signals in the presence of oncoming vehicles at the junctions. No other significant interactions were observed.

Table 3 Mean period of turn signal activation (s) by type of hazards and gender

| | Presence of other vehicles at junction (hazards) | | | | | | | |
| | No traffic | | Oncoming vehicle | | Crossing vehicle | | Presence of both | |
	M	SE	M	SE	M	SE	M	SE
Male	11.7	0.876	17.3	1.196	12.9	2.072	14.6	2.171
Female	14.9	1.231	13.4	1.261	15.3	0.882	14.8	1.853
Overall	11.7	0.793	15.4	1.160	11.3	1.436	13.7	1.961

Table 4 Mean period of turn signal activation (s) by type of hazards and age group

| | Presence of other vehicles at junction (hazards) | | | | | | | |
| | No traffic | | Oncoming vehicle | | Crossing vehicle | | Presence of both | |
	M	SE	M	SE	M	SE	M	SE
Young	13.0	0.838	16.0	1.016	11.7	1.549	14.0	1.615
Old	13.5	1.443	14.8	2.194	20.3	3.844	19.0	1.000
Overall	11.7	0.793	15.4	1.160	11.3	1.436	13.7	1.961

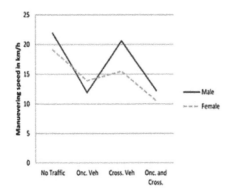
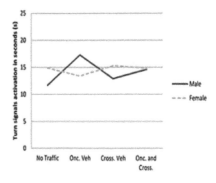

Figure 2 Mean maneuvering speeds and period of turn signal activation by gender and types of hazards.

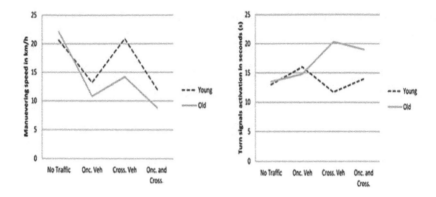

Figure 3 Mean maneuvering speeds and period of turn signal activation by age group and types of hazards.

4   CONCLUSIONS

The goal of the present study was to measure the hazard perception skills of novice motorcyclists in responding to hazards at junctions. The results revealed that age and gender were significant factors in predicting accident risk of novice motorcyclists at junctions. Female and older motorcyclists were more at-risk on both performances measured. No significant differences were observed in the maneuvering speed and turn signal use of female and older motorcyclists in the presence of all types of hazards at junctions. No particular patterns of turn signal use were found in the female group. For instance, female motorcyclists used less signal in the presence of oncoming vehicle compared to when there was no traffic at junctions. The lack of these patterns and the performances of female and older motorcyclists could indicate that the current rider training program is inadequate in preparing the novice motorcyclists to deal with hazards on the road effectively.

Another important finding in the current study was that although male or younger motorcyclists responded to oncoming vehicles substantially, no significant effects of the crossing vehicles were observed on their riding performances. They did not significantly reduce their speed nor give longer turn signals when only crossing vehicles were present at the junctions. This observation has led us to believe that the hazard perception skills of all novice motorcyclists in our sample were not entirely improved during the training. One of the crucial improvements to the current rider training program is to include exposure to actual road traffic during the practical training. Although the safety of the learner motorcyclists during the training is a major concern, consideration and proper planning should be made to include this exposure. A possible solution could be having a closed section of the actual roads with limited traffic as a training ground.

Studies have found that experience is an important factor in the improvement of drivers' skill in gap acceptance (Cavallo and Laurent, 1988; Clarke et al., 2006). For example, Clarke et al. (2006) analyzed accidents cases involving young drivers (17 – 25) in the UK and found that accidents while making right turn across oncoming traffics (cross-flow turn) were affected the most by the years of driving experience. They concluded that in cases where young drivers were fully or partially at fault, the biggest skill improvement was found in numbers of cross-flow turn accidents. Novice drivers' lack of skill in gap acceptance was also found by Mitsopoulos-Rubens et al. (2002) who concluded that novice drivers accepted more gaps and were less adept compared to experienced drivers.

One limitation of this study is the small number of older participants (17) available in the sample. Nonetheless, this was expected because most of the novice motorcyclists applying for riding license are of younger age. To summarize, the different levels of skill improvement found in this study coupled with the effect of experience in skill improvement suggest that stricter measures and riding restrictions should be considered before a full license is given to the novice motorcyclists.

ACKNOWLEDGEMENTS

This study was funded by Malaysian Institute of Road Safety Research and the Malaysian ministry of transport through a research grant awarded to the first author.

REFERENCES

Alexander, J., Barham, P., Black, I. 2002. Factors influencing the probability of an incident at a junction: results from an interactive driving simulator. *Accident Analysis and Prevention* 34: 779–792.

Crundall, D., Crundall, E., Clarke, D., Shahar, A. 2012. Why do car drivers fail to give way to motorcycles at t-junctions? *Accident Analysis and Prevention* 44(1): 88-96.

Cheng, A.S.K., Nga, T.C.K., Lee, H.C. 2011. A comparison of the hazard perception ability of accident-involved and accident-free motorcycle riders. *Accident Analysis and Prevention* 43: 1464–1471.

Crundall, D., Humphrey, K., Clarke, D. 2008. Perception and appraisal of approaching motorcycles at junctions. *Transportation Research Part F: Traffic Psychology and Behaviour* 11(3): 159-167.

Clarke, D.D., Ward, P., Bartle, C., Truman, W. 2006. Young driver accidents in the UK: the influence of age, experience, and time of day. *Accident Analysis and Prevention* 38: 871–878.

Clarke, D.D., Forsyth, R., Wright, R. 1999. Junction road accidents during cross-flow turns: a sequence analysis of police case files. *Accident Analysis and Prevention* 31: 31–43.

Clarke, D.D., Ward, P.J., Jones, J. 1998. Overtaking road-accidents: differences in manoeuvre as a function of driver age. *Accident Analysis and Prevention* 30(4): 455–467.

Cavallo, V. and Laurent, M. 1988. Visual information and skill level in time-to-collision estimation. *Perception* 17(5): 623-632.

Mitsopoulos-Rubens, E., Triggs, T., Regan, M. 2002. Comparing the gap acceptance and turn time patterns of novice with experienced drivers for turns across traffic. *Proceedings of*

428

the Fifth International Driving Symposium on Human Factors in Driver Assessment, Training and Vehicle Design. Big Sky, Montana, USA.

Pai, C.-W. 2009. Motorcyclist injury severity in angle crashes at T-junctions: Identifying significant factors and analysing what made motorists fail to yield to motorcycles. *Safety Science* 47(8): 1097-1106.

Pai, C.-W., Hwang, K.P., Saleh, W. 2009. A mixed logit analysis of motorists' right-of-way violation in motorcycle accidents at priority T-junctions. *Accident Analysis and Prevention* 41(3): 565-573.

Pai, C.-W., Saleh, W. 2008. Exploring motorcyclist injury severity in approach-turn collisions at T-junctions: Focusing on the effects of driver's failure to yield and junction control measures. *Accident Analysis and Prevention* 40(2): 479-486.

Royal Malaysian Police (RMP). 2010. *Statistical report of road accidents in Malaysia.* Kuala Lumpur: Bukit Aman Traffic Branch.

Spek, A.C.E., Wieringa, P.A., Janssen, W.H. 2006. Intersection approach speed and accident probability. *Transportation Research Part F* 9: 155-171.

CHAPTER 44

The Influence of Presentation Mode and Hazard Ambiguity on the Domain Specificity of Hazard Perception Tasks

Sarah Malone, Antje Biermann and Roland Brünken

Saarland University
Saarbrücken, Germany
s.malone@mx.uni-saarland.de

ABSTRACT

The improvement of the validity of the driving test can be seen as a possibility to reduce the novice drivers` high accident risk. One empirical indicator of test validity is a clear difference between expert and novice test performance. Expertise research in general has revealed that experts outperform novices more clearly with increasing domain specificity of the given task–it's similarity to the characteristic demands of a defined domain–(Glaser & Chi, 1988), and that expert knowledge is better connected and thus more flexible than that of novices in a certain domain (Spiro, Feltovich, Jacobson, & Coulson, 1991). Expertise research in the domain of road safety tries to identify quantifiable aspects of driving skill that can explain the increased accident liability of novice drivers (Horswill & McKenna, 2004). Hazard perception which is the driver's ability to identify emerging hazards and potential dangerous situations has been identified as one such relevant skill (McKenna & Crick, 1994). The aim of the present study was to investigate whether the domain specificity of hazard perception tasks can be enhanced by the implementation of animated instead of static pictures of traffic scenarios and whether the ambiguity of the presented hazards has any impact on this effect.

According to expertise research we assumed, that experts outperform novices more clearly if dynamic presentations are involved, because they are more specific for the driving domain than static pictures. This effect could be even more obvious if the given tasks require the use of driving specific knowledge if rather implicit than explicit hazards are presented.

In an experimental study 63 learner drivers (novices) and 70 individuals with some driving experience (experts) were to react on potential hazards within 31 different driving scenarios presented on a computer screen. Whereas previous studies into hazard perception assessment included either videos (e.g. Sagberg & Bjornskau, 2006) or static pictures (e.g. Biermann et al., 2008) of traffic situations, we compared animations and pictures of exactly the same traffic scenarios. We used a 2 by 2 by 2 factorial mixed design with repeated measures on the factor hazard ambiguity (explicit vs. implicit hazard) and the two between subjects factors expertise (experts vs. novices) and presentation mode (animated vs. static).

The results indicated that experts in general identified more breaking hints ($\eta_p^2 = .20$) and by tendency were faster in reacting than novices ($\eta_p^2 = .03$). This confirms that the reaction task represents a valid measurement of driving expertise. There was no simple interaction between the factors expertise and presentation mode but a three way interaction ($\eta_p^2 = .08$). The interaction indicated that animation does not generally improve the assessment of hazard perception but for those scenarios in which the use of knowledge about traffic rules is required (implicit hazards). This reveals that the decision to choose animations in order to assess hazard perception should depend on content aspects of a task. This allows the conclusion that the most specific and valid reaction time task for the driving domain includes the animated presentation of traffic scenarios with implicit hazards.

Keywords: hazard perception, driving assessment, expertise

1. INTRODUCTION

In Germany the number of fatal traffic accidents is declining continuously. But there is still an increased risk for novice drivers to die in traffic crashes in comparison to more experienced drivers. Possibilities to face this problem could be seen for example in an optimized driving test. In Germany the theoretical driving test has been integrated in computer based testing systems in 2008. Principally there has been an adaptation from the former paper pencil test that mainly consisted of multiple choice items with static pictures of traffic situations to a computerized version. But besides that, computer based testing offers the possibility to integrate innovative presentation (e.g. animations) and response formats (e.g. reaction time measures) in the theoretical driving test, too. The aim of the present study was to investigate whether specific features of computer based information presentation can improve the validity of those tests and consequently improve its ability to sort out those individuals that are not yet competent enough to drive safely.

The computer based presentation feature under investigation in the present study was the possibility to present visual information in a dynamic rather than a static way, which should affect the assessment of skills that have been acquired in a moving sphere like driving competence. Therefore we combined findings from instructional psychology, traffic psychology and expertise research to examine whether the use of dynamic presentations increase the validity of the expertise assessment of novice vs. expert drivers.

2. THEORETICAL BACKROUND

Expertise research usually focuses on extraordinary performance of individuals in a certain domain. Ericsson and Lehmann (1996) emphasized the meaning of expertise development for the understanding of adaptive processes that individuals perform in order to accomplish everyday tasks.

Experts outperform novices in a specific domain: they are faster in specific problem solving tasks and they make fewer mistakes (Glaser & Chi, 1988). Expertise researchers agree about these facts. In most of the theories concerning the development of expertise the possibility to make experiences in the domain is concerned as critical requirement of the acquisition of expertise. The fact that young novice drivers are overrepresented in traffic accidents (e.g. Mayhew, Simpson, & Pak, 2003) supports the hypothesis that driving expertise evolves through experience.

Researchers in the domain of road safety try to identify some quantifiable aspects of driving expertise that can explain the increased accident liability of some drivers (Horswill & McKenna, 2004). Knowing about skills that are relevant for driving security can serve as a basis for all efforts made in order to improve driver training and assessment. Research has revealed that one skill in which driving novices perform inferior to experienced drivers is hazard perception (e.g. McKenna & Crick, 1994). In addition to that, individuals with fewer accidents in the past performed better in hazard perception tasks then those with more accidents (e.g. McKenna & Horswill, 1999; Quimby, Maycock, Carter, Dixon, & Wall, 1986; Quimby & Watts, 1981). These results show that hazard perception tests are valid measures of one aspect of driving competence.

Hazard perception refers to the driver's ability to identify emerging hazards and potential dangerous situations (McKenna & Crick, 1994). There are different technology based methods to assess this ability. Studies into hazard perception assessment usually include the measurement of the time taken to detect hazards in either videos or animations (e.g. Sagberg & Bjornskau, 2006) or static pictures (e.g. Biermann et al., 2008) of traffic situations. However, none of these studies involved exactly the same traffic situations in an animated and in a static version in order to compare the two presentation modes. We assumed that technology based assessment offers the opportunity to enhance validity of measures by integrating animations. One finding of expertise research guides our hypothesis, that animation will help to differentiate between experts and novices: Experts outperform novices more clearly with increasing domain specificity of the task (Glaser & Chi, 1988). Animated presentations of traffic scenarios should be more realistic than static presentations and therefore more specific for the driving domain. Animations offer nearly the same visual impressions of traffic scenarios as driving in real traffic does: Movement can be perceived directly. In addition the task itself becomes more demanding if dynamic presentations are used. Individuals have to react in time to avoid an accident just as in real traffic.

Another stable finding of expertise research is that expert knowledge is better connected and thus more flexible than the knowledge of novices in a certain domain (Spiro, Feltovich, Jacobson, & Coulson, 1991). This flexibility can be attributed to the experts repeated learning and application of that knowledge in different situations. Therefore the need for use of domain specific knowledge for task solution should increase the advantage of the experts. In none of the studies researching hazard perception, scenarios including explicit hazards, that can be identified without any knowledge of traffic rules, and scenarios including implicit hazards, that can only be identified with knowledge of traffic rules, were used in order to compare them.

3. HYPOTHESES

Expertise research has revealed that experts outperform novices in domain specific tasks. One of the specific tasks in the driving domain is hazard perception. According to prior research we postulated that driving experts would outperform driving novices in both, the static and the animated version of the materials. More specific, experts were supposed to identify more hazards (hypothesis 1) and react faster to them than novices (hypothesis 2).

The hypothesis that follows from the concept of domain specificity is that the animated version of the reaction time task is better qualified to differentiate driving experts from driving novices than the static version because reacting on time to hazards within animations corresponds to the demands of real driving. Experts should outperform novices more clearly when the hazard perception test includes rather dynamic than static presentations concerning the number of correctly identified traffic hazards (hypothesis 3) and reaction times (hypothesis 4).

This interaction should be even more obvious if the given tasks require the use of driving specific knowledge (implicit hazards). This applies to the number of identified hazards (hypothesis 5) and reaction times (hypothesis 6). The last hypotheses arise from the assumption, that a task should be more specific for a certain domain if it requires the application of knowledge that is acquired while expertise development in that domain.

4. METHOD

Sixty-four learner drivers (novices; 39% male) and 76 individuals who owned their driving licence for more than two years (experts; 29% male) took part in the experiment. The participants were between 16 and 58 years old ($M = 22.32$; $SD = 6.15$).

The subjects were shown 31 different driving scenarios from the driver's perspective in a random order. The scenarios were presented either in an animated version or in a static version on a computer screen. Individuals were randomly assigned to the different experimental conditions. Half of the novices and 43 % of the experts worked with the static version of the testing materials. The participants' task was to react as soon as they became aware of a hint that signed that the driver should slow down or even break within every driving scenario. There was at maximum one hazard or signal included in every scenario that indicated the need for a reduced velocity. Six Scenarios were distractors without any breaking hints. There were two types of hints in the 25 scenarios left. On the one hand explicit hazards emerged in 18 scenarios – for example a child crossing the road - that could be recognized without having any driving specific knowledge. On the other hand an implicit breaking hint could be for example a traffic sign that indicates that the driver has to give way to oncoming traffic (7 scenarios). To react adequately to this type of situation the participants had to use their knowledge about traffic rules and signs. Figure 1and 2 show examples of traffic scenarios taken from the static version of the testing material.

Figure 1 Computer-generated image of traffic situation. The image shows an explicit hazard. Both cars try to drive on the middle lane.

Figure 2 The image demonstrates an implicit hazard that could emerge because vehicles coming from the right have right-of-way at this intersection.

The design of the experiment was a 2 by 2 by 2 factorial mixed design with repeated measures the factor hazard ambiguity (explicit hazards vs. implicit hazards). The two between subjects factors were expertise (experts vs. novices) and presentation mode (static vs. dynamic). Two dependent variables were logged for each scenario. First it was recorded if the participant reacted appropriate. That means that he pressed a button on the keyboard when there was a hazard and didn't press the button when there was no hazard within the scenario. Additionally the time from the first appearance of the hint until the reaction by the participant was recorded. In case of too early or no response at all, no reaction time was logged for the respective item.

5. RESULTS

Results are presented separate for the two dependent variables: correctly identified traffic hazards and reaction time. Four Novices had to be excluded from the analysis because their reaction time has not been recorded because of a technical error. Table 1 shows the mean results for proportion of correctly identified hazards.

Table 1 Results for correctly identified hazards

	Experts				Novices			
	Static		Dynamic		Static		Dynamic	
	$n = 33$		$n = 43$		$n = 30$		$n = 30$	
	M	SD	M	SD	M	SD	M	SD
Implicit hazard	.57	.20	.58	.16	.46	.25	.41	.21
Explicit hazard	.81	.12	.58	.13	.71	17	.52	.16

An analysis of variance indicated that experts were able to identify more breaking hints correctly within the scenarios $(F (1, 132) = 20.68; p < .001; \eta_p^2 = .14)$. In addition the participants performed better when they worked with the static version than with the dynamic version $(F (1, 132) = 21.76; p < .001; \eta_p^2 = .14)$. The participants identified more breaking hints correctly when there were rather explicit than implicit hazards $(F (1, 132) = 61.02; p < .001, \eta_p^2 = .32)$. The analysis did not show an interaction between expertise and presentation mode $(F (1, 132) = .001; p = .97)$ but an interaction between expertise and hazard ambiguity by tendency $(F (1, 132) = 3.10; p = .08)$. However, there was an interaction between presentation mode and hazard ambiguity $(F (1, 132) = 27.80; p < .001; \eta_p^2 = .17)$. The results indicate that the overall effect of presentation mode is limited. Only for the scenarios with explicit hazards the static version was easier than the dynamic version. In the case of scenarios with implicit hazards the static version was as difficult as the dynamic version. A triple interaction could be found by tendency $(F (1, 132) = 2.93; p = .09)$.

Table 2 shows the results for reaction times.

Table 2 Results for reaction times

	Experts				Novices			
	Static		Dynamic		Static		Dynamic	
	n = 33		n = 43		n = 30		n = 30	
	z	SD	z	SD	z	SD	z	SD
Implicit hazard	-.15	.78	-.06	.30	-.19	.88	.35	.56
Explicit hazard	-.02	.55	-.20	.26	.21	.84	-.07	.33

Note. Z-scores were calculated including mean reaction times for the static and the dynamic version for each scenario.

Experts were faster in the reaction to breaking hints then novices by tendency (F (1, 132) = 3.61; p = .06; η_p^2 = .03). There was neither a main effect of presentation mode (F (1, 132) = .82; p = .37) nor for hazard ambiguity (F (1, 132) = .05; p = .82). No interactions between expertise and presentation mode (F (1,132) = .96; p = .33) or the factors expertise and hazard ambiguity (F (1, 132) = .01; p = .91) could be found. Analysis revealed an interaction between presentation mode and hazard ambiguity (F (1, 132) = 94.4; p < .001; η_p^2 = .42). In addition, there was a triple interaction between expertise, presentation mode, and hazard ambiguity (F (1,132) = 9.90; p < .05; η_p^2 = .07). The triple interaction is visualized in Figure 3.

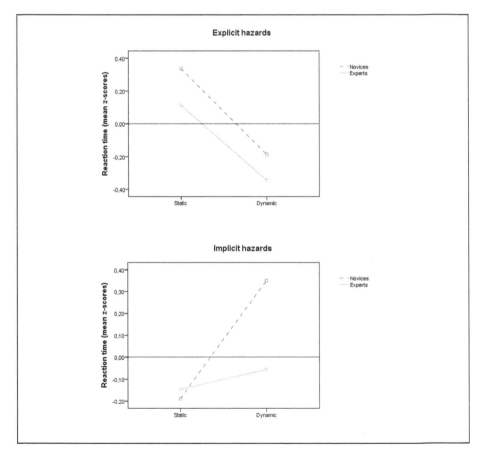

Figure 3 Visualization of the triple interaction between expertise, presentation mode, and hazard ambiguity (dependent variable: reaction time).

The results show that only for those scenarios which contain implicit hazards, the presentation mode has an impact on reaction times dependent on expertise of the subject. Experts reacted faster than novices when these scenarios were presented in a dynamic way. There was no difference in reaction times of experts and novices when the testing material included only static pictures. When specific traffic knowledge was not required, experts outperformed novices to a similar amount with static and with dynamic presentations.

6. DISCUSSION

The main aim of the study was to investigate whether the validity of a hazard perception test can be optimized through the integration of animation. As assumed before experts recognized more traffic hazards correctly and responded faster to them by tendency. Hypotheses 1 and 2 are supported by the results. These results are consistent with the findings of previous research concerning hazard perception. Therefore we conclude that the reaction task and the chosen traffic scenarios are adequate to differentiate between real experts and real novices, which can be concerned as an indication of validity.

In addition, it was assumed that the difference between experts and novices concerning the number of correctly identified hazards (hypothesis 3) and reaction times (hypothesis 4) would be more remarkable when the hazard perception test included rather dynamic than static presentations of the traffic scenarios. This interaction could not been shown, which means that both hypotheses have to be rejected. The inclusion of dynamic presentations impedes the reaction task for both, experts and novices. This result can be ascribed to the transient character of dynamic presentations. There is more information presented in animations than in pictures that could have distracted the subjects or even made them react too early when the critical hint was not visible yet.

Reflections about the domain specificity of expert performance led to the hypothesis, that the superiority of the experts in hazard perception tasks would be higher if the participants had to use specific knowledge, such as traffic rules, to solve a task. This interaction could be shown by tendency for the number of correctly identified hazards but not for reaction times. Hypothesis 5 is supported by tendency whereas hypothesis 6 is not supported.

Results concerning reaction times indicated that only for those scenarios with implicit hazards, in which the use of traffic rules was required for task solution, experts outperform novices more clearly in the animated version. Expertise shows off best, when domain specificity of the task is maximized: dynamic presentation reflects the perception in real driving, the need for a reaction at the right moment reflects the challenges of the real driving task, and the need for specific driving knowledge is given along with it. This result leads to the assumption that domain specific knowledge is internalised by experience. Therefore the decision whether to choose animated or static presentations to assess a skill like driving should depend on content aspects of a task.

Hazard perception testing has been proven as a valid measure of one component of driving competence that develops with driving experience. Therefore it should be a gain to include a hazard perception test in the German driving test. We

438

recommend the enrichment with animations of at least those scenarios that require traffic knowledge for a correct reaction. Further research is needed to investigate whether the validity of the test can be developed to the maximum if domain specificity is optimized even more. In order to realize more domain specificity driving simulators can be used.

ACKNOWLEDGEMENT

The present study was conducted in the frame of a research project funded by the German Federal Highway Research Institute (Research Grant FE 82.326).

REFERENCES

Biermann, A., Skottke, E.-M., Anders, S., Brünken, R., Debus, G., & Leutner, D. (2008). Entwicklung und Überprüfung eines Wirkungsmodells: Eine Quer- und Längsschnittstudie. In R. Brünken, G. Debus & D. Leutner (Eds.), *Wirkungsanalyse und Bewertung der neuen Regelungen im Rahmen der Fahrerlaubnis auf Probe (Berichte der BASt, Heft M 194)* (pp. 46-111). Bremerhaven: Wirtschaftsverlag NW.

Ericsson, K. A., & Lehmann, A. C. (1996). Expert and exceptional performance: Evidence of maximal adaptation to task constraints. *Annual Review of Psychology, 47*(1), 273–305.

Glaser, R., & Chi, M. T. H. (1988). Overview. In M. T. H. Chi, R. Glaser & M. J. Farr (Eds.), *The nature of expertise* (pp. xv-xxviii). Hillsdale: NJ: Erlbaum.

Horswill, M. S., & McKenna, F. P. (2004). Drivers' hazard perception ability: Situation awareness on the road. In S. Banbury & S. Tremblay (Eds.), *A cognitive approach to situation awareness: Theory and application* (pp. 155-175). Aldershot: Ashgate.

Mayhew, D. R., Simpson, H. M., & Pak, A. (2003). Changes in collision rates among novice drivers during the first months of driving. *Accident Analysis and Prevention, 35*, 683-691.

McKenna, F. P., & Crick, J. L. (1994). *Developments in hazard perception. Final Report*: Department of Transport (UK).

McKenna, F. P., & Horswill, M. S. (1999). Hazard perception and its relevance for driver licensing. *IATSS Research, 23*(1), 36-41.

Quimby, A. R., Maycock, G., Carter, I. D., Dixon, R., & Wall, J. G. (1986). *Perceptual abilities of accident involved drivers* (TRL Research Report 27). Crowthorne, Berkshire TRL Limited.

Quimby, A. R., & Watts, G. R. (1981). *Human factors and driving performance* (TRL Report LR1004). Crowthorne, Berkshire: Transport and Road Research Laboratory.

Sagberg, F., & Bjornskau, T. (2006). Hazard perception and driving experience among novice drivers. *Accident Analysis and Prevention, 38*, 407-414.

Spiro, R. J., Feltovich, P. J., Jacobson, M. J., & Coulson, R. L. (1991). Cognitive flexibility, constructivism, and hypertext: Random access instruction for advanced knowledge acquisition in ill-structured domains. *Educational Technology, 31*(5), 24-33.

CHAPTER 45

Towards a Stochastic Model of Driving Behavior under Adverse Conditions

R.G. Hoogendoorn, B. van Arem, S.P. Hoogendoorn

K.A. Brookhuis and R. Happee

Delft University of Technology
r.g.hoogendoorn@tudelft.nl

ABSTRACT

Adverse conditions have been shown to have a substantial impact on traffic flow operations following substantial adaptation effects in driving behavior. In order to determine whether driver support systems are effective under these circumstances, it is crucial to capture stochasticity due to human factors in mathematical models of driving behavior. To this end in this contribution we introduce a new stochastic car-following model based on psycho-spacing theory using a Bayesian network modeling approach with parameter learning. We show that this model yields a relative adequate prediction of longitudinal driving behavior in case of adverse conditions through several examples. The paper is concluded with a discussion section and recommendations for future research.

Keywords: adverse conditions, human factors, longitudinal driving behavior, driver heterogeneity.

1 INTRODUCTION

Adverse conditions have been shown to have a substantial impact on traffic flow operations. Examples of adverse conditions, defined as conditions following an unplanned event with a high impact and a low probability of occurring, are emergency situations due to man-made or naturally occurring disasters, adverse weather conditions (e.g. fog) and freeway incidents (e.g. car-accidents). Human factors can be assumed to have a substantial influence on driving behavior in case of adverse conditions. Human factors (e.g. personal characteristics, mental workload) are however complex and diverse, and lead to differences between drivers (inter driver heterogeneity) as well as within drivers (intra driver heterogeneity).

In order to determine whether systems supporting the driver in case of adverse conditions, such as lane reversal, are effective, it is crucial to be able to capture this stochasticity in driving behavior due to inter and inter driver heterogeneity in mathematical models. A complicating factor is however that in the past driving behavior has predominantly been modeled through microscopic deterministic models. These models are called microscopic as they capture traffic flow on the level of individual vehicles. Therefore and in contrast with macroscopic models they are by definition built on driving behavior specifications. In these models inter and intra driver heterogeneity is considered an error and therefore do not sufficiently consider human factors (Boer, 1999).

The vehicle interaction task in driving behavior can be divided into a longitudinal and a lateral vehicle interaction subtask. Longitudinal vehicle interaction subtasks consist of acceleration, deceleration, synchronization of the speed with the speed of the lead vehicle and maintaining a desired distance to the lead vehicle. Lateral vehicle interaction subtasks consist of lane changing, merging and overtaking.

In the longitudinal vehicle interaction subtask, two main regimes can be distinguished, namely: free flow and congested driving. In the congested driving regime behavior is assumed to be determined by car-following. Several mathematical microscopic models have been developed aiming to mimic car-following behavior under a wide range of conditions and to use them in microscopic simulation software packages (Brackstone & MacDonald, 1999).

However, in spite of the vast number of available car-following models, most of these models do not incorporate human factors. For example, the only human element in most of these models is a finite reaction time. Furthermore, most models assume that that drivers can adequately perceive lead vehicle related stimuli (no matter how small) as well as that drivers are able to adequately evaluate situations and respond appropriately. Behavioral scientists have quite rightly noticed that this field of knowledge has been underexploited by traffic flow theorists (Boer, 1999; Ranney, 1999; Van Winsum, 1999).

A good step towards the incorporation of human factors in mathematical models of car-following behavior was achieved through the development of psycho-spacing models. However, although car-following patterns closely resemble the ones predicted by psycho-spacing theory, these models are also deterministic in character. These models are deterministic as it is assumed that each driver will always change its acceleration on crossing a perceptual thresholds. In reality however, perceptual thresholds vary due to workload, distraction etc. (Brackstone et al., 2002; Hoogendoorn et al., 2011a; Hoogendoorn et al., 2012; Wagner, 2005).

It is therefore crucial to develop a stochastic model based on the principles of psycho-spacing theory. This model should be able to capture intra as well as inter driver heterogeneity due to human factors. This incorporation of heterogeneity could possibly lead to a more adequate evaluation of the efficacy of measures aimed at mitigating the impact adverse conditions have on traffic flow operations.

To this end, in this contribution a new stochastic psycho-spacing model using a Bayesian network modeling approach is introduced. In this context, in the next section we discuss stochasticity in psycho-spacing models. We continue with setting up a simple Bayesian net aimed at predicting so-called action points in the relative speed – spacing (Δv, s) plane as well as acceleration in case of adverse conditions. Next, conditional probabilities were learned for the nodes through a Maximum Likelihood Estimation (MLE) approach using empirical data collected with a helicopter near the Dutch cities of Rotterdam and Apeldoorn as well as data collected through two driving simulator studies. Action points were estimated from this data using a new data analysis technique (Hoogendoorn et al., 2011b).

We will show that this approach yields an adequate prediction of action points in the relative speed – spacing (*Δv,s*) plane as well as acceleration at the action points, while also accounting for stochasticity. We also show that the model yields substantial differences in the distribution of joint probabilities of action points and acceleration at the action points in case of adverse conditions.

2 STOCHASTICITY IN PSYCHO-SPACING MODELS

The differing versions of psycho-spacing models have independently been developed by a number of researchers since the 1960s. Perhaps the earliest version of these models was developed by Michaels (1963) and Todosiev (1963). In psycho-spacing models car-following behavior is described on a relative speed - spacing (Δv, *s)* plane and controlled by perceptual thresholds.

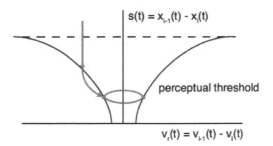

Figure 1 Basic psycho-spacing model

These thresholds, based on changes in Δv and *s* delineate a central area in which a driver of a following vehicle is unable to notice any change in his dynamic conditions (Brackstone et al., 2002). On crossing one of these thresholds, a driver will perceive that an unacceptable situation has arisen and will adjust his longitudinal driving behavior through a change in the sign of his acceleration, typically in the order of 0.2m/s^2 (Montroll, 1959). In the remainder of this contribution these points are referred to as so-called 'action points'. The driver will maintain this acceleration until the other threshold is crossed producing the typical 'spirals' (Figure 1).

From the available research (e.g. Evans et al., 1977) it can be observed that car-following patterns closely resemble the ones predicted by psycho-spacing theory. However, research has shown that the assumption of deterministic perceptual thresholds does not hold in reality as large differences can be expected between and also within drivers. Furthermore, in the original formulation of the model (e.g. Montroll, 1959) it is assumed that drivers will only change the sign of their acceleration. It can be conjectured that this acceleration change at the action points is dependent on Δv and s and will also show a large degree of stochasticity. The aforementioned is supported by Brackstone et al. (2002). In their research using empirical data of UK motorways, they established that large variations could be observed in the position of action points in the $(\Delta v, s)$ plane. The aforementioned is also supported by Wagner (2005) who compared actual action points to the action points which would be expected in case of deterministic perceptual thresholds. Here it was shown that again large differences exist between the position of action points in the Δv, $s)$ plane. Also adverse conditions been shown to lead to substantial differences in the position of action points as well as differences in acceleration at the action points. From research reported in Hoogendoorn et al. (2011a) it followed for example that adverse weather conditions led to a larger scatter in the position of action points compared to normal driving conditions. Also acceleration at the action points changed substantially, as was shown through a Multivariate Regression Analysis. The error (being the sum of the residuals) was substantial indicating a large degree of stochasticity. Furthermore in Hoogendoorn et al. (2011c) it was shown that distracting freeway incidents on the opposite lane have a substantial influence on the position of action points and acceleration as well. In this regard empirical trajectory data was used in order to estimate the position of action points. Finally emergency situations have been shown to lead to substantial differences in the position of action points as well as to differences in acceleration at the action points (Hoogendoorn et al., 2012). In case of emergency situations action points are much less scattered in the $(\Delta v, s)$ plane compared to normal driving conditions. Again, also differences in acceleration at the action points were observed.

It can therefore be concluded that the deterministic perceptual thresholds in the original formulation of the psycho-spacing models do not hold in reality. Large differences in the position of action points can be observed. These differences are dependent on external circumstances, such as adverse conditions. It is therefore crucial to develop a stochastic psycho- spacing model, able to adequately capture differences in the positions of action points as well as acceleration at the action points.

3 SETTING UP THE BAYESIAN NETWORK

We started with setting up a simple Bayesian net. The net consists of nine discrete nodes. In the net, nodes representing lead vehicle related stimuli, i.e. positive relative speed Δv_{pos} (approach), negative relative speed Δv_{neg} (separation) and spacing s are connected to action points Ap as well as to acceleration increases

a_{inc} and reductions a_{red} at the action points. Three discrete parent nodes were added to the net representing emergency situations, adverse weather conditions and freeway incidents. The constructed Bayesian net is depicted in Figure 2.

The numbers of classes within the nodes for s, Δv_{pos}, Δv_{neg} are based on the maximum values observed in the datasets discussed in the ensuing. These classes therefore represent different classes of spacing and speed differences with the lead vehicle. The node representing action points consists of two classes (1=no action point, 2=action point), while the nodes representing acceleration increases and reductions contain six classes. Finally, the parent nodes representing the adverse conditions contain two classes (1=no adverse condition, 2=adverse condition).

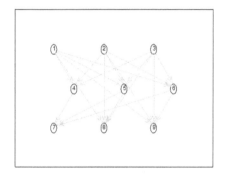

Figure 2   Designed extended Bayesian network. In this network the following nodes are incorporated: Emergency situations (1), Adverse weather conditions (2), Freeway incidents (3), Spacing(4), Positive relative speed (5), Negative relative speed (6), Action point(7), Acceleration increases (8) and Acceleration reductions (9)

We continued with constructing conditional probability tables (CPT's) using random conditional probabilities and making inferences. These inferences were based on a ground state, i.e. no evidence was fed to the network for a specific adverse condition, value of spacing s, positive relative speed Δv_{pos} or negative relative speed Δv_{neg}.

4   LEARNING PARAMETERS

In order to acquire a realistic prediction of action points in the relative speed – spacing $(\Delta v, s)$ plane as well as acceleration at action points, we trained the network using empirical trajectory data. The data used in this contribution firstly consists of two vehicle trajectory datasets collected through the data collection approach proposed in Hoogendoorn and Schreuder (2005). The first dataset covered 500m of roadway stretch yielding 935 trajectories. The data was collected during the afternoon peak hour at the three-lane A15 motorway to the South of the Dutch city of Rotterdam. The second dataset collected through this approach included an incident on the A1 motorway. This dataset consists of 199 trajectories. At the incident a van rolled over and ended in the median strip.

Also data collected through two driving simulator studies described in Hoogendoorn et al. (2011a) and Hoogendoorn et al. (2012) was used. In sum, 4 different datasets were used, including normal driving conditions, an emergency situation, adverse weather conditions and a freeway incident. Action points were estimated from the empirical trajectory data. To this end we used the data analysis technique proposed in Hoogendoorn et al. (2011b). This resulted in the distribution of action points in the relative speed ($\Delta v,s$) plane illustrated in Figure 3. We continued by creating a tabula rasa through again filling the CPT's with random parameters. In this regard initial parameters do not matter as we are able to find the globally optimal Maximum Likelihood Estimates (MLE) independent of where we start.

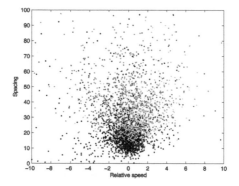

Figure 3 Distribution of action points in the ($\Delta v,s$) plane. Action points under normal driving conditions are presented by blue points. The black, red and green points represent respectively action points in case of emergency situations, adverse weather conditions and incidents.

First it was assumed that the examples in the training data set are drawn independently from the underlying distribution. In other words: examples are conditionally independent given the parameters of the graphical model. Furthermore, in computing the parameter estimators in the discrete cases, we additionally assumed that all counts corresponding to the parameters in the Bayesian network are positive. In this regard the maximum likelihood estimators for the parameters are given by (Niculescu et al., 2006):

$$\hat{\theta} = \frac{N_i}{k_i N}$$

The problem of maximizing the data-log-likelihood subject to the parameter sharing can be broken down into sub problems, one for each probability distribution. One such sub problem can be restated as:

$$P = \arg\max\left\{h(\theta)\big|g(\theta)=0\right\} \tag{1}$$

where:

$$h(\theta) = \sum_i N_i \log \theta_i \tag{2}$$

and:

$$g(\theta) = \left(\sum_i h_i \theta_i\right) - 1 = 0 \tag{3}$$

When all the counts are positive, it can be easily proven that P has a global optimum which is achieved in the interior of the region.

4 MARGINAL AND JOINT PROBABILITIES OF ACTION POINTS THROUGH PARAMETER LEARNING

After learning the parameters in the nodes using the method described in the previous section, we calculated the marginal probabilities for all the nodes in the Bayesian net, given the evidence of respectively no adverse condition, an emergency situation, an adverse weather condition and a freeway incident. After calculating the marginal probabilities, we calculated the joint probabilities $P([s, \Delta v_{pos}, \Delta v_{neg}],Ap)$ using the learned parameters. This is the probability of an action point occurring given the probabilities of a certain value of s, Δv_{pos} and Δv_{neg}. We smoothed the joint probabilities twice using a 3-by-3 convolution kernel. Next we fed evidence to the network of respectively an emergency situation, adverse weather conditions and a freeway incident. Again we calculated the marginal probabilities and the joint probabilities $P([s, \Delta v_{pos}, \Delta v_{neg}],Ap)$ using the learned parameters. In Figure 4 we present the joint probabilities given evidence of the three adverse conditions.

When comparing the joint probabilities of the ground state to the joint probabilities given evidence of an emergency situation, an adverse weather condition and a freeway incident it can be observed that in all four cases the highest joint probabilities are located at smaller values of s, Δv_{pos} and Δv_{neg}. However, substantial differences between the joint probabilities in the ground state and given evidence of the three adverse conditions can be observed. For example, given the evidence of an emergency situation, high joint probabilities of an action point occurring are located at much smaller values of s than in the ground state. Furthermore, when comparing the location of the joint probabilities in Figure 4 to the position of action points in Figure 3 it can be observed that they are relatively similar. We continued with calculating the joint probabilities of acceleration reductions a_{red} and acceleration increases a_{inc}. As an example, the results for the ground state and given evidence of an emergency situation are presented in Figure 5. In order to provide a more insightful figure, only joint probabilities are shown given the evidence $s=20$.

446

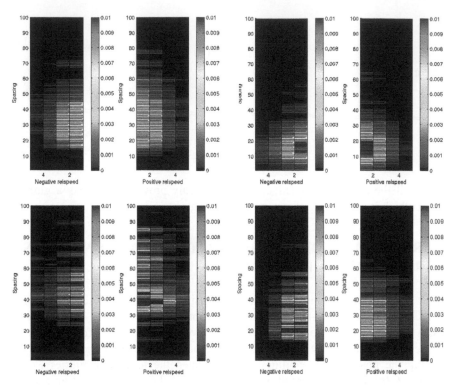

Figure 4 Joint probabilities $P([s, \Delta v_{pos}, \Delta v_{neg}], AP)$ smoothed twice using a 3-by-3 convolution kernel of the ground state (top left) and given evidence of an emergency situation (top right), an adverse weather condition (bottom left) and a freeway incident (bottom right).

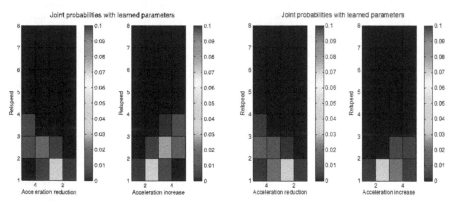

Figure 5 Joint probabilities $P([s, \Delta v_{pos}, \Delta v_{neg}, Ap], A_{red})$ and $P([s, \Delta v_{pos}, \Delta v_{neg}, AP], A_{inc})$ of the ground state (left) and given evidence of an emergency situation (right).

Figure 5 shows that in the ground state the joint probabilities are the highest at smaller values of relative speed Δv and acceleration a. This means for example that the probability of larger accelerations at small speed differences with the lead vehicle is smaller than the probability of smaller acceleration at small speed differences with the lead vehicle. When comparing the joint probabilities of the ground state to the situation in which evidence of an emergency situation was fed to the network, some slight differences can be observed. It can be observed from the figure that the joint probability of larger accelerations at smaller values of relative speed Δv are slightly higher given the evidence of an emergency situation than in the ground state. This is supports the findings reported in Hamdar and Mahmassani (2008) and also Hoogendoorn et al. (2012).

5. DISCUSSION

Adverse conditions have been shown to have a substantial impact on traffic flow operations following substantial changes in driving behavior. Human factors can be assumed to play a substantial role in driving behavior under adverse conditions leading to intra and inter driver heterogeneity. In order to determine whether solution approaches are effective, it is crucial that mathematical models of driving behavior capture this stochasticity. A complicating factor is however that in the past driving behavior has predominantly been modeled through deterministic models. In order to capture stochasticity in driving behavior in this contribution we took a first step in modeling car-following behavior through a Bayesian network modeling approach based on psycho-spacing theory. We set up a Bayesian network with discrete nodes and connected them through links. Parameters in the conditional probability tables were learned through an MLE approach using actual data. Action points were estimated using a new data analysis technique. From the marginal as well as the joint probabilities it followed that the Bayesian network performs relatively well in predicting action points in the $(\Delta v, s)$ plane under normal driving conditions and adverse conditions. The network also provides a reasonably adequate prediction of acceleration at the action points.

However, in order to acquire a more representative prediction of action points and acceleration at the action points more data is needed. Furthermore, the used Bayesian network was static and consisted of discrete nodes. In order to acquire a more realistic representation of the longitudinal driving task it is recommended to transform the current net into a dynamic continuous Bayesian network. Further-more, it is recommended to extend the network with nodes representing human factors. Examples of these nodes are age, driving experience and mental workload.

ACKNOWLEDGMENTS

The research presented is part of the research program "Traffic and Travel behavior in case of Exceptional Events", sponsored by the Dutch Foundation of Scientific Research MaGW-NWO.

REFERENCES

Boer, E. (1999). Car following from the drivers perspective. Transportation Research Part F: Traffic Psychology and Behavior, 2(4), 201206.

Brackstone, M., and M. McDonald (1999). Car-following: a historical review. Transportation Research Part F: Traffic Psychology and Behavior, 2(4), 181196.

Brackstone, M., B. Sultan and M. McDonald (2002). Motor Driver behavior: studies on car-following. Transportation Research Part F: Traffic Psychology and Behavior, 5, pp. 31-46.

Evans, L., and R. Rothery. Perceptual thresholds in car following: a recent comparison. Transportation Science, 11, 1977, pp. 60-72.

Hamdar, S., and H. Mahmassani (2008). From existing accident-free car-following models to colliding vehicles: Exploration and assessment. Transportation Research Record: Journal of the Transportation Research Board, 2088(-1), 45–56.

Hoogendoorn, R.G., S.P. Hoogendoorn, K.A. Brookhuis, and W. Daamen. (2011a). Adaptation Effects in Longitudinal Driving Behavior, Mental Workload and Psycho-spacing Modeling in case of Fog. Transportation Research Records (in press).

Hoogendoorn, S.P., R.G. Hoogendoorn and W. Daamen (2011b). Wiedemann Revisited: A New Trajectory Filtering Technique and its Implications for Car-following Modeling, Transportation Research Records, (in press)

Hoogendoorn, R.G. S.P. Hoogendoorn, K.A. Brookhuis and W. Daamen (2011c). Psycho-spacing models and adverse conditions: A close look at incidents. 2011 International Conference on Networking, Sensing and Control, ICNSC 2011, art. no. 5874915, pp. 329-334

Hoogendoorn, R.G., S.P. Hoogendoorn, and K.A. Brookhuis (2012). Driving Behavior in case of Emergency Situations: A Psycho-spacing Modeling Approach. Proceedings of the Annual Meeting of the Transportation Research Board 2012, Washington DC.

Hoogendoorn, S.P. and M. Schreuder. (2005). Tracing congestion dynamics with remote sensing: towards a robust method for microscopic data collection. Proceedings of the Annual Meeting of the Transportation Research Board 2005, Washington DC.

Michaels, R. M. (1963). Perceptual factors in car following. Proceedings of the second international symposium on the theory of road traffic flow, pp. 4459.

Montroll, E. W.(1959). Acceleration and clustering tendency of vehicular traffic. Proceeding symposium on theory of traffic flow (pp. 147157). Research Laboratories, General Motors Corporation, Detroit, USA.

Niculescu, R., T. Mitchell, and R. Rao. (2006). Bayesian network learning with parameter constraints. The Journal of Machine Learning Research, 7, 13571383.

Ranney, T. (1999). Psychological factors that influence car-following and car- following model development. Transportation Research Part F: Traffic Psychology and Behavior, 2(4), 213219.

Todosiev, E.P. (1963). The action point model of driver vehicle system. Engineering Experiment Station, The Ohio State University, Columbus, Ohio, Report 202 A-3.

Van Winsum, W. (1999). The human element in car following models. Transportation Research Part F: Traffic Psychology and Behavior, 2(4), 207211.

Wagner, P. (2005). Empirical Description of Car-Following. Proceeding of Traffic and Granular Flow '03. pp 15-27

CHAPTER 46

The Effect of Low Cost Engineering Measures and Enforcement on Driver Behaviour and Safety on Single Carriageway Interurban Trunk Roads

João Lourenço Cardoso

Laboratório Nacional de Engenharia Civil (LNEC)
Av. Brasil, 101, 1700-066 Lisboa, PORTUGAL
joao.cardoso@lnec.pt

ABSTRACT

The application of low-cost road and traffic engineering measures (LCEM) is a cost-effective method for reducing accidents and their consequences. Empirical evidence shows that enforcement contributes to improvements on driving behaviour and road safety; however, results from reducing enforcement are seldom presented.

In this paper a presentation is made of the impact LCEM and changes in enforcement intensity had on selected driving behaviour variables and safety levels on a 170 km single carriageway trunk road. LCEM were implemented on the road, followed a year later by the commitment of exceptionally intense and severe law enforcement and, after two years, its relaxation. The impacts of these safety interventions were evaluated through observational before-after studies. The expected number of injury accidents was reduced by 41% (less 75% fatalities), when considering the combined effect of LCEM and enforcement; suppression of strict enforcement was related to a 20% increase in the number of fatalities.

Keywords: safety effects, low cost measures, enforcement, driving behaviour

1   INTRODUCTION

Road safety interventions are efforts deliberately aiming at improving the safety performance of road transport systems (Wilpert, B. and Fahlbruch, 2002). These interventions may focus on any element of the system (human, infrastructure and vehicle), include any of the phases of the accident (pre-crash, crash and post-collision) or be designed to mitigate one of the contributing factors to the phenomenon of accident (exposure, risk, severity of injuries and injuries sustained).

There is much empirical evidence showing that the effective use of legal measures and their enforcement may contribute for a quick improvement of both driving behaviour and road safety (Elvik et al, 2009; and Elliot and Broughton, 2005). Experience shows that the full realisation of the benefits of effective enforcement of road user behaviour legislation depends largely on the definition of a set of measurable compliance targets and on the permanence of police monitoring activity. Enforcement rigor and tolerance margins are also known to impact on enforcement effectiveness (De Waard and Rooijers, 1994).

Safety belt use, driving under the influence of alcohol and drugs, speed limit violations and red light running are the main analysed driving behaviour issues targeted in published studies. According to some estimates, increasing the seat belt wearing rates in the European Union countries to 100% would result in a reduction of 7000 fatalities annually (Christ et al, 1999).

Low-cost engineering measures (LCEM) are physical road safety interventions on the infrastructure that have a low capital cost and can be implemented quickly (ETSC, 1996).

LCEM are specifically designed to prevent the accidents occurring on selected high risk zones of the road network, and reduce their consequences; they may also be directed at specific safety problems in a given area. Consequently, LCEM have favourable benefit to cost ratios, and their systematic application is a cost-effective method for achieving quick reductions in accident frequencies and their consequences. Experience has shown that benefit/cost ratios between three and 24 may be achieved, in countries where moderate to high monetary values are attached to the prevention of death and injuries resulting from road accidents (FHWA, 1993).

LCEM can be applied on road sites, road sections or small areas, located on urban, suburban or rural roads. Due to the low-cost requirement, application of LCEM is usually constrained to the existing road area; therefore, no expensive and lengthy land acquisition and environmental impact evaluation processes are needed prior to their application. LCEM include several types of interventions, including: minor changes in the layout of a road, namely at junctions; improvements in road signing and marking; improvement of pavement surface characteristics; removal of roadside obstacles; changes in junction operation and traffic channelization; pedestrian crossings; and physical measures to reduce speed.

The road safety performance of a traffic system may be expressed in several ways. One may measure the social acceptability of road danger in a community, by classifying the sense of danger of its members, thus obtaining what is known as subjective safety. Nominal safety refers to the level of compliance to technical standards and legal or formal rules set for designing and using the road system. Objective safety is quantitatively expressed by the long term expected number of accidents, casualties or damage likely to result from traffic operation; additionally, objective safety may be descriptively analysed using historical accident severity data (AASHTO, 2010).

In this paper, objective safety is used to evaluate and describe the impact that LCEM and changes in enforcement intensity had on selected driving behaviour variables and safety levels on IP 5 road, a 170 km single carriageway two lane Portuguese trunk road (Figure 1). The analysis covers a ten year time period.

Figure 1 Overall aspect of the IP 5 road

2   IP 5 DESIGN CHARACTERISTICS AND IMPLEMENTED SAFETY INTERVENTIONS

The analysed road section is one of the main routes connecting Portugal's central and northern maritime ports to Spanish hinterland and Europe. It crosses a mountainous area, between the coastal plains and the central Iberian plateau. The IP 5 route is used by a significant percentage of the international road traffic originating or arriving to Portugal. Authorized traffic in this road is restricted to motorized vehicles.

At the time of the first intervention, IP5 was a single carriageway two lane road with climbing lanes in selected grades (corresponding to over a third of the total road length). Average daily traffic volumes (AADT) were very high, between 4400 and 10000 vehicles. Heavy goods vehicle (HGV) traffic was also intense, both in absolute numbers and expressed as a percentage of the total number of passing vehicles: between 1700 and 3450 daily vehicles, or 17% to 32% of total AADT. IP 5 safety performance was especially poor, at the time of the first interventions: 35 fatalities and 37 serious injuries being registered annually, as a consequence of 508 accidents.

The road had 3.75 m wide lanes and 2.5 m wide paved shoulders. Selected geometric layout data are presented in Table 1.

Table 1 Main geometric design characteristics of analysed IP 5

Geometric characteristic	Road section			
	Albergaria Viseu	Viseu bypass	Viseu Guarda	Guarda V. Formoso
Length (km)	53	17	63	37
Design speed (km/h)	80	100	100	100
Minimum horizontal radius (m)	300	450	300	600
Length of tangents (%)	38	38	43	58
Length with additional climbing lane (%)	45	40	49	33

The most frequent types of accidents occurring in IP 5 (head-on collisions and run-off-the road, namely on curves, on approach zones to interchanges, and on zones with additional climbing lanes) were mostly related to speeding and irregular overtaking (this resulting, in part, from the unforeseen high growth in traffic volumes).

The analysed safety interventions consisted of the implementation of LCEM on the trunk road; followed a year later by the commitment of exceptionally intense and severe law enforcement; and, subsequently (after two years), its relaxation. The sequential application of these safety interventions and the planned monitoring of resulting developments allowed for the evaluation of their individual impacts through observational before-after studies.

Three sets of LCEM were implemented (Cardoso and Roque, 2000): improvement of traffic operations on sections with climbing lanes, namely by application of traffic regulations enhancing the number of passing opportunities for cars; increase in visibility conditions and operation predictability for traffic leaving and entering IP 5, at interchanges; and changes in road environment (carriageway and roadside area) intended to influence driving behaviour.

This last set of measures comprised several LCEM: measures to improve surface water drainage, traffic sign visibility and overall visibility; mandatory use of day-light running lights in the road section, through the installation of appropriate vertical signs; setting a 90 km/h speed limit on the whole route section; the installation of edge rumble strips along the entire road (and the corresponding warning signs); the installation of new no-passing zones on selected dangerous sites; the repositioning of reflecting road studs at the road axis and at all new no-passing zones; and signing of horizontal curves according to consistency criteria. The obligation to use day-light running lights (still not widespread in the Portuguese road network) was intended to improve long distance vehicle conspicuity and to suggest drivers the sense of being on a special road that required extra driving precaution.

Plastic position marker posts (Figure 2) were installed at the road axis in the vicinity of interchanges, as an additional warning to drivers and to impede late passing manoeuvres from intruding the beginning of no-passing zones. New arrow markings were provided in acceleration and deceleration lanes.

Implementation of this engineering intervention took less than six months and its total costs amounted to 840,000 US dollars (at 1998 prices), less than 5,000 US dollars per kilometre (Cardoso and Roque, 2000).

Figure 2 Position marker posts at the road axis at the approach to an interchange

Approximately six months after the LCEM implementation, a special enforcement campaign started on the IP 5 route. Under the motto '*Maximum safety - zero tolerance*', this campaign (MSZT) was subject to widespread media coverage, including the personal intervention of high ranking government officials.

Characteristics of the enforcement activity were changed in two ways. Firstly, tolerance levels were eliminated (for instance in the case of prohibited manoeuvres or if the lights were not switched on), or reduced to the minimum technically allowed by the measuring devices (radars and alcohol tests). This approach, based on the belief that IP 5 was so dangerous that no exceptions to ruled driving behaviour could be granted to normal drivers, was a major change from the more traditional approach to driving law enforcement. As an example, on normal roads, drivers expected that speeding up to 20 km/h over the speed limit would be punished only under very 'unfortunate' conditions. Furthermore, moderate severity was usually employed by the police for other offences; under normal circumstances, there was a moderate chance that a violation could be granted a simple warning, instead of a ticket.

Secondly, the overall activity of the traffic police was increased by more than 75% in the first four weeks of the campaign and by 25% in the following 24 months. Growth in police activity was achieved by increasing the available types of enforcement actions and by raising the number of simultaneous traditional patrols. Traditional patrol activity (with marked cars and motorcycles) was complemented with helicopter patrols, automatic photo-radar devices and 'camouflaged' vehicles equipped with video and radar. The number of police patrols on the road was raised from the original 9 patrols (8 hours shifts) per day, to 16 patrols per day in the first four weeks and to 11, since then, during two years.

Police officers were instructed to broaden the focus of their enforcement activity, to include issues such as speed limits, blood alcohol, no-passing zones, day-light running lights, and HGV weight limits and tachograph regulations (rest and driving periods).

A special traffic sign was installed along the IP 5 road, to remind drivers that the road was subject to extraordinary enforcement activity.

3   ANALYSIS METHOD

Safety related developments following the referred interventions on IP5 were monitored by LNEC, through an agreement with the Portuguese main roads administration. The impacts on driving behaviour were analysed using an observational before-after study; developments in AADT were considered, as well (Cardoso and Roque, 2000).

Selected driving behaviour variables were measured at three instances of the project life cycle:

- Immediately before the implementation of the corrective measures ('Before');
- Three months after the execution of the planned corrections ('A_LCEM'), before the MSZT campaign;
- One year after the start of the enforcement campaign ('A_MSZT').

Traffic behaviour variables included the following: hourly traffic flows; spot speeds; headways; distance between front wheels (right and left) and the right edge line; and the lights switch position (on/off).

Furthermore, two years after the end of the enforcement campaign, hourly traffic flow and spot speed measurements were carried out on the tested road ("A_E_MSZT").

Seat belt compliance was not evaluated, since previous yearly country wide measurements had resulted in front seat belt compliance rates above 94%.

Safety impacts were analysed using an observational before-after study with control sections. The expected number of accidents and the observed number of fatalities and severe injuries were used as safety performance variables. The multivariate regression empirical Bayes method described by Hauer (1998) was used in the analysis of developments in the expected number of accidents.

Several roads similar to IP 5, in the Portuguese NRN, were used as a control group, to account for the overall effect of safety developments in the country during the analysed periods, assuming that the effect of the disturbing variables was identical in both the treated IP 5 site and the rest of IP roads. The control group consisted of 201 IP road sections, corresponding to 625 km.

Impacts of traffic volume developments on accident frequencies were taken into account using previously developed safety performance functions. Earlier studies showed that accident frequencies on IP roads do not vary linearly with average annual daily traffic changes: they are related to a power slightly above the unit:

$$E = 4.483 \times 10^{-4} \times L \times AADT^{1.022} \tag{1}$$

Where:
E = Annual number of accidents in a road section;
L = Length of the road section (km);
AADT = Average annual daily traffic (vehicles).

The dispersion parameter of this safety performance function is equal to 0.519 (Cardoso, 2007).

Due to the absence of safety performance functions for the number of fatalities and killed and serious injuries, only descriptive analyses of developments in these variables were made. For these variables, a linear variation with AADT was assumed.

4   EFFECTS ON DRIVING BEHAVIOR AND SAFETY

4.1   Driving behaviour

Following the application of the LECM, significant reductions in desired speeds (-11 km/h to -5 km/h) were observed on curves, especially in the inside lane. On tangents, no significant change in desired speeds of cars was noticed, due to LCEM.

The enforcement campaign originated reductions in desired speeds on long tangents, resulting in significantly lower speeds on the IP 5 when compared with other similar Portuguese IP routes. This is highlighted in Figure 3, which contains the results of average and 85th percentile speed measurements on IP 5 and four other IP roads, during MSZT campaign and two years after its end.

456

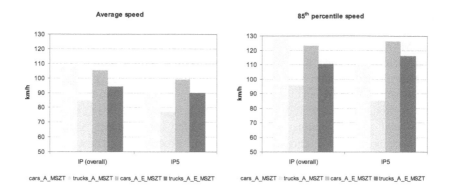

Figure 3 Comparison between speed characteristics during the enforcement campaign and after its end, on IP 5 and other IP roads

Rumble markings also resulted in a better compliance of cars and trucks to the defined lane space: inside shoulder encroachments were substantially reduced (from 20%~50% to 5%~35%) and the distance to the right edge line for vehicles in the inside lane significantly increased by 25 cm (to 35~87 cm).

Only a very small percentage (between 0% and 11%) of drivers used daylight running lights in the before period. After installation of the corresponding signs (LCEM), this percentage increased significantly to values above 90%. Neither the enforcement campaign nor its end had any significant impact on these high compliance rates.

Changes in truck passing regulations and the introduction of signs reminding the minimum legal distance between consecutive trucks on the descending lanes in road sections with additional climbing lanes did not have noteworthy influence on queuing trucks, namely as regards minimum headways and lane choice. Overall, there is a perception that compliance to these types of regulations is not enforced very strictly, even under the full effect of the MSZT. Nevertheless, due to these violations there were less passing opportunities and bigger queues; factors which affected negatively traffic capacity and safety performance.

4.2   Road safety

As mentioned before, LCEM and MSZT were not introduced simultaneously; however, the time interval between the application of LCEM and the start of MSZT (less than a year) was too short to allow for an assessment of the isolated impact of LCEM. Three triennial periods were considered in the assessment of road safety impacts: prior to the combined application of LCEM and MSZT; under the influence of both types of intervention; and after the suppression of MSZT. Injury accident data occurred on IP5 and on the control group IPs were collected from official statistics.

Table 2 summarizes the calculated safety parameters for the three analysed periods. In the first line estimated expected numbers of injury accidents are presented. For both after periods (with and without MSZT campaign) two values are provided: the observed values (estimated in the case of the expected value); and the values which were calculated taking into account only the changes in AADT and in the control group safety (taking the previous period as a reference). Therefore, the latter numbers express the amount of occurrences that would have been produced if no interventions had taken place.

Table 2 Safety developments on the IP 5

Type of safety indicator	Before	LCEM+MSZT		Suppression of MSZT	
		Observed	Calculated	Observed	Calculated
Expected number of accidents	331	367	624	377	428
Observed number of accidents	335	374	608	385	438
Number of fatalities	85	43	199	52	43
Number of killed and seriously injured victims	188	107	380	125	107

The implementation of both MEBC and SMTZ contributed to a significant decrease in the expected number injury accidents (-41%); roughly, this corresponds to less 75% fatalities and less 70% killed and seriously injured victims (less 273 victims in three years). Additionally, it may be concluded that the suspension of the enforcement campaign was accompanied by a 20% increase in the number of fatalities and a 17% increment in the number of killed and seriously injured victims (more 18 victims in three years); however, the expected number of injury accidents on the IP 5 continued to decrease at a greater rate than in the rest of the IP road network.

5   CONCLUDING REMARKS

The application of MEBC resulted in rapid improvement of some aspects of driving behaviour, such as lower variability in speed and path choice and lower night time speeds in dangerous curves. The implementation of the intense and rigorous enforcement campaign fostered further improvements, most noticeably as regards speed choice under normal conditions. The behavioural changes had a direct influence in reducing injury accident frequency; even stronger reductions were obtained in the number of fatalities and seriously injured victims.

458

The absence of a long distance alternate route to IP 5 contributed to the success of the enforcement campaign.

Suppression of MSTZ campaign was accompanied by increased speeds and severity of accidents, resulting in higher numbers of serious injuries and fatalities.

ACKNOWLEDGMENTS

The presented results were obtained in a study developed by LNEC for *Estradas de Portugal, S.A.*, the National roads administration (then JAE), which provided AADT data. Information on the infrastructure characteristics and driving behaviour were collected *in situ* by LNEC; enforcement data was provided by the Portuguese police (BT-GNR) and the National Road Safety Authority (then DGV) supplied accident data.

Currently, the IP 5 route has been substituted by a motorway; however, the safety benefits of this last improvement are not analysed in this paper.

REFERENCES

AASHTO, 2010. *Highway Safety Manual*. Washington, D.C.: American Association of State Highway and Transportation Officials

Cardoso, J.L.; Roque, C.A. Low cost engineering measures and stricter enforcement. A successful combination to improve road safety on a dangerous rural route. Paper presented at the 11th International Conference "Traffic Safety on Three Continents", Pretoria, South Africa, 2000.

Cardoso, J. L. 1998. *Definição e detecção de zonas de acumulação de acidentes na rede rodoviária nacional.(Definition and detection of high accident risk sites – in Portuguese, with English summary)*. Lisboa, Portugal: Laboratório Nacional de Engenharia Civil.

Christ, R., Delhomme, P., Kaba, A., Makinen, T., Sagberg, F., Schulze, H., Siegrist, S, 1999. *GADGET Final Report - Investigations on influences upon driver behaviour. Safety approaches in comparison and combination*. Vienna: KfV Austrian Road Safety Board.

De Waard, D. and Rooijers, T. 1994. An experimental study to evaluate the effectiveness of different methods and intensities of law enforcement on driving speed on motorways. *Accident Analysis and Prevention* 26: 751-765.

Elliott, M. and Broughton, J., 2005. *How methods and levels of policing affect road casualty rates. TRL Report TRL637*. Wokingham, UK: TRL Limited

Elvik, R., Høye, A., Vaa, T., and Sorensen, M., 2009. *The handbook of road safety measures. Second edition*. UK: Emerald Press

ETSC, 1999. *Police Enforcement Strategies to Reduce Traffic Casualties in Europe*. Brussels: European Transport Safety Council.

FHWA, 1993. *The 1993 annual report on highway safety improvement programs*. Washington, D.C.: Federal Highway Administration.

Hauer, E. 1997. *Observational Before-After Studies in Road Safety*. Tarrytown, New York: Pergamon/Elsevier Science, Inc.

Wilpert, B. and Fahlbruch, B. (Eds.) 2002. *System safety. Challenges and pitfals of intervention*. Oxford, UK: Pergamon, Elsevier.

CHAPTER 47

The Vision of Roundabout by Elderly Drivers

Gianfranco Fancello, Claudia Pinna, Paolo Fadda

DICAAR – Dept. Of Civil Eng. Land use Arch. -University of Cagliari, Italy

fancello@unica.it , claudia.pinna@centralabs.it , fadda@unica.it

1   INTRODUCTION

Nowadays we frequently call third age those 65 to 75-80 year olds who are still active and self sufficient and fourth age the over 80's who instead tend to spend less time outside the home. The issue with these population increases is that elderly prefer their own car to fulfill their mobility needs; as everybody else does. The accident rates are higher for older drivers and increase exponentially for drivers over the age of 75 years [8]; some authors argue that risk of accident for elderly would be overestimated influenced by reduced mileage travelled [10]: infact an analysis of accident data has revealed that the number of accidents caused by elderly drivers is, on average and per unit distance driven, actually lower than other age groups. It should be noted though that there are contrasting standpoints in the literature on whether elderly drivers are to be considered high risk. One side claims that they are not any worse than any other drivers for the following reasons:

1. in absolute terms older drivers do not represent a high risk group; by contrast as they recognize their own limitations and slower reactions, they tend to drive more cautiously and observe the rules of the road [9]; they also use less the car in difficult situations such as on highways or in bad weather and poor visibility [12];

2. in spite of their responsible behavior behind the wheel, there is an increase in the number of road accidents involving injuries to the elderly: this is to be attributed not, as mistakenly believed, to the higher accident rate, but to the greater physical fragility of over 65 year olds who are more likely to sustain injury in the event of an accident; for 80 years old drivers, the probability of dying if involved in an accident is five times higher than drivers between 30 and 59 years old [11];

3. they don't drive under the influence of alcohol or drugs or falling asleep; similarly, very few older drivers are involved in single vehicle accidents [9];

The other side notes that elderly drivers are indeed a high risk group and this is due to the smaller number of miles driven, chiefly for reasons associated with their mental and physical faculties. The reasons for this claim are follows:

1. Analysis of age, gender and cohorts, indicated that middle age drivers are safer than younger drivers who, in turn, are safer than older drivers. Older male drivers are safer than older female drivers and more recent cohorts of older drivers are safer than more distant cohorts;

2. Elderly drivers are a high risk population because of the type of accidents that older drivers are usually involved. In fact, side impact crashes are twice as likely to occur with older drivers rather than younger drivers. Due to the severity of these types of crashes, it is generally harder for an older person to recover than a younger driver;

3. There is a higher incidence of elderly drivers involved in accidents occurring in more complex and mentally demanding situations, such as for instance at junctions or slip roads, where the driver has to carry out several tasks simultaneously (merging into traffic flow, pedestrian crossings etc.). Recent studies have shown that high risk maneuvers include merging from a slip road into mainstream traffic, multi-lane driving and turning across traffic at unsignalized intersections [7].

Several studies have shown that mental and physical conditions, driving habits and behavior when performing certain maneuvers differ with age. In particular, the mental and physical faculties of over 65-year olds as well as their ability to concentrate behind the wheel deteriorate more rapidly, with the result that there is a greater likelihood of them having an accident caused by human error [7][9].

Over the last tenyears, a great deal of attention has been focused on driver perception: since the vision is more affected by age, to see well even in darkness, a person of 60 years old needs eight times more light than a 20 year old. The eye is more sensitive to dazzling light and movements that occur in the area at the end of the visual field are not perceived clearly: there is a reduction of the visual field [2]. The focus for different distances is adjusted slowly [1] and with age also increases the minimum duration of the gaze needed to identify the signal details [4]. Car speed and the evolution in real-time traffic affect the risk of the accident for which the mental processing time plays a fundamental role. The elderly tend to reduce their driving speed to be able to successfully manage the several driving situations; when this is not possible, they may be induced to perform traffic maneuvers under pressure, within time intervals not compatible with the their ability. Some physical abilities improve with increasing age: Kline showed that elderly drivers are able to distinguish signals degraded better than youngs, because they are used to perceive better ordered and clear scenes. The accidents caused by elderly drivers occur early in the morning or late evening, under specific conditions, because of reduced visibility or in the presence of intersections less visible [7]. In particular, older drivers find it difficult to identify the traffic signals in the presence of distracter items: for this reason, the intersections are difficult to manage. Crossing an intersection requires performing certain operations, if these are concentrated in a short time, the driver may have a heavy workload and the decline of his mental and physical abilities. The changes that occur with aging must be recognized and addressed with appropriate interventions, taking into account the needs of elderly drivers.

2   METHODOLOGY

The study involved 11 elderly drivers: two more young drivers (Subject A and B) were tested to compare results and evaluate how aging affects the visual perception while driving. The selected drivers had age between 67 and 82 years; only 2 women were tested: in fact generally older women tend to not renew a driver's license once they retire and 65 years of age reached; 54.54% of the drivers have eye problems, but only 45.45% regularly wears glasses. 45.4% of drivers suffer from hypertension, while 18% have problems with high cholesterol and adult-onset diabetes (a disease with direct implications on the view), only 18% do not have any general health problem. The drivers are all resident in the same city (Cagliari) and moreover already knew the path used in the test phase. Regard to the test: people drive on a suburban location 8 km long, in a single line road: roundabout for the test was localized at km No. 4. It 'a four-arm roundabout with a diameter of 40 meters, with a traffic level of service not less than B; a suburban roundabout was chosen to avoid disturbing effects stemming from congestion, or the presence of pedestrians that might influence the perception of the geometry of the roundabout; the tests were conducted on weekdays, because the roundabout is located in an area that during the holidays is affected by significant tourism flow; the tests were conducted on different days, always at the same time to avoid that the visual perception was affected by different light conditions; also during the 13 days of testing we have registered the same weather conditions (sunny day without clouds). Each driver has used own car so that there was a strong familiarity with the vehicle, so distractions have been minimized.

Tests of visual perception have been made using a "eye tracker" portable with two cameras synchronized: the first camera points the eye, allowing you to view the pupil, by recording its movements; the other camera looks the external environment observed by the driver: synchronization allows you to see instantly what the eye sees and how long. This tool uses eye tracking technology called "Dark Pupil Tracking". The data obtained from the registration were analyzed with software "Gaze Tracker" which identifies the points or areas observed by the driver, and calculates the duration of the look: also allows to define the movements of the eyes to different points of view, thus allowing the reconstruction eye movements.

The visual field was divided into six areas to evaluate the "look zones" of drivers: road and other vehicles, dashboard, vertical road signals, left side mirror, rear view mirror, other elements. For each driver were recorded: the observed points, with and without saccadic eye correlated movements , the points of gaze fixation and the look zones. As shown in the results' paragraph, it was not possible to carry out the look areas analysis within the roundabout: in fact, during the maneuver, the drivers adjust own posture, changing completely the point of view and then altering the fixed points: so it's no possible to divide the scene and the screen in fixed looks areas. Therefore, we decided to change the analysis of perception, developing a matrix as done: in each line (i) there is the subdivision of the overall travel time of the roundabout, second by second; in each column (j), there is the indication of new points of interest on which could focus the attention of the driver while driving in the roundabout.

These points are: previous vehicle, arm of the roundabout located to the left of the driver, arm of the roundabout located to the right of the driver, arm of the roundabout located in front of the driver, road signs, center island, traffic inside the roundabout, exit arm of the roundabout.

In the matrix, the element $a_{i,j}$ is: 1 if during i^{th} second when driver is inside the roundabout, he looks the jth element; 0, otherwise.

In this way, the travel time of the roundabout has been transformed from continuous to discrete, thus defining the observation time for each zone.

To compare the performance of several drivers who were tested, the observation times were expressed as a percentage of the total travel time of the roundabout.

2.1 Perception index

In 1964 Neisser [13] defined a synthetic indicator to define the average time T that a man needs to find a target signal within the visual scene: this time depends on the total number of observations (n) and by the average look time (t). For Neisser, the time needed to acquire the stimulus is:

$$T = t * \frac{n}{2}$$

By analyzing this formula, we note that do not include either the number of looks, or the duration of each look. So we decided to change the Neisser formula: we propose a new indicator that uses the gaze duration and the complexity of perception. Compared to Neisser index, in this new formulation were considered the same variables as, for example, the total number of gazes and the duration of each fixation, but, in this case, we consider not the average time value but the real time of fixation. Moreover, for real time of fixation (t), we choose to use an exponential structure:

$$C(t) = \begin{cases} \sum_{t=1}^{t_{max}} (\alpha n_t)^t & t \geq 1 \; ; \; \alpha > 1 \\ 0 & t < 1 \end{cases}$$

This new index provides a measure of the difficulty of perception of the same stimulus for different subjects, or it may be representative of the difficulties encountered by a single person in the observation of different stimuli.

The index can then be used to compare the ability of perception of the various subjects or to assess the complexity of a stimulus than others, for the same subject.

For this reason we think that this index may work better than Neisser Index for the case study, namely the perception of part of a road (roundabout) by elderly drivers. The advantages are: real time has been considered, not the average fixation time; the length of the single observation is introduced; the new index penalizes those who have more difficulty in perceiving a signal and maintains the view on this for more time, averting his eyes from the road: so it is more realistic. The exponential form is better for perception in parallel [15], that drivers use to search for stimuli in complex scenes as the road intersections.

3   DATA ANALYSIS

All collected data during the experiment, were analyzed: we have distinguished the data observed in an unconscious way (where the gaze doesn't linger) from data about points of fixation of the gaze, during which the driver deliberately focuses on that stimulus to collect the meaning.

3.1   Look zones analysis

Visual field of drivers has been studied in detail, through "look zones" analysis.

Initially, the visual field was divided in different areas that the driver observes while driving; for each one, the observation time has been calculated , compared to the total lenght of the test. The table shows the percentage of observation of each area.

Table 1 Percentage of observation of the various zones

Drivers	Road And Vehicles	Vertical Road Signs	Dashboard	Rear view Mirror	Left Side Mirror	Other Elements
Subject 1	81,64%	10,92%	6,12%	1,13%	0,00%	0,00%
Subject 2	93,55%	6,30%	0,12%	0,04%	0,00%	0,00%
Subject 3	77,84%	12,15%	6,53%	0,00%	0,00%	0,78%
Subject 4	99,13%	0,87%	0,00%	0,00%	0,00%	0,00%
Subject 5	87,31%	2,19%	0,50%	0,00%	0,00%	0,00%
Subject 6	92,87%	5,50%	0,30%	0,75%	0,00%	0,57%
Subject 7	62,38%	25,03%	0,00%	0,00%	12,59%	0,00%
Subject 8	98,70%	0,22%	0,87%	0,21%	0,00%	0,00%
Subject 9	56,16%	18,79%	12,53%	0,00%	12,53%	0,00%
Subject 10	94,55%	5,54%	0,00%	0,00%	0,00%	0,00%
Subject 11	89,87%	0,67%	8,81%	0,00%	0,65%	0,00%
Mean	84,91%	8,02%	3,25%	0,19%	2,34%	0,12%
Subject A	76,89%	21,28%	0,10%	1,24%	0,00%	0,00%
Subject B	88,09%	6,15%	5,07%	0,57%	0,12%	0,00%
Mean	82,49%	13,72%	2,59%	0,91%	0,06%	0,00%

The analysis of the zones look reveals that elderly drivers while driving observe more elements as the route, road markings and the vehicle in front, all observed on average for 84,91% of the test lenght. The 63.64% of drivers have observed the road for more than 85% of driving time: only two drivers (Subjects 7 and 8) have observed the way for a short time (62,38%, and 56,16%), because they keep more attention on road signs and on mirrors. Most of older drivers (81.82%) have kept short attention to road signs: in fact the signals are often perceived as a background, since they are not read, but seen as a "road outline". So, elderly drivers consider them not useful and then don't observe them. On the contrary, young drivers have observed the road signs more than elderly (5,7%). Regarding dashboard, it has been observed for an average of 3,25% of the test lenght. In details, four drivers have paid considerable attention to the dashboard, observed it for 8,5% of driving time; on the contrary the 72,73% of drivers tested has paid little attention to this element (0,24%) and three have not ever it watched. However, older drivers have observed the dashboard more than the youngs. Car instrument panel and dashboard are often distracting: infact sometimes they are not easy to legible and the driver often must to linger the gaze too long, diverting attention from the road. To avoid it , in man-machine interface design should be adopted some measures to improve the drivers perception and to minimize the time required to gather information.

During the test all drivers, older and young, use less the mirrors (both left side mirror and rear view mirror): this is probably due to the presence of low volumes of traffic and few overtaking maneuvers. The rear view mirror was not used by 63,64% of older drivers, while 4 drivers have observed it only for 0,53% of the test time. For young drivers the average time of observation of this mirror is equal to 0,91%. Only three elderly users have observed the left mirror (27% of the sample): specifically, subjects 7 and 8 have observed it for 12,5% of the time, but they never looked the other mirror (rearview). About young drivers, only the subject B has observed the left mirror (0,12% of time test), while the subject A doesn't use the left mirror, using mostly the rearview mirror. Drivers are often distracted by the presence of some external elements, such as advertising signals and the surrounding landscape. To find out how much the drivers have observed them, all points of fixation that did not fit into the above categories have been included in "other".

During test, only 2 older drivers have observed other than those necessary for driving though for short time (0,68% on average test time). The young drivers have never diverted attention from their task. This difference can be ascribed to a decreased ability to concentrate by the elderly.

3.2   The roundabout analysis

Only in 1989 the first roundabout has been made in Italy: for it still now many motorists, especially the elderly, have too short experience with this geometric road element. The way how elderly drivers approach the roundabout and then they cross it shows their lack of training in this specific maneuver, bringing out the different techniques used while driving. After twenty years, today the roundabouts are a widespread design choice, but despite this, some older drivers are not familiar with this type of intersection. One of the tested drivers (n°9) had never used this type of intersection, making some driving mistakes: for example, being inside of the roundabout, she slowed her vehicle to allow cars coming from the right arm of the roundabout to pass; so not knowing to have priority, she has complied with the ordinary rules that regulate traffic in free intersections (priority to vehicle that coming from right side). Moreover, even the n°9 driver is not able to make U-turn in the roundabout, but she has taken the arm of the roundabout earlier than that one recommended. The table shows the observation times for the eight selected items, expressed as a rate of the total journey time of the roundabout.

Table 2 Percentage of observation of the different zones in the roundabout

Drivers	Previous vehicle	Arm to the left	Arm to the right	Arm in front	Road Signs	Center Island	Traffic inside	Exit arm
Subject 1	35,00%	10,00%	10,00%	5,00%	0,00%	25,00%	5,00%	10,00%
Subject 2	34,62%	19,23%	11,54%	0,00%	0,00%	15,38%	3,85%	15,38%
Subject 3	44,44%	5,56%	11,11%	0,00%	0,00%	16,67%	0,00%	22,22%
Subject 4	42,31%	7,69%	3,85%	15,38%	0,00%	15,38%	7,69%	7,69%
Subject 5	41,67%	8,33%	29,17%	4,17%	0,00%	8,33%	0,00%	8,33%
Subject 6	54,29%	8,57%	5,71%	5,71%	0,00%	20,00%	0,00%	5,71%
Subject 7	22,73%	9,09%	18,18%	4,55%	0,00%	31,82%	0,00%	13,64%
Subject 8	15,00%	20,00%	35,00%	15,00%	0,00%	5,00%	0,00%	10,00%
Subject 9	32,00%	8,00%	16,00%	0,00%	0,00%	24,00%	8,00%	12,00%
Subject 10	47,62%	14,29%	9,52%	4,76%	0,00%	14,29%	4,76%	4,76%
Subject 11	41,67%	4,17%	8,33%	12,50%	8,33%	8,33%	12,50%	4,17%
Mean	37,39%	10,45%	14,40%	6,10%	0,76%	16,75%	3,80%	10,36%
Subject A	35,00%	5,00%	5,00%	10,00%	0,00%	20,00%	10,00%	15,00%
Subject B	40,00%	8,00%	20,00%	8,00%	0,00%	8,00%	8,00%	8,00%
Mean	37,50%	6,50%	12,50%	9,00%	0,00%	14,00%	9,00%	11,50%

Analysis of data shows that the element more observed is the vehicle in front, and that it was seen by older drivers for 37,39% and by young people for 37,50% of total travel time of the roundabout. The older drivers, running through the roundabout, often observe the center island to set own vehicle path ; in fact center island was seen by older drivers for 16,75% of the time and by the youngs for 14% of the time of fixation. All drivers have looked more the arm of the roundabout located at own right side than one located at own left, although they should give priority to traffic coming from the left side. The older drivers have looked the arm of the roundabout located at right side for 14,40% of the total time and the arm on the left side for 10,45%; the youngs have less observed both arms, for 12,50% (right side) and for 6,50% (left side) of the total time of running across the roundabout.

The young drivers have looked the arm of the roundabout which is on the opposite side for 9% of the time, but the elderly only for 6,10% of the time (three old drivers have never looked it). The elders have observed the arm to go out from the roundabout for 10,36% of the time, while young people only for 11,5%. The traffic flow inside the roundabout has been looked by older drivers or 3,80% of the time, and by youngs for 9%: three elderly drivers have never observed vehicles circulating within. Only one old driver has seen the road signs in the roundabout, to understand how go out by the roundabout. Road signs were not observed by young drivers. This means that the road signs are not watched by drivers who do not consider them as driver assistance, as also shown by previous analysis regarding the whole route. That is because within complex scenes as intersections (particularly roundabouts), there are too many stimuli that can't be understood togheter, therefore older drivers exclude some of them such as road signs.

3.3   Roundabout perception index

The new perception index was calculated using data obtained from the "Eye tracker" movies, referring to only the roundabout in the path: in particular the analysis has focused on two elements which, as just mentioned, were the most watched by drivers: the vehicles (the previous one and the other present) and the arm of the roundabout to the right side. The analysis of movies has allowed us to calculate, for each driver tested, both the number of gaze's fixations which are necessary to perceive the roundabout elements, and the duration of each fixation. The following tables show the values of t and n_t, calculated for all drivers, about the observation of the two elements mentioned above. T is the total time that needs the driver to run through the roundabout. In this case, the perception index has been calculated assuming $\alpha = 2$. First of all, we calculate the new index about perception the previous vehicle and other vehicles: if the index value is high, it means that the driver has found difficulty to perceive that stimulus or road element.

Table 3 Observation data of the previous vehicle and other vehicles

	T	t_1	n_1	t_2	n_2	t_3	n_3	t_4	n_4	t_7	n_7	$\sum(\alpha n)^t$
Subject 1	21	1	5	2	1							14
Subject 2	26	1	3	2	3							42
Subject 3	20	1	1	2	2	3	1					26
Subject 4	25	1	3	2	4							70
Subject 5	25	1	4	2	1			4	1			28
Subject 6	35	1	3	2	3	3	1			7	1	178
Subject 7	22	1	5									10
Subject 8	20	1	3									6
Subject 9	25	1	4	2	2							24
Subject 10	26	1	4	2	3							44
Subject 11	24	1	2					4	1			20
Subject A	20	1	4			3	1					16
Subject B	24	1	4			3	2					72

The average index value for the whole sample, is equal to 42. only three drivers tested have a value greater than average: two of which are elderly (Subject 6 and Subject 10) and one young (Subject B); the driver n°6 has difficulty in perceiving easily the previous vehicle and other vehicles, so the perception index is equal to 178: he is slower to drive inside the roundabout (35 seconds, 10 seconds longer than the average time of the sample) and moreover he watches the vehicles 8 times, one of which had a duration of 7 seconds; the subject n°8 shows an high perception of external stimuli, with a complex index equal to 6 (even less than of young drivers): the subject has observed other vehicles only three times (each time for one second), then acquiring the stimulus in a short time, despite his age. The subject n°11 has a reduced index value (20) despite he has looked the previous car for 4 seconds: in fact, he observes the vehicles only 3 times and in two of them the duration of fixation is short (1 second). Then, we calculate the new index about the perception of the arm of the roundabout located to the right side of drivers.

Table 4 Observation data of the arm of the roundabout

	T	t_1	n_1	t_2	n_2	t_3	n_3	$\sum(\alpha n)^i$
Subject 1	21			2	1			4
Subject 2	26	1	3					6
Subject 3	20			2	1			4
Subject 4	25	1	1					2
Subject 5	25	1	2	2	1	3	1	16
Subject 6	35	1	2					4
Subject 7	22	1	2	2	1			8
Subject 8	20	1	2	2	2	3	1	28
Subject 9	25	1	4					8
Subject 10	26	1	2					4
Subject 11	24			2	1			4
Subject A	20	1	4			3	1	16
Subject B	24	1	5					10

In this case the values of the index of perception are generally lower than those previously calculated for the perception of the vehicles: it means that, as a rule, this element has an understanding easier for all the drivers tested. It may be due to specific characteristics of this element observed, because the vehicles are a dynamic stimulus, then the driver needs to check them continuously for knowing instantly their position and their speed. Instead the right arm of the roundabout is a static element: the driver observes it to assess the position of the vehicles that come from this, although , before entering into the roundabout, they are required to give way to traffic. therefore the information about the right arm of the roundabout are for drivers then hierarchically less useful than those about traffic flow. The subject n°6, which in the previous case had had a big complexity in the perception of the stimulus, in this case has acquired the information quickly, by observing the stimulus 1 once and for 2 seconds: his perception index is the lowest for the whole sample, equal to 4. Even the Subject n°1, n°10 and n°11 were given the same difficulty of perception, even if the Subject 10 has observed the stimulus and only 2 times for 1 second each one. In this case the Subject n°8 has had more difficulty to perceive the right arm of the roundabout (index = 28), while he hasn't the same difficulty for vehicles (previous and others): infact his index, in that case, is = 6. It means that difficulty of perception often depends to subjective aspects.

4   CONCLUSIONS

The roundabout, for its geometrical layout, allows the reduction of the points of conflict, and then the number of stimuli to be seized for the driver is less than a conventional intersection. So in theory elderly drivers should be facilitated in these situations; but it does not happen because of their unfamiliarity and familiarity with this type of intersection. With increasing age there is a reduction of the visual field: the older drivers must make frequent rotations of the neck, to perceive the stimuli that come from opposite ends of the visual scene. This study found that older drivers perform a rotation of the neck also to observe the left mirror. This process, if done repeatedly, can cause fatigue. Observing the behavior of the elderly when they drive inside a roundabout, we found that they have a limited knowledge of traffic rules for this intersection and so they adapt their driving behavior to other vehicles, as confirmed by the high rate observation (37.39%). Again, older drivers have looked less the road signs. Within complex scenes such as roundabouts, elderly neglects them to observe the other stimuli. therefore would be preferable to locate roundabouts in homogeneous areas (altimetrically and planimetrically) , to help drivers to distinguish even from a distance and to facilitate their perception. The new index provides a measure of the complexity of perception, which depends on the number and duration of fixations. Through this index we have tried to propose an objective measure about the use of road signs, in such a way that they represent a guide for the drivers and not an additional risk factor. In the future, starting from this study, it will be possible to implement the data so as to include additional variables, which, as stimuli, characterize the behavior of the driver (age and visual capacity) and that influence the driving perception.

5   REFERENCES

[1] K.M. Butler, R.T. Zacks, J.M.Henderson, 1999. Suppression of reflexive saccades in younger and olderadults: age comparison on an antisaccade task. *Memory Cognit,* 27(4): 584-591.

[2] B. Crassini, B.Brown, K. Bowman, 1988. Age related changes in contrast sensitivity in central and peripheral retina. Perception, 17: 315-332.

[3] G. Daigneault, P. Joly, J.Y. Frigon, 2002. Previous convictions or accidents and the risk of subsequent accidents of older drivers, *Accident Analysis and Prevention,* 34, 257-261.

[4] R.E. Dewar, D.W. Kline, H.A. Swanson, 1994. Age differences in the comprehension of traffic sign symbols. Transp. Res. Rec., 1456:1-10.

[5] R.E. Dewar, D.W. Kline, F. Shieberg, H.A. Swanson, 1994. Symbol signing design for older drivers. Final Report. Federal Highway Administration.

[6] G. Fancello, E.Pani, P. Fadda, 2003. Road safety on elderly drivers: an experimental human factors analysis, XXIInd PIARC World Road Congress Proceedings.

[7] G. Fancello, Stamatiadis N., Pani E., Fadda P.,Wilkinson, 2008. Are Older Drivers different in the Us And Italy? Urban Trasport XIV – Wit Press, Southampton.

[8] J.H. Guerrier, P. Manivannan, S. Nair, 1999. The role of working memory, field dependence, visual search, and reaction time in the left turn performance of older female drivers. *Applied Ergonomics*, 30, 109–119.

[9] Hakamies-Blomqvist L., 2004. Older drivers – a review, VTI rapport 497a.

[10] J. Langford, R. Methorst, L. Hakamies-Blomqvist, 2006. Older drivers do not have a high crash risk-a replication of low mileage bias. *Accident Analysis and Prevention*, 38: 574-578.

[11] H.C. Lee, D. Cameron, A.H. Lee, 2003. Assessing the driving performance of older adult drivers: on-road versus simulated driving, *Accident Analysis and Prevention*, 35, 797-803.

[12] R. B.Naumann, A.M. Dellinger, M. Kresnow, 2011. Driving self-restriction in high-risk conditions: how do older drivers compare to others?. *Journal of safety research* 42: 67-71.

[13] U. Neisser, R.Lazar, 1964. Searching for novel targets. *Perceptual and motor skills*,19: 427-432.

[14] C. Owsley, B.T. Stalvey, J.M. Phillips, 2003. The efficacy of an educational intervention in promoting self-regulation among high-risk older drivers, *Accident Analysis and Prevention*, 35, 393-400.

[15] T. Ueno, 1968. Visual search time based on stochastic serial and parallel processings. *Perception & Psychophysics*, Vol. 3.

CHAPTER 48

The Effect on Passenger Cars' Meeting Margins When Overtaking 30 Meter Trucks on Real Roads

Sandin, J., Renner, L. and Andersson, J.

VTI, the Swedish National Road and Transport Research Institute
Linköping, Sweden
jesper.sandin@vti.se

ABSTRACT

The purpose was to study the effect of vehicle length on meeting margins during overtaking maneuvers. A field study video-recorded overtaking maneuvers of a 30 m and a 24 m truck on a two-lane road. The difference in average meeting margins between the trucks was not statistically significant. An ocular assessment of the video material revealed a few critical situations during the overtaking maneuvers of the 30 m truck; all with meeting margins less than 3 s. Although these results should be interpreted with great caution as the number of analyzed overtaking maneuvers was limited, two previous studies describe similar findings. The conflict technique is discussed as a tool in the assessment of critical meeting margins. It is concluded that more field studies and data are needed to estimate the risks when overtaking Longer Combination Vehicles.

Keywords: overtaking maneuver, meeting margin, safety margin, time gap, TTC, crash risk, conflict technique, truck, longer combination vehicle

1   BACKGROUND

Longer Combination Vehicles (LCVs) ranging from 25 m to 53 m in length have received increased interest in recent years as they are expected to increase transport efficiency and thereby decrease costs as well as carbon dioxide emissions

(Knight et al. 2008; Mellin and Ståhle 2010; Grislis 2010). However, before they are introduced in the traffic system on a larger scale, their impact on traffic safety and other road users' behavior has to be investigated.

Because the time required to overtake a vehicle increases with the length of the overtaken vehicle (Vierth et al. 2008), overtaking-related crashes are often supposed to increase with LCVs (Knight et al. 2008). Vierth et al (2008) analyzed police-reported accidents involving heavy vehicles in Sweden during the years 2003 to 2005. The authors found no statistical evidence in that material indicating that overtaking accidents were more frequent for combination vehicles up to 25.25 m than for rigid 18 m vehicles. According to (Knight et al. 2008), there is a lack of studies that quantify the risks of overtaking maneuvers in terms of accidents rates.

Hammarström (1976) suggested using the concept of meeting margins as an indirect risk measure for overtaking maneuvers of longer vehicles. Meeting margin is defined as the time elapsed from the conclusion of an overtaking maneuver to the moment the overtaking vehicles front meets the front of an oncoming vehicle in the opposite lane. In Hammarströms study, an overtaking was considered concluded when the rear of an overtaking vehicle was 10 m ahead from the front of the test vehicle. Hammarström (1976) compared meeting margins extracted from video-recorded overtaking maneuvers of an 18 m and a 24 m long test vehicle kept at the speed of 70 km/h. Video data were collected on two-lane roads classified as wider or narrower than 10.5 m in total width. The width of the narrow roads was mostly 7-9 m. The speed limit ranged from 70 to 90 km/h. Hammarström analyzed meeting margins up to 7 s for the two road widths separately, and found that the meeting margins were significantly smaller on the wider roads. The results showed that the difference in meeting margins between the two test vehicles was very small on the narrow road; 4.5 s for the 18 m vehicle and 4.3 s for the 24 m vehicle. However, the differences were not statistically significant. The average speeds were 97 km/h and 114 km/h for flying overtakings of the 18 m and 24 m vehicle respectively.

Troutbeck (1981) carried out a similar study of overtaking behavior extracted from video-recorded overtaking maneuvers of test vehicles of various lengths (5, 16, 18, 20 and 21 m) driving at speeds ranging from 60 to 80 km/h. Video data was collected on two-lane rural highways in Australia. The speed limit on the highways was 110 km/h with pavement widths of about 7.4 m and shoulder widths between 2.4 and 3 m. Instead of measuring meeting margins, Troutbeck measured safety margins which have a slightly different definition of the conclusion of an overtaking maneuver. For safety margins, an overtaking maneuver was considered concluded when the overtaking vehicle was completely back in its own lane. Safety margins up to 30 s were analyzed. The results showed that the cumulative frequency of safety margins was best described with a cubic spline function where the most frequent safety margin was 4.5 s with an average value of 10 s for accelerating overtaking maneuvers. The results indicated that the distribution of safety margins was independent of test vehicle length, when this length exceeded 16 m. The distributions of safety margins were largely dependent on traffic volumes. If the overtaken vehicle was traveling at 70 km/h, then the median relative speed during overtaking was 26.5 km/h for accelerative overtakings by cars and 38.1 km/h for flying overtakings by cars. As the speed of the overtaken vehicle increased, the relative speed decreased.

In order to compare the cumulative distribution of safety margins with Hammarströms (1976) meeting margins, Troutbeck (1981) recalculated the safety margins up to 7 s into meeting margins. The data used for comparison related to overtakings with visible oncoming cars only and to the narrow roads in Hammarströms study. Because no significant effect of vehicle length could be discerned in the analysis of safety margins or meeting margins, Troutbeck decided to pool the data from each study regardless of the overtaken vehicle length. Troutbeck found that the two distributions from each study were significantly different. The distributions are shown in Figure 2 – in the discussion.

With reference to LCVs, the field studies of Hammarström (1976) and Troutbeck (1981) were carried out more than 30 years ago, and with vehicles that are of standard length today. A more recent field study of overtaking maneuvers of 30 m LCVs was conducted by Barton and Morrall (1998). However, they did not estimate meeting margins or safety margins.

Consequently, and in relation to the previous studies, the purpose of the current study was to examine whether overtaking maneuvers is affected by a 30 m LCV by meeting margins as an indirect measure of risk.

2   METHOD

Two trucks were used in the field study. One, in Sweden, ordinary timber truck with a length of 24 m and total weight of 60 ton used as a reference, and one Longer Combination Vehicle for timber transport with a length of 30 m and total weight of 90 ton (Skogforsk 2012). The 30 m truck was equipped with a speed limiter of 80 km/h, and the drivers of the 24 m truck were instructed not to drive faster than 80 km/h. Each truck was equipped with four digital video cameras and overtaking maneuvers were video-recorded together with GPS position, truck speed and time on a Video VBOX (Racelogic 2012). The data collection took place during 6 months on a 50 km two-lane road in the north of Sweden. The two-lane road was 7-8 m wide and without shoulders, with one lane in each direction. The speed limit was 90 km/h. The 30 m truck was granted an exemption to operate on the road.

During the data reduction the times t_A, t_B and t_C were extracted manually from the video data (Figure 1). The definitions of the times t_A, t_B and t_C were as follows:

t_A: was registered when three centerline road markings were seen in one of the rearward-facing cameras up to the front of the overtaking vehicle – this corresponds to a distance of approximately 37.5 m from the front of the truck.

t_B: was registered at the time when the whole overtaking vehicle was first seen in the video from one of the forward-facing cameras which corresponds to a relative distance of approximately 2 m between the front of the truck and the rear of the overtaking vehicle.

t_C: was registered at the time when the overtaking vehicle is front to front with an oncoming vehicle in the opposite lane.

The speed of the overtaken truck was registered at the times t_A, t_B and t_C. The time differences between t_A and t_B were used to calculate the mean speeds of the overtaking vehicles based on the truck speeds at position A and B.

Meeting margin is defined as the number of seconds elapsed between the passing of the overtaken truck and the time when the overtaking vehicle is front to front to an oncoming vehicle in the opposite lane (Figure 1). For each overtaking maneuver, the meeting margin was calculated as $t_C - t_B$

All the overtaking maneuvers were registered during the data reduction. However, the times t_A, t_B and t_C were only extracted if the overtaking vehicle was a passenger car, and if the meeting margin $t_C - t_B$ was less than 10 s.

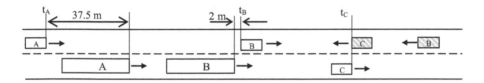

Figure 1 Illustration of the times t_A, t_B and t_C that were extracted from the video data of overtaking maneuvers on the two-lane road. Letters indicate the relative positions of the involved vehicles in a specific stage of the overtaking maneuver. The diagonally striped vehicle indicates an oncoming vehicle in the opposite lane.

3 RESULTS

The descriptive analysis shows that the number of registered overtaking maneuvers of the 30 m truck differs from the registered number of overtaking maneuvers of the 24 m reference timber truck. The 30 m truck has been driven more than the reference truck, but exact information of how much both trucks have been driven is not available. The ratio of the total number of overtaking maneuvers is 1:5, i.e. 1078 and 265 overtaking maneuvers of the 30 m truck and the 24 m truck respectively. The ratio of the number of overtaking maneuvers with meeting margins less than 10 s is also about 1:5, i.e. 88 and 19 overtaking maneuvers of the 30 m truck and the 24 m truck respectively.

The overtaking maneuvers do not differ unsystematically in distribution over the time of day or the spread of location over the studied distance. In total, the result of the descriptive analysis therefore shows that the data is valid even though possible effects have to be interpreted very carefully due to the difference in number of overtaking maneuvers between the trucks.

The average speeds during overtaking maneuvers of the trucks are much higher than the speed limit. The speed limit of the two-lane highway was 90 km/h while the overtaking speeds are 117±11 km/h of the 30 m truck and 115±14 km/h of the reference truck. There is no significant difference in these average overtaking speeds of the trucks (independent t-test; p>0.05).

The average meeting margins on the two-lane highway is 6.7±1.8 s for the 30 m truck and 6.9±1.7 s for the reference truck —with the median of 7.0 and 7.2. There is no significant difference in the meeting margins between the trucks (independent t-test; p>0.05).

An ocular assessment of the video material reveals a few critical situations during the overtaking maneuvers of the 30 m truck. At meeting margins less than 3 s, the oncoming vehicle is required to considerably maneuver to the right and/or require the truck driver to brake to avoid collision or a near incident. While for meeting margins of more than 3 s the same behavior is not found. The ocular assessment therefore concludes that the limit for a critical situation is just under 3 s on the two-lane highway. Three critical situations are found under this limit during the overtaking maneuvers of the 30 m truck, see the cursive numbers in Table 1.

Table 1 The share of overtaking maneuvers in different meeting margin intervals on the two-lane road for both trucks, in cumulative percentage (the number of overtaking in brackets; critical situations in cursive)

Meeting margin interval (s)	The share of overtaking maneuvers	
	30 m	24 m (ref)
< 2	*0%*	
< 3	*3,4% (3)*	*0%*
< 4	5,7% (5)	5,3% (1)
< 5	19,3% (17)	21,1% (4)
< 6	35,2% (31)	26,3% (5)
< 7	50,0% (44)	42,1% (8)
< 8	71,9% (63)	73,7% (14)
< 9	92,0% (81)	84,2% (16)
< 10	100% (88)	100% (19)

4   DISCUSSION

The purpose of the current study was to examine whether meeting margins during overtaking maneuvers is affected by vehicle length. Meeting margin was used as an indirect measure of risk. The difference in average meeting margins between the 30 m and the 24 m truck used as a reference was not statistically significant -nor was the difference in average overtaking speed. These results should be interpreted with caution as the number of overtaking maneuvers with meeting margins less than 10 s are limited, especially for the 24 m reference truck for which merely 19 overtaking maneuvers were analyzed.

The view of the authors of the current paper is that meeting margins is a better measure than safety margins because the conclusion of an overtaking maneuver is defined at a fixed distance. The value of a safety margin is calculated from the time where the overtaking vehicle is completely back in its own lane, and this time is likely to vary from one overtaking maneuver to another. Although the definitons of meeting margin in the present study differs from the definition by Hammarström, we agree with Hammarström (1976, p.17) who states that: "As long as calculated meeting margins are compared with calculated meeting margins only, the calculated margins exact agreement with the actual measured meeting margins is of less importance."

In order to compare the cumulative distribution of safety margins with Hammarström's meeting margins, Troutbeck (1981) recalculated the safety margins into meeting margins up to 7 s. The distributions are shown in Figure 2. In sake of comparison, Figure 2 also describes the distribution of the meeting margins up to7 s from the present study when recalculated to fit Hammarström's definition.

In short, the reduction of the meeting margiΔt between the conclusion of an overtaking maneuver of 2 m ahead in the present study and the previous definition of 10 m ahead from the truck can be expressed as $\Delta t = 8/(v_{car} - v_{truck})$, where v_{car} is the speed of an overtaking car and v_{truck} is the speed of the overtaken truck expressed in m/s. Note that the recalculated meeting margins are estimations based on the average overtaking speeds.

Because no significant difference in meeting margins could be discerned between the two trucks in the present study, the data sets were pooled. An analysis of the pooled meeting margins showed that the recalculated average meeting margin became 4.9 s for both vehicles, and the average Δt became 0.9 s.

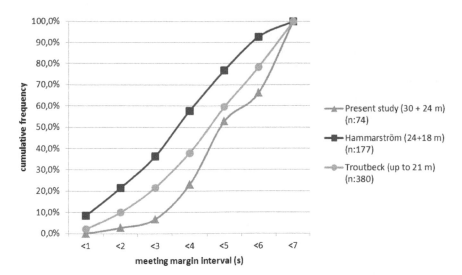

Figure 2 The share of overtaking maneuvers in pooled meeting margin interval by Hammarström's definition for each of the three studies, in cumulative percentage. The length of the vehicles in each respective study is indicated in the label to the left together with the number of overtaking maneuvers on which the distributions are based.

As can be seen in Figure 2, there are clear differences between the graphs, However, one cannot conclude that meeting margins are reduced with the length of the overtaken vehicles when comparing the distributions from each respective study. If that was the case, then the meeting margin distribution from the present study would have been distributed towards the smallest meeting margins, followed by the distributions of Hammarström and Troutbeck. Thus, one simple interpretation of Figure 2 could be that the distributions are not influenced by vehicle length at all. Such an interpretation is supported by Troutbecks results which indicated that the distribution of safety margins was independent of test vehicle length, when this length exceeded 16 m. While this interpretation requires careful consideration, there are possibly also other factors that influence the meeting margin distributions, like traffic volume and road width.

Troutbeck found that discerned differences in safety margins, and therefore meeting margins, were predominantly affected by the traffic volume in the opposite lane. Troutbeck states that as traffic volumes increases - the gap available for overtaking in the opposing traffic decreases, and most motorists are prepared to accept a shorter gap and compensate for this deficiency by overtaking more quickly – and thereby reducing the safety margins. Hammarström do not report whether traffic volume affected the meeting margins. Neither have the present study registered the traffic volume during the overtaking maneuvers. However, the road at which the present study was carried out has very low traffic volumes, which is likely to explain why the distribution is furthest to the right in Figure 2.

Hammarströms results showed that the meeting margins were clearly reduced with the width of the road. The reason is that on wider roads, where three vehicles can fit in laterally, overtaking maneuvers allows for meeting margins to be shorter or even negative. The present study collected data from the most narrow road (7-8 m and without shoulders) compared to the total road widths of Hammarström (7-9 m) and Troutbeck (9.8-10.4 m). This may be an additional explanation to why the distribution of the present study is furthest to the right in Figure 2.

A full understanding of the differences between the three meeting-margin distributions requires a more detailed analysis, which is beyond the scope of the present paper. However, the question is if comparing cumulative distributions is the best way to estimate risk by means of meeting margins. For example, Troutbeck (1981, p.81) reports that "...most safety margins [and therefore meeting margins] were in excess of the minimum required by most motorists. Thus, the mean safety margin is not important; neither is the standard deviation. What is important is the distribution of small safety margins in the lower tails of the distributions for overtakings from the various data groups." Furthermore, Hammarström (1976, p. 14) writes that "In order to compare hazardous overtaking maneuvers, a clear definition is needed of how small a meeting margin can be in order to be labeled hazardous. Meeting margins the size of 0 s are of course hazardous [for narrow roads in particular]. However, how much larger than 0 s can a meeting margin be in order to still be labeled hazardous? Such a limit is impossible to exactly determine scientifically."

With reference to the last sentence, it is nevertheless noteworthy that all three studies describe observations of critical situations among the distributions of smaller meeting margins below 2-3 s. Troutbecks analyses indicated that if an overtaking around a truck of 16 m or longer is required to be completed with a small safety margins of about 2 s, then the time travelled by the overtaking vehicle at the end of the maneuver is largely reduced. This result is in accordance with the ocular assessment in the present study which revealed that at meeting margins less than 3 s (appr. 2 s according to Hammarströms definition), the oncoming vehicle was required to considerably maneuver to the right and/or required the driver of the 30 m vehicle to brake to avoid collision or a near incident. Hammarström observed that the relative shares of meeting margins below 1 s are larger for the 24m vehicle compared to the 18 m vehicle. However, the differences were not statistically significant.

The observed relation between small meeting margins and criticality at the end of an overtaking maneuver is reminiscent of the reasoning behind the concept of the conflict technique (Almqvist, 2006; Ekman, 1996; Hydén, 1987; Svensson, 1998). The conflict technique is a method for studying traffic safety, which demonstrates how serious conflicts are related to actual traffic accidents. Conflict technique studies are performed by observers who register and classify critical traffic situations according to collision course, time and speed. If the time for preventing maneuvers is less than a limit value the event is classified as a serious conflict.

In the present study, the average speeds during the overtaking maneuvers were 115 and 117 km/h for the 24 m truck and the 30 m truck respectively. According to the conflict technique, a "time to accident" shorter than about five seconds would be considered a serious conflict in these speeds. However, the conflict technique cannot be directly applied to the concept of meeting margins associated with overtaking maneuvers. Firstly, the conflict technique is primarily developed for urban traffic. Secondly, time to accident is measured different than the meeting margin measure though time to accident is measured from when a driver starts a preventing maneuver. However, the end definition of the measurement period is the same. These differences make it difficult to directly compare the field study data to the conflict technique limit of a serious conflict. However, the comparison is possible in principle.

When illustrating the field study data for time and speed, i.e. meeting margin and overtaking mean speed, one can see that the identified critical situations from the ocular assessment differ from the other data. This difference can be described as a gap with a similar gradient as the conflict technique definition of serious conflicts, see Figure 3.

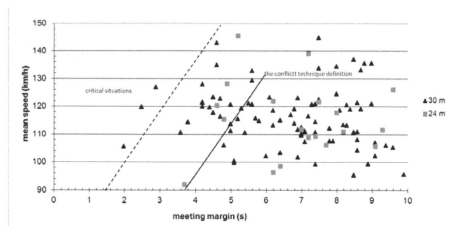

Figure 3   Illustrating the overtaking maneuver data from the two-lane highway (The dotted line shows the gap between the critical situations and the other data. The solid line illustrates the conflict technique definition of serious and non-serious conflicts *if* comparable to the field study data)

A serious conflict is an unwanted traffic event and drivers do not knowingly enter such a conflict since it is perceived as extremely unpleasant. Instead, most overtaking maneuvers are performed with a good time margin, not only to avoid accidents but also to such a degree that one avoids near accidents, i.e. serious conflicts. The gap between the critical situations and the other data might illustrate this. However, this is a suggestion and should only be considered as an idea for further studies. Undeniable such a definition of serious conflicts for overtaking maneuvers would be highly usable.

5   CONCLUSIONS

The purpose of the present study was to examine whether overtaking maneuvers is affected by a 30 m Longer Combination Vehicle by meeting margins as an indirect measure of risk. The difference in average meeting margins between the 30 m and the 24 m truck used as a reference was not statistically significant - nor was the difference in average overtaking speeds. However, an ocular assessment of the video material revealed a few critical situations during the overtaking maneuvers of the 30 m truck; all with meeting margins less than 3 s. These results should be interpreted with great caution as the number of analyzed overtaking maneuvers was limited, especially for the reference truck. It is concluded that more field studies and data are needed to estimate the risks when overtaking Longer Combination Vehicles.

ACKNOWLEDGMENTS

The authors would like to acknowledge the people involved in the ETT-project and Bjälmsjö Transport AB for excellent collaboration when carrying out the field study, and the Swedish Transport Administration for financial support.

REFERENCES

Almqvist, S. 2006. *Loyal speed adaptation: Speed limitation by means of an active accelerator and its possible impact in built-up areas.* Department of Technology and Society Traffic Engineering, Lund University, Lund, Sweden.

Barton, R.A., and J. Morrall. 1998. Study of long combination vehicles on two-lane highways. *Transportation Research Record* 1613: 43-49.

Ekman, L. 1996. *On the treatment of flow in traffic safety analysis – a non-parametric approach applied on vulnerable road users. Bulletin 136.* Department of Traffic Planning and Engineering, Lund University, Lund, Sweden.

Grislis A. 2010. Longer combination vehicles and road safety. *Transport* 25(3): 226-343.

Hammarström, U. 1976. *Overtakings of long combination vehicles – Study of meeting margins. [Omkörningar av långa fordonskombinationer. Studie av mötesmarginaler.]* VTI report 103, National Road and Traffic Research Institute (VTI), Linköping, 1976.

Hydén, C. 1987. *The development of a method for traffic safety evaluation: The Swedish Traffic Conflict technique.* Department of Traffic Planning and Engineering, Lund University, Lund, Sweden.

Knight, I., W. Newton, P.A. McKinnon, T. Barlow, I. McCrae, M. Dodd, G. Couper, H. Davies, A. Daly, W. McMahon, E. Cook, V. Ramdas, and N. Tylor. 2008. *Longer and/or Longer and Heavier Goods Vehicles (LHVs) – a Study of the Likely Effects if Permitted in the UK.* Final Report, Published project report 285, TRL, 2008.

Mellin A. and J. Ståhle. 2010. *Situational and future analysis – Longer and heavier road and rail vehicles. Subproject 1 of the Co-modality project.* VTI report 676, Swedish National Road and Transport Research Institute (VTI), Linköping, 2010.

Racelogic 2012 Accessed February 2012, http://www.videovbox.co.uk/

Skogforsk 2012. "ETT - Modular System for Timber Haulage." Accessed February 2012, http://www.skogforsk.se/en/Research/Logistics/ETT/

Svensson, A. 1998. *A method for analyzing the traffic process in a safety perspective.* Department of Traffic Planning and Engineering, Lund University, Lund, Sweden.

Troutbeck, R.J. 1981. *Overtaking Behaviour on Australian Two-lane Rural Highways. Australian Road Research Board.* Special Report, SR No. 20, Vermont South, Victoria, Australia, 1981.

Vierth, I., H. Berrel, J. McDaniel, M. Haraldsson, U. Hammarström, M.-R. Yahya, G. Lindberg, A. Carlsson, M. Ögren, and U. Björketun. 2008. *The effects of long and heavy trucks on the transport system. Report on a government assignment.* VTI report 605, Swedish National Road and Transport Research Institute (VTI), Linköping, 2008.

CHAPTER 49

The Effect of Text Messaging on Young Drivers' Simulated Driving Performance

Qian Zhang, Zuhua Jiang

Department of Industrial Engineering and Logistics Management
Shanghai Jiao Tong University, Shanghai, China
Irenezhang333@gmail.com

ABSTRACT

This study compared the driving performance of two groups of Chinese young drivers using a low–cost fixed based driving simulator: driving with concurrent text messaging tasks (Text messaging group, n=30) and driving without text messaging tasks (Control group, n=30). Driving performance suffered during text messaging tasks on participants' Vehicle speeds, Driving maintenance, Reaction time (RT) and Time–to–collision (TTC), and Accidents. Results demonstrated that young drivers' driving performance was negatively influenced by concurrent text messaging tasks. Participants in Text messaging group had significantly lower mean speed, much larger lane variation, and committed more accidents than drivers in Control group. Besides, their RTs were nearly twice as much as drivers' in Control group while TTCs were only about one third of drivers' in Control group. The findings lend further support of the dangers of young drivers being distracted by text messaging and suggest that the access to text messaging should be curtailed.

Keywords: text messaging, young drivers, driving performance, simulation

1   INTRODUCTION

The emergence of new electronic devices in vehicles has increased the number of distractions to which drivers are potentially exposed. Among the various in–

vehicle information systems and entertainment systems, one of the most popular devices used while driving is cell phone. During the last decade, cell phones have transitioned from a luxury enjoyed by the few, to a must–have item enjoyed by most people in China. As of July 2011, the number of cell phone subscribers in China exceeded 920 million, nearly eleven times more than that in December 2000. Meanwhile, the number of young drivers in China has also increased dramatically. There were about 23 million young drivers aged between 18 and 25 years in 2010, accounted for about 15% of the total car drivers (National Bureau of Statistics of China 2011). The rapid growth in both cell phone use and young driver numbers has led to concerns regarding the impact of cell phone use on young drivers' driving performance, and a great deal of research has been done on this subject. Because of more willing to try new technologies and being overconfident with their driving skills, young drivers are more likely to engage in cell phone use while driving (National Highway Traffic Safety Administration [NHTSA] 2009) and suffer the effects of distraction than older drivers (Lee 2007; Hosking et al. 2009).

Although the detrimental effects of cell phone use on driving performance have been well documented by a lot of studies (e.g. Strayer and Johnston 2001; Horrey and Wickens 2006; Wang et all. 2010), most of the studies have focused on the verbal communication function of the cell phones, whereas the effects of text messaging on driving performance are underrepresented in the research literature. In 2008, the Royal Automobile Club (RAC) Foundation asked 2002 members on Facebook to self–report on whether they text while driving. Alarmingly, 45% admitted doing so. Gras et al. (2006) found that 19.1% of the Spanish drivers admitted texting on highways and 22.5% on rural roads at least once a month. In China, people are more willing to send text messages since the cost of texting is cheaper than cell phone conversation, and more than 830 billion text messages were sent by Chinese people in 2010 (Ministry of Industry and Information Technology of China 2011). A survey found that among Chinese drivers, 20% admitted texting while driving at least once a week while this proportion increased to one third among younger drivers between 18 and 30 years (Zhuang 2007).

Kircher et al. (2004) studied the effects of receiving a text message on the performance of ten experienced drivers, and found it significantly increased reaction times in a peripheral detection task and generally reduced drivers' speed. Hosking et al. (2006) measured the effect of texting, both receiving and sending text messages, among twenty young drivers aged between 18 and 21 years, and found that when texting, young drivers spent more time looking away from the road environment, had lower ability to maintain lane position, and failed to see signs instructing them to change lane more frequently during driving. However, few studies have been conducted on the effect of text messaging among Chinese drivers, especially young drivers. Therefore, this study aims to provide additional insight into the effect of text messaging on young drivers' simulated driving performance. It was hypothesized that when engaging in text messaging, young drivers would display reduced speed, poorer lateral lane control, increased reaction time (RT) and decreased time–to–collision (TTC), and more accidents.

2   METHOD

2.1   Participants

Sixty young drivers aged between 18 and 25 years were recruited from Shanghai Jiao Tong University via notices placed in the university bulletin boards. Participants who described themselves as regular users of text messaging were selected to ensure that any performance effects seen are not due to drivers being unfamiliar with text messaging itself. All the participants were randomly assigned to two groups: Control group without text messaging tasks (n=30) and Text messaging group (n=30). For the text messaging group, participants were required to use their own cell phones to ensure the familiarity with phone operation. Moreover, only cell phones with a standard alphanumeric keypad were allowed, participants whose personal cell phones had touch–screens or QWERTY keyboards were excluded. All participants were required to hold a valid Chinese driver license, be in good physical and mental health, and drive more than 2000 km annually. Each participant received a gift worth ￥50 for their participation.

2.2   Apparatus

The driving simulation program used in this study was developed by our research team based on software developing platform—Multigen^{TM} Creator, Vega and Visual C++ .Net. Multigen Creator is a comprehensive software toolset for real–time 3D modeling, which was used to build 3D models in the simulated driving scenario. Vega provides a software environment for the creation and deployment of virtual reality through which the real–time visual and audio simulation of the driving scenario can be realized. The driving simulation program was written in C++ programming language by calling Vega API functions.

The simulated driving experiment was conducted in Human Factors Laboratory of Shanghai Jiao Tong University. A low–cost, fixed base driving simulator was used for this session. The simulation program was operated on a standard desktop computer with an Intel^{TM} Core i7–920 processor and Nvidia^{TM} GeForce GTS450 graphic card. Participants were asked to sit in a stationary chair at a large desk. A 150–inch projector screen was located about 2 meters in front of the driver's seat. The laboratory was equipped with a large black curtain to minimize the environmental distractions. A Logitech^{TM} momo steering wheel was mounted to the front of the desk while accelerator and brake pedals were fixed on the floor. Two speakers and a subwoofer were used to present the realistic engine and road noises realized by the audio module in the program. For each participant, the C++ program recorded data of the current time, speed, gear, steering wheel angle, driving distance, lane position (x, y, and z coordinate value), and accidents frequency at every 0.2 second, then output all these data as a ".dat" document. A digital video camera was mounted in the laboratory to record the whole simulation process.

2.3   Simulation scenario

The simulated road created for this study was an accurate representation of a two–lane road in Shanghai. The overall length was about 8 km, containing a 6 km–

long round–city highway and a 2 km–long city's road. The lane widths and other road engineering characteristics were all built according to related Chinese national standard (GB 5768–2009: Road traffic signs and markings). The speed limit was 100 km/h for round–city highway and 60 km/h for city's road respectively.

Two traffic hazards were contained in the simulated road, as shown in Figure 1. The first hazard located at 2.5 km, which was a road works site preceded by a road works hazard warning sign (250 m prior), a right lane closed sign (50m prior), and a temporary 40 km/h speed restriction sign (20 m prior). The road works lasted 100 m and terminated with a "works end–100km/h" sign. Approximately 1 km later, the participants reached the second hazard, a car pulling out of a curb parking area as the participant approached. When the participant's vehicle was 100 m away, the parked car began to blink its left indicator light. When the participant was 20 m away the parked car pulled into the participant's lane, accelerating to 80 km/h.

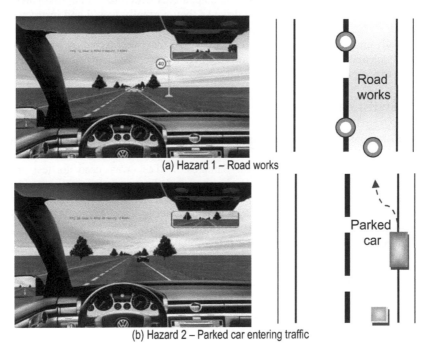

(a) Hazard 1 – Road works

(b) Hazard 2 – Parked car entering traffic

Figure 1 Overview of the two hazards in driving scenario

2.4   Measures

A short driving history questionnaire was developed to gather information on the participants' demographics and driving background. Participants in Control group drove along the same route as the Texting drive but without additional texting tasks, whereas participants in Text messaging group were required to keep on writing text messages all through the course of drive. The instructions as to what to write and the message recipient were provided via a notice on the desk.

Driving performance was measured by four categories: Vehicle speeds (mean speed, mean maximum speed, and mean speeds at hazards), Driving maintenance (speed variation, lane position, and lane variation), Reaction time (RT) and Time–to–collision (TTC), and Accidents (an accident was recorded when the car collided with any of the objects during driving).

2.5 Procedures

Each participant was scheduled for the experiment one by one in advance. Upon arrival, they were given a general overview of the whole experiment and assigned randomly to either Control group or Text messaging group. Before the simulation session, participants were asked to complete an informed consent agreement, a short driving history questionnaire. Then each of them was given a short practice until they felt comfortable operating the simulated apparatus which lasted approximately 10 minutes. Participants were asked to drive as they normally would and instructed to obey all traffic rules. When the participants reached about 8 km, the scenario automatically ended, their simulation data and video files were saved by a research assistant immediately. The average time required for each participant to complete the whole experiment was about 30 minutes.

2.6 Statistical Analyses

The Kolmogorov–Smirnov test was used to assess whether all baseline date met the normal distribution. The data for simulated driving performance were normally distributed and therefore Independent–Samples T Test was used to test for significance differences between Control group and Text messaging group. The distributions of several driving history variables and accidents differed significantly from a normal distribution and therefore non–parametric tests (Mann–Whitney U and Chi–square test) were performed. All statistical analysis was conducted using IBM^{TM} SPSS Statistics 19, with an alpha level of 0.05.

3 RESULTS

3.1 Demographics and driving history

Table 1 Demographics and driving history information

	Control group (n=30)	Text messaging group (n=30)
Mean age (S.D.)	22.50 (1.59)	21.80 (1.63)
Mean license holding time (S.D.) /month	19.83 (11.82)	18.57 (11.52)
Mean annual mileage (S.D.) /km	3393.33 (1491.34)	3710.00 (2169.12)
Males /%	47	50
Accident involvement in previous 1 year /%	23	33

Demographics and driving history information are shown in Table 1. No significant differences were found on age (t=1.684; p=0.098), license holding time (t=0.42; p=0.676), annual mileage (t=−0.659; p=0.513), male proportion (Chi–square=0.067, *df*=1, p=0.796), and accident involvement in previous 1 year (Chi–square=0.739, *df*=1, p=0.390) between the two groups.

3.2 Vehicle speeds

3.2.1 Mean speed

Figure 2 shows the mean speed values of the two groups, significantly lower speeds were observed in Text messaging group (Control: 69.1 km/h vs. Texting: 57.3 km/h; t=4.957; p=0.000).

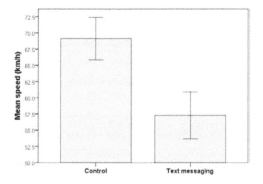

Figure 2 The mean speed values of the two groups through the course of drive

3.2.2 Mean maximum speed

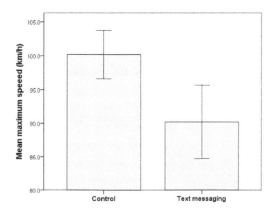

Figure 3 The mean maximum speed values of the two groups through the course of drive

Figure 3 shows the mean maximum speed values of the two groups over the course of the trial. The maximum speed was significantly lower in Text messaging group (Control: 100.2 km/h vs. Texting: 90.2 km/h; t=3.128; p=0.003). Five people in Control group exceeded the maximum speed limit (100 km/h) while one people in Text messaging group did so.

3.2.3 Mean speed at hazards

Shown in Figure 4 are the mean speeds for the two groups as they approached and passed through the two hazard sites. As can be seen in the figure, participants in Control group reduced their speeds greatly as they approached each hazard point, whereas the average speeds of drivers in Text messaging groups decreased less. Statistically significant difference were found between the groups at both Hazard 1 (Control: 28.55 km/h vs. Texting: 40.68 km/h; t=−3.154; p=0.003) and Hazard 2 (Control: 47.88 km/h vs. Texting: 60.12 km/h; t=−3.262; p=0.002).

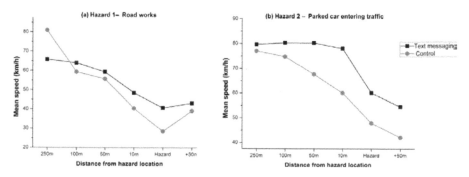

Figure 4 Participants' mean speeds through each of the two hazards

3.3   Driving maintenance

Participants in Text messaging group had a significantly higher lane variation (t=−4.274; p=0.000), which was almost twice as much as that in Control group, as shown in Table 2. In addition, participants showed a slightly lower speed variation and preferred driving near the centerline when engaging in text messaging tasks.

Table 2 Means and standard deviations for driving maintenance behavior

	Control group (n=30)	Text messaging group (n=30)
	Mean (S.D.)	Mean (S.D.)
Speed variation, *km/h*	20.67 (2.12)	19.88 (5.03)
Lane position, m^*	2.09 (0.54)	2.50 (0.90)
Lane variation, m^{***}	2.47 (1.71)	4.93 (2.65)

Note. *** p<0.001, *p<0.05.

3.4 Reaction time and time–to–collision

Two measures of the drivers' deceleration reactions were examined at each hazard location: deceleration reaction time (RT) was measured in seconds from a point 250m prior to each hazard; and deceleration time–to–collision (TTC) was measured in seconds to reach the hazard at the current speed. During the approach to Hazard 1, 86.7% (26/30) of the drivers in Control group registered a deceleration response (release their foot off accelerator to the brake pedal), whereas 73.3% (22/30) of drivers in Text messaging group did so. As for approaching Hazard 2, 83.3% (25/30) of the drivers in Control group, and 63.3% (19/30) of drivers in Text messaging group registered a deceleration response. A comparison of the two groups' performance on the two deceleration measures across the two hazard sites indicated significant group differences, as shown in Figure 5.

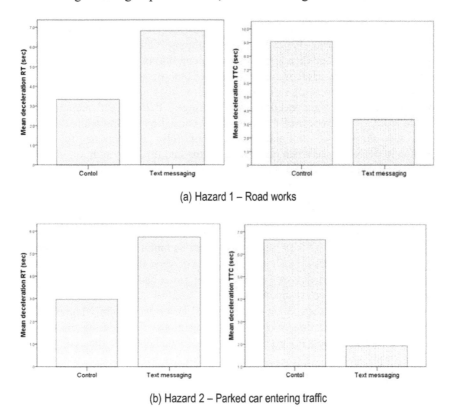

(a) Hazard 1 – Road works

(b) Hazard 2 – Parked car entering traffic

Figure 5 Participants' deceleration reactions at each of the two hazards

Hazard 1 is shown in the top panels of Figure 5. Independent–Samples T Test revealed that the Control group was significantly different from the Text messaging group on both RT (Control: 3.33s vs. Texting: 6.82s; t=−3.103; p=0.003) and TTC (Control: 9.07s vs. Texting: 3.35s; t=5.085; p=0.000). The bottom panels of Figure 5 show the participants' reactions at Hazard 2. Statistically significant difference were also found between the groups on both RT (Control: 2.97s vs. Texting: 5.74s; t=−2.995; p=0.005) and TTC (Control: 6.65s vs. Texting: 1.92s; t=4.205; p=0.000).

3.5   Accidents

Because of the relatively infrequent occurrence of the accidents in driving simulation, total accident frequencies were summed for analysis. Participants in Text messaging group committed more accidents than drivers in Control group, significant differences were observed between the two groups on accidents (Mann–Whitney U test, Z=−2.183, p=0.029).

4   DISCUSSIONS

This study was an initial explore on the impairment to driving performance caused by concurrent text messaging tasks among Chinese young drivers. Participants tended to reduce their speed in texting conditions, their mean speed and mean maximum speed were all significantly lower than drivers' in Control group. This corresponds with the results of Kircher et al. (2004) who found that drivers tended to reduce speed when receiving text messages. Participants' reduction in speed indicated their awareness of the impairment caused by texting while driving. However, this attempt to mitigate risk cannot fully compensate for their deterioration in performance when attempting to text and drive. In addition, drivers in Control group enjoyed significantly lower speeds at both hazard points than participants in Text messaging group.

Across the two hazard sites, drivers in Control group had a significantly faster RTs and longer TTCs, releasing the accelerator earlier and further away from the hazards than drivers in Text messaging group. When engaged in concurrent text messaging tasks, participants' RTs were nearly twice as much as drivers' in Control group while their TTCs were only about one third of drivers' in Control group.

As hypothesized, participants in Text messaging group had a significantly larger lane variation, nearly twice as much as that in Control group. However, it is unanticipated that the speed variation of the control group was slightly higher (not significantly different) than that of Text messaging group. Considering the lower mean speed of the Text messaging group, it is understandable that they enjoyed a lower speed variation.

The results demonstrated that young drivers' driving performance was negatively influenced by concurrent text messaging tasks, and suggested that the access to text messaging should be curtailed.

5   LIMITATIONS

Since this is an initial explorative study on the effect of text messaging on driving performance among Chinese young drivers, a number of limitations should be taken into account when generalizing the results from experiment studies to real–life situations. First, all the participants in this study were college students which couldn't represent the whole demographic distribution of young drivers in China. Second, since the simulated driving experiment was conducted on a low–cost fixed based driving simulator, the results from this experimental study should be further verified by more studies using high fidelity driving simulators or field test. Finally, the factor of texting experience were not taken into account in this study, therefore, it is unclear that whether different texting abilities would contribute to the variance on driving performance, further studies should be conducted.

ACKNOWLEDGMENTS

The authors would like to give thanks to both the Participation in Research Program of Shanghai Jiao Tong University for financially supporting this research under Contract No.T020PRP20041 and all the participants' efforts in this study.

REFERENCES

Charlton, S.G. 2009. Driving while conversing: cell phones that distract and passengers who react. *Accident Analysis and Prevention* 41:160-173.

Gras, M. E., Cunill, M., and Sullman, M. J. M. et al. 2007. Mobile phone use while driving in a sample of Spanish university workers. *Accident Analysis and Prevention* 39:347-355.

Horrey, W. J. and Wickens, C. D. 2006. Examining the impact of cell phone conversations on driving using meta-analytic techniques. *Human Factors* 48:196-205.

Hosking, S., Young, K., and Regan, M. 2006. The effects of text messaging on young novice driver performance. Monash University Accident Research Centre, Report No.246.

Kircher, A., Vogel, K., and Bolling, A. et al. 2004. Mobile telephone simulator study. Swedish National Road and Transport Research Institute, Linkoping, Sweden.

Lee, J. D. 2007. Technology and teen drivers. *Journal of Safety Research* 38:203–213.

Ministry of Industry and Information Technology of China. Accessed Feburary 1, 2011. http://www.miit.gov.cn.

National Highway Traffic Safety Administration. 2009. Driver Electronic Device Use in 2008. Washington, DC:US Department of Transportation. Report no. DOT HS-811-184.

National Bureau of Statistics. Accessed 2011. http://www.stats.gov.cn.

Reed, N. and Robbins, R. 2008. The effect of text messaging on driver behaviour.UK. Transport Research Laboratory. Report no. PPR 367.

Standardization Administration of China. Accessed 2009. *GB5768−2009: Road traffic signs and markings*. China standards press. http://www.spc.net.cn.

Strayer, D. L. and Johnston, W. A. 2001. Driven to distraction: dual-task studies of simulated driving and conversing on cellular telephone. *Psychological Science* 12:462-466.

Wang, Y., Zhang, W., and Bryan, R. 2010. The effect of feedback on attitudes toward cellular phone use while driving: a comparison between novice and experienced drivers. *Traffic Injury Prevention* 11:471-477.

Zhuang, K.M., Bai, H.F., and Xie, X.F. 2007. A Study on Risky Driving Behavior and Related Factors. *Acta Scientiarum Naturalium Universitatis Pekinensis* 5:1–8.

CHAPTER 50

Situational Awareness in Road Design

Dr Guy Walker and Ipshita Chowdhury

Institute for Infrastructure and the Environment
Herriot-Watt University
Edinburgh
g.h.walker@hw.ac.uk

ABSTRACT

This article places theories of SA into contact with the issue of Self Explaining Roads. Twelve drivers took part in an on-road study and performed a verbal commentary as they drove around a defined test route. The verbal transcripts were partitioned into six road types, and driver SA was modeled using semantic networks. The content and structure of these networks was analyzed and cognitively salient endemic road features were extracted. These were then compared with aspects of driver behavior. The findings show that SA is highly contingent on road type, and that motorways/freeways are the most cognitively compatible with drivers. Cognitive incompatibilities grow rapidly as road types become increasingly minor and less overtly 'designed'.

Keywords: Situation Awareness, Semantic Networks, Verbal Protocols, Naturalistic Study

1   INTRODUCTION

1.1   Self Explaining Roads

Certain psychological parameters have been embedded in road design guidance for a considerable number of years (AASHO, 2004; Highways Agency, 2011). As a result, the most modern form of road, the high speed, multi-lane, limited access motorway (the UK term for freeway, autostrada, autobahn etc.), provides drivers with what Vanderbilt (2008) describes rather fancifully as a 'toddlers view of the world':

"we make the driving environment as simple as possible, with smooth, wide roads marked by enormous signs and white lines that are purposely placed far apart to trick us into thinking we are not moving as fast as we are. [..] a landscape of outsized, brightly colored objects and flashing lights, with harnesses and safety barriers that protect us as we exceed our own underdeveloped capabilities" (p. 90).

Motorways represent an unusual form of 'total environment' yet the behavioral outcomes of this environment are undeniable. In most countries in the world motorways are the safest types of road to travel on despite carrying by far the largest volumes of traffic at the greatest speeds (e.g. dft, 2008). Motorways, therefore, are an excellent example of a Self Explaining Road, one that does not "need any additional explanation or learning process to know what it means and what to expect" (Stelling-Konczak et al., 2011).

A powerful mapping exists between the objective state of the built environment and the perceived state of that environment on the part of the individual within it. This concept is a familiar one. Norman (1990) refers to affordances and 'gulfs of evaluation', both of which describe a person's attempts to make sense of their context and how it matches their expectations and intentions. Because it is not possible to create motorway-like 'total environments' in all situations, significant gulfs of evaluation can begin to occur on other, non-motorway types of road. SER research is about reducing such gulfs (Theeuwes & Godthelp, 1995) and the results so far are encouraging. For example, Charlton et al. (2010) present a study in which SER derived changes to road infrastructure, most notably changes to the visual environment, resulted in significant reductions in speed.

For the purposes of this paper it is interesting to note that while the SER approach makes reference to a number of cognitive antecedents as explanations for these desirable behavioral adaptations, including affordances (e.g. Weller et al., 2008), schemas (e.g. Charlton et al., 2010), expectation and prediction (e.g. Stelling-Konczak et al., 2011), one concept that is not cited, and which could serve to unify a number of ideas already in use, is SA.

1.2   Situational Awareness

The concept of SA explains how drivers use information from the world to combine long-term goals (like reaching a destination) with short-term goals (such as avoiding collisions) in real time (Sukthankar, 1997). Drivers are required to keep track of a number of critical variables in a dynamic and changeable environment, including their route, their position, their speed, the position and speed of other vehicles, road and weather conditions and the behavior of their own vehicle. Drivers also need to be able to predict how these variables will change in the near future in order to anticipate how to adapt their own driving (Gugerty, 1997).

SA is an important factor in driving safety. Gugerty (1997) points out that "errors in maintaining situation awareness are the most frequent cause of errors in real-time tasks such as driving" (p.498) and that poor SA can be attributed to more accidents than improper speed or technique. Endsley (1995) proposes a set of generic SA design guidelines which now appear highly relevant to the topic of Self Explaining Roads. These principles are:

1. Reduce the requirement for [drivers] to make calculations
2. Present data in a manner that makes understanding and prediction easier.
3. Organize information in a manner that is consistent with the [driver's] goals.
4. Indicators of the current mode or status of the [driver – vehicle – road] system can help to cue the appropriate situational awareness.
5. Critical cues should be provided to capture attention during critical events.
6. Global situational awareness is supported by providing an overview of the situation across the goals of the [driver].
7. [driver – vehicle - road] system-generated support for projection of future events and states will support SA.
8. [driver – vehicle - road] system design should be multi-modal and present data from different sources together rather than individually.

* Text in square brackets added by authors

In SER literature, the goal to change the characteristics of roads in order to influence driver behavior is met with the identification of so-called endemic road features (Charlton et al., 2010). These include a wide range of entities and artifacts, from road width and lane markings through to landscaping and roadside furniture. What the review of SA shows us is that care should be exercised in how the identification of these features is undertaken. Out of all the possible features in the built environment, which ones are the more 'cognitively salient'? The identification of these features in SER literature has proceeded along various lines and a good summary is provided in Charlton et al. (2010). Examples include several recent papers which use a form of 'picture sorting' task to discover what road features distinguish different road types (e.g. Stelling-Konczak et al., 2011). Other methods include questionnaires (e.g. Goldenbeld & Van Schagen, 2007) and driving simulator studies (e.g. Aarts & Davidse, 2007). Curiously, very few studies take place in real road environments, and no studies have been identified which make meaningful reference to, or use of, SA as a concept. This article will attempt to make progress on these issues using a network based approach to SA representation, and an empirical paradigm that employs naturalistic driving.

3   METHOD

3.1   Design

The experiment is based upon 'real-world' driving in which individuals use their own vehicles around a defined course on public roads and were required to provide a verbal commentary as they drove. The commentary required drivers to 'speak out loud' about what information they were taking from the environment, what they intended to do with it, and to explain their driving actions. The transcript of this commentary was analyzed using Leximancer, a tool for automatically creating semantic networks from text data (Smith, 2003). These outputs were dependent upon one independent variable: road type. This had six levels; 1) motorway(freeway), 2) major A/B classification road, 3) country road, 4) urban road, 5) junction and 6) residential road. This is an initial study and a relatively homogenous sample was used. Driving style, speed and time to complete the course was measured and outlying participants were excluded from analysis. All experimental trials took place at defined times in order to avoid peak traffic conditions in the study area and to offer some control over traffic density. All runs took place in dry weather conditions with good visibility.

3.2 Participants

The primary purpose of the study is to test the concept of SER, SA and semantic networks, and to do this effectively in the context of a first study required a relatively homogenous sample. This comprised of twelve male drivers, ranging in age from 17 years to over 50. The modal age category was 21-25 and mean driving experience was 13.5 years, ranging over 3 to 44 years. All drivers held a full UK driving license with no recent major endorsements and reported that they drove approximately average mileages per year of 10 to 12 thousand miles. All drivers in the sample have, therefore, been exposed to many hundreds of hours of driving task performance. Participants were members of the public recruited through mail shots and adverts but in order to ensure the degree of homogeneity sought, all drivers were screened using the Driving Style Questionnaire (DSQ; West et al., 1992).

3.3 Materials

Twelve cars of mixed type (from sports coupes to people carriers/MPVs) were used in the study. Car drivers were audio recorded whilst they drove using a microphone and laptop computer. The on-road route was contained within the West London area of Surrey and Berkshire and was 14.5 miles in length not including an initial three mile stretch used to warm up participants. The route is comprised of one motorway section (70 mph speed limit for 2 miles), seven stretches of major road (50/60 mph speed limits for 6 miles), two stretches of country road (60 mph speed limit for 3 miles), three stretches of urban roads (40 mph limit for 1 miles), one residential section (30 mph limit for 1 mile), and fifteen junctions (>30 mph speeds for 1 mile).

3.4 Procedure

Formal ethical consent was obtained from all participants before the study commenced with particular emphasis placed on control of the vehicle and safety of other road users remaining the participants' responsibility at all times. An instruction sheet on how to perform a verbal commentary was read by the

494

participant, which described that they should drive as they normally would but provide a constant commentary about why they were performing current actions, what information they were taking from the environment, how the vehicle was behaving and what actions they plan to take. Drivers were instructed to keep talking even if it appeared to them that what they were saying did not make obvious sense. The experimenter provided examples of the desired form and content that it should take.

There then followed a warm-up phase. A three mile approach to the start of the test route enabled the participants to be practiced and advised on how to perform a suitable verbal commentary. This involved providing suggestions and guidance from the passenger seat and pulling over to review the audio transcript and advise where necessary. All participants were able to readily engage in this activity with a mean word per minute rate in excess of 30. Minimal advice was needed. The verbal commentary acted as a form of secondary task and the high rate of verbalizations seem to indicate spare capacity and little grounds to suspect interference with the primary driving task.

During the data collection phase the experimenter remained silent aside from offering route guidance and monitoring the audio capture process. The drivers, meanwhile, provided a constant verbal commentary as previously instructed. Drivers were de-briefed upon return to the start location.

3.5   Data Analysis

An approach to the measurement and representation of SA that is compatible with several on-road data collection methodologies, is one based on semantic networks. Semantic networks are an established way of representing knowledge (e.g. Collins & Loftus, 1975). Creating networks to represent driver SA involves extracting information elements from drivers (called nodes or vertices in semantic networks) and establishing links (or edges) between them. The result is that when elements become temporally, spatially, causally or semantically interlinked they begin to form 'concepts'. It is claimed that "one can produce dictionary-like definitions of concepts" and that a definition of any situation (or state of knowledge) can be represented (Ogden, 1987).

A novel development reported in this paper is the use of a sophisticated software tool called Leximancer™ which automates the creation of semantic networks, and does so with complete repeatability. Leximancer™ uses text representations of natural language in order to create themes, concepts and links. This is achieved by algorithms which refer to an in-built thesaurus and to features of text such as word proximity, quantity and salience. Leximancer™ has been used extensively in previous studies (e.g. Rooney et al., 2010; Hewett et al., 2009; Cretchley et al., 2010), however, the application of this technique to a more intensive form of analysis based on verbal commentaries in real-life transport contexts is a novel one.

4 RESULTS AND DISCUSSION

4.1 Semantic Networks

The twelve drivers contributed a total of 27,225 words into the analysis (Mean = 4537; SD = 2343, Minimum word count = 1759, Maximum = 7531). Leximancer™ subjects this raw textual data to six main stages in order to create the semantic networks:

1. Conversion of raw text data (definition of sentence and paragraph boundaries etc.).
2. Automatic concept identification (keyword extraction based on proximity, frequency and other grammatical parameters).
3. Thesaurus learning (the extent to which collections of concepts 'travel together' through the text is quantified and clusters formed).
4. Concept location (blocks of text are tagged with the names of concepts which they may contain).
5. Mapping (a visual representation of the semantic network is produced showing how concepts link to each other).
6. Network analysis (this stage is not a part of the Leximancer™ package but was carried out as an additional step to characterize the structural properties of the semantic networks).

A total of six networks are derived from this process, one for each road type. The networks provide a representation of the knowledge extant in the driver's working memory as they encountered each environment, and a proxy for their SA. They have been produced automatically with complete repeatability from raw verbal transcripts: no manipulation of the data or the process of network creation was undertaken. An example network is shown below in Fig 1.

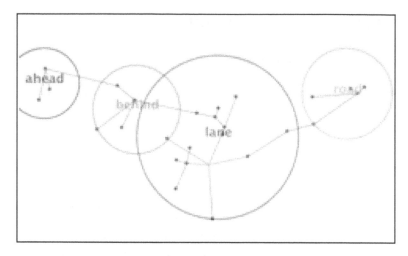

Figure 1   A semantic network produced by Leximancer™ referring to the verbal transcripts obtained during motorway (freeway) driving.

In Leximancer™ the nodes in the networks are extracted from the verbal transcripts and are referred to as 'concepts'. Each one is ascribed a relevance value from 0 to 100%, which is a value derived from the number of times the concept occurs as a proportion of the most frequently occurring concept (Smith, 2003). In total, 174 concepts were extracted from the six semantic networks. In order to reduce the data to the highest scoring concepts and to avoid the inevitable idiosyncrasies of low scoring, highly personal and infrequently occurring concepts, those which scored lower than two standard deviations of the mean in each individual road-type category were excluded. This gives rise to a high scoring subset of 25 which in turn represent directly, or refer to endemic road features (as shown in Table 1). What is clear is that marked changes occur in the content of driver SA as they enter and exit different road environments, to such an extent that no key concepts are common across them. Road type is thus a powerful contingency factor in subsequent driver SA, a finding which accords with the varied literature on Self Explaining Roads.

4.2   Hazard Incident Rate and Speed Adaptation

Hazards are defined as any object or entity that has the potential to cause a driver to change direction or speed (Coyne, 2000). When drivers are asked to provide a verbal commentary they are, in effect, talking about hazards. Hazard Incident Rate (HRI) is a term employed in Police and advanced driver training and is a reflection of the temporal nature and pacing of hazards. Based on the analysis above it becomes possible to calculate a crude Hazard Incident Rate (HIR) based on individually encoded items related to the external environment. HIR is measured in hazards per hour.

Table 1 List of endemic road features crossed with road type.

Concept	Motorway	A/B Roads	Urban Roads	Country Roads	Residential Roads	Junctions
Ahead	■					
Behind	■				■	
Bend				■		
Braking						
Car	■		■		■	
Check		■				■
Clear						■
Coming	■					
Corner				■		
Doing		■				
Fourth		■				
Front	■					
Gear				■		
Hill						
Indicating						■
Lane	■	■				
Mirrors						■
MPH		■				
Parked			■			
Pull		■			■	
Road	■		■			
Around				■		
Slow		■		■		
Take						
Third						

Figure 2 shows an interesting relationship between speed adaptations and Hazard Incident Rate (HIR). Broadly speaking, the hazard incident rate increases across different non-motorway road types. Similarly, as the hazards per hour increase, speeds adapt in a downward direction. Thus for motorways (478 hazards per hour) the speed is 69.23mph whereas for residential roads (1564 hazards per hour) the speed is 23.38mph. The overall pattern suggests that drivers are using their SA to interpret the road context and adapt their behavior to suit. However, whilst the overall pattern suggests a favorable downward adaptation, the speed versus hazard incident rate for the most behaviorally compatible road type, motorways, suggests that speeds should, if anything, be more negatively adapted. For example, for the hazard incident rate for residential roads to match that of motorways, the downward adaptation in speed would have to be greater than the on-road data suggests: 7.14mph instead of 23.38mph. Drivers seem best able to manage this in respect to major A/B classification roads, where the difference in actual and 'ideal' speed is only 4.22mph (or 9.33%). This rises dramatically as the roads reduce in status, reaching a difference between actual vs. 'ideal' for residential roads of 69.43%. This is a very crude indicator, with many extraneous and other variables to take into consideration so there is a need to not overstate the case. Nevertheless, when this difference between actual and 'ideal' speeds is considered as a whole a powerful trend emerges, one that looks like a progressive uncoupling of the perceived state of the situation (i.e. SA) from the actual state. The challenge is to identify and understand the endemic features that may help driver SA to become more aligned to given road situations.

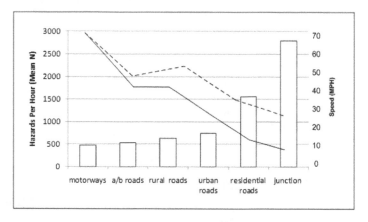

Figure 2 Hazard Incident Rate (hazards per hour) overlain with the average speeds actually attained on the on road course (dotted line) and the speeds that would need to be attained in order to match the HIR for motorways (solid line).

5 CONCLUSIONS

This paper has attempted to build on the maturing field of SA within the road safety domain, and to put the concept in touch with the practical engineering problem of safer road infrastructure design. An argument has been put forward that SA could be a valuable adjunct to the developing strand of work in Self Explaining Roads and the results gained are broadly consistent with this. The study is exploratory in nature, but it succeeds in applying a sophisticated approach to SA in a real-world setting, unearthing cognitively salient features of road infrastructure, and tentatively exploring whether the SA that drivers develop is linked to appropriate driving behaviors. As an exploratory study it is right to reflect on some of the inherent limitations, principle among which are those around a small sample size and the efficacy of verbal protocols to elicit tacit knowledge. Clearly, the work reported in this paper is not mature enough in itself to directly inform engineering interventions on a large scale, but they are sufficiently well developed to enable new hypotheses to be developed and for further experiments to be conducted.

REFERENCES

Aarts, L. T, & Davidse, R. J. (2007). Herkenbare vormgeving van wegen. (Rep. No. R-2006-18). Leidschendam: SWOV.

AASHTO, 2004. A policy on geometric design of highways and streets. Washington: Amer Assn of State Hwy.

Charlton, S. G., Mackie, H. W., Baas, P. H., Hay, K., Menezes, M & Dixon, C. (2010). Using endemic road features to create self-explaining roads and reduce vehicle speeds. Accident Analysis and Prevention, 42, 1989 – 1998.

Collins, A. M. & Loftus, E. F. (1975). A spreading-activation theory of semantic processing. Psychological Review 8(6): 407-428.

Coyne, P. (2000). Roadcraft The Police Drivers Manual. London: HMSO.

Cretchley, J., Rooney, D., & Gallois, C. (2010). Mapping a 40-year history with Leximancer: Themes and concepts in the Journal of Cross-Cultural Psychology. Journal of Cross-Cultural Psychology, 41 (3), 318-328.

Dft, 2008. Road statistics 2008: traffic, speeds and congestion. Available at: http://data.gov.uk/dataset/road_statistics_-_traffic_speeds_and_congestion

Endsley M. R. (1995). Toward a theory of situation awareness in dynamic systems. Human Factors, 37, 32-64.

Goldenbeld, C. & Van Schagen, I. (2007). The credibility of speed limits on 80km/h rural roads: the effects of road and person(ality) characteristics. Accident Analysis and Prevention, 39(6), 1121 – 1130.

Gugerty, L. J., 1997. Situation awareness during driving: explicit and implicit knowledge in dynamic spatial memory. Journal of Experimental Psychology: Applied, 3, (1), 42-66.

Hewett, D. G., Watson, B. M., Gallois, C., Ward, M., Leggett, B. A. (2009). Intergroup communication between hospital doctors: Implications for quality of patient care. Social Science and Medicine, 69, 1732–1740.

Highways Agency, 2011. Design manual for roads and bridges. Available at: http://dft.gov.uk/ha/standards/dmrb/index.htm

Norman, D. A. (1990). The design of everyday things. New York: Doubleday.

Ogden, G.C. (1987). Concept, knowledge and thought. Annual Review of Psychology, 38 203-227.

Rooney, D., Paulsen, N., Callan, V.J., Brabant, M., Gallois, C., and Jones, E. (2010). A New Role for Place Identity in Managing Organizational Change . Management Communication Quarterly. 24(1), pp. 104-121

Smith, A. E. (2003). Automatic extraction of semantic networks from text using leximancer. Proceedings of HLT-NAACL, Edmonton, May-June.

Stelling-Konczak, A., Aarts, L., Duivenvoorden, K & Goldenbeld, C. (2011). Supporting drivers in forming correct expectations about transitions between rural road categories. Accident Analysis and Prevention, 43, 101-111.

Sukthankar, R., 1997. Situation awareness for tactical driving. Unpublished doctoral dissertation, Carnegie Mellon University, Pittsburgh.

Theeuwes, J. & Godthelp, H. (1995). Self explaining roads. Safety Science, 19, 217-225.

Vanderbilt, T. (2008). Traffic: why we drive the way we do (and what it says about us). London: Penguin.

West, R., Elander, J., French, D., 1992. Decision making, personality and driving style as correlates of individual accident risk: Contractor report 309. Transport Research Laboratory, Crowthorne.

CHAPTER 51

Driver's Behavior Assessment as an Indicator of Accident Involvement

Nikolaos Katsianis, Nikolaos Eliou, Evdokia Iliadou

University of Thessaly
Volos, Greece
neliou@uth.gr

ABSTRACT

The present paper examines the correlation between driver's behavior and accident involvement. Driver's behavior is being described by measurable driving parameters, which are acquired by the recording of the driving pattern of a sample of drivers while running through specific road sections. After the field tests, each driver fills out a questionnaire regarding driving experience and traffic accident involvement. The outcome of the data processing is the formation of two types of parameters, driving parameters and accident parameters. The correlation of these types of parameters is being examined with the use of Pearson Correlation Coefficient. Finally, it is suggested that the driving parameters with the higher coefficient value could be the strongest traffic accident predictors.

Keywords: driver's behavior, accident involvement, accident predictors

1   DRIVER'S BEHAVIOR

Driver's behavior studies have pointed out that there are several motion relevant parameters that can reflect each one's driving pattern. The most popular parameter is speed, which appears to be connected with risky driving behavior. Speed defines the time available to react in front of serious traffic conflicts and also affects the seriousness of the consequences of a crash [Rothengatter, 1988]. Furthermore it's commonly accepted that excessive speed is responsible for the largest amount of accidents. Studies have shown that accidents frequency is increased exponentially in relation to speed [Taylor, 2002].

However, it can't be argued that speed is the most representative of one's driving pattern, as there is no free choice of speed value by the drivers, either due to speed limits or due to traffic flow reasons. [Young et al., 2011]. Moreover, it can be argued that higher speed values do not lead to high accident frequency, for example motorways do not exhibit as many accidents as their high speed values should indicate. Thus, it can be concluded that the major problem is either the inappropriate speed selection for a given road and weather conditions, or the extended speed variation of different vehicles using the same road, as smaller speed differentiation decreases collision points and accident likelihood [Rothengatter, 1988, Aarts and van Schagen, 2006].

Acceleration value is shown to be more representative of driving pattern, as there are no limits by the law and drivers are not aware of the acceleration value, so they can't adjust their performance to the desired level. However, researches have shown that acceleration can have similar values at both risky and safe traffic situations [Bagdadi & Varhelyi, 2011]. Thus, many researchers have used acceleration-related parameters, which derive from the processing of acceleration values.

Robertson, Winnet & Herrod (1992) used an index called "Equivalent Vector Acceleration", defined as the sum of all transverse and longitudinal accelerations, as a measure of driving behavior. The former index was found in subsequent studies to correlate strongly with the number of accidents, thus verifying the claim that "safe" drivers can react better on the road avoiding heavy accelerations and decelerations, which may increase the risk of losing vehicle control [Lajunen, Karola & Summala, 1997]. In addition, a comparative assessment regarding speed (average speed, max speed etc) and an acceleration variable (celeration variable) concluded that the number of accidents correlates strongly with the acceleration variable, which was measured as the total sum of speed variations in the longitudinal direction [af Wahberg, 2006].

The rate of change of acceleration is another derivative of acceleration that has been used as a parameter of driving behavior, with some of its values characterized critical for safe driving [Nygard, 1999, Bagdadi & Varhelyi, 2011]. Additional parameters that can characterize driving behavior are travel time percentage of acceleration, deceleration or steady speed [af Wahlberg 2007].

Finally, the distance kept from the front vehicle is claimed to be an important safety critical parameter, as it seems to be indicative of the risk that the driver is willing to take [Evans & Wasiliewski, 1981 και 1982, Brackstone et al, 2008]

2   ACCIDENTS

The number of accidents is the most profound way to identify a risky driving behavior. However, it can be argued that risky driving behavior could not be defined by the total number of accidents one has been involved in. Additional parameters should be taken into consideration, like exposure (e.g. mileage), usual road network (e.g. urban network, motorways or rural roads) and driving conditions (e.g. night driving). The classification of the accident types is also considered to be necessary, regarding culpability, accident severity and number of involved vehicles [af Wahlberg, 2009]. Since driver's behavior studies aim to define (or explain) which behavioral characteristics are responsible for accidents, it is rational that the research should include only the culpable accidents.

The accident data sources used can be summarized to the following; police records [Evans & Wasielewski,1981], insurance companies' records [Dalziel & Job, 1997, etc], fleet records and self reports. It can be said that the majority of these sources is not very reliable, as not all the accidents are being reported to the police and the insurance companies, plus researches have shown that people tend to forget some of their accidents, especially when they have many of those, while they also forget the exact time of their accidents, placing them much earlier from their actual time [af Wahlberg, 2009]. Fleet records seem to be the most reliable source, as long as the sample consists of drivers occupied in a certain business or organization.

3   METHOD

The methodology procedure consisted of two phases. On phase one, Field Operation Tests were performed, which included a sample of 10 drivers, driving along a specified route, which consisted of a two way urban road stretch and a freeway stretch, part of the former National Road Network. The urban section is a two way urban arterial with one lane for each direction, stopped controlled intersections and a speed limit of 50 km/h. The time of the tests was chosen in order to avoid heavy traffic situations, which are quite common for the given road. The freeway section is a non-separated two way road with one lane for each direction plus an auxiliary lane and a shoulder, where free flow conditions are a typical situation. Special equipment based on differential GPS combined with a data logger was placed in each participant's vehicle in order to record the movement dynamic characteristics for every run. The parameters used in this study are speed and velocity, which have been recorded in a frequency of 20 Hz. After the data were collected, they were processed properly in order to form the parameters chosen as drivers' behavior indexes, which include:

- Average speed
- Acceleration index (also known as celeration) [af Wahlberg, 2006 and 2007], defined as the sum of all speed variations (in absolute values) towards the number of total measurements (for N=20Hz, there are 20 values per second)
- Travel time percentages for acceleration, deceleration and relevant steady speed
- Travel time percentage when the rate of change of acceleration is below the critical value of -9.9 m/sec3, which has been used as an indicator of safety-critical driving behavior [Nygard, 1999, Bagdadi & Varhelyi, 2011].

The second phase included the collection of accident and mileage data from the drivers, via questionnaires. The method used was self-reports as it was considered to be the most reliable and practical data source from the ones available. Drivers were asked about their total mileage, while they noted their mileage during the past five years, which is considered to be a time zone that their driving style hasn't change. Moreover, data about road conditions and vehicles used were gathered. However, these data weren't used in the present paper, due to the rather small sample and the consequent lack of the necessary data to perform a reliable statistical analysis.

Regarding accident classification, given that this study's objective is to correlate driving behavior with accident involvement, participants were asked to identify the culpability of the accident by giving a short description of the incidents. In some cases where the culpability wasn't clearly defined by the description, the solution was given by the question if there had been any way of avoiding the accident by the driver. Furthermore, data regarding the severity of the accidents were collected, based on the drivers' estimation of the material damage. The classification resulted from all the above is the following:

- Culpable accidents per 10000km of mileage
- Culpable accidents during the past 5 years per 10000km of mileage.

Although the objective of the research is to study the accidents resulting from driving behavior, additional parameters were examined regarding the total number of all accidents, regardless culpability, such as:

- Total accidents per 10000km of mileage
- Total accidents during the past 5 years per 10000km of mileage

Accidents per 10000km were selected rather than accidents per km for a better and more comprehensible data comparison.

Pearson correlation coefficient was used to analyze the correlation among driving parameters and accident data.

4   RESULTS

The results from the two phases, (phase one: driving parameters, phase 2: accident parameters) are presented on the following tables. Tables 1 and 2 contain the respective results regarding the urban road section and the freeway section. The results from the accident data categorization and the respective parameters are presented on Tables 3 and 4.

Finally, Tables 5 and 6 contain the Pearson Correlation Coefficient values for every possible combination between driving parameters and accident parameters. The accident parameters that were examined are culpable accidents for the past 5 years and culpable accidents for total mileage (columns 1 and 2). However, it is interesting to see how the correlations work with the parameter "all accidents" as well (columns 3 and 4).

504

Table 1 Driving Parameters on urban road section

Driver s/n	Speed (km/h)	Celeration (m/sec2)	Time spent on accelerating	Time spent on decelerating	Time spent on even speed	Frequency of abrupt change of acceleration
1	34,177	0,423	50,44%	40,10%	9,46%	1,08%
2	34,930	0,495	44,32%	49,13%	6,54%	1,26%
3	37,259	0,417	43,38%	48,68%	7,94%	0,42%
4	37,358	0,381	47,43%	42,50%	10,07%	1,27%
5	41,665	0,352	46,52%	43,38%	10,09%	1,20%
6	38,562	0,632	47,80%	46,62%	5,58%	3,38%
7	36,838	0,446	45,59%	44,98%	9,43%	2,52%
8	29,660	0,761	46,18%	48,85%	4,97%	11,98%
9	37,969	0,595	49,22%	45,63%	5,15%	1,26%
10	37,579	0,535	44,90%	48,45%	6,65%	1,37%

Table 2 Driving Parameters on freeway

Driver s/n	Speed (km/h)	Celeration (m/sec2)	Time spent on accelerating	Time spent on decelerating	Time spent on even speed	Frequency of abrupt change of acceleration
1	57,248	0,380	41,57%	49,14%	9,29%	0,39%
2	56,278	0,365	46,78%	41,12%	12,09%	0,09%
3	66,101	0,401	37,39%	39,93%	22,68%	0,04%
4	72,668	0,432	45,46%	45,68%	8,85%	0,43%
5	65,780	0,391	55,54%	35,97%	8,49%	0,26%
6	78,168	0,735	45,98%	51,61%	2,40%	1,76%
7	54,818	0,393	49,86%	37,39%	12,75%	0,20%
8	50,954	0,598	53,43%	38,81%	7,76%	0,95%
9	70,972	0,402	40,71%	44,81%	14,48%	0,49%
10	70,838	0,514	52,18%	40,83%	6,99%	0,22%

Table 3 Mileage and Accident Data main categorization

driver s/n	Mileage	Years of driving	All accidents	Culpable accidents	Mileage for the past 5 years	All accidents last 5 years	Culpable accidents last 5 years
1	75000	11	1	0	50000	1	0
2	100000	14	0	0	40000	0	0
3	22000	4	0	0	22000	0	0
4	250000	12	3	1	80000	0	0
5	100000	12	2	2	50000	0	0
6	150000	15	2	1	60000	1	1
7	600000	40	4	2	35000	1	1
8	120000	5	1	1	120000	1	1
9	70000	7	3	1	50000	3	1
10	150000	12	0	0	40000	0	0

Table 4 Accident Parameters

driver s/n	Culpable accidents per 10000 km driven for the past 5 years	All accidents per 10000 km driven for the past 5 years	Culpable accidents per 10000 km driven	All accidents per 10000 km driven
1	0,000	0,200	0,000	0,133
2	0,000	0,000	0,000	0,000
3	0,000	0,000	0,000	0,000
4	0,000	0,000	0,040	0,120
5	0,000	0,000	0,200	0,200
6	0,167	0,167	0,067	0,133
7	0,286	0,286	0,033	0,067
8	0,083	0,083	0,083	0,083
9	0,200	0,600	0,143	0,429
10	0,000	0,000	0,000	0,000

Table 5 Pearson Correlation Coefficients for the urban road section

	Culpable accidents per 10000 km drived for the past 5 years	Total of culpable accidents per 10000 km drived	All accidents per 10000 km drived for the past 5 years	All accidents per 10000 km drived
Speed	0,030	0,409	0,028	0,291
Celeration	0,385	0,077	0,285	0,121
Time spent on accelerating	0,197	0,306	0,586	0,708
Time spent on decelerating	0,034	-0,197	-0,209	-0,402
Time spent on even speed	-0,260	-0,030	-0,310	-0,150
Frequency of abrupt change of acceleration	0,195	0,153	-0,031	-0,077

Table 6 Pearson Correlation Coefficients for the freeway section

	Culpable accidents per 10000 km drived for the past 5 years	Total of culpable accidents per 10000 km drived	All accidents per 10000 km drived for the past 5 years	All accidents per 10000 km drived
Speed	-0,027	0,185	0,062	0,308
Celeration	0,225	0,052	-0,066	-0,072
Time spent on accelerating	0,001	0,376	-0,343	-0,172
Time spent on decelerating	0,039	-0,232	0,271	0,250
Time spent on even speed	-0,038	-0,190	0,117	-0,048
Frequency of abrupt change of acceleration	0,349	0,217	0,174	0,218

4   CONCLUSIONS

The initial hypothesis of the present research is that drivers' behavior characteristics affect the possibility of accident involvement. The physical meaning of the values on the correlation table is that driving parameters with high correlation coefficient value tend to affect stronger the possibility of accident involvement, and thus could be characterized as accident predictors. It is noted though that the rather small sample (N=10) does not allow the drawing of general conclusions, as the reliability of the statistical method is questionable. However, the correlation tables show that there is an actual connection between specific driving parameters and culpable accidents. Moreover, the two different road types seem to give similar correlation values.

On the urban road section, the parameter having the strongest correlation with accidents is celeration (0,385), having an also high correlation value on the freeway too (0,225). The same goes for the frequency of abrupt change of acceleration, having high values both on the freeway section (0,349) and the urban section (0,195).

The parameter "time spent on accelerating" seems to affect accident possibility only in urban areas (0,197), while the correlation seems to be totally insignificant on the freeway. The hypothesis that safe drivers can "read" the road properly and avoid accelerations and decelerations seems to be verified, since the parameter "time spent on even speed" has a negative correlation with accident frequency on the urban road section (-0,260). Speed and time spent on decelerating do not seem to correlate significantly with accident involvement for both urban and freeway sections.

It is expected that the correlations regarding the total number of culpable accidents would be higher than the ones regarding the 5 year period, although this has not been noticed in all of the cases. It is noted that the 5 year period was selected in order to get a significant amount of accident data, which would be doubtful had the period been shorter.

Regarding the parameters of all accidents regardless of culpability, it is noted that the correlations do not seem to follow any general rule, as some correlations may be strongly positive for the urban section and significantly negative for the freeway.

One of the shortcomings of the present research is definitely the small size of the sample and also the rather short routes (approximately 5 km each). Moreover, the sample was quite homogeneous, meaning that most of the drivers had similar accident records (only a few if not any), which also limited the amount of accident data. Had the amount of data been a lot larger, it would be possible to categorize the accidents according to the road network where they occurred and then examine the respective correlations. For example, the driving parameters of the urban road section would be correlated with the culpable accidents occurred on urban areas.

Another shortcoming of this methodology is the fact that people tend to alter their driving style after accidents [Wahlberg, 2007], so there is a possibility that drivers with recent accidents, when being part of such a research, do not exhibit a risky driving pattern.

508

Finally, some suggestions for further research should probably include a large sample of drivers, many hours of recording when driving along long routes with various characteristics and perhaps the recording of additional driving parameters such as the headway distance drivers keep from the vehicle in front, in order to obtain a more accurate picture of their driving behavior.

REFERENCES

Aarts, L. and van Schagen, I. 2006. Driving speed and the risk of road crashes: A review, *Accident Analysis & Prevention*, 38: 215-224.

Arthur, W. Jr., and Doverspike, D. 2001. Predicting motor vehicle crash involvement from a personality measure and a driving knowledge test. *Journal of Prevention and Intervention in the Community*, 22: 35–42.

Bagdadi, O., Varhelyi, A. 2011. Jerky driving – An indicator of accident proneness? *Accident Analysis and Prevention*, 43: 1359–1363.

Brackstone, M., Waterson, B., McDonald, M. 2009. Determinants of following headway in congested traffic. *Transportation Research Part F*, 12: 131–142.

Dalziel, J. R., and Job, R. F. 1997. Motor vehicle accidents, fatigue and optimism bias in taxi drivers. *Accident Analysis and Prevention*, 29: 489–494.

Evans,L. and Wasielewski,P. 1981. Do accident-involved drivers exhibit riskier everyday driving behavior? *Accident Analysis & Prevention*, 14 (1): 57–64.

Evans,L. and Wasielewski,P. 1982. Risky driving related to driver and vehicle characterestics. *Accident Analysis & Prevention*, 15 (2): 151–136.

Lajunen, T., Karola, J., and Summala, H. 1997. Speed and acceleration as measures of driving style in young male drivers. *Perceptual and Motor Skills*, 85: 3–16.

Nygard, M. 1999. A Method for Analysing Traffic Safety with Help of Speed Profiles. Tampere University of Technology.

Robertson, D. I., Winnet, M. I., & Herrod, R. T. 1992. Acceleration signatures. *Traffic Engineering & Control*, 33: 485–491.

Rothengatter, T. 1988. Risk and the absence of pleasure: a motivational approach to modeling road user behavior. *Ergonomics*, 31(4): 599–607.

Taylor, M., Baruya, A., Kennedy, J. 2002. *The Relationship between Speed and Accidents on Rural Single-carriageway Roads*. TRL, Wokingham (Report No 511).

af Wahlberg, A.E. 2006. Speed choice versus celeration behavior as traffic accident predictor. *Journal of Safety Research*, 37: 43–51.

af Wahlberg, A.E. 2007. Aggregation of driver celeration behavior data: Effects on stability and accident prediction. *Journal of Safety Science*, 45: 487–500.

af Wahlberg, A.E. 2007. Long-term effects of training in economical driving: Fuel consumption, accidents, driver acceleration behavior and technical feedback. *International Journal of Industrial Ergonomics*, 37: 333–343.

af Wahlberg, A.E. 2009. Driver behaviour and accident research methodology : unresolved problems. Ashgate Publishing Limited, Farnham, England, 43-46.

Young, M., Birrell, S., Stanton, N. 2011. Safe driving in a green world: A review of driver performance benchmarks and technologies to support 'smart' driving. *Applied Ergonomics*, 42: 533–539.

CHAPTER 52

Study on Young Novice Drivers' Unsafe Driving Behavior

Zuhua Jiang, Dong Man and Qian Zhang

Department of Industrial Engineering and Logistics Management,
Shanghai Jiao Tong University, Shanghai, China
Zhjiang@sjtu.edu.cn

ABSTRACT

Nowadays, the high crash involvement of the increasing young novice drivers has become a great concern in China. This study aims to investigate the potential factors influencing young novice drivers and also how these factors affect their driving behavior through simulated driving experiment. Both a Driver Behavior Questionnaire (DBQ) and a simulated driving experiment were conducted in 73 Chinese young novice drivers. Unsafe driving behavior was defined by the DBQ, including errors, lapses, ordinary violations and aggressive violations. The effects of driving potential, objective characteristics and subjective characteristics on young novice drivers' unsafe driving behavior were evaluated based on the simulated driving platform. The results show that subjective characteristics have the greatest impact on young novice drivers' unsafe driving behavior. The findings provide further reference for young novice drivers' driving safety study and driving training design in simulated driving environment.

Keywords: young novice drivers; unsafe driving behavior; DBQ; simulation

1   INTRODUCTION

By the end of 2009, car ownership in China has increased to about 76.2 million with an annual growth rate of 17.8%. Meanwhile, the number of car drivers has increased to 138 million with an annual growth rate of 13.2%, among which drivers with driving experience less than 3 years accounted for 34.8%. Moreover, Chinese urban mortality rate is about 50 people per million vehicles, which is 26.5 times that of Japan, and 17.8 times that of the United States.

A great deal of foreign research has explained why young drivers are over–represented in accidents. Fisher and Pollatsek (2006) suggested that "lack of skills and driving experience" was the most important reason for young drivers' accident involvement. Besides, over–confident was considered as another important factor of the high accident involvement rate among young driver. Lucidi (2000) found that compared to drivers of other age, young drivers often overestimated their driving skills and adopted unreasonable driving strategies to address the potential and encountered driving dangers. This study aims to investigate the potential factors influencing young novice drivers and also how these factors affect their driving behavior through simulated driving experiment. A total of 73 young novice drivers were chose as the sample; both a driving potential test and a driving simulation platform were used to assess the potential factors influencing young novice drivers' driving behavior by potential variables, basic variables, and psychological variables.

2 PARTICIPANTS

Seventy–three students, 56 male and 17 female, at Shanghai Jiao Tong University served as participants. Participants' age ranged from 19 to 24 years with a median of 26.2 years. All participants owned a valid Grade C driver's license for a median of 36.8 months. Since the data of five participants were regarded as invalid data, finally there were 68 valid data (17 female and 51 male test data).

3 DRIVING POTENTIAL TEST

3.1 Test Conditions

In this study, the driving potential test included 4 tests: attention breadth test, time reaction test, depth perception test and speed perception test. Participants' performance through the test were then analyzed and classified, which reflected the objective differences of different personalities, ages and driving experience among the young novice drivers. The driving potential test aimed to explore the physical condition and psychological factors of the participants, and whether they have the potential to become an excellent driver.

3.2 Attention Breadth Test Result

This experiment was tested in three groups of different time: 1.5s, 1.0s and 0.5s, and each set of test will be randomly repeated 12 times. In the end, the instrument records 50% probability point which means the attention span value. Results are shown in Table 1 and Figure 1.

Table 1 Attention span test results

attention span test	Male		Female		Experience Group		Inexperienced Group	
	Mean	Var	Mean	Var	Mean	Var	Mean	Var
0.5s	8.39*	1.87	7.75*	2.29	8.05*	1.84	8.44*	2.15
1.0s	8.91*	2.31	9.59*	1.94	8.44*	2.24	9.88*	1.98
1.5s	9.43*	2.87	8.88*	2.47	8.39*	2.30	10.4*	2.93

Note: *indicates P> 0.05; ** indicates P <0.05; *** P <0.01.

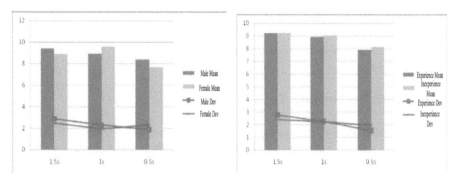

Figure 1 Data analysis diagram of attention span test

3.3 Time Reaction Test

This experimental test was divided into two parts: voice response test and color response test. During the voice response section, the instrument's sound timing was random and automatically, participants need to response of the sound signal as fast as possible. This session repeated 10 times and the average response time was recorded; during the color reaction section, the instrument displayed four colors randomly, and the participants need to response to the color as soon as possible, and the average response time was recorded. Results are shown in Table 2 and Figure 2.

Table 2 Time reaction test results

Time reaction test	Male		Female		Experience Group		Inexperience Group	
	Mean	Var	Mean	Var	Mean	Var	Mean	Var
voice	0.23*	0.08	0.23*	0.08	0.20**	0.04	0.24**	0.10
color	0.60*	0.17	0.58*	0.19	0.60*	0.14	0.60*	0.20

Note: *indicates P> 0.05; ** indicates P <0.05; *** P <0.01.

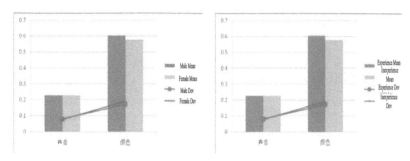

Figure 2 Data analysis diagram of time reaction test

Results showed that there are no significant differences between male and female group in time reaction ability. Significant differences were found between the experienced and inexperienced groups in voice response ability, but not in color response test ability.

3.4 Depth Perception Test

In depth perception test, participants were asked to sitting two meters away from the instrument, and try their best to move the three rods into one plane by four buttons (forward, backward, fast forward, and fast backward). This session was measured in three times, the difference between the actual coordinates and the ideal coordinates was recorded. Results are shown in Table 3 and Figure 3.

Table 3 Depth perception test results

Depth perceptio n test	Male		Female		Experience Grpup		Inexperienced Group	
	Mean	Var	Mean	Var	Mean	Var	Mean	Var
depth	4.99*	5.45	7.62*	7.89	4.09*	2.51	6.22*	7.56

Note: *indicates P> 0.05; ** indicates P <0.05; *** P <0.01.

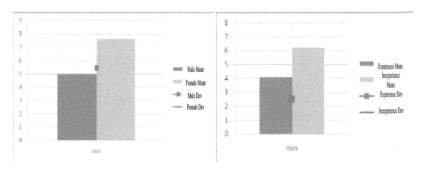

Figure 3 Data analysis diagram of depth perception test

Results showed that there are no significant differences between male and female group in depth perception ability, The same pattern was also found among the experienced group and inexperienced group.

3.5   Speed Perception Test

Table 4 Speed perception test results

Speed perception test	Male		Female		Experience Grpup		Inexperienced Group	
	Mean	Var	Mean	Var	Mean	Var	Mean	Var
Low speed	1.62*	3.28	1.82*	5.29	1.19*	2.87	0.94*	3.62
Middle speed	0.17**	0.74	0.90**	1.14	0.07**	0.50	0.46**	0.90
Fast speed	0.24*	0.89	0.33*	0.52	0.35*	1.33	0.20*	0.41

Note: *indicates P> 0.05; ** indicates P <0.05; *** P <0.01.

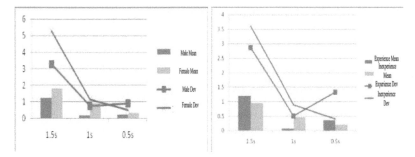

Figure 4   Data analysis diagram of speed perception test

This experiment was tested in three different speeds: low speed, middle speed and fast speed. Participants must anticipate when the moving cursor would reach the designated point at current speed. At the end of the experiment, the reaction time and standard time differences were recorded by the instrument (absolute value). Results are shown in Table 4 and Figure 4.

Results showed that there are no significant differences between male/female group and experience/inexperienced group in low speed perception and fast speed, while significant differences were found in the middle speed.

4 STIMULATION AND APPARATUS

4.1 Apparatus

The laboratory hardware platform consisted of a desktop simulator, projector and sound, as shown in Figure 5. Desktop simulator consists of a computer and a Logitech MOMO driving simulator. Participants manipulate the steering wheel, throttle, brake, handle, and buttons to control the simulation vehicle. Real–time visual feedback via a projector or computer monitor presented to the user with sound playback through the audio equipment.

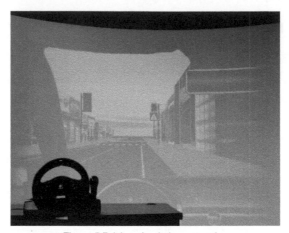

Figure 5 Driving simulation scenarios

Driving simulation data acquisition consisted of two parts. One part of the data was from the score recorded by assistants including four parts: violations, attention lapses, reaction time and accident. Violations recorded including speeding, running red lights and stop signs. Attention lapses were defined when participants failed to stop at the stop license, fail to notice other vehicles and almost crash with them. Reaction time reflected their reaction time after red light turning into green light. The accident recorded the number of accidents and a detailed description of the accident. Driving record score information recorded by four laboratory assistants. After the experiment, repeated comparisons to determine the accuracy of the recorded score with the video was performed. The other part of the data obtained by the C++ programming language (i.e., data of every frame, including real–time location information of the current simulation time, steering angle and accelerator pedal, steering wheel angle ranged from –120 to 120 degree, and values of the accelerator pedal ranged from 0–100). The data was an important basis for driving stability performance which was saved as a text format automatically.

4.2 Procedure

Upon arrival, the participants were given an overview of the activities involved and the time required for the experiment. After completion of a brief questionnaire, each participant was given a short session and allowed to drive until they felt comfortable operating the simulator. Drivers were given general instructions about the simulated road and asked to drive in the simulator as they would normally drive. At the same time, the assistant should be carefully observed participants' driving performance, recorded unsafe behavior and accidents.

4.3 Driving simulation data analysis

4.3.1 Driving stability curve

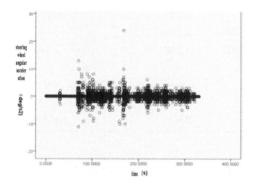

Figure 6 The acceleration plot chart of steering wheel angle

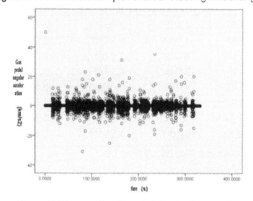

Figure 7 The acceleration plot chart of gas pedal

Figure 6 and Figure 7 is a steering angle and the accelerator pedal curve of a participant in experienced group, the more concentrated the point cloud is, the better the driving stability and control.

4.3.2 Driver's score data analysis

The means and variance of each variable were shown in table 5.

Two separate dimensions of the factors were extracted from potential conditions. Driving potential dimensions were distinguished according to the four driving potential factors (attention span, reaction time, depth perception, speed of perception); basic variable dimensions were distinguished according to three factors (gender, age, driving mileage). The influence of the three independent dimensions (driving potential, the basic variables, and psychological variables) on unsafe behavior was shown in Table 6 and Table 7.

516

Table 5 Potential conditions variables

Dimension variables	Mean	Var
Potential variables		
1.attention span test (point)	9.069	2.232
2.time reaction test (s)	0.227	0.079
3.depth perception test (mm)	5.605	6.144
4.speed perception test (s)	1.546	2.841
Basic variables		
1.gender	-----	-----
2.age	26.24	5.394
3.driving mileage	23614	41669

Table 6 Results of driving potential dimension variables

driving potential	unsafe behaviors			
	Error driving	Attention loss	General violation	Serious violation
attention span(<9.096)	4.435*	3.652**	0.739*	1.130**
attention span(>9.096)	4.000*	4.318**	0.568*	0.796**
time reaction(<0.227)	4.040*	3.920*	0.920***	0.800*
time reaction(>0.227)	3.944*	4.167*	0.444***	1.000*
depth perception(<5.605)	4.158*	3.860**	0.579**	0.860*
depth perception(>5.605)	3.667*	4.833**	1.000**	1.000*
speed perception(<1.546)	3.917*	2.500***	0.500*	0.917*
speed perception(>1.546)	4.105*	4.351***	0.684*	0.877*

Note: *indicates P> 0.05; ** indicates P <0.05; *** P <0.01.

Table 7 Results of basic driving variables

driving potential	unsafe behaviors			
	Error driving	Attention loss	General violation	Serious violation
gender (male)	3.539**	3.558**	0.769***	0.942*
gender(female)	5.706**	5.471**	0.294***	0.706*
age(<28)	4.368*	4.763**	0.605*	0.842*
age(>28)	3.710*	3.130**	0.710*	0.936*
Driving miliage(<5000)	4.833**	4.639**	0.444**	0.861*
Driving miliage(>5000)	3.094**	3.250**	0.813**	0.844*

Note: *indicates P> 0.05; ** indicates P <0.05; *** P <0.01.

5   EXPERIMENTAL RESULTS AND DISCUSSION

Participants who performed better than the average level were generally correspond to lower number of unsafe behaviors than the control group in the driving simulation, but the result did not show significant difference. The experimental results also showed that young drivers would sometimes overestimate their driving potential, which easily led to unreasonable driving behavior. This conclusion is closer to Lucidi (2000). Participants who perform better in the attention breadth test show greater attention loss in the driving simulation, which probably because participants with better attention breadth would be distracted more often by the driving environment. Participants who perform better in reaction time test show greater number of violations than the control group, the explanation maybe that fast speed response time lead to a slight decline to the driving stability. Though every single test showed that no significant differences were found between the groups, the simulation performance of these groups appeared significant differences, which indicated that though individual capacity has little impact on people's driving behavior, but when all these capacities were taken into account together, they would have significantly impact on driving behavior. The more comprehensive situation was, the greater difference between participants appeared. Overall, the driving potential did influence the driving performance among young novice drivers, though not playing a leading role.

In the basic variable dimension, the male group performed better than the female subjects in error and attention loss test, which showed that male drivers' driving proficiency and their confidence level were better than the female drivers; on the contrary, during the test of general violations and serious violations, the female drivers were generally performing better than the male participants. Female drivers who have lower driving proficiency were often drive more carefully, sometimes they chose maintaining a lower speed during the driving simulation even in order to avoid accidents. Participants older than 28 years old and with driving mileage more than 5000 km showed better simulation test results than others which indicated that the driving experience was a key factor to distinguish driving performance between young drivers.

6   CONCLUSION

This study aimed to analyze the potential factors that influence young novice drivers' performance. The degree of influence of driving potential dimension, basic attributes dimension and psychological attributes dimension on young novice drivers' unsafe behaviors was discussed. Results showed that psychological attributes have the greatest impact on young drivers' unsafe behavior. The three independent dimensions variables in different groups were compared, results showed how different variables convert to unsafe behaviors under a variety of factors. When the accumulation of unsafe behavior exceeds a certain threshold, traffic accidents would be caused. This study provided an important reference to the

518

emerging young novice drivers' safety issues, and it also provided reference for the design of the training program under the virtual simulation environment.

REFERENCES

Fabio, L., Anna, M. G., and Roberto, S. 2010. Young novice driver subtypes: Relationship to driving violations, errors and lapses. *Accident Analysis and Prevention* 42:1689–1696.

James, R. 2000. Human error: models and management. *British Medical Journal* 320:768-770.

Lucidi, F., et al. 2010. Young novice driver subtypes: Relationship to driving violations, errors and lapses. *Accident Analysis and Prevention* 42: 1689–1696.

Parker, D., Reason, J. T., and Manstead, A. S. R., et al. 1995. Driving errors, driving violations and accident involvement. *Ergonomics* 38:1036—1048.

Verschuur, W. L., Hurts, K. 2008. Modeling safe and unsafe driving behavior. *Accident Analysis and Prevention* 40:644—656.

Reason, J., Manstead, A., Stradling, S., et al. 1990. Errors and Violations on the Roads: a Real Distinction. *Ergonomics* 33:1315—1332.

Reason, J. 1990. Human Error. Cambridge University Press. London.

Parker, D., Reason, J. T., and Manstead, A. S., et al. 1995. Driving Errors, Driving Violations and Accident Involvement. *Ergonomics* 38:1036—1048.

CHAPTER 53

A Psycho-ergonomics Approach to Adaptive Behavior of Vulnerable Road Users

Marin-Lamellet Claude*, Marquié Jean-Claude**

* French institute of science and technology for transport, development and networks (IFSTTAR) Laboratory of Ergonomics and Cognitive Sciences applied to Transport (LESCOT)
25 Avenue François Mitterrand case 24 69675 Bron– France
claude.marin-lamellet@ifsttar.fr
** CLLE (Cognition, Language, Ergonomics), UMR 5263 CNRS-EPHE, University of Toulouse (Toulouse II)
5 allée Antonio Machado, 31058 Toulouse Cedex 9, FRANCE.
marquie@univ-tlse2.fr

ABSTRACT

In this paper, the framework proposed by Marquié (1993) to formalize adaptive behavior mechanisms from a psycho-ergonomics perspective is presented and applied to different transport contexts (driving and travelling by public transports) and categories of vulnerable road users (older, brain injured, blind). This framework proposes that to adjust his performance level, the individual can differently mobilize his/her basic cognitive processes and knowledge, or implement accommodation processes, namely strategic compromise related to either the goals (changing the goal structure), or the means (mobilizing new technical or human resources). Components of the framework are illustrated by recent results obtained with vulnerable road users (older or people with disabilities) through structured and instrumented observations of actual behavior in situations of moving and through laboratory studies, some of them using a driving simulator. Possible ways of improving the framework are discussed.

Keywords: aging, disability, adaptive strategies, mobility, transportation

INTRODUCTION

The displacement activity is subject to a need for regulation or adaptation, that is to say, a "feedback control which maintains the balance of an organized structure or organization under construction" (Piaget, 1967). This activity involves individual (skills) and socio-organizational components (rules related to the presence of others in travel environment). The displacement activity is a complex dynamic process in which the main actor (the person who moves) intends to keep the control on it. The individual adapts his/her behavior to maintain the control of the situation, that is to say maintaining an acceptable performance level given the amount of available resources (Hoc & Amalberti, 2007).

Behavioral adaptation process can be performed both in the short term i.e. in real-time during the course of the activity, and in the long-term through the capitalization and the consolidation of the expertise of the person.

Older people, persons who are affected by neurodegenerative diseases, and people with disabilities (blind or wheelchair users), can be considered as vulnerable road users. For them, the adaptation process could be seen as a corrective response made to a temporary or permanent deprivation of a functional resource, in order to minimize the negative impact on displacement activity performance. In the case of "normal" aging, this means to adapt the behavior to progressive losses, while in the case of traumatic events, it means to face with a sudden and often irreversible loss of functional resources. Finally, in the case of a congenital disability, the reference is the norm of the group people are affiliated to because of the absence of an original situation.

Behavioral adaptation can be initiated by the individual or "imposed" from outside. An Adaptation under the control of the individual is based on a mechanism of detection of a malfunction (the deviation from the norm), on sensitivity to signals coming from outside or from his own current experience. In the population of vulnerable road users, this detection can be disturbed by brain injury consequences, illness or psychological disorder: this is the case, for example, with elderly drivers suffering from Alzheimer's disease or drivers with traumatic brain injury who remain insensitive to the reactions of other drivers. In some cases, people may be aware of their difficulties but deny them, as is often the case for people with traumatic brain injury. A national policy, through appropriate measures can also be used to compel a person to adapt his/her behaviour in a displacement activity, such as driving. Thus, in some countries, drivers with neurological sequelae are only allowed to drive in certain contexts (low traffic, no highway, short distance from home ...).

The nature and conditions of implementation of adaptive mechanisms should be considered as a major research subject for the psycho-ergonomics community, especially for populations that are limited in their functional abilities. In addition, adaptive mechanisms are submitted to cognitive control of the individual and their triggering conditions may also vary among populations with disability resulting from diseases (e.g., Alzheimer's disease), neurological sequelae (due to Traumatic Brain injury – TBI) or congenital problem (blind).

Few frameworks in the current litterature allow a better understanding of these adaptive mechanisms and account for their diversity. The objective of this presentation is to propose a framework from an ergonomics and psychology perspective and to explore it in the context of the mobility of vulnerable road users.

A FRAMEWORK FOR ADAPTIVE BEHAVIOR MECHANISMS

When performing an activity such as moving in a transport context, the individual mobilizes basic cognitive processes (memory, attention, executive functions) and knowledge (mental representations and more generally all knowledge) in relation to requirements of the task. The comparison made by the individual between the task goals, available personal (sensory, cognitive ...) and external resources, and the predicted or observed activity performance leads to anticipated adjustments or corrections. The framework proposed by Marquié (1993 ; Marquié & Isingrini, 2001 ; see also, for some components, Baltes, Dittmann-Kohli, Dixon, 1984 ; Dixon & Bäckman, 1995 ; Salthouse, 1990) suggests that to adapt himself to the situation, the individual can mobilize differently his cognitive or knowledge based processes, or trigger a process of accommodation that will involve either a strategic level by a reorientation of goals, or a more operational level by mobilizing a technical resource or using a different working organization (see Figure 1).

A first level of adaptation takes place in the cognitive system of the individual. Several mechanisms can, at this stage, play an important role in maintaining performance despite the usual decline with age that affects the cognitive processes:

- *Preservation*: This mechanism refers to the idea that the efficiency of some basic cognitive abilities are maintained during normal aging because of their regular use in daily life (Marquié, 1997). This mechanism, classically described by the adage "use-it or lose-it", is certainly attractive but is very difficult to confirm empirically, mainly because of its strong interaction with socioeconomic factors.
- *Compilation*: It describes the fact that, with practice, segments of behaviour become automatic and, as a result, no longer require (or little) the contribution of basic cognitive resources. Controlled processes are becoming less necessary because most of the activity is managed by procedural knowledge.
- *Compensation*: This is an old concept in psychology. It is often referred to in a nonspecific, general sense of behavioral response to the occurrence of a decline, a deficit or an increase in task requirements, in order to reduce the gap between capacity and task demand "(Marquié & Isingrini, 2001). The concept should however be used only to describe the sollicitation of nonspecific latent skills or the use of specific know-how resulting from extensive practice of the task, in response to declines of the basic cognitive processes involved in the task.
- *Response bias*: It refers to decision criteria. Decision criteria are assumed

522

to vary with age. For example the elderly are known to adopt in some tasks more cautious strategies, such as increased proactive or retroactive visual control, more conservative decisions under uncertainty, greater focus on accuracy than on response speed. Response bias has been studied using the signal detection paradigm. Marquié & Baracat (2000 ; see also Baracat & Marquié, 1992) in their review of the literature, concluded that response bias is also strongly influenced by variables such as gender or education level. Future research should also examine this in relation with several disabilities.

- *Remediation*: Cognitive remediation is a technique that is similar to methods used in clinical rehabilitation. Its objective is to increase the level of cognitive efficiency in daily living activitities, either by training the basic processes involved in the impaired task or by developping alternative skills.

The second level of behavioral adaptation in the Marquié's framework, called "*goal-mediated accommodation* ", refers to changes in the hierarchy of goals as a response to the individual's difficulty in maintaining an acceptable performance level in a given context (e.g., no longer drive at night for an elderly driver or avoid a complex subway station for a person who is blind).

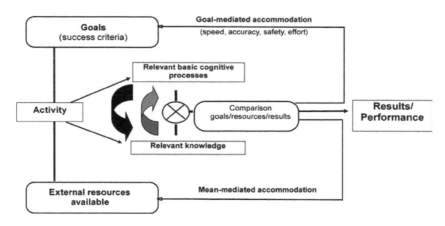

Figure 1: A model of adaptive behaviour mechanisms (from Marquié, 1993; Marquié & Isingrini, 2001)

The third level concerns " *mean-mediated accommodation* ", that is to say, the resort to an external assistance, either technical or human, to maintain the level of performance or to minimize its decline (for example a driving assistance system, a vocal guidance system for people with visual impairments, a technical device enabling access to the train for wheelchair users, or a more efficient organization of work).

Illustration in the driving context

Older drivers are a interesting population for the study of driving because they have a great driving experience. Long practice in performing a particular activity is a powerful resource that can be used to limit the deleterious effects of aging on cognitive functioning.

The form of behavioral adaptation frequently encountered in this population is *goal-mediated accommodation*, that is to say the action "to balance the goals in a different way." For example, older drivers reduce distances traveled, avoid driving at night or complex maneuvers, as in some intersections (Gabaude, Marquié, & Obriot-Claudel, 2010). Older drivers are redefining the priorities of their goal, downgrading the "speed criterion" in the order of priorities. Due to a slowdown in information processing, the time constraint becomes a major difficulty for older drivers, especially with the increasing density of traffic that requires making decisions quickly and increases the time pressure. In all our studies in real driving conditions, we have observed that older drivers adopted lower speed than younger subjects.

This type of adaptation is certainly efficient, but it does not solve everything. When engaged in a driving situation, older drivers must implement in real time corrective processes at the cognitive level to exploit the maximum of their available capacities. Observations in real driving situations showed that in terms of decision making, normal aging is associated with a tendency to focus on cautious decisions and to use less risky criteria (Delorme & Marin-Lamellet 1998). In this respect, a pathology effect was observed in the case of Alzheimer's disease, presumably due to an alteration of risk perception and a metacognitive related regulation which could be explained by the fact that this pathology often impairs the front part of the cortex (Marin-Lamellet, Lafont, Paire-Ficout, Laurent, Thomas-Antérion & Fabrigoule 2008). Curiously, this has not been observed in TBI patients when placed in a driving situation. If their driving performance is degraded, it is essentially in situations like the roundabout due to a lack of anticipation and reduction of visual scanning of the environment. The TBI population is interesting because it confirms what was observed in the older driver. Indeed, brain injured drivers were mostly young drivers with a limited driving experience. As they were observed in their driving when they left the rehabilitation center, they were not yet aware of the impact of their new functional state on their driving performance and goal-mediated accommodation was not observed (Marin-Lamellet, Etienne & Bedoin 2007).

Another dimension proposed by the framework is remediation. Recent advances in functional cerebral imaging techniques have confirmed the plasticity of the human brain, even after 60 years. An important question concerning remediation is whether it should focus on the specific task in which deficits are identified, or target the cognitive processes that are involved in this task. If the content of remediation must eventually be adapted to each person, it seems that in the perspective of an application to the transport context, both approaches could be taken into account and are not incompatible. In the doctoral work of Marjolaine Masson (Masson, Marin-Lamellet, Colliot and Boisson 2009), the hypothesis was that persistent solicitation of attentional functions would be more beneficial for the control of the driving task when it is done on a driving simulator compared to computerized exercises. Sixteen older people entered a three-week nine-hour training program in which they were randomly allocated to either a simulator-based training or a computer-based training. Off-road (driving simulator and neuropsychological testing) and on-road evaluations were used to assess the ability of driving participants pre-and post-training. Both groups improved in many neuropsychological evaluations. Significant improvements in driving simulator test were found in favor of the simulator-based training group. However, improvements were greater for the computer-based training group regarding the on-road test.

Mean-mediated accommodation is another mode of possible behavioral adaptation for vulnerable road users. It refers to the use by the individual of external means in a new kind of "instrumental, material or human compensation." In the driving context, Driving assistances (ITS -Intelligent technical systems- or ADAS -Advanced driving assistive technologies-) have been shown to have potential for aiding older drivers in different situations (Simoes & Marin-Lamellet 2002):

- collision warning systems at intersections that have the ability to alert drivers to oncoming hazards;
- automated lane changing and merging systems to assist older drivers with difficulties in passing zones;
- blind spot and obstacle detection systems that may assist drivers with limited sight;
- driver condition monitoring to measure fluctuations in sleepiness, cognitive load and factors associated with medications.

However, potential conflicts between driving assistance system and driving adaptive behaviour of older drivers have been pointed out by Caird (2004), who also noted "*Systematic analysis of how, when and why these adaptations are adopted by older drivers has not yet received adequate research attention*".

Illustration in the Public transport context

Public transport is another area of research where various adaptive behaviour mechanisms can be observed and interpreted in the light of the framework described above. If the activity of travel by public transport is different from driving, it can still be considered as a dynamic situation soliciting cognitive processes. It is also characterized by a strong interaction between the individual and his social and political environment.

In the field of public transport, the process of *goal-mediated accommodation* has been demonstrated with the persons with visual impairments through a survey in all public transport systems: bus, subway, Suburban Train, local and national train (Marin-Lamellet, Pachiaudi, & Le Breton-Gadegbeku, 2001). A questionnaire administered by phone to 174 persons with visual impairments, showed that they favored the subway (over 60% of them) when it was available because it was considered more reliable than the bus. In addition, the fact that subways systematically stop at all stations was presented as a positive factor, as this decreases the attention load during the travel. The bus was favored mainly by the older persons with visual impairments or with limited autonomy because of the possibility for them to get help from the driver. Moreover, the survey showed that persons with visual impairments planned more travel time than necessary when using the bus or train, and they avoided dangerous areas (infrastructure configuration, crossing streets ...) based on personal experiences or experiences from others people who are blind or visually impaired.

Little work has been done on the cognitive adaptation mechanisms used by persons with visual impairments and cognitive control made while traveling. We conducted an experiment whose objective was to observe persons with visual impairments when using a vocal guidance system and a guidance tactile pavement (Marin-Lamellet & Aymond, 2008). This study illustrates the process of *mean mediated accommodation*. Participants (N=172) made a journey in real conditions in a train station or subway station, and were accompanied by an observer and an mobility instructor for safety.

Based on the different criteria used to characterize the performance of persons with visual impairments (travel time, number of stops, number of requests for assistance to the mobility instructor, trajectory errors), the results showed an improvement in the efficiency of the displacement and of the guidance process. The participants who were blind showed in two distinct situations, vocal guidance only and vocal guidance associated with tactile guidance pavement, time to complete the trip higher than the participants who were partially sighted and they made more stops in both situations. On the other side, the proportion of time spent stopped evolved differently between the two situations. Indeed, in the situation with vocal guidance system alone, the participants who were partially sighted spent more time stopped than the participants who were blind, while in the situation vocal guidance associated with tactile guidance pavement, the proportion of time spent stopped was about the same order. This was mainly due to the fact that the participants who were blind were doing more short stops (0-10 sec) than the participants who were partially sighted. It was conceivable that the shorter stops made by the participants who were blind were connected to a more frequent listening to vocal messages, however, when we isolated the stops made in the areas of reception of the vocal system, the participants who were blind were doing almost as many short stops than the participants who were partially sighted. Given that the average rate of trajectory error was higher for the participants who were blind than for the participants who were partially sighted, the observed difference in the percentage of time spent stopped could be interpreted as the fact that the participants who were partially sighted tended to stop a little longer than the participants who were blind, but walked faster between stops. These longer stops from the participants who were partially sighted can be interpreted as a strategy of visual information search in an attemp to use visual information available in the station.

How to improve the framework ?

The few previous examples illustrate the interest of the behavioral adaptive framework proposed by Marquié and in the following of this presentation, we will discuss two possible improvements of this framework:

- Adding a social dimension as a trigger or a regulation of behavioral adaptation;
- Adding a temporal aspect in the organization of this framework: if all forms of accommodation are available, which one will be primarily favored and why?

The idea of behavioral adaptation in psycho-ergonomics is that the individual fails to maintain a satisfactory cost-performance balance and therefore must regulate its activity to find an acceptable compromise. The trigger for the regulation or the factor that determines the choice of the behavioral adaptation type can be safety or the pressure from the social context within which the individual evolves. Thus, in the case of an older driver, reducing speed can cause problems by triggering aggressive behaviors from other drivers (use the horn, insults) and cause the adoption of a new strategy for avoiding certain driving situations. We were also surprised in our experiments on vocal guidance systems for people with visual impairments, that many of these participants asked to "see" (that is to say, touch) the information system before performing the study because they did not want to wear in public a device which could stigmatize or devalue them (acceptability). This population may refuse a technical device that would present this drawback, even losing thereby an opportunity to maintain or improve its mobility performance.

In everyday life, when an individual is in a situation where performance becomes insufficient, he/she implements a behavioral adaptation process, thus rising the following questions: what mechanism will be selected first? Does a typical sequence of different mechanisms selection exist? For example, is a *mean-mediated accommodation* always tried before a *goal mediated accommodation*? We could assume that this latter form of behavioral adaptation is implemented when all other forms of adaptation have failed to maintain the level of performance. It is expected that there exists a high between-individual variability related to the characteristics and features of the situation (available resources, rigidity of goals).

It is also likely that metacognition plays an important role in the choice of behavioral adaptation mechanisms. It is conceivable that individuals with low awareness of their capabilities and low self confidence will seek to quickly implement a *goal-mediated accommodation* process without trying to implement other strategies. We observed this phenomenon in the verbal comments collected at the end of our experiments with persons with visual impairment. Some of them said: "Finally, it also allowed me to realize that, with a small effort, I could take the subway", thus showing that they initially considered themselves as not being able to perform this activity. The use of human help, which refers to *mean-mediated accommodation* in Marquié's framework, is also frequent among these people who tend to look for simplicity instead of implementing more sophisticated strategies. Here again, we collected illustrative verbal comments: "In fact, I'm not looking too much, I keep up to the entrance of the station and I expect that another passenger proposes to accompany me, and generally this works well, people are nice, but you cannot do that at night, there is no longer enough people. " Such mechanisms will be unacceptable to other people who will claim complete autonomy and will make a point of honor to fend for themselves. Another difficulty is that this does not allow one to accumulate practice and to elaborate skilled expertise.

In working situations, which are generally standardized, we can consider that operators have everything they need to do their work. Thus, when they have to implement an adaptive behavioral process, they have a set of possibilities in their immediate environment. This is not necessary the case in transport situations like driving or travelling by bus. Indeed, if an older driver is in a difficult position to find his/her way with a conventional map, he/she cannot use a navigation system if the vehicle is not equipped. In an experiment conducted on a driving simulator, we showed that when they have at their disposal several methods of control of an on board information system (keyboard, touchpad, voice command), older drivers modify their control strategy of the information system based on the complexity of the situation; when the latter required more of control of the vehicle and attentional resources, older participants stopped to use the keyboard interface to use the voice command (Kamp, Marin-Lamellet, Forzy and Causeur 2001). However, beyond the provision of technical means, the decision to acquire a system for driving assistance depends on economic factors and the familiarity of the person with communication technologies.

CONCLUSION

This presentation was an attempt to illustrate, with a psycho-ergonomics perspective, some of the possible processes that vulnerable road users (older or people with disabilities) can implement in adaptive behavior. The framework proposed for such mechanisms postulated that the individual could mobilize differently his cognitive or knowledge based processes, or trigger a process of accommodation that will involve either a strategic level by a reorientation of goals (goal-mediated accommodation), or at a level more operational by mobilizing a technical resource or use a different organization (mean-mediated accommodation).

Technological trends in recent years coupled with a growing requirement for mobility in the vulnerable road users population, will provide an opportunity for researchers to study more situations in which these persons could have a set of possible solutions in terms of behavioral adaptation.

REFERENCES

Baltes, P. B., Dittmann-Kohli, F., & Dixon, R. A. (1984). New perspectives on the development of intelligence in adulthood: toward a dual-process conception and a model of selective optimization with compensation. In P. B. Baltes & O. G. Brim, Jr (Eds.), Life-span development and behavior Vol. 6. NY: Academic Press, pp. 33-76.

Baracat, B. & Marquié, J. C. (1992). Age related changes in sensitivity, response bias, and reaction time in a visual discrimination task. Experimental Aging Research, 18(2), 59-66.

528

Caird, J. K. (2004). In-vehicle intelligent transportation systems: Safety and mobility of older drivers. Transport Research Board; TRB Conference Proceedings CP 27, 236-255.

Delorme, D., & Marin-Lamellet, C. (1998). Age-related effects on cognitive processes: Application to decision majing under uncertainty and time pressure. In J. Graafmans, V. Taiple & N. Charness (Eds.), Gerontechnology (pp. 124-127). Burke: VA: IOS Press.

Dixon, R.A., & Bäckman, L. (1995). Compensating for psychological deficits and declines. Managing losses and promoting gains. Mahwah, New Jersey: LEA, 346 p.

Gabaude, C., Marquié, J.-C., & Obriot-Claudel, F. (2010). Self-regulatory driving behaviour in the elderly: relationsips with aberrant driving behaviours and perceived abilities. Le Travail Humain, 73(1), 31-52.

Hoc, J.-M., & Amalberti, R. (2007). Cognitive Control Dynamics for Reaching a Satisficing Performance in Complex Dynamic Situations. Journal of Cognitive Engineering and Decision Making, 1(1), 22-55.

Kamp, J.F, Marin-Lamellet, C., Forzy, J.F. and Causeur, D. (2001) HMI aspects of the usability of internet services with an in-car terminal on a driving simulator; IATSS Research, vol25, N°2, 29-39.

Marin-Lamellet, C., Pachiaudi, G., Le Breton-Gadegbeku, B. (2001) Information and Orientation Needs in Public Transport for Persons Who are Blind or Partially Sighted: the BIOVAM Project. Transportation Research Record, n°1779, 203-208.

Marin-Lamellet, C., Etienne, V., & Bedoin, N. (2007). Compétence de conduite et traumatisme crânien. Paper presented at the 11th International Conference on Mobility and Transport for Elderly and Disabled Persons (TRANSED), Montréal, Canada.

Marin-Lamellet, C., Lafont, S., Paire-Ficout, L., Laurent, B., Thomas-Antérion, C., & Fabrigoule, C. (2008). The impact of attentional and executive impairments on driving abilities in normal aging and Alzheimer disease. 4th ICTTP, Washington DC.

Marin-Lamellet, C., & Aymond, P. (2008) Combining verbal information and a tactile guidance surface: the most efficient way to guide people with visual impairments in transport stations? British Journal of visual impairment, vol26(1), 63-81.

Marquié, J.-C. (1993). Vieillissement cognitif, expérience, contraintes de l'environnement. Perspectives théoriques et ergonomiques. Unpublished Thèse d'Etat, Université Paul Sabatier, Toulouse.

Marquié, J. C. (1997). Vieillissement cognitif & expérience : l'hypothèse de la préservation. Psychologie Française, 42(4), 333-344.

Marquié, J.-C., & Baracat, B. (2000). Effects of age, educational level and gender on response bias in a recognition task. Journal of gerontology: Psychological sciences, 55B, 266-272.

Marquié, J.-C., & Isingrini, M. (2001). Aspects cognitifs du vieillissement normal. In E. Aubert & J.-M. Albaret (Eds.), Vieillissement et Psychomotricité. Marseille: Solal Editeur, 77-113.

Masson, M., Marin-Lamellet, C., Colliot, P. and Boisson, D. (2009). Cognitive rehabilitation of traumatic brain injury drivers: a driving simulator approach. 21th ITMA World Congress, The Hague, Netherlands, 26-29 April 2009, p81.

Piaget, J. (Ed.). (1967). Biologie et connaissance. Paris: Gallimard.

Salthouse, T. A. (1990). Influence of experience on age differences in cognitive functioning. Human Factors, 32(5), 551-569.

Simoes, A., Marin-Lamellet, C. (2002) Road users who are elderly: drivers and pedestrians; in Human factors for the highway engineer, Fuller.R. & Santos.J.A. eds, Elsevier Science Ltd, Oxford, UK, 255-275.

CHAPTER 54

The Effects of Short Time Headways within Automated Vehicle Platoons on Other Drivers

Magali Gouy[1], Cyriel Diels[2], Nick Reed[1], Alan Stevens[1], Gary Burnett[3]

TRL[1]
Wokingham, United Kingdom
mgouy@trl.co.uk

Jaguar Land Rover research[2]
Coventry, United Kingdom
cdiels@jaguarlandrover.com

University of Nottingham[3]
Nottingham, United Kingdom
Gary.Burnett@nottingham.ac.uk

ABSTRACT

The implementation of platoons of electronically coupled vehicles on European roads is anticipated to bring about significant changes in transportation, with potential benefits such as improved fuel efficiency, increased road capacities, safety and driver comfort. However, before automated driving systems can be implemented, they must be thoroughly tested to ensure that their operation causes no undesirable safety issues. This paper investigates possible "contagion" effects, whereby the short time headways (THW) within automated platoons may lead drivers in the vicinity of the platoon also to adopt a shorter THW. In a medium-fidelity driving simulator, 30 participants were asked to follow a lead vehicle on a three-lane highway whilst being confronted with two different platoons having different THWs (1.4 sec. and 0.3 sec.) and a baseline condition without any platoon. Results showed a significant effect of the platoon on drivers' THW. Specifically when driving adjacent to the platoon with the shortest THW, participants' average and minimum THW were significantly smaller than when driving adjacent to the platoon with the longer THW. The limitation of these common parameters of THW and the potential to examine THW over time are discussed.

1   BACKGROUND

The driving task is developing towards increasing levels of automation (Walker et al., 2001). Automated driving is generally anticipated to be accompanied with improvement in fuel consumption, traffic flow, safety and convenience of the driving task. Hence there is an increasing interest in developing and implementing automated driving. Projects sponsored by Defense Advance Research Project Agency (DARPA) (Buehler et al., 2007) in the United States and EUREKA (Prometheus project) (Williams, 1988) as well as the EC's Framework research program (e.g. Chauffeur and SARTRE projects) in Europe are main actors for the development of automated drive. One form of automated driving, called 'platooning', entails the electronic coupling of vehicles by means of driver assistance systems on regular unmodified highways. Within a platoon, the first vehicle is driven manually while the others follow automatically maintaining a very short headway (Kunze et al., 2009).

Because the technical feasibility of such systems has already been shown, the focus is increasingly put on human factors issues associated with the system (Lank et al., 2011). Thus far, research has focused mainly on the interaction between drivers and automated systems and has largely neglected the interaction between automated systems and other traffic participants. When automated systems emerge on normal highways, they will have to interact with manually driven vehicles. Therefore, it is of paramount importance to understand how the manually driven and automated vehicles in a platoon will interact and make sure that this interaction is safe. Robinson et al. (2010) identified a range of critical scenarios to tackle before the introduction of platoons. One scenario considered the perspective of non-platoon drivers, specifically a non-platoon driver entering a platoon. Two simulator studies analyzed subjective data from participants that interacted with a platoon to investigate the acceptance of the system (Lank et al., 2011) and to determine platoon requirements, such as comfortable intra-platoon gap and platoon length (Larburu et al., 2010). A study investigating changes in behavior of non-platoon drivers analyzed, through a field study, the speed and overtaking time of non-platoon drivers (Lank et al., 2011) and results show no difference in behavior while overtaking between the platoon vehicles maintaining short distances (distance= 11 yd) and the sample case with larger distances between vehicles (distance= 54 yd). Despite the attention that has been already given to the interaction between platoon and non-platoon drivers, further work is needed to cover the whole range of factors influencing non-platoon drivers in their interaction with a platoon and thereby cover all drivers' variables.

In recent years, concerns have been raised that drivers can react in unexpected ways to the introduction of safety measures, a phenomenon labeled "behavioral adaptation" (OECD, 1990). Skottke (2007) demonstrated a behavioral adaptation to the small time headways (THW) adopted when driving within a platoon in the form of reduced THWs in the subsequent driving period after having left a platoon. Results provide support for the idea of a shift in the frame of reference causing behavioral adaptation. Thus, after a drive with small THW a "normal" THW appears larger and drivers will therefore tend to keep a shorter "normal" THW during non-automated manual driving after leaving a platoon. Since drivers are not alone on the road, social psychological processes can be expected to influence them (De Pelsmacker and Janssens, 2007). Therefore it is conceivable that a behavioral change of platoon drivers might affect non-platoon drivers that in turn show a similar behavior.

The aim of the present study is therefore to verify the hypothesis that drivers exhibit behavioral adaptation to small THWs in platoons when driving in the vicinity of platoons, as reflected in a subsequent reduction in their adopted THWs.

However, THW is influenced by many factors and there is still a lack of detailed understanding about how these factors operate on THW (Brackstone and McDonald, 2007). Generally, they can be divided into two categories: individual and situational factors. Individual factors are stable over time and related to the personality and skills of drivers. Several studies show that THW is consistent within a driver and that each driver has a preferred THW but preferred THW differs substantially across drivers (e.g. Van Winsum and Heino, 1996). Van Winsum and Heino (1996) showed that drivers who preferred to follow the lead vehicle more closely were more efficient in the control of braking compared with drivers who preferred to follow at a longer distance. Additionally, personality factors such as sensation seeking appear to influence the choice of THW (Jonah et al., 2001). Finally, several situational factors temporarily affect THW such as conditions of reduced visibility (Van der Hulst et al., 1998), traffic (de Waard et al., 2008), time on task (Fuller, 1981) and intoxication (Smiley et al., 1985).

2   METHOD DESCRIPTION

2.1   The driving simulator

The study was conducted in the medium-fidelity TRL driving simulator. The simulator comprises a Honda Civic family hatchback right-hand drive car with a five-speed manual gearbox. The driving environment is projected at a resolution of up to 1920×1457 onto three forward screens to give the driver a 210° horizontal forward field of view. A rear screen provides a 60° rearward field of view, thus enabling normal use of all mirrors. Its engine and major mechanical systems have been replaced by an electric motion system that drives rams attached to the axles underneath each wheel. These impart limited motion in three axes (heave, pitch, and roll). A stereo sound system provides simulated engine, road, and traffic noise. The driving simulation is generated by SCANeR Studio 1.1 software (OKTAL). The driving performance data are recorded at a frequency of 20 Hz throughout each participant's drive. THW is calculated as follow: distance to the next vehicle [m] / speed [m/s], the distance to the next vehicle is measured along the road from the front of the "Ego" vehicle to the rear of the lead one.

2.2 Experimental design and procedure

Participants were asked to follow a lead vehicle on a UK's three-lane highway and specifically to stay in the same lane and not to lose track of the lead vehicle. Participants were given written instructions about the experiment at the beginning. The inside lane (left lane as driven in the UK) was dedicated to the platoons. There were four platoons in total, each platoon was driving at a smaller velocity than the lead vehicle (90 kph = 55.92 mph). Participants had not been informed about the system platoon of electronically linked vehicles to reduce the likelihood of drivers forming their own ideas about the purpose of the trial. Since the driven vehicle was required to follow the lead vehicle, participants drove in the middle lane and thereby drove in the lane adjacent to the platoons (when present). The lead car maintained a constant velocity (93 kph = 57.79 mph) - thus variations in THW can exclusively be attributed to the participants' behavior. At the beginning of the trial, the lead vehicle slowly accelerated ($1 m/s^2$) so that participants could easily catch up and start the car following task. The outside lane was occupied by fast traffic (140 kph = 87 mph) in order to make participants realize that other cars could move into the gap ahead if this was to be left too large.

The experimental design involved two factors: the time headway adopted in platoons and the treatment order. The first factor had two levels: in one condition (THW03) the time headway adopted in platoons was 0.3 sec; and in the other (THW14) the time headway was 1.4 sec. The second factor was determined by the treatment order, as a carry-over effect from one condition to the other was suspected. Participants were randomly assigned to one group. Before starting with the treatment conditions, participants were asked to accomplish a baseline (BL) in which there was no platoon on the left lane (see Table 1).

Table 1 Study design

	1st Drive	2nd Drive	3rd Drive
Group 1	Baseline	THW 03	THW 14
Group 2		THW 14	THW 03
	16 minutes	16 minutes	16 minutes

Participants drove for 16 minutes in each condition. The first minute was not included in the data analysis because it was considered as a run-in into the trial. The procedure is illustrated in Figure 1.

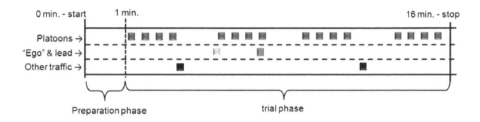

Figure 1 Schema of the study design taking the condition THW14 as example

2.3 Participants

A total of 30 drivers participated in the experiment, all of whom had previous experience with the driving simulator. The group consisted of 15 males and 15 females. Participants all held a valid driver's license for at least one year (M=20.93, SD=13.88). The age of the participants ranged between 20 and 63 (M=40.53, SD=14.06). The mileage ranged between 2000 and 56000 miles a year (M=11965.52, SD=9753.14). Participants were compensated for their time and expenses incurred by taking part in the study.

3 RESULTS AND DISCUSSION

The average THW was calculated for the three driving conditions. As can be seen in figure 2, THW was clearly higher during the baseline drive. Participants showed on average a smaller time headway in the platoon condition THW03 (M=1.90, SD=.95) than in the condition THW14 (M=2.02, SD=.89). A mixed ANOVA revealed a main effect of the driving conditions on mean THW [$F(1.14, 32.06)$=63.18, p<.000, η^2p=.69] and contrasts showed a significant difference between the two platoon conditions [$F(1, 28)$=5.85, p=.022, η^2p=.17]. Similarly, average min. THW was higher in the Baseline and lower in THW03 (M=1.00, SD=.57) than THW04 (M=1.16, SD=.72). There was a main effect of driving condition on average min. THW [$F(1.32, 37.15)$=34.74, p<.000, η^2p=.55] and contrasts showed a significant difference between the two platoon conditions [$F(1, 28)$=5.77, p=.023, η^2p=.17]. Results support the idea of an influence of traffic on drivers' THW and suggest a shift of THW toward the lead vehicle in THW03.

Concerning the factor group, figure 2 shows a negligible difference between the two groups in the Baseline demonstrating a similarity of the two groups. In mean THW, it seems that group 1, who started with THW03 keep a shorter THW than group 2 in THW14. However, an ANOVA revealed no significant differences between the two groups in mean THW and average min THW.

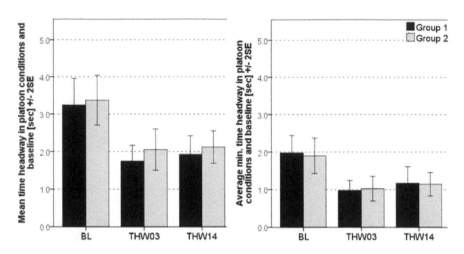

Figure 2 mean THW and average min. THW in platoon conditions and Baseline (BL) [sec.] +/- 2SE.

Plotted over time, THW demonstrates a sinusoidal variation (Figure 3). Thus, drivers' THW fluctuated even if the speed of the lead vehicle was constant. This is apparently a manifestation of the acknowledged fact that drivers can be considered to behave as an operator in a complex monitoring and control task (Michon, 1985).

Consequently, sinusoidal frequencies within the THW data become relevant for analysis, and were subsequently investigated using the Fourier transform procedure in Matlab (Stearns and Hush, 1990). When the drive starts, the participants gradually entered a loop control based on external input (e.g. environment) and outputs of the vehicle. As it takes a while for the loop control to stabilize, data at the beginning of the trial were not included in the Fourier transform. Similarly, data at the end of the trial have not been included either, as fatigue might be expected to also influence the control loop.

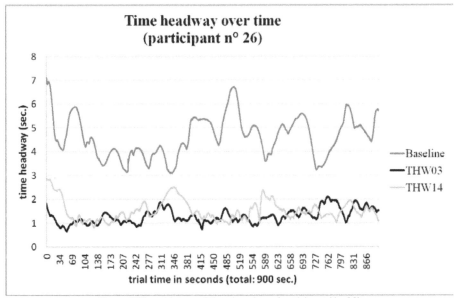

Figure 3 Time history of time headway (participant Nr. 26)

Results of the Fourier analysis show that the THW function incorporates different cluster of frequencies that are surmised to arise from two different processes:

- An open loop control represented by low THW frequencies. This control is occurring over long periods of time and is triggered by external inputs under a driver's conscious control.
- A closed loop control represented by high THW frequencies. Very small and automatic adjustments are represented here, which are expected to be unconsciously achieved by a driver.

These two loop controls can be, but are not necessarily, present in the same driving task. Figure 4 show two different frequency spectra as an example to illustrate this phenomenon. Figure a) shows that the open loop control is predominantly operating in this drive by this specific driver as low frequencies have greater amplitude. In contrast in b) there is a less clear predominance of the low frequencies, as the closed loop control seems to operate here.

Results show that there is variability between drivers concerning the loop control and there is also variability in the frequencies of the THW function between drivers and in different driving conditions.

536

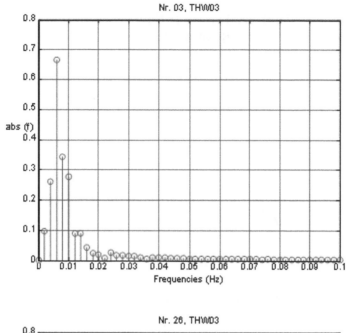

a)

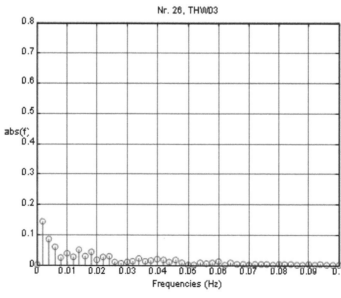

b)

Figure 4 Frequencies spectrum (x= frequencies in Hz; y= amplitudes as relative value) for participant Nr. 3 (a) and Nr. 26 (b) in THW03.

4   CONCLUSIONS

In an initial attempt to investigate effects of short THWs in platoons on the non-platoon drivers we found an effect on the mean and average minimum THW. Nevertheless, we also found that decomposing the THW function into its constituent frequencies by means of Fourier analysis provides richer information about dynamic aspects of THW and offers other perspectives than the static parameter mean THW. Indeed a driver, whose THW over time is a function composed mainly of high amplitude and small frequencies may have the same mean THW when compared to a driver, whose THW over time is a function composed mainly of small amplitudes and high frequencies. In summary, there is undeniably a masking of information if only the mean THW is considered in the analysis. Thus, two parameters emerge from a Fourier analysis of THW: the frequency spectrum; and the type of control exercised (open loop control vs. close loop control).

Results clearly show a substantial variability between drivers and the different driving situations when these more subtle parameters are considered. Further work is therefore needed to investigate how these parameters are influenced by individual and situational factors.

ACKNOWLEDGEMENTS

The research leading to these results has received funding from the European Community's Seventh Framework Program (FP7/2007-2013) under grant agreement n°238833/ ADAPTATION project. www.adaptation-itn.eu

REFERENCES

BRACKSTONE, M. & MCDONALD, M. 2007. Driver headway: How close is too close on a motorway? *Ergonomics,* 50, 1183-1195.
BUEHLER, M., IAGNEMMA, K. & SINGH, S. 2007. *The 2005 darpa grand challenge: The great robot race,* Springer Verlag.
DE PELSMACKER, P. & JANSSENS, W. 2007. The effect of norms, attitudes and habits on speeding behavior: Scale development and model building and estimation. *Accident Analysis & Prevention,* 39, 6-15.
DE WAARD, D., KRUIZINGAA, A. & BROOKHUIS, K. A. 2008. The consequences of an increase in heavy goods vehicles for passenger car drivers' mental workload and behaviour: A simulator study. *Accident Analysis and Prevention,* 40, 818–828.
FULLER, R. 1981. Determinants of time headway adopted by truck drivers. *Ergonomics,* 24, 463-474.

JONAH, B., THIESSEN, R. & AU-YEUNG, E. 2001. Sensation seeking, risky driving and behavioral adaptation. *Accident Analysis & Prevention,* 33, 679-684.

KUNZE, R., RAMAKERS, R., HENNING, K. & JESCHKE, S. 2009. Organization and Operation of Electronically Coupled Truck Platoons on German Motorways. *Intelligent Robotics and Applications,* 135-146.

LANK, C., HABERSTROH, M. & WILLE, M. 2011. Interaction of Human, Machine, and Environment in Automated Driving Systems. *Transportation Research Record: Journal of the Transportation Research Board,* 2243, 138-145.

LARBURU, M., SANCHEZ, J. & RODRIGUEZ, D. G. 2010. SAFE ROAD TRAINS FOR ENVIRONMENT: Human factors' aspects in dual mode transport systems. *ITS.* Busan.

MICHON, J. 1985. A critical view of driver behavior models: What do we know, what should we do. *Human behavior and traffic safety,* 485–520.

OECD 1990. Behavioural adaptations to changes in the road transport system. Paris.

ROBINSON, T., CHAN, E. & COELINGH, E. 2010. Operating Platoons On Public Motorways: An Introduction To The SARTRE Platooning Programme. *ITS.* Busan.

SKOTTKE, E.-M. 2007. *Automatisierter Kolonnenverkehr und adaptiertes Fahrverhalten. Untersuchung des Abstandsverhaltens zur Bewertung moeglicher kuenftiger Verkehrsszenarien.,* Hamburg, Dr. Kovac.

SMILEY, A., MOSKOWITZ, H. M. & ZIEDMAN, K. 1985. Effects of drugs on driving: Driving simulator tests of secobarbital, diazepam, marijuana, and alcohol. *In:* WALSH, J. M. (ed.) *Clinical and Behavioral Pharmacology Research Report.* Rockville: U.S. Department of Health and Human Services.

STEARNS, S. D. & HUSH, D. R. 1990. Digital signal analysis. *Englewood Cliffs, NJ, Prentice Hall, 1990, 460 p.,* 1.

VAN DER HULST, M., ROTHENGATTER, T. & MEIJMAN, T. 1998. Strategic adaptations to lack of preview in driving. *Transportation Research Part F: Psychology and Behaviour,* 1, 59-75.

VAN WINSUM, W. & HEINO, A. 1996. Choice of time-headway in car-following and the role of time-to-collision information in braking. *Ergonomics,* 39, 579-592.

WALKER, G. H., STANTON, N. A. & YOUNG, M. S. 2001. Where is Computing Driving Cars? A Technology Trajectory of Vehicle Design. *International Journal of Human Computer Interaction,* 13, 203-229.

WILLIAMS, M. Year. PROMETHEUS-The European research programme for optimising the Road Transport System in Europe. *In:* IEE Colloquium on Driver Information, 1988 London. IET, 1.

CHAPTER 55

Exploring Drivers' Compensatory Behavior when Conversing on a Mobile Device

Gregory M. Fitch & Richard J. Hanowski

Virginia Tech Transportation Institute
Blacksburg, Virginia USA
gfitch@vtti.vt.edu

ABSTRACT

We present initial results pertaining to a research program focused on investigating drivers' compensatory behavior when conversing on a mobile device. The research seeks to connect the empirical research finding that driving performance degrades when conversing on a mobile device to the naturalistic driving research finding that conversing on a mobile device is associated with a decreased risk of a safety-critical event (SCE). The current study investigated the mean speed of Commercial Motor Vehicles (CMVs) when drivers were either conversing, or not conversing, on a mobile device. The mean speed when drivers engaged in other mobile device subtasks was also explored. Speed across device use was investigated in low, moderate, and high driving task demands using selection criteria developed in Fitch & Hanowski (Fitch & Hanowski, 2011). We found that CMV drivers' mean speed does not decrease when engaged in any mobile device use. Furthermore, drivers' mean speed when conversing on a cell phone increased by 4 km/h in low driving task demands, and by 2 km/h in moderate driving task demands. CMV drivers may not decrease their speed because they are subject to greater economic pressures to reach their destination on time. They do regulate their mobile device use, however. Mobile device use was significantly less in high driving task demands.

Keywords: Distraction, Cell Phone, Portable Electronic Device, Adaptation

1   INTRODUCTION

Fatalities that arise from drivers distracted by a mobile device are tragic. Research seeks to identify the dangers of mobile device use while driving with the ultimate goal of preventing crashes. What aspects of mobile device use increase drivers' risk have received tremendous focus, primarily because there have been contradictory findings.

Early epidemiological research found that using a cell phone – be it handheld or hands-free – is associated with a quadrupling of the risk of injury and property damage crashes (McEvoy et al., 2005; Redelmeier & Tibshirani, 1997). These alarming risk estimates drew significant concern and have been used to equate mobile device use to driving with a blood alcohol concentration of 0.08 percent (Strayer, Drews, & Crouch, 2006). However, a series of Naturalistic Driving Studies (NDSs), which recorded video of the driver behavior leading up to a safety-critical event (SCE) (e.g., a crash, near-crash, crash-relevant conflict, or unintentional lane deviation), not only investigated the SCE risk associated with mobile device use, but also examined the SCE risk of specific mobile device subtasks. These studies unanimously found that SCE risk was associated with complex subtasks such as texting and dialing – because they require glances away from the forward roadway for an extended period of time - while conversing on a cell phone was not associated with an increased risk (Hickman, Hanowski, & Bocanegra, 2010; Klauer, Dingus, Neale, Sudweeks, & Ramsey, 2006; Olson, Hanowski, Hickman, & Bocanegra, 2009). These results were found for both Commercial Motor Vehicle (CMV) as well as light vehicle (LV) drivers.

The results from the NDSs and epidemiological studies have fostered debate for the following reasons. First, the SCE risk associated with mobile device use computed in the NDSs was not found to be near a fourfold increase like it was found in the epidemiological research. Dingus, Hanowski, & Klauer (2011) comment that this may be because the epidemiological studies were unable to precisely know what the driver was doing just prior to the crash. Because the timing of the cell phone records was imprecise, the driver could have completed his cell interaction up to 10 minutes prior to having to react to a crash circumstance. Furthermore, the driver could have been doing a number of tasks with the cell phone, only one of which is a conversation. Young et al. (2011) sought to resolve this discrepancy by examining the exposure metric used in the crash risk estimate. After developing a driving consistency index that measures the percentage overlap in driving times from one day to the next, Young et al. (2011) estimate that the two epidemiological studies mentioned above may have actually had less driving time in their control window on the day before the crash, than in the window just before the crash, a bias that may have overestimated the relative risk for cell phone use while driving. In applying the driving consistency index (DCI) to the relative risk estimates made in the epidemiological studies, Young et al. (2011) found that the relative risk for cell phone use lowered to about one, aligning the estimates with the findings from the NDSs.

Second, the NDSs found that just conversing on either a hand-held cell phone, hands-free cell phone, or CB radio was not associated with an increased risk of a

SCE. In some cases, it was found to be associated with a decreased risk (Hickman, Hanowski, & Bocanegra, 2010; Olson, Hanowski, Hickman, & Bocanegra, 2009). In contrast, numerous empirical studies have shown that driving performance degrades when concurrently conversing on a cell phone. The National Safety Council synthesized these findings in a white paper titled "Understanding the Distracted Brain" (National Safety Council, 2010), which highlights that conversing on a mobile device: 1) increases drivers' ratings of workload (Horrey, Lesch, & Garabet, 2009), 2) leads to missed signals and slower reaction times (Strayer & Johnston, 2001), 3) leads to poor speed maintenance and headway distance (Rakauskas, Gugerty, & Ward, 2004), 4) makes drivers look but fail to remember seeing objects (Strayer, Drews, & Johnston, 2003), 5) reduces the area that drivers scan (Atchley & Dressel, 2004; Maples, DeRosier, Hoenes, Bendure, & Moore, 2008), 6) increases reaction time to unexpected events (Caird, Willness, Steel, & Scialfa, 2008; Horrey & Wickens, 2006), 7) decreases travel speed (Cooper, Vladisavljevic, Strayer, & Martin, 2008; Young, Regan, & Hammer, 2003), 9) increases following distance (Cooper et al., 2008), 10) makes drivers less likely to change lanes (Cooper et al., 2008), 11) leads to missed navigational signage (Drews, Pasupathi, & Strayer, 2004), and 12) increases stop light violations when in an intersection dilemma zone (Horrey, Lesch, & Garabet, 2008).

To date, it has been unclear why the NDS research finding that cell phone conversation does not increase SCE risk is incongruent with the above empirical research. One possible explanation is that there is a subjectively reported impression that conversing on a cell phone improves alertness in monotonous driving conditions. Jellentrup, Metz, & Rothe (2011) investigated this issue by performing a field study and found that drivers that received regularly scheduled phone calls were more alert and awake during the conversation and up to twenty minutes afterwards. Perhaps drivers conversing on a mobile device in the NDSs benefited from this raised alertness and thus were involved in fewer SCEs. Another explanation, however, is that the NDSs did not isolate cases when the driving task demands were high. As such, the SCE risk of conversing in challenging driving conditions may have been outweighed by the lower risk of such activity in less demanding conditions. Fitch and Hanowski (2011) sought to address this issue by investigating the risk of mobile device use as a function of driving task demands. They divided existing NDS datasets into low, moderate, and high driving task demand subsets based on exogenous parameters identified in the driver workload literature (Hulse, Dingus, Fischer, & Wierwille, 1989; Nygren, 1995; Piechulla, Mayser, Gehrke, & Koenig, 2003; Schweitzer & Green, 2007; Zhang, Smith, & Witt, 2009). Table 1 presents the selection criteria used. The odds ratio and 95% confidence interval for specific mobile device subtasks was then computed separately in the low, moderate, and high driving task demand subsets. Two revealing results were found: 1) conversing on a mobile device was not associated with an increased risk in the low, moderate, or high driving task demands examined (Table 2), and 2) CMV drivers conversed on their mobile device less frequently in high driving task demands (Figure 1).

Table 1 Selection Criteria used to Create Low, Moderate, and High Driving Task Demand Subsets

Driving Task Demands	Selection Criteria
Low	• All SCEs and baseline epochs that occurred on straight, level, and dry roads during daylight conditions with no adverse weather, at Level of Service (LOS) A and that were not junction-related
Moderate	• All SCEs and baseline epochs that did not meet the "high" or "low" driving task demand selection criteria
High	• All SCEs and baseline epochs that were LOS C or greater, or • All SCEs and baseline epochs that occurred at an intersection, were intersection related, occurred on an entrance/exit ramp, a driveway, or parking lot

Table 2 Odds Ratios for CMV Drivers' Mobile Device Use in Low, Moderate,and High Driving Task Demands

Task	Low OR	Low 95% C. I.	Moderate OR	Moderate 95% C. I.	High OR	High 95% C. I.
Cell phone use (Collapsed)	0.86	0.49 - 1.53	1.25*	1.10 - 1.43	0.65*	0.46 - 0.92
Text	-	-	35.19*	12.39 – 99.96	-	-
Dial	1.83	0.44 - 7.65	9.98*	7.23 - 13.66	0.67	0.24 - 1.84
Converse on hand-held cell phone	1.37	0.69 - 2.72	1.17	0.98 - 1.41	0.86	0.51 - 1.44
Converse on hands-free cell phone	0.26	0.06 - 1.05	0.42*	0.32 - 0.55	0.51*	0.31 - 0.82
Converse on CB radio	1.5	0.47 - 4.81	0.37*	0.25 - 0.55	1.02	0.50 - 2.09

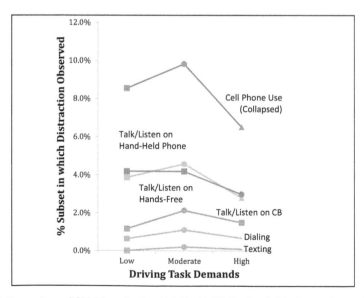

Figure 1 Percentage of CMV Samples in which the Mobile Device Subtask was observed in Low, Moderate, and High Driving Task Demands

Why would conversing on a hands-free cell phone be associated with a decreased risk in high driving task demands? Although this result is perplexing, perhaps drivers were at a decreased risk because they changed the way they drove *for the better*, and not *for the worse*. That is, the performance decrements observed in the empirical studies, such as slowing down, increasing headway, failing to scan mirrors, and not changing lanes, is evidence of drivers increasing their safety margins (Kristie, Michael, & Lee, 2008). Slowing down, increasing their range, looking forward more often, and staying in the same lane may increase the time available to respond to unfolding conflicts and avoid involvement in an SCE. Kristie et al. (2008) discuss the role of self-regulation in moderating the risk of a crash when conversing on a cell phone. They comment that many distraction studies treat drivers as passive receivers and processors of distracting information. They go on to delineate multiple ways in which drivers adjust their driving behavior in response to changing or competing task demands to maintain an adequate level of safe driving.

The objective of the current study was to determine whether the CMV drivers in the NDSs changed how they drove while conversing on a mobile device in a manner similar to the drivers observed in the empirical studies. If such changes can be observed, then a connection between the NDSs and empirical research would exist. Because this research is ongoing, this paper presents the results of the first step taken by exploring whether the drivers analyzed in Fitch & Hanowski (2011) were at a decreased SCE risk when conversing on a mobile device because they decreased their speed.

2   METHOD

This study consisted of a re-analysis of the low, moderate, and high driving task demand subsets pertaining to CMV drivers that were analyzed in Fitch & Hanowski (2011). The datasets were produced by installing video cameras (Figure 1) and vehicle sensors (e.g., accelerometers, range-sensing radars, etc.) in participants' vehicles and continuously recording their driving performance without an experimenter in the vehicle. SCEs that occurred during the 3 million vehicle miles collected were detected by algorithms and validated by data analysts who visually inspected a 6-second window of video and sensor data for each potential event. These datasets provided an instant replay of the driver, vehicle, and environment in the seconds leading to an SCE. Detailed methods pertaining to how the data were collected and originally analyzed are presented in Olson et al.'s (2009) case-control study of CMV driver distraction. It comprised 4,451 SCEs (i.e., crashes, near-crashes, crash-relevant conflicts, and unintentional lane deviations) as well as 19,888 baseline epochs that were randomly sampled based on the time driven by each driver. Flags were created whenever specific mobile device subtasks were observed in the video. Flags were created for texting on a cell phone, dialing on a cell phone, talking/listing on a hand-held cell phone, and talking/listing on a hands-free cell phone. If a flag was true for any of these subtasks, then a flag for "cell phone use (collapsed across subtasks)" was also created (so that "using a cell

544

phone" could be analyzed as a dichotomous variable). A flag was also created for talking/listing on a CB radio. Odds ratios and the respective 95% Confidence Intervals (C.I.s) were then computed for mobile device use and each subtask to investigate the association between their presence and the occurrence of an SCE. These odds ratios were computed using the samples in the low driving task demand subset, the moderate driving task demand subset, and again with the high driving task demand subset.

The speed of the subject vehicle in each event was then queried. T-tests were performed to investigate whether the mean speed when conversing on a mobile device (hand-held, hands-free, or CB radio) differed from the mean speed when not conversing on a mobile device.

3   RESULTS

Table 3 presents the mean speed of the CMVs when drivers were either engaged, or not engaged, in a specific mobile device subtask. In low driving task demands, drivers conversing on a hands-free cell phone travelled at 97.9 km/h (SE = 1.0 km/h, n = 246) while drivers not conversing on a hands-free cell phone travelled at 94.1 km/h (SE = 0.2 km/h, n = 5,623), $p = 0.0005$. Drivers engaged in cell phone use (collapsed across subtask) travelled at 96.5 km/h (SE = 0.7 km/h, n = 501) while drivers not engaged in cell phone use (collapsed) travelled at 94.1 km/h (SE = 0.2 km/h, n = 5,368), $p = 0.0008$.

In moderate driving task demands, drivers conversing on a hands-free cell phone travelled at 96.1 km/h (SE = 0.6 km/h, n = 671) while drivers not conversing on a hands-free cell phone travelled at 93.9 km/h (SE = 0.1 km/h, n = 15,556), $p = 0.0002$. Drivers conversing on a hand-held cell phone travelled at 95.2 km/h (SE = 0.5 km/h, n = 743) while drivers not conversing on a hand-held cell phone travelled at 93.9 km/h (SE = 0.1 km/h, n = 15,484), $p = 0.0164$. Drivers engaged in cell phone use (collapsed across subtask) travelled at 95.6 km/h (SE = 0.4 km/h, n = 1,586) while drivers not engaged in cell phone use (collapsed) travelled at 93.8 km/h (SE = 0.1 km/h, n = 14,641), $p < 0.0001$.

In high driving task demands, no significant differences in mean speed were observed when drivers engaged in any of the mobile device subtasks.

Table 3 Mean Truck Speed in km/h with and without Driver Performing a
Mobile Device Subtask

Driving Task Demands	Subtask	Mean Speed w/o Subtask (km/h)	Mean Speed w/ Subtask (km/h)	t-test
Low	Cell Phone Use (Collapsed)	94.1 (SE = 0.2, n = 5368)	96.5 (SE = 0.7, n = 501)	t(617.4) = -3.37, p = 0.0008*
	Texting	94.3 (SE = 0.2, n = 5868)	101.4 (SE = - , n = 1)	Not Computed
	Dialing	94.3 (SE = 0.2, n = 5831)	94 (SE = 2.4, n = 38)	t(5867) = 0.1, p = 0.9202
	Talking/Listening HH	94.3 (SE = 0.2, n = 5642)	95.3 (SE = 1, n = 227)	t(5867) = -0.95, p = 0.342
	Talking/Listening HF	94.1 (SE = 0.2, n = 5623)	97.9 (SE = 1, n = 246)	t(5867) = -3.5, p = 0.0005*
	CB Radio	94.3 (SE = 0.2, n = 5800)	94.8 (SE = 2.3, n = 69)	t(69.2) = -0.22, p = 0.8272
Moderate	Cell Phone Use (Collapsed)	93.8 (SE = 0.1, n = 14641)	95.6 (SE = 0.4, n = 1586)	t(2059.4) = -4.53, p <.0001*
	Texting	94 (SE = 0.1, n = 16193)	93.5 (SE = 2, n = 34)	t(33.3) = 0.25, p = 0.8075
	Dialing	94 (SE = 0.1, n = 16046)	94.8 (SE = 1, n = 181)	t(186.3) = -0.85, p = 0.3943
	Talking/Listening HH	93.9 (SE = 0.1, n = 15484)	95.2 (SE = 0.5, n = 743)	t(836.5) = -2.4, p = 0.0164*
	Talking/Listening HF	93.9 (SE = 0.1, n = 15556)	96.1 (SE = 0.6, n = 671)	t(741.6) = -3.72, p = 0.0002*
	CB Radio	94 (SE = 0.1, n = 15880)	94.6 (SE = 0.9, n = 347)	t(16225) = -0.75, p = 0.456
High	Cell Phone Use (Collapsed)	59.3 (SE = 0.6, n = 2064)	61.4 (SE = 2.3, n = 143)	t(2205) = -0.88, p = 0.3786
	Texting	59.4 (SE = 0.6, n = 2205)	96.2 (SE = 2, n = 2)	Not Computed
	Dialing	59.4 (SE = 0.6, n = 2192)	63.1 (SE = 6.7, n = 15)	t(2205) = -0.51, p = 0.6126
	Talking/Listening HH	59.3 (SE = 0.6, n = 2145)	61.9 (SE = 3.6, n = 62)	t(2205) = -0.72, p = 0.4736
	Talking/Listening HF	59.4 (SE = 0.6, n = 2142)	59 (SE = 3.4, n = 65)	t(2205) = 0.11, p = 0.9137
	CB Radio	59.3 (SE = 0.6, n = 2174)	64.9 (SE = 5.2, n = 33)	t(2205) = -1.13, p = 0.2597

4   DISCUSSION

This study reports the initial results of a research program focused on investigating drivers' compensatory behavior when conversing on a mobile device. Drivers' mean speed when conversing on a mobile device was compared to drivers' mean speed when not conversing on a mobile device. Comparisons were made for CMV drivers operating a heavy vehicle in low, moderate, and high driving task demands.

We found that drivers significantly increased their mean speed when conversing on a mobile device. However, this increase had an upper bound of 4 km/h in low driving task demands, and 2 km/h in moderate driving task demands. The practical implication of a 4 km/h speed increase when driving on a straight, dry, non-junction road with a low traffic density is believed to be inconsequential (Hanowski, 2011). The same might be said for a 2 km/h increase in mean speed in moderate driving task demands. Overall, the results suggest that CMV drivers do not slow down when conversing on a mobile device in low or moderate driving task demands.

Significant speed differences were not observed in high driving task demands. This was likely because there were few observations of drivers engaging in mobile device use in these conditions (Fitch & Hanowski, 2011), making inferential tests data limited. CMV drivers may have conversed less frequently in high driving task demands as a way of regulating their workload.

An explanation for why drivers did not slow down, and in specific cases sped up, when conversing on a mobile device could be that they looked less frequently at the speedometer and thus made fewer speed corrections. Another explanation could be that CMV drivers do not slow down because they are subject to economic pressures to reach their destination on time. It is worth noting that Sayer, Devonshire, and Flannagan (2007) did not observe LV drivers slow down when using a cell phone in a large-scale field operational test of integrated vehicle-based safety systems. Our results align with this finding.

The failure to observe drivers slowing down when conversing on a mobile device does not discount the compensatory behavior hypothesis. First, CMV drivers are professionals who are paid to reach a destination at a specific point in time. Driving is their primary task. The finding that they converse less often in high driving task demands suggests that they regulate their secondary task engagement when the driving conditions become challenging (Fitch & Hanowski, 2011).

Second, there is already an indication that drivers observed in NDSs look forward more readily when conversing on a cell phone and thus become more likely to detect an unfolding conflict (Hanowski, Olson, Hickman, & Bocanegra, in press; Sayer et al., 2007; Victor & Dozza, 2011). Hanowski et al. (in press) report that the same CMV drivers examined in the current study had a shorter mean eyes-off-road time when conversing on a hands-free phone compared to when not conversing on a hands-free phone. Sayer et al. (2007) found that glance durations away from the forward roadway were their shortest when LV drivers used a cellular phone. Likewise, Victor and Dozza (2011) investigated LV drivers' eye glances to the forward roadway using the 100-Car NDS data. They found that drivers engaged in cell phone conversation looked forward on average 0.05 percent more often than drivers not engaged in a cell phone conversation (mean Eyes Off Forward Roadway = 0.12% when talking, while mean Eyes Off Forward Roadway = 0.17% when not talking). Because driving is primarily a visual task, the finding that conversing on a mobile device is associated with a decreased risk can be better understood when the increase in visual attention to the forward roadway is considered.

Finally, it is possible that drivers increased their headway to the lead vehicle. Headway was not analyzed in the current study, or any other NDS to the authors' knowledge, but it will be the focus of the next step in the research program. An increased headway to a lead vehicle when conversing on a mobile device would increase drivers' safety margin and help explain the decreased SCE risk associated with conversing on a mobile device.

5 CONCLUSIONS

CMV drivers, who were observed through a large-scale NDS, were not found to slow down when conversing on a mobile device. Rather, they were observed to significantly increase their speed by 4 km/h in low driving task demands, and by 2 km/h in moderate driving task demands. Although, these speed increases are not believed to implicate driver safety in the context that they were observed. Overall, the findings indicate that CMV drivers keep speed, regardless of mobile device use, when operating a CMV in order to reach their destination on time. Regulating how much they use their mobile device may therefore be how they compensate for the increased workload.

6 LIMITATIONS

This study was exploratory and faces the following limitations. First, the driving context was binned according to broad classifications of driving task demands. The precision of the results could be increased if more specific driving contexts were isolated. Second, although the CMV subject pool is relatively homogeneous, the odds ratios reported in Fitch and Hanowski (2011) do not control for driver demographics. Finally, this study did not perform a within subject comparison of drivers' mean speed when engaged, and not engaged, in a conversation. Such tests would increase the precision of the speed estimates.

ACKNOWLEDGMENTS

This research was made possible by funding provided by the National Surface Transportation Safety Center for Excellence located at the Virginia Tech Transportation Institute in Blacksburg, Virginia. Additionally, the Federal Motor Carrier Safety Administration and the National Highway Traffic Safety Administration sponsored the collection and reduction of the CMV driver naturalistic driving data that was analyzed in this study.

REFERENCES

Atchley, P. & Dressel, J. (2004). Conversation limits the functional field of view. Human Factors: The Journal of the Human Factors and Ergonomics Society 46(4), 664-673.
Caird, J. K., Willness, C. R., Steel, P., & Scialfa, C. (2008). A meta-analysis of the effects of cell phones on driver performance. Accident Analysis & Prevention, 40(4), 1282-1293.

Cooper, J. M., Vladisavljevic, I., Strayer, D. L., & Martin, P. T. (2008). Drivers' lane changing behavior while conversing on a cell phone in a variable density simulated highway environment. Proceedings of the 87th Annual Meeting of Transportation Research Board.

Dingus, T. A., Hanowski, R. J., & Klauer, S. G. (2011). Estimating crash risk. Ergonomics in Design: The Quarterly of Human Factors Applications, 19(8), 8-12.

Drews, F. A., Pasupathi, M., & Strayer, D. L. (2004). Passenger and Cell-Phone Conversations in Simulated Driving. Proceedings of the Human Factors and Ergonomics Society 48th Annual Meeting 48, 2210-2212.

Fitch, G. M. & Hanowski, R. J. (2011). The risk of a safety-critical event associated with mobile device use as a function of driving task demands. Proceedings of the 2nd International Conference on Driver Distraction and Inattention.

Hanowski, R. J. (2011). The Naturalistic Study of Distracted Driving: Moving from Research to Practice. SAE International Journal of Commercial Vehicles, 4(1), 286-319.

Hanowski, R. J., Olson, R. L., Hickman, J. S., & Bocanegra, J. (in press). Driver distraction in commercial motor vehicle operations. In M. Regan, T. Victor & J. Lee (Eds.), Driver distraction and inattention: advances in research and countermeasures: Ashgate.

Hickman, J. S., Hanowski, R. J., & Bocanegra, J. (2010). Distraction in Commercial Trucks and Buses: Assessing Prevalence and Risk in Conjunction with Crashes and Near-Crashes. Report No. FMCSA-RRR-10-049. Washington, DC: Federal Motor Carrier Safety Administration.

Horrey, W. J., Lesch, M. F., & Garabet, A. (2008). Assessing the awareness of performance decrements in distracted drivers. Accident Analysis & Prevention, 40(2), 675-682.

Horrey, W. J., Lesch, M. F., & Garabet, A. (2009). Dissociation between driving performance and drivers' subjective estimates of performance and workload in dual-task conditions. Journal of Safety Research, 40(1), 7-12.

Horrey, W. J. & Wickens, C. D. (2006). Examining the impact of cell phone conversations on driving using meta-analytic techniques. Human Factors: The Journal of the Human Factors and Ergonomics Society, 48(1), 196-205.

Hulse, M. C., Dingus, T. A., Fischer, T., & Wierwille, W. W. (1989). The Influence of Roadway Parameters on Driver Perception of Attentional Demand Advances in Industrial Ergonomics and Safety (pp. 451-456). New-York: Taylor and Francis.

Jellentrup, N., Metz, B., & Rothe, S. (2011). Can talking on the phone keep the driver awake? Results of a field-study using telephoning as a countermeasure against fatigue while driving. Proceedings of the 2nd International Conference on Driver Distraction and Inattention.

Kristie, Y., Michael, R., & Lee, J. (2008). Factors Moderating the Impact of Distraction on Driving Performance and Safety Driver Distraction (pp. 335-351): CRC Press.

Maples, W. C., DeRosier, W., Hoenes, R., Bendure, R., & Moore, S. (2008). The effects of cell phone use on peripheral vision. Optometry - Journal of the American Optometric Association, 79(1), 36-42.

National Safety Council. (2010). Understanding the distracted brain: Why driving while using hands-free cell phones is risky behavior. from http://www.nsc.org/safety_road/Distracted_Driving/Documents/Dstrct_Drvng_White_Paper_Fnl(5-25-10).pdf

Nygren, T. E. (1995). A conjoint analysis of five factors influencing heavy vehicle drivers' perceptions of workload. Proceedings of the Human Factors and Ergonomics Society 39th Annual Meeting. 1102-1106.

Olson, R. L., Hanowski, R. J., Hickman, J. S., & Bocanegra, J. (2009). Driver Distraction in Commercial Vehicle Operations: Final Report. Contract DTMC75-07-D-00006, Task Order 3. Washington, D.C.: Federal Motor Carrier Safety Administration.

Piechulla, W., Mayser, C., Gehrke, H., & Koenig, W. (2003). Reducing drivers mental workload by means of an adaptive man–machine interface. Transportation Research - Part F, 6, 233-248.

Rakauskas, M. E., Gugerty, L. J., & Ward, N. J. (2004). Effects of naturalistic cell phone conversations on driving performance. Journal of Safety Research, 35(4), 453-464.

Sayer, J. R., Devonshire, J. M., & Flannagan, C. A. (2007). Naturalistic driving performance during secondary tasks. Proceedings of the Fourth International Driving Symposium on Human Factors in Driver Assessment, Training and Vehicle Design.

Schweitzer, J. & Green, P. A. (2007). Task Acceptability and Workload of Driving City Streets, Rural Roads, and Expressways: Ratings from Video Clips. Report No. UMTRI-2006-6. Ann Arbor, Michigan: The University of Michigan Transportation Research Institute (UMTRI).

Strayer, D. L., Drews, F. A., & Crouch, D. J. (2006). A comparison of the cell phone driver and the drunk driver. Human Factors: The Journal of the Human Factors and Ergonomics Society, 48(2), 381-391.

Strayer, D. L., Drews, F. A., & Johnston, W. A. (2003). Cell phone-induced failures of visual attention during simulated driving. Journal of Experimental Psychology: Applied, 9(1), 23 - 32.

Strayer, D. L. & Johnston, W. A. (2001). Driven to distraction: Dual-task studies of simulated driving and conversing on a cellular telephone. Psychological Science(12), 462–466.

Victor, T. & Dozza, M. (2011). Timing matters: Visual Behavior and Crash Risk in the 100-Car On-Line Data. Paper presented at the Second International Conference on Driver Distraction and Inattention. Paper # 96-0.

Young, K., Regan, M., & Hammer, M. (2003). Driver Distraction: A Review of the Literature. Report No. 206. Victoria: Monash University Accident Research Centre.

Zhang, H., Smith, M. R. H., & Witt, G. J. (2009). Driving Task Demand-Based Distraction Mitigation. In M. A. Regan, J. D. Lee & K. L. Young (Eds.), Driver Distraction: Theory Effects, and Mitigation. Boca Raton, Florida: CRC Press.

CHAPTER 56

A Safety Adapted Car Following Model for Traffic Safety Studies

Kaveh Bevrani[1], Edward Chung[2]

[1,2] Smart Transport Research Centre, Queensland University of Technology
Brisbane, Australia
k.bevrani@ qut.edu.au
edward.chung@qut.edu.au

ABSTRACT

Traffic safety studies demand more than what current micro-simulation models can provide as they presume that all drivers exhibit safe behaviors. All the microscopic traffic simulation models include a car following model. This paper highlights the limitations of the Gipps car following model ability to emulate driver behavior for safety study purposes. A safety adapted car following model based on the Gipps car following model is proposed to simulate unsafe vehicle movements, with safety indicators below critical thresholds. The modifications are based on the observations of driver behavior in real data and also psychophysical notions.

NGSIM vehicle trajectory data is used to evaluate the new model and short following headways and Time To Collision are employed to assess critical safety events within traffic flow. Risky events are extracted from available NGSIM data to evaluate the modified model against them. The results from simulation tests illustrate that the proposed model can predict the safety metrics better than the generic Gipps model. The outcome of this paper can potentially facilitate assessing and predicting traffic safety using microscopic simulation.

Keywords: Car following, safety metrics, behavioural modelling, simulation

1   INTRODUCTION

Since the development of the first car following (CF), more than 60 years ago, CF is a major sub-model for every microscopic simulation model. Early models,

Gazis–Herman–Rothery (GHR) (Gazis et al., 1959, Herman et al., 1959, Rothery, 1997) and linear models (Pipes, 1967) only model following phase. Later safety distance or collision avoidance models set up a safety distance. The Gipps model (1981), the most successful collision avoidance model, is able to switch between free flow and following states. Psychophysical models simulate driver performance as sequential control reacting to a few thresholds.

CF entails the interaction of nearby vehicles in the same lane, and so has a major role in traffic safety studies. CF model has been used in new technologies, such as Advance Vehicle Control Systems, to mimic driver actions (Brackstone and McDonald, 1999).The potential of micro-simulation to evaluate safety related factors has been recognized (Bonsall et al., 2005, Hamdar and Mahmassani, 2008, Bevrani and Chung, 2011). Though there has been small advancement in applying these models to analyze traffic safety. Some safety studies using microscopic simulation have been undertaken particularly at intersections. Archer (2005) used micro-simulation within signalized and unsignalized urban intersections, calculating some safety indicators based on the concept of "conflicts". According to the research that authors of this paper has carried out which is under review for publication, and also Lee and Peng (2005) in terms of the precise traffic metrics such as headway and speed deviation the Gipps model performs better than other CF models such as Psychophysical, Cellular Automata, GHR, and IDM. As a result the Gipps model is chosen to be further modified. The Gipps CF model is examined for its abilities for safety study purposes. The Gipps CF model is broadly used in microscopic software. Lee and Peng (2005) stated that the Gipps CF model best presents driver behavior, and simplicity of the model is another advantage.

Two of the most common types of motorways crashes are rear-end and sideswipe crashes. In real traffic situations, frequent near crash events called "surrogate safety measures" or "safety indicators" can be used instead of real crashes. Traditional statistical models use accident histories to predict current safety conditions and their disadvantages highlight that they occur relatively rarely (Bevrani and Chung, 2011). Rear-end crashes are the interest of this paper, and therefore CF will be examined rather than Lane changing models. Hence, in order to assess safety in microscopic simulation models, Time to Collision (TTC) and Short following headways are preferred, which directly show the crash risk metrics.

The following section explains the methodology of the research including; the used real data, the Gipps CF model structure and calibration process of the simulation models. The third Section demonstrates the Gipps model performance and specifically analyzes Headway and TTC reproduction. In fourth section based on the observation of Gipps model's results, modifications are proposed which lead to the development of the Safety Adapted Car Following model (SACF). The SACF model performance indicators are compared with the generic Gipps model. Additionally the improvements of the safety indicators reproduction using the SACF are illustrated. Section 5 presents the summaries and conclusions and highlights the robustness of the proposed model.

2 MODELLING AND REAL DATA

2.1 Gipps CF Model

The base CF model in this paper is Gipps CF model which proposed speed according to Equation 1. Gipps (1981) assumes that the follower car can estimate all the parameters with the exception of bn-1. He proposed to use $\hat{b}$ instead of bn-1, and did not explain what this amount in reality can be.

Equation 1:
$$v_n(t + \tau) = \min\left\{v_n(t) + 2.5a_n\tau(1 - v_n(t)/V_n)\sqrt{\frac{0.025 + v_n(t)}{V_n}}, b_n\tau + \sqrt{b_n^2\tau^2 - b_n[2[x_{n-1}(t) - s_{n-1} - x_n(t)] - v_n(t)\tau - v_{n-1}(t)^2/\hat{b}]}\right\}$$

an: Maximum acceleration(m/s²); bn Maximum deceleration(m/s²); sn: vehicle size(m); Vn: Desired speed(m/s); xn(t): vehicle location t; vn(t): the speed of vehicle n at time t((m/s); τ: the apparent reaction time(sec), $\hat{b}$: the driver of vehicle n estimation about bn-1. Gipps implements an extra safety margin to calculate a safe distance θ= τ /2(sec).

2.2 Real Data for Calibration And Validation

NGSIM program collected comprehensive vehicle trajectory information on southbound US 101, Hollywood Motorway, in Los Angeles, CA, on June 15th, 2005. The area covered was roughly 640 meters in length and included five mainline lanes. In this paper 15 minutes of data are used: from 7:50 to 8:05 a.m. including around 3000 vehicles. The major focus of this research is CF behavior, among the available trajectories, those either themselves or their leader experience lane changes were omitted. 251 trajectories remain and the rest of study is applied to these trajectories. All these 251 vehicles should have at least one TTC below 3 seconds. Vehicles should have headways less than 1.50 seconds.

The sum of UTheil's Inequality Coefficient of speed as objective function is minimized. A Genetic Algorithm (GA) is implemented to search for global solutions for minimizing the objective variables for each trajectory. By this way the Gipps model parameters are calibrated. At first instant the location of the follower and leader is reset to the real position and then by taking the leader trajectory and applying the CF model, simulation is done.

In this work main safety indicators are headway= $\frac{\Delta x}{v}$ and TTC= $\frac{\Delta x}{\Delta v}$. It is important to understand the relationship between the two. Short headways occur more frequently than short TTCs. However short headway does not necessarily end a crash, because it has to simultaneously happen with instability in the traffic where driver cannot react on time. On the other hand TTC includes both a short headway and instability. In other words a short headway eventually needs to end to a short TTC to cause any serious risk. Therefore short headways can show a potential risk of a crash in case of any instability in future, while TTC shows a present risk of crash that has occurred. Any model that can present both of these indicators with a higher level of accuracy can be a better model to be used for safety analysis.

3   GIPPS MODEL SAFETY INDICATORS REPRODUCTION

To examine the safety indicators of TTC or short headway created by the Gipps model, individual real and simulated trajectories must be looked at. For this reason a pair of vehicles in NGSIM data as sample is modeled. To determine real behavior of the Gipps model, the simulation is implemented (FIGURE 1). The question is what is the smallest TTC if the Gipps model is applied?

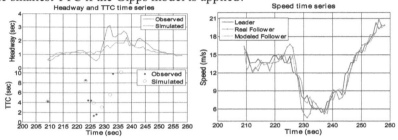

FIGURE 1 Headway, TTC and speed profile for a pair of following vehicles in the NGSIM

To explore how the Gipps model creates TTCs, a leader and a follower are artificially modeled. Leader speed is fixed to 13 m/s, while in each scenario follower speed is changed from 16 to 9 m/s. FIGURE 2-a illustrates one example of these tests in which the follower has a speed of 16 m/s. Speed is calculated simply from Equation 1 by changing spacing. It shows that the Gipps model until spacing of 20 meters, always predicts a lower speed for follower than leader to protect the safe distance. After 20 meters spacing, a higher speed is given to the follower which causes short TTCs. The minimum TTC is 6. In other words once the Gipps model is applied, the minimum feasible TTC in simulation is about 6 seconds, which is not obviously a dangerous safety event. Based on this results, apart from the gratitude of TTC it also is expected that the frequency of TTCs to be much less than reality.

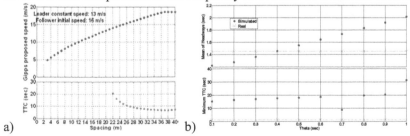

FIGURE 2 a) Speed Vs spacing and TTC b) the effects of θ on headway and TTC

FIGURE 2-b demonstrates another sensitivity analysis on a pair of following vehicles in the NGSIM data, showing the effect of the θ in the Gipps model on headways and TTC. The results indicate that θ has linear effects on the headways, while a clear relationship between θ and TTC values cannot be observed. This shows that by adjusting θ small headways can be reproduced in the Gipps model.

554

However it also shows that simply changing the safe distance parameters θ is not likely to reproduce real TTC.

The Gipps CF model is implemented for the chosen NGSIM data (FIGURE 3). Different simulation steps are tested. In simulation step 0.5 second the distribution of TTCs is most similar to a real distribution. However the frequency of TTC events is about 50% (FIGURE 3) of observed. This is in the meantime that parameter θ is set to zero. This result is far from real frequency of the risky events in the real data. FIGURE 3-b illustrates a good agreement with reality distribution. By adjusting θ to zero and simulation step 0.5 second the headway distribution is well matched. In the next step it is expected that applying a modification especially according to driver behavior, can improve the Gipps model's manner in producing TTCs.

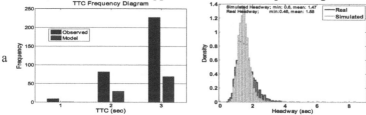

FIGURE 3 Real Vs simulated TTC frequency and headway distribution

4   THE PROPOSED MODIFIED CF MODEL

This paper serves to mimic traffic safety indicators. Applying three modifications to the generic Gipps model has led to the development of a modified CF model adapted for safety study purposes. This Safety Adapted CF Model (SACF) framework is presented in FIGURE 4.

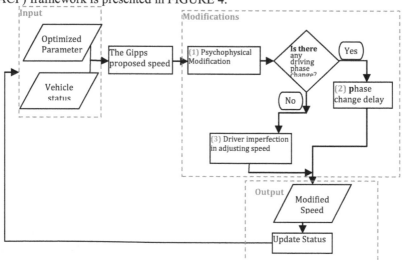

FIGURE 4 the Safety Adapted CF Model (SACF)

The details of each modification are discussed in this section. Modification one and three are global and applied to the model in every simulation step, while modification 2 is local and effects specific points of each trajectory, and effective on safety metrics reproduction. These three modifications enhance the ability to reproduce safety metrics that is close to reality. Despite Hamdar and Mahmassani (2008) work, where they modified the Gipps model by adding a random risk value to observe the number of crashes, here we do not intend to create crashes. Therefore to achieve the better reproduction of the chosen safety metrics the barriers of the Gipps model are eliminated and based on the observation of real data the adjustments are proposed. The modifications also are supported by logical notions either reported in the literature or based on reasonable assumptions. The modifications include applying human perception limitation, driver reaction delay within driving phase changes and at last Driver imperfection in adjusting speed.

4.1   First Modification: Human Perception Limitation

The modification to the Gipps model is a psychophysical concept. Examining individual trajectories (FIGURE 1) highlight that the real follower does not react to every small action of its leader. There are little fluctuations in the real follower while follower in the Gipps model reacts to every small action of its leader. Humans cannot detect any small changes of speed or spacing under certain thresholds. For this reason in this section a psychophysical notion saying humans cannot perceive speed differences under a specific threshold is applied. This modification to the Gipps model means the driver does not adjust their safe distance unless they perceive a recognizable change in speed. As a result Equation 2 is applied:

Equation 2: $v_n(t) = \begin{cases} v_n^{Gipps}(t) & if\ \frac{\Delta v_n(t) - \Delta v_n(t-1)}{\Delta v_n(t-1)} \geq 0.2 \\ v_n(t-1) & otherwise \end{cases}$

This means that if the driver's speed difference exceeds the driver cognition threshold, 0.2, Gipps should be applied to calculate the new speed otherwise the previous speed should be kept.   The modification obviously decreases the unrealistic noises in the speed profile (Figure 5) and helps to create more critical TTCs, however it improves the speed and space profile. Trajectories from Figure 5 illustrates that the modified model has fewer fluctuations than the Gipps model.

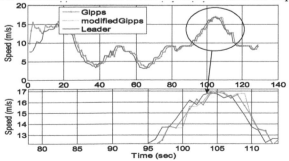

Figure 5 The Gipps model vs. the psychophysically modified Gipps CF model

4.2 Second Modification: Driver Reaction Delay

Examining the number of trajectories highlights that there is an extra delay for the follower driver when they stay in a stable driving phase for a while (either in a constant speed or in an acceleration phase) and then in a sudden change needs to turn to a deceleration phase. The follower's extra delays are illustrated In FIGURE 6 by circles. This notion is commonsense. Humans placed in a fixed situation for a period of time are not ready to react in their minimum reaction time. As a result of these extra delays the follower speed becomes much higher than leader, with the exception of some points where the driver can look a few cars downstream and can anticipate the deceleration phase earlier as indicated by the last rightmost circle in FIGURE 6. The anticipation ability is also introduced to the model by assuming that drivers are able to anticipate in a uniform distribution of ¼ of the occasions. In contrast with the real follower the Gipps CF model does not differentiate between different maneuvers and driving phases. This discussed concept is applied to the Gipps model Equation 3.

Equation 3:   If: $\Delta v_n(t) < \frac{\sum_{i=t-q}^{t-1} \delta v_n(i)}{q} - \frac{\vartheta}{1/\lambda}$

Then: Until T delay → $v_n(t) = v_n(t-1) + 0.95 \times \left(\frac{\sum_{i=t-q}^{t-1} \Delta v_n(i)}{q} \right)$

$\delta v_n(i)$: Current follower speed trend = $v_n(t) - v_n(t-1)$; and $\frac{\sum_{i=t-q}^{t-1} \Delta v_n(i)}{q}$: Average speed changes of the follower vehicle in the last few seconds; ϑ= the speed criteria which can determines if the vehicle is in deceleration time. For simulation step of half a second, $\vartheta=1$ (m/s); λ = simulation step per second; q: Number of simulation steps that the average speed changes are calculated in that period and it is equal to $\frac{T}{\lambda} * I$, where "I" is the time that represents driver previous stable status. Here it is assumed that I= 4 (sec); T: reaction time, it is assumed to be 1 second

This command causes the driver to keep the same speed status as the past status and react after one second delay. It does not apply the Gipps equations before that delay time. The commands are written independent of simulation steps and as a result in any other simulation steps the modification behaves appropriately. The impact of this command is showed in FIGURE 6. The simulated trajectory gets closer to the real follower trajectory in the time of phase changes.

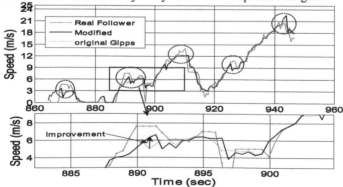

FIGURE 6 The Gipps model Vs the modified driver reaction time Gipps model

4.3   Third Modification: Driver Speed Adjustments

Driver imperfection in speed adjustment is modeled using the real NGSIM data to determine the relation between speed and spacing, and the speed a driver will chose. Yang and Peng (2010) used naturalistic CF to show that the distribution of accelerations drivers use in CF situations depending on the space gap with the vehicle in front, is different. For this reason it is assumed that: $v(t) = f(\Delta x)$, or the driver speed is a function of spacing.

Equation 4: $v^{modified}(t) = v^{Gipps}(t) + normal\ random\ number(\mu, \sigma)$

$\mu = ax^3 + bx^2 + cx^1 + d$ and $\sigma = ex^3 + fx^2 + gx^1 + h$

μ: Model's mean value, σ: Standard deviation, a, b, c ..,h: Speed variation model's coefficients, x: Spacing, The random number is diver speed adjustment imperfection factor

To model deviation of speed as a function of spacing and speed magnitude, we benefit from the NGSIM data. Five models are created for the mean values and 5 models for standard deviation values for different range of speed (0-25 m/s). In each span the speed difference between leader and follower is calculated in NGSIM data and mean and standard deviation is counted as one point. Cubic equations were found to have the minimum residuals from the real data. An example of these models can be seen in FIGURE 7-a. These 10 cubic formulas create a matrix of coefficients. The individual trajectories (FIGURE 1) identify that the Gipps model closely emulates its leader speed profile by keeping a safe distance from the leader. At an individual level FIGURE 7-b indicates that the modified model speed profile keeps a higher and more realistic difference than leader speed profile, while the generic Gipps model follows the similar speed profile as the leader.

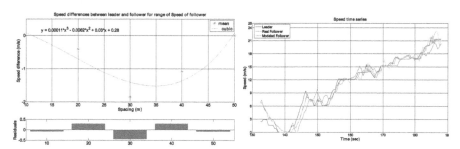

FIGURE 7 a) an example of the mean model of speed differences b) following vehicle trajectories after applying the third modification

4.4   Outcome

The simulation of NGSIM data using the new modified model showed a significant improvement in the reproduction of headways and TTCs. TABLE 1 identifies that the frequency of the critical headways and TTCs of the proposed model are very similar to reality. It also shows that the TTC estimation errors are decreased, speed profile slightly improved and remaining error metrics slightly worsens. To clearly show the robustness of the proposed SACFM, the distribution diagrams are presented at FIGURE 8-b.

TABLE 1 Root Mean Square Error and Aggregate Frequency Results

Model	RMSE		Simulated / Real Frequency	
	TTC	Headway	TTC <3sec	headway <1 sec
Generic Gipps model	1.42	0.13	0.31	0.50
SACF model	1.06	0.14	0.91	1.07

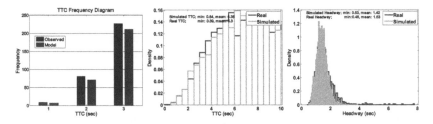

FIGURE 8 The results of the SACFM

8   CONCLUSION

Of the different CF models the Gipps CF model represents driver behavior in a fairly good manner. However, the Gipps model is still far from being functional in terms of the required level of accuracy for traffic safety studies. The Gipps CF model was examined by implementing sensitivity tests on the model parameters and tracking the safety indicators within the model structure. As a result few ideas developed to explore how the model can be improved for safety study purposes.

This research therefore modified the Gipps CF model and established a unique framework called Safety Adapted CF Model improves the Gipps CF model by applying human imperfection in information, delay in reaction and action, which can be used for investigating safety measures within motorways. The modifications specifically included: applying a human perception limitation, driver extra delay in driving phase changes and driver imperfection in adjusting speed. Additionally this paper highlighted the points that any model should address to be able to simulate Driver more realistically for the purpose of safety studies.

Simulations in aggregate and disaggregate level were used to support the validity of the SACF model. The SACF model demonstrated better capabilities to simulate unsafe vehicle movements with short TTCs and Headways. The model makes the simulated models potentially able to be used to specifically evaluate *near rear-end crashes* events. The outcomes of this research assist to proactively evaluate safety in motorways via microscopic simulation models like Aimsun (TSS-Transport Simulation Systems, 1997) which uses Gipps CF model. Eventually as the future work of this research, it should be stressed that the SACF model should be tested in a platoon of vehicles and its ability to evaluate traffic control plans needs to be examined.

ACKNOWLEDGEMENTS

Thanks to the contribution of the Academic Strategic Transport Research Alliance (ASTRA) and also to Karyn Gonano for her language advice.

REFERENCES

Archer, J. (2005) Indicators for traffic safety assessment and prediction and their application in micro-simulation modelling: A study of urban and suburban intersections. Department Of Infrastructure, Centre for Traffic Research. Stockholm, Sweden, Royal Institute of Technology.

Bevrani, K. & Chung, E. (2011) An Examination of the Microscopic Simulation Models to Identify Traffic Safety Indicators. International Journal of Intelligent Transportation Systems Research, 1-16.

Bonsall, P., Liu, R. & Young, W. (2005) Modelling safety-related driving behaviour - Impact of parameter values. Transportation Research Part A: Policy and Practice, 39, 425-444.

Brackstone, M. & Mcdonald, M. (1999) Car-following: a historical review. Transportation Research Part F: Traffic Psychology and Behaviour, 2, 181-196.

Gazis, D. C., Herman, R. & Potts, R. B. (1959) Car-Following Theory of Steady-State Traffic Flow. Operations Research, 7, 499-505.

Gipps, P. G. (1981) A behavioural car-following model for computer simulation. Transportation Research Part B: Methodological, 15, 105-111.

Hamdar, S. & Mahmassani, H. (2008) From Existing Accident-Free Car-Following Models to Colliding Vehicles: Exploration and Assessment. Transportation Research Record: Journal of the Transportation Research Board, 2088, 45-56.

Herman, R., Montroll, E. W., Potts, R. B. & Rothery, R. W. (1959) Traffic Dynamics: Analysis of Stability in Car Following. Operations Research, 7, 86-106.

Lee, K. & Peng, H. (2005) Evaluation of automotive forward collision warning and collision avoidance algorithms. Vehicle System Dynamics: International Journal of Vehicle Mechanics and Mobility, 43, 735 - 751.

Pipes, L. A. (1967) Car following models and the fundamental diagram of road traffic. Transportation Research, 1, 21-29.

Rothery, W. (1997) Car Following Models. Traffic Flow Theory.

Tss-Transport Simulation Systems (1997) Aimsun. IN BARCELÓ, J. (Ed. Barcelona.

Yang, H. H. & Peng, H. (2010) Development of an errorable car-following driver model. Vehicle System Dynamics: International Journal of Vehicle Mechanics and Mobility, 48, 751 - 773.

CHAPTER 57

Linking Behavioural Indicators to Safety: What is Safe Driving and What Is Not?

Marieke H. Martens, Roald J. van Loon, Rino F.T. Brouwer

TNO
Soesterberg, the Netherlands
roald.vanloon@tno.nl

ABSTRACT

Safety is often quantified by the number of traffic accidents and the severity of those accidents. A significant part of the causes of traffic accidents can be traced back to the driver. In recent years, there were several efforts to describe driver behaviour and its relation with risk, by evaluating the output of driving behaviour (i.e. behavioural indicators) using measures like speed, time headway, time to collision and lateral position. But what determines a situation to be critical or unsafe? This is the central research question in a new TNO project. Although there are established functions that describe the relationship between some of these indicators and the level of risk, they are still widely based on historical accident data, are usually not linked to their surroundings or to other behavioural indicators, seldom include temporal factors, and researchers have almost never reached consensus concerning critical values.

Keywords: driver behaviour, individual level, surrogate measure, safety criteria

Intuitively, everybody has an idea about what can be defined as unsafe driving. The reality however is that the limit to safe driving, especially at the individual level, is barely known. This paper will present an overview of the current links between driving behaviour and safety, and calls for joining forces and sharing expertise in order to come to a full understanding of safe and 'healthy' driving.

1   BEHAVIOURAL MODELS OF DRIVING

The idea that the driver is an important factor in driving and that the behaviour of that driver could be theoretically analysed has been around since at least the first half of the previous century (Gibson & Crooks, 1938). At least 90% of the causes of traffic accidents can be traced back to the driver (Smiley & Brookhuis, 1987), so although the relevance is clear the development of a comprehensive model of driving behaviour has progressed at a slow pace (cf. Michon, 1985). One of the proposed reasons for this lack of progress is the preoccupation in the traffic safety field with accidents (Ranney, 1994) with a focus on developing models that explain accident-causing behaviour specifically, or everyday driving, or maybe both, consequently resulting in a diluted focus altogether.

Early work on the link between driver behaviour and traffic safety constituted mainly of researching individual's accident proneness, with an emphasis on identifying common traits related with above-average crash risk. This so-called selective (Barrett, Alexander, & Forbes, 1973) or differential accident involvement (McKenna, 1982) approach largely uses postdictive research designs to develop crash predictors, a method that appears to be unique to the field of accident analysis (Barett et al., 1973 ?) and already early-on criticized due to its lack of concern for the psychological processes underlying the drivers' behaviour (Rabbitt, 1981). The problems with using accidents as a criterion for describing driving behaviour are not only methodological. For example, accident involvement lacks stability over time (e.g. Burg, 1970), which could cause statistical power problems as accidents are infrequent and hence analysis over a longer period of time is required.

Contrary to describing traffic safety as the proneness of an individual to accidents without regarding any dynamic relations would be to include the dynamic relations of the driver with the surroundings and model driving behaviour more cognitively. Michon (1985) distinguished the former (taxonomic models) from the latter (functional models) which yield more potential in understanding the complexity of the driving task. Functional models include motivational models (driver as an active decision maker) such as the risk-avoidance model (Fuller, 1988), information-processing models (driver as a passive responder) with behaviour as a consequence of automaticity, and second-generation motivational models where the driver makes decisions on either an operational, tactical or strategic level of information processing (Michon, 1985).

Interesting to mention is the distinction between the driver as a passive responder in information-processing models, and the driver as an active decision maker in motivational models, which shows similarities with the distinction between what the driver is capable of doing, driving performance, and what the driver chooses to do, driving behaviour (Shinar, 1978; Evans, 1991). Where motivational models describe what the driver actually does (e.g. driving behaviour), both the taxonomic and information-processing models focus almost entirely on the limitations of the driver's capabilities (e.g. driving performance).

2 ACCIDENTS AND TRAFFIC SAFETY

Commonly, traffic safety is defined by the number of accidents and the severity of those accidents. Traffic safety engineers use this measurement to be able to study the factors affecting the likelihood of an accident (Lord & Mannering, 2010), and ultimately qualifying transportation infrastructure and changes in that infrastructure as either safe or unsafe. The advantages of using this type of measurement are straight-forward: it is useful to identify locations in the infrastructure where a relatively high number of accidents occur in comparison to similar locations in that same infrastructure, it allows for post-hoc evaluation of design changes implemented to enhance the level of safety, and it helps us to monitor trends regarding traffic safety over time.

Besides the obvious advantages, this type of safety assessment does have its limitations. One of the larger disadvantages is intuitively the fact that this approach is reactive and does not allow for ex-ante evaluations; we need a number of accidents recorded before we can assess the level of safety. On top of that, accidents are rare so waiting for even a single accident to occur can take quite some time. Another disadvantage of this approach is the apparent impossibility to find the actual cause of the accident, despite the improvements in the field of accident analysis. For example, it is still difficult to establish the role of behavioural factors like vigilance or other cognitive impairments just prior to the accident as drivers do not easily admit to errors they make themselves using driver questionnaires (Lajunen & Summala, 2003), and eye witness testimonies are not always reliable (Memon, Mastroberardino, & Fraser, 2008). Understanding these causes is precisely what would be required to prevent an accident from happening at all.

2.1 Traffic beyond accidents

Traffic is characterized by a much broader set of events than accidents alone. Just before an accident occurs, the course to an accident could be changed somehow, for instance using an avoidance manoeuvre by one of the drivers involved. These 'near accidents' represent some drop in the level of safety but are excluded when we define traffic safety purely as a function of the number of accidents. Also, actual accidents could be subdivided based on severity, allowing for a more fine grained safety analysis. Hydén (1987) described the traffic process as a continuum of all possible events and outcomes, ranging from undisturbed passages up to fatal accidents. The advantage of analysing traffic safety using this continuum is the ability to relate more events to traffic safety besides accidents and thus a shorter period of observation is required. Next to that, it could allow traffic safety to be defined as a more quantitative measure instead of a mere qualitative safe (no accident)/unsafe (accident) distinction. By looking at the temporal and spatial proximity characteristics of all unsafe events in Hydén's continuum, we could shift from a reactive approach using accident statistics to an ex-ante prediction of the number of accidents. Svensson (1992) even argues that sometimes accident rates are predicted better by proximal characteristics than by statistics.

2.2 Surrogates for safety-related incidents

As stated before, traffic accidents are rare and it is advantageous to use traffic events beyond accidents as surrogates (Hydén, 1987; Svensson, 1992; Tarko et al., 2009). These events become even statistically rarer when controlling for confounding factors like weather, road and traffic conditions (Songchitruska & Tarko, 2006). Especially in naturalistic driving studies where high resolution data is collected but usually few accidents under similar conditions are observed, surrogates like high acceleration/deceleration, separation time between vehicles (Hayward, 1972), or lane departures (LeBlanc et al., 2006) could be useful. These surrogate measures for safety-related incidents are capable of detecting changes in driving behaviour before an accident actually occurs.

With the help of the technological advancements in the last decade, the car manufacturing industry has started equipping cars with devices to help people drive more safely (navigation systems, sensors, adaptive cruise control), encourage safer driving styles (collision avoidance warnings), and even notify drivers about specific undesirable driving events like the occurrence of sharp turns and aggressive accelerations (Lee, 2007). Besides being a real-time feedback application for drivers, these systems are a source of information helpful in studying driving behaviour continuously and trying to link specific behavioural changes to safety-related events.

3   BEHAVIOURAL INDICATORS AND TRAFFIC SAFETY

Functional models go beyond taxonomical analysis of statistics and take the cognitive driver in account as an integral part of the traffic process. Despite the evidence that other criteria for safety besides "an accident has occurred" seems more promising, taxonomic behavioural models of driving do have a link to traffic safety because it is based on actual unsafe situations (accidents), and functional models appear to be lacking such a connection.

To overcome this issue, surrogate measures (or indicators) for behavioural output are developed in an effort to distinguish unsafe from safe driving. Based on the relationship of each of these surrogate measures with accident and conflict data, and with the help of the increasing availability of research data from technological advancements, driving studies describe safety effects by reporting substantial changes in these measures. In this chapter, we shall describe some of these relationships.

3.1 Longitudinal indicators

Probably the most commonly used behavioural metric linked to safety is speed. Nilsson (1982, 1997) analysed changes in the number of accidents in Swedish traffic situations where speed regimes changed during the 70's and 80's, and developed speed-risk functions per severity accordingly. Nilsson did however include a limited number of urban speed regime changes, and Cameron & Elvik (2010) raised doubts on the applicability of the model in those lower speed zones. Based on the raw data of a meta-analysis conducted earlier by Elvik and colleagues (2004) but this time categorising by road type, they found lower predictive power estimates for the number casualties in urban areas. Consequently, they suggested that for urban roads the speed behavioural indicator alone was not sufficient to predict safety in terms of number of casualties. The relation between speed and risk is still under debate though (e.g., Hauer, 2009), and partly because of some of the methodological flaws in using accidents as a criterion discussed earlier. For instance, accidents are intuitively more severe if speed increases, making it more likely that the accident will be reported and hence analysed altogether. Also, the severity of an accident is more dependent on the difference in speed upon collision, than on the speed of either vehicle alone.

Other than speed per se, variation in speed is also linked to safety. Salusjärvi (1990) initially defined the changes in accident risk as a function of the changes in speed variability. However, Kloeden and colleagues (2001) differentiated the speed variability-risk function according to road type and found that speed variability when driving on urban roads (60 km/h) is less predictive for risk changes than it is for rural roads (80-120 km/h). Apparently, the behavioural indicator of speed variability shows similar differences in predictive power with regards to road type as speed; statistically speaking again a moderating effect of road type.

Next to indicators related to speed, a temporal indicator for driving behaviour has also been proposed. This indicator, the time-to-collision (TTC) refers to the time left before two following vehicles collide provided they continue on the same course and speed (Hayward, 1972). Although the TTC concept has been known for decades, there is no real consensus on it's critical values yet. Van der Horst and Godthelp (1989) proposed that a TTC of less than 1.5s should be considered critical; an idea backed by Svensson (1992) for urban areas but not by Maretzke and Jacob (1992) who proposed 5s as a critical cut-off value.

A different indicator related to vehicle separation time which requires less parameters to calculate is the time headway (TH), which is defined as the time that elapses between the front of the lead vehicle passing a point on the roadway and the front of the following vehicle passing that same point (e.g. Vogel, 2003). Similar to the TTC, no real consensus is established regarding critical values of TH. Evans and Wasielewski (1982) first found that drivers who maintain a TH shorter than 1s have a considerable increased change of being involved in an accident, but reported a year later that they could not establish a relationship between preferred headway and accident involvement (Evans & Wasielewski, 1983). This again could for instance be explained by a moderating effect; drivers preferring a short TH could be more alert (and hence respond faster to prevent an accident), while other (older) drivers prefer a long TH to compensate for higher response times.

In a comparison of the TH and TTC measurements, Vogel (2003) concluded that both measurements are suitable for different purposes; whereas the TTC could be used to evaluate the actual level of safety of a situation, the TH values only indicates that critical TTC values could occur.

3.2 Lateral indicators

Besides indicators measuring behaviour regarding forward movement and separation time (e.g. longitudinal indicators), we can also measure behaviour by looking at lane keeping performance (e.g. lateral indicators). The general rationale behind these indicators is that increased lane swerving and/or increased number of lane exceedances indicates reduced vehicle control and hence a reduced level of safety.

Among the most common used lateral indicators are indicators describing lane position and lane exceedances. By looking at variances in distributions of observed lane positions (e.g. standard deviation of lateral position, SDLP), we can estimate the probability of the vehicle leaving its lane (O'Hanlon et al., 1982). The number of times a vehicle actually leaves its lane or the time spent outside of its lane also tells us something about the level of risk (Wierwille et al, 1996; Östlund et al., 2004).

Another indicator called the time-to-line-crossing (TLC) differentiates between levels of risk preceding lane exceedances and describes the time it takes to reach the lane marking assuming constant steering angle and speed (Godthelp & Konings, 1981; Godthelp et al., 1984). Although a small TLC value indicates that a lane exceedance is likely to occur within a short time frame, it does not imply reduced lateral control. After all, the TLC value is recalculated with every change in speed and/or steering angle hence providing us with a limited perspective on overall lane keeping performance.

4   MEASURING INDIVIDUAL DRIVING BEHAVIOUR

Starting from the early differential accident involvement research on, focus has almost always been on evaluating groups of drivers (the elderly, the young, grouped by annual mileage, and even by marital status and socio-economic status; e.g., Arthur, Barret, & Alexander, 1991). However, driving behaviour related research questions are seldomly asked with regard to the individual driver. Although it is valid to state that for example an increase in speed with 3 km/h significantly decreases the level of safety with 17% (e.g., Nilsson function), it is still problematic to translate that behavioural change back to any changes in the safety level of an individual's traffic situation with regard to road width, speed of other vehicles in the vicinity, other behavioural changes like lateral position, and even the individual's driving performance.

In this section, we focus on differences in driving behaviour between individual drivers, and why there probably is no model of individual driving behaviour with regard to safety yet. Also, we will discuss how we could apply the usage of behavioural indicators to current modelling attempts, and what efforts already has been put into a) combining those behavioural indicators and b) defining criteria for safe driving behaviour.

4.1 The good driver and the bad driver

Many studies describe the phenomenon that a small number of drivers contribute to a large number of safety events, e.g. the good driver bad driver phenomenon. In the "100 car study", a naturalistic driving study conducted by the United States National Highway Traffic Safety Administration, a mere 10% of the drivers caused almost 35% of all events (Dingus et al., 2006).

It is based on this notion that the actuarial sciences have a long and interesting history of identifying factors predicting high levels of individual driving risk and that it has become a standard to directly relate insurance rates to specific individual risk factors (Venezian, 1981; Walters, 1981). However, those individual risk factors usually do not go much further than demographic variables and mileage. Beyond that, it is the historical accident data of the individual driver to quantify accident risk and set insurance rates. Methodologies to establish insurance rates are therefore based on accident proneness, and although the insurance industry quantifies individual risk in order to classify safety risks, they are actually taxonomically and postdictively assessing the safety risks of (groups of) individuals without regard to the driving behaviour of those individuals.

Current methods are arguably advantageous for the insurance industry as they are simple and still explain surprisingly well (e.g., Venezian, 1981). However, the same transition as we have seen in the evolution of driving behaviour models, where we have moved beyond models of accident proneness and try to incorporate the dynamic properties of driving behaviour to better predict traffic safety, would result in an opportunity for the insurance industry to set rates according to actual unsafe driving of that individual instead of likely unsafe driving in general.

4.2 Driving as a satisficing task

One of the greater issues regarding the lack of development of a comprehensive model of 'safe individual driving' is probably one of a historical nature: almost all early modellers of driving behaviour were physicists and had an engineering background. As Hancock (1999) pointed out in a comment on an earlier review of car-following models (Brackstone & McDonald, 1999a), there is a tendency to derive equations from the physical world and to try to fit these to behavioural response data. The result is that a large number of models frame the question of driving behaviour around the idea of optimal performance, consequently explaining performance variation as being some sort of noise. However, drivers primarily drive well enough to accomplish their task but do not seek continual improvement towards optimal performance (Hancock, 1999). E.g. driving is, with few exceptions, a satisficing task (Hancock & Scallen, 1999) which might explain why a large number of behavioural models have proved disappointing (for a review see Ranney, 1994) as they assume the driving task can be modelled easily and deterministically (Brackstone & McDonald, 1999a).

Hancock (1999) also argues that behavioural indicators like TTC, which are used often in car following models, are primarily perceptual signals to trigger behavioural changes for instance in order to avoid collision with the leading vehicle and that avoidance systems should be based upon replicating the human response to threat (Hancock, 1993; for the generic ecological approach see Gibson & Crooks, 1938). Hence, successfully relating one behavioural indicator (in this case TTC values) to traffic safety is only feasible if it is done with regard to other behavioural indicators as it is the combination of all of these indicators that represent human behavioural output.

4.3 Combining indicators and linking them to safety

It has become clear that different behavioural indicators can be linked to risk. The question at hand is how these behavioural indicators relate to each other, what their combined relationship with risk is, and how we can define safety criteria for them. In an effort to come up with a single estimate to assess a change in risk, researchers in the European AIDE project (adaptive integrated driver-vehicle interface) integrated some of the earlier mentioned behavioural indicators (Janssen et al, 2008). They concluded that based on the current literature the only way to obtain a single estimate was to assume independent measurements across different indicators, so that the single risk estimate was the product of a multiplication of those indicators. Although the approach of Janssen and colleagues (2008) can be lauded for its simplicity, the assumption of independence across indicators is, of course, questionable.

Even with a solid understanding of the relationship between behavioural indicators, it is still difficult to relate their integral to traffic safety. After all, for instance for driver support systems it is crucial to have a definition of 'safety' and have consensus on some sort of a cut-off value at which the system will support (or at least start to support) the driver. The same requirement stands when assessing whether driving under the influence of a specific drug is acceptable.

Brookhuis and colleagues (Brookhuis, Waard, & Fairclough, 2003; Brookhuis, 1995) reported an attempt to identify levels of driver impairment in other driving studies. For each of the different behavioural indicators, they defined two types of criteria; a) a relative cut-off value indicating a significant change in the individual driver's performance, and b) an absolute cut-off value which defined impaired driving for any driver. However, in a study to investigate drowsy driving, Brouwer and colleagues (2005) could not identify drowsy (and hence impaired) driving based on the criteria set by Brookhuis and colleagues (2003). Furthermore, they found that for different individual drivers, different behavioural indicators were better predictors for 'unsafe' driving than others.

5   WHAT IS SAFE DRIVING AND WHAT IS NOT?

We discussed different behavioural models to describe traffic safety, and discussed the concerns about using accidents as a criterion for describing driving behaviour. We also mentioned how we can assess traffic safety by looking at traffic events beyond accidents, and that behavioural output measures (behavioural indicators) have been developed in an effort to distinguish unsafe from safe driving on a macro-level. However, it appears to be difficult to link safety to driving behaviour on an individual level.

In the end, the question boils down to: what is safe driving and what is not? In the past, criteria were set in countries for a legal blood alcohol level behind the wheel, but they were not straightforward and international, since every country uses their own national limit. Apparently, clear and unambiguous criteria (e.g. cut-off values) need to be set for individual driving behaviour and, with the technological advancements in naturalistic driving research, behavioural indicators are promising in measuring precisely that. This is the central question in a recently started TNO project.

Although some effort is already taken in this direction, there is a fair deal of issues that remain under-addressed. First of all, successfully relating one behavioural indicator to traffic safety is only feasible if it is done with regard to other behavioural indicators (e.g. an increase in speed could very well be counteracted by a decrease in speed variation). Combining them in one model however might be a problem as for different individual drivers, some behavioural indicators appear to be more predictive for safety than others. Also, when trying to use behavioural output measures to qualify safe driving as an alternative to accident statistics, we need to be careful not to end up in a paradox. Current efforts to set criteria for safe driving are based on relationships linking behavioural indicators to risk: relationships often based on a function of accidents statistics to begin with.

Even though based on accident risk, we have also seen that some behavioural indicators tend to be dependent on road type in explaining safety. This issue is probably not limited to road type. Low standard deviation lateral position (SDLP) within a narrow lane may be less safe than a higher SDLP in a wide lane. Low speed in fog may not be safer than a high speed with good preview. So even if we are able to combine different behavioural indicators in to one risk factor, we still need to link it to its surroundings.

5.1 The next steps

Gaining a better understanding of individual driver variability and being able to link individual driving behaviour to safety might be the next challenge in developing some overall and coherent model for 'safe driving'. Besides the obvious scientific relevance of this model, it could help us to establish international standards for safety criteria. These criteria could have great applicability in the field of driver support systems, as these systems could be improved up to where they act specifically upon the individual driver the vehicle. The same knowledge could also be used in so-called pay-how-you-drive insurance methods, as an improvement over the current (taxonomic) models.

A lot of work still needs to be done to accomplish this. First of all, the definition of safety needs to be (re-)addressed with regard to individual driving and the usability for accident statistics in that context. We might also need to start addressing specific levels of behavioural indicators as a cue for other behavioural changes (e.g. avoidance behaviour) and also use the latter to link the former with safety. The concept that driving is a satisficing task and that we are not continuously aiming for optimal performance might find similarities in psychological research like performing under stress or habituation; it could very well be that we need to search for ranges of safe values instead of one cut-off values beyond which driving becomes unsafe.

The step from safety at a macro-level (accident statistics) towards safety at the individual level could also prove to be too big. Framing our research questions in an 'intermediate' context first could be helpful, for instance where we start looking at variances in behaviour among different groups of impaired drivers and trying to see coherences between groups at different levels of impairment. The next challenge would then be to translate those findings from the intermediate level back to the individual.

To be able to set international standards for behavioural indicators, (international) collaboration is needed between researchers either willing to share their existing data sets or their thoughts and ideas in an effort to tackle this issue. Only then could we come to a better understanding of safe and healthy driving.

REFERENCES

Arthur, W., Barret, G. V., & Alexander, R. A. (1991). Prediction of Vehicular Accident Involvement: A Meta-Analysis. *Human Performance, 4*(2), 89–105.

Barrett, G. V., Alexander, R. A., & Forbes, B. (1973). *Analysis and performance requirements for driving decision making in emergency situations.* (No. DOT HS-800 867). Washington, DC: NHTSA.

Brackstone, M., & McDonald, M. (1999a). What is the answer? And come to that, what are the questions? *Transportation Research Part F: Traffic Psychology and Behaviour, 2*(4), 221–224.

Brackstone, M., & McDonald, M. (1999b). Car-following: a historical review. *Transportation Research Part F: Traffic Psychology and Behaviour, 2*(4), 181–196.

Brookhuis, K. A. (1995). Driver Impairment Monitoring System. In M. Vallet & S. Khardi (Eds.), *Vigilance et Transports: Aspects fondamentaux, dégradation et préventation* (pp. 287–297). Lyon: Presses Universitaires de Lyon.

Brookhuis, K. A., De Waard, D., & Fairclough, S. H. (2003). Criteria for driver impairment. *Ergonomics, 46*(5), 433–445.

Brouwer, R. F. T., Duistermaat, M., Hogema, J. H., Waard, D. de, Brookhuis, K. A., & Wilschut, E. S. (2005). *Detecting drowsiness under different highway scenarios using a simple hypovigilance diagnosis system.* TNO Report TNO-DV3 2005-D009. Soesterberg: the Netherlands.

Burg, A. (1970). The stability of driving record over time. *Accident Analysis & Prevention, 2,* 57–65.

Cameron, M. H., & Elvik, R. (2010). Nilsson's Power Model connecting speed and road trauma: Applicability by road type and alternative models for urban roads. *Accident Analysis & Prevention, 42*(6), 1908–1915.

Dingus, T. A., Klauer, S. G., Neale, V. L., Petersen, A., Lee, S. E., Sudweeks, J., Perez, M. A., et al. (2006). *The 100-Car Naturalistic Driving Study: Phase II - Results of the 100-Car Field Experiment* (No. DOT HS 810 593). Washington, DC: NHTSA.

Elvik, R., Christensen, P., & Amundsen, A. (2004). *Speed and road accidents: An evaluation of the Power Model* (TOI Report No. 740/2004). Norway.

Evans, L. A. (1991). *Traffic safety and the driver*. New York, NY: Van Nostrand Reinhold.

Evans, L., & Wasielewski, P. (1982). Do accident-involved drivers exhibit riskier everyday driving behavior? *Accident Analysis & Prevention, 14*(1), 57–64.

Evans, L., & Wasielewski, P. (1983). Risky driving related to driver and vehicle characteristics. *Accident Analysis & Prevention, 15*(2), 121–136.

Fuller, R. (1988). On learning to make risky decisions. *Ergonomics, 31*, 519–526.

Gibson, J. J., & Crooks, L. E. (1938). A theoretical field-analysis of automobile driving. *American Journal of Psychology, 51*(3), 453–471.

Godthelp, H., Milgram, P., & Blaauw, G. (1984). The Development of a Time-Related Measure to Describe Driving Strategy. *Human Factors: The Journal of the Human Factors and Ergonomics Society, 26*(3), 257–268.

Godthelp, J., & Konings, H. (1981). Levels of Steering Control; Some notes on the Time-To-Line Crossing concept. Presented at the 1st European Annual Conference on Human Decision and Manual Control, Delft: Technical University Delft.

Hancock, P. . (1993). Evaluating in-vehicle collision avoidance warning systems for IVHS. *Evaluating in-vehicle collision avoidance warning systems for IVHS* (pp. 947–958). Berlin: Springer.

Hancock, P. . (1999). Is car following the real question – are equations the answer? *Transportation Research Part F: Traffic Psychology and Behaviour, 2*(4), 197–199.

Hancock, P. A., & Scallen, S. F. (1999). The Driving Question. *Transportation Human Factors, 1*(1), 47–55. doi:10.1207/sthf0101_4

Hauer, E. (2009). Speed and Safety. *Transportation Research Record: Journal of the Transportation Research Board, 2103*(-1), 10–17.

Hayward, J. (1972). Near-miss Determination Through Use of a Scale of Danger. *Highway Research Record, 384*, 22–34.

Hydén, C. (1987). The development of a method for traffic safety evaluation: The Swedish Traffic Conflicts Technique. *Bulletin* (Vol. 70). Lund: Institute för Trafikteknik.

Janssen, W., Nodari, E., Brouwer, R., Plaza, J., Ostlund, J., Keinath, A., Tofetti, A., et al. (2008). *Specification of AIDE methodology*. AIDE IST-1-507674-IP, Deliverable 2.1.4.

Kloeden, C. N., Ponte, G., & McLean, A. J. (2001). *Travelling Speed and the Risk of Crash Involvement on Rural Roads* (No. CR 204). Australian Transport Safety Bureau.

Lajunen, T., & Summala, H. (2003). Can we trust self-reports of driving? Effects of impression management on driver behaviour questionnaire responses. *Transportation Research Part F: Traffic Psychology and Behaviour, 6*(2), 97–107.

LeBlanc, D., Sayer, J., Winkler, C., Ervin, R., Bogard, S., Devonshire, J., Mefford, M., et al. (2006). *Road Departure Crash Warning System Field Operation Test: Methodology and Results* (Technical Report No. DTFH61-01-X-00053). Ann Arbor, MI: UMTRI.

Lee, J. D. (2007). Technology and teen drivers. *Journal of Safety Research, 38*(2), 203–213.

Lord, D., & Mannering, F. (2010). The statistical analysis of crash-frequency data: A review and assessment of methodological alternatives. *Transportation Research Part A: Policy and Practice, 44*(5), 291–305.

Maretzke, J., & Jacob, U. (1992). Distance warning and control as a means of increasing road safety (pp. 105–114). Presented at FISITA '92: Safety, the vehicle and the road.

McKenna, F. P. (1982). The human factor in driving accidents: An overview of approaches and problems. *Ergonomics, 25*, 867–877.

Memon, A., Mastroberardino, S., & Fraser, J. (2008). Münsterberg's legacy: What does eyewitness research tell us about the reliability of eyewitness testimony? *Applied Cognitive Psychology, 22*(6), 841–851.

Michon, J. A. (1985). A critical review of driver behavior models: What do we know, what should we do? In R. Schwing & L. A. Evans (Eds.), *Human behavior and traffic safety* (pp. 487–525). New York: Plenum Press.

Nilsson, G. (1982). The effects of speed limits on traffic accidents in Sweden. Presented at the Intern. symp. on the effects of speed limits on accidents and fuel consumption, Dublin, Paris: OECD.

Nilsson, G. (1997). *Speed management in Sweden.* Linkoping, Sweden: Swedish national road and transport research institute.

O'Hanlon, J., Haak, T., Blaauw, G., & Riemersma, J. (1982). Diazepam impairs lateral position control in highway driving. *Science, 217*(4554), 79–81. doi:10.1126/science.7089544

Östlund, J., Peters, B., Thorslund, B., Engström, J., Markkula, G., Keinath, A., Horst, D., et al. (2004). *Driving performance assessment methods and metrics* (Deliverable 2.2.5 of the AIDE project No. IST-1-507674-IP, SP2).

Rabbitt, P. M. A. (1981). Cognitive psychology needs models for changes in performance with old age. In J. Long & A. Daddeley (Eds.), *Attention and Performance IX.* Hillsdale, NJ: Lawrence Erlbaum Associates.

Ranney, T. A. (1994). Models of driving behavior: A review of their evolution. *Accident Analysis & Prevention, 26*, 733–750.

Salusjarvi, M. (1990). Finland. In G. Nilsson (Ed.), *Speed and safety: research results from the Nordic countries.* Linkoping: VTI.

Shinar, D. (1978). *Psychology on the road: The human factor in traffic safety.* New York, NY: John Wiley & Sons.

Smiley, A., & Brookhuis, K. A. (1987). Alcohol, drugs, and traffic safety. In J. A. Rothengatter & R. A. de Bruin (Eds.), *Road Users and Traffic Safety* (pp. 83–105).

Songchitruksa, P., & Tarko, A. P. (2006). The extreme value theory approach to safety estimation. *Accident Analysis & Prevention, 38*(4), 811–822.

Svensson, A. (1992). *Further development and validation of the Swedish traffic conflicts technique.* Lund: Lund University.

Tarko, A., Davis, G., Saunier, N., Sayed, T., & Washington, S. (2009). *Surrogate Measure of Safety: White Paper.* Transportation Research Board ANB20(3) Subcommittee on Surrogate Measures of Safety.

Van der Horst, R., & Godthelp, H. (1989). Measuring road user behaviour with an instrumented car and an outside-the-vehicle video observation technique. *Transportation Research Record, 1213*, 72–81.

Venezian, E. (1981). Good and Bad drivers, a Markov model of Accident Proneness (pp. 86–90). Presented at the 86st Casualty Actuarial Society Casualty Actuarial Society.

Vogel, K. (2003). A comparison of headway and time to collision as safety indicators. *Accident Analysis & Prevention, 35*(3), 427–433.

Walters, M. A. (1981). Risk Classification Standards (pp. 1–18). Presented at the 86st Casualty Actuarial Society Casualty Actuarial Society.

Wierwille, W., Tijerina, L., Kiger, S., Rockwell, T., Lauber, E., & Bittner, A. J. (1996). *Task 4: Review of Workload and Related Research* (No. DOT HS 808 467 (4)). Heavy Vehicle Driver Workload Assessment. US Department of Transportation, NHTSA.

CHAPTER 58

Incident Investigation Training Needs for the Australasian Rail Industry

Herbert C. Biggs, Tamara D. Banks, Nathan Dovan

CARRS-Q, Queensland University of Technology and the CRC for Rail Innovation
Brisbane, Australia
h.biggs@qut.edu.au

ABSTRACT

A core component for the prevention of re-occurring incidents within the rail industry is rail safety investigations. Within the current Australasian rail industry, the nature of incident investigations varies considerably between organisations. As it stands, most of the investigations are conducted by the various State Rail Operators and Regulators, with the more major investigations in Australia being conducted or overseen by the Australian Transport Safety Bureau (ATSB). Because of the varying nature of these investigations, the current training methods for rail incident investigators also vary widely. While there are several commonly accepted training courses available to investigators in Australasia, none appear to offer the breadth of development needed for a comprehensive pathway. Furthermore, it appears that no single training course covers the entire breadth of competencies required by the industry. These courses range in duration between a few days to several years, and some were run in-house while others are run by external consultants or registered training organisations. Through consultations with rail operators and regulators in Australasia, this paper will identify capabilities required for rail incident investigation and explore the current training options available for rail incident investigators.

Keywords: incident investigation, training needs analysis, rail

1   RAIL INCIDENT INVESTIGATION IN AUSTRALIA

Although the industry agrees that rail safety investigations are a core component for preventing the reoccurrence of incidents (Watson, 2004), the current approach to the training and development of investigators varies considerably between Australian organisations. A mixture of informal learning opportunities and formal courses are being implemented to train investigators. Some of the larger organisations have taken the initiative of creating their own in-house training courses by adapting varying training programs from overseas institutions and/or Australian safety agencies. Other organisations are outsourcing their training and encouraging their staff to participate in short professional development courses. While several courses are readily available in the training market, the content and quality of these courses is varied with only a few courses being aligned to recognised higher qualifications under the Australian Qualification Framework (AQF). Furthermore, it is has been recognised that none of the current courses offer the breadth of development required for a comprehensive career pathway in incident investigation (Short, Kains and Harris, 2010).

The level of experience and relevant qualifications obtained varies considerably between investigators. This relates to the severity level of the incident that they may need to investigate. More specifically, the depth of knowledge, skills and abilities required to conduct investigations is related to the level of investigation required for each incident. The rail industry typically classifies incidents in relation to five levels. Levels 1 and 2 represent severe incidents, for example when a fatality occurs. These incidents are investigated by specialist organisations such as the Australian Transport Safety Bureau and the Independent Transport Safety Regulators. These organisations have dedicated teams of investigators who specialise in conducting serious incident investigations. These investigators have extensive knowledge, skills and abilities in the area of investigation. Levels 3-5 incidents are less severe, for example a near hit may be classified as a level 5 incident. These lower level incidents are typically investigated internally by rail organisations. Some of these organisations treat incident investigation as an add-on task and require all line managers and supervisors to be able conduct basic investigations as part of their job on an as needs basis.

In the past rail organisations have devised their own training agendas to satisfy their individual needs. Historically this may have been considered appropriate as the application of safety policies and legislative acts varied between the Australian States and Territories. However with the harmonisation of the workplace health and safety acts and the appointment of a national rail safety regulator in 2012, there is currently strong support within the rail industry for the development of a unified approach to investigator training. It is believed within the industry that a national or structured learning framework would ensure that investigators develop a unified and comprehensive skill range.

With the goal of facilitating a collaborative approach to the development of a national competency framework for rail incident investigations, the Australian CRC for Rail Innovation commissioned an in-depth training needs analysis. The information presented in this Chapter details some of the findings from this analysis. In conducting this analysis the research team undertook consultations with Rail Operators and Regulators in Queensland, New South Wales, Victoria, Western Australia, South Australia and also New Zealand, to identify rail incident investigator training needs and explore current training options.

2   TRAINING NEEDS

A preliminary training needs analysis was conducted to determine what training is necessary to perform the job of rail incident investigator. As the data from this analysis was to be used to guide the development of a national training framework for rail incident investigators, the analysis was conducted at a task level. A modified-Delphi method was utilised involving a combination of qualitative and quantitative methods with a panel of subject matter experts. This modified-Delphi methodology was selected as it provided a structured technique to gain consensus from a panel of experts (Hsu and Sandford, 2007; Keeney, Hasson and McKenna, 2001; Linstone and Turoff, 1975). Three rounds of data gathering and analysis were conducted. In each round of data collection, the findings from the previous round were presented back to the subject matter experts. The subject matter experts were provided with the opportunity to confirm, add, delete or amend their responses in light of the group data. An advantage of this technique is that it allowed subject matter experts to remain anonymous to one another, thereby reducing the potential for influence or bias throughout the rounds.

In the first round, exploratory data was gathered through informal consultations with subject matter experts from six Australian organisations interested in maintaining rail safety. These organisations comprised rail transport service providers, transport investigators and statutory safety bodies. The purpose of these initial consultations was to identify what employee capabilities are needed to perform the job of rail incident investigator. A thematic analysis was conducted on the interview data to generate a list of employee capability requirements that subject matter experts perceived should be incorporated into a national training framework for rail incident investigators.

To confirm the integrity of the list generated in round one, a structured interview and focus group approach was implemented in round two of the training needs analysis. This involved presenting the preliminary list to managers and actual incident investigators from 17 organisations from Australia interested in maintaining rail safety. Subject matter experts from four of the six organisations that participated in stage one data collection also participated in the stage two data collection. Senior investigators from two rail organisations in New Zealand also expressed an interest in participating in this training needs analysis. Interviews were conducted with subject matter experts from these two additional organisations to increase the generalisabilty of the findings to include investigative training needs in both Australia and New Zealand. The participating organisations comprised: rail transport service providers; transport investigators; statutory bodies; safety boards;

regulators; transport authorities; and private companies that operate trains to transport their stock. In these interviews, subject matter experts were asked to review the list of employee capability requirements that was generated from round one and to confirm or amend each item based on what they perceived was essential attributes for rail incident investigators.

The findings from round two were then collated and several duplications were removed to yield a list of 71 employee capability requirements. This list was then piloted to enhance understanding of industry specific terms by two rail investigation experts. Based on the feedback from the pilot study, several minor terminology adjustments were implemented. The third round of data collection then involved presenting the refined list of 71 employee capability requirements via an online questionnaire asking managers and incident investigators, from the same 19 organisations that participated in round two, to weight the list. Based on information gathered in the first two rounds of data collection, the researchers believed that differences may exist between the attributes required to conduct investigations of minor incidents as compared to major incidents, therefore the researchers requested that participants complete the survey twice.

The first survey directed participants to rate the list based on perceived capability requirements for low level investigations of minor incidents. The second survey directed participants to rate the list based on perceived capability requirements for high level investigations of serious incidents. Participants rated each item using a Likert scale ranging from one representing 'not required at a minimal level' to seven representing 'full understanding/competence required'. A further option of selecting 'unsure' was also provided for each item to reduce the risk of respondents selecting a central rating if they were unfamiliar with some of the listed capabilities. Examples of the wording of items in the survey include 'interpersonal communication' and 'time management'. The questionnaire was conducted via an online survey hosted by the researchers and was accessed by participants via a unique email link. To maintain anonymity, demographic details pertaining to age and gender were not collected. Fifty-two respondents rated the list of attributes required to conduct investigations of minor incidents and 47 respondents rated the list for major incidents.

Overall the mean rating of perceived need across the 71 capabilities was greater for higher level investigations 5.9 than lower level investigations 5.3. In reviewing the rankings of the 71 capabilities it is interesting to note that the generic requirements such as being objective, using critical thinking and maintaining safety tended to be rated higher than the industry specific requirements such as knowledge of heritage railway operations and rail vehicle crash dynamics. The capabilities most perceived to require full understanding/competence for higher level as compared to low level investigations can be viewed in Table 1. In comparison the capabilities ranked the lowest, with a perceived need of only partial understanding/competence can be viewed in Table 2. The finding that generic capabilities are perceived to be more important than industry specific capabilities is consistent with previous research conducted in the Health industry by Biesma (2008).

Table 1 Means and standard deviations for capabilities most perceived to require full understanding/competence

High Level Investigation	Mean (SD)	Low level Investigation	Mean (SD)
Thorough	6.60 (.53)	Site Safety	6.30 (.91)
Objective	6.58 (.61)	Personal Safety	6.28 (.90)
Critical/Analytical Thinking	6.56 (.70)	Understanding Personal Limitations (use SME)	6.22 (.89)
Managing Investigations	6.52 (.64)	Confidentiality	6.21 (.95)
Attention to Detail	6.52 (.70)	Attention to detail	6.11 (1.07)
Confidentiality	6.52 (.75)	Objective	6.02 (1.00)
Site Safety	6.48 (.70)	Thorough	6.00 (1.06)
Problem Solving	6.48 (.70)	Recommendations	5.94 (1.44)
Compliance	6.42 (.96)	Problem Solving	5.94 (1.15)
Personal Safety	6.42 (.67)	Critical/Analytical Thinking	5.89 (1.11)

Table 2 Means and standard deviations for capabilities least perceived to require full understanding/competence

High Level Investigation	Mean (SD)	Low level Investigation	Mean (SD)
Rolling Stock Fundamentals	5.27 (1.27)	Signaling system and electrical fundamentals	4.62 (1.65)
Signaling system and electrical fundamentals	5.24 (1.38)	Statistical Analysis	4.59 (1.77)
Rail vehicle occupant safety	5.12 (1.52)	Rail ergonomics	4.53 (1.67)
Rail vehicle crash dynamics	5.12 (1.53)	Rail resource management	4.45 (1.61)
Rail resource management	5.06 (1.52)	Rail vehicle occupant safety	4.38 (1.91)
Wheel rail interface	5.06 (1.52)	Infrastructure construction and maintenance	4.36 (1.65)
Rail ergonomics	5.04 (1.36)	Level Crossings	4.34 (1.70)
Infrastructure construction and maintenance	5.02 (1.42)	Wheel rail interface	4.25 (1.69)
Level Crossings	4.94 (1.36)	Rail vehicle crash dynamics	4.00 (1.96)
Heritage Operations	4.29 (1.82)	Heritage Operations	3.66 (1.94)

Examination of the 71 variables found 49 significant differences (p<.05), with means greater on high level investigations compared to low level investigations across all requirements. This finding supports the commonly reported perception in the industry that the training provided for investigators should cover the same content regardless of whether the investigators role is assessing high or low level incidents. It also supports the industries preference for the use of a spiral framework whereby the training content is covered in increasing depth as the investigator courses become more advanced. For example when asked about training content for the differing levels, one investigator stated "the training topics covered should be the same, just more in-depth" for senior investigators.

3   CURRENT AUSTRALASIAN APPROACH TO TRAINING

Rail incident investigator training in Australasia is currently a combination of formal courses and informal on-the-job learning arrangements. All of the 17 Australian organisations and the two New Zealand organisations, described above, provided some study leave and payment of course fees for appropriate training courses to up-skill their rail investigators. In addition to providing training through formal courses, many organisations provided informal training through mentoring arrangements with experienced investigators and internal job rotations. For example, one Australian organisation requires new safety investigators to rotate through a four week orientation where they are exposed to various roles within the organisation to obtain a general overview of how rail operations are performed in their organisation. A senior investigator from this organisation reported that it was critical to "up-skill new investigators on the technical side of rail through on-the-job training." In regards to the importance of on-the-job training, a senior investigator from another organisation described how they considered members recruited from within the rail industry to also be inexperienced investigators. They suggested that regardless of whether new investigators were recruited from within or outside of the rail industry, mentoring and placements within rail investigator teams would provide valuable, relevant and practical learning experiences.

The degree to which organisations' currently train and develop their rail incident investigators through formal courses varied widely both within and between organisations. In regards to variance within organisations, it was frequently reported that the majority of incident investigations pertained to low level investigations. In response to this, most organisations had a team of several investigators capable of investigating low level investigations. Typically organisations would only have one or two senior investigators that were responsible for investigating the more serious high level investigations. To meet the different levels of knowledge and skills required for investigating low versus high level investigations, most organisations reported offering a range of training options to their staff. For example one rail organisation, through their in-house registered training organisation, provided a fundamentals training course designed to equip their staff with the basic knowledge and skills required for participating in level 5 investigations. Course attendance was compulsory for all their line managers and supervisors. Training for investigators required to undertake level 4 investigations was outsourced to an Australian University that offered a modified Diploma in criminal investigations. As this organisation did not have any staff responsible for investigating higher level incidents, they did not offer any more advanced investigator training. In comparison, a transport safety organisation that had safety officers who were responsible for investigating higher level incidents encouraged their safety officers to complete a graduate certificate course in Safety Science at an Australian University.

In regards to variance between organisations, it was observed that there is no standard approach to the level of training required or the content covered. The rail organisations appear to be operating in silos. Many registered training organisations are attempting to sell specialised investigation courses and several of the larger rail organisations have developed in-house training programs that they only offer internally to their own staff. Most courses on offer are available as short courses, for example the two day course on technical derailment or the 10 day Certificate IV in Government (Investigations) which covers evidence collection, warrants, acts and legislation. More advanced training options include, but are not limited to, the Australian Transport Safety Bureau's Diploma in Transport Safety Investigations; the CQ Australia University's Bachelors in Accident Forensic Investigations and the CQ Australia University's Masters in Accident Investigation. Across the Australian training providers, a range of courses are available to cover many aspects of incident investigation. Some of these content areas comprise: risk management; human factors; compliance investigations; systems auditing; derailment cause analysis; evidence collection; safety culture; ICAM; warrants; acts; regulations; and occupational health and safety legislation. Investigator training has also been outsourced to some international training providers. For example Australian and New Zealand investigators have completed some of the professional development short courses offered through Cranfield University in the United Kingdom.

As there is currently no nationally accepted standard or qualification framework relating to rail incident investigators, the selection and training of investigators is perceived to be challenging in the rail industry. For example one senior safety manager reported that he perceived there to be a lack of suitable training options for new or potential investigators. He commented that when he was recently recruiting for an investigator, he obtained sufficient interest in the position with 14 applications being received. However, only one of the applicants had completed a satisfactory level of training. He commented that he would like to see future training options be available to provide aspiring investigators with sufficient rail industry knowledge and investigative experience to be able to perform the role of rail incident investigator.

The desire for a national training framework was echoed across many organisations. For example a General Manager of Safety from another organisation stated "having a standardised training program would increase the pool of applicants we can recruit from. It would also allow us to outsource our investigations to an external investigator" and a senior investigator from another organisation stated "I would like to see accident investigators have more career opportunities and be more homogenous in their approach to investigations. The problem lies in getting a curriculum up and running that can help facilitate this."

4 DEVELOPMENT OF A TRAINING FRAMEWORK

The findings from this training needs analysis indicate strong industry support for the development of a national training program and capability framework. A collaborative approach to the development of industry accepted training standards for rail incident investigators has potential benefits for both rail organisations and individual investigators. In regards to organisational benefits, the development of a standard competency framework for investigators can increase efficiency and reduce training costs. The time taken for rail organisations to independently develop training curricula and resources can be very high. Smaller organisations therefore may not have the resources to train rail investigators in the same way as their larger counterparts. Furthermore the lack of clarity and choice in current training options makes it difficult for organisations to provide appropriate training for their staff. Additionally, the development of a standard competency framework would allow organisations to share their resources from a larger pool of qualified professionals. In regards to benefits for the individual investigators, the development of a nation wide capability framework would allow nationally recognised career pathways to be articulated. This may assist investigators who desire to move between organisations or across State boundaries as their prior learning can recognised at a National level. The introduction of a professional standard for railway investigators also has the potential to provide an assurance of higher consistency and standards with regards to job performance and increased credibility in the industry.

The Australian rail industry appears to understand the importance of having a uniform approach and has begun moving towards a standardised competency framework and training. Feedback obtained in the current training needs analysis indicates that the industry perceives that it is essential for training investigators to obtain training in generic capability requirements such as being objective, using critical thinking and maintaining safety. Although the rail industry did value organisation/rail specific experience, they placed less emphasis on the obtainment of these industry specific capability requirements, such as knowledge of heritage railway operations and rail vehicle crash dynamics, through course based training. The finding that generic capabilities are perceived to be more important than industry specific capabilities is consistent with previous research conducted in the United Kingdom (Biesma, 2008). It also suggests that in developing a national competency framework for rail investigators that it may be possible to utilise some generic competency units that have previously been established and accredited under the Australian Quality Framework for training. The current training needs analysis also identified a common perception in the industry that the training provided for investigators should cover the same content regardless of whether the investigators role is assessing high or low level incidents. It is suggested that a spiral model be applied where by the depth of training content increases in line with course advancement.

Given that the Australasian rail industry stands to gain substantial benefits from taking a more collaborative approach to the development of rail incident investigations, it is recommended that further training needs analysis continue to be conducted. Future research should aim to map the job capabilities identified in the current research to job competencies and articulate the relevant knowledge and skill requirements. It is recommended that industry representatives are consulted to determine: how important the competencies are to the successful performance of incident investigation and current competence levels in investigative teams. This information could be used to build on the current research findings and further guide the development of competency areas for training for current and future incident investigators.

ACKNOWLEDGMENTS

The authors are grateful to the CRC for Rail Innovation (established and supported under the Australian Government's Cooperative Research Centres program) for the funding of this research. Project No. P4.113, Project Title: Rail Incident Investigator Training and Capability Framework. The authors would also like to acknowledge the numerous investigators and senior safety personnel who generously shared their time and knowledge to increase our understanding of rail incident investigator training requirements.

REFERENCES

Biesma R.G., M. Pavlova., R. Vaatstra., G.G. Van Merode., K. Czabanowska., T. Smith, and W. Groot. 2008. Generic Versus Specific Competencies of Entry-Level Public Health Graduates: Employers' Perceptions in Poland, the UK and the Netherlands. *Advances in Health Sciences Education,* 13: 325-343.

Braithwaite, G. "Re-inventing (with wheels, wings and sails) - A new look at transport accident investigator training." Paper presented at ISASI, Gold Coast, Australia, 2004.

Hsu C, and B. A. Sandford. 2007. The Delphi technique: Making sense of consensus. *Practical Assessment, Research and Evaluation,* 12(10): 1-7.

Keeney S., F. Hasson, and H.P. McKenna. 2001. A critical review of the Delphi technique as a research methodology for nursing. *International Journal of Nursing Studies,* 38: 195-200.

Linstone H. A, and M. Turoff. 1975. *The Delphi Method: Techniques and Applications.* Addison-Wesley: Reading, MA.

Short, T., M. Kains, and R. Harris. 2010. Scoping a National Rail Incident Investigation Qualification. Brisbane: Cooperative Research Centre for Rail Innovation.

Watson S. 2004. Training rail accident investigators in UK. *Journal of Hazardous Materials,* 111: 123-129.

CHAPTER 59

Control Loss Following a Simulated Tire Tread Belt Detachment

Lee Carr, Robert Rucoba, Robert Liebbe, Amanda Duran

Carr Engineering, Inc.
Houston, USA
amanda@ceimail.com

ABSTRACT

In 2002, the University of Iowa conducted a study of driver responses to a simulated rear tread belt detachment using the National Advanced Driving Simulator (NADS). The researchers evaluated the ability of test subjects to maintain control of simulated vehicles with this simulated tire failure. They also evaluated the hypothesis that a vehicle's linear range understeer gradient would affect that ability. The vast majority of subjects that "drove" the simulated vehicle that was closest to a real-world vehicle had no difficulty maintaining control using steering and braking, a result consistent with real-world findings in other studies. Contrary to real world findings in other studies, the majority of Ranney NADS subjects failed to realize that a tire failure had occurred and they displayed no appropriate response to the event itself. Further, those subjects that did lose control did so as a result of extremely small steering commands inconsistent with the recorded remaining cornering capacities of real vehicles. In order to resolve those apparent conflicts among studies, the available data underlying the Ranney NADS study were obtained with the cooperation of the National Highway Traffic Safety Administration including drivers' steering, braking, and acceleration inputs before, during and after the simulated tread belt detachment. In this paper, these data as well as available information regarding the NADS protocols and the NADS computer modeling, including vehicle and tire parameters, were analyzed to identify sources of the conflicts when compared to real world studies. This independent analysis concludes that the conflicts arise because the Ranney NADS study methodology is significantly flawed and because its underlying data do not support the conclusions stated by its authors and should not be used as a basis for a human factors study.

1   INTRODUCTION

The National Highway Traffic Safety Administration (NHTSA) sponsored a study entitled "An Investigation of Driver Reactions to Tread Separation Scenarios in the National Advanced Driving Simulator (NADS)" in 2002 [i]. Three vehicle simulation models were created to evaluate the ability of test subjects to maintain control in the event of a rear tire tread belt detachment and to examine the influence of linear range understeer gradient on that ability. The current paper studies factors that affect all three of the vehicles and therefore does not deal with varying understeer gradient which is better suited for a vehicle dynamics forum rather than a human factors forum. The original authors of the study reached the following basic conclusions:

- *"When drivers had prior knowledge of the imminent tread separation, they were significantly less likely to sustain loss of vehicle control following the tread separation."*
- *"Driver age, the tendency to react with steering input, and higher vehicle speeds both at tread separation and more importantly at the time of initial control response, affected drivers' abilities to control the vehicle following tread separation."*

Although not a specifically stated opinion of the Ranney NADS authors, certain results of the Ranney NADS experiments have been cited by others as evidence that tread belt detachment are events likely to result in control loss.

There have been numerous evaluations conducted by other researchers of these issues because of concerns for public safety and to assist in the identification of countermeasures to minimize injury risk that may arise out of a tire failure related crash. Among these are [ii, iii, iv, v, vi, vii, viii, ix, x, xi, xii, xiii, xiv, xv, xvi, xvii, xviii, & xix]. Each of the reports and/or their underlying data, with the exception of Firestone [xiii], are in conflict with the conclusions of the NADS authors cited above and they are in direct opposition to the notion that drivers are likely to have control problems when a rear tire tread belt detachment occurs. Indeed, the NHTSA finding [xii] specifically analyzed the same NADS authors' hypotheses and subsequently rejected them. In an attempt to resolve these apparent conflicts, the underlying data and available information regarding the NADS protocols and the NADS computer modeling were obtained through the NHTSA and other sources so that an independent technical analysis could be conducted. This study presents a more complete and in depth analysis of the Ranney NADS study than the authors' previous publication [xx].

2   THE RANNEY NADS STUDY

The Ranney study used the National Advanced Driving Simulator (NADS) located at the University of Iowa. The NADS simulator is a 24 foot dome capable of replicating vehicle kinetic and dynamic motion within six degrees of freedom that contains a full vehicle cab surrounded by video monitors. The cab portion of a

1997 Jeep Cherokee was installed on the simulator for all experiments in the Ranney Study. A driver view of a roadway was provided and the results were recorded using four video cameras to record driver reaction. The simulator operated using NADS software and a modified mathematical model of the Cherokee. A soundtrack was provided to simulate the noise of a tire failure. The performance of the vehicle was based on an existing simulation model of a 1997 Jeep Cherokee that had been developed for use in past non-tire failure related studies at lower speeds. A model was developed to represent the failed tire's force and moment capacities. Data were recorded for dynamic responses including yaw rates, lateral and longitudinal accelerations, speed, brake pedal force, throttle position and steering. Data were requested from the NHTSA which maintained a partial record and those materials were provided for this analysis. NHTSA provided data in the form of two Excel spreadsheets and a video for each of the Ranney NADS drivers. Based upon our review of the Ranney report, additional materials were created in the Ranney NADS study including data and video that were not made available by these authors' request to the NHTSA and we have initiated another request to the NHTSA to find other available materials. As of this date, no additional materials have been made available.

36 subject "drivers" described as being representative of the driving population including factors for gender and age was assigned to "baseline" Vehicle 1. Each subject drove their assigned "vehicle" for approximately 5 minutes during a practice drive for purposes of orientation. Each driver participated in two additional simulation runs where they experienced an "unexpected" rear tire failure and an "expected" rear tire failure at approximately 75 miles per hour.

Tabulated datasets containing the original driver commands and vehicle dynamic response are not available at this time. Each video file contains the three simulation runs for a single subject: the practice drive, the unexpected tread belt detachment scenario, and the expected belt detachment scenario. The video is a composite of camera views including exterior of the NADS dome, forward view of the roadway and an over the shoulder angle in-cab behind the driver. The video also displayed a data overlay with pertinent information about each run, including test subject number, tread belt detachment position (right or left rear), speed/velocity, accelerator pedal position, steering wheel angle and applied brake forces. Because no other data record is available to us at this time, all data for this analysis is based on the video record.

The videos and data from "Vehicle 1" were analyzed to better understand the tire failure simulation and the drivers' reactions to the unexpected and expected tread belt detachment event. Verbal signals and body motion (looking around) were utilized to determine if and how drivers reacted during the experiments. For all subjects that lost control of "Vehicle 1" following the tread belt detachment, an examination of each video frame was performed to record all driver commands. Graphical representations of these data for selected simulations of "Vehicle 1" are contained in available upon request. At this time, only limited analysis of "Vehicle 2" and "Vehicle 3" video were performed in part because these authors have determined these vehicle models to be even less representative of any real-world

584

vehicle than the "baseline" Vehicle 1. As an added note, these authors have determined that the Vehicle 1 model was also unrepresentative of a real world vehicle and the results of the study using the Vehicle 1 model are also highly suspect. This will be discussed in greater detail below.

Review of the videos showed that the majority of drivers input little to no steering during the tread belt detachment event to maintain their intended path on the roadway. As some of the drivers looked around in response to the noise, they unintentionally input 2° to 3° of steering, likely only because they were momentarily distracted and not as an intentional control input to maintain path.

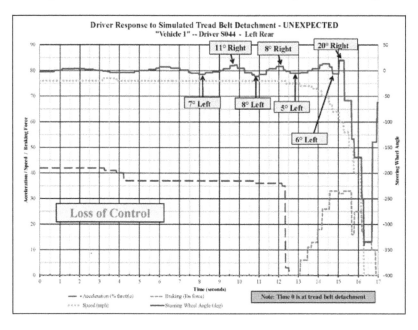

Figure 1 - Driver S044 Data for an Unexpected Tread Belt Detachment

In the "baseline" Vehicle 1 group, 36 subjects experienced the simulated tire failure as an unexpected event. Only one subject of the 36 made control demands, without observer prompting, which resulted in a "control loss". Data for this single event is shown in Figure 1. The researchers defined control loss as occurring when the "vehicle" achieved a yaw rate greater than 15 °/second. Indeed, all but this one subject continued to "drive" after the unexpected failure with no problem nor significant change in behavior until an observer advised the subjects that a failure had occurred. The simulated "failure" event itself was insufficient to instigate a significant driver reaction.

Additional data was plotted and available upon request. For example, the simulated tread belt detachment occurred at time 0 for driver S021 in "Vehicle 1", and the vehicle's speed is indicated to be approximately 76 mph which is maintained for more than 12 seconds reflecting no significant change in throttle position and no significant use of the brake. Further, the video record confirms there is no control issue with this subject for more than 12 seconds after the simulated tire failure. In that time interval an oscillatory motion of the steering wheel is observed where at least six changes in steering wheel direction are observed that are 11° or less. There is then a steering input to the right of 20° followed by an obvious counter steer to the left as the vehicle spun. During the last three steering wheel oscillations, the driver applied the brake with a substantial pedal force. These data suggest that very small steering wheel inputs of 20° or less cause the simulated vehicle to slide or spin in an uncontrolled way prompting the driver to apply the brakes.

A similar pattern is evident for the five other subjects in this group who had no problem due to the simulated tread belt detachment but did have a problem after being prompted by an observer. As is portrayed in the data, the observer waited approximately 15 seconds after the simulated "failure" to notify the subject of a tire problem. These data for the five subjects who experienced a control loss only after notification clearly confirm that the "failure" event itself was insufficient to instigate a significant driver reaction.

All of these data confirm that the "baseline" model Vehicle 1 displayed steering sensitivity, yaw damping and cornering capacity characteristics that would not accommodate even the most minor of steering and/or braking inputs. In spite of that, the majority of the subjects still did not have a problem controlling the "baseline" model Vehicle 1, despite its inability to accommodate even minor disturbances. The data further confirm that the simulated tread belt detachment event did not create forces in and of itself that caused any subject to lose control. It is noted that there are no bases offered by the Ranney NADS authors for the statement that results "...generalize to real-world driving."

Table 1 - Control loss for "Vehicle 1" for simulated tread belt detachment

	Unexpected Event				**Expected Event**			
	No Control Loss	**Loss of Control**			No Control Loss	**Loss of Control**		
		Before Notification	After Notification	Total		Before Notification	After Notification	Total
"Vehicle 1"	30	1	5	6	35	0	1	1

3   REAL WORLD EXPERIMENTS AND DATA

As stated in the Introduction, extensive real world analyses have been performed to determine the conditions, circumstances, causes and results of tire failure related events including tread belt detachments and many of these have been reported in the open literature and other public records. These have uniformly found that specific events occur. The first phase involves progressive deterioration of the tire and there is noise and vibration feedback to the driver because the tire is out of round and out of balance. Some steering correction is required because the separating tire creates drag on one vehicle side. After the separating portion becomes large enough, interaction of its end with the pavement and with the vehicle causes noise and vibration that are commonly described by occupants as an explosion or other unmistakable event. This real world circumstance calls into question the Ranney NADS result where most subjects failed to recognize the "failure's" occurrence.

In the next phase, centrifugal force causes the fragments to detach from the tire and interact with the vehicle. After this occurs, the failed tire has diminished traction. This results in diminished cornering capacity because forces and moments can no longer be balanced in response to a steering command that exceeds the reduced tractive capacity of the failed tire. These authors have conducted several hundred investigations of incidents involving tread belt detachments. Under certain conditions, passenger cars, utility vehicles, vans, light trucks, heavy trucks and buses have entered into uncontrolled slides following the tire failure. Numerous case studies are available to demonstrate how the final slide of various passenger cars from the roadway into a median where it overturned.

Detailed reconstructions confirm that often the tread belt detachment event occurs very quickly and that it occurs at a point relatively close to the point of initiation of the final left turn. Often, reliable witnesses and physical evidence confirm that the vehicle first moved somewhat to the direction of the side where the tread belt detachment occurred. This frequently observed pattern of events bears no similarity to the results observed in the Ranney NADS study. Full and partial tread separation experiments conducted on a flat track machine found that "the main effect of the delamination appears to be the effect on the tractive force, Fx, and the vertical force, Fz" [27]. It is the offset longitudinal drag (tractive force) that creates a moment which causes the vehicle to yaw slightly in the direction of the side experiencing the delamination. In addition, the author in [27] studied the tire force modeling used in the NADS simulations and concluded that "it is therefore expected that the delamination forces in actual separation events would be significantly higher than those measured in" the cited NADS references and in the simulation itself. The video records in the Ranney NADS study show that no such event occurs, confirming that the simulation of the event is not correct.

As described above, the Ranney "baseline" model Vehicle 1 results in limit sliding and oscillation with very small steering commands. This calls into question whether or not cornering capacity of the simulated vehicle faithfully reflects cornering capacity of a real vehicle with a rear tire that has experienced a tread belt detachment. Many researchers have conducted tests to measure this cornering capacity [ii, iii, iv, v, vi, vii, viii, ix, x, xi, xvi & xvii] for a wide range of vehicles. Data following SAE Recommended Practice J266 for a representative sample of utility vehicles with intact tires show a positive understeer gradient at increasing lateral accelerations. Available test data demonstrate the common choice by highway vehicle designers to provide a steering design that requires increasing steer angle to follow a given path as cornering severity increases. This increase (understeer gradient) is small at low to moderate lateral accelerations. At higher lateral accelerations, those designs typically provide a higher and increasing gradient.

Available data on the effect of a rear tire tread belt detachment demonstrate the gradient or slope of the data plot for each vehicle is smaller than with intact tires at low to moderate lateral accelerations and, in practical terms, each has approximately zero slope (neutral steer). None display a true "linear range" of gradient in this condition. The cornering limit is also reduced so that at lateral acceleration demands in the 0.4-0.6 G range, that limit may be reached. At this limit, the gradient is sharply negative for all vehicles regardless of their linear range gradient. This is true for a real Jeep Cherokee substantially the same as the Ranney NADS "baseline" Vehicle 1 model. For all tested vehicles this is a capacity that is sufficiently high that it will accommodate the demands of normal driving and will provide a margin beyond that to accommodate even unusual maneuvering [xxi]. Each vehicle will achieve a relatively high yaw rate and "spin" if the driver persists in a steering demand beyond the traction capacity of the failed rear tire regardless of the sign or magnitude of their linear gradient. These authors are unaware of any controversy that this is the mechanism by which control loss occurs in real world crashes although there are various hypotheses regarding the cause of the rear wheel's diminished traction.

All available data confirm that a vehicle, including a real Jeep Cherokee substantially the same as the Ranney NADS "baseline" Vehicle 1 model, with a tread belt detached rear tire retains a cornering capacity in the range of 0.4-0.6 g's on dry pavement surfaces. All available data confirm that a steering demand much greater than those observed in the Ranney NADS study, can be accommodated without problem. This result calls into question the fidelity of even the Ranney NADS "baseline" Vehicle 1 model where control loss is observed as very small steering angles.

4   LIKELY CAUSES OF CONFLICTS

There are a variety of anomalous results that arise from the Ranney NADS study compared to other available studies of the same hypotheses. This independent review of the Ranney NADS data and the conditions under which the study was conducted has led to the following major conclusions regarding the likely causes of those conflicts.

The signals provided to the "drivers" were not faithful reproductions of those available in a real world tread belt detachment. As a tire deteriorates before detachment, noise and vibration are accompanied by drag that requires a near constant steering input as compensation. While these may not be of sufficient magnitude to provide an imperative signal to a driver, the Ranney NADS steering data show that there was no consideration given to the steering compensation required in a likely real world event pre-detachment. When the detachment event occurs, unmistakable audible and tactile signals are created as the tire fragment strikes the vehicle and the road and its mass and shape create longitudinal and vertical accelerations unlike any common occurrence. These take the form of noise and shaking and cause at least momentary drag that requires steering correction to maintain path. The fact that most drivers failed to recognize the noise and accelerations and took no significant steering corrections to maintain a path comparable to those recorded in real detachment test events confirm that the signals were not representative.

The instruction given to prompt the "drivers" who displayed no problematic post-detachment response (35 of 36 subjects in the "baseline" Vehicle 1 unexpected event) was not one consistent with a real world event. The observers waited approximately fifteen seconds and then stated: *"You have just experienced a tread separation. Drive as you would normally with a tread separation."*

Indeed, the video shows each subject had been doing just that for approximately 15 uneventful seconds after the "failure" as simulated. Further, there is nothing in the record to suggest that any of the subjects had a basis to decide what it means to "Drive as you would normally with a tread separation" - or to have any experience whatsoever with the rare event of a tread belt detachment to guide their response.

The cornering capacity of the simulated vehicles was so small and unrealistic that very minor steering wheel demands resulted in control loss for some "drivers". Steering data for those few "drivers" of the "baseline" Vehicle 1 model who had a control loss imply that the problem occurred with very small steering angles of 10 degrees or less. This conclusion is based on tabulation of the steering data as a function of time. Tests of real world vehicles with a detached rear tire confirm that they will accommodate the steering demands that are documented in the Ranney NADS control loss events without any risk of control loss. Available data confirm that this problem likely arises because the Ranney authors chose incorrect characteristics of the failed tire in their computer modeling. These data show that the simulated tire from the Ranney NADS study would cause extreme steering sensitivity and would create a cornering limit approximately 1/3 of a real detreaded tire. With so little traction, even minor steering could result in high yaw rate or "spin". It is noted that the majority of the "drivers" of the "baseline" Vehicle 1 had no problem even with this "vehicle", a result consistent with real world findings that most drivers do not have a problem bringing their vehicle to a safe stop in the event of a tire failure. The unrealistic low modeled traction is common to all three of the simulated "vehicles" in the Ranney NADS study and is one of the likely causes of conflicting results between the Ranney NADS study and observed real world vehicle and driver performance.

The NADS "baseline" Vehicle 1 model does not represent a real world vehicle. In addition to inaccuracies in the tire normal model and tire detreaded tire model, there exists a lack of proper Vehicle 1 model validation. Limited information about the dynamic runs used to validate the Vehicle 1 model is presented in published literature [i,[xxii],[xxiii]]. No information is publicly available about validating the Vehicle 1 model at the speeds used in the Ranney NADS or at the limits of tire traction. Vehicle dynamic performance begins to change significantly beyond the linear range of understeer, and it appears that no real-world testing was conducted to validate the Vehicle 1 model at higher speeds or outside the linear range. Prior publications by NHTSA engineers provide excellent guidelines on proper protocol for validating vehicle dynamics simulations. "Validation is defined as showing that, within some specified operating range of the vehicle, a simulation's predictions of a vehicle's responses agree with the actual measured vehicle's responses."[xxiv] "A computerized, mathematical model of a physical system, such as a vehicle stability and control simulation, will be considered to be valid if, within some specific operating range of a system, a simulation's predictions of a system's responses of interest to specified input(s) agree with the actual physical system's responses to the same input(s) to within some specified level of accuracy."[xxiv] "A computerized, mathematical model of a physical system, such as a handling and control vehicle dynamics simulation, will be considered to be valid if, within some specified operating range of the physical system, a simulation's predictions of the system's responses of interest to specified input(s) agree with the actual physical system's responses to the same input(s) to within some specified level of accuracy." [xxv] Without proper documentation, testing, or validation of the Vehicle 1 model, it is impossible to evaluate the error or analyze the accuracy of the overall Ranney NADS study other than to state that the Vehicle 1 model is not likely to represent a real world vehicle and is a likely source of error in the conclusions reached in the Ranney NADS study.

The fact that "Vehicle 2" and "Vehicle 3" were purely "hypothetical in the eyes of the Ranney NADS study authors and are derivatives of a vehicle model that is clearly not representative of a real world vehicle, renders the results derived from the "Vehicle 2" and "Vehicle 3" experiments highly unrealistic as well.

In the event of a real world tire failure, the risk is presented as the cornering capacity of the failed tire is reached and exceeded. Studies, including that performed by the NHTSA of the Ranney study hypotheses [xii], conclude that limit performance is by definition non-linear. This non-linear performance is observed when driver steering commands result in a high yaw rate or "spin". The Ranney NADS "baseline" Vehicle 1 model displayed control loss with steering commands that are only a fraction of those required to reach the cornering limit of real world vehicles. The Ranney data for the 1 of 36 subjects in the "baseline" Vehicle 1 that lost control without prompting displays an oscillation in the steering consistent with a modeling instability not reflective of the real world. All available tests of tread belt detachments on real vehicles confirm that after the tread and belt have detached, all vehicles tend to resume a straight line path that requires no steering correction.

5   CONCLUSION

Researchers utilizing models or simulations in all fields must ensure that such models or simulations be as accurate and realistic as possible to achieve accurate and reliable results. The Ranney study using the NADS was a novel use of that technology and represents a promising method for observing driver response in what may be a dangerous test situation depending on that driver's choices or the nature of the problem being studied. However, the Ranney study contains flaws in its methodology and flaws in the modeling of a tire failure and subsequent vehicle handling characteristics that created an unrealistic vehicle yaw response that prevents drawing any valid conclusions regarding driver likelihood of control loss in the event of a tread belt detachment.

REFERENCES

i. Ranney, T. A, Heydinger, G. Watson, G, Salaani, K, Mazzae, E. N., and Grygier, P., "Investigation of Driver Reactions to Tread Separation Scenarios in the National Advanced Driving Simulator (NADS)", National Highway Traffic Safety Administration. Report No. DOT HS 809 523, 2003.

ii. Dickerson, C.P., Arndt, M.W., and Arndt, S. M., "Vehicle Handling with Tire Tread Separation" SAE Technical Paper 1999-01-0120, 1999. doi: 10.4271/1999-01-0120

iii. Arndt, M.W., Thorne, M. and Dickerson, C.P., "Properties of Passenger Car Tires with Tread Detachment", SAE Technical Paper 2000-01-0697, 2000. doi: 10.4271/2000-01-0697

iv. Arndt, S.M, and Arndt, M.W., "The Influence of a Rear Tire Tread Separation on a Vehicle's Stability and Control.", SAE Technical Paper 2001-06-0145, 2001.

v. Fay, R.J., Robinette, R.D., Smith, J., Flood, T., and Bolden, G., "Drag and Steering Effects from Tire Tread Belt Separation and Loss", SAE Technical Paper 1999-01-0447, 1999. doi: 10.4271/1999-01-0447

vi. Gardner, J., "The Role of Tread/Belt Detachment in Accident Causation" presented at the International Tire Exhibition and Conference 1998, Paper 27A. doi: 10.4271/2007-01-0846

vii. Klein, E.Z. and Black, T.L., "Anatomy of Accidents Following Tire Disablements", SAE Technical Paper 1999-01-0446, 1999. doi: 10.4271/1999-01-0446

viii. Tandy, D. F, Tandy, K.T., Durisek, N.J., Granat, K.J., Pascaralla, R.J., Carr, L., and Liebbe, R. L., "An Analysis of Yaw Inducing Drag Forces Imparted During Tire Tread Belt Detachments", SAE Technical Paper 2007-01-0836, 2007. doi: 10.4271/2007-01-0836

ix. Ford Motor Company, Firestone Tire Root Cause Update and Explorer Vehicle Dynamics Presentation, NHTSA/Ford Motor Company Meeting, personal communication, March 28-29, 2001.

x. Ford Motor Company, Vehicle Dynamics Update, NHTSA/Ford Motor Company Meeting, personal communication, June 13, 2001

xi. Ford Motor Company, Vehicle Dynamics Update, NHTSA/Ford Motor Company, MRA Meeting, personal communication, June 28, 2001.

xii. National Highway & Traffic Safety Administration Press Release "NHTSA Denies Firestone Request for Ford Explorer Investigation", http://www.dot.gov/affairs/nhtsa01102.htm, February 12, 2002, NHTSA 11-02.

xiii. Guenther, Dr. D. A., FTI/SEA Consulting, "Engineering Evaluation Of Explorer Directional Control", Bridgestone/Firestone submission to the NHTSA, personal communication, August 2001.

xiv. Baker, J. S., "Tire Disablements and Accidents on a High-Speed Road", Traffic Institute, Northwestern University, Evanston, Illinois, 1968.

xv. Arndt, M. W., Rosenfield, M., Arndt, S. M., and Stevens, D.C., "Analysis of Causes of Unintended Rollover During a Tread Separation Event Test", presented at Icrash 2006, Anthens, Greece, July 4-7, 2006.

xvi. Arndt, M. W., Rosenfield, M., and Arndt, S. M., "Measurement of Changes to Vehicle Handling Due To Tread-Separation-Induced Axle Tramp", SAE Technical Paper, 2006-01-1680, 2006. doi: 10.4271/2006-01-1680

xvii. Arndt, S. M., Arndt, M. W, and Rosenfield, M., "Effectiveness of Electronic Stability Control on Maintaining Yaw Stability When an SUV Has a Rear Tire Tread Separation", SAE Technical Paper, 2009-01-0436, 2009. doi: 10.427/2009-01-0436

xviii. Tandy, D. F., Neal, J., Pascarella, R., and Kalis, E., "A Technical Analysis of a Proposed Theory on Tire Tread Belt Separation-Induced Axle Tramp", SAE Technical Paper, 2011-01-0967, 2011. doi: 10.427/2011-01-0967

xix. Gilbert, M., Mueller, T., and Nirvelli, J., "The Effect of Tread-Separation on Vehicle Controllability", presented at the HIFI Tire Tech 2010 Conference, Houston, Texas, August 13, 2010

xx. Rucoba, R., Carr, L., Liebbe, R., Duran, A., "An Analysis of Driver Reactions To Tire Failures Simulated With The National Advanced Driving Simulator (NADS), presented at the Sixth International Driving Symposium on Human Factors in Driver Assessment, Training and Vehicle Design, Lake Tahoe, CA, June 27-30, 2001.

xxi. Carr, L, Liebbe, III, R., Crimeni, J., and Johnston, M., "Motor Vehicle Driver Characteristics – Crash Avoidance Behavior", SAE Technical Paper 2007-01-0449, 2007. doi: 10.4271/2007-01-0449

xxii. Salaani, M.K. and Heydinger, G.J, "Model Validation of the 1997 Jeep Cherokee for the National Advanced Driving Simulator", SAE Technical Paper 2000-01-0700, 2000. doi: 10.427/2000-01-0700

xxiii. Salaani, M.K., Guenther, D.A., and Heydinger, G.J., "Vehicle Dynamics Modeling for the National Advanced Driving Simulator of a 1997 Jeep Cherokee", SAE Technical Paper 1999-01-0121, 1999. doi: 10.4271/1999-01-0121

xxiv. Heydinger, G.J., Garrott, W.R., Chrstos, J.P., and Guenther, D.A., "A Methodology for Validating Vehicle Dynamics Simulations", SAE Technical Paper 900128, 1990. doi: 10.4271/900128

xxv. Garrott, W.R., Grygier, P.A., Chrstos, J.P., Heydinger, G.J., Salaani, K., Howe, J.G., and Guenther, D.A., "Methodology for Validating the National Advanced Driving Simulator's Vehicle Dynamics (NADSdyna)", SAE Technical Paper 970562, 1997. doi: 10.4271/970562

²7. Daws, J.W., "Force Characteristics of Tire Tread Delamination", Tire Society, Akron, OH, September 23-24, 2003

Section VII

Methods and Standards

CHAPTER 60

Textual Data Collection and Analysis for Human Factors

Michael R Neal

Fielding Graduate University
Arvada, CO USA
mneal@javakats.com

ABSTRACT

Researchers can use novel data collection and analysis methods in research and development to provide additional flexibility and increase insight. Text-based data, in particular, provide an excellent method for the collection of observations about human factors. Voice transcripts, survey comments, and several forms of media can be distilled into textual summary notes or via computer and manual transcription. Surveys can add additional comment sections to allow free-form text input. Additionally, researcher notes, comments, and detailed observations can be captured in free-form text. However, the analysis and understanding of these comments is often difficult to perform. This is especially true when the text data are voluminous and unstructured.

This paper will examine two text datasets as examples of collection and analysis methods. The first is spontaneous conversation snippets from UK lorry drivers. The second data set is a contrived questionnaire concerning a new vehicle dashboard display. Graphics, quantitative measurements, and qualitative analysis were generated, and a sentiment lens was enabled to identify positive and negative commentary. Tables were included to illustrate specific insights from the data concerning human factors issues. Finally, notes and commentary describe how to conduct textual data collection and analysis with the automated text analysis software Leximancer.

Keywords: human factors, analysis, text data, survey, conversation, Leximancer

596

1 TEXT DATA IN RESEARCH

Qualitative research involves the study of unstructured text, audio, and video data. It requires the acquisition, organization, processing, and interpretation of these data. Qualitative data are unique and can offer insights into information because the data can preserve chronology, consequences of specific events, and often detailed explanations (Miles & Huberman, 1994). Textual data are traditionally rich with context and ancillary metadata and, specifically, Straus and Corbin (1998) observed, "The analyses of these data in the social sciences is broad-based and is about 'persons' lives, lived experiences, behaviors, emotions, and feelings as well as about organizational functioning, social movements, cultural phenomena, and interactions ..." (p. 11). Clearly, using text data collected by a variety of methods could provide insight in all domains including human factors for transportation studies.

The use of software that supports qualitative methods can be an invaluable resource for researchers. Manual and computer assisted techniques for analysis of textual data have existed for decades, and transportation studies have been conducted on a wide variety of topics. Several recent studies focusing on text as a data source, which are relevant to transportation industry research, have been conducted. In *Cognitive Compatibility of Motorcyclists and Car Drivers*, Walker, Stanton, and Salmon (2011) examined how motorcyclists and car drivers interpret road situations differently. Grech, Horberry, and Smith (2002) examined accident reports in *Human Error In Maritime Operations: Analyses Of Accident Reports Using The Leximancer Tool*. Plant and Stanton (2011) performed transcript analysis manually and with an automated tool to determine effectiveness of a computer assisted approach in *A Critical Incident in the Cockpit: Analysis of a Critical Incident Interview Using the Leximancer Tool*.

Media sources often can provide data for issues. Larkin, Previte, and Luck (2008) examined press reports and citizen attitudes towards elderly drivers. Crofts Bisman (2010) conducted a literature review of journal articles seeking to interrogate accountability in accounting practices. These two studies might not exactly match studies transportation industry researchers are pursuing, but they help illustrate possibilities in publicly available text collections. Finally, since social media is increasingly becoming part of consumer discourse and can be considered a text collection, this source is worth considering for many new studies.

2 DATA COLLECTION AND DESIGN

In all industry sectors, including transportation, descriptive data collection has often been a difficult process. Surveys with specific questions can be complex to design and difficult to create sufficient interest from prospective participants. Interviews in a user experience laboratory can provide more flexibility but introduce potential researcher bias. To deal with these issues, researchers could explore innovative and ad hoc approaches to data gathering. Voice recordings, researcher notes, and informal observations can potentially be converted into text data and analyzed. Once a researcher or human factors laboratory has established techniques and acquired tools, fast and efficient processing of these data can be accomplished with regularity.

Data collection methods need to be expanded. For example, a one-question survey for an area of research is an example where free-form comments are entered into a single text entry box. Recordings of commentary can be made and transcripts extracted. Both of these collection methods provide freedom of expression and flexibility of researchers' and users' vocabulary to describe experiences.

Rapid collection of focused unstructured text data can lead to more cost-effective studies and faster turn around, provided analysis methods are sufficient. One powerful use of this technique is to perform preliminary studies on phenomena where time and cost restraints limit planning and resource allocation. These preliminary studies can give insights to guide researchers to more detailed and deliberate follow-up studies.

3   TECHNIQUES FOR FREE-TEXT DATA ANALYSIS

One consideration for the analysis of any text is the process to extract meaning. With free-form text, this can be even more challenging, as a lack of structure may obfuscate relationships and strength of data. In text analysis, manual or computer assisted coding has traditionally been directed to look for specific content. This process itself can take days or weeks depending on the complexity and fidelity of the coding dictionary required. This has been a barrier to quick analyses and for widespread adoption by many disciplines and studies. If processes and tools are developed, then more options exist and become both practical and cost-effective to pursue.

3.1 Text Analytics Software

There are several new and more automated text analysis software packages now available. For these studies, the Leximancer text analytics software program was used. Leximancer is a statistics-based tool that provides sophisticated quantitative analysis and data to assist the researcher's qualitative analysis of free-form text. Leximancer also creates its own dictionary of terms and does automatic coding. This significantly reduces researcher bias in the initial steps of a text analysis, because Leximancer serves as a discovery agent for the text dataset. The concepts generated from the text compose themes for further explorations.

4.0 DATA ANALYSIS EXAMPLES

Two different datasets have been chosen for examples of text analysis. The first is UK lorry driver Shout Out dataset, which captured observations from lorry drivers while they were driving. The second dataset is a contrived survey for human factors review of a new dashboard layout for a vehicle. The survey responses were organized by age and years driving experience, providing additional structure and demographic data.

4.1 Data Set 1: Shout Out Data

The first dataset contains data collected in the United Kingdom from "shout outs" by commercial truck drivers. The data is described as follows (Walker, personal communication, February 2012):

The experiment is based upon 'real-world' driving in which individuals use their own vehicles around a defined course on public roads and were required to provide a verbal commentary as they drove. The commentary required drivers to 'speak out loud' about what information they were taking from the environment, what they intended to do with it, and to explain their driving actions.

This dataset was selected as an example since the data are particularly unorganized, unstructured, and challenging to interpret. For example, there is no specific or structured meta-data describing the person, location, situation, time of day, or other contextual factors. The data are raw text segments converted from audio recordings.

In the case of this shout out data, Table 1 provides a few examples for illustration.

Table 1 Examples from UK Shout Out dataset

	Examples from Shout Out Data
1	surrounded by lorries
2	um, I'm in second lane
3	ur, it's all clear
4	indicate again and pull across. ok mirrors, two mirrors check inside lane [something about] the car ahead just gonna sit back now to watch this Astra make sure no cars in the second lane are gonna pull across
5	there againàyou

This is particularly challenging data to analyze for several reasons. First, the syntax, spelling, voice-to-text translation errors (e.g., Example 5), and vernacular make for sparse and disjointed data. Second, there does not appear to be a clear context or continuity between statements. Traditional qualitative methods would require significant researcher interpretation, and theme detection would prove difficult if the dataset were larger. However, there is inherent value in this type of unfiltered data. Additionally if data this sparse and unclear can be used by researchers to add value and understanding to a situation, particularly human issues, it could suggest a number of possibilities for quick and illuminating studies previously not considered.

The Shout Out transcribed text was placed in a .txt file and loaded into

Leximancer. There were 331 statements composed of 2,266 words for an average of 6.8 words per statement. The shortest statements were single words (e.g., *indicating*, *accelerating*), and the longest statement was 17 words. There was no attempt to combine multiline entries that may have been interpreted to form a single statement. No data cleansing was performed and terms such as *um*, *ur* were not added to the list of ignored words. It is common for researchers to perform some data cleansing such as removing translation errors, misspellings. However, this data was not altered for demonstration purposes to show an extreme case.

The Shout Out dataset is an excellent example for researchers to consider given its characteristics. While a detailed study attempting to draw reproducible conclusions concerning certain phenomenon may require more structured data collection and processing methods, the ability to take such an unstructured collection of text comments enables new potential applications in research.

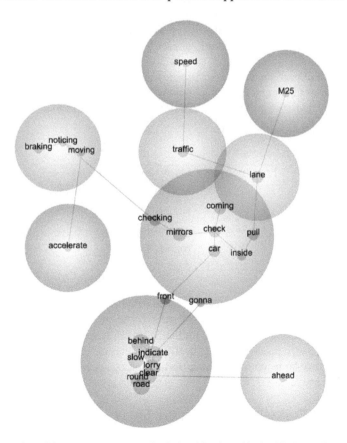

Figure 1 Illustration of themes, concepts, and relationships found in the UK Shout Out dataset

Figure 1 shows a Leximancer concept map of the UK Shout Out dataset. The large circles denote themes, or clusters, of related concepts. The black text with dots represents actual concepts. These concepts have been found automatically by

Leximancer and illustrate major topics of discussion. Concepts mentioned together often in the text are located near each other on the map, often within the same theme circle.

The bottom left theme circle shows a tight cluster of concepts to do with safely maneuvering around obstacles (i.e., *behind, indicate, slow, lorry, clear, round, road*). The larger theme circle directly to the upper right includes concepts that denote discussion around visibility and mirrors (i.e., *coming, checking, mirrors, car, pull, inside*). This type of visual representation is excellent as a first pass view into the data to provide a general indication of topics, how they relate, and what is discussed most. Researchers can then use this information to explore and review the original source text segments efficiently.

While there are numerous quantitative reports produced by a software tool such as Leximancer, one interesting and typically meaningful measure is the concept frequency ranking. This is not a word frequency, as a Leximancer concept is composed of terms that travel near each other throughout the text. This means that the software may code concepts in comments where the keyword is not present as long as sufficient indicative terminology exists there. Thus, concept counts usually outnumber simple word frequency counts and are more meaningful. Leximancer uses co-occurrences of the concepts as a measure of relationship between them.

Table 2 shows the top five concept counts for the UK Shout Out dataset, as well as the most related concept for each. Recall that the data are sparse, so these counts are low.

Table 2 Top concept counts for UK Shout Out dataset

Concept	Frequency	Most Related Concept
lane	30	M25 (3)
behind	26	Round (9)
car	24	mirrors (4)
road	20	round (9)
front	19	road (9)

While this dataset was small, interesting observations can be made. The major topics of conversation from the lorry drivers relate to the two largest themes of maneuvering between lanes (perhaps on the M25). However, since Leximancer is a statistical tool, large datasets will typically be more suited for analysis. Both the facility to analyze large amounts of data and increased accuracy from larger datasets should encourage researchers to collect as much data as possible.

4.2 Data Set 2: Dashboard Display Survey

The second data set is a contrived survey of drivers asked to comment on a new dashboard display for their vehicles. This display is code-named the ALD12. The survey was open-ended, and respondents were asked to enter comments on what they liked and disliked about the new dashboard. Demographic data was collected, including the drivers' age and years of experience. This researcher fabricated the responses, ranging from "looks fine" to multiple sentence responses containing specific feedback. There were one hundred responses created with a variety of ages and years experiences. The years of experience was roughly correlated with a starting age of 25 years for a driver.

The data was spellchecked to improve quality, although that is not required for a tool such as Leximancer. The data were compiled in an Excel spreadsheet with a column each for Age, Years Experience, and Comments. Table 3 shows five example comments from this data.

Table 3 Example fictitious data from dashboard survey

Age	Yrs Exp.	Comment
25	1	Looks fine
30	4	emergency marker location excellent. The colors are clear and provide a good differentiation.
44	10	emergency marker location excellent. ALD12 is much better than HUD4. HUD4 would often break and not properly display the emergency light indicators. ALD12 should be implemented immediately. I did not have a chance to look at the HUD5 but my mates said it was pretty poor.
55	18	I liked the old display. The colors are too hard to see in daylight on ALD12. I guess the emergency light is okay but the fuel indicator light is awful
64	38	Looks bad. Older drivers should not be forced to learn new displays. Our driving records are good and safety is learned. Why was this changed? Also, I cannot even read the fuel indicator light.

The data are meant to represent a variety of potential comments from dashboard display evaluators. Note that both point of view (e.g., "I") and personality emerge in these data (e.g., "mates said it was pretty poor"). While reading a small number of comments or determining a sampling strategy to extract data might provide insight, using a sophisticated text analytics tool to process an entire dataset will provide more useful results.

For this analysis, a sentiment lens was enabled which created codes to look for comments expressing positive and negative tone. For example, a detractor, or one providing negative feedback might use words such as *poor*, *bad*, and *awful*. A promoter might use words such as *good*, *excellent*, and *great*. Note that comments often can and do contain both types of sentiment, so it is useful to know about the proximity or co-occurrence of sentiment concepts.

A compound concept was created during the analysis combining the concepts *emergency*, *indicator*, and *light*. This decision was made because these three terms referred to a single entity. It is also reasonable to assume that a researcher may be examining this survey data to find sentiment and concepts around a specific entity, such as an emergency indicator light.

Figure 2 shows a Leximancer concept map of the dashboard survey. As in Figure 1, the large circles denote themes, or clusters, of related concepts. The black text represents concepts. Segmentation tags shown in red text have been clustered among the topical concepts to represent the ranges of driver experience. These tags were drawn from the structured data associated with each comment in the data. The groupings were as follows: 1-5 years, 6-15 years, and over 15 years. In this case, researchers wanted to compare the opinions of drivers within these ranges. Another approach would have been to examine comments and attitudes by age.

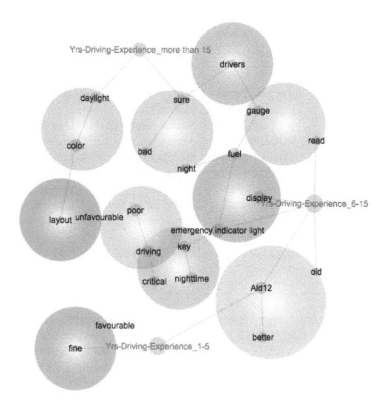

Figure 2 Illustration of themes, concepts, and relationships in the fictitious dashboard survey

This concept map shows a clear separation of issues according to the drivers' years of experience. The favorable sentiment tag is near the younger drivers, indicating that they tended to be more positive in their comments. Notice in the middle of the display that unfavorable feedback centers on dashboard layout. Automatic sentiment analysis provides a powerful tool to gain insight from a human factors survey or even from observational data.

The Leximancer tool automatically created a thesaurus for each concept. This thesaurus was composed of words associated with the concept throughout the data. This facility allows concepts to be coded even when the actual keyword does not appear in the text. For sentiment, the thesaurus was initially comprised of a standard positive and negative word list. These lists were adapted to the data, however, so that irrelevant and misleading words were removed, and some new (domain-specific) sentiment words were included. In Table 4, the sentiment terms found by Leximancer are shown, and their origins are explained.

Table 4 Thesaurus terms for unfavorable sentiment

Word	Score	Origin
poor	3.91	Standard
bad	3.41	Standard
weather	3.32	Leximancer Discovered
critical	3.21	Standard
difficult	2.66	Standard
useless	2.44	Researcher Added
arrangement	2.15	Leximancer Discovered
choices	2.15	Leximancer Discovered
combinations	2.15	Leximancer Discovered
problem	2.15	Standard

Table 4 shows the top 10 ranked terms that provide evidence of unfavorable sentiment. The score is a measure of strength of evidence in indicating negative tone. It is a standardized probability score.

The Origin column shows how the word made it into the definition for unfavorable tone. First, Leximancer provided a standard list of words generally indicating unfavorable tone. Words such as *bad*, *awful*, and *poor* were in this list. Second, the researcher added several terms specific to the data, such as *useless*. Third, Leximancer determined additional negative terms using word association information. This discovery of new words to indicate negative tone demonstrates one benefit of moving beyond manual, keyword coding by using a sophisticated software tool.

5 CONCLUSIONS

In the transportation industry, a thorough analysis of observation data can provide insights. For example, frequent concepts, specific word usage, sentiment, and word relationships can aid researchers in identifying connections between comments and specific functionality and applications. However, in all industry sector domains, including transportation, descriptive data collection and analysis has often been a difficult process. Several potential approaches include a one-question survey form for each area of research interest where free-form comments are entered into a single text entry box. Recordings of commentary with transcripts extracted are another potentially rich source of information. Media and social media text collections are yet another source of data for user interaction with systems. All of these collection methods provide a freedom of expression and flexibility of researcher and user vocabulary to describe experiences.

This paper used real-world UK Shout Out data and a contrived survey to describe a method for deriving meaning from the data. Graphics, quantitative measurements, and qualitative analysis were demonstrated for those data. A sentiment lens was enabled and positive and negative commentary identified. Several display techniques were used including a concept map and tabular data of concept counts and related concepts as evidence of results for meaning making.

ACKNOWLEDGMENTS

The author would like to acknowledge Dr. Guy Walker of Heriot-Watt University in Edinburgh for providing the U.K. Shout Out dataset. Also, Julia Cretchley from Leximancer support and Dr. Andrew Smith of the University of Queensland provided a review of technical facts and provided input on techniques and observations for the use of Leximancer in transportation industry studies.

REFERENCES

Crofts, K. and J. Bisman. 2010. Interrogating accountability: An illustration of the use of Leximancer software for qualitative data analysis, *Qualitative Research in Accounting & Management* 7(2): 180–207.

Grech, M., T. Horberry, and A. Smith. 2002. Sept 30-Oct 4. Human error in maritime operations: Analyses of accident reports using the Leximancer tool. Paper presented at the *Proceedings of the 4th Annual Meeting of the Human Factors and Ergonomics Society*, Baltimore, U.S.

Larkin, I. K., J. A. Previte, and E. M. Luck. 2008. Get off our roads Magoo: Are elderly drivers entitled to drive on our roads?. In *Proceedings Australian and New Zealand Marketing Academic Conference (ANZMAC)*, Sydney, Australia.

Miles, M. B. and A. M. Hubermam. 1994. *Qualitative data analysis*. Thousand Oaks CA: Sage Publications.

Plant, K. L and N. A. Stanton. 2011. A critical incident in the cockpit: Analysis of a critical incident interview using the Leximancer tool. Paper presented at 3rd CEAS Air&Space Conference, Venice, Italy.

Strauss, A. and J. Corbin. 1998. Basics of qualitative research: Techniques and procedures for developing grounded theory. London, UK: Sage Publications.

Walker, G. H., N. A. Stanton, and P. M. Salmon. 2011. Cognitive compatibility of motorcyclists and car drivers. *Accident Analysis and Prevention* 43: 878–888.

CHAPTER 61

Automotive HMI International Standards

Christian Heinrich

Daimler AG
Sindelfingen, Germany
christian.heinrich@daimler.com

ABSTRACT

More and more information technology has been implemented into cars over the last years. The HMI for these new applications has to comply with several requirements for automotive HMI. The purpose of the following article is to show what has been done in the field of design rules and methods for automotive HMI in the past and to show the opportunities of a closer cooperation between the designers of automotive and information HMI.

Keywords: HMI, automotive, standardization

1 HISTORY

Since the author was not with the group from the beginning the following is based on information Anders Hallen, a charter member, shared with us at the 10th anniversary of the group.

It all started with PROMETHEUS (**PRO**gra**M**me for a European Traffic with **H**ighest **E**fficiency and **U**nprecedented **S**afety)

PROMETHEUS was initiated in 1985 with the purpose to develop the European traffic scenario of the future with improved safety, environment and efficiency as a goal and to encourage development of the European electronics and supplier industries. One of the working groups in PROMETHEUS was created to tackle the Human Factors and HMI questions.

Well into the program, the need for standardisation was realised. Within CEN (Comité Européen de Normalisation) the technical committee CEN TC278 was formed in 1991 for this purpose. CEN 278 held 13 working groups (WG) of which 12 were trying to build systems with new technology and one, WG10, which was entrusted with the task of using new technologies to solve the problems of human machine interaction.

Discussions of lifting the CEN work to an international, ISO, level started early 1993, since it became clear that it is inefficient to have local standards in the automotive business. Eventually, ISO TC204 (Transport Information and control systems) was established to mirror the work of CEN278 on an international level. TC204/WG13 was given the task of standardising road-side HMI while in-vehicle HMI should be handled within TC22/SC13. WG8 was formed for this purpose and held its first official meeting in Paris in November of 1994. CEN TC 278 WG 10 still exists and consists of the European members of ISO WG8. This is important to give the standardisation of automotive HMI a voice on European level.

The working group developed the following standards and regularly reviews them.

2   Design Standards

Design standards define certain properties of an HMI and follow continuously technological and methodological progress. In many cases they represent an important basis for regional recommendations and memoranda of understanding.

2.1   Dialog management principles

The Standard ISO 15005 describes general principles of driver system interaction and the respective compliance procedures:

- The driver must be able to override any intervention of the system towards driving functions
- Systems that are not intended to be used while driving must be deactivated or the manual must contain an appropriate warning
- A TICS dialog shall regulate the flow of information so that it can be easily perceived
- Glances of 1.5 seconds shall be sufficient to gather relevant information
- System reaction time should not exceed 250 ms.

This standard was the basis both of the European Statement of Principles (ESoP 2006) and of the AAM Guidelines (Alliance of Automotive Manufacturers) (AAM 2006).

2.2   Visual presentation of information

The Standard ISO 15008 first published in 2003 and revised in 2009 defines parameters that are relevant for the safe and quick recognition of visual information.
- The required contrast is defined for different light condition: Day, night, twilight and direct sunlight
- Depending on the importance of information different font sizes are required, measured as angle from the driver's point of view
- To ensure readability certain properties of the used fonts are prescribed (Stroke width, aspect ratio, spacing)

2.3   Auditory presentation

The standard ISO 15006 gives recommendations and requirements for auditory signals:
- The sound level must be loud enough to be well perceived but shall not startle the driver
- The timing shall be appropriate for the type of information
- To ensure good audibility also in case of age related hearing loss a frequency range from 400 Hz to 2000 Hz is recommended
- For important warnings redundant visual information is required.

3. PERFORMANCE STANDARDS

This specific type of standards defines metrics for assessing driver workload. They specify the application of a given method, the necessary equipment and data treatment. They not set specific values e.g. for an acceptable workload.

3.1   Occlusion method to assess visual distraction

ISO 16673 describes a method to determine workload with a special focus on chunkability (Foley, 2009).

The test subject wears goggles which can switch between a transparent and an opaque state. During the operation of the device the time with transparent shutter is measured (TSOT, Total Shutter Open Time). This value is compared with task duration of unobstructed view.

608

Figure 1 shows the occlusion goggle in the opaque state (Baumann et al, 2004)

The ratio metric R is a measure for the chunkability of the task. If a task is chunkable it is easier for the driver to split the task in short parts, which leads to shorter glances away from the road. Also Total shutter open time (TSOT) can be used as a metric for driver workload and is specified in the document.

3.2   Simulated lane change test

ISO 26022 uses a simplified driving situation (Mattes and Hallen, 2009). It can be easily applied in any driving simulator, both simple or sophisticated. The road has three lanes and the driver has to change lanes according to signs appearing beside the road while also operating a device.

Figure 2 shows the driving scene. In this example the driver should change to the right lane.

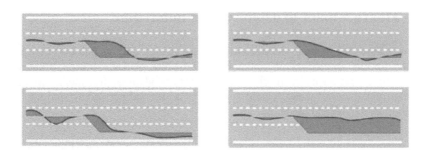

Figure 3 shows different types of errors during the lane change: Too late, too slow, bad lane keeping, no reaction.

The deviation (marked in red in Fig. 3) from the ideal pathway is a measure of the driving quality and thus for the distraction from the device.

Because of simplicity the results are highly reproducible (Bengler, Mattes, Hamm and Hensel, 2010).

3.3 Detection Response Task

The basic concept of ISO NWI 17588 is to use the subject's reaction to a stimulus and measure whether and how fast the test subject is able to react while operating a device or secondary task. The miss rate and the reaction delay time are measures of the mental and cognitive load. The tests can be performed with or without additional driving task. The purpose of the standard is to define the parameters of the experiment so that results from different labs can be compared. (van der Horst. and. Martens, 2010; Engström, 2010).

In the NWI the following stimuli are considered (examples see Fig 4):
- Visual stimuli on fixed external (remote) locations
- Visual stimuli fixed on the head
- Haptic stimuli

Figure 4 shows different stimuli: 1) visual stimuli as LED in the cluster instrument 2) a visual stimulus fixed on the head 3) A small vibrator mounted on the shoulder of the test subject

610

3.4 Measurement of driver visual behaviour

The standard ISO 15007 defines different measures and techniques that are important for the evaluation of visual behaviour like glance duration, fixation, transition time and saccade. It gives also recommendations for mounting and calibration of video and measurement systems. The second part gives also advice for experiment design and planning.

This standard is currently under revision to include automated eye tracking and automated data analysis. It is covered more detailed in a separate chapter of these proceedings.

3.5 Calibration Task

The purpose of ISO TS 14198 is to define tasks that can be reproduced very easily and precisely to allow comparison of results from different labs. SURT (Surrogate Reference Task) and CCT (Critical Tracking Task) were chosen for that purpose (Bengler, et al., 2010).

With the SURT the test person has to find the circle which is bigger than the others and then move the cursor field in such a way that it covers this very circle. The difficulty is scalable both for the cognitive and the manual demand: The difference in size of the circles can be varied, so that a small difference leads to a greater cognitive demand. With the size of the cursor field the manual demand can be changed.

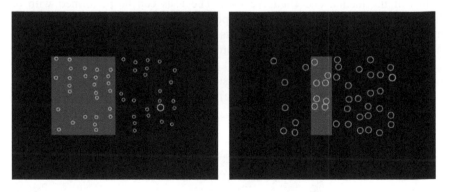

Figure 5 shows an easy and a difficult version of SURT

In the CTT a target bar which moves randomly up and down is presented on a display (see Figure 6). The test subject has to keep this target bar as close as possible to a horizontal center line (dashed line in Figure 5) by means of two keys. The speed of the random movement of the bar can be changed to make the workload scalable.

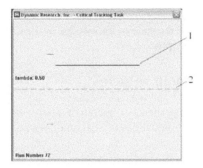

Figure 6: Typical screen of CTT with target bar (1) above the center line (2)

The application of these artificial calibration tasks in combination with performance measurement methods allows to check the correctness of equipment, processes instructions and subjects in the evaluation environment.

4 PROCESS STANDARDS

4.1 Procedure for assessing suitability for use while driving

ISO 17287 describes a process oriented approach; the following items must be documented:
- Intended use in the respective context
- Operations that are not intended while driving
- Measures that have been taken to prohibit these operations
- Measures that have been taken avoid foreseeable misuse
- Form in which failures of the system are presented to the driver
- Information about usability studies including the results. These studies should consider intended use and context

4.2 Procedure for determining priority of messages

ISO TS 16951 describes to methods two determine the priority of warnings, the Index Method and the Matrix Method.

With the Index Method each warning is rated regarding criticality and urgency by specialists on a scale of four values. The priority is then calculated by the mean of these values.

With the Matrix Method the relative priority for each pair of warnings is determined by experts. The results are documented in a matrix were each warning is represented by a row and a column.

5. TECHNICAL REPORTS

Technical reports represent the state of knowledge at the time they are published and are not subject to a review process.

5.1 Warning system in vehicles

ISO TR 16352 is a comprehensive literature study on warnings (so called Mutschler report) covering all modalities (visual, auditory and tactile) with a special focus on assistance systems.

5.2 Integration of safety-critical and time-critical warning signals

The work ISO TR 12204 started with the aim to develop a standard. But based on the respective research (Chiang, Llaneras and Foley, 2006), (Sato and Akamatsu, 2008), (Abe. and Itoh, 2008), (DOT 2008) it became clear that the matter is quite complicated and it was not possible to define simple rules that found consensus among the experts.

The following methods are described in the report:
• Procedure for assessing integration needs
• Timely comprehension methodology
• Appropriate response methodology

This report can be used as basis for further development.

6. THE WORKING GROUP

The Working Group 8 consists of 35 members from 11 countries. They come from universities, research institutes and automotive industry. They are also members of their national standardization organizations. The group meets twice a year for three days. In the first two days the taskforces of the different projects meet, often in parallel sessions. In the meeting of the working group on the third day all decisions are finalized.

If somebody is engaged in automotive HMI topics, either in research or as part of the business, and is interested in contributing to this standardization effort there are three ways to do so:
• If nominated by the national body one can participate in the meetings
• It is also possible to contribute to the work of the task forces by mail and conference calls
• As with all international standardization committees national mirror committees exist which also deal with the above mentioned topics

7. CONCLUSIONS

The development of high quality standards on automotive HMI means a high workload and often takes quite a time to ensure applicability and validity. But the importance of having international standards to accompany the innovation process justifies the effort the WG members undertake.

ACKNOWLEDGMENTS

The author would like to acknowledge the contributions of all members of ISO TC 22 SC13 WG 8. Special thanks go to Klaus Bengler, Jörg Breuer and Stefan Mattes for valuable comments on this document.

REFERENCES

AAM 2006 "Statement of Principles, Criteria and Verification Procedures on Driver Interactions with Advanced In-Vehicle Information and Communication Systems" accessed February 9, 2012 http://www.autoalliance.org/files/DriverFocus.pdf

Abe, G. and M. Itoh: "How Drivers Respond to Alarms Adapted to Their Braking Behaviour", Journal of Mechanical Systems for Transportation and Logistics, 1(3), pp. 331-342, 2008 (11).

Baumann M, A. Keinath, JF. Krems, K. Bengler K: „Evaluation of In-vehicle HMI using occlusion techniques: Experimental Results and Practical Implications". Applied Ergonomics. 2004; 35: 197-205.

Bengler, K, S. Mattes, O. Hamm and M. Hensel (2010) Lane change Test: Preliminary Results of a Multilaboratory Calibration Study. In G. L. Rupp, Performance Metrics for Assessing Driver Distraction chapt. 14 (pp 243 – 253), SAE International, Warrendale, Pensylvenia

Chiang, D., E. Llaneras and J. Foley, Driving Simulator Investigation of Multiple Collision Alarm Interference Issues, Proceedings of Driving Simulator Conference Asia/Pacific 2006, CD-ROM, 2006

DOT HS 810 905: Integrated Vehicle-Based Safety Systems (IVBSS): Human Factors and Driver-Vehicle Interface (DVI) Summary Report, February 2008

Engström, J(2010) The Taktile Detection Task as a Method for Assessing Drivers'Cognitive Workload. In G. L. Rupp, Performance Metrics for Assessing Driver Distraction (pp 90 – 103), SAE International,, Warrendale, Pensylvenia

ESoP 2006, European Statement of Principles on the Design of Human Machine Interaction 2006, accessed February 9, 2012. http://eur-lex.europa.eu/LexUriServ/site/en/oj/2007/l_032 /l_03220070206en02000241.pdf

Foley, J. P. (2009). Now You See It, now you Don't: Visual Occlusion as a Surrogate Distraction Measurement Technique. In M. A. Regan, J. D. Lee, and K. L. Young (Eds.), Driver Distraction: Theory, Effects, and Mitigation (pp. 107-121). Boca Raton, Fla.: CRC Press.

Mattes, S., and A Hallen, . (2009). Surrogate Distraction Measurement Techniques: The Lane Change Test. In M. A. Regan, J. D. Lee, and K. L. Young (Eds.), Driver Distraction: Theory, Effects, and Mitigation (pp. 107-121). Boca Raton, Fla.: CRC Press.

Sato, T and M. Akamatsu, Preliminary Study on Driver Acceptance of Multiple Warnings while Driving on Highway, Proceedings of SICE Annual Conference 2008, 872-877, 2008

van der Horst, R. A. and M. H. Martens (2010) The Peripheral Detection Task (PDT): On-line Measurement of Driver Cognitive Workload and Selective Attention. In G. L. Rupp, Performance Metrics for Assessing Driver Distraction (pp 73 – 89), SAE International, , Warrendale, Pensylvenia

CHAPTER 62

Evaluation of Automotive HMI Using Eye-Tracking - Revision of the EN ISO 15007-1 & ISO TS 15007-2

[1]Christian Lange

[1]Ergoneers GmbH,
85077 Manching, Mitterstraße 12
Germany

ABSTRACT

This paper presents the revision of documents EN ISO 15007-1 and ISO/TS 15007-2 which was done by the ISO TC22SC13WG8 working group consisting of eye-tracking specialists worldwide. Both ISO documents were published in 1999. Since then many research studies were conducted, which lead to an increasing level of knowledge about eye-movement behavior. In parallel to that, eye-tracking technology developed enabling fully automated data analysis.

Due to that both standards were revised in the ISO TC22SC13WG8 working group to include latest findings in eye-movement behavior and latest developments in eye-tracking technology.

Keywords: Eye-Tracking, Experiments, Data Analysis, Driver Information Systems, EN ISO 15007-1, ISO/TS 15007-2

INTRODUCTION

An increasing number of driver assistance and driver information systems can be found in modern cars. Those systems might cause driver distraction which can be measured using eye-tracking technology.

To enable developers and research to conduct eye-tracking experiments to measure driver visual behavior with respect to transport information and control systems and to gather valid and comparable results EN ISO 15007-1 & ISO TS 15007-2 were developed.

EN ISO 15007-1

EN ISO 15007-1 has been developed to give guidance on the terms and measurements relating to the collection and analysis of driver visual behavior data. This approach aims to assess how drivers respond to vehicle design, the road environment, or other driver-related tasks in both real and simulated road conditions. It is based on the assumption that efficient processing of visual information is essential to the performance of the driving task.

Regarding latest findings in eye-movement behavior the sections "Terms & Definitions", "Basic Measures" and "Glance Metrics" were revised and completed. This now leads to a full set of validated glance metrics which allow fully description and understanding of driver's visual behavior. In addition to that it's clearly described and documented with formulas how to calculate those metrics. Thereby it's ensured that all user of eye-tracking technology are calculating the metrics in the same way, what leads to valid and comparable results.

One of the most important basic measures is the so called "glance duration" because all the glance metrics are based on the glance duration. Therefore a clear definition for the glance duration is given:

"glance duration
time from the moment at which the direction of gaze moves towards an area of interest (e.g., the interior mirror) to the moment it moves away from it."

Furthermore a figure (see figure 1) visualizes the explanation and helps to understand it correctly.

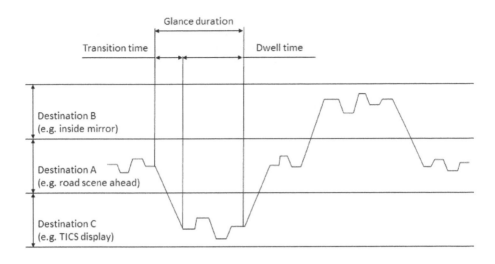

FIGURE 1 Visualization of a glance duration to a TICS display (e.g. navigation displays) (from EN ISO 15007-1)

The existing set of glance metrics was revised and new glance metrics that were found in the last years were included. To enable an easy and standardized usage of the metrics a clear explanation of the metrics is given, which is combined with a formula of how to calculate the metrics. The metrics listed below are examples taken from EN ISO 15007-1 to show how the metrics are presented:

"number of glances
count of glances to an area of interest (or set of related Areas of Interest) during a condition, task, subtask or sub-subtask. Unit [glances].
      EXAMPLE   9 glances

total glance time
summation of all glance durations to an area of interest (or set of related Areas of Interest) during a condition, task, subtask or sub-subtask; Total glance time = $\sum$(glance duration1, glance duration2, ..., glance duration n). Unit [s]
      EXAMPLE   17,88 s

mean glance duration
mean duration of all glance durations to an area of interest (or set of related Areas of Interest) during a condition, task, subtask or sub-subtask; Mean glance duration = (Total Glance Time)/(Number of glances) during a condition, task or subtask; unit [s]
      NOTE   1,28 s

Area of Interest attention ratio
Ratio representing the percent of time glances are within an area of interest (or set of related Areas of Interest) during a condition, task, subtask or sub-subtask; Area of Interest attention ration = $\sum$(glance duration1, glance duration2, ..., glance duration n)/(duration of condition, task, subtask or sub-subtask) * 100%; unit [%]
EXAMPLE   53,47 %"

ISO/TS 15007-2

ISO/TS 15007-2 gives guidelines on equipment and procedures for analyzing driver visual behavior, intended to enable assessors of transport information and control systems (TICS) to
- plan evaluation trials;
- specify (and install) data capture equipment, and;
- validate, analyse, interpret and report visual-behavior metrics (standards of measurement).

It is applicable to both road trials and simulated driving environments.

ISO/TS 15007-2 now takes into account the developments in eye-tracking technology in the past 10 years and describes the currently existing eye-tracking methodologies:
- Head-mounted eye-tracking systems
- Remote eye-tracking systems

To give an idea of how the methods are explained, the description for the head-mounted eye-tracking technology is shown below:

"Head-mounted eye-tracking systems
With head-mounted eye-tracking systems, the subject wears components of the eye-tracking system directly on the head. The components necessary for the eye-tracking are mounted on a helmet, a cap or on a device similar to glasses. Head-mounted eye-trackers may consist of the following components:
- Scene camera: this camera records what the subject can see
- Eye camera: this camera records at least one eye
  NOTE 1 The eye can be recorded directly or via an IR reflective mirror;
  NOTE 2 Calibration of eye camera to scene camera is necessary to transform the x- and y-coordinates from the eye camera coordinate system to the scene camera coordinate system. Thereby the head-mounted eye-tracking system is able to indicate in the scene camera view where the subject is looking.
- Infrared LED: the infrared LED typically makes the eye visible in the infrared spectrum. Thereby the system becomes more independent and robust from the surrounding lighting conditions."

It is also described and clarified to which challenges, automated data analysis with novel eye-tracking systems might lead and how one has to treat them. Therefore the revised ISO standard contains sections about "Quality of Eye-Tracking data" and "Artefacts" taking into account topics like check of calibration, validity and validation of data and pupil detection errors.

A very important point is that the eye-tracking system is calibrated correctly. Otherwise it indicates a wrong glance location what would lead to fully wrong and invalid results. Therefore the calibration has to be treated very carefully. Below is shown how the revised ISO standard introduces a procedure how to check the calibration before and after the experiment.

"The calibration check makes ensure that the eye-tracking system has calculated the point of fixation correctly after the calibration and that there is no drift in calibration during the experiment.

The required procedure is as follows:

- At the beginning of each participant session a two-step check should be used:
    o Use the calibration procedure provided by the eye-tracking system in use (e.g. matrix of 4, 9 or 16 dots)
    o Once calibrated have participant glance to each area of interest important to the study and verify that each glance is properly displayed/calculated
- At the end of the experiment a single step check should be used:
    o Have participant glance to each area of interest important to the study and verify that each glance is still properly displayed/calculated. If it is not, it is a sign that there's a drift in calibration that occurred. The experimenter needs to determine whether the shift is still acceptable or not.

To evaluate whether too much drift in calibration has occurred the table below may be helpful:

TABLE 1 — Criteria for calibration quality

Good	Glance is inside AOI
Satisfactory	Glance is near border of intended AOI and not in another AOI
Poor	Glance is far from intended AOI and/or in adjacent AOI

If the shift in calibration is poor there are two options:
    o Re-calibrate at the moment of drift
    o Discard the data of the participant and replace with a new participant"

Furthermore, ISO/TS 15007-2 now provides a section with interpretations of the commonly applied glance metrics, what helps users of the eye-tracking methodology better understand their findings. The example below for the glance metric "AOI attention ratio" shows how this is done.

"AOI attention ratio
The AOI attention ratio describes the visual demand of an area of interest (especially when operating a TICS). A high AOI attention ratio combined with a long mean glance duration while operating a TICS may indicate that the task's design does not allow it to be visually interrupted and resumed easily without loss of information. Design improvements to this aspect of the task may reduce its visual demand.

NOTE: The above mentioned glance metrics should be interpreted in common for the TICS evaluation and not isolated from each other, because this may lead to wrong conclusions.

E.g.1: glance frequency and mean glance duration may be traded off within a fixed sample interval. That is, very long glance durations (indicative of high workload demand) may be associated with fewer rather than more glances. Thus it is important to consider the two measures together, especially if the sample interval is fixed rather than allowed to reflect task completion time.

E.g.2: When comparing the speed gauge in a HUD with a conventional speedometer in the instrumental cluster one may find that subjects have a higher number of glances and a higher glance frequency to the speed gauge in the HUD than to the speedometer in the instrumental cluster. When only taking into account those two metrics this might lead to the conclusion that the HUD is more distracting. When also taking into account that the mean glance duration to the HUD is shorter than to the instrumental cluster and that the total glance time to both areas of interest for the same time interval is the same one will draw the conclusion that subjects control their speed more often with the HUD without being more distracted which leads to an increase in safety.

E.g.3: The HUD also offers an opportunity to highlight another example of where multiple metrics should be examined. Because the HUD is typically located high in the field of view and closer to the driver's line of sight to the forward roadway, transitions to nearby areas of interest on the roadway are shorter – and the probability of noticing events in nearby regions (such as unexpected roadway events like pedestrians or braking vehicles) is increased, with response times to them facilitated (see Kiefer, R.J. (1996a). A review of driver performance with head-up displays. Proceedings of the Third Annual World Congress on Intelligent Transport Systems. Orlando, FL, USA)."

OUTLOOK

The revised ISO documents will probably lead to a more valid and comparable usage of eye-tracking technology to measure driver visual behavior with respect to transport information and control systems.

In a next step, both ISO documents will be merged in one document which provides researchers and developers all necessary information and guidance for their daily work.

REFERENCES

EN ISO 15007-1: Road vehicles — Measurement of driver visual behavior with respect to transport information and control systems - Part 1: Definitions and parameters (under ballot)

ISO/TS 15007-2: Road vehicles - Measurement of driver visual behavior with respect to transport information and control systems - Part 2: Equipment and procedures (under ballot)

CHAPTER 63

Designing the Non-linearity of Human Motion Perception for Dynamic Driving Simulators

Anca M. Stratulat[1], Vincent Roussarie[1],

Jean-Louis Vercher[2], Christophe Bourdin[2]

1PSA Peugeot Citroën
Vélizy-Villacoublay, FRANCE
2Aix-Marseille University,
Marseille, FRANCE
anca.stratulat@gmail.com

ABSTRACT

Driving simulators allow the exploration of research fields that are difficult to reach in normal conditions, like testing driving assistance systems during dangerous situations. However, there are some perceptive aspects of the simulations that still need to be improved. This is due to their mechanical limitations that do not allow us to produce accelerations or braking at their physical value. As a consequence, driving simulators' motion algorithm is using tilt in combination with linear translations to produce a sensation of linear acceleration. The present study show that motion perception depends on the manner tilt and translation are used together to provide a unified percept of linear acceleration. Our results show that there is a large variability on how humans perceive accelerations and decelerations and it is not advisable to use motion algorithms without taking into account non-linearity of human perception.

Keywords: driving simulator, motion perception, multisensory integration.

1   INTRODUCTION

Driving simulators can be used to simulate special driving situations, like driving into fog, or even take into account dangerous scenarios produced by the state of the driver (drowsiness, fatigue etc). All these scenarios can help car manufacturers to study the acceptance of new security systems built on cars (e.g. Advanced Driving Assistance System - ADAS) and to reduce the number of accidents caused by these typical situations. The present article is devoted to the study of motion perception, more specifically to the perception of acceleration and braking on a dynamic driving simulator.

The development of driving simulators was inspired from flight simulators, on which the motion is not simulated only by audio-visual scenes (static simulator), but also by mechanical motion that follows a specific algorithm (dynamic simulator). All dynamic driving simulators are based on the same basic algorithm, called *washout algorithm*, which is set on tilt-translation ambiguity. The tilt-translation ambiguity is based on the incapacity of the vestibular system to distinguish between acceleration and tilt under certain conditions. It means that a tilt, rather than a translation, may be interpreted as acceleration if other sensory inputs (e.g. visual, auditory) are presenting acceleration. Tilt-coordination technique in a dynamic simulator implies that the mechanical tilt is synchronized with visual linear translation in order to produce a sensation of linear acceleration (Figure 1).

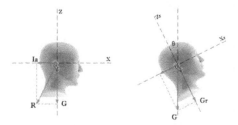

Figure 1 Perceptually equivalent situations, as described in Holly and McCollum (Holly, 1996), are the basis of tilt-coordination technique. The inertial force due to linear acceleration (Ia) together with gravity (G) gives the resulting vector GIA (gravito-inertial acceleration) or R (a). The vestibular system cannot distinguish between this situation and a tilt of the head, produced below angular velocity threshold, so that the gravity corresponds to GIA (b). Therefore, a linear acceleration can be simulated by either linear translation or tilt (or a combination of both).

Recent studies on tilt-coordination for the simulation of linear acceleration have shown the effectiveness of this technique (Groen and Bles, 2004, Berger, Schulte-Pelkum and Bülthoff, 2010). Berger and collaborators used a Stewart platform to produce backward pitch tilt in combination with visual forward accelerations. According to their study, subjects perceived the tilt as linear acceleration. However, all the studies on tilt-coordination technique show that its effectiveness is limited by the sensitivity of the semicircular canals, and suggests keeping the rotational velocity under 2 deg/s (Groen and Bles, 2004). To go further this limitation, it has been proposed that tilt-coordination could be used in combination with physical translations (linear accelerations). Groen and colleagues investigated the effects of short forward accelerations combined with backward pitch (tilt-coordination technique) in order to evaluate the perceived realism of take-off accelerations simulated on an advanced flight simulator and they observed that the best perception of acceleration is obtained when the motion is simulated at 20% on translational motion and at 60% for tilt-coordination (Groen, Valenti and Hosman, 2000). Therefore the addition of physical motion to visual accelerations may consistently improve the sensation of overall motion. And, as most of the modern driving simulators possess motion-based platforms that allow longer surge translations, and therefore stronger linear accelerations capabilities, some questions arise concerning the use of tilt and translation for the simulation of longitudinal motion: *Can the perception of linear self-motion be improved by the use of larger translations (up to 10 m) in combination with tilt-coordination? If so, what is the best tilt/translation ratio to simulate a realistic acceleration or deceleration? Does this ratio vary with the level of acceleration or deceleration? Is this ratio perceived differently for deceleration or acceleration?*

In order to answer to these questions, we developed a series of three experiments, carried on a dynamic driving simulator that allows us to produce a large range of motions (up to 10 m on longitudinal translations).

2   GENERAL METHODS

1.2   Experimental device – dynamic driving simulator

The general device was a dynamic driving simulator, composed a cell (where the car is placed), a hexapod and an X-Y motion platform (Figure 2). For a detailed description, see (Chapron and Colinot, 2007).

Figure 1 Dynamic driving simulator SHERPA by PSA Peugeot-Citroën. The translational movements of the hexapod are limited to ±5m, ±2.75m and ±20cm, on X, Y and Z respectively. The rotational movements are limited to ±18deg, ±18deg and ±23deg, on pitch, roll and yaw respectively. The X-Y motion platform can reproduce linear movements of 10 and 5 meters.

1.3   Experiment 1 - Perception of linear breaking

The aim of this first experiment was to study whether the perception of a deceleration of -0.8 m/s² remains the same regardless of tilt/translation ratio, or whether it changes with this ratio. 14 volunteers participated in this study. The visual stimulation consisted of a straight one-lane road surrounded by an empty green field, and bordered by trees at 18 m intervals, up to a white line traced on the floor. A red wall was positioned 23 m after the white line. To avoid spatial positioning of the subjects, no granular textures were used on the road or on the field (uniform color) and no trees were bordering it after the white line. The dynamic stimulation, which controlled the motion of the cell, consisted in simulating braking through tilt and translation. We used 5 different tilt/translation ratios to simulate the same final deceleration, following a cosine curve with a peak of -0.8 m/s² (Figure 3). The ratios were composed of inverse-proportional percentages of tilt and translation (condition 1: 100/0% tilt/translation, condition 2: 75/25% tilt/translation, condition 3: 50/50% tilt/translation, condition 4: 25/75% tilt/translation, condition 5: 0/100% tilt/translation). A 6th condition was added, with only visual simulation of the braking (no dynamic stimulation). This condition was used to determine the effects of pure visual stimulation compared to the other conditions.

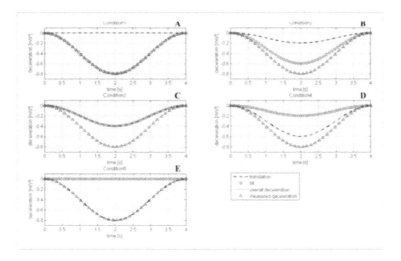

Figure 3 Sinusoidal profiles of deceleration simulated by translation (dash-dot line) and of equivalent deceleration simulated by tilt (circles), represented for each of five conditions. The sum of the two signals is always the same, with a peak of 0.8 m/s² (solid line). The latter is compared to the signals measured by the three-dimensional accelerometer (triangles).

The participants were seated in the driver's seat, but had no access to controls during the simulation (passive simulation). The car drove towards the red wall, accelerating for about 6sec before reaching a constant speed of 50 km/h. Once it crossed the white line, the car automatically started to brake and the red wall instantly disappeared. The braking lasted 4 sec before the car came to a full stop. The car always stopped in the same position, which is with its bumper touching the invisible wall. However, subjects were not informed about this invariability. We used a 2AFC paradigm to ask the subjects to evaluate the intensity of the braking. They were asked to answer the following question "Did the car stop BEFORE or AFTER the red wall?" and to rate their answer with a certainty level from 1 to 6, with 1 representing the lowest certainty, and 6 representing the highest certainty.

The performance of the task was analyzed in terms of braking perception errors. A score of -1 was attributed to any BEFORE answer, and a score of 1 to any AFTER answer, and these scores were multiplied by the corresponding level of certainty.

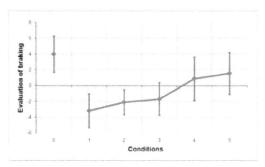

Figure 4 Mean values of perception errors for each condition. Negative values for perception error are equivalent to an overestimation of the braking (participants considered the car stopping before the wall). Positive values are equivalent to an underestimation of the braking (participants considered the car stopping after the wall).

The results, presented in figure 4, suggest that the perception of braking is non-linear and depends on the tilt/translation ratio. If more tilt is used, the braking is overestimated. Inversely, if more translation is used, the braking is underestimated. By using a Probit function, we determined the best tilt/translation ratio (the ratio which produces the most realistic perception of braking) at 35/65% tilt/translation, meaning 0.28 m/s² in tilt and 0.52 m/s² in translation (figure 5).

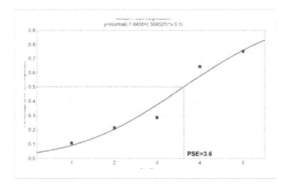

Figure 5 The Probit regression calculated for conditions 1 to 5. The point of subjective equality (PSE) corresponds to the condition for which participants answered AFTER in 50% of the case and BEFORE for the other 50%. In this study, the PSE value is 3.6.

This first study shows that the perception does not remain constant across the change of tilt/translation ratio. If the deceleration is mainly produced by tilt, braking is perceived as stronger than when it is mainly simulated through translation. An explanation should probably be sought in the functionality of the vestibular system and the interactions among all the sensory cues involved in the process of multi-sensory integration during the braking.

Therefore, the differences in perception for the different ratios may be due to proprioceptive cues. Because the head of the participants was free to move, it is possible that movements of the head during braking varied from one ratio to another, depending on the amplitude and the velocity of the tilt. This motion will produce not only neck proprioceptive signals, but also extra-vestibular ones. Then, even though the level of rotational velocity of the cell was kept below the threshold of perception of semi-circular canals, we cannot exclude that the head movements could have been added to the tilt of the cell, increasing the level of vestibular stimulation and/or generating a proprioceptive stimulation at neck level.

Given the non-linearity of the sensory systems, we inquire if the best tilt/translation ratio will remain constant for other levels of deceleration.

1.4 Experiment 2 - Variation of braking perception with the level of deceleration

In order to answer this question, a second experiment was conducted. We used the same experimental design and protocol, but tested 3 different levels of deceleration (-0.6, -1.0 and -1.4 m/s²). Moreover, if the head perception is indeed the cause of the extra-weighting for vestibular and proprioceptive cues, then a perceptive difference should be observed between free-head and fixed-head conditions. Therefore, 24 subjects participated to the free-head condition and 9 to the fixed-head condition.

Braking distance and initial constant speed were adapted to the level of deceleration, in order to have the same braking time (4 sec). Each of the 3 decelerations was dynamically simulated on 5 tilt/translation ratios. As in experiment 1, the subjects' task was to answer the following question "Did the car stop BEFORE or AFTER the red wall?" by using the certainty levels.

In the case of free-head condition, the results show an increased overestimation of braking proportional with the quantity of tilt (figure 6). In order to determine the most realistic tilt/translation ratio, a psychometric function was computed for each level of deceleration. The points of subjective equality or PSEs were determined for each levels of deceleration as follows: $PSE_{-0.6}=8.85$, $PSE_{-1.0}=4.99$, $PSE_{-1.4}=3.34$. The results suggest that the PSE depends on the level of deceleration (figure 7).

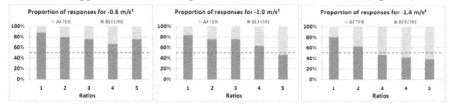

Figure 6 The number participants answering BEFORE or AFTER for each condition (free-head condition).

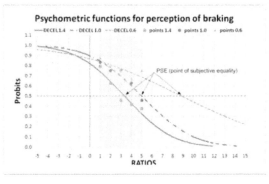

Figure 7 The psychometric functions calculated for each deceleration level in free-head condition.

For fixed-head condition, the subjects detected differences in braking intensity between trials for the same level of deceleration (figure 8). Nonetheless, they found that the fixed head was an uncomfortable position and totally unnatural. Interestingly, the results show a non-linear evaluation of the braking in relation to the quantity of tilt or of translation. For -0.6 m/s², there is an overwhelming underestimation of the braking when only tilt is used, but this underestimation fluctuated with the quantity of translation used. For -1.0 m/s², there is a slight overall underestimation, while for -1.4 m/s², there is a clear underestimation for pure translation (Figure 9).

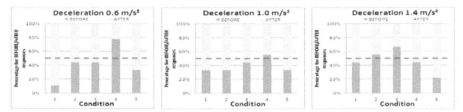

Figure 2 - Mean values of certainty levels for each level of acceleration for fixed-head condition. The maximum value (6) represents a 100% certainty, while the minimum value (1) represents a 0% certainty.

The results of this second experiment suggest that braking intensity is perceived as directly proportional to the quantity of tilt used to simulate the deceleration, and inversely proportional to the amount of translation when the head is not fixed to the headset. Thus, the most realistic tilt/translation ratio depends on the level of simulated deceleration. A plausible explanation for this difference in the perception of different ratios of the same level of deceleration might be the presence of proprioceptive cues (motion of the head, contact of the seatbelt with the body, cervical proprioception, neck proprioception etc.). This explanation is also sustained by the results of the fixed-head condition that suggest that the movement of the head modifies the final perception of deceleration. Therefore, the movement of the head plays an important role in the perception of deceleration on a dynamic driving simulator. This movement increases the weight of proprioceptive (neck) and vestibular (semicircular canals) cues.

Studies on perception of acceleration/deceleration or forward/backward motion (Bringoux, Nougier, Barraud, Marin and Raphel, 2003) showed that, even if the two opposite motions are identical from the physical point of view (except direction), there are external factors that intervene in the final perception of motion like familiarity (Holly and McCollum, 2008). Thus, we question whether the effects of tilt/translation ratios described for various deceleration levels on motion perception are also found for acceleration (forward motion).

1.4   Experiment 3 - Perception of linear acceleration

A third experiment conducted on the same dynamic simulator, was using a similar experimental design and protocol, with passive subjects. 30 volunteers participated in this study. The visual scenario consisted in a straight two-lane road with granular texture surrounded by an empty green field, with grass texture. The two lanes were separated by a continuous white line. A finish line was positioned at a given distance from the departing point of the car. As we tested 3 different levels of acceleration, the distance between the starting point and the finish line was adjustable. The car was advancing at constant speed of 50 km/h, on the left lane. On the right lane, at a certain distance in front of the subject's car, another car was running at constant speed of 50 km/h. Both cars were heading for the finish line. At the given distance in front of the finish line, the subject's car started to accelerate and the second car instantly disappeared. The acceleration took exactly 3 sec. The moment the subjects' car passes the finish line, it starts to brake until full stop. The acceleration of the subject's car was strong enough to reach the other car and to pass the finish line in the same time. The subject was informed about the constancy in speed of the second car, but not about the synchronous arrival of the cars. The subjects were submitted to acceleration of 1.0, 1.5 or 2.0 m/s^2 (maximal peak). It was dynamically simulated by a combination of backward tilt and forward translation, with the same 5 tilt/translation ratios as used in previous experiments. In order to evaluate perception of acceleration, we used a 2AFC paradigm. When the subject's car passed the finish line, the subject had to answer to the following question: "Who crossed the finish line first?" by using the certainty levels.

The results show no differences between the 5 tilt/translation conditions for the same level of acceleration. However, the results show an increased overestimation of the acceleration proportional to the level of acceleration (Figures 10 and 11).

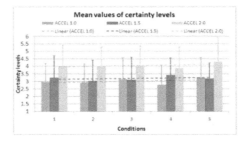

Figure 10 Mean values of certainty levels for each level of acceleration. The maximum value (6) represents a 100% certainty, while the minimum value (1) represents a 0% certainty.

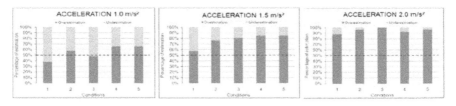

Figure 11 Number of subjects that overestimated (blue) or underestimated (orange) the acceleration for each level of acceleration and for each condition. The red line represents the level for accurate perception of acceleration.

Our data suggest that one level of acceleration can be simulated by any of the 5 tilt/translation ratios, without influencing the final perception of acceleration.

Because of the high levels of acceleration studied, most of the conditions including tilt had an angular velocity greater than the semi-circular canals' threshold of 3.7 deg/s, according to (Benson, 1990). However, it seems that the detection of tilt did not influence the final perception of linear acceleration. The constancy of perceived linear acceleration along the tilt/translation ratios could be due to the optic flow produced by the texture of the field and road. The optic flow presented to the participants the same displacement speed and traveled distances for one level of acceleration, independently of the ratio used. Our results suggest that the perception of linear acceleration could be then influenced by the non-inertial cues, like visual cues, which determine the CNS to interpret all motions as linear acceleration. Viewed from the perspective of Bayesian framework (Zupan, Merfeld and Darlot, 2002), the non-inertial cues could change the reliability of the inertial cues.

2   GENERAL DISCUSSION

The results of these studies show that the accelerations and decelerations are not perceived in the same manner and therefore they should be simulated differently on dynamic driving simulators. Moreover, the case of breaking seems to be more complex from the perception point of view, given that the best tilt/translation ratio depends on the level of deceleration.

From perceptive point of view, these results suggest that, in the final perception of motion, the reliability of each sensory cue changes relative to the multisensory interaction and context. This may be explained by optimal cue integration theory, which represents a weighted linear combination of perceptual estimates from the individual cues (Angelaki, Gu and DeAngelis, 2009) or Bayesian framework. This theory is considered to be particularly well-suited for the interaction of vestibular cues with other sensory information and especially for self-motion perception. The Bayesian approach states that the contribution of each sensory cue carries a certain weight, representing a probability density function conditioned by "*a priori*" information (Zupan, Merfeld and Darlot, 2002). In our case, the prior information may be represented by the context (acceleration/deceleration, fixed/free head). The perception of linear accelerations, positive or negative, seems to be based on the integration of different sensory cues: mainly vestibular, but also proprioceptive and visual, even if they were not always present in our experiments. The weight of each sensory cue is based on its reliability, which changes from one experiment to another, depending on the prior information.

The present work suggests that human self-motion perception is non-linear and that the CNS combines vestibular cues with visual and proprioceptive information, but also with prior information represented by the driving situation, in order to produce a coherent final perception of linear acceleration. Moreover, this combination follows an optimal probabilistic manner.

3   CONCLUSIONS

Our overall results suggest that the perceived intensity of braking depends on the tilt/translation ratio used to simulate the level of deceleration, while the perceived intensity of acceleration does not depend on this ratio. Therefore, the algorithms used on dynamic driving simulators have to take into consideration the characteristics of multisensory integration. Using classical washout algorithms seems to limit the realism of the simulations. As a conclusion, we propose to integrate the results of the present studies in the definition of washout algorithms in order to take into consideration the non-linearity of human sensory systems.

REFERENCES

Angelaki, D. E., Gu, Y., and DeAngelis, G. C., 2009. Multisensory integration: psychophysics, neurophysiology, and computation. *Current Opinion in Neurobiology*, 19, 452-458.

Benson, A., 1990. Sensory functions and limitations of the vestibular system. In *Perception and control of self motion*. L. E. Ass. (Ed.). R. Warren and A.H. Wertheim.

Berger, D. R., Schulte-Pelkum, J., and Bülthoff, H. H., 2010. Simulating believable forward accelerations on a stewart motion platform. *ACM Transactions on Applied Perception*, 7.

Bringoux, L., Nougier, V., Barraud, P.-A., Marin, L., and Raphel, C., 2003. Contribution of somesthetic information to the perception of body orientation in the pitch dimension. *Quaterly Journal of Experimental Psychology Section A - Human Experimental Psychology*, 56(5), 909-923.

Chapron, T., and Colinot, J.-P., 2007. The new PSA Peugeot-Citroën Advanced Driving Simulator. Overall design and motion cue algorithm. *Driving Simulation Conference.*

Groen, E. L., and Bles, W., 2004. How to use body tilt for the simulation of linear self motion. *Journal of Vestibular Research*, 14(5), 375-385.

Groen, E., Valenti, C. M., & Hosman, R., 2000. Psychophysical thresholds associated with the simulation of linear acceleration. *AIAA Modeling and Simulation Technologies Conference.*

Holly, J. E., and McCollum, G., 1996. The shape of self-motion perception-I. Equivalence classification for sustained motions. *Neuroscience*, 70(2), 461-486.

Holly, J. E., and McCollum, G., 2008. Constructive perception of self-motion. *Journal of Vestibular Research*, 18(5-6), 249-266.

Schlack, A., Krekelberg, B., and Albright, T. D., 2008. Speed perception during acceleration and deceleration. *Journal of Vision*, 8(8), 9.1-11.

Zupan, L. H., Merfeld, D. M., and Darlot, C., 2002. Using sensory weighting to model the influence of canal, otolith and visual cues on spatial orientation and eye movements. *Biological Cybernetics*, 86(3), 209-230.

CHAPTER 64

Inferring Cognitive State from Observed Behaviour

Sarah Sharples[1], Tamsyn Edwards[1], Nora Balfe[2] & John R. Wilson[1,2]

[1] Human Factors Research Group
Faculty of Engineering
University of Nottingham
University Park, Nottingham
NG7 2RD, UK

[2] Network Rail
Ergonomics Team
40 Melton Street
London NW1 2EE, UK

ABSTRACT

The discipline of human factors has traditionally been informed by a variety of different research approaches, spanning controlled experimental studies to research 'in the wild'. There has however been little work investigating the relationship between observable behaviours and subjective reports. This paper examines the potential for applying a structured approach to collection of data in a range of contexts to enable increased use of observed behaviours in informing human factors theory. Two cases are presented from the domains of rail control and air traffic control. Some evidence for relationships between observed behaviour and cognitive state is presented along with suggestions that individuals differ in the extent to which their cognitive states are 'observable'.

Keywords
Observation, Workload, Fatigue, Stress, Control tasks, Human Factors Methods.

1. INTRODUCTION

The discipline of human factors has traditionally been informed by a variety of different research approaches, spanning controlled experimental studies to research

634

'in the wild'. Traditionally laboratory studies have been used effectively for demonstrating causal links between independent and dependent variables, but have also been criticised for their lack of ecological validity. On the other hand, field studies have been demonstrated to yield rich data that produces detailed descriptions of work activities but have not always been effective in yielding generalisable findings.

Subjective methods have been applied in a number of contexts to enable collection of data in the field regarding user reports of cognitive state. Typical measures might include measures of workload (e.g. Hart & Staveland, 1988), situation awareness (Mogford, 1997), stress (Martinussen & Richardsen,, 2006) or fatigue (Schroeder, Rosa & Witt, 1998). Although such methods have been well validated and their value in the field context is high, they have suffered from criticism due to the need to normally apply them retrospectively to avoid disrupting or changing work behaviours, as well as their reliance on the ability of the respondent to accurately report their cognitive state.

Anecdotally, many industries and work activities have reported that it is possible to 'see' how well a person is coping with their work. This is particularly seen in control environments, where a control manager might observe an operator for clues that they might be experiencing stress, or high workload. Such 'clues' might be a change in the way an operator communicates, or even non-verbal behaviour changes, such as postural changes. Observation is also a commonly used technique in understanding human factors of work and activity in context (e.g. Farrington-Darby, Wilson, Norris & Clarke, 2006).

There has however been little work in the field of human factors investigating the nature of such observable behaviours in a work context, and in particular, looking at the relationship between observable behaviours and subjective reports. The concept of 'behavioural markers' is accepted within crew resource management (Flin & Martin, 2009) and observation of medical practice (Carthey, de Leval, Wright, Farewell, Reason, 2003) as well as being used in support of diagnosis of some clinical conditions (e.g. Bellassen, Igloi, Cruz de Sousa, Dubois & Rondi-Reig, 2012).

This paper examines the potential for applying a structured approach to collection of data in a range of contexts to enable increased use of observed behaviours in informing human factors theory. Two cases are presented from the domains of rail control and air traffic control.

2. CASE 1: WORKLOAD IN RAIL SIGNALLING OPERATIONS

Case 1 was led by the third author and had the overall aim of investigating the impact of automation on rail signalling performance. As part of this work, different methods to measure the impact of automation were applied, including structured observation of performance and subjective reporting of workload.

Balfe et al (2008) derived a structured approach for observing rail signalling operations, identifying interaction, planning, monitoring, quiet time and communications as five broad categories of behaviour that formed a mutually exclusive and exhaustive set, with detailed sub-categories for further behavioural analysis (see Sharples et al., 2011 for more detailed description of observation framework). This approach is now used widely within the UK rail industry and has been demonstrated to yield high inter-rater reliability (76%-95%, Balfe, 2010). A broad view of workload can to a certain extent be inferred by examining the distribution of the behaviours amongst the different categories (i.e. a person with high workload can be assumed to have less quiet time for example) but in other cases the relationship between observed behaviour and experienced workload may be less clear cut- for example it is known that an individual who is experiencing higher workload may be working harder to maintain the same level of performance – therefore a high level of workload may be reflected in a higher level of interaction, or there may be no increase to the level of interaction but instead an increase in active monitoring time.

In addition, a dedicated tool has been developed to measure workload in rail signallers – the Integrated Workload Scale (IWS) (Pickup, 2005). This scale has been used extensively within the rail environment to analyse operator behaviour and inform design of future rail signalling environments. The study therefore applied both the IWS and Balfe et al's (2008) observational framework to obtain data to examine more closely the relationship between subjectively reported experiences of workload and observed interaction in rail signalling.

2. 1. Case study 1 method

The study used a simulator to examine the performance and behaviour of six rail signallers. The simulator asked participants to complete a 30 minute scenario three times with different levels of signalling automation in use. Participants were asked to report their subjective level of workload using the IWS (which yields a univariate workload score of between 1-9) every two minutes and the observation classification scheme was applied throughout the 30 minute period. Data from the condition where the highest level of automation was present (Automatic Route Setting) is presented here. A period of disruption was introduced half way through each session to introduce variability in the types of actions required and perceived level of workload.

2. 2. Case study 1 results

The detailed results of this study are reported in Balfe (2010) – this paper specifically examines the relationship between the observed behaviour and focuses on the condition with the highest level of automation.

Figure 1. shows the pattern of workload reported by participants during the 30 minutes period of work. It can be seen that in general workload increased after the period of disruption was introduced into the task, half way through the experimental period. In addition, it is noted that the SDs represent variability within participants, suggesting that even though the conditions were quite closely controlled there was still a difference in the absolute level of workload reported (consistent with typical subjective reporting methods that have more relative than absolute validity).

Figure 2 shows the distribution of different activities during the 30 minute period of observation. It can be seen from the graph that during the first period of work the dominant activity was passive monitoring, and that after the period of disruption there was an increased level of active monitoring and intervention.

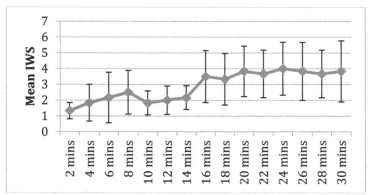

Figure 1. Mean & SD of self-reported workload over 30 minutes for participants using Automatic Route Setting

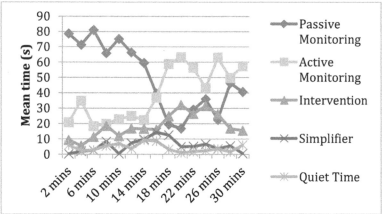

Figure 2. Mean seconds spent in different types of activities over 30 minute period for signallers using automatic route setting.

Statistical analysis reveals that overall, there were significant correlations between observed passive monitoring, active monitoring and intervention (Passive, r=-0.526, df=87, p<0.001; Active: r=0.444, df=87, p<0.001; Intervention: r=0.342, p<0.001, df=87). However, when data from individual signallers was examined it was clear that for some signallers there was a very strong relationship between their observable behaviour and their reported workload, whereas for others there was no obvious relationship. Of the six signallers observed, four had strong relationships between observed monitoring behaviour (passive or active) with correlation coefficients ranging from 0.62 to 0.84, but two had very low correlations ranging from 0 (due to no variation in reported subjective workload levels) to 0.13. This appears to indicate that there may be two types of people – those whose subjective workload states are 'observable' and those whose states are not observable. Figure 3 shows examples of signallers who a) had a high level of correlation between observed behaviour and subjective workload and b) who did not demonstrate relationship between observed behaviour and subjective workload.

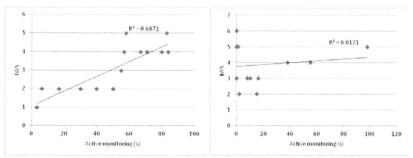

Figure 3. Example of signaller a) with strong correlation between observed behaviour and subjective report of workload and b) with weak correlation between observed behaviour and subjective report of workload

3. CASE 2: PERFORMANCE, BEHAVIOURS AND SUBJECTIVE REPORTS IN AIR TRAFFIC CONTROL

Case 2 was led by the second author and had the overall aim of examining interactions between different influencing factors on air traffic control performance. This study also used a simulated environment to elicit a range of actions and experiences. Participants were students who had been trained in ATC principles during a dedicated training session and who had demonstrated their knowledge of these principles in a practice task. The study was part of a larger programme of work to investigate interacting effects of different factors on ATC performance, therefore a number of measures were applied including physiological measures, eye tracking, a range of subjective measures, performance analysis and observation of a series of specific 'behavioural markers'. This paper just considers the results from the subjective measures of workload, fatigue and stress, measurements of posture and provides some comments on initial results from analysis of behavioural markers. This study did not have a structured observation framework to the same extent as the work in the rail control context, but instead examined whether there were any potential inferences from looking at posture alone, specifically angle of a person's back when seated at the control desk. Posture was therefore felt to be a useful indicator of 'observable behaviour' and has some analogy with the difference between active and passive monitoring in study 1 (where passive monitoring is usually reflected in a more reclined posture than active monitoring).

3. 1. Case study 2 method

Case study 2 applied a unidimensional workload tool developed for the ATC environment – the Instantaneous Self-Assessment tool (ISA). Stress was measured by the Stress Arousal Checklist (Mackay et al., 1978) and fatigue by a visual analogue scale (i.e. Lee, Hicks, Nino-Murcia 1991). The workload measure (ISA) was applied every 4 minutes, whilst measures of fatigue at stress were applied alternately every 11 or 12 minutes. This time variance was necessary to correspond with specific break points in the simulation program. Posture was measured continuously and average posture (in terms of degrees of back from vertical) was calculated for every 4 minutes, corresponding with measure periodicity of workload. The task completed by the participants required acceptance and routing of aircraft and two periods of high work demand were introduced into the task to elicit variability in performance and experience data.

3. 2. Case study 2 results

Figures 4, 6 & 7 shows the mean and SD of subjective reports of workload, fatigue and stress over the two hour period of task completion. It can be seen that all three of these graphs show two 'peaks' which correspond to the manipulated high levels of demand. The posture graph shows as similar pattern with participants in general leaning further forward at the times when high demand was experienced and higher levels of workload, fatigue and stress were reported.

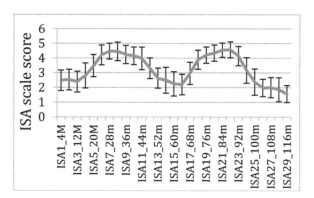

Figure 4. Mean and SD of self-reported workload over 116 minutes for participants completing simulated ATC task

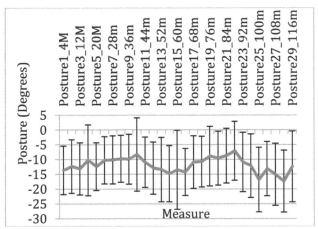

Figure 5. Mean and SD. Posture over task period

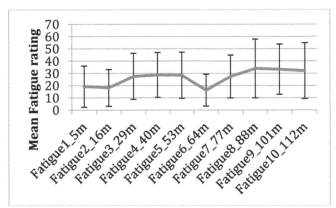

Figure 6. Mean and SD of self-reported fatigue over task period

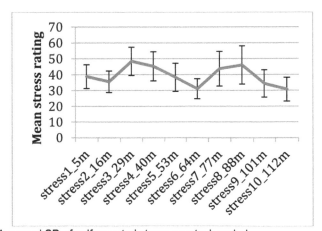

Figure 7. Mean and SD of self reported stress over task period

Overall correlation analyses reveal that there were significant correlations over the entire time period between posture and workload (r_s=0.17, df=839 p<0.0005) and posture and stress (r_s=0.21, df=288, p<0.005) but that there was not a significant correlation between posture and fatigue (posture and fatigue r_s=0.1, df=288, p>0.05).

Analysis was also completed to investigate whether, as in case study 1, there appeared to be different levels of 'observability' for different participants. Figure 8 demonstrates the range of correlation coefficients obtained for individual participants for workload and posture (i.e. a single bar on the chart represents the correlation coefficient obtained for all data for that individual participant when comparing workload and posture) This graphical representation suggests that in fact there may be three different levels of 'observability' of different participants as represented by the ovals on the graph – those whose correlations are above 0.59, those between 0.59 and 0.34, and those below 0.3.

The posture analysis, like the observation framework applied in case study 2, applied a mutually exclusive and exhaustive categorisation of behaviour. However, in this study, specific instances of behavioural markers were also observed (calculated as number of times a behaviour was observed during each 20 minute period of the task). Initial analysis reveals very strong correlations between instances of discrete events, such as sitting back, (r_s=-0.98, df=3 p<0.005) sat forward towards screen (r_s=0.9, df=3 p<0.05) and slump down in seat (r_s=-0.98, df=3 p<0.005) and self report of workload.

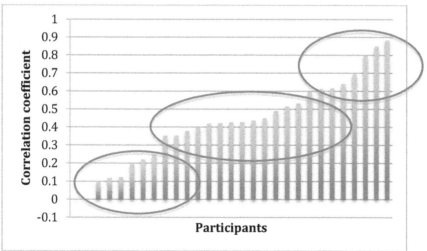

Figure 8. Representation of correlation coefficients (between posture and workload) for individual participants, suggesting the possible existence of three different levels of 'observability' of participants

4. DISCUSSION

This paper has presented two examples where subjective report data appears to have an association with observable behaviour. This is valuable, as it provides support for the instinctive belief that meaningful insight can be gained from detailed observation of behaviour, which is particularly valuable when we wish to analyse safety critical complex work situations, where anything that disturbs operators, such as completion of self-report measures, may not be practical.

However, it is clear that observation alone is not sufficient to indicate cognitive states such as workload, fatigue or stress. For some individuals there appears to be a strong relationship between observed behaviour and cognitive state, but for others the relationship is not so clear. Therefore, if observation is to be used either for human factors analysis or for real-time observation of performance in operational contexts (e.g. a control manager wishing to observe whether staff are coping with a situation or experiencing a high level of workload) it is critical that methods are found to identify which individuals are highly 'observable' and which ones are not.

The application of multiple methods demonstrated in this paper also helps to shed light on the classical issue of understanding the relationship between workload, effort and behaviour. It has been known for some time that some operators work harder to maintain performance, implying that it is not appropriate to measure workload using primary task performance measures alone; the data here reinforces that view but suggests that performance combined with observations of characteristics of how that performance is achieved (e.g. by changing effort towards intervention and active monitoring and thus reducing passive monitoring time) may provide useful indications of workload for some participants. Physiological measures may provide further opportunity to collect data unobtrusively in real-world situations to further examine this issue.

A mixed methods approach offers continued opportunity to develop understanding of the way in which individuals complete work, and thus how future work systems should be designed. However, it is important to realise that individuals may vary in the extent to which their observable behaviour is related to their subjective experience of work.

ACKNOWLEDGEMENTS

The authors would like to thank the Network Rail Ergonomics Team, Theresa Clarke, Barry Kirwan, Eurocontrol and Laura Millen for their support of and input to the work presented in this paper.

REFERENCES

Balfe N. (2010) Appropriate Automation of Rail Signalling Systems: A human factors study. PhD thesis, University of Nottingham

642

Balfe, N., Wilson, J. R., Sharples, S. & Clarke, T. (2008). Structured Observations of Automation Use. In P. D. Bust (Ed.) *Contemporary Ergonomics 2008*, London: Taylor & Francis, 552-557.

Bellassen V, Igloi K, Cruz de Sousa L, Dubois B, Rondi-Reig L (2012) Temporal Order Memory Assessed during Spatiotemporal Navigation As a Behavioral Cognitive Marker for Differential Alzheimer's Disease Diagnosis. *J. Neuroscience* **32**(6): 1942-1952

Carthey, J., de Levala, M.R., Wright, D.J., Farewell, V.T. & Reason, J.T. (2003) Behavioural markers of surgical excellence. *Safety Science* **41**(5), 409–425.

Farrington-Darby, T., Wilson, J.R., Norris, B.J. & Clarke, T. (2006): A naturalistic study of railway controllers, *Ergonomics*, **49**(12-13), 1370-1394.

Flin, R. & Martin, L. (2001): Behavioral Markers for Crew Resource Management: A Review of Current Practice, *The International Journal of Aviation Psychology*, **11**(1), 95-118

Hart, S. G. & Staveland, L. E. (1988) Development of NASA-TLX (Task Load Index): results of empirical and theoretical research. In P. A. Hancock and N. Meshkati (eds), *Human Mental Workload* Amsterdam: North-Holland, 139-183.

Lee, K.A., Hicks, G. & Nino-Murcia, G. (1991) Validity and reliability of a scale to assess fatigue. *Psychiatry Research*, **36**(3), 291–298

Mackay, C., Cox, T., Burrows, G. & Lazzerini, T. (1978) An inventory for the measurement of self-reported stress and arousal. *British Journal of Social & Clinical Psychology*, **17**(3), 283-284.

Martinussen M. & Richardsen A. (2006) Air traffic controller burnout: survey responses regarding job demands, job resources, and health. *Aviation Space & Environmental Medicine*. **77**, 422– 8.

Mogford, R.H. (1997): Mental Models and Situation Awareness in Air Traffic Control. *The International Journal of Aviation Psychology*, **7**(4), 331-341.

Pickup, L., Wilson, J.R., Norris, B.J., Mitchell, L., & Morrisroe, G. (2005). The integrated workload scale (IWS): A new self-report tool to assess railway signaller workload. Applied Ergonomics, 36(6), 681-693

Schroeder, D.J., Rosa, R. & Witt, A. (1998) Some effects of 8- vs. 10-hour work schedules on the test performance/alertness of air traffic control specialists. *International Journal of Industrial Ergonomics* **21**, 307-321.

Sharples, S., Millen, L., Golightly, D. & Balfe, N. (2011) The impact of automation in rail signalling operations. Journal of Rail and Rapid Transit. Proceedings of IMechE, Part F 225(2) 179-191

CHAPTER 65

Head EAST at the Next Intersection? A Systems Approach to the Evaluation of Road Performance

Paul M. Salmon, Miranda Cornelissen, Kristie L. Young

Human Factors Group, Monash University Accident Research Centre, Building 70, Clayton Campus, Monash University, Victoria 3800, Australia

ABSTRACT

The safety of road users at intersections currently represents a key road safety issue. This paper argues that current reductionist, road user centric approaches to studying behavior at intersections are limited, and that a holistic approach, focusing on all road users, their vehicles and the road infrastructure is required. The Event Analysis of Systemic Teamwork (EAST) is put forward as a suitable approach for studying intersection systems and is demonstrated through a test analysis of performance at three major intersections in Melbourne, Australia. The analysis shows that the EAST method is able to identify key differences across different road users, including the tasks they perform, the knowledge that they use and also their interactions with the road system and infrastructure. In closing, the benefits of using EAST for road safety applications are discussed, along with the implications of the findings for enhancing safety at intersections.

INTRODUCTION

The safety of different road users at intersections currently represents a key road safety issue. In Australia, for example, the majority of urban crashes and a substantial proportion of rural crashes occur at intersections (McLean et al., 2010). Further, in Victoria, Australia, between 2001 and 2005, 47% of car crashes, 47% of pedestrian crashes, 58% of cyclist crashes and 38% of motorcycle crashes occurred at intersections (VicRoads, 2011).

In recent times researchers have made a strong case for the 'systems' approach when considering road user behavior and safety (e.g. Larsson et al, 2010). Under this approach safety is an emergent property arising from the interactions of all parts of the system; thus there is a need for research to consider the entire system, as opposed to component parts in isolation such as individual road users (e.g. drivers). Under this philosophy, intersection conflicts are seen not to arise solely from adverse road user behaviors; rather they are a product of the interaction between various factors, such as the road environment and intersection infrastructure, road user training, road rules and regulations, environmental conditions, and the behavior of all road users (e.g. drivers, pedestrians, riders). Previous studies of road user behavior at intersections, however, have largely been road user centric, focusing on road users in isolation and factors such as driving errors (e.g. Gstalter and Fastenmeier, 2010; Sandin, 2009).

Interventions designed to improve safety will be better informed through considering the entire intersection 'system' as opposed to merely individual road user behavior. This is challenging, not least because existing approaches, such as simulation (e.g. de Winter et al, 2009) and questionnaires (e.g. Reason et al, 1990) typically focus on one component only (e.g. 'driving' simulators, 'driver' behavior questionnaire). To support the systems approach in road safety, new approaches to the assessment of road transport 'system' performance, that consider all road users along with the road environment, are required.

This paper presents a novel application of a popular framework of ergonomics methods to the analysis of intersection 'system' performance. The analysis moves towards systems analyses in road transport by considering the intersection 'system', comprising different road users (e.g. drivers, riders, cyclists, and pedestrians), vehicles, and the road environment. Specifically, using data derived from a semi-naturalistic on-road study, the Event Analysis of Systemic Teamwork framework (EAST; Stanton et al, 2005) was used to evaluate performance at 3 major intersections. The aim was to test the utility of the framework approach for understanding road transport system performance.

EVENT ANALYSIS OF SYSTEMIC TEAMWORK

The EAST framework (Stanton et al, 2005) encompasses a suite of ergonomics methods for analyzing performance within complex sociotechnical systems. To date the approach has been applied for this purpose in a diverse set of domains, including the military, aviation, emergency services, energy distribution, and air traffic control. Underpinning the framework is the notion that complex sociotechnical system performance can be meaningfully described via a 'network of networks' approach. Specifically, performance is analyzed via three different but interlinked perspectives: task, social and knowledge networks. Task networks represent a summary of the goals and subsequent tasks being performed within a system, social networks analyze the organization of the system and the communications taking place between agents (both human and non-human) and knowledge networks describe the information that agents use and share in order to perform activities. This so-called 'network of networks' approach is represented in Figure 1.

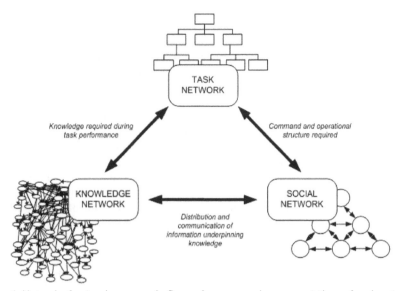

Figure 1. Network of networks approach; figure shows example representations of each network.

EAST uses a combination of ergonomics methods to produce and interrogate these networks. Initially, task analysis (HTA; Stanton, 2006) is used to construct task networks, social network analysis (SNA; Houghton et al, 2006) is used to analyze the social networks and propositional networks (Salmon et al, 2008) are used to construct and analyze knowledge networks. If required, a deeper level of analysis is provided through the application of further approaches, including Operation Sequence Diagrams (OSDs) and cognitive task analysis interview.

INTERSECTION CASE STUDY

Methodology

EAST was applied to data derived from a semi-naturalistic on-road study of driver, rider, cyclist and pedestrian behaviour (see Salmon et al, In press, for full description of study). Five participants from each road user group negotiated a 15km urban test route in the south-east suburbs of Melbourne, Australia (pedestrian participants negotiated the three intersections focused on in this article only). All participants provided concurrent verbal protocols whilst negotiating the route and camera's were used to record road user behaviour and the road scene.

RESULTS

Task networks

'Negotiate intersection' task models were constructed for each road user group based on verbal protocol data and a review of the road scene video data (see Figure 2 for example driver and rider task networks). The task models highlight the

646

differences, in terms of high level goals and tasks performed, between the four road user groups when negotiating intersections. The task models demonstrate that the only task common across the four road user groups is that of 'maintaining situation awareness', which of course entails different things for each form of road user. Pedestrians ostensibly have the simplest goals to pursue in approaching and checking the intersection, determining their path through the intersection, activating the pedestrian crossing controls, waiting for the green man and then crossing the intersection. All three forms of vehicle-based road users have to continuously operate their vehicle whilst maintaining situation awareness and a safe position on the road. They first have to determine their route through the intersection, following which car drivers take the appropriate lane and either follow traffic or lead traffic through the intersection when the light is green. Cyclists and riders, on the other hand can either filter to the front of the traffic queue or take their place in the traffic queue (which may or may not entail crossing lanes in the path of other traffic). Upon receiving a green light they then proceed through the intersection.

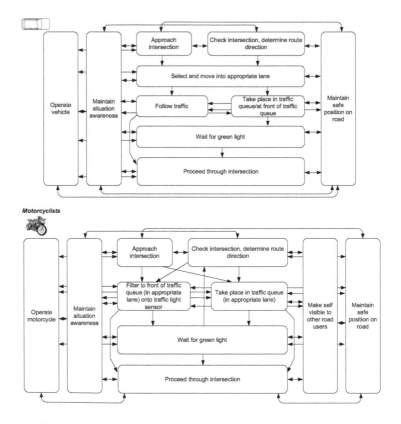

Figure 2. ⮡Negotiate intersection⮡ task networks for drivers and riders.

Although presented at a high level of granularity, the task models confirm that there are key differences in the goals being pursued and tasks being performed across the different road users. Moreover, these differences create the potential for conflicts between road users (e.g. riders and cyclists filtering in conjunction with drivers changing lanes and following traffic ahead). Finally, the analysis suggests that the level of cognitive workload across road users will be different; riders and cyclists, for example, have additional tasks such as making themselves visible to other road users and determining whether or not to filter up through traffic queues.

Social networks

Social networks were constructed for each road user group at the three intersections. In this case the networks comprised human agents (e.g. road users), vehicle agents, and infrastructure agents (e.g. traffic lights, road markings). Links between participants and other agents (e.g. other traffic, traffic lights, and lane markings) were established through analysis of VPA data (i.e. participant's reference to 'checking traffic lights' represents a link between participant and traffic light). Frequency counts within each group were then used to calculate the total number of links present per intersection across the five participants within each group. The social networks for the second intersection are presented in Figure 3.

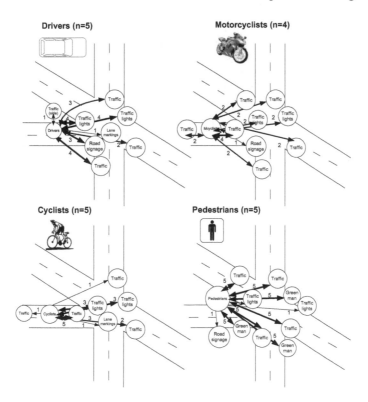

Figure 3. Social network analysis for intersection 2.

The social networks revealed some key differences across the different road user groups. For example, at intersection 1 drivers had less interactions with the road environment and other road users, focusing mainly on the traffic, traffic lights, and lane markings and speed cameras directly in-front of them. Riders and cyclists, on the other hand, had a much richer interaction, focusing on the traffic and traffic lights from all directions as well as the speed cameras and lane markings. This was indicative of the anticipatory nature of motorcycling and cycling through intersections (e.g. anticipating traffic lights, other traffics movements). Driving through intersections, on the other hand, seems to be more characteristic of a 'wait and follow' task in that drivers choose their route, situate themselves in the lane, and follow other traffic through.

Interactions with traffic behind the vehicle were also different across the road users. At all three intersections, riders and cyclists included a focus on the traffic following behind them, whereas drivers did not. Other differences revealed by the social networks were that cyclists checked the condition of the road surface (e.g. for bumps, debris etc) and the obvious difference that pedestrians had a heavy focus on the red and green man crossing indication lights.

Knowledge networks

An intersection knowledge network was constructed for each participant based on content analysis of their verbal transcripts. Example driver, rider, cyclist and pedestrian knowledge networks are presented in Figure 4. Within Figure 4 the nodes represent concepts and the lines represent relationships between concepts. Overall each network provides a representation of each road user's situation awareness when negotiating the three intersections.

The structure and content of the networks was analysed quanitatively and qualitatively. Network structure was analysed using the network density and diamater metrics. Network density represents the level of interconnectivity of the network in terms of links between concepts and is expressed as a value between 0 and 1, with 0 representing a network with no connections between concepts, and 1 representing a network in which every concept is connected to every other concept (Kakimoto et al, 2006; cited in Walker et al, 2011). Higher levels of interconnectivity indicate an enhanced, richer level of situation awareness with more linkages between concepts. Less rich situation awareness is embodied by a lower level of interconnectivity, since the concepts underpinning situation awareness are not well integrated. Diameter is used to analyse the connections between concepts within networks and also the paths between the concepts (Walker et al, 2011). Greater diameter values are indicative of more concepts per pathway through the network (Walker et al, 2011). Denser networks therefore have smaller values since the routes through the network are shorter and more direct. Lower diameter scores are therefore indicative of better awareness, since the holder is able to generate awareness through the linkage of concepts, whereas higher diameter scores are indicative of a model of the situation comprising more concepts but with less links present between them. The mean density and diameters values for participants' networks across the three intersections are presented in Figure 5.

649

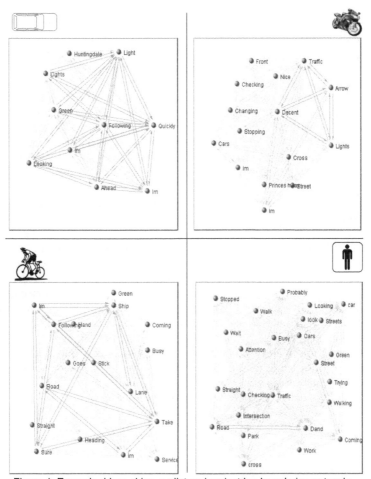

Figure 4. Example driver, rider, cyclist and pedestrian knowledge networks.

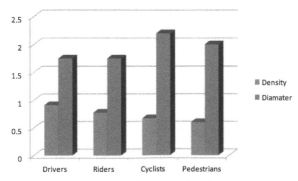

Figure 5. Mean density and diameter values for road user groups.

Figure 5 show that the density of drivers' networks (0.91) was greater than the other three groups when negotiating the intersections. The rider group had the next greatest mean density score (0.77) followed by the cyclists (0.67) and pedestrians (0.60) respectively. These results suggest that driver situation awareness, when examined through network analysis, was more interconnected whilst negotiating the intersections.

The cyclist group had a greater mean diameter (2.2), followed by pedestrians (2) and then the drivers and riders, who achieved the same mean diameter value (1.75). These results suggest that cyclist networks comprised more concepts but with fewer connections between them compared to pedestrians, drivers and riders. This suggests that the concepts underpinning cyclist situation awareness were less well integrated than in the pedestrians, drivers' and cyclists' situation awareness networks.

Analysis of network content involved a qualitative interrogation of the networks. This involved creating a 'master' network for each road user group by combining all concepts within each road user group together. Unique concepts (i.e. present only in one road user groups master network) and common concepts (i.e. that were present in all of the four road user groups' master networks) were then identified. Figure 6 shows the total number of concepts within each master network along with the number of concepts unique to each road user groups. Figure 6 demonstrates that situation awareness in each road user group was different, both in terms of the nature and number of concepts underpinning it. Drivers had a total of 21 concepts within their master network, with 3 of them being unique to drivers. Riders had a total of 24 concepts, with 9 of them unique. The cyclist master network comprised 39 concepts, with 16 being unique. The pedestrian master network comprised 54 concepts, 30 of which were unique to pedestrians. Finally, only 3 concepts (lights, cars, and lane) were common across all four road user groups.

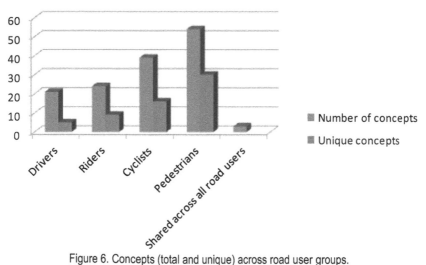

Figure 6. Concepts (total and unique) across road user groups.

DISCUSSION

The aim of this paper was to test the utility of the EAST framework for assessing road transport system performance. The analysis demonstrates that, based on on-road study data, the tripartite network analysis-based approach is suitable for examining road transport performance and is capable of identifying differences across road users relating to the tasks being performed, the interactions with road objects and infrastructure, and the knowledge being used. Specifically, differences were found in the task, social and knowledge networks of drivers, riders, cyclists, and pedestrians when negotiating the same three intersections.

The analysis demonstrates then that, even when faced with the same road situation, the tasks, interactions with road environment, and knowledge used by distinct road users is different. This might not be particularly groundbreaking; however, identifying a structured methodology that is capable of revealing these differences is. Further, the framework used moves towards the much heralded use of systems analysis techniques rather than reductionist, individual road user techniques for road safety-related analyses. Notably, although a systems approach has been called for by many (e.g. Larsson et al, 2010), little has been put forward in the way of methods for supporting such an approach in road safety. Further applications of the EAST framework in the road transport context are therefore urged.

The analysis itself raises questions regarding how well intersections currently cope with distinct road users. If, as the present analysis suggests, tasks, interactions, and knowledge are different across road users, then two pressing lines of inquiry are raised. First, whether these differences create conflicts between road users, and second, the extent to which road system design (e.g. road environment design, training, road rules and regulations) caters for, and supports these differences. Combining the three network representations together reveals both potential conflicts and instances where the road system could better support all road users. For example, riders/cyclists decision to filter up through the traffic queue and enactment of the filtering task occur when drivers are heavily focused on lane selection and the traffic and traffic lights directly in-front of them. Both road users are performing very different tasks, interacting with different parts of the intersection system and using different knowledge, and yet there may be instances where one road user's (driver) tasks, interactions, and knowledge do not incorporate the other (rider/cyclist). Moreover, intersections do not appear to be supporting this interaction: there are no filtering lanes, there are no road signs warning drivers of the likelihood of riders and cyclists filtering, driver training has no focus on cyclists and riders point of view at intersections. An in-depth analysis of intersections, entailing the application of EAST to identify conflicts between road users and inadequate intersection design therefore seems to be an important line of inquiry for enhancing safety and reducing trauma.

As an exploratory proof of concept study this research did have some limitations. First, the study used a small participant sample size which prevented the use of statistical tests on the networks produced. Future studies to be undertaken by the authors will incorporate larger sample sizes amenable to statistical analysis of findings. Second, the road users negotiated the same intersections under similar traffic conditions, but not at the same time as one another (i.e. they were not negotiating the intersections together at the same time). Analysis of intersection system performance when road users are interacting together at the same intersection at the same time would provide more valid data on the potential conflicts between them.

ACKNOWLEDGEMENTS

Dr Paul Salmon's contribution to this research was funded through his Australian National Health and Medical Research Council Post doctoral training fellowship.

REFERENCES

de Winter, J. C. F., de Groot, S., Mulder, M., Wieringa, P. A., Dankelman, J. & Mulder, J. A. (2009). Relationships between driving simulator performance and driving test results. *Ergonomics*, 52:2, pp. 137-153

Houghton, R. J., Baber, C., McMaster, R., Stanton, N. A., Salmon, P. M., Stewart, R., Walker, G. H. (2006). Command and control in emergency services operations: a social network analysis. *Ergonomics*, Vol 49, pp. 1204 – 1225.

Gstalter, H., & Fastenmeier, W. (2010). Reliability of drivers in urban intersections. *Accident Analysis & Prevention*, 42:1, pp. 225-234.

Larsson, P., Dekker, S. W. A., Tingvall, C. (2010). The need for a systems theory approach to road safety. *Safety Science*, 48:9, pp. 1167-1174

McLean, J., Croft, P., Elazar, N., & Roper, P. (2010). Safe Intersection Approach Treatments and Safer Speeds Through Intersections: Final Report, Phase 1, AP–R363/10, Austroads, Sydney, NSW.

Salmon, P.M., Stanton, N.A., Walker, G.H., Jenkins, D.P., Baber, C., & McMaster, R. (2008). Representing situation awareness in collaborative systems: a case study in the energy distribution domain. *Ergonomics*, Vol 51, 3, pp.367 – 384.

Reason, J., Manstead, A., Stradling, S., Baxter, J., & Campbell, K. (1990). Errors and violations on the roads: a real distinction? *Ergonomics*, 33 (10-11), 1315-1332.

Sandin, J. (2009). An analysis of common patterns in aggregated causation charts from intersection crashes. *Accident Analysis & Prevention*, 41(3), pp. 624-632.

Stanton, N. A. (2006). Hierarchical task analysis: Developments, applications and extensions. *Applied Ergonomics*, 37, pp. 55-79.

Stanton, N. A., Salmon, P. M., Baber, C., Walker, G. (2005). Human factors methods: A practical guide for engineering and design. Ashgate, Aldershot, UK.

Vicroads. (2011). CrashStats. Retrieved 17/02/2011, from http://www.vicroads.vic.gov.au/Home/SafetyAndRules/AboutRoadSafety/StatisticsAndResearch/CrashStats.htm

Human Factors Now Showing in 3D: Presenting Human Factors Analyses in Time and Space

Miranda Cornelissen*, Paul M. Salmon*, Neville A. Stanton#, Roderick McClure*

* Monash Injury Research Institute, Monash University, Melbourne, Australia
# University of Southampton, Southampton, United Kingdom
miranda.cornelissen@monash.edu

ABSTRACT

Human centered design requires that designers and analysts apply human factors analyses to real world contexts. However, the outputs of most human factors methods are abstract representations and their relation to the real world context remains absent or implicit, forcing the analyst to infer a link. A review of those Human Factors methods that have some element of space and time in the representation of their outcomes could potentially make the link between abstract outcomes and real world contexts explicit. The aim of this paper was to explore the ability of these selected methods for representing analyses of road user behavior in real world driving contexts. The methods were applied to right hand turns at intersections and the outputs were mapped onto an intersection schematic. The results highlighted strengths and weaknesses of each method and showed that link analysis and network analysis-based methodologies were best able to represent the complexity and interaction of the example task best. It is concluded that these are the most promising approaches for linking abstract outcomes and real world representations of Human Factors analysis when studying complex sociotechnical systems.

Keywords: Human Factors methods, road user behavior, Operation Sequence Diagrams, Link analysis, propositional networks, intersections

654

1   INTRODUCTION

Human Factors methods provide insight into behavior of actors within complex sociotechnical systems. Task Analysis methods are amongst the most popular Human Factors methods. These include Hierarchical Task Analysis (HTA; Annett, 2004), Cognitive Work Analysis (CWA;Rasmussen, Pejtersen, & Goodstein, 1994; Vicente, 1999) and the Critical Decision method (CDM; Klein, Calderwood, & MacGregor, 1989).

Despite the widespread application of these methods, analysis outcomes are often restricted to abstract representations, e.g. a hierarchy of goals (HTA) or qualitative descriptions of decision making (CDM). The methods subsequently fail to place descriptions of behavior in the real world context and force analysts to infer that link. For example, a HTA depicts a hierarchical representation of goals, but fails to indicate where in the real world these goals are pursued.

Addressing the lack of real world representation is important because the lack of such representation is an obstacle for involving Human Factors methods during the early phases of the system or product design life cycle. Without real world representation of analyses outcomes, it is difficult for designers to visualize analysis outputs in the context of real world performance. Context analysis is key to successful task analysis (Diaper & Stanton, 2004). The implications of Human Factors analyses are thus often lost and not being incorporated into designs. Without real world representations designers of the end user system often fail to observe the link between the outputs and the context of their real world problem and will consequently discount the Human Factors input to their design problem.

Some Human Factors methods do represent tasks in time and space, e.g. Operation Sequence Diagrams (OSD; Kirwan & Ainsworth, 1992). While these methods were originally developed to represent temporal aspects of a task or the interaction of users with elements of the system, these could potentially make the link between abstract outcomes and real world application explicit. The aim of this paper is to evaluate Human Factors methods that represent their outcomes in time and/or space to determine whether they can provide the link between abstract representations of some of Human Factors' most popular methods and real world applications based on the above criteria.

1.1   The need for time and space representations in road transport

Road transport represents a domain in which Human Factors based designs are seemingly underrepresented in the early phases of the design process, and often get involved only at the latter stages of the design process to evaluate usability of an engineering based design. In addition, Human Factors research in road transport has mainly concentrated on single road users (c.f. Baldwin, 2009), isolated components such as infrastructure design (c.f. Dill, Monsere, & McNeil, 2012) and technology systems (c.f. Birrell, et al., 2011). From a systems theoretic viewpoint road design is better supported through an understanding of how all road users act together and with the environment in the road transport context (Larsson, Dekker, & Tingvall, 2010; Salmon & Lenné, 2009; Salmon, Stanton, & Young, 2011). Without the involvement of Human Factors methods at the front end of the road design process, cognitive processes of road users, as well as the complex interaction of road users amongst each other and the environment will not be adequately supported by the resulting road design.

The use of Human Factors methods to inform road design could be aided by representing the results of these methods in the road design context. Such methods would have to: first, represent both physical and cognitive processes; second, represent the interaction of road users with each other and the environment; third, allow subsequent application of human centered design principles; fourth, allow a non Human Factors expert to relate the outcomes to their real world context intuitively.

2 HUMAN FACTORS METHODS THAT PRESENT RESULTS IN TIME AND SPACE

A review of Human Factors methods handbooks (Kirwan & Ainsworth, 1992; Salmon et al., 2011; Stanton, Hedge, Brookhuis, Salas, & Hendrick, 2004; Stanton, Salmon, Walker, Baber, & Jenkins, 2005) revealed the following three well established methods that represent their results in time and/or space: Operational sequence diagrams, link analysis, and network analysis methods. These are discussed below.

2.1 Operation sequence diagrams

Process charting methods are used to represent a task sequence or process graphically using standardized symbols. These methods display a breakdown of task step components, the sequential flow of tasks, and the temporal aspects of the activity. For each component task a process chart indicates who will perform it, what collaboration is necessary between which actors, and also what technological artifacts are used to perform the task (Kirwan & Ainsworth, 1992).

Operation Sequence Diagrams (OSD; Kirwan & Ainsworth, 1992) were selected because, of the process charting methods available, it was specifically developed to deal with complex activities and interaction between actors and teams. It was therefore expected to be able to deal with the complexity of the interactions of road users with each other and the environment in the road system. OSDs represent tasks as linear and parallel processes, attempting to resemble the real world temporal aspects of the task. The output of an OSD depicts a task process, the task steps performed and the interaction between agents and technological artifacts or agents over time during the task performance (Kirwan & Ainsworth, 1992).

OSDs allow modeling of the temporal aspects of task analysis results as well as the interaction of actors in the system with each other and the system. The resulting template can then be mapped onto a real world scenario.

2.2   Link Analysis

Interface evaluation methods are used to assess the usability of an interface. These methods aim to improve the design by understanding or predicting user interaction with it (Stanton & Young, 1999). Examples of such methods include heuristic analysis, link analysis and layout analysis.

One method that explicitly represents analyses of human behavior in the real world is link analysis (Drury, 1990). This approach is used to identify and represent the associations between elements of a system during task performance, for example, eye fixations, communication and physical interactions (e.g. button presses) (Stanton & Young, 1999).

A link analysis represents task analysis results as interactions of the different actors with each other and the various elements of the system mapped onto a real world scenario. This will depict patterns in the interactions and will reveal any gaps or conflicts in the requirements for interaction with the system.

2.3   Network methodologies

Network representations represent systems as collections of nodes in a network structure. The meaning of the connections or nodes depends on the theoretical framework of the method. Examples of such methods are Social Network Analysis (SNA; Driskell & Mullen, 2004) and propositional networks (Salmon, Stanton, Walker, & Jenkins, 2009). Whereas these methods do not explicitly map the results in space, network structures have spatial characteristics that could be of benefit to representing results in a real world context.

Propositional networks (Salmon, et al., 2009) are used for modeling distributed situation awareness in collaborative systems. This method depicts a system's awareness as a network of information elements and the relationships between them. These elements are derived from task performance data, often involving verbal transcript or interview recordings of the task execution. Subsequently, in a content analysis, information elements and the relationship between these elements are identified and drafted in a network.

The propositional networks will represent knowledge and information, which can subsequently be mapped onto a real world scenario, relating to the objects in the world it refers to; either static or changing while proceeding through the intersection.

3   APPLYING METHODS TO INTERSECTION SCENARIO

The three methods described above represent Human Factors results in time or space in different ways, but show the potential for representing analyses of different road user behavior in intersection contexts.

To evaluate their utility for this purpose, the three methods are applied to the analysis of intersection behaviors from a real world scenario to compare their representations. To methods are evaluated using the following criteria:

1. The method must represent physical as well as cognitive processes
2. represent the interaction of road users with each other and the environment
3. facilitate human centered design principles to be applied subsequently
4. facilitate non-Human Factors experts to intuitively relate the outcomes to their real world context

The task under analysis will be a right hand turn at a signalized intersection. In Australia, road users travel on the left hand side of the road and right hand turns therefore represent a relatively complex interaction with an intersection and an opportunity to interact with many different road user groups. A right hand turn for traffic on the road (e.g. drivers, motorcycle riders and cyclists) means crossing the intersection and continuing on the intersecting road. Traffic using the footpath (e.g. pedestrians and cyclists) travel to the far right corner, as viewed from their approach.

Data was gathered during an on road study of road user behavior at intersections (Cornelissen, Salmon, & Lenné, 2011). The data from the on road study included driver, motorcycle rider, cyclist and pedestrians Verbal Protocol Analysis (VPA), provided as they negotiated a route comprising seven intersections, and post drive Critical Decision Method (CDM; Klein, et al., 1989) interviews focusing on the three right hand turns conducted at major intersections encountered during the route. Both the VPAs and CDMs were transcribed verbatim.

3.1 Operation Sequence Diagrams

OSDs of a right hand turn at a signalized intersection were constructed for all road user groups using the OSD template. An OSD for drivers is depicted in Figure 1. The template was populated with data from a Task Analysis of the transcripts detailing a right hand turn. A description of this Task Analysis is presented elsewhere (Cornelissen, Salmon, & Young, under revision). The main part of the diagram describes the task steps needed to execute a right hand turn at a signalized intersection and aims to reflect the temporal aspect of such task. The numbers 2.1 – 3.0 represent tasks that can or will be executed continuously, with the exception of maintaining position (2.2), which is not executed during changing lanes (1.2). An OSD usually contains a row for each agent (human and technological) involved in the task. However, the OSDs failed to comprehensively depict all road users and the technological agents in one diagram due to the interactive nature of the task and the multiple relations between task steps, e.g. directional control (3.0) is used for avoiding collisions with objects (2.6). Therefore the diagrams were constructed per road user. Standardized symbols are used to represent the task steps, with symbols representing operational activities, receiving information, and decisions. This analysis focuses on the operations (circle) and decisions (diamond). Again the interactive nature and the continuous use of information from the environment by road users, which is only partially restricted by time and space, lead to a level of complexity that the OSD failed to represent meaningfully.

658

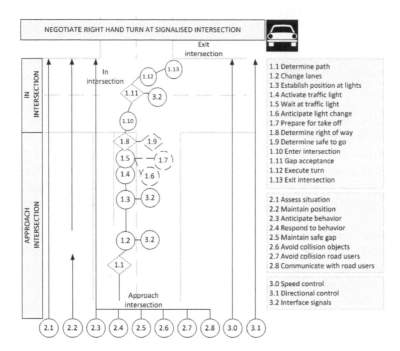

Figure 1. Operation Sequence Diagram for drivers

3.2 Link analysis

A link analysis was constructed based on the analysis of road users' transcripts of the VPA and CDM interviews. Interactions of road users with objects and actors in the system were derived from these transcripts and represented in spatial diagrams. The frequency of interactions with the elements is represented in Figure 2 by the thickness of the lines in the link analysis diagram (i.e. the more interactions the thicker the link).

While the analysis showed differences in the diagrams between all road users, the main difference existed between road users on the road and those using the footpath. Figure 2 presents the results for both drivers and pedestrians. The analysis showed a difference in the elements of the environment that drivers and pedestrians interact with. For example, pedestrians mainly attend to the facilities provided, such as traffic lights, as well as the different potential stages of their crossing. While drivers also focus on the facilities provided, e.g. traffic lights, they attend to more objects all together and tend to focus more on traffic at the intersection, other traffic's traffic lights, traffic in front, as well as the road infrastructure including lane markings, cats eyes, median strips and traffic islands.

659

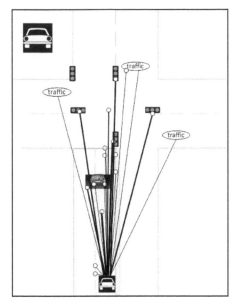 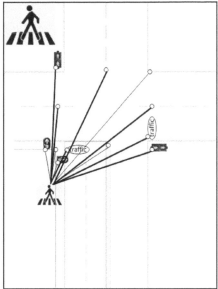

Figure 2. Link Analysis for Driver (left) and Pedestrian (right)

3.3 Propositional networks

The propositional networks were constructed using the VPA verbal transcripts analyzed with the Leximancer™ content analysis software. Leximancer™ automates the content analysis procedure by processing verbal transcript data through five stages: conversion of raw text data, concept identification, thesaurus learning, concept location and mapping (i.e. creation of a network). This led to the creation of a network representing situation awareness for each road user group. Each network therefore aggregates data from the on road study of three right hand turns for each participant in a particular road user group. The concept mappings produced by this software, were exported to Microsoft Visio and subsequently mapped onto the intersection representation. If nodes, such as traffic lights, had a particular physical presence in the intersection scenario, nodes were placed at that position in the intersection. Nodes related to the road users thinking process, such as 'I am' or 'because' were placed close to the road user. Furthermore, nodes related to actions of road users in relation to objects in the environment, such as following or looking, were place on the edge of the scenario in between the road user and the objects. In addition, if a node was represented more than once in the intersection space, inferred from the transcripts and networks, the node was represented multiple times. For example, traffic can be present at the four road sections of the intersection, representing traffic ahead, side and oncoming traffic, and is therefore presented four times in the representation of the intersection. The propositional networks for drivers and motorcycle riders are presented in Figure 3

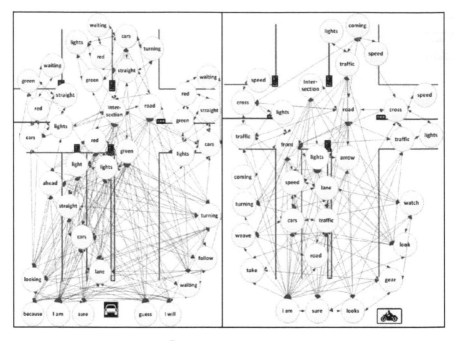

Figure 3. Propositional network for drivers (left) and motorcycle riders (right)

The networks depict a representation of the road user's use of information and knowledge during the task. It shows information derived from objects in the environment, e.g. 'lights' and 'lane', and concepts related to task execution, e.g. 'looking' and 'follow'. The networks demonstrate that although drivers and motorcycle riders have the same goal (execute a right hand turn at a signalized intersection) both use different information and knowledge in pursuing this goal. The driver is mainly occupied by the status of the lights, both the ones facing the driver as those at other sections, and waiting before the turn can be executed. While the motorcycle rider also engages with the traffic light, the overall focus is much more on the interaction with other traffic. This is represented by the nodes depicting the weaving or filtering in between traffic in their own lane, as well as concerns with cross and oncoming traffic interfering with their path.

DISCUSSION

The aim of this paper was to establish whether Human Factors methods that represent their results in time and space can bridge the gap between the abstract representation of results, and real world application, the end goal being to facilitate the integration of Human Factors analyses outputs into the system design lifecycle. Three different methods were applied to an intersection scenario: OSDs, link analysis, and propositional networks. The analysis demonstrated the different approaches to representing tasks in time and or space as well as strengths and weaknesses of these methods in their capacity to bridge the gap between abstract representations and real world applications. The results are summarized in Table 1.

Table 1. Summary table

	Operation Sequence Diagram	Link Analysis	Propositional networks
Represent turning right at intersection in time	✗ Linear process fails to represent continuous and interactive aspect of task	✗ Method uses spatial representation, although changes over time in spatial diagram can be depicted	✗ Method uses spatial representation, although changes over time in spatial diagram can be depicted
Represent turning right at intersection in space	✗ Linear process fails to represent continuous execution of tasks in space	✓ Links between road users and elements are depicted in intersection space	✓ Knowledge networks are overlaid on elements in space to which they refer
Represent physical and cognitive processes	✓	✓	✓
Represent interaction of road users with each other and environment	✗ Linear and parallel representation of tasks makes comprehensive integration with links between task steps, environment and road users impossible	✓ Represented by links between elements in environment and road users representing gazes, communication physical movements etc.	✓ Represented by information elements and description of relationships between these elements
Application of Human centered design principles	N/A Failed to represent task	✓ Information can be grouped or placed differently based on results Conflicts between road users can be anticipated	✓ Improve information provided based on needs road users Conflicts between road users can be anticipated
Intuitive relation to real world context	N/A Failed to represent task	✓ Links represent gazes, communication and physical movements	✓ Nodes and links describe road users and objects as well as their characteristics and interactions with each other

It is concluded that link analysis and propositional networks provide the most promising means of representing road transport behavior in the context of real world applications. These methods were however not originally intended to be used for this function, and both have a different approach to representing tasks in space. Therefore further application of these methods as the bridging method between abstract representations and real world context is encouraged. It should be verified how the methods deal with input from different types of Task Analysis methods and how they scale up for more complex scenarios. It should also be established whether the above methods could be combined in order to represent multiple aspects of the real world context. Subsequently, the methods should be tested to verify whether the representation in a real world context aids distinguishing between a good and bad design and whether it will identify requirements for and consequences of modifications to designs.

ACKNOWLEDGEMENT

The work is part of Miranda Cornelissen's PhD, funded by a Monash Graduate Scholarship and a Monash International Postgraduate Research Scholarship.

REFERENCES

Annett, J. (2004). Hierarchical task analysis. In D. Diaper & N. A. Stanton (Eds.), The handbook of task analysis for human-computer interaction (pp. 67 - 82). Mahwah, NJ: Lawrence Erlbaum Associates.

Baldwin, C. L. (2009). Individual differences in navigational strategy: implications for display design. Theoretical Issues in Ergonomics Science, 10(5), 443-458. Birrell, S. A., Young, M. S., Jenkins, D. P., & Stanton, N. A. (2011). Cognitive Work Analysis for safe and efficient driving. Theoretical Issues in Ergonomics Science, iFirst.

Cornelissen, M., Salmon, P. M., & Lenné, M. G. (2011). Understanding road user behavior using a system based approach: modeling intersection negotiation using cognitive work analysis. Paper presented at the AAAM, Paris, France.

Cornelissen, M., Salmon, P. M., & Young, K. L. (under revision). Same but different? Understanding road user behaviour at intersections using cognitive work analysis.

Diaper, D., & Stanton, N. A. (2004). Handbook of Task Analysis in Human-Computer Interaction. London: CRC Press.

Dill, J., Monsere, C. M., & McNeil, N. (2012). Evaluation of bike boxes at signalized intersections. Accident Analysis & Prevention, 44(1), 126-134.

Driskell, J. E., & Mullen, B. (2004). Social Network Analysis. In N. A. Stanton, A. Hedge, K. A. Brookhuis, E. Salas & H. Hendrick (Eds.), Handbook of Human Factors and Ergonomics Methods (pp. 58.51-58.56). Boca Raton, USA: CRC Press.

Drury, C. G. (1990). Methods for direct observation of performance. In J. R. Wilson & E. N. Corlett (Eds.), Evaluation of human work: A practical ergonomics methodology (2nd ed.). London: Taylor and Francis.

Kirwan, B., & Ainsworth, L. K. (1992). A guide to task analysis. London: Taylor & Francis.

Klein, G. A., Calderwood, R., & MacGregor, D. (1989). Critical decision method for eliciting knowledge. IEEE Transactions on systems, man and cybernetics, 19(3), 462-472.

Larsson, P., Dekker, S. W. A., & Tingvall, C. (2010). The need for a systems theory approach to road safety. Safety Science, 48(9), 1167-1174.

Rasmussen, J., Pejtersen, A. M., & Goodstein, L. P. (1994). Cognitive systems engineering. New York: Wiley.

Salmon, P. M., & Lenné, M. G. (2009). Putting the 'system' into safe system frameworks. Australasian College of Road Safety Journal, 30(3), 21-22.

Salmon, P. M., Stanton, N. A., Lenné, M. G., Jenkins, D. P., Rafferty, L. A., & Walker, G. H. (2011). Human factors methods and accident analysis: practical guidance and case study applications. Surrey, England: Ashgate Publishing Limited.

Salmon, P. M., Stanton, N. A., Walker, G. H., & Jenkins, D. P. (2009). Distributed Situation Awareness. Aldershot, UK: Ashgate Publishing.

Salmon, P. M., Stanton, N. A., & Young, K. L. (2011). Situation awareness on the road: review, theoretical and methodological issues, and future directions. Theoretical Issues in Ergonomics Science, iFirst.

Stanton, N. A., Hedge, A., Brookhuis, K. A., Salas, E., & Hendrick, H. (2004). Handbook of Human Factors and Ergonomics methods. Boca Raton, FL: CRC Press.

Stanton, N. A., Salmon, P. M., Walker, G. H., Baber, C., & Jenkins, D. P. (2005). Human factors methods: a practical guide for engineering. Aldershot: Ashgate Publishing Limited.

Stanton, N. A., & Young, M. S. (1999). A guide to methodology in ergonomics: Designing for human use. London: Taylor & Francis.

Vicente, K. J. (1999). Cognitive work analysis: Toward safe, productive and healthy computer-based work. Mahwah, New Jersey: Lawrence Erlbaum Associates, Inc

CHAPTER 67

An Attempt to Predict Drowsiness by Bayesian Estimation

Atsuo MURATA, Takehito Hayami, Yusuke Matsuda and Makoto Moriwaka

Graduate School of Natural Science and Technology, Okayama University
Okayama, Japan
murata@iims.sys.okayama-u.ac.jp

ABSTRACT

EEG (*EEG-MPF*, *EEG-α/β*), heart rate variability (*RRV3*), tracking error and subjective rating of fatigue (drowsiness) while performing a simulated driving task were measured to predict drowsiness on the basis of Bayesian estimation method. The relation between these measurements and drowsiness was analyzed. As a result, *EEG-MPF* tended to decrease with the increase of drowsiness. It tended that *EEG-α/β*, *RRV3* and tracking error increased with the increase of drowsiness. Then, we have proposed a method that can predict drowsiness by applying Bayesian estimation method to physiological measurements such as *EEG-MPF*, *EEG-α/β*, *RRV3*. Bayesian estimation carries out a statistical inference using some kind of evidences or observations and calculating the probability that a hypothesis is true. As a result, we confirmed that the proposed method could to some extent predict the symptom of decreased consciousness (drowsiness).

Keywords: EEG, HRV, drowsiness, Bayesian estimation.

1   INTRODUCTION

Importance of monitoring drowsiness during driving has been paid more and more attention. The development of the system that can monitor drivers' arousal level and warn drivers of a risk of falling asleep and causing a traffic accident is essential for the assurance of safety during driving. However, effective methods for warning drivers of the risk of causing a traffic accident have not been established.

Brookhuis et al., 1993, carried out an on-road experiment to assess driver status using measures such as Electro- encephalography (EEG) and Electrocardiography (ECG). They found that changes in EEG and ECG reflected changes in driver status. Kecklund et al., 1993, recorded EEG continuously during a night or evening drive for eighteen truck drivers. They showed that during a night drive a significant intra-individual correlation was observed between subjective sleepiness and the EEG alpha burst activity. End-of the-drive subjective sleepiness and the EEG alpha burst activity were significantly correlated with total work hours. As a result of a regression analysis, total work hours and total break time predicted about 66% of the variance of EEG alpha burst activity during the end of drive. Galley, 1993, overcame a few disadvantages of EOG in the measurement of gaze behavior by using on-line computer identification of saccades and additional keyboard masking of relevant gazes by the experimenter. As EOG, especially saccades and blinks, is regarded as one of useful measures to evaluate drivers' drowsiness, such an improvement might be useful to detect the low arousal state of drivers. Wright et al., 2001, investigated sleepiness in aircrew during long-haul flights, and showed that EEG and EOG are potentially promising measures on which to base an alarm system. Skipper et al., 1986, made an attempt to detect drowsiness of driver using discrimination analysis, and showed that the false alarm or miss would occur in such an attempt.

Many studies used measures such as blink, EEG, saccade, and heart rate to assess fatigue (Fukui et al., 1971, Fumio, 1998, Milosevic, 1978, Piccoli et al., 2001, Sharma, 2006, Tejero et al., 2002). McGregor et al., 1996, suggested caution in interpreting saccade velocity change as an index of fatigue since most of the reduction in average saccade velocity might be secondary to increase in blink frequency. No measures alone can be used reliably to assess drowsiness, because each has advantages and disadvantages. The results of these studies must be integrated and effectively applied to the prevention of drowsy driving. To prevent drivers from driving under drowsy state and causing a disastrous traffic accident, not the gross tendency of reduced arousal level but more accurate identification of timing when the drowsy state occurs is necessary. It is not until such accurate measures to identify drowsiness and predict the timing of drowsy driving is established that we apply this to the development of ITS which can surely and reliably avoid unsafe and unintentional driving under drowsy and low arousal state.

Although the studies above made an attempt to evaluate drowsiness (or sleepiness) on the basis of psychophysiological measures, Landstrom et al., 1999, examined the effectiveness of sound exposure as a measure against driver drowsiness. They used twelve lorry drivers in a total of 110 tests of a waking (alarm) sound system. The effectiveness of the waking sound system was verified through subjective ratings by lorry drivers. This system is used by a driver when he or she feels that their arousal level is becoming lower, and there is a risk of falling asleep. The disadvantage of this system is that one must intentionally and spontaneously use the waking alarm system by monitoring their drowsiness by oneself. Eventually, it is necessary for automotives in future to detect the arousal level of a driver automatically and warn drivers of the drowsy state by using some effective measures such as a waking sound system. Therefore, Murata and Hiramatsu, 2008, and Murata and Nishijima, 2008, have made an attempt to objectively evaluate the drowsiness of drivers using EEG or HRV measures.

Murata and Hiramatsu, 2008, and Murata and Nishijima, 2008, succeeded in clarifying the decrease of EEG-MPF or the increase of RRV3 when the participant's arousal level is low. However, it was not possible to predict the drowsiness on the basis of the time series of EEG-MPF or RRV3. Although detecting the arousal level automatically and warn drivers of the drowsy state is an ultimate goal in such studies, it is, at present, impossible to develop such a system unless such studies (Murata and Hiramatsu, 2008 and Murata and Nishijima, 2008) are further enhanced and the prediction technique on the basis of some useful methodology is established. Few studies made an attempt to predict the arousal level systematically on the basis of physiological measures.

The aim of this study was to propose a useful prediction method of drowsy state. The final and future goal was to apply the results to the development of ITS (Intelligent Transportation System) that can warn drivers of their low arousal state and to prevent driving under low arousal level from occurring. The EEG and ECG during a monotonous task were measured, and it was investigated how these measures changed under the low arousal (drowsy) state. The EEG measurement was adopted in order to evaluate the arousal level accurately as a baseline. The time series of mean power frequency of EEG was plotted on X-bar control charts. Heart rate variability (HRV) measure *RRV3* were derived on the basis of R-R intervals (inter-beat intervals) obtained from ECG. Using a Bayesian estimation method, an attempt was made to predict the timing when the participant actually felt drowsy.

2   METHOD

2.1   Participants

Nine male graduate or undergraduates (from 21 to 26 years old) participated in the experiment. They were all healthy and had no orthopedic or neurological diseases.

2.2 Apparatus

EEG and ECG activities were acquired with measurement equipment. An A/D instrument PowerLab 8/30 and a bio-amplifier ML132 were used. Surface EEG was recorded using A/D instrument silver/silver chloride surface electrodes (MLAWBT9), and sampled with a sampling frequency of 1 kHz. According to international 10-20 standard, EEGs were led from O1 and O2.

2.3 Design and Procedure

The participants sat on an automobile seat, and were required to carry out a simulated driving task. They were also asked to evaluate their arousal level every 1 min according to the following category: 1: arousal, 2: a little bit drowsy, 3: drowsy.

EEG (*EEG-MPF*, *EEG-α/β*) and heart rate variability (*RRV3*) while performing a simulated driving task were derived to evaluate drowsiness. The relation between these measurements and drowsiness was analyzed. As well as the physiological measures above, the tracking error during a simulated driving task was recorded. The psychological rating of drowsiness reported every 1 min was used as a baseline of change of drowsiness with time.

2.4 Data Analysis

FFT was carried out every 1024 data (1.024s). Before the EEG data were entered into an FFT program, the data were passed through a cosine taper window. In such a way, the mean power frequency was calculated. This was plotted as an X-bar control chart. Using the X-bar control chart, the judgment of drowsiness of participants was carried out. The ECG was led from V5 using BiolaoDL-2000(S&ME). On the basis of ECG waveform, R-R intervals (inter-beat intervals) were obtained. Heart rate variability (HRV) measure *RRV3* was derived as follows. The moving average per ten inter-beta intervals was calculated. Variance of past three inter-beat intervals was calculated as *RRV3*, which is regarded to represent the functions of para-sympathetic nervous systems.

In order to apply Bayesian estimation method, the likelihood $P(x_j|H_i)$ was calculated as a ratio of the number of judgment H_i to the total number of judgments (30 judgments). First, an interval of 30 s was selected. When more than 10% of data exceeded the threshold, we judged that the arousal level was low. The 30s-interval was moved by 1s, and the judgment of arousal level was carried out for the whole analysis interval. The procedure for calculating the likelihood is summarized in Figure 1.

Using the following Bayesian theorem, $P(H_1|x_j)$ and $P(H_2|x_j)$ were calculated

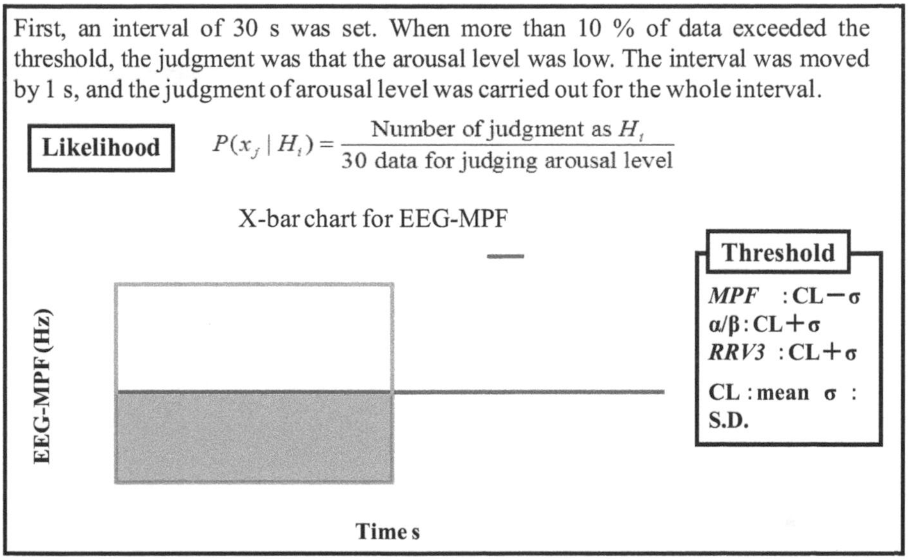

First, an interval of 30 s was set. When more than 10 % of data exceeded the threshold, the judgment was that the arousal level was low. The interval was moved by 1 s, and the judgment of arousal level was carried out for the whole interval.
Likelihood
$P(x_j \mid H_i) = \frac{\text{Number of judgment as } H_i}{\text{30 data for judging arousal level}}$
X-bar chart for EEG-MPF
EEG-MPF (Hz)
Time s
Threshold
MPF : CL−σ
α/β : CL+σ
RRV3 : CL+σ
CL : mean σ : S.D.

Figure 1 Procedure for calculation of likelihood.

Integration of Bayesian judgment
H_1 : drowsy H_2 : arousal
Measurement
x_1: MPF
x_2: α/β
x_3: RRV3
Data
Hypothesis
H_1
H_2
x_1 0.90 0.10
x_2 0.25 0.75
x_3 0.70 0.30
$P(H_i|X)$ 0.875 0.125
Maximum
Bayesian theorem
$P(H_i \mid x_j) = P(H_i)P(x_j \mid H_i)/P(x_j)$
Caluculation of
$P(H_1|x_j)$, $P(H_2|x_j)$
Integration formula
$P(H_i \mid X) = \prod_{j=1}^{M} P(H_i \mid x_j) / P(H_i)^{M-1} \sum_{i=1}^{N} \frac{\prod_{j=1}^{M} P(H_i \mid x_j)}{P(H_i)^{M-1}}$
Data vector $X = (x_1, x_2, \ldots x_j, \ldots x_M)$

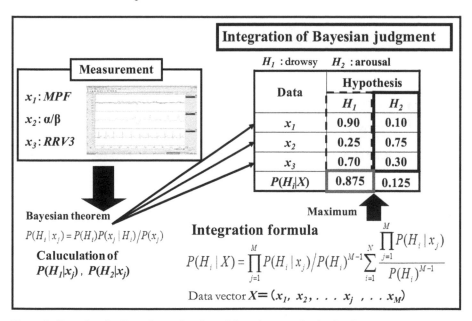

Figure 2 Procedure for Bayesian estimation of drowsiness on the basis of measurements x_1, x_2, and x_3.

using the following formula.

$$P(H_i|x_j) = P(H_i)\, P(x_j|H_i) \,/\, P(x_j)$$

668

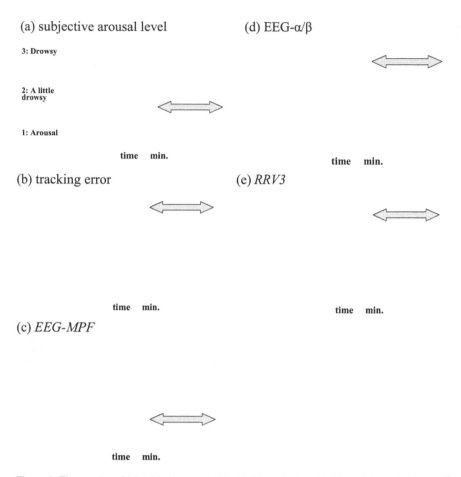

Figure 3 Time series of (a) subjective arousal level, (b) tracking error, (c) *EEG-MPF*, (d) *EEG-α/β*, and (e) *RRV3*.

According to the procedure shown in Figure 2 (using Eq.(1) below), the estimation of drowsiness by Bayesian estimation was carried out. In an example of the numerical calculation, H_1 is estimated as true.

3   RESULTS

3.1   Time series of biological Information

As a whole, *EEG-MPF* tended to decrease with the increase of psychologically evaluated drowsiness. It also tended that *EEG-α/β*, *RRV3* and tracking error increased with the increase of psychologically evaluated

(a) tracking error

(d) *RRV3*

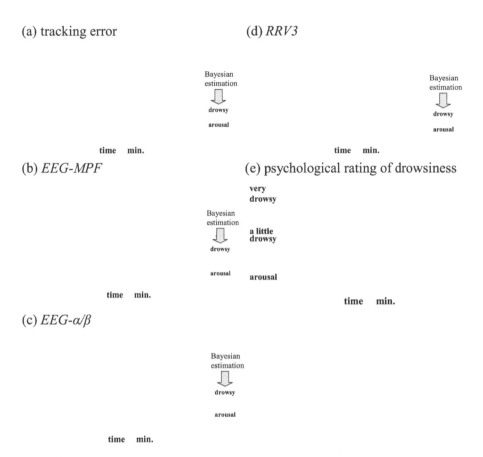

(b) *EEG-MPF*

(e) psychological rating of drowsiness

(c) *EEG-α/β*

Figure 4 Change of Bayesian estimation of drowsiness (participant A). (a) tracking error, (b) *EEG-MPF*, (c) *EEG-α/β* and (d) *RRV3*, and (e) psychological rating of drowsiness.

drowsiness. Such tendency has also been clarified by Brookhuis et al., 1993, Murata and Hiramatsu, 2008 and Murata and Nishijima, 2008. It is important to predict drowsiness using such evaluation measures. The changes of (a) subjective arousal level, (b) tracking error, (c) *EEG-MPF*, (d) *EEG-α/β*, and (e) *RRV3* are depicted in Figure 3.

3.2 Application of Bayesian estimation and prediction of drowsiness

Thus, a method for predicting drowsiness was proposed by applying Bayesian estimation to physiological measurements. Bayesian estimation carries out a statistical inference using some kind of evidences or observations and calculating the probability that a hypothesis is true. The following estimation method was proposed. The conditional probability $P(H_i|X)$ was derived using Eq.(1).

$$P(H_i|X) = \frac{\displaystyle\prod_{j=1}^{M} P(H_i|x_j)}{P(H_i)^{M-1} * \displaystyle\sum_{i=1}^{N} \frac{\displaystyle\prod_{j=1}^{M} P(H_i|x_j)}{P(H_i)^{M-1}}} \quad (1)$$

Data Vector $X = (x_1, x_2, \ldots, x_j, \ldots, x_M)$

In this case, M and N are equal to 3 and 2, respectively. x_1, x_2 and x_3 correspond to EEG-MPF, EEG-α/β, and RRV3, respectively. The data vector X corresponds to (x_1, x_2, x_3). H_1 and H_2 represent the drowsy and the arousal states, respectively. H_1 included the categories 2: a little bit drowsy and 3: drowsy. H_2 corresponded to the category 1: arousal. On the basis of the experimental data, $P(H_1|x_1)$, $P(H_1|x_2)$, $P(H_1|x_3)$, $P(H_2|x_1)$, $P(H_2|x_2)$, and $P(H_2|x_3)$ were calculated. Using Eq.(1), the probabilities $P(H_1|X)$ and $P(H_2|X)$ were calculated every 1s. The estimation was carried out as $\max(P(H_1|X), P(H_2|X))$.

The time series of biological information such as *EEG- MPF, EEG-α/β*, and *RRV3* during the simulated driving task was observed together with the results of proposed Bayesian estimation. As a result, it was suggested that the proposed method can predict the symptom of decreased consciousness (drowsiness). Change of (a) tracking error, (b) *EEG-MPF*, (c) *EEG-α/β*, (d) *RRV3* (participant A), and (e) psychological rating of drowsiness is shown together with the results of Bayesian estimation in Figure 4. When we judged that H_1(drowsy) was likely on the basis of Bayesian estimation, the corresponding physiological measures (Figure 4 (b)-(d)) and tracking error (Figure 4(a)) tended to change in accordance with this.

4   DISCUSSION

As a result of predicting drowsiness by applying Bayesian estimation to biological information, 300-600 s after the continuation of H_1 estimation led to the increasing tendency of tracking error. The results indicate that applying Bayesian estimation to biological information (c) *EEG-MPF*, (d) *EEG-α/β*, and (e) *RRV3* can potentially be used to warn drivers of drowsiness or low arousal state.

Although it was observed that the tracking error increased after 300-600s of the continuation of H1 estimation, it must be explored when the warning must be presented to drivers. If the warning is presented to drivers frequently and unnecessarily, drivers must feel distracted. In future research, it must be discussed when the warning should be presented to drivers so that they don't feel distracted and the timing of warning presentation can be properly and timely presented to drivers so that drowsy driving state can be surely avoided.

The following probability was calculated on the basis of the analytical data. When the judgment H_1 (drowsy state) continued for more than 1 min, the ratio of interval where the judgment of low arousal level (drowsiness) appeared to the total 600 s was calculated using the tracking error data. The mean ratio for all participants (nine participants) was 0.962 (S.D.(Standard Deviation): 0.059). This demonstrates that the Bayesian estimation can predict the appearance of drowsiness with higher accuracy of more than 0.95.

It must be noted that measurement EEG and ECG of drivers and simultaneously carry out the estimation on the basis of the algorithm proposed is impossible. This must be overcome in future to use the proposed in real-world driving environment.

5   CONCLUSIONS

In this study, a prediction method of drowsiness using Bayesian estimation was proposed, and the effectiveness of this method was verified using biological information. As a result, the proposed method led to higher prediction accuracy. In future work, the elaboration of calculation of posterior probability $P(H_1|x_1)$, $P(H_1|x_2)$, $P(H_1|x_3)$ should be accomplished to enhance the prediction accuracy of the proposed method. The algorithm for the faster calculation of posterior probability should also be necessary so that the prediction systems is put to practical use.

REFERENCES

Brookhuis,K.A. and Waard,D., 1993. The use of psychophysiology to assess driver status, *Ergonomics* 36: 1099-1110.

Fukui,T. and Morioka,T., 1971. The blink method as an assessment of fatigue, *Ergonomics* 14: 23-30.

Fumio,Y., 1998. Frontal midline theta rhythm and eyeblinking activity during a VDT task and a video game; useful tools for psychophysiology in ergonomics, *Ergonomics* 41: 678-688.

Galley,N., 1993. The evaluation of the electrooculogram as a psychophysiological measuring instrument in the driver study of driver behavior, *Ergonomics* 36: 1063-1070.

Hershhman,R.L., 1971. A rule for the integration of Bayesian options, *Human Factors* 13(3): 255-259.

Kecklund,G. and Akersted,T., 1993. Sleepiness in long distance truck driving: An ambulatory EEG study of night driving, *Ergonomics* 36: 1007-1017.

Landstrom,U., Englund,K., Nordstrom,B., and Astrom,A., 1999. Sound exposure as a measure against driver drowsiness", *Ergonomics* 42: 927-9379.

McGregor, D.K. and Stern, J.A., 1996. "Time on task and blink effects on saccade duration", *Ergonomics* 39: 649-660.

Milosevic,S., 1978. Vigilance performance and amplitude of EEG activity, *Ergonomics* 21: 887-894.

Murata,A. and Hiramatsu,Y., 2008. Evaluation of Drowsiness by HRV Measures -Basic Study for Drowsy Driver Detection-, *Proc. of IWCIA2008*: 99-102.

Murata,A. and Nishijima,K., 2008. Evaluation of Drowsiness by EEG analysis - Basic Study on ITS Development for the Prevention of Drowsy Driving-, *Proc. of IWCIA2008*: 95-98.

Piccoli,B., D'orso,M., Zambelli.P.L., Troiano,P., and Assint,R., 2001. Observation distance and blinking rate measurement during on-site investigation: new electronic equipment, *Ergonomics* 44: 668-676.

Sharma,S., 2006. Linear temporal characteristics of heart interbeat interval as an index of the pilot's perceived risk, *Ergonomics* 49: 874-884.

Skipper,J.H. and Wierwillie,W., 1986. Drowsy driver detection using discrimination analysis, *Human Factors* 28: 527-540, 1986.

Tejero,P. and Choliz,M., 2002. Driving on the motorway: the effect of alternating speed on driver's activation level and mental effort, *Ergonomics* 45: 605-618.

Wright,N. and McGown,A., 2001. Vigilance on the civil flight deck: incidence of sleepiness and sleep during long-haul flights and associated changes in physiological parameters, *Ergonomics* 44: 82-106.

CHAPTER 68

Effects of Visual, Cognitive and Haptic Tasks on Driving Performance Indicators

Christer Ahlström[1], Katja Kircher[1], Annie Rydström[2],
Arne Nåbo[3], Susanne Almgren[3], Daniel Ricknäs[4]

[1]Swedish National Road and Transport Research Institute
Linköping, Sweden
christer.ahlstrom@vti.se
[2]Volvo Car Corporation
Gothenburg, Sweden
[3]Saab Automobile
Trollhättan, Sweden
[4]Scania
Södertälje, Sweden

ABSTRACT

A driving simulator study was conducted by using the same setup in two driving simulators, one with a moving base and one with a fixed base. The aim of the study was to investigate a selection of commonly used performance indicators (PIs) for their sensitivity to secondary tasks loading on different modalities and levels of difficulty, and to evaluate their robustness across simulator platforms.

The results showed that, *across platforms*, the longitudinal PIs behaved similarly whereas the lateral control and eye movement based performance indicators differed. For *modality*, there were considerable effects on lateral, longitudinal as well as eye movement PIs. However, there were only limited differences between the baseline and the cognitive and haptic tasks. For *difficulty*, clear effects on PIs related to lateral control and eye movements were shown. Additionally, it should be noted that there were large individual differences for several of the PIs.

In conclusion, many of the most commonly used PIs are susceptible to individual differences, and, especially the PIs for lateral control, to the platform and environment where they are acquired, which is why generalizations should be made with caution.

Keywords: driver distraction, performance indicators, workload, simulator

1   INTRODUCTION

Road vehicle manufacturers are now prompted to make sure that equipment in the vehicle (provided as standard or option) does not impose unacceptable workload during driving. For example, the European Commission has established guidelines stating that "the system should be designed so as not to distract or visually entertain the driver" (EsoP Expert Group). Since the guidelines do not prescribe in a testable form what this actually means, the ultimate objective of this study is to design a practically feasible test for secondary task workload.

Attempts have been made to operationalize distraction and to standardise tests for secondary task workload. Examples include the 15-seconds rule (SAE, 2000), the occlusion technique (ISO, 2007), the lane change test (Mattes, 2003), different detection response tasks (van Winsum, Martens, & Herland, 1999), and measures of visual behaviour (ISO, 2002, under development).

The validity of tests that are performed in a laboratory setting, such as the 15-second-rule, has been questioned (eg. Reed-Jones, Trick, & Matthews, 2008). It may also be questioned if secondary task load should be assessed by adding another (third) task, as in the detection response tasks. In this project, validity and safety will be achieved by performing the test in high-fidelity driving simulators (with and without moving base). Further, distraction assessment will not rely on any imposed third task. Instead, the evaluation will be conducted under rather natural conditions. The test will rely on performance indicators (PIs) that are sensitive to different task modalities, to different task difficulty levels and at the same time insensitive to traffic conditions and individual differences.

The main aim of this study was to verify a selection of commonly used PIs that have been reported to be sensitive to distraction and workload. More specifically, the PIs were tested on their ability to quantify different task loads in terms of task modality (visual, cognitive and haptic), task difficulty (easy, medium and hard), traffic (with/without lead vehicle) and simulator type (moving base vs. fixed base).

2   METHOD

The experiment was conducted in the VTI Driving Simulator III, an advanced moving base simulator, and in the Volvo Car Corporation (VCC) driving simulator, an advanced fixed base simulator. Both simulators are equipped with SmartEye Pro eye tracking systems. The moving base simulator is equipped with a visual system providing a 120° forward field of view while the fixed base simulator has 180° field of view. Both simulators use displays for the rear view mirrors.

The simulated road was a rural road (lane width 3.2 m) made up of alternating left and right curves with a radius of 1000 m and with an arc length of 90°. The posted speed limit was 90 km/h.

2.1   Participants

In the moving base simulator 24 drivers participated, and in the fixed base simulator 12 drivers participated. The inclusion criteria were; 30 – 45 years of age, no glasses, and experience with using modern technology while driving.

For the moving base simulator test, 9 women and 15 men were randomly selected (mean age 42 ± 5 years, driver licenses for 23 ± 6 years). For the fixed base simulator test, 4 women and 8 men were recruited amongst VCC employees (mean age 34 ± 7 years, licenses for 17 ± 7 years).

2.2   Task modality

A visual, a cognitive and a haptic secondary task were chosen to test the ability of the PIs to distinguish between different task modalities. The *visual task* was a modification of the arrows task (Östlund et al., 2004). It is a visual search task with the goal to determine whether at least one arrow is pointing upwards. The easiest level consisted of a 4x4 matrix where all the arrows pointed to the left or to the right, except for the possible target arrow. The second level consisted of a 6x6 matrix where all the arrows pointed downwards and the most difficult level consisted of a 6x6 matrix where all the arrows were aimed randomly.

The *cognitive task* was a slightly modified version of the Paced Serial Addition Task (Gronwall, 1977; Sampson, 1956). Single digit numbers were read from tape, with a new number being read every third second. The participants were to respond with the sum of the last two numbers read. The first level consisted of the addition of ones and twos. The second level consisted of additions where the sum was below ten and the third level consisted of additions where the sum of the terms was in the range from ten to twenty.

The goal of the *haptic task* was to insert wooden bricks of three different shapes into matching holes in a wooden box. The box was placed between the front seats, within easy reach of the driver. To prevent the driver from looking down at the bricks the box was covered with a black cloth. There were three cylindrical, three triangular and three cubic wooden bricks to choose from. On the easiest level the drivers were asked to only insert the cylindrical bricks, on the medium level they were asked to insert cubic and/or triangular bricks in any order, and on the hardest level the drivers should alternate between cubic and triangular bricks.

2.3   Design and Procedure

An identical within-subjects design was run in both simulators. For driving performance assessment the factor *modality* was varied on four levels (baseline, visual, cognitive, haptic), the factor *traffic* was varied on two levels (lead car, no

676

lead car), and the factor task *difficulty* was varied on three levels (easy, medium, hard). In addition, the factor *simulator* was tested between subjects.

Each participant completed three blocks where each block focused on one of the three secondary tasks. The order of the blocks was balanced between participants. Within each block, there were always three sections where the participant performed the secondary task; (i) without driving, (ii) driving without a lead vehicle, and (iii), driving with a lead vehicle ahead. The order of the three sections within each block was balanced across participants.

In each block, the participant first drove for 30 s without performing the task. Then the reference task was executed on Level 1 for 30 s. This was followed by another break of 30 s. Then the reference task was executed on Level 2 for 30 s. This was followed by another 30 s break, the execution of Level 3 of the reference task, and finally another break of 30 s.

2.4 Performance Indicators

The projects HASTE (Östlund, et al., 2004) and CAMP (Angell et al., 2006) have made a substantial contribution in investigating the suitability of different PIs for driver state assessment. The PIs used here builds heavily on the work done in these projects.

Each PI was calculated based on the 30-second segments described in the previous section. During baseline, the segments were extracted at corresponding locations along the road. In total, 13 PIs were derived. Eye behaviour PIs were only derived if the gaze quality was above 0.2 for at least 70 % of the data in the segment (gaze quality is a measure from 0 – 1 provided by the eye tracking manufacturer).

Longitudinal PIs were the average speed, the standard deviation of speed, speeding index and minimum time headway (THW). The speeding index basically quantifies how much the driver is violating the speed limit (Hjälmdahl, Dukic, & Pettersson, 2009).

Lateral PIs were the mean lane position, the modified standard deviation of lane position, the number of lane crossings, the steering wheel reversal rate (SWRR), the lane departure distance and the lane departure area. The modified standard deviation is calculated by filtering the lane position time series with a 0.1 Hz high-pass filter before the standard deviation was calculated (Östlund et al., 2005). The SWRR was determined as the number of steering wheel reversals per minute larger than 1°. The lane departure distance was the distance travelled with any part of the vehicle outside the lane boundaries, and, similarly, the lane departure area was determined as the total area that any part of the vehicle was outside the lane boundaries.

Eye movement PIs were percent road centre (PRC), longest glance away from the road and total eyes-off-road time. PRC was defined as the percentage of gaze data that lies within a circle with 8° radius (Victor, Harbluk, & Engström, 2005), the longest glance was the time duration of the longest consecutive sequence of gaze data that was directed outside a circle with 20° radius, and eyes-off-road time was the percentage of time that the gaze was directed outside a circle with 20° radius.

3   RESULTS

There were only minor differences between the two simulator types for PIs related to *longitudinal PIs* (Table 1). There was no overall offset, and the main difference was that the visual task led to stronger speed reductions for the moving base simulator. This had effects both on the mean speed and the speeding index. The mean speed lay close to the posted 90 km/h speed limit when no lead car was present, and was reduced to about 85 km/h for the visual task in the moving base simulator (see Figure 1). In the traffic condition with a lead vehicle present, the mean speed was limited by the lead vehicle and lay at around 87 km/h. Again, the speed reduction could be observed for the visual task for the moving base simulator. The minimum THW was similar for both simulators and all modality types, except that it increased with increasing task difficulty level for the visual and the haptic task in the moving base simulator. However, this increase was due to one outlier who reduced the speed massively and therefore increased the THW so much that the mean of all participants got affected. As can be seen in Figure 1, the values for the individual drivers vary substantially around the mean, especially in the visual condition and with a lead vehicle present. This is not particular to mean speed, but rather the rule for most of the PI tested here.

The *lateral PIs* show a different picture. Here, major differences between the two simulator types can be found, both as a general offset, and for those factor combinations that include modality and task difficulty level. With the visual and the haptic task active, the drivers positioned themselves further to the left and the SDLP was higher in the fixed base simulator. The number of lane crossings did not differ considerably between simulators, but the lane departure area was higher on average in the fixed base simulator, again, influenced by modality, with the highest lane departure areas for the visual task, and in the case of the static simulator, for the haptic task as well. Interestingly, this difference is only evident for the medium task difficulty level, which is a phenomenon that to some extent could be observed for all lateral PIs, for both simulators, especially for the visual and the haptic tasks. The only exception is the steering wheel reversal rate (SWRR), which is based on the driver's input to the car, and not on the car's position in relation to the road. This PI lay at around 15-18 reversals per minute for all modalities, including baseline, for the fixed base simulator, with a tendency for slightly higher values for the most difficult task level. In the moving base simulator the SWRR was significantly lower during baseline, with 10-12 reversals per minute, and the reversal rate did not increase for the more difficult task levels.

Different *eye movement related PIs* were computed for the comparison of the visual behaviour in the two simulators. Most of the PIs had a general offset between the simulators. The drivers in the fixed base simulators accumulated more time with their eyes off the road, especially during the visual and the cognitive task, while the drivers in the moving base simulator cast longer single glances at the display. The only eye movement based PIs that did not differ between simulators was PRC (Figure 3), which reacted in similar ways to modality, interacting with the task difficulty level. The more difficult the visual task, the lower was the percentage of glances to the forward road. The cognitive and haptic task tended to lead to an increased PRC as compared to baseline driving, in the case of the cognitive task with an increasing tendency for more difficult task levels.

Table 1. F-values and significances for all PIs for main effects and 2-way interactions of the within-subjects analyses. Within each row the upper values represent the moving base simulator and the lower values represent the fixed base simulator. A grey box illustrates a significant difference (*p < .05, **p < 0.01) between the two simulator types.

	PI	modality	Level	lead car	modality x level	modality x lead car	level x lead car
Longitudinal	mean speed	17.0**		29.0** 6.2*			
	std speed	3.3* 7.7**		5.9* 9.7*			4.9*
	speeding index	16.1**		31.8** 6.5*			
	min THW		4.8*	n/a		n/a	n/a
Lateral	mean lat pos	3.5* 9.2**	19.7** 48.2**		6.0*		
	mod std lat pos	49.9** 14.5**	9.1** 9.0**	12.7**	8.3** 8.6**		
	# lane crossings	20.3** 5.4**	22.9** 7.2*		3.9*		
	lane dept. area	11.6** 12.2**	19.0**		17.8**		
	lane dept. distance	15.9** 8.9**	11.4** 40.7**		16.3**		
	stw reversal rate	30.3** 40.5**	110.8**	4.7*	14.8**		
Visual	PRC 8 deg radius	67.6** 70.8**	3.9*	8.7*	6.0** 12.4**		
	longest glance	277.8** 8.8*	12.6** 4.9*		13.9**		
	eyes-off-road time	148.6** 43.6**	44.5**		45.8** 21.6**		

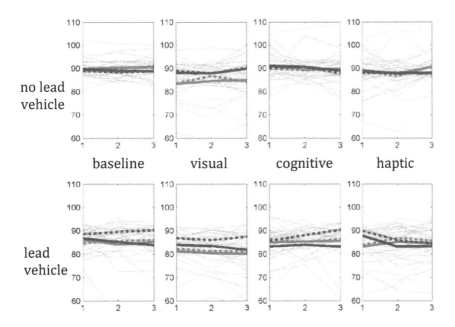

Figure 1. Mean speed per modality, traffic condition and level of task difficulty for individual drivers in the moving base (red) and fixed base (blue) drivers, with mean (bold solid line) and median (bold dashed line) values.

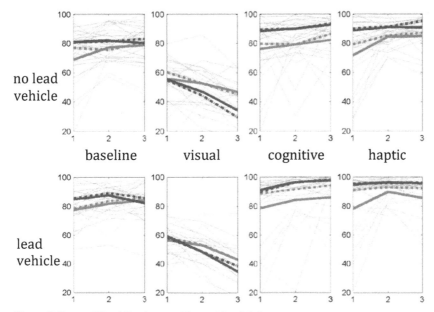

Figure 2. Percent Road Centre, see Figure 1 for details.

4   DISCUSSION

The goal of this study was to investigate commonly used PIs for their sensitivity to different levels of difficulty of secondary tasks loading on different modalities, and to evaluate their robustness across platforms.

The same participants could not be used in both simulators due to the large physical distance between the simulators. This resulted in a confounding between the participants and platform. This is aggravated by different recruiting procedures; the participants driving the moving base simulator were recruited from the general public, while the participants in the fixed base simulator were recruited in-house and consisted of a group of engineers.

Generally speaking, the longitudinal PIs behaved more similarly across platforms, while the lateral control was much more impaired in the fixed base simulator. This finding supports the results from Greenberg et al. (2003), that motion cueing may have a strong impact on lateral driving PIs when disturbances are present. Caution should be exerted in the interpretation of outcomes based on lateral PIs in fixed base simulators. It can be assumed that the lane tracking part of driving is relatively automated for experienced drivers (Land, 2006), therefore it is unlikely that tasks that do not require extended glances away from the forward roadway should lead to decreased lateral control of the extent observed here.

It has become common to bundle modalities into "visual/haptic" and "talking/listening". While it is true that haptic tasks often involve eye glances to aid the hand movements, the results shown here do not fully support a split along those lines. Here the haptic task was conducted without visual support, the eyes could therefore remain directed at the road. Most PIs showed more similarities between the cognitive and the haptic task than the visual and the haptic task. Interestingly, the PIs that were affected most similarly by the visual and the cognitive load were the lateral PIs in the fixed base simulator, whose validity possibly is most questionable. Therefore, it is suggested that haptic tasks without visual component be treated separately, and if they prove not to be very detrimental their increased deployment in HMI solutions should be investigated.

There are some concerns with the results that need to be highlighted. It is unexpected that the lateral PIs should show the most extreme values for the medium level of task difficulty, especially as this could also be observed for the baseline condition, in which no task at all was conducted. For this condition the log data from the corresponding time window were sampled, while the drivers did nothing but to drive throughout the section. One possible explanation might be that the curvature of the road differed in the second segment. For future work of the same kind it is recommended to keep the curve radius and direction constant during secondary task activity, to avoid possible confoundings with road layout. A further concern is the large intraindividual differences found across most of the investigated PIs. In some cases the behaviour is quantitatively different, with some drivers having much higher values than others, but with the general trend being equal, while for other PI the differences appear to be of a qualitative nature. An example is lane departure distance – some drivers do not leave their lane at all, while some drive outside of their lane for a substantial amount of time. This behaviour could be recoded into "manages the task" vs. "does not manage the task". When the goal is to investigate whether a system leads to potentially unsafe behaviour, the PI should possibly be the percentage of drivers who did not manage the task, instead of reporting the average lane departure distance, for example. Using mean values can be very misleading when dealing with distributions that have two or more peaks, or are otherwise not suitable to be represented by a mean value.

As in many studies in the literature, the major part of the PIs used here describes the control level of driver behaviour (Michon, 1985). When evaluating the safety impact of a new system, however, what does it mean if drivers increase their SDLP significantly with a few centimetres? It may be more important to know whether a new system will make drivers commit more violations, or whether a behaviour surfaces that definitely is safety critical. For this type of indicator it may be better to look at the tactical level of driving behaviour, where interactions with other road users play an important role. On this level it is also less likely that behaviour is highly automated, such that a negative impact might be discovered earlier.

To conclude, assessing the effects of secondary tasks on traffic safety via PIs that are based on the control level of driving behaviour is difficult. Results vary widely between participants, and the implications of changes in some PI on safety are not always clear. It is therefore suggested to investigate the possibility of using PIs on the tactical level of driving behaviour, and, instead of reporting mean values, that potentially are not very expressive, it may be worth exploring the notion that a certain safety envelope may not be breached by any or x % of the participants. This may make the interpretation of results much more straightforward.

ACKNOWLEDGEMENTS

We are very grateful to the competence centre ViP for funding the study.

REFERENCES

Angell, L., Auflick, J., Austria, P. A., Kochhar, D., Tijerina, L., Biever, W., et al. (2006). Driver Workload Metrics Task 2 Final Report.

EsoP Expert Group. (2007). Commission recommendation of 22 December 2006 on safe and efficient in-vehicle information and communication systems: update of the European Statement of Principles on human machine interface. Official Journal of the European Union, L 32/200.

Greenberg, J., Artz, B., & Cathey, L. (2003). The effect of lateral motion cues during simulated driving. Paper presented at the DSC North America.

Gronwall, D. M. (1977). Paced auditory serial-addition task: a measure of recovery from concussion. Percept Mot Skills, 44(2), 367-373.

Hjälmdahl, M., Dukic, T., & Pettersson, J. (2009). Uppåt. Uppföljning av åkeriers trafiksäkerhetsarbete (VTI Report No. R658). Linköping: Swedish National Road and Transport Research Institute (VTI).

682

ISO. (2002). Road vehicles - measurement of driver visual behaviour with respect to transport information and control systems, Part 1: Definitions and parameters. ISO 15007-1:2002.

ISO. (2007). Road vehicles - ergonomic aspects of transport information and control systems - occlusion method to assess visual demand due to the use of in-vehicle systems. ISO 16673:2007: International Organisation for Standardisation.

ISO. (under development). Road vehicles - measurement of driver visual behaviour with respect to transport information and control systems, Part 2: Equipment and procedures. ISO/DTS 15007-2.

Land, M. F. (2006). Eye movements and the control of actions in everyday life. Prog Retin Eye Res, 25(3), 296-324.

Mattes, S. (2003). The lane change task as a tool for driver distraction evaluation. In H. Strasser, K. Kluth, H. Rausch & H. Bubb (Eds.), Quality of work and products in enterprises of the future (pp. 57-60). Stuttgart, Germany: Ergonomia Verlag.

Michon, J. A. (1985). A critical review of driver behaviour models: What do we know? What should we do? In L. A. Evans & R. C. Schwing (Eds.), Human behaviour and traffic safety (pp. 487-525). NY: Plenum Press.

Reed-Jones, J., Trick, L. M., & Matthews, M. (2008). Testing assumptions implicit in the use of the 15-second rule as an early predictor of whether an in-vehicle device produces unacceptable levels of distraction. Accident Analysis & Prevention, 40(2), 628-634.

SAE. (2000). Navigation and route guidance function accessibility while driving (SAE Recommended Practice J2364, draft of January 20, 2000). Warrendale, PA: Society of Automotive Engineers.

Sampson, H. (1956). Pacing and performance on a serial addition task. Can J Psychol, 10(4), 219-225.

van Winsum, W., Martens, M. H., & Herland, L. (1999). The effect of speech versus tactile driver support messages on workoad, driver behaviour and user acceptance (No. TNO-Report TM-99-C043). Soesterberg, Netherlands: TNO.

Victor, T. W., Harbluk, J. L., & Engström, J. A. (2005). Sensitivity of eye-movement measures to in-vehicle task difficulty. Transportation Research Part F: Traffic Psychology and Behaviour, 8(2), 167-190.

Östlund, J., Nilsson, L., Carsten, O., Merat, N., Jamson, H., Jamson, S., et al. (2004). HMI and safety-related driver performance (HASTE Deliverable No. 2): European Commission.

Östlund, J., Peters, B., Thorslund, B., Engström, J., Markkula, G., Keinath, A., et al. (2005). Driving performance assessment - methods and metrics (EU Deliverable, Adaptive Integrated Driver-Vehicle Interface Project (AIDE) No. D2.2.5).

CHAPTER 69

Assessment of Different Experiences Related to the Automobile

Caio Márcio Almeida e Silva[1], Maria Lúcia Okimoto[2], Fabrício Benites Bernardes[3], Sônia Isoldi Marty Gama Müller[2]

[1]Universidade Federal do Paraná |
Instituto de Tecnologia para o Desenvolvimento - Lactec
Curitiba, Brasil
caiomarcio1001@yahoo.com.br

[2] Universidade Federal do Paraná
Curitiba, Brasil
lucia.demec@ufpr.br
soniaisoldi@ufpr.br

[3] Porto Alegre, Brasil
fabrício_benites@hotmail.com

ABSTRACT

The paper presents a study that had for objective to verify the varied people's evaluation in different levels of experience with products. To the whole, they announced 30 people divided in two groups of fifteen: one composed by people that had the physical interaction for the first time with the product, and other composition for people with more than two years of use. As result, it was verified that the only category that presented statistical difference was the technology.

Keywords: design, experience, automobile

1   INTRODUCTION

This paper treats of the different types of experiences that we can have with a product. Hekkert (2008) conceptualized the term "Product Experience" as a set of consequences caused by the interaction between a person and a product. This set includes all our senses involved in interaction, product-related meanings, and the resulting feelings and emotions. However, there are some more specific approaches, such as: Karana, Hekkert e Kandachar (2009); Ludden, Schifferstein e Hekkert (2009); Fenko, Schifferstein, Huang e Hekkert (2009); Mugge, Schifferstein e Schoormans (2010); Schifferstein, Otten, Thoolen, Hekkert (2010); Desmet, Nicolás e Shoormans (2008); Okudan e Mohammed (2008); Boztepe (2007); Popovic, Blackler e Mahar (2003); Schifferstein e Cleiren (2005) and Silva (2012).

The experience with the product is differentiated of the usability by some authors. Bevan (2009) for instance, it points that that difference is in the development of the study, as well as in their objectives. In the experience with the product, the focus is to improve all the experience with certain workmanship and in their different levels: expectation, interaction, and powder-interaction.

And in those three interaction levels, we can identify some present components. Ribeiro (2006) it presents five of those components that should be considered in the experience. They are them: sensorial, sentimental, of use, social, and of motivation.

The experience with the product was also worked starting from the intuitive use by Silva (2012). The author considers the importance of the cognitive processing, the sensorial incentives and of the aspects for the intuitive use; for an experience of intuitive interaction. Besides, the author verifies that, sometimes, the expectation of the future user's experience is shown different from the real lived experience.

This paper presents an exploratory experimental study on the experience with the product in different modes of interaction. From the fundamentals of "Product Experience" we propose the development of a study that aims to establish a comparison involving two types of experiences related to a car sedan, that is in the brazilian market.

2   METHODOLOGY

This paper has as objective to establish a comparison involving two types of experiences related to a present automobile in the brazilian market. The experiences are: starting from the first physical contact with the automobile and starting from the users' familiarity with more than two years of use. The tests were accomplished in field and it counted with the aid of: boards with images of the automobile, a video, a notebook STi 15 inches, the automobile, protocol and pen. As techniques, a questionnaire and a scale were used. As metric, we have the metric ones based in specific subjects of the product and the metric of solemnity-report (TULLIS and ALBERT, 2008).

To the whole, they announced 30 people voluntarily. You gave 30, fifteen were subject users of the automobile with more than two years of use (group B); and fifteen were people that had the first physical contact with the interior and external of the automobile (group A). The approach had the ethical aspects assisting the Deontology Code of the Ergonomist Certified - NORMA ERG BR 1002 (ABERGO, 2003), with the signature of a Term of Free and Illustrious Consent for all the participants. The experiment lasted, on average, 15 minutes and it was initiate with the signature of the Term of Free and Illustrious Consent. Soon afterwards the participants were well educated for the moderator of all of the procedures of the experiment.

The approach to the people of the group was accomplished in located dealership in João Pessoa, Paraíba, Brazil. There, the researcher was infiltrated in the team of "after-sales". Like this, the reception of the subjects for the participation of the research was made by one of the atendentes to the people that agendaram services in the workshop. That practice assured the time of the participants' use (confirmed in the register), as well as the random participation. The application of the questionnaires of the group happened in the interval of time between the 26 and January 31, 2012. In that interval of time, they were used four days for such application.

The application for the people of the group B happened from January 26 to February 25, 2012. The place of the experiment varied. In that stage of the research, the moderator drove the automobile until the participant. Like this, the same interacted with the automobile and, soon afterwards, he/she answered to the questionnaire.

The participants evaluated the product, starting from the experience that they had, starting from eight different categories: I comfort, safety, use easiness, speed, technology, stability, pleasure and differentiation. The evaluation was made through a scale of ten centimeters, without graduation (figure 1). The scale had in the left end the denial" "indicative and in the right end the "indicative intensity". As example, we have for the category safety the indicative "is not safe" in the end left, and "very safe" in the right end.

Figure 1: Example of the scale.
Source: The authors (2012)

3   RESULTS

The results were organized in tables starting from the relationship between the participants and the categories:

Table 1: Results of the relationship among the participants the categories, starting from the experience of physical interaction with the product for the first time (I Group A)

Participants	Confort	Safety	Usability	Speed	Technology	Stability	Pleasure	Differentiation
1	9.4	8.9	7.4	9.3	9.5	9.1	9.4	8.6
2	8.3	7.4	7.1	8.8	8.3	8.3	7.5	7.0
3	7.2	6.5	8.2	8.5	8.2	8.9	7.1	5.7
4	9.6	9.1	7.1	8.9	9.3	9.2	6.7	5.5
5	8.6	9.6	9.5	9.7	9.4	9.4	9.4	8.2
6	9.5	9.6	9.6	9.9	9.9	9.8	10	9.9
7	10	10	10	10	10	10	10	10
8	10	10	10	10	10	10	10	10
9	8.6	9.4	9.6	9.7	9.8	9.6	9.7	9.7
10	8.6	9.8	9.8	8.5	9.7	8.5	9.7	9.5
11	7.6	8.2	8.3	8.2	7.9	7.5	8.0	8.8
12	9.3	9.4	8.2	8.4	9.0	9.2	9.5	5.3
13	9.0	7.3	8.7	7.0	8.8	6.6	7.7	2.1
14	9.8	9.6	8.2	3.9	5.1	8.3	8.0	5.9
15	10	9.4	10	10	7.8	9.3	4.9	5.0
Medium	9.03	8.95	8.78	8.72	8.85	8.91	8.50	7.41

For the group A, we had the following ranking of appraised categories:

1. Confort
2. Safety
3. Stability
4. Tecnology
5. Usability
6. Speed
7. Pleasure
8. Diferentiation

Table 2: Results of the relationship among the participants the categories, for users with more than two years of experience with the product (Group B)

Participants	Confort	Safety	Usability	Speed	Technology	Stability	Pleasure	Differentiation
1	8.8	8.7	8.9	8.9	5.2	9.0	8.8	9.2
2	10	10	10	10	3.6	10	10	10
3	5.7	7.8	9.3	8.1	2.8	5.5	5.2	2.5
4	8.6	2.0	8.7	8.2	8.9	9.2	8.8	9.3
5	9.9	9.9	9.9	6.8	1.5	9.8	9.6	5.5
6	8.6	9.9	9.7	9.7	5.7	9.7	10	9.6
7	8.9	9.6	9.9	9.8	9.6	9.9	9.8	4.8
8	9.7	9.8	9.8	9.8	8.8	10	9.9	10
9	7.3	9.5	9.8	7.4	5.7	9.6	9.5	5.8
10	9.4	7.5	9.7	9.9	9.5	9.7	9.7	9.4
11	9.7	9.2	9.9	9.8	9.6	9.7	9.7	9.6
12	9.4	9.7	9.7	9.8	9.9	9.9	8.7	9.8
13	9.1	9.1	9.2	9.2	9.4	9.4	9.6	9.3
14	9.3	10	8.3	10	8.3	10	10	9.9
15	10	10	10	10	8.4	10	10	10
Medium	8.96	8.85	9.52	9.16	7.13	9.43	9.29	8.31

For the group B, had the following ranking of appraised categories:

1. Usability
2. Stability
3. Pleasure
4. Speed
5. Confort
6. Safety
7. Diferentiation
8. Technology

Starting from the tabulation of the data, a statistical test of average comparison was accomplished (test-t) using a level of significance of 5%. Of the eight appraised categories, seven possess averages same statistically. They are them: I comfort (with "P" Valor same to 0,844), Safety (with "P" Valor same to 0,868), use Easiness (with "P" Valor same to 0,026), Speed (with "P" Valor same to 0,377), Stability (with "P" Valor same to 0,187), Pleasure (with "P" Valor same to 0,132) and Differentiation (with "P" Valor same to 0,313).

According to the test, the only category that presented average different statistically it was "technology". That reveals us that the evaluation that the participants had to do on the technology of the automobile him/it for the first time is different from the evaluation on users' that possess more than two years of use technology.

As it was presented by Bevan (2009), the experience can be divided in three levels: expectation, interaction and powder-interaction. In the case of the seven categories that you/they obtained an average same statistically, the level of the expectation was shown equal to the of interaction and powder-interaction. Already in the category "technology", the evaluation among those levels of experience was shown different.

That difference presented in the category "technology" might have been influenced by some factors. Among them, we detached two: experiences previous with similar products and expectation for never to have interacted with the physical product before. In that category, it was confirmed what Silva (2012) highlighted: the difference among the expectation of the experience and to real experience. However, in that case, that was verified starting from different people's groups.

Ribeiro (2006) it presented some components of the experience. They are them: sensorial, sentimental, of uses, social and motivational. For the group A, with the experience starting from the first physical contact, identified mainly the following components: sensorial, social and motivational. Already for the group B, with the experience starting from more than two years of use, identified mainly the following components: sentimental, of use.

4   CONSIDERATIONS

The article presented a study that had for objective to verify the varied people's evaluation in different levels of experience with products. To the whole, they announced 30 people divided in two groups of fifteen: one composed by people that had the physical interaction for the first time with the product, and other composition for people with more than two years of use.

When comparing the equal of the categories, it was verified that the category "technology" presented difference proven statistically. In the other categories, the statistically averages stayed same.

We recommended for future works that the participants' sample is enlarged. Another recommendation, is that she can compare cars different with participants also different. Finally, an unfolding of that research for the design of the information is to apply the methodological procedure of that article to evaluate the experience with the product starting from the video and images; and to compare them with the experiences obtained starting from the use and of the first physical contact.

ACKNOWLEDGMENTS

To Capes and Fundação Araucária for the financial support.

REFERENCES

A. V. Cardello and P. M. Wise. Taste, smell and chemesthesis in product experience, In: Product Experience, Oxford: Elsevier (2008).

B. G. Rutter, A. M. Becka and D. A. Jenkins. 'User-centered approach to ergonomic seating: a case study' *Design Management Journal* Vol Spring (1997) 27–33.

B. Thomas and M. Van-Leeuwen. 'The user interface design of the fizz and spark GSM telephones human factors in product design' in W S Green and P W Jordan (eds) *Current Practice and Future Trends*, Taylor & Francis, London (1999) pp 103–112.

C. M. A. e Silva and M. L. Okimoto. Considerando a intuição no uso de produtos. In: Anais do 11º Congresso Internacional de Ergonomia e Usabilidade de interfaces humano-tecnologia: produtos, informações, ambiente construído e transporte. Manaus, 2011.

C. M. A. e Silva. Experiência com o produto a partir do uso intuitivo. Dissertação de mestrado. Progama de pós-Graduação em Design, Universidade Federal do Paraná, Curitiba, 2012.

D. A. Norman. O Design do dia-a-dia. Rio de janeiro, Rocco, 2006.

D. A. Dondis. Sintaxe da linguagem visual. 2ª edição. São Paulo, Martins Fontes, 1997.

H. C. Okoye. Metaphor mental model approach to intuitive graphical user interface design, College of Business Administration thesis, Cleveland State University, Cleveland (1998).

H. N J. Schipperstein and P. Hekkert. *Product Experience*. Elsevier, 2008.

H. N J. Schipperstein; M. P. H. D. Cleiren. *Capturing product experiences: a split-modality approach*. In: Acta Psychologica 118. Elsevier, p. 293–318, 2005.

H. T. Neefs. On the visual appearance of objects. In: *Product Experience*. Oxford, Elsevier, 2008.

J. Nielsen. Usability engineering. Boston, Academic Press, 1993.

K. Krippendorff. The semantic turn. Boca Raton, Taylor & Francis Group, 2006.

L. Lidwell, K. Holden. B. Jill. Princípios Universais do Design. Porto Alegre, Bookman, 2010.

Csikszentmihaly, M. Flow: The Psychology of Optimal Experience. New York: Harper Perennial, 1991.

M. B. P. Ribeiro. Design experiencial em ambientes digitais: um estudo do uso de experiências em web sites e junto a designers e usuários de internet. Dissertação (Mestrado em Design), UFPE, 2006.

M. H. Sonneveld and H. N. J. e SCHIFFERSTEIN, H. N. J. The tactual experience of objects. In: Product Experience. Oxford: Elsevier, 2008.

M. Van Hout. (2004). Interactive Products and User Emotions. Dissertação de mestrado. Twente.

M. Van Hout. Interactive Products and User Emotions. Dissertação de mestrado. Twente.

M. Wong. Princípios de Forma e Desenho. São Paulo: Editora Martins Fontes.

N. Bevan. What is the difference between the purpose of usability and user experience methods? Proceedings of the Workshop UXEM'09 (INTERACT '09). Uppsala: ACM Press, 2009.

R. Van EgmondVAN. The experience of product sounds. In: Product Experience, Oxford: Elsevier, (2008).

R. Arnheim. Arte e percepção visual, Nova versão, São Paulo, Pioneira, 2005.

T. Frank and A. Cushcieri. 'Prehensile atraumatic grasper with intuitive ergonomics' *Surgical Endoscopy* Vol 11 (1997) 1036–1039.

W. Cybis and A. Bertiol. Ergonomia e usabilidade: conhecimentos, métodos e aplicações. São Paulo, Novatec Editora, 2007.

CHAPTER 70

An Evaluation of the Low-Event Task Subjective Situation Awareness (LETSSA) Technique

Rose, J.A*., [2]Bearman, C., [3]Dorrian, J.

[1,3]University of South Australia
Adelaide, Australia
*rosja005@mymail.unisa.edu.au

[2]Central Queensland University
Adelaide, South Australia

ABSTRACT

Introduction: Accurate situation awareness is an important part of driving a train and it is important to have effective measurement techniques to assess drivers' levels of situation awareness. The most widely used subjective measure of situation awareness is the Situation Awareness Rating Technique (SART) (Taylor, 1990). Although SART was developed using information from military flight crews, Endsley, Selcon, Hardiman and Croft (1998) state that it is suitable for any domain without need for customisation, although there appears to be little or no research confirming this. Based on comparisons between military aviation operations and long-haul train driving and on the researchers' experience of using SART in long-haul train driving, it is argued that a new measure needs to be developed to assess situation awareness for low-event tasks. Based on a task analysis of the train driving task produced by Rose and Bearman (2012) and on Endsley's three stages of situation awareness, a new measure of low-event task subjective situation awareness was developed (LETSSA). This paper describes an initial study that seeks to provide a basic evaluation of the new LETSSA technique. **Method:** To test the new measure, simulator experiments were conducted using participants with no train driving experience. Twenty-three volunteers (20 males, 3 females, aged 22-70y), attended two sessions in a full-cab, high-fidelity train simulator. In the first session, information provided to assist situation awareness was low and in the second session, information provided was high. Measures included a summary measure of train driving violations, subjective performance ratings, NASA Task Load Index (NASA-TLX) workload measure, Situation Awareness Global Assessment Technique (SAGAT) and LETSSA. **Results:** Wilcoxon rank-sum tests revealed significant differences in scores of subjective and actual performance, LETSSA, SAGAT, and workload between the first and second sessions (p<.01). Investigating session-to-session change for each individual participant revealed higher consistency between LETSSA (22 indicated improvement) and actual performance (23 improved) than between SAGAT (only 19 indicated improvement) and actual performance. Correlational analysis of the differences between sessions (high awareness score minus low awareness score) for LETSSA and objective performance, and SAGAT and objective performance found low correlations that were not significant. **Discussion:** The study was designed to produce an increase in situation awareness in the second, compared to the first experimental session. Overall, this was clearly reflected by all measures, including SAGAT and LETSSA. Inspection of change in individual participants revealed that the direction of change was more consistent in LETSSA than in SAGAT (22 versus 19 participants), suggesting that LETSSA may be a better measure in this instance. The lack of correlations with performance in terms of the change between sessions suggests that the magnitude of the change was not well-reflected by either measure. These findings represent the first step in a program of studies that will investigate use of LETSSA in more detail, and compare these results in novices with those in experienced train drivers.

1 INTRODUCTION

The level of awareness that a person has of their current situation (i.e. their situation awareness) is an important part of making decisions, especially when operating within complex, dynamic systems (Endsley, 1995). Accurate situation awareness has been linked to effective decision-making and successful task performance in several domains, including: flight crews (Orasanu, 1995), air traffic controllers (Mogford, 1997), motor vehicle drivers (Gugerty, 1997), and even chess players (Chi, 2006). For example, Durso, Bleckley and Dattel (2006) found that situation awareness was predictive of air traffic controllers' task performance over and above other cognitive tests, such as intelligence, working memory and spatial memory. This suggests that accurate situation awareness is a particularly important element in cognitively-oriented tasks (Durso, Bleckley & Dattel, 2006).

Situation awareness can be defined as the knowledge pertaining to specific situations or environments (Endsley, 1995). According to Endsley, there are three levels of situation awareness: perception, comprehension, and projection. At the first level, the operator needs to perceive elements of the information available in their environment that relate to the tasks they are engaged in. At the second level, the operator needs to understand the significance of these aspects to form an accurate mental picture of their environment. At the third level, the operator predicts what will occur in the future based on their mental picture of the environment.

Situation awareness research to date has been examined mainly in high-event situations characterised by constantly changing information and high mental workload, such as vehicle driving and the take-off and landing phases in aviation (Luther, Livingston, Gipson and Grimes, 2005). In contrast, long-haul train driving is a low-event situation where the driver is monitoring gauges and the external environment for the large majority of time. Due to differences in the cognitive demands of the two types of situations, it may be the case that findings from research in high-event situations may not be transferrable to low-event situations. While information gained from other domains can be used as a starting point for examining situation awareness in long-haul train driving, the validity of such information needs to be examined.

SAGAT (Endsley, 1995) is a method used in simulators to obtain an objective measure of situation awareness (i.e. as part of the scoring system, subjective responses are compared with actual awareness indicators) and has been widely used in many domains (e.g. Endsley & Rodgers, 1996; Matthews, Pleban, Endsley & Strater, 2000). Data is collected by way of questionnaires administered at various freezepoints, or pauses, in a simulation. Questions relate to Endsley's three levels of situation awareness and capture an operator's situation awareness at specific points in time. The questions are not the same for all tasks but must be developed for each specific task. One of the disadvantages of SAGAT is that it cannot be used in the real world as it relies on freezing a simulation to collect data. Measurement of situation awareness in the real world is usually collected via subjective measures, which provide retrospective assessments of situation awareness (as opposed to almost instantaneous measurement as in SAGAT).

Subjective measures of situation awareness are commonly used for pilots and air traffic controllers, with the most widely used measure being the Situation Awareness Rating Technique (SART) developed by Taylor (1990). SART was developed using information from military flight crews and it contains ten dimensions that fall into three clusters: demands on attentional resources (instability, complexity, and variability of the situation); supply of attentional resources (arousal, concentration, division of attention, and spare capacity); and understanding of the situation (information quantity, information quality, and familiarity) (Taylor, 1990). SART can be administered as a 10-dimension measure using a 7-point rating scale for each of the ten constructs or as a simpler 3-dimension scale with ratings for each dimension to be made on a 100 millimetre line from low to high. SART can be administered at the end of a simulation or in naturalistic settings when operators take a break from their task.

There has been much discussion about whether SART does, in fact, measure situation awareness. According to Endsley (1993), two of the three components of SART are measures of workload (supply and demand of resources) and only one component (understanding the situation) is aligned with other definitions of situation awareness. Endsley argues that situation awareness and workload are flip sides of the same coin, with situation awareness resulting from workload as well as generating the need for workload. However, there are other factors that influence situation awareness, such as decision-making and system capabilities, thus it is possible to have high workload with poor situation awareness, low workload with poor situation awareness, low workload with high situation awareness and high workload with high situation awareness (Endsley, 1993). SART ratings have been found to be significantly correlated with subjective workload ratings (Selcon, Taylor and Koritsas, 1991), suggesting support for the argument that components of SART measure workload rather than situation awareness.

Endsley, Selcon, Hardiman and Croft (1998) state that SART is suitable for any domain without need for customisation, however, there appears to be little or no research confirming this. An exploratory study conducted by the authors with ten long-haul train drivers raised concerns regarding the efficacy of SART as a measure of situation awareness for this type of task. When asked to complete the SART questionnaire, several drivers were unsure of the meaning of some of the questions, even though the questions had been modified to make them more appropriate to this participant group. The responses also suggest that variability in some of the dimensions is very low which could affect the overall situation awareness score. For example, nine out of ten experienced drivers rated their familiarity with the situation as very high and the only driver who rated familiarity as low had never driven that section of track before. This suggests then that this element of the SART questionnaire is not sensitive enough to discriminate between drivers on familiar sections of track, which is the vast majority of driving for this group.

There are considerable differences between the cognitive demands of driving long-haul trains and high-event situations, such as piloting a fighter aircraft. In contrast to a fighter pilot for whom every mission is likely to be different, a long-haul train driver usually travels the exact same route on every trip, with the only variations generally being temporary speed restrictions, signal aspects (including crossings), weather (affecting traction and visibility in particular) and occasionally debris/people on or near the tracks. In addition, freight trains will often have variations in loading, number of carriages, and placement of engines, while other types of long-haul trains (such as coal trains) usually have the same number of carriages on every trip (with the occasional variation), either fully loaded or empty. Decision-making for fighter pilots relates to mission objectives, operational risk, aircraft handling and weapons deployment. In long-haul train driving, decision-making relates mostly to increasing speed or applying brakes in order to manage speed appropriately for the track, and to manage buff and draft forces along the length of the train (Biemans, Swaak &Wibbels, 2006).

In terms of planning ahead, fighter pilots' plans may need to be altered rapidly because of external events and multiple plans may be formulated. For long-haul train drivers, the terrain remains the same and drivers frequently travel along the same track, so much of the train driver's task relies on remembering when to begin braking and when to speed up. While there are many factors that must be taken into consideration, such as train and track quality, weather, and visibility, the plan is unlikely to change drastically. In interviews conducted by the authors, the drivers stated that they use various landmarks to know where they are and when to begin braking or accelerating. Thus the planning ahead component is very similar from one day to the next with the exception of temporary speed restrictions, and variations in factors such as weather, and track and train quality. It should be noted that while the task of long-haul train driving may be low-event that does not mean it is simple. Fatigue is a common problem for train drivers, as with other shiftworkers, and the resultant reduction in alertness can lead to errors and an increased risk of accidents (Dorrian, Roach, Fletcher & Dawson, 2007).

One of the questions in SART refers to whether the participant was concentrating on many parts of a situation or focussed on only one. With its focus on high-event domains, where the operator needs to be aware of numerous elements of their environment, it is assumed in SART that if an operator is focussing on only one element they have a lower level of situation awareness than someone focussing on several elements. However, in long-haul train driving, there are sections of track and occasions when the driver should be focussing on only one or two elements of the environment and focussing on too many elements may, in fact, reduce their situation awareness relevant to their current goal. Therefore a low score on this question may not necessarily be a true indication of the driver's awareness.

These concerns with current subjective measures of situation awareness suggest that a subjective measure of situation awareness that is more specific to the task of driving long-haul trains is required.

Following a method similar to that used by Wilson et al (2001), a task-specific, subjective, self-rating measure of situation awareness was developed based on a task analysis of the train driving task compiled by Rose and Bearman (2012) and based on Endsley's (1995) three-level model of situation awareness. The aim was to develop a measure that could combine some of the construct advantages of SAGAT with the logistic benefits of SART. Like SART, the new measure is subjective, can be applied at the end of a simulation and can therefore be used in the real world. However, unlike SART which asks very general questions that purport to be appropriate for all tasks, the new measure is task specific. In this way, it is similar to SAGAT. As a first step in assessing this new measure, it was tested in a pilot study using a full-cab, high-fidelity train simulator. Specifically, this study aimed to evaluate the efficacy of a low-event task, subjective, self-rating measure of situation awareness (LETSSA).

2 METHOD

2.1 Participants

Participants were recruited from the general population via posters in model railway shops and on noticeboards in shopping centres around the Adelaide (South Australia) suburban area. Participants had no prior experience driving trains, held a current motor vehicle driver's licence, had good vision, and were aged between 22 and 70 (M=43.7, SD=12.9). In total, there were 25 participants, two of whom were excluded because they only attended one session. Of the 23 participants who completed both sections of the study, three were female. Participation was voluntary and no remuneration was given.

2.2 Measures

The train simulator is a high-fidelity, full-cab simulator with 26-L braking and a generic console as shown in the picture below. The simulated train used in the test runs was a 594-metre, 4800-tonne, fully loaded grain train with 50 carriages being pulled by three locos.

A number of measures of participant situation awareness, performance and workload were collected during the study. Subjective situation awareness was measured using the new low-event task subjective situation awareness (LETSSA) measure. This measure asks the participant to rate their agreement from 1 to 10 about 10 statements relating to the three levels of situation awareness. For example:

- I knew my speed and whether I was speeding, driving too slowly or about right
- I knew the upcoming track profile and was able to plan ahead to minimise buff and draft

The range of possible scores was 10-100, with higher scores indicating higher situation awareness, and the questionnaire took around two minutes to complete.

Objective situation awareness was measured using the Situation Awareness Global Assessment Technique (SAGAT) (Endsley, 1995). SAGAT is a widely used objective measure of situation awareness, with numerous studies supporting its validity (e.g. Fracker, 1990; Gugerty, 1997) and reliability (e.g. Collier & Folleso, 1995; Gugerty, 1997), thus it was considered an appropriate measure for the purpose of comparing LETSSA with an objective situation awareness measure. In the SAGAT procedure, simulations are paused at one or more locations, at which point operators respond to questions based on Endsley's three levels of situation awareness. Questions may also be asked at the finish of the simulation. SAGAT questions for this study were derived from a task analysis of train driving (Rose & Bearman, 2012), with a combination of multiple choice answers and questions requiring a written response. For example:

- Is the front of your train: going uphill/on level ground/going downhill/don't know
- What is the current speed limit?

Responses to these questions are then marked against the correct answers, providing objectivity. There was a total of 12 questions in each of two questionnaires (one for the midway pause and one for the end of the simulation). The range of possible scores was 0-12 on each questionnaire (0-24 combined total), with higher scores indicating higher situation awareness, and the questionnaires took approximately two minutes each to complete.

An objective measure of performance was calculated based on demerit points such that low scores signified good performance and high scores signified poor performance. Demerit points were given for speed infringements (1 demerit point for every 5 seconds more than 1km above the speed limit), failing to sound the whistle (1 demerit point per instance), excessive buff and draft (1 demerit point for every 100m travelled with forces over 700kN), 3 demerit points for stopping unnecessarily (usually indicating excessive use of air brakes), 5 demerit points for a signal passed at danger, and 2 demerit points for stopping before the red signal is sighted. The lowest possible score is zero and the upper limit depends on individual track design and driver behaviour, and higher scores indicate better performance. The scores were calculated by the researcher's observations during the simulation test runs and data gathered from the simulator's data analyser.

A subjective measure of performance was compiled based on the participant's perceptions of the elements of the objective measure. This measure contained 8 questions requiring multiple choice responses. For example:

- How many times did you enter a temporary speed restriction zone or crossover at higher than the posted speed if at all?
- How often did you fail to sound the whistle or were late to sound the whistle when required if at all?

The responses were either: never, rarely, sometimes, often, or constantly, with scores of 4, 3, 2, 1 and 0 respectively. The range of possible total scores was 0 to 32, with higher scores indicating higher subjective performance, and the questionnaire took approximately two minutes to complete.

Workload was measured using the NASA-TLX questionnaire (Hart & Staveland, 1988), which has been widely used across numerous disciplines and has been found to be a reliable and valid measure of subjective workload (e.g. Xiao,

Wang, Wang and Lan, 2005). There are six questions requiring a response ranging from 0 to 10 on each question, and a range of possible total scores of 0 to 60, with higher scores indicating higher workload.

2.3 Procedure and Scenarios

The procedure used in the study was designed to ensure a difference in situation awareness between the two test simulations. Situation awareness was manipulated through practice and the availability of the train performance display.

Participants attended the train simulator for approximately 2½ -3 hours on two occasions, exactly one week apart.

1-week apart, same time of day		
Session 1: Low SA		
ACTIVITY		**TPD**
Instructions, familiarisation (15 mins)		
Easy practice Run (15 mins)		YES
-BREAK-		
Difficult Practice Run1 (25 mins)		YES
Difficult Practice Run2 (25 mins)		NO
TEST RUN 1		**NO**

Session 2: High SA		
ACTIVITY		**TPD**
Difficult Practice Run3 (25 mins)		YES
Difficult Practice Run4 (25 mins)		YES
-BREAK-		
Difficult Practice Run5 (25 mins)		YES
Difficult Practice Run6 (25 mins)		YES
TEST RUN 2		**YES**

Figure 1. Protocol Diagram – Two Testing Sessions, low and high Situation Awareness (SA). TPD stands for Train Performance Display. Data analysed in the results section comes from Test Run 1 and Test Run 2.

Session 1: Low Situation Awareness (SA) - In the first session, participants were given instructions on driving the simulator and drove a practice run with an easy scenario and a short passenger train set-up. The easy scenario had very little gradient change, no road crossings, no whistle-boards, no people/animals, and very few variations in speed. The initial practice run was followed by two further practice runs on a more difficult scenario, which had steep gradients, crossings, numerous speed limit variations, whistle-boards, and people/animals. The more difficult scenario used a longer, heavier train set-up, which simulated a 594-metre long, 4800-tonne fully-loaded grain train with 50 wagons being pulled by three locomotives. The distance of the track was approximately 13 kilometres, which took approximately 25 minutes to traverse. The scenarios used for the two difficult practice runs used the same section of track but with some variations such as signal aspects, location of animals, and position of track workers. Participants then drove the first test run using a scenario with the long, heavy train and with the difficult track but with changes from the practice scenario in signal aspects, placement of temporary speed restriction/workers, people and animals, and the addition of a car waiting at a crossing. The same order of scenarios was used for all participants.

During the easy practice scenario and the first difficult scenario, a train performance display (TPD) was available, displaying curvature and gradient of track, upcoming signals, speed signs, levels of buff and draft, and current speed limit. This display was turned off for the second practice of the difficult scenario and for the first test simulation. This was done to ensure a relatively low level of situation awareness in the first test run due to the lack of information required to maintain situation awareness.

Session 2: High Situation Awareness (SA) - On their return exactly one week later (at the same time and day to avoid circadian influences [Van Dongen & Dinges, 2001]), participants practiced the test scenario four times with the use of the train performance display. During practice, they were instructed on the best way to drive the simulator to ensure optimum performance including adherence to speed limits, avoiding excessive use of brakes, and minimising buff and draft. The second test simulation was then run, with the train performance display available to provide information required to maintain situation awareness.

In order to ensure that a difference did in fact exist and in order to make comparisons with the new task specific self-rating measure, the Situation Awareness Global Assessment Technique (SAGAT) (Endsley, 1995) was used to measure objective situation awareness. The simulation was paused approximately halfway through the scenario, at which point SAGAT was administered. Immediately following the end of the simulation, SAGAT was again administered, followed by all the other measures, in random order. The method of freezing simulations has been used in numerous studies (for example, Endsley & Bolstad, 1994; Mogford, 1997). Endsley (1995) found that freezing simulations for thirty seconds, one minute or two minutes does not adversely affect performance. LETSSA was administered at the end of each simulation run only.

3 RESULTS

Figure 2 indicates that, overall, there was an improvement in objective and subjective performance, situation awareness as measured using LETSSA and SAGAT measures, and a decrease in workload during the high information compared to the low information session. Wilcoxon rank-sum tests (also shown in Figure 2) indicated that these differences were significant ($p < 0.01$).

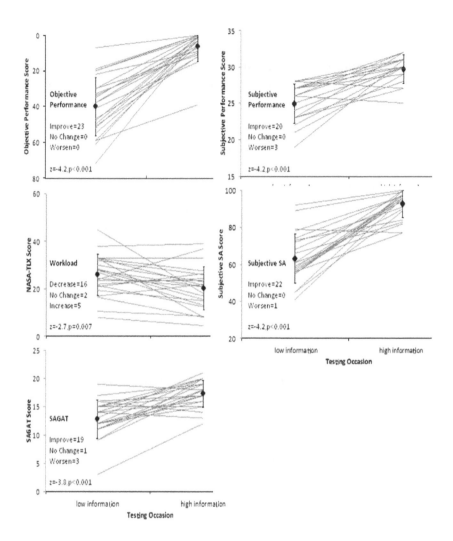

Figure 2. Grey lines indicate scores for each individual (n=23) on objective performance (note the reversed direction of the y-axis), subjective performance, subjective workload, subjective situational awareness (SA) and SAGAT. Diamonds show group means with standard error bars. Number of participants showing improvement, no change or worsening for objective and subjective performance, subjective SA and SAGAT, and a decrease, no change or increase for subjective workload is indicated. Results of Wilcoxon rank-sum test results are given (z, p).

Improvement in objective performance was consistent across all participants and all but one participant showed an improvement in LETSSA. Scores on SAGAT were less consistent, with one "no change" and three "worsen" across the two sessions. There was some variability in subjective performance, and workload had the most varied scores with 16 "decrease", two "no change", and five "increase."

In order to investigate the relationships between the magnitude of the changes between the low and the high information sessions, differences between scores were calculated and correlational analyses of these changes were conducted between measures. All Spearman Rho Rank correlations were low (<=0.38) and not significant (p>0.05, Table 1).

Table 1. Spearman Rho Correlation Matrix.

Obj. Performance	Subj. Performance	NASA-TLX	LETSSA	SAGAT
1.000	0.345	-0.377	-0.218	0.213
0.345	1.000	0.160	0.298	0.019
-0.377	0.160	1.000	0.323	0.175
-0.218	0.298	0.323	1.000	0.101
0.213	0.019	0.175	0.101	1.000

4 DISCUSSION

This study has aimed to provide a basic evaluation of the new LETSSA technique for measuring subjective situation awareness. The LETSSA technique was developed because of concerns with the applicability of SART (the most common measure of subjective situation awareness) to low-event domains. These concerns were based on comparisons between the high-event domains used to develop SART and low-event domains, such as long-haul train driving; and on the researchers' experience of using SART in long-haul train driving. LETSSA presents a more tailored approach to measuring subjective situation awareness, being based on Endsley's (1993) three-level model of situation awareness and a task analysis of train driving (Rose & Bearman, 2012).

The results show that LETSSA can capture expected differences in situation awareness. Both LETSSA and SAGAT were sensitive to the manipulation of situation awareness in this study, which was also reflected in the subjective and objective performance measures. This is an important first evaluation of LETSSA and shows that this measure is an effective measure of differences in situation awareness. Comparison of the individual participants' scores on SAGAT and LETSSA (in Figure 2) shows that the direction of change was relatively consistent across participants on both measures, with LETSSA showing slightly more consistent change than SAGAT. Surprisingly, changes in LETSSA between the two sessions did not appear to be correlated with SAGAT. While LETSSA is designed to measure subjective situation awareness and SAGAT is designed to measure objective situation awareness it would be expected that the change in the two measures would be correlated, particularly as both measures were developed based on the same task analysis (by Rose and Bearman, 2012). It has also been found that SAGAT and SART are not highly correlated. Snow and Reising (2000) suggest that the lack of correlation between SAGAT and SART is because they are measuring different concepts or are based on differing definitions of situation awareness. In the current study, LETSSA and SAGAT appear to similarly reflect the direction of the change in situation awareness, but not the magnitude of this change. The low correlations in the present study should however be interpreted with caution because of the low numbers of participants.

One of the criticisms of SART is that it is correlated with workload (Selcon, Taylor & Koritsas, 1991). The correlation between LETSSA and workload in this study was non-significant, but found to be 0.323. This is higher than expected. It should be noted that workload significantly decreased between the two test runs and subjective situation awareness (as measured by LETSSA) significantly increased. This, combined with the inconsistent changes seen in workload (16 decrease, 2 no change and 5 increase), seems to suggest that the measures are independent. Further research with larger numbers of participants is required to determine the relationship between workload and LETSSA.

A primary consideration for the current study is the use of novices to investigate use of LETSSA. Clearly, use of the new measure with experienced train drivers will likely yield different results. Indeed, this study represents a first stage in a program, which will involve experienced train drivers, in order to specifically compare changes in situation awareness captured by this measure, and the way in which this differs based on task experience. It should also be noted that the questions used in this study were to some extent guided by the limitations of the simulator. Thus, it was not possible to include the various communication tasks required of train drivers, for example. While the current version of LETSSA is appropriate for simulator use, its use in naturalistic settings would be limited. As the LETSSA technique develops it will be necessary to develop a version that encompasses the additional tasks of the train driver (such as communication) so that it better captures the real world situation.

It is clear then, that while LETSSA shows promise as a measure of subjective situation awareness, further research is required to determine its relationship to other measures of situation awareness and measures of workload and performance. Further research is also required to examine whether LETSSA can distinguish between smaller differences in situation awareness, such as may be expected for train driving experts. These studies which form the next stage of the evaluation process are currently being run.

ACKNOWLEDGEMENTS

The authors are grateful to the CRC for Rail Innovation (established and supported under the Australian Government's Cooperative Research Centres program) for the funding of this research. Project No. R2.102, Human Factors Analytic Tools.

REFERENCES

Biemans, M., J. Swaak, and M. Wibbels. 2006. Cabin information support for enhanced decision-making of train drivers. *Proceedings of the 16th World Congress on Ergonomics*, 10-14 July 2006.

Chi, M.T.H. 2006. Laboratory Methods for Assessing Experts' and Novices' Knowledge. In *The Cambridge Handbook of Expertise and Expert Performance*, eds. K.A. Ericsson, N. Charness, P.J. Feltovich and R.R. Hoffman, pp.167-184. Cambridge: Cambridge University Press.

Collier, S.G. and K. Folleso. 1995. SACRI: A measure of situation awareness of nuclear power plant control rooms. In *Experimental analysis and measurement of situation awareness,* eds. D.J. Garland and M.R. Endsley, pp. 115-122. Daytona Beach, Florida: Embry-Riddle University Press.

Dorrian, J., G.D. Roach, A. Fletcher, and D. Dawson. 2007. Simulated train driving: Fatigue, self-awareness and cognitive disengagement. *Applied Ergonomics*, 38, 155-166.

Durso, F.T., M.K. Bleckley, and A.R. Dattel. 2006. Does Situation Awareness Add to the Validity of Cognitive Tests? *Human Factors,* 48, 721-733.

Endsley, M.R. 1993. Situation Awareness and Workload: Flip Sides of the Same Coin. *Proceedings of the 7th International Symposium on Aviation Psychology.*

Endsley, M.R. 1995. Toward a Theory of Situation Awareness in Dynamic Systems. *Human Factors,* 37, 32-64.

Endsley, M.R. and C.A. Bolstad, 1994. Individual Differences in Pilot Situation Awareness. *The International Journal of Aviation Psychology,* 4, 241-264.

Endsley, M.R. and M.D. Rodgers. 1996. Attention distribution and situation awareness in air traffic control. *Proceedings of the 40th Annual Meeting of the Human Factors and Ergonomics Society, Santa Monica, CA,* pp. 82-85.

Endsley, M.R., S.J. Selcon, T.D. Hardiman, and D.G. Croft. 1998. A comparative analysis of SAGAT and SART for evaluations of situation awareness. *Proceedings of the Human Factors and Ergonomics Society 42nd Annual Meeting.*

Fracker, M.L. 1990. Attention gradients in situation awareness. In *Situational Awareness in Aerospace Operations (AGARD-CP-478)*, Conference Proceedings no. 478, pp. 6/1-6/10. Neuilly Sur Seine, France: NATO-AGARD.

Gugerty, L.J. 1997. Situation Awareness During Driving: Explicit and Implicit Knowledge in Dynamic Spatial Memory. *Journal of Experimental Psychology: Applied,* 3, 42-66.

Hart, S.G. and L.E. Staveland. 1988. Development of NASA-TLX (Task Load Index): Results of Empirical and Theoretical Research. In *Human Mental Workload,* eds. P.A. Hancock and N. Meshkati, pp. 139-183. Amsterdam: North Holland Press.

Luther, R., H. Livingston, T. Gipson, and E. Grimes. 2005. Methodologies for the Application of Non-Rail Specific Knowledge to the Rail Industry. *Second European Conference on Rail Human Factors, York.*

Matthews, M.D., R.J. Pleban, M.R. Endsley, and L.D. Strater. 2000. Measures of infantry situation awareness for a virtual MOUT environment. *Proceedings of the Human Performance, Situation Awareness and Automation: User Centered Design for the New Millennium Conference, October 2000.*

Mogford, R.H. 1997. Mental Models and Situation Awareness in Air Traffic Control. *The International Journal of Aviation Psychology,* 7, 331-341.

Orasanu, J.M. 1995. Situation Awareness: Its Role in Flight Crew Decision Making. *Proceedings of the Eighth International Symposium on Aviation Psychology,* pp.734-739, April 24-27, 1995, Columbus, Ohio.

Rose, J.A. and C. Bearman. 2012. Making effective use of task analysis to identify human factors issues in new rail technology. *Applied Ergonomics,* 43, 614-624.

Selcon, S.J., R.M. Taylor, and E. Koritsas. 1991. Workload or situational awareness?: TLX vs SART for aerospace systems design evaluation. *Proceedings of the Human Factors Society 35th Annual Meeting,* 62-66. Santa Monica, CA: Human Factors Society.

Snow, M.P. and J.M. Reising. 2000. Comparison of Two Situation Awareness Metrics: SAGAT and SA-SWORD. *Proceedings of the IEA 2000/HFES 2000 Congress.*

Taylor, R.M. 1990. Situational awareness rating technique (SART): The development of a tool for aircrew systems design. *Situational Awareness in Aerospace Operations,* AGARD-CP-478, 3/1-3/17. Neuilly Sur Seine, France: NATO-AGARD.

Van Dongen, H.P.A. and D.F. Dinges. 2001. Circadian rhythms in fatigue, alertness and performance. In *Principles and Practice of Sleep Medicine* (3rd ed.), eds. M.H. Kryher, T. Roth & W.C. Dement, pp. 391-399. Philadelphia, Pennsylvania: W.B. Saunders.

Wilson, J.R., L. Cordiner,, S. Nichols, L. Norton, N. Bristol, T. Clarke, and S. Roberts. 2001. On the Right Track: Systematic Implementation of Ergonomics in Railway Network Control. *Cognition, Technology & Work,* 3, 238-252.

Xiao, Y.M., Z.M. Wang, M.Z. Wang, and Y.J. Lan. 2005. The appraisal of reliability and validity of subjective workload assessment technique and NASA-task load index. *PubMed, 23,* 178-181.

CHAPTER 71

Designing Developmentally Tailored Driving Assessment Tasks

Erik Roelofs[1], Marieke van Onna[1], Karel Brookhuis[2],

Maarten Marsman[1], Leo de Penning[3]

[1] Cito, National Institute for Educational Measurement, The Netherlands
[2] Delft University of Technology, Groningen University, The Netherlands
[3] Netherlands Organization for Applied Scientific Research TNO
Erik.Roelofs@cito.nl

ABSTRACT

Mislevy's evidence centered design model for assessment was used in order to develop a procedure by which driving task scenarios can be derived as a basis for developmentally tailored driving assessments. Borrowing from recent theories on driving and driving errors, task environment attributes were defined which may complicate the sub processes of driving and thus may result in varying task difficulty. A universe of assessment tasks was defined by combining basic driving tasks and critical task environment attributes. A collection of 55 critical driving task scenarios was selected from 39 video recorded driving lessons, throughout different stages of driving education. Results of a difficulty rating study pertaining to these scenarios including experienced driving instructors show that the scenarios discriminate well between beginning and advanced learner drivers. A contrasting comparison between easy and difficult scenarios revealed that the degree of time pressure, the number of necessary maneuvers to be carried out consciously in a given time-span may have been a discriminating factor in addition to the specified task environment attributes.

Successful scenario solution can be predicted by using an IRT function, where solution probability is a function of driver ability and task difficulty. Implications for assessment design activities are discussed.

Keywords: simulator, assessment for learning, driving, evidence-based design

1 INTRODUCTION: DEVELOPMENTS IN DRIVER TRAINING AND DRIVER ASSESSMENT

During the last decade a shift towards competence-oriented driver training has taken place. Until recently the drivers' task was conceived of as a set of elementary driving tasks pertaining to vehicle control at a low, automated level, and to maneuvering in interaction with other traffic, applying traffic rules. Nowadays, driving is considered as a broad domain of competence, in which the driver is expected to make decisions, taking into account the task environment, combining his interests and those of other traffic participants. Higher order aspects of driving are considered crucial in this process: risk tolerance, reflection on one's own driving behavior and hazard perception. The increased emphasis on integration of lower and higher order aspects of driving is reflected in goals for driver education (the GDE-matrix; Hatakka et al., 2002). Moreover, competence frameworks are being elaborated which specify driver roles and the accompanying knowledge and skills needed for driving. Initial driver training programs increasingly build on these frameworks. As a result, driver training has evolved into a hierarchical learning sequence, particularly stressing the mental processes involved in driving. Driver training is provided by professional driving instructors, who supply tasks in daily traffic. In some European countries a two-phased driver training program, including a pre-license and post-license phase, is put into practice. In addition, initiatives for permanent (ongoing) road safety education have emerged. In sum, acquisition and maintenance of driving competence can be considered an ongoing or even a life-long process.

An indispensable part of driver training and learning is the use of formative assessments, which are used to inform and support the (learner) driver about the levels of driving competence, and the underlying performance aspects that need further attention. Using this information subsequent training may be tailored to the drivers' specific needs (Stiggins, 2002). However, to facilitate higher order level learning, formative assessments should go beyond isolated testing of knowledge about traffic rules and technical driving skills. Fitting in with a competence oriented approach, formative assessments need to: a) address critical elements of the driving task and its circumstances as they appear in daily driving and throughout the driving career; b) fit in with the stage of development of the (learner) driver, c) be informative about mental processes.

To tailor formative assessments to different stages of driver competence it is vital to identify driving assessment tasks that discriminate between drivers from different stages. The aim of this study was to develop a systematic procedure by which to arrive at such task scenarios.

The following research questions were addressed:
- Is it possible to identify a collection of driving tasks of varying complexity and difficulty levels using the evidence centered design model?

- To which extent do driver instructors reliably rate the levels of task difficulty related to these tasks, when they are applied to learner drivers in different training stages?

2   METHOD

The evidence centered design model for assessments as described by Mislevy, Steinberg, and Almond (2002) was used to arrive at a collection of assessment tasks for educational purposes. This model for assessment design helps to sort out the relationships among attributes of a candidate's competence, observations which prove competence and situations which elicit relevant driver performance. The design model is composed of six interrelated sub-models: the Student Model, the Evidence Model, the Task Model, the Assembly Model, the Presentation Model, and the Delivery Model. The first three models mentioned generally constitute the basis for assessment design activities.

In this study, task scenarios with different task demands were collected while the task difficulty levels were estimated by driving instructors. The categorization of the tasks and the rating variable were based on an elaboration of the evidence centered design model applied to the assessment of driving.

2.1 A student and evidence model as a basis for collecting task scenarios

For the purposes mentioned above a Student Model was elaborated that describes the processes that take place during driving. Competent driving is defined as the ability to carry out driving tasks as they exist within a universe of traffic situations, varying in complexity. The quality of the mental processes must be expressed in terms of performance measures if we wish to make inferences about driving competence. In the current study measures are used that are related to the following outcomes of driving (cf. Roelofs, Van Onna & Vissers, 2010), i.e. 1) safety: the driver's awareness of the situation, resulting in correct timing of actions, adapting speed and using "space cushions"; 2) facilitating traffic flow, which implies not impeding the progress of other road users, and using the road efficiently; 3) consideration with other road users, which means giving others opportunities to fulfill their tasks, or adapting to their mistakes; 4) controlled driving, referring to steering and controlling the car smoothly, without stutters and jerks or departures of smooth lines. The four measures may correlate substantially.

As stated in modern models of driving, the driver is confronted with different levels of task demands which can be close, below or above the ability to solve the concomitant problems (Fuller, 2005). The odds of being in control of the traffic situation are dependent on the balance between task demands and driver abilities. In cases of very high task demands even experienced drivers may encounter difficulties in solving traffic situations.

In a third part of the Student Model the characteristics of the task environment that influence task complexity were described and explained. More specifically, critical task environmental attributes that hamper or facilitate the driving process were identified from error-taxonomy studies (Reason, 1990; Stanton and Salmon, 2009). Errors that occur under specific circumstances can be seen as indications of a mismatch between the complexity level of the task and the available level of competence. The environmental attributes can be categorized with respect to the driving process they hamper, yielding the features of the task environment that may influence task difficulty (see Table 1).

Table 1. Critical attributes of the driving task environment that may hinder driving task processes

Critical attributes of the driving task environment	Which process is being complicated?		
	Perception	Decision making	Action execution
Sight obstruction	X		
Hearing obstruction	X		
Discontinuous traffic environment	X	X	
Other participants arrive at scene at the same time		X	
Reduced space to carry out actions		X	X
Inferior road conditions			X
Weather conditions hindering lateral vehicle control			X
Road characteristics hindering lateral vehicle control			X

In the Evidence Model of this study, each performance measure in the student model is seen as an ability that can have a broad range of values. Item response theory (IRT) models explicitly balance the (driver) ability and task difficulty in predicting the odds of a successful solution of the task by the driver. The evidence model in this study relies on IRT. For each assessment item, an item response function gives the probability that a person with a given ability level (expressed by the parameter θ), will respond correctly on a task. Drivers with low ability have a low probability while drivers with high ability are very likely to respond correctly. Each task may have a difficulty level β. A basic IRT model is the Rasch model (Rasch, 1960). The probability of a successful solution of task j (X_j=1), is a function of the difference between ability θ and task difficulty β_j:

$$P(X_j = 1 \mid \theta) = \frac{\exp(\theta - \beta_j)}{1 + \exp(\theta - \beta_j)}$$

(1)

IRT offers the possibility to select test items that are tuned to the level of ability of the learner. In the current project we aimed to identify a collection of driving tasks that differ in terms of task demands. However, IRT analyses are useless without sound item construction, based on elaborated ideas of task difficulty. Without those ideas, the range of task difficulties may be too narrow, or the average task difficulty may be off target. The Student model described above specifies these ideas on task difficulty.

2.2 Collection of driving task scenarios

The task universe was defined by the following attributes: a) the basic subtask of driving as mentioned in the Dutch driving curriculum: turning, merging, longitudinal driving, stopping, crossing, passing, lane changing; b) all relevant defining situational characteristics of the task, referring to e.g. road conditions, road type, weather conditions, traffic intensity, types of other road users; c) the mental processes that are possibly hampered in the task situation; d) the performance criteria for which the scenario is most critical.

Driving lessons from 13 instructors enrolling 39 learner drivers were recorded, using three digital video cameras: one was directed at the driver, one at the road ahead, one at the road behind. All instructors used the training method of Driver Training Stepwise (DTS; Nägele, & Vissers, 2003) which consists of four consecutive learning stages: (1) vehicle control (2) driving in simple traffic situations; (3) driving in complex traffic situations (4) independent driving. To ensure a variety in task complexity, lessons throughout all stages were filmed.

The mixed footage (front shield, drivers' face and rear window) was analyzed for the occurrence of situations in which the learner driver committed an error or a near error, as indicated by verbal comments or pedal interference. The analysis resulted in a collection of 55 scenarios. Each of them represented one specific driving task and involved the complication of one or more sub processes of driving. The greater part of the scenarios related to turning and cruising (both 15 scenarios, 27.3%). A smaller number of scenarios pertained to merging (8, 14.5%), crossing (5; 9.1%), passing (8; 14.5%) and lane changing (4; 7.3%). Out of the 55 scenarios 15 (27.3%) referred to a complicated perception process, 31 (56.4%) to a complicated decision process, and 9 (16.4%) to a complicated action execution.

2.3 Subjects

Eight experienced driving instructors (mean age 45.7 years (SD=7.0) and 15 years of experience on average (SD=7.0)) participated. None of them had taken part in the recorded lessons. However, the instructors themselves used the DTS method during their regular training.

2.4 The rating procedure

All scenarios were rated in terms of complexity using the following procedure.

During two three-hour group sessions driver instructors were asked to view the video scenes which were projected on a screen (60 by 80 inches). The drivers' intention was mentioned by the session moderator; e.g. "the driver intends to pass a row of parked cars". After that, the actual execution of the task was displayed on the screen twice. The instructors were asked to respond to the following question: 'In how many cases out of 10 will the learner driver at the end of DTS stage 2 and 4 respectively solve this traffic task safely, efficiently and independently?' Safe and efficient were explained as: the learner driver or other road users do not need to reduce speed strongly, wait for long times, change their course or evade, touch other road users or objects or to cross road lines. 'Independently' was explained as referring to a situation in which no verbal or operational instructor intervention is needed to have the learner driver fulfill the task.

Two subjects missed the second session, whereas one subject did not attend the first session.

2.5 Method of data analysis

Each instructor rating was expressed in terms of a probability, e.g. 0.70 (7 out of 10). The rater agreement was computed by calculating the inter rater reliability coefficient and the Gower similarity index. In case of sufficient agreement the ratings would be pooled. Using the Rasch IRT model, the instructors' ratings can be conceived as the success probability of the learner driver with ability θ on an item with difficulty β, as stated in a rewritten variant of Equation 1:

$$\ln\left[\frac{P(X_j = 1 \mid \theta)}{1 - P(X_j = 1 \mid \theta)}\right] = \theta - \beta \tag{2}$$

This means that a logit transformation on the average instructor probabilities allows for linear modeling of abilities of drivers and item difficulties. In this study, only two levels of driver ability, corresponding to DTS stages 2 and 4, were considered. In addition, 55 item difficulties were involved. These parameters were estimated by analysis of variance with only main effects.

3. RESULTS

Since some of the instructors either missed the first or the second session, the dataset was split into three overlapping subsets that were analyzed separately. Version 1 contained the ratings of five instructors who rated all scenarios, version 2 the ratings of seven instructors who rated the first 28 scenarios, version 3 the ratings of six instructors who rated the last 28 scenarios.

The inter-rater reliability of ratings for learner drivers at DTS 2 was good for all data subsets ($r>= 0.82$). The inter-rater reliability of the ratings for learner drivers at DTS4 was sufficient for all data subsets ($r>=0.69$). The Gower indexes were all larger than 0.85. These levels of agreement indicate that the estimates of the different driving instructors could be pooled for further analysis with the Rasch model. All individual ratings for a given task scenario were averaged across instructors to obtain pooled estimates of the probability of a successful solution of task j, $P(Xj=1|\theta)$. These pooled probabilities averaged about 0.19 for DTS stage 2, and about 0.74 DTS stage 4.

A restricted ANOVA with only main effects was run on the logit transformations of the pooled estimates of $P(Xj=1|\theta)$. Error in the logits accounted for only 5.5% of the total variance. The larger part of the variance in the logits was accounted for by the DTS stages (80.7%); the levels of scenario difficulty accounted for 13.7% of the total variance. Note that the scenarios had a relatively low variation in difficulty levels. Only two scenarios were significantly easier, and three were significantly more difficult than the average item difficulty.

In Figure 1 the predicted values of $P(Xj=1|\theta)$ given the estimated values of βj, are plotted for four of the tasks j. Task scenario 34 was relatively easy, resulting in relatively large probabilities of a successful solution. Scenarios 13 and 40 were of average difficulty and task scenario 55 was relatively difficult. The predicted values of $P(Xj=1|\theta)$ at the ability levels corresponding to DTS stages 2 and 4, differed somewhat from the observed pooled estimates. Of the predicted probabilities 19% diverged more than 0.1, but none of them diverged more than 0.2. This seems to indicate a reasonable fit of the Rasch model.

When contrasting the scenarios estimated as most and least difficult some observations can be made, which provides some insight into the relevance of specific scenario characteristics for scenario difficulty. The three scenarios rated as most difficult involved a combination of tasks within a short time span and under conditions with limited visibility of other participants. In addition, traffic participants from different directions entered into the scenario at the same time, leaving little time and space to make decisions and carry out the relevant maneuver. The most difficult scenario took place during evening darkness. In this scenario the driver drove on a parallel road, next to the main road. The starting location is referred to as 'L1' in figure 2. The driver intends to change to the far outer lane of the main road to make a turn left at oncoming traffic lights. This location is referred to as 'L2' in figure 2. To do so, two lanes: a bike lane and a lane for traffic heading straight forward or turning to the right had to be crossed.

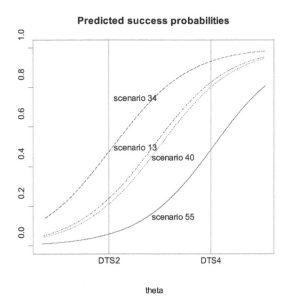

Figure 1   Predicted probabilities of a successful solution for four task scenarios as a function of ability

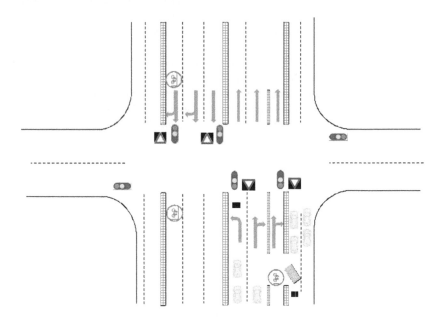

Figure 2. Scenario rated as most complex

The tasks rated as easiest pertained to situations in which right of way was clearly regulated by means of traffic lights; there was only one maneuver to carry out without time pressure. There was only one other road user at the scene, and this road user did not drive in a conflicting direction.

4. CONCLUSIONS

This study aimed to develop a procedure by which a collection of driving assessment tasks could be developed. This study illustrates the usefulness of the evidence centered design model. Borrowing from Fuller's theory driving is seen as the dynamic interaction between the determinants of task demands and driver capability, involving various sub-processes. Using insights from error studies the task environment attributes were derived which may complicate these sub-processes. By combining basic driving tasks with critical environment attributes the universe of assessment tasks was defined.

A collection of informative driving assessment task scenarios was selected out of videotaped driving lessons throughout different stages of driving acquisition. Experienced driving instructors viewed the video scenarios and estimated solution probabilities for each of them, taking into consideration beginning and advanced learner drivers. The estimated solution probabilities differed meaningfully between the two envisioned target groups DTS stage 2 and DTS stage 4. However, most of the evaluated driving tasks were considered difficult for beginning drivers. Regarding the intended formative purposes, these tasks would not fit to the needs of this target group. The high average difficulty level suggests that easier tasks should have been identified.

Encouraging were the high levels of inter-rater reliability among driver instructors, suggesting that the procedure of estimating task difficulty is feasible in practice. The Rasch model could be used as part of the evidence model for assessment. The data seemed to support a reasonable fit of this model. The desirable measurement properties of the Rasch model include the fact that the number-correct test score is a sufficient statistic for the ability level. Also, the Rasch model offers the possibility to model adaptive testing to meet a variety of (learner) drivers.

Within the limited scope of this study we could not systematically sort out the effects of the complicating factors within the task scenario on estimated task difficulty. Maybe this is because each complicating source can vary in the degree of complication it causes. Possibly other un-specified factors would account for the difference between relatively easy and difficult task scenarios. A contrasting comparison between easy and difficult items revealed that the degree of time pressure, the number of necessary maneuvers to be carried out consciously in a given time-span may have been a discriminating factor in addition to the specified factors.

These latter two factors (time pressure and conscious versus automatic processing) appear as important in a recent task analyses method presented by Fastenmeier and Gstalter (2007), referred to as SAFE (Situative

Anforderungsanalyse von Fahraufgaben). This method allows the identification of which elements of the task lead to more or less complexity and objective risk in any driving task and also arrives at quantitative measurements of these dimensions. A suggestion for follow-up study would be to further elaborate driving tasks further and rate their difficulty, building on the research by Fastenmeier and Gstalter.

Another aspect that is mentioned in Fastenmeier and Gstalter's model is determining the level of risk associated with the task at hand. The search for critical scenarios to be employed in assessments for learning task complexity per se is not sufficient. If one is to choose what to include in driving assessments, possible safety risks make up another inclusion criterion.

In a follow-up study simulator task scenarios will be transformed into simulator assessment tasks. In addition, a cognitive assessor model will be developed, which will automatically score drivers' performances, based on human expert scoring rules, at critical moments in the scenarios. In general, the idea of validity by design, as proposed by Mislevy (2007) needs to be tested in a more thorough and rigorous setting, enabling an evidentiary argument, an argument from what we observe learner drivers do in a few particular circumstances, to inferences about what they can do more generally. These inferences can in turn be the basis for tailoring subsequent training to the specific needs of the driver.

REFERENCES

Fastenmeier, W., and Gstalter, H. (2007). Driving task analysis as a tool in traffic safety research and practice. *Safety Science, 45*, 952–979.

Fuller, R. (2005). Towards a general theory of driver behaviour. *Accident Analysis and Prevention, 37*, 461-472.

Hatakka, M., Keskinen, E., Baughan, C., Goldenbeld, Ch., Gregersen, N.P., Groot, H., Siegrist, S. Willmes-Lenz, G. and Winkelbauer, M. (2003) *Basic driver training: New models*. Turku, University of Turku.

Mislevy, R. J., Steinberg, L. S., & Almond, R. G. (2002). On the roles of task model variables in assessment design. In S. Irvine & P. Kyllonen (Eds.), *Generating items for cognitive tests: Theory and practice*. Mahwah, NJ: Erlbaum.

Nägele, R.C & Vissers, J.A.M.M. (2003). *'Driver Training Stepwise (DTS). Evaluation of the follow-up in the province of Gelderland*. Veenendaal, Traffic Test.

Rasch, G. (1960). *Probabilistic models for some intelligence and attainment tests*. Copenhagen: Danish Institute for Educational Research.

Roelofs, E.C., Onna, M. van, Vissers, J. (2010). Development of the Driver Performance Assessment: Informing Learner Drivers of their Driving Progress. In L. Dorn (Ed.) *Driver behavior and training, volume IV* (pp. 37-50). Hampshire: Ashgate Publishing Limited.

Reason, J., 1990. *Human Error*. Cambridge University Press, Cambridge.

Stanton, N.A., & Salmon, P.M. (2009).Human error taxonomies applied to driving: A generic driver error taxonomy and its implications for intelligent transport systems. *Safety Science, 47*, 227–237

Stiggins, R. J. (2002). Assessment crisis: The absence of assessment FOR learning. *Phi Delta Kappan, 83* (10), 758-765.

CHAPTER 72

Development of a Coaching Program for Young Drivers in Their First Period of Solo Driving

Erik Roelofs[1], Jan Vissers[2], Marieke van Onna[1], Gerard Kern[3]

[1] Cito, National Institute for Educational Measurement, The Netherlands
[2] DHV group, Amersfoort, The Netherlands Regional
[3] Agency for Traffic Safety Gelderland, The Netherlands
Erik.Roelofs@cito.nl

ABSTRACT

During the first period of solo driving, young novice drivers go through the most dangerous phase of their driving career. In line with European developments, a Dutch second phase coaching program, referred to as the 'Drive Xperience (DX)', has been developed for young novice drivers. The program combines the use of web-based driver assessment for learning with a full day of training and coaching. In this chapter the design principles of the program are highlighted. The empirical study focused on a description of the participating group of young drivers (n=2257) as compared to a reference group of young drivers (n=345) and a group of lease car drivers (1129), of whom the latter were involved in similar advanced driver training programs. Comparative analyses were carried out on web-based assessment data pertaining to driving history, specific risk factors, and self-images related to safe driving. Results show that the DX program attracts young drivers that show a more risky profile than average young drivers and lease car drivers in terms of speed violations, irritated driving and the number of fines. The rather positive self-image at the same time calls for attention in terms of program design. Implications of program fit to the target group are discussed.

Keywords: young drivers, coaching, driving behavior, self-image, risk factors

1 INTRODUCTION

During the first six to twelve months of solo driving, young novice drivers experience the most dangerous phase of their driving career: their accident risk reaches a peak by then. This is why several European countries have introduced second phase training programs. A recent evaluation of the Austrian model (Gatscha and Brandstaetter, 2008) showed that the number of personal injury accidents among 18-year-old novice drivers was reduced with 28 per cent in all regions of Austria after the introduction of the obligatory second phase system.

Several studies stress that the content of second phase driving programs should not be restricted to lower level vehicle control skills. Moreover, concentrating on the lower vehicle control skills can even be counterproductive (Glad, 1988). Instead, second phase driving programs should rather address driving style variables which refer to the way that people choose to drive (Elander, West and French, 1993; McKenna, 2009), which can be explained by the way drivers make decisions on the level of life tasks and on the strategic, tactical and operational levels of driving tasks.

In a literature review Helman, Grayson and Parkes (2010) mention driving style variables on the strategic and participation level of driving tasks that are associated with collision risk are reviewed: speed choice, close following of vehicles in front, high-risk overtaking, violation of traffic laws and engaging in distracting activities, such as speaking on mobile phones while driving. Factors on the lifestyle level, such as fatigue-related factors (Groeger, 2006; McKenna, 2009), and alcohol use, themselves fall outside the driving domain, but are associated with collision risk. The awareness of these factors, the readiness and the ability to take them into consideration before and during driving involve higher order skills on the part of the young driver, as elaborated in the Goals of Driver Education matrix (Hatakka, et al., 2002). Among these skills are self-reflection and awareness of emotions during driving. The purpose of second phase training programs is to further develop higher order driving skills that appear to play an important role in the reduction of one´s accident risk.

The scholars mentioned above stress that development of complex higher order skills should involve increased levels of self-regulation. Although varying approaches may stimulate self-regulation on the part of the learner, there is consensus about the idea that some form of coaching is likely to be effective. Recently, within the field of driver training, principles of driver coaching have been defined (Bartl, et al., 2010). Coaching:

- puts the learner in an active role;
- builds on the prior knowledge and experience of the learner;
- encourages the learner to identify his/her goals and to meet these goals;
- raises the awareness, responsibility and self-acceptance of the learner;
- raises awareness of the learners' values, goals, motives and attitudes as well as his sensations and emotions, knowledge, skills and habits;
- addresses the learner's internal obstacles to change.

The current study focused on a Dutch second phase coaching program that was developed as part of the EU-project 'Evaluation of post-license training schemes for novice drivers' (NovEv; Sanders and Keskinen, 2004; Vissers, 2006).

In this study, attention was focused on a description of the participating group of young drivers. For the program to be successful in addressing the target group, it aims to attract a representative group of young drivers, who are in need of various forms of extra insights into their driving styles. Another issue pertains to the potential program fit with the target group. In order to explore this fit, driving characteristics of participating young drivers in 'The Drive Xperience Program' were compared with a Dutch reference group of young drivers, and with experienced drivers who participate in advanced driver training programs. The following research questions will be addressed:

- What are the driving characteristics of young drivers enrolled in a program for coached driving in terms of their driving behavior, their personal risk factors, and their self-perceptions regarding driving proficiency?
- In what respects do participating young drivers differ from the average young Dutch driver?
- In what respects do young drivers differ from experienced drivers, who are enrolled in advanced driver training programs?

2 METHOD

2.1 Design characteristics of the second phase coaching program 'Drive Xperience'

The design principles and content of the Drive Xperience (DX) program second phase coaching program were drawn from different sources of research, which are described below.

First, throughout the training program the GDE matrix (Hatakka et al, 2002) is used as a conceptual framework for all learning experiences. During discussion about driving styles, it is emphasized that the actual quality of participation in traffic situations is influenced by decisions on higher level tasks. These include life tasks, e.g. socializing with peers, recreating, working, and strategic tasks, e.g. choice of transportation mode, and route decisions.

Second, in the coaching program coaches consider driving as a cognitive/affective decision making process, during which the driver (ideally verbally) needs to reflect on his/her own decisions, actions and consequences.

Third, one of the problems with young drivers is that they often fail to safely adapt their choice of traffic situations to their level of proficiency (Fuller, 2005; De Craen, 2010). In the coaching program young drivers are supported to search for a balance between their own level of proficiency and the complexity of the traffic situations they engage in.

Fourth, the ultimate goal of the coaching program is that young drivers improve their quality of driving at all GDE task levels. To determine the quality of driving behavior, coaches are encouraged to pay attention to five inter-related criteria of driving performance (Roelofs, et al., 2010): 1) Safety: The driver's ability to solve traffic situations in such a way that time and space are available for all participants to carry out driving activities without conflicting or colliding with objects or others; 2) Facilitating traffic flow: The ability to drive in a way that does not impede the progress of other road users, and helps an optimal flow of traffic; 3) Consideration with other road users: The ability to give others opportunities to fulfill their tasks, or to adapt to their mistakes, without showing aggression or irritation; 4) Controlled driving: Steering and controlling the car smoothly, without stutters and jerks or departures from smooth lines. 5) Environmentally-responsible driving: Driving in such a way that emissions of harmful gases and noise levels are kept to a minimum and that optimal use of fuel is achieved.

During a pre-test period of four weeks before the start of the training program the participants were asked to complete two web-based assessments: The Driver Risk Assessment and the Driver Self-Assessment. These assessments result in an individual driver profile, accessible for both participants and coaches and are used for tailoring the training content to individual learning needs and for reflective purposes on the part of the young drivers.

The one day driver coaching program consists of three parts:

- Coached trip: The main objective of the coached trip is to present the driver with feedback about his 'everyday' driving performance.
- Track experience: The main objective of the track experience is for participants to experience the limits of their skills in vehicle control and to share these experiences with other group members.
- Group discussion: The main objective of the group discussion is to stimulate recognition of potentially hazardous situations in rather 'normal' driving situations. Risks of alcohol use and other risky driving behaviors are discussed.

2.2 Instrumentation

In this study data from two assessments were employed. The data were used to give a detailed description of the participants. As the instruments have also been administered to experienced drivers (driving in leased cars) enrolled in driver training programs, comparisons can be made between the former and the latter group.

First, the Driver Risk Assessment, a web-based questionnaire consisting of 119 questions regarding the driver history and the drivers' behavioral risk factors was administered. Personal background variables pertain to age, gender, and years of driving experience. Driving history pertains to mileage, the kind and variety of traffic situations to which the driver is exposed, the number of active and passive accidents the driver was involved in during the past three years and the number of fines received for various reasons the past year. Questions regarding behavioral risk factors pertain to the following topics: speed choice under various conditions, lane preference on motorways during various traffic conditions, alcohol use and driving, the use of adversary alcohol strategies in combination with driving, anger in reaction to others participants' violations, distraction and fatigue. Most of the behavioral questions ask for a response on a Likert scale, where each scale point represents a behavioral option.

Second, The Driver Self-Assessment, a web-based questionnaire consisting of 45 illustrated questions was administered. It comprises a self-assessment of driving ability addressing the five criteria for driving competence described above. The questions pertain to strategic, tactical and operational behavior choices on different road types. Each question is accompanied by an illustrative picture, showing the essential features of the traffic situation to which the question refers. Participants respond by using a four-point Likert-scale.

2.3   Participants

For this study data were used from two groups: young drivers enrolled on the DX-coaching program (referred to as 'DX group') and lease car drivers enrolled on advanced driver training programs (referred to as 'LC group'). The DX group (n=2257) participated in a total of 81 training groups, distributed across 45 municipalities. The training programs took place in the period between August 2010 and January 2012 and enrolled 1238 male and 1019 female participants. The participants had held their driving license between six months and two years. They were aged between 18 and 24 years. Their average mileage amounts to 7725 kilometers, which is almost equal to the average Dutch young driver.

The LC group (n=1129) came from 36 companies that sent their associates to advanced driver training in small groups. The LC group differs in terms of age and gender composition from the average Dutch population of drivers. The former group is relatively younger (75 per cent is younger than 45 versus 53.4 per cent for the population), has fewer years of driving experience (28.8 per cent less than 10 years versus 22.5 per cent) and contains more men (81 per cent versus 53 per cent).

To enable comparison with a reference group of young drivers (referred to as 'YD-Ref group') several analyses were performed on a database containing data from a large-scale survey using representative samples of the Dutch population. This survey has been carried out almost yearly since 1990 (Eversdijk et al., 2000) until 2005. For purposes of comparison only the most recent data (2005) were used (n=345), since traffic safety figures have changed considerably since 1990.

3   RESULTS

DX participants regularly drive during rush hour and on motorways (means 2.9, 3.4, see Table 1). They relatively often drive on roads outside built-up areas which have no separated lanes (mean 3.8), and which serve participants with different speeds and vehicle masses. In addition, DX participants regularly drive in the town center of a big city (mean 3.0). Finally, they sometimes drive during weekend nights (mean 2.4; about once a month). LC group drivers drive (significantly) more often in the situations mentioned (t-values 21.5, 18.3, 3.4, 17.0 and 5.5 respectively; df= 1712; all significant at the .001 level).

Table 1. frequency of driving in different traffic situations

	DX group (n=1165)		LC group (n=549)	
Frequency of driving:	M	SD	M	SD
During rush hour	2.9	1.2	4.2	1.0
On motorways	3.4	0.9	4.3	0.8
On roads outside built-up areas	3.8	0.9	4.0	1.0
In the town center of a big city	3.0	1.1	4.0	0.9
During weekend nights*	2.4	1.0	2.7	0.9

Note 1= almost never; 5= every day; * 1= almost never; 4= every weekend

Table 2. Average number of accidents per million kilometers

Group and age	Mileage*	Active accidents	Passive accidents	Fines
DX group (n=1165)	7,725	22.5	10.8	78.1
YD-Ref group (n=289)	11,611	21.9	11.4	48.7
LC group (n=549)	26,734	7.2	4.7	78.1

Note: * Kilometers per year

Table 2 shows the accident risk expressed in the average number of accidents and fines per one million kilometers.

The average mileage for the DX group was significantly lower than for the young driver reference group (YD-Ref group; t=-6.1, df=1452, $p<.001$). LC group drivers had a higher average mileage than both young driver groups. Both young driver groups had higher accidents risks than the LC group, both for accidents where the driver was legally liable (active accidents) and for accidents where the other driver involved was legally liable (passive accidents). The risk for fines was equally high for the DX group and the LC group (78.1 per million kilometers).

Since the data concerned frequency data on a fixed number of kilometers that were skewed to the right, a Poisson model was appropriate for testing the null-hypothesis that the active accident rates in both groups are equal. This hypothesis was rejected ($\chi2$= 3.7, df = 1, p=.06). The DX group had a significantly lower passive accident rate than the YD-Ref group ($\chi2$= 6.9, df=1, p=.01). The DX group had a much higher fine rate than the YD-Ref group ($\chi2$= 3996,0, df=1, $p<.001$). Their fine rate equals the one of LC group drivers.

Table 3. Self-reported behavior regarding risk factors

	DX group (n=1165)		LC group (n=565)		YD-ref Group (n=345)	
	M	SD	M	SD	M	SD
Alcohol use (2 items, α=.54)	.38	.23	.42	.18	-	-
Adversary alcohol strategies (11 items, α=.92)	.06	.17	.32	.33	-	-
Concentration loss (13 items, α=.81)	.10	.09	.08	.08	-	-
Seriously angry towards others (2 items, α=.64)	.03	.14	.02	.11	.01	.13
Irritated towards others (4 items, α=.61)	.19	.26	.17	.24	.13	.23
Withholds anger towards others (4 items, α=.59)	.57	.30	.56	.31	.43	.26
Stays calm towards others (2 items, α=.45)	.75	.35	.78	.32	-	-
Violation of speed limits on various roads and circumstances (9 items, α=.93)	.15	.13	.16	.14	-	-
Violation of speed limits under favorable circumstances (3 items, α=.65)	.24	.18	.25	.18	.20	.19
Driving on outer lanes (5 items, α=.78)	.35	.24	.38	.27	-	-
Fatigue during driving (5 items, α=.73)	.10	.11	.11	.11	-	-

Note: 0.-.16: rarely; .17-.33: occasionally; .34-.50: sometimes; .51-.67: rather often; .68-.84: very often; .85-1.0: most of the times

Table 3 shows results regarding behavioral risk factors as measured by means of the Driver Risk Assessment. The DX participants sometimes drink alcohol (mean .38) and they do not differ in this respect from the lease car driver (mean = .42). They score low on the scale 'use of adversary alcohol strategies' (mean: = .06). This scale consists of questions about well-known but adversary strategies to minimize the effects of alcohol once the participants find themselves at places where alcohol is served, e.g. pretending to drink alcohol, wait a while before driving home. LC group drivers occasionally combine alcohol and driving, including the use of the adversary strategies (mean = .33).

A next risk factor, loss of concentration rarely occurs with DX participants (mean: = .10). This scale refers to situations of unconscious driving and distractions.

The anger subscales referred to the drivers' reaction on other participants' violations or deviating driving behavior. In terms of anger DX participants very often stay calm (mean: = .75). Reactions of withheld anger occur rather often (mean = .57). Irritated reactions towards other participants occur occasionally (mean = .19). Seriously angry reactions occur very rarely (mean: = .03). Although small, the differences with the LC group are, except withheld anger (t=0.3), significant (t= -2,1, 2,1 and 2,2 respectively).

Available data from the YD-ref group show that average young drivers react less frequently with irritation (t = 4.2; df = 1508; p <.01), withheld anger (t = 8.5; df = 1508; p <.01) and serious anger (t=2.5, df = 1508; p <.01) than the DX participants.

A fourth risk factor is speed and lane preference. On average DX participants claim they rarely violate the limits (mean = .15) and even less frequently than lease car drivers (mean = .16; t = -2.0; df = 1712; p < .05). Under favorable weather and road conditions DX participants occasionally violate speed limits (mean: = .24). The difference with lease car drivers (mean: = .25) is not significant (t = -1.0; df = 1712; ns). Under favorable conditions young drivers more often violate the speed limits than the YD-ref group (means: = .24 and .20 respectively; (t = 2.1; df = 1508; p <.05). Regarding outer lane choice, both DX participants and LC drivers sometimes prefer the outer lanes on the motorways (means: = .35 and .38 respectively). LC drivers do this more often (t = -2.3, df = 1712; p < .05) than DX participants.

Finally, the scores for fatigue suggest that on average DX participants rarely experience fatigue while driving (mean: = .10). LC drivers experience fatigue a little more frequently (mean: = .11, t = -2.0, df = 1712, p < .05).

Table 4. driving proficiency as reported on the Driver Self-Assessment

	DX group (n=2109)		LC group (n=719)	
	Mean	SD	Mean	SD
Committing driving errors (12 items, α =.74)	.21	.12	.17	.13
Driving in a hurry (6 items, α =.71)	.45	.17	.36	.18
Perceived quality of one's own driving behavior (43 items, α =.80)	.73	.08	.76	.09

Note: 0.-.16: rarely; .17-.33: occasionally; .34-.50: sometimes; .51-.67: rather often; .68-0.84: very often; .85-.1most of the times

Table 4 shows the results on the Driver Self-Assessment. The first scale consists of 12 items addressing errors that affect traffic safety (e.g. following too close, braking too late), traffic flow (e.g. causing others to wait), and vehicle control (e.g. errors when braking, using gears). On average, DX participants occasionally commit errors (mean: = .21) that affect these criteria. However, they do so significantly more frequently than LC drivers (mean: .17; t=7.6, df = 2826, p<.001)

The second scale, 'driving in a hurry' refers to the tendency to drive faster than the speed limits and to get irritated by those who drive slower than that. DX drivers sometimes drive in a hurry (mean: = .45) and do so significantly more often than LC drivers (mean: = .36; t=11.7, df = 2826, p<.001). The overall scale 'Perceived quality of one's own driving behavior' consists of 43 items, covering the degree to which the driver reports a safe, flow-aiding, social, controlled and environmentally considerate way of driving. On average DX drivers report a high frequency of this driving behavior (mean: = .73). However, they score significantly lower than LC drivers (mean: = .76; t=8.6, df = 2826, p<.001).

4   CONCLUSIONS

Looking at the driving history in terms of their mileage, the relative accident risk and the number of fines it becomes clear that DX participants form at least an average group of young drivers. The organizing agency feared that the participants would be well-educated and good drivers, but this does not seem to be the case. On the contrary, there are indications that the relative risks of (active) accidents and fines are higher than they are in a representative Dutch reference group of young drivers. In terms of the available comparable risk factors studies, the DX participants score less favorably than the average young driver: the former reports irritation towards other drivers, making errors and violating the speed limits more often than the latter group.

The second main question was to what extent the design and content of the coaching program had addressed the driving characteristics of the participating young drivers?

Reflecting on the question whether the DX program address specific young driver needs, the following observations can be made. First, DX participants find themselves in risky traffic situations, i.e. on roads outside built-up areas and in the town center of a large city. While doing so, they occasionally violate the speed limits. They sometimes choose outer lanes when driving on motorways. Taken together with the reported relatively high accident involvement, this calls for attention to the risks of speed as is the case in the DX coaching program.

Second, a relatively favorable picture emerges regarding the combination of alcohol use and driving. DX participants mostly refrain from drinking when they have to drive. Alcohol use seems more of an issue for the lease car driver involved in this study.

Third, DX participants sometimes find themselves hurrying and irritated through the traffic. At the same time they hold a rather positive self-image of their own driving behavior. Their self-image does not differ to a large extent from that of the more experienced lease car drivers.

The tendency to drive at higher speeds than is permitted, to react with irritation towards other drivers combined with a positive self-image may result in overestimation of the own driving proficiency. In turn this overestimation may result in imbalances between task demands and real proficiency, causing increased risks (Fuller, 2005; De Craen, 2010). The average driving profile of DX participants matches well with the design and content of the coaching program. During the coached trip the participants are encouraged to form realistic self-images, by means of non-judgmental peer feedback during the trip. Drivers become more conscious about their driving and the immediate consequences for other traffic participants.

In a follow-up study two issues need to be addressed. First, attention will be paid to possible short and long term effects of the DX program on driving behavior, as recently recommended by other authors (cf. Mynttinen et al., 2010). Second, more focus is needed on the features of the coaching program, which contribute to or impede changes in attitudes and driving behaviors of the participants. To those ends, more intrusive forms of data collection will be employed. Direct observations, interviews and performance assessments will be employed to sort out the effective ingredients of coaching on young drivers, even if the effects may be indirect (McKenna, 2010).

REFERENCES

Bartl, G., et al. (2010). High impact approach for Enhancing Road safety through More Effective communication Skills In the context of category B driver training. EU HERMES Project Final Report.

De Craen, S., (2010). The X-factor. A longitudinal study of calibration in young novice drivers. Dissertation. Leidschendam: Stichting Wetenschappelijk Onderzoek Verkeersveiligheid SWOV

Elander J., et al. (1993). Behavioral correlates of individual differences in road traffic crash risk: an examination of methods and findings. Psychological Bulletin, 113 (2) 279–294.

Eversdijk, J.J.C., et al. (2000). PROV 1999 Periodiek Regionaal Onderzoek Verkeersveiligheid [PROV 1999: periodic Dutch regional research into traffic safety]. Report number: TT00-66. Veenendaal: Traffic Test BV.

Fuller, R., (2005). Towards a general theory of driver behavior. Accident Analysis and Prevention, 37, 461-472.

Gatscha, M., and Brandstaetter, C., (2008) Evaluation der zweiten Ausbildungsphase in Österreich [Evaluation of the second phase system in Austria]. Forschungsarbeiten aus dem Verkehrswesen. Vol. 173. Austrian Federal Ministry of Transport, Innovation and Technology. Vienna, Austria.

Glad, A., (1988) Fase 2 I foreoplaringen. Effect pa ultkkes riskoen. [Driver Education's second phase. Its effect to the accident risk.] Report No. 0015. Oslo, Transportokonomiskt institut.

Groeger J A., (2006). Youthfulness, inexperience, and sleep loss: the problems young drivers face and those they pose for us. Injury Prevention, 12 19–24.

Hatakka, M., et al. (2002) From control of the vehicle to personal self-control; broadening the perspectives to driver education. Transportation Research Part F, 5, 201-215.

Helman, S., et al. (2010). A review of the effects of experience, training and limiting exposure on the collision risk of new drivers. TRL Insight Report INS005. Bracknell (UK): Transport Research Laboratory.

McKenna, F. P., (2009). Can we predict driver behavior from a person's sleep habits? Behavioral Research in Road Safety, 19th Seminar, 30 March–1 April 2009. London: Department for Transport.

McKenna, F. P., (2010). Education in Road Safety Are we getting it right? London: RAC Foundation.

Mynttinen, S., et al. (2010). Two-phase driver education models applied in Finland and in Austria – Do we have evidence to support the two phase models? Transportation Research Part F: Traffic Psychology and Behavior, 13(1), 63-70.

Roelofs, E.C., et al. (2008). Development of multimedia tests for responsive driving. In L. Dorn (ed.) Driver Behavior and Training, Volume III (pp 251-264), Hampshire: Ashgate Publishing Limited.

Roelofs, E.C., at al. (2010). Development of the Driver Performance Assessment: Informing Learner Drivers of their Driving Progress. In L. Dorn (Ed.) Driver behavior and training, volume IV (pp. 37-50). Hampshire: Ashgate Publishing Limited.

724

Sanders, N., and Keskinnen, E., (eds.) (2004). EU NovEV project; Evaluation of post-license training schemes for novice drivers. Final Report. International Commission of Driver Testing Authorities CIECA, Rijswijk, The Netherlands.

Vissers, J.A.M.M., (2006). Evaluatie Tweede Fase Opleidingsprogramma Gelderland 2006 [Evaluation second phase driver training program]. Amersfoort, DHV.

CHAPTER 73

Field Operational Trials in the UK

Ruth Welsh, Andrew Morris, Steven Reed, James Lenard

Transport Safety Research Centre
Loughborough University, UK
r.h.welsh@lboro.ac.uk

Stewart Birrell
MIRA Ltd, UK

ABSTRACT

Significant research and development in Europe in recent years has been focused on Intelligent Transport Systems (ITS). Many ITS functions are available on portable navigators and Smartphones and their market penetration is increasing considerably. Nevertheless no standards directly related to the use of Aftermarket and Nomadic Devices in vehicles exist and there is little published knowledge about their impact on the driver behavior. Field Operational Trials (FOTs) being undertaken within Europe, funded under the EU TeleFOT project, are assessing the impact that such devices have focusing upon 4 domains; Safety, Mobility, Efficiency and the Environment. User uptake and acceptance is also being considered. This paper describes the UK FOTs and presents examples of analyses using the different data sources available at the time of writing.

Keywords: driver behavior, field operational trials

1   BACKGROUND

Significant research and development in Europe in recent years has been focused on Intelligent Transport Systems (ITS). Many ITS functions are available on portable navigators and Smartphones and their market penetration is increasing considerably. Nevertheless no standards directly related to the use of Aftermarket and Nomadic Devices in vehicles exist and there is little published knowledge about their impact on the driver behavior and the user acceptance.

TeleFOT (Field Operational Tests of Aftermarket and Nomadic Devices in Vehicles), involves a large scale pan-European field trial aiming to assess the impact of functions provided by nomadic devices on the driving task, as well as on the transportation process as whole. A Field Operational Test (FOT) is a relatively new method, especially in Europe, for studying the impacts of functions on transportation, i.e. on driving, traffic and transport. From a technical and methodological perspective, the FESTA Handbook (2008) defines an FOT as "A study undertaken to evaluate a function, or functions, under normal operating conditions in environments typically encountered by the participants using quasi-experimental methods."

Therefore TeleFOT aims to compare the impacts that the use of the function has on travel and behavior compared with a baseline condition during which the function is not operating. In order to achieve this, the drivers' control over or interaction with the function(s) has to be manipulated by the research team. "Normal operating conditions" implies that the drivers use the vehicles during their daily routines, that data logging works autonomously and that the drivers do not receive special instructions about how and where to drive. Except for some specific occasions, there is no experimenter in the vehicle, and typically the study period extends over several months.

The FOTs in TeleFOT are organized in three test communities based in Northern (Finland, Sweden), Central (Germany, UK, France) and Southern (Greece, Italy, Spain) Europe. This paper describes in detail the FOTs in the UK and illustrates the type of analyses that are being undertaken with the various data sources available.

2   FIELD OPERATINAL TRIAL DESIGN

FOTs have been undertaken by two separate organizations in the UK, Loughborough University and MIRA Ltd. Data have been collected using a combination of a large scale FOT (LFOT) and 2 detailed FOTs (DFOT). The LFOTs aim to investigate normal, everyday use of the functions provided by the nomadic device over a long period of time using a naturalistic driving context with the participants using their own vehicles. Basic data are logged concerning the vehicle's movement using GPS and accelerometer readings integral to the nomadic device. The DFOTs are more experimental in nature; selected participants drive a predefined route in a highly instrumented vehicle capable of precisely recording driver behavior through video observations and also capable of logging parameters such as lane departure, time headway and fuel consumption. Typically, DFOT data are being collected in order to answer research questions where the level of data being collected in the LFOTs is insufficient due to the need for specialised data collection equipment.

The LFOT assessed the impact of static navigation and speed alert provided by an N-Drive Navigation device. DFOT1 was designed to complement this LFOT and considered the navigation function within the N-Drive device. DFOT2 assessed the Green Driving Support, Forward Collision Warning and Lane Departure Warning functions provided by a prototype in-vehicle Smart driving system; Foot-LITE.

2.1 Study design

Table 1 below provides an overview of the study design for each FOT.

Table 1 Field Operational Trial Study Design

Parameter	LFOT	DFOT1	DFOT2
Number of participants	80	40	40
Design	Within subjects 1 month control 8 months experiment	Between Subject 50% 1 hour no function /50% 1 hour with 1st use of function then 1 hour as experienced user	Within Subjects 2 conditions (1 experimental, 1 control)
Driving conditions	Free choice	Controlled route Accompanied	Controlled route Accompanied
Functions	Static Navigation Speed Alert	Static Navigation	Green Driving Forward Collision Lane Departure

2.2 Data Acquisition

2.2.1 LFOT data

Three main types of data have been collected in the LFOT; Questionnaire Data, Travel diary data and logged data. Participants completed a background questionnaire at the start of the study together with an initial User Uptake questionnaire. Subsequent User Uptake questionnaires were also completed at the study mid-point and upon completion. Participants were also asked to complete a one week travel diary at 3 monthly intervals during the study, with the fist being during the first week of the study. The N-drive navigation device with speed alert acted as the data logger in the LFOT. This recorded GPS positioning at 1Hz, date and time and when the device gave information to the driver.

2.2.2 DFOT 1 data

Two data acquisition systems have been used in DFOT1; Race Technology DL1 data logger and Seeing Machine's 'FaceLAB' equipment.

Race Technology DL1 is a standalone motorsport/industrial data logger and video system which can store data up to a maximum of 100Hz from a number of sources including its built in high accuracy 20Hz GPS and 3g, 3 axis accelerometers. Analogue inputs can be configured through a range of sensors equating to 13 additional data sources such as wheel speeds, shaft speeds, engine speeds, temperatures, pressures etc. The DL1 system can be fully integrated with a four channel video system allowing data overlay and analysis of multiple video channels. Up to 50 hours of logging can be achieved with a 1GB compact flash card.

FaceLAB uses state of the art non-contact and unobtrusive camera based technologies to accurately detect and track both eye and face movements. The system uses Infrared (IR) reflection to detect eye location and identify the pupil; from this all other measures are derived. Data are generated for head location (X, Y and Z, and roll, pitch and yaw) plus for more detailed analysis of the eye behavior such as eye movement and blink rate. These measures allow distraction measures to be recorded and analyzed. Analysis of gaze and head movement can be significantly improved by using FaceLAB world model software. This allows a 3D 'world' to be created modeling the environment around the subject, from which gaze intersections with model objects can be plotted. In DFOT1 models of interior components of the trial vehicle were constructed, such as mirrors, instrument panel and position of the navigation device, and eye gaze measures analyzed for intersections with these objects.

2.2.3 DFOT 2 data

DFOT 2 also uses the Race Technology DL1 described above. In addition data are also collected from the OBDII port and an adapted lane departure warning camera incorporated with the Foot-LITE device at approximately 5-7 Hz. This allows instantaneous & average fuel economy (mpg), current & ideal gear position, engine speed (rpm) & load (%), throttle position (%), number of lane deviations & coasting warnings, headway and idling time, vehicle position in lane and accelerator pedal & speed smoothness to be calculated and assessed.

In addition, both DFOTs used the NASA-TLX Task Load Index in order for participants to rate their concentrations levels during the course of each DFOT (Hart 1988)

3   EXAMPLE ANALYSES

This section provides examples of the type of analysis that is being carried out on different types of data being collected. Within the TeleFOT project around 50 key research questions (RQ) with associated hypotheses are being analyzed across the following impact assessment domains; Safety, Mobility, Efficiency the Environment and User Uptake. Thus, it is only feasible to present a snapshot of some early results here. The derivation for all of the RQs for each domain can be referenced in the analysis plans available on the TeleFOT website (http://www.telefot.eu/). At the time of writing, travel diary data and logged data from the LFOTs within the TeleFOT project are still in the post processing phase and hence unavailable for analysis. Questionnaire data from the LFOTs and data from the DFOTs can however be analyzed.

3.1   Questionnaire data LFOT

Questionnaire data is typically being used to identify participant's perceptions and the way that these change over a period of time using the TeleFOT function. For example, one question relates to the participants perception of safety offered by the function. Participants were asked in the initial user uptake questionnaires, prior to use of the functions, whether they thought their 'safety' during driving would change following installation of the navigation device to be tested.

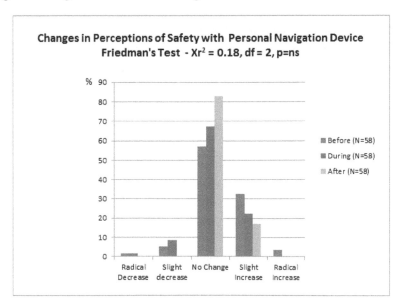

Figure 1 Changes in perceptions of safety with use of Navigation Function

The participants were asked to rank the likely changes in perception of safety on a 5-point scale where '1' represents radical decrease in safety and '5' represents radical increase. They also responded to this question during their trial and subsequently upon completion of the trial. Data, currently available for 58 participants, are presented in figure 1 above.

A Friedman's Test has been used to determine whether there are differences in the participants' responses during the three phases of the study. The p value is non-significant and hence it can be concluded that in general, the use of navigation did not significantly change the participants perception of safety whilst driving, with the majority reporting that they perceived no safety benefit.

3.2   DFOT data

The UK DFOT1 focussed on looking at how risk might affect visual behaviour in traffic situations, in particular when using a Personal Navigation Device. Drivers were asked to drive a prescribed route involving 27 separate intersections/junctions and were asked to rate their perceived level of concentration required by them to safely negotiate the intersection/junction on a 1 to 4 scale whereby 1 equalled relatively low level of concentration required to safely negotiate the junction/intersection and 4 = very high level. The intention of determining this rating by the participants was that it could be used as a surrogate for driver 'risk' during junction negotiation.

Tables 2 and 3 below highlight the types of junctions with the mean highest and lowest concentration ratings based upon responses from 20 participants.

Table 2 - Junctions with Highest Concentration Ratings

Junction number	Description	Average Concentration Rating
7	Sharp right turn on busy local street with pedestrians present	2.57
25	Merge on a slip-road from minor road to major A-Road	2.44
27	Right turn on busy roundabout with heavy traffic flow	2.41
16	Entry onto and exit from roundabout on a minor road but with a road crossing through	2.38
3	Entry onto and exit from roundabout on a minor road but with a road crossing through	2.11

Table 3- Junctions with Lowest Concentration Ratings

Junction number	Description	Average Concentration Rating
10	Left turn but with small deviation (more of a 'merge')	1.3
20	Right turn from minor road to minor road	1.33
19	Left turn from minor road to minor road	1.38
11	Left turn from major road to minor road	1.41
12	Right turn then left turn but both on minor roads	1.57

The junctions with the lowest mean 'concentration ratings' as judged by the subjects comprised junctions mainly on minor roads involving a left-turn which are situations where no gap-choice is expected and where traffic volumes are normally relatively light. On the other hand, junctions with the highest concentration ratings usually involved busier roads with higher volumes of traffic and with pedestrians present.

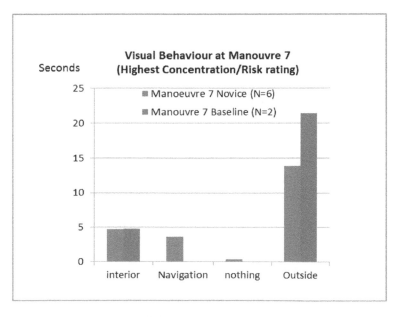

Figure 2: Visual Behaviour at Intersections with Highest Mean Concentration Ratings

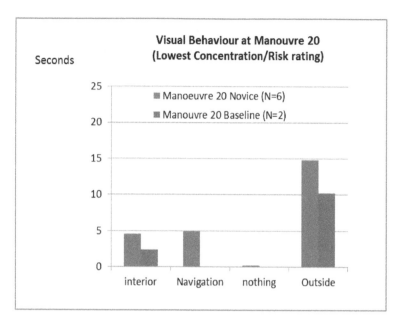

Figure 3: Visual Behaviour at Intersections with Lowest Mean Concentration Ratings

FaceLAB recordings were made of the drivers' visual behaviour looking in particular at duration of glances to the road, the vehicle interior and (where applicable) the function being tested (static navigation in the case of DFOT1). An assessment was made of the difference in visual behaviour between a 'base-line' condition (in which no navigation system was present) and the experimental condition in which the subjects used the navigation device for the first time. At all junctions/intersections

When all data are processed, the analysis will involve a full evaluation of the comparison of visual behaviour at each junction by the subjects in each of the three conditions used in the trials. Figures 2 and 3 above illustrate an example of the analysis that will be conducted.

Data collected from DFOT 2 showed that when using the Smart driving advisor fuel efficiency for a mixed driving route lasting approximately 1 hour increased by 4.1% (figure 4), this difference was statistically significant ($F_{(1,67)} = 16.0$, p < 0.001). The increase in fuel efficiency observed was achieved without an impact on journey time or average speed, which are common misconceptions with adopting an eco-driving style. Whilst an average increase of just over 4% is not as large as the 5-15% cited as being possible with the use of eco-driving assistance systems as proposed by Klunder et al (2009), intra-participant increases were often between 13 and 17%. Results from DFOT 2 also showed that participants perceived workload during the driving task (as measured with TLX) as being higher, on average, when using the Smart driving advisor at 30.2 (S.D. = 18.7) verses 25.8 (16.8) in the control condition. This difference was predominately made up of increases in

Mental Demand and Effort; however, interestingly a very small improvement in self-determined driving performance was also observed when using the Smart driving feedback and could be a result of positive feedback being offered by the Foot-LITE system (e.g. 'Good Speed Consistency).

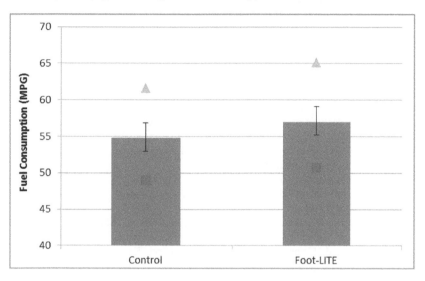

Figure 4: Mean fuel consumption for the entire journey in DFOT 2. Error bars represent standard deviation of data, red squares minimum and green triangles maximum.

4   CONCLUSIONS

This paper has described the nature of Field Operational Trials that have been undertaken in the UK as part of the EU funded TeleFOT project. At the time of writing, the data collected were still being processed and hence the data available for presentation of result are somewhat limited. The paper does however give an indication of the methods for data collection that are becoming more widespread within the transport research community. It also demonstrates ways in which the data can be used to answer questions of particular relevance to policy makers in respect of safety, mobility, efficiency, the environment and user uptake.

The TeleFOT project has collected a vast amount of data from Field Operational Trials across Europe. The data have great potential to better inform the research community, industry and policy makers and also to provide a rich source for furthering analytical methods such as video data analysis for behavioral scientists.

The authors hope, along with other colleagues in the TeleFOT project, to present more fully the results of the data analysis being undertaken in future publications.

734

ACKNOWLEDGEMENTS

The research leading to work presented in this paper has received funding from the European Commission Seventh Framework Programme (FP7/2007-2013) under grant agreement n° FP7-ICT-2007-2, Project TeleFOT (Field Operational Tests of Aftermarket and Nomadic Devices in Vehicles). The authors would like to specially thank the TeleFOT consortium that has supported the development of this research.

REFERENCES

Alvarez-Filip, L., N. K. Dulvy, and J. A. Gill, et al. 2009. Flattening of Caribbean coral reefs: Region-wide declines in architectural complexity. *Proceedings of the Royal Society, B* 276: 3019–3025.

EU project FESTA (2008). 'Deliverable n. 6.4: FESTA Handbook', p. 1. *Available at* http://www.its.leeds.ac.uk/festa/index.php

Hart, S. G. and Staveland, L. E. 1988 Development of NASA-TLX (Task Load Index): Results of empirical and theoretical research. *P. A. Hancock and N. Meshkati (Eds.) Human Mental Workload. Amsterdam: North Holland Press.*

Klunder, G., Malone, K., Mak, J., Wilmink, I., Schirokoff, A., Sihvola, N.,Holmén, C., Berger, A., de Lange, R., Roeterdink, W. and Kosmatopoulos, E. 2009. Impact of Information and Communication Technologies on Energy Efficiency in Road Transport: Final Report. TNO report for the European Commission, 1-126

CHAPTER 74

An Assessment of Cognitive Workload Using Detection Response Tasks

Conti, A.S.[1], Dlugosch, C.[1], Vilimek, R.[2], Keinath, A.[2] & Bengler, K.[1]
[1] Technische Universität München - Institute of Ergonomics
Boltzmannstraße 15, 85747 Garching, Germany
[2] BMW Group
80788 Munich, Germany
conti@lfe.mw.tum.de

ABSTRACT

In-vehicle technologies are becoming less visual manual and more cognitively oriented. This trend includes technologies such as speech interaction, which alleviates visual-manual demand shifting the focus to cognitive demand. For our purposes, cognitive workload is defined as the load incurred by a human operator by a given task(s). Because of this technological shift to the cognitive, it follows that the ways to assess cognitive workload should be evaluated in more detail. This is especially important for industries needing to quantify, predict, and test how cognitively demanding a system is in order to derive distraction or to optimize conditions for the safest in-vehicle interaction. Detection response tasks (DRTs) have been suggested as sensitive to cognitive workload and therefore a good technique to employ to measure cognitive workload. The specific aim of the current study was to assess how sensitive DRTs are to cognitive workload. The DRTs evaluated are the remote DRT, head mounted DRT and tactile DRT. The DRTs were tested under dual-task/static and triple-task/dynamic settings. In addition to the DRTs, five cognitively loading tasks, artificial and naturalistic, were used to manipulate cognitive workload in two difficulty levels: easy and difficult. Results reveal how sensitive the DRTs are to cognitive workload and to variations thereof, as well as how DRTs perform in different settings. Implications and future directions will be discussed.

Keywords: Cognitive workload, Detection Response Tasks, Dual task, N-back.

1   INTRODUCTION

In-vehicle devices such as navigation or phoning apparatus were originally developed based on an interaction between the device and the user, requiring visual search and manual input. Due to a high overlap between device usage and the driving task, developers are looking in to other possibilities to alleviate an already inundated visual-manual domain. As an alternative, speech-interaction, for example, has been proposed as a way to by-pass visual-manual tasks. This technology, however, is not without potential fault and introduces yet another issue, namely that of cognitive workload. The basic mission is to assess the size of these effects and their relevance for traffic safety. In light of this, detection-response tasks (DRTs) have been proposed as the steadfast, economical, and easy to use way of assessing cognitive workload online (Merat, N. & Jamson, A.H., 2008; Jahn, G., Oehme, A., Krems & Gelau, 2005). Therefore, finding a sensitive, valid, and reliable way to measure cognitive workload is important not only for the assessment of in-vehicle devices, but also of additional tasks currently considered acceptable while driving, such as listening to an audiobook or conversing. The current research evaluates the DRT methodology with the aim of gaining a better understanding of how it can be used in applied settings.

Cognitive or mental workload is the dynamic between the amount of mental resources available and those required by a task (Hart & Staveland, 1988). The feeling of being cognitively loaded can depend on the ability (Gonzalez, 2005) and motivation of a person (Vidulich & Wickens, 1986). Additionally, cognitive workload can vary across individuals; such differences can even be seen at a neurological level in varying cortical activation patterns (Jaeggi, Buschkuel, Etinne, Ozdoba, Perrig & Nirkko, 2007). Due to its non-physical nature, cognitive workload is something which must be inferred based on task performance or other overt parametric, rather than directly observed (Casali & Wierwille, 1984). Because of this, techniques such as secondary task methodology have been developed to assess cognitive workload. The ideology behind the secondary task paradigm is that a secondary task is operationalized as the measure of resources not being used in the performance of the primary task (Wickens & Hollands, 2000). Therefore, it is assumed that, provided the primary task is being performed following a suitable instruction, the secondary task performance reflects the residual capacity left over.

There are many questions surrounding cognitive workload, the measurement thereof, and its applicability. In an attempt to narrow down these questions, the current experiment, a joint effort between BMW and the Department of Ergonomics (TU Munich), is to be considered the first step in our research endeavor to fully understand the power of the DRT as a cognitive workload assessment tool. The specific research question addressed in this publication is how the different DRT settings compare to each other, especially in terms of their sensitivity to cognitive workload. Our experiment tested three different DRTs: head-mounted, tactile, and peripheral, in dual-task/static and dynamic/triple-task settings, in order to evaluate their ability to measure cognitive workload. These modalities were specifically focused on as a result of current ISO discussions over each of their efficiencies as measuring tools. The cognitively loading additional or "secondary tasks" (ISO terminology) were implemented to induce overall cognitive workload and to test the effects of different types of cognitive workload on DRT task performance. These additional tasks induced cognitive workload in two ways: easy, the simpler variant with less task demand, and difficult, the more complex variant with an increased task demand. The aim of this was to test whether the DRTs were, in addition to being able to measure cognitive workload, sensitive enough to detect variations of such workload.

2   METHODS

2.1   Participants

All participants were required to have obtained their driver's licenses. Eighteen participants (9 female) took part in this experiment. The age range of the participants was between 19 and 27 years old, with a mean of 22.33 (SD = 2.50). All participants were right handed and had normal or corrected-to-normal vision.

2.2   Tasks and Procedure

Prior to experimentation, participants were guided by the experimenter through a brief familiarization period where all tasks and response methods were explained to the participants. To minimize task switching effects, trials were arranged in the following hierarchy: DRT or baseline, driving condition, and secondary task, and were thus blocked together accordingly. Baselines, DRT, secondary task, and difficulty levels were randomized across participants. Non-driving blocks were always performed prior to their respective driving blocks, with the exception of baseline measurements. Participants were instructed to drive safely, when driving was part of their task set, and to perform the tasks as fast and accurately as possible.

The driving task was a simple simulated two-lane highway scene. This simulated driving task required participants to simply drive along the highway scene, maintaining a constant speed of 80km per hour.

"Secondary tasks" were those that were neither driving nor DRTs. These tasks were used to induce different levels and types of cognitive workload. Each task had both an easy and difficult variant, which were deciphered via cognitive task analysis. In this analysis, tasks were broken down in to their fundamental cognitive subparts. The research team then assessed and agreed upon the best way, per task, to manipulate difficulty. Five secondary tasks in two difficulty levels were used in this experiment: n-back (n levels 0 & 2; Mehler, Reimer, Coughlin & Dusek, 2009), counting task (modeled after Bengler, Kohlmann & Lange, 2012), control command task, sentences/dialogue task (modeled after an experiment reported in Baddeley & Hitch, 1974), and surrogate reference task (SuRT; as per ISO Calibration document draft- ISO/DTS 14198). All secondary tasks were user-paced with the exception of the n-back, which was system paced. The SuRT and n-back tasks were introduced to this experiment as the common barometer used by all ISO collaborating laboratories.

Three detection response tasks were used in this experiment: head-mounted (HDRT), tactile (TDRT), and peripheral or remote (RDRT). All detection response tasks used in this experiment required participants to detect a stimulus and then to respond to its presence via button press. The interstimulus interval randomly oscillated between 3000 – 5000 ms. Stimuli were presented for 1000 ms in duration. Responses were always given with the left hand via button press. The RDRT was set at an 80cm viewing distance. Four red LEDs were arranged horizontally and spread symmetrically 17° and 32° from the center point. The HDRT consisted of 1 LED being mounted to a baseball-type cap. The viewing distance was kept constant across participants at 18cm (measured along the center line of the hat's brim, starting from where the brim is joined to the cap). For the TDRT, a small electric vibrating node was placed on the left shoulder of the participant.

2.3 Equipment

All experiments were carried out in a fixed simulator at the Technical University Munich. The driver's seat was positioned centrally in this mockup with an active steering wheel (Wittenstein). The large LCD monitor to the front of the driver was used to display the simulated highway driving scenario (SILAB; Veitshöchheim, Germany). During the driving simulation, no gear shifting was required. DRT responses were given by pressing a button fastened to the left index finger. A screen to the right of the driver was used to display the visual-manual task (SuRT) whereby responses were input by a small keypad located underneath the screen. Speech segments used in the experiment were produced by Text Speaker 3.19 (DeskShare Inc., 2000) with the voice character "Hans" (IVONA Software). Responses to speech based tasks were given verbally by the participant.

3 RESULTS

The current study was performed in order to evaluate the DRT as a valid cognitive workload assessment technique; each DRT's sensitivity to variations in cognitive workload and a comparative look at the DRTs were of particular interest.

In figure 1, mean RTs for each DRT can be seen (all mean DRT RTs discussed are in milliseconds unless otherwise specified). A total of six repeated-measures ANOVAs were run on the current data set; one per DRT, per static/dynamic scenario. Triple-task situations generally yielded longer RTs than those for the dual-task situations. Across all DRTs, the dual-task and triple-task scenario mostly mirror each other's trend, with a few exceptions. DRT RTs when the n-back task was also performed was among the best in terms of performance, often showing the lowest RT; concurrent SuRT task performance, on the other hand, lead to some of the longest DRT RTs.

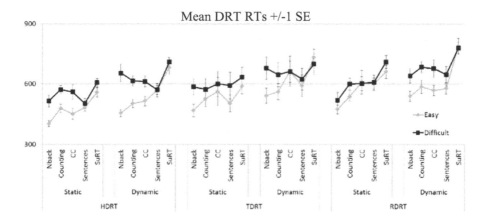

Figure 1: Mean DRT RTs (ms) for easy and difficult secondary task variants as a function of secondary task, setting, and DRT.

For the HDRT static scenario, significant main effects of secondary task, $F(4,68) = 10.49, p < .001, \eta2 = .328$ and difficulty were found, $F(1,17) = 28.80, p < .001, \eta2 = .629$. The Bonferroni post hoc test revealed that the DRT RTs during the SuRT task ($M = 581.90, SD = 77.88$) were significantly higher than when the n-back task ($M = 458.69, SD = 80.65$), counting task ($M = 524.71, SD = 82.25$), cc task ($M = 505.53, SD = 118.64$), and sentences task ($M = 493.32, SD = 65.30$) (all $ps < .05$), were performed. Additionally, DRT RTs when the n-back was performed were significantly lower than when the counting task, $p < .05$, was performed. For difficulty as a main effect, DRT RTs when easy ($M = 474.67, SD = 68.12$) secondary tasks were performed were significantly lower than when the secondary task was the difficult ($M = 550.99, SD = 19.12$), $p < .001$.

In the HDRT dynamic scenario, secondary task, $F(2.64,44.87) = 12.66, p < .001, \eta2 = .427$ (Greenhouse-Geisser), and difficulty, $F(1,17) = 27.66, p < .001, \eta2 = .619$, were found as significant main effects. The Bonferroni post hoc test revealed that the DRT RTs were significantly longer when the SuRT ($M = 694.79, SD = 103.76$) task was performed than when the n-back ($M = 554.42, SD = 116.65$), counting ($M = 558.80, SD = 87.81$), cc ($M = 562.79, SD = 84.74$) and sentences tasks ($M = 569.02, SD = 103.88$), (all $ps < .05$), were performed. Secondary task difficulty showed the same effect as per the static scenario; DRT RTs when the easy ($M = 544.11, SD = 79.92$) tasks were being performed were significantly lower than when difficult ($M = 631.82, SD = 89.25$), $p < .001$, tasks were performed. There was a significant interaction effect between secondary task and difficulty level, $F(4,68) = 7.23, p < .001, \eta2 = .298$. This indicated that the difficulty level had different effects on the mean DRT RTs depending on which secondary task was being performed concurrently. For the most part, difficult task settings caused an increase in RT. This was most strongly seen for the n-back and counting tasks. For the sentences task, very little difference was observed across task difficulty levels.

For the TDRT static scenario, a significant main effect of secondary task difficulty, $F(1,17) = 11.38$, $p < .01$, $\eta2 = .401$, was found. The Bonferroni post hoc test revealed that the DRT RTs when the easy ($M = 529.57$, $SD = 163.86$) tasks were being performed were significantly lower than when difficult ($M = 597.00$, $SD = 200.64$), $p < .01$, tasks were performed.

In the TDRT dynamic scenario, significant main effects of secondary task $F(4,68) = 5.88$, $p < .001$, $\eta2 = .257$, and difficulty, $F(1,17) = 12.35$, $p < .01$, $\eta2 = .421$, were found. The Bonferroni post hoc test revealed that the DRT RTs were significantly longer when the SuRT ($M = 715.40$, $SD = 184.84$) task was performed than when the n-back ($M = 609.88$, $SD = 187.25$), counting ($M = 603.60$, $SD = 207.30$) and sentences ($M = 607.46$, $SD = 220.18$), (all $ps < .05$), tasks were performed. Secondary task difficulty results showed that the DRT RTs when the easy ($M = 617.80$, $SD = 186.28$) tasks were being performed were significantly lower than when difficult ($M = 661.90$, $SD = 212.72$), $p < .01$, tasks were performed. There was a significant interaction effect between secondary task and difficulty level, $F(4,68) = 3.27$, $p < .05$, $\eta2 = .161$,. This indicated that the difficulty level had different effects on the mean DRT RTs depending on which secondary task was being performed concurrently. As seen in the HDRT, there was a strong mean DRT RT difference across the easy and difficult performances of the n-back and counting tasks. However, this was not the case with the cc and SuRT tasks, which produced effects such that the easy variant performed better (this observation was strongest for the SuRT task).

For the RDRT static scenario, a significant main effect of secondary task, $F(4,68) = 9.04$, $p < .001$, $\eta2 = .347$, was found. The Bonferroni post hoc test revealed that the DRT RTs during the n-back task ($M = 495.98$, $SD = 109.93$) were significantly lower than when the counting ($M = 568.63$, $SD = 88.83$) and SuRT ($M = 685.59$, $SD = 109.97$) (both $ps < .05$), tasks were performed. Additionally, DRT RTs when the counting task was performed were also significantly lower than when the SuRT task, $p < .05$, was performed.

In the RDRT dynamic scenario, significant main effects of secondary task $F(4,68) = 16.50$, $p < .001$, $\eta2 = .492$, and difficulty, $F(1,17) = 12.85$, $p < .01$, $\eta2 = .431$, were found. The Bonferroni post hoc test revealed that the DRT RTs were significantly longer when the SuRT ($M = 784.50$, $SD = 120.31$) task was performed than when the n-back ($M = 590.07$, $SD = 115.12$), counting ($M = 635.53$, $SD = 104.52$), cc ($M = 622.58$, $SD = 117.70$), and dialogue ($M = 612.72$, $SD = 125.93$), (all $ps < .01$), were performed. In terms of secondary task difficulty, DRT RTs when the easy ($M = 612.18$, $SD = 87.01$) tasks were being performed were significantly lower than when difficult ($M = 686.04$, $SD = 114.51$), $p < .01$, tasks were performed.

4   DISCUSSION

The aim of the current study was to take the first steps in evaluating the different DRTs and to compare their cognitive workload measuring potential. Participants performed several cognitively loading tasks, both artificial and naturalistic, with each of the DRTs under examination: HDRT, TDRT, RDRT. All experimental combinations were performed in both a static/dual-task and dynamic/triple-task setting to additionally assess differences in sensitivity and performance.

As to be expected, the triple task (dynamic) scenario generally yielded longer RTs and higher standard errors. This finding is completely in accordance with what is typically observed when many tasks are concurrently performed. In two situations, this was not the case (i.e. RDRT for CC and sentences). This discrepancy could reflect the overall difference observed between DRT performances while artificial tasks versus the naturalistic tasks are being performed. All DRTs were discriminative to cognitive workload induced by the artificial tasks (i.e. n-back, counting); this is the case with the exception of the SuRT task. In line with the idea that DRTs are able to measure cognitive workload, it makes sense that a task such as the SuRT, which is more visual-manual than cognitive, does not fit with the overall data trend. In terms of naturalistic tasks (i.e. control command, sentences), different sensitivities to cognitive workload are revealed for each DRT.

The HDRT was observed to be the most sensitive to variations in cognitive workload in both the dual and triple task situations. This finding confirms that cognitive workload was induced across all secondary tasks and the HDRT proved sensitive enough to detect this. An additional interaction effect between these two variables was found in the triple task condition, such that the task difficulty seemed to cause an inflation of the mean DRT RT, for all secondary tasks except for the dialogue condition where only a slight increase was found for difficult relative to easy.

For the RDRT and TDRT, the triple task scenario was much more sensitive than its respective dual-task variant. In the RDRT dual-task setting, a significant main effect of secondary task difficulty was found, while in the triple-task setting main effects of secondary task, difficulty, and an interaction effect between them, were found. The TDRT also followed this trend of increased sensitivity to cognitive workload given an increased task load. In the TDRT dual-task setting, mean DRT RTs were significantly affected by the secondary task performed and in the triple, difficulty of the secondary task was also a main effect. Though it is not clear, at this point, why exactly the DRTs are more sensitive in the triple task, a possible explanation could be that the dual-task situation was simply too easy for some participants and therefore participants were not aroused enough to perform the given tasks (Yerkes-Dodson law, 1908). This, as we saw, was not the case for the HDRT.

5   CONCLUSION

The current experiment represents the first steps at systematically evaluating the DRT tasks and their ability to measure cognitive workload. Though all DRTs did show sensitivity to cognitive workload, this sensitivity varied according to whether the task situation was static or dynamic. It seems that the HDRT is the most sensitive to variations in cognitive workload for the dual-task/static scenario. Though the RDRT and TDRT were also somewhat sensitive, this sensitivity varied and even increased along with an increasing task demand (i.e. triple-task/dynamic setting). Further analyses will incorporate a deeper look at the secondary tasks' manipulation of cognitive workload as well as secondary task and driving performance. Such analyses will lend a perhaps clearer picture of exactly which type of cognitive workload can be measured by which DRT. Future work should direct itself towards investigating the effect of task instruction on performance as well as validating the induced workload via physiological methodologies.

REFERENCES

Baddeley, A.D., & Hitch, G.J. (1974). Working memory. In G.A. Bower (Ed.), *Recent advances in learning and motivation* (Vol. 8, pp. 47–90). New York: Academic Press.

Bengler, K., Kohlmann, M., & Lange, C. (2012). Assessment of cognitive workload of in-vehicle systems using a visual peripheral and tactile detection task setting. In: Work: A Journal of Prevention, Assessment and Rehabilitation 41 (Supplement 1/2012), S. 4919–4923.

Casali, J. G., & Wierwille, W. W. (1984). On the measurement of pilot perceptual workload: A comparison of assessment techniques addressing sensitivity and intrusion issues. *Ergonomics, 27,* 1033-1050.

Gonzalez, C. (2005). Task workload and cognitive abilities in dynamic decision making. *Human Factors, 47* (1). Accessed from: Carnegie Mellon University, Department of Social and Decision Sciences. Paper 32. http://repository.cmu.edu/sds/32

Hart, S. G., & Staveland, L. E. (1988). Development of NASA-TLX (Task Load Index): Results of experimental and theoretical research. In P. A. Hancock & N. Meshkati (Eds.), Human Mental Workload . Amsterdam: North Holland.

ISO/DTS 14198 Road vehicles — Ergonomic aspects of transport information and control systems — Calibration tasks for methods which assess demand due to the use of in-vehicle systems. Revised Draft Version (2011).

Jaeggi, S.M., Buschkuehl, M., Etienne, A., Ozdoba, C., Perrig, W., & Nirkko, A.C. (2007). On how high performers keep cool brains in situations of cognitive overload. *Cognitive Affective & Behavioral Neuroscience,* 7, 75-89.

Jahn, G., Oehme, A., Krems, J., & Gelau, C. (2005). Peripheral detection as a workload measure in driving: Effects of traffic complexity and route guidance system use in a driving study. *Transportation Research Part F: Traffic Psychology and Behaviour,* 8(3), 255-275.

Mehler, B., Reimer, B., Coughlin, J. F., & Dusek, J.A. (2009). Impact of Incremental Increases in Cognitive Workload on Physiological Arousal and Performance in Young Adult Drivers. *Transportation Research Record: Journal of the Transportation Research Board, 2138,* 6-12.

Merat, N., Jamson, A. H., & Kingdom, U. (2008). The Effect of Stimulus Modality on Signal Detection: Implications for Assessing the Safety of In-Vehicle Technology. *Human Factors, 50*(1), 145-158.

Vidulich, M. A., & Wickens, C. D. (1986). Causes of dissociation between subjective workload measures and performance; Caveats for the use of subjective assessments. *Applied ergonomics, 17*(4), 291-6.

Wickens, C. D., & Hollands, J. (2000). Engineering Psychology and Human Performance, 3rd edn. (Upper Saddle River, NJ: Prentice Hall).

Yerkes R.M., & Dodson J.D. (1908). The relation of strength of stimulus to rapidity of habit-formation. *Journal of Comparative Neurology and Psychology,* 18, 459–482.

CHAPTER 75

Statistical Methods for Building Robust Spoken Dialogue Systems in an Automobile

Pirros Tsiakoulis, Milica Gasic, Matthew Henderson, Joaquin Planells-Lerma, Jorge Prombonas, Blaise Thomson, Kai Yu, Steve Young

Cambridge University Engineering Department
{pt344, mg436, mh521, jp566, jlp54, brmt2, ky219, sjy}@cam.ac.uk

Eli Tzirkel

General Motors Advanced Technical Center – Israel
eli.tzirkel@gm.com

ABSTRACT

We investigate the potential of statistical techniques for spoken dialogue systems in an automotive environment. Specifically, we focus on partially observable Markov decision processes (POMDPs), which have recently been proposed as a statistical framework for building dialogue managers (DMs). These statistical DMs have explicit models of uncertainty, which allow alternative recognition hypotheses to be exploited, and dialogue management policies that can be optimised automatically using reinforcement learning. This paper presents a voice-based in-car system for providing information about local amenities (e.g. restaurants). A user trial is described which compares performance of a trained statistical dialogue manager with a conventional handcrafted system. The results demonstrate the differing behaviours of the two systems and the performance advantage obtained when using the statistical approach.

Keywords: spoken dialogue systems, statistical dialogue management, POMDP

1   INTRODUCTION

Spoken dialogue has many potential advantages over, or in combination with, other modalities in an automobile. This has led to the deployment of various in-car systems integrating spoken dialogue, with applications such as telephony, entertainment, and navigation.

However, the user experience of these systems is often extremely poor. Part of the problem is that the noisy conditions can cause a drastic deterioration of automatic speech recognition performance. This results in systems repeatedly asking the same question, which in turn causes users to become frustrated.

This paper discusses how a change to the dialogue manager can result in improved overall performance, even in the context of poor recognition accuracy. Current dialogue managers are typically designed as deterministic decision networks. Since system designers will usually include a confidence threshold before accepting a given piece of information, the system ends up asking the same question over and over. This decision network structure also makes it difficult for system designers to use the information in N-best lists of speech-recognition hypotheses. The result is typically a system that is not robust to understanding errors.

Recently, a statistical approach to building dialogue managers has been developed based on partially observable Markov decision processes (POMDPs). These statistical DMs have explicit models of uncertainty which allow alternative recognition hypotheses to be exploited, and dialogue management policies that can be optimised automatically using reinforcement learning. This paper will present a voice-based in-car system for providing information about local amenities (e.g. restaurants). A user trial is described which compares performance of a trained statistical dialogue manager with a conventional handcrafted system.

2   SYSTEM OVERVIEW

The system architecture is shown in figure 1. It consists of 5 distinct modules: automatic speech recognition (ASR), semantic decoding or spoken language understanding (SLU), dialogue management (DM), natural language generation (NLG) and speech synthesis or text-to-speech (TTS).

The user speaks to the system, and the acoustic signal y is converted by the speech recognizer into a set of sentence hypotheses w, which represents a probability distribution over all possible things that he might have said. In practice, w is represented by an N-best list. The sentence hypotheses are converted into an N-best list of dialogue acts v by a semantic decoder. The dialogue manager represents the dialogue state s in a factored form via the triple $<u,g,h>$ (see also figure 2) where u is the actual (but unknown) user utterance, g is the assumed user goal and h represents the history (Williams and Young, 2007). Since this state cannot be directly observed the system maintains a probability distribution b over this state, which is called the belief state. The belief state is updated at every user turn using Bayesian inference treating the input v as evidence.

746

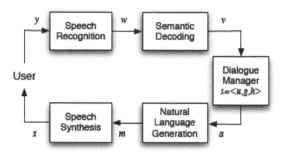

Figure 1 Architecture of the Spoken Dialogue System

Based on b, the optimal system response a is selected using a mapping $\pi:b{\rightarrow}a$ which is called a policy and which is trained automatically using reinforcement learning. The abstract system dialogue act a is converted to an appropriate utterance m by a natural language generator and then converted to speech x by a HMM-based speech synthesizer.

The spoken dialogue system is implemented using the ATK platform (http://mi.eng.cam.ac.uk/research/dialogue/atk_home.html). ATK is an API designed to facilitate building experimental applications for systems trained using the HTK speech recognition toolkit (see http://htk.eng.cam.ac.uk). It consists of a C++ layer sitting on top of the standard HTK libraries. ATK is multi-threaded and supports real-time speech recognition. ATK allows a variety of components to be connected together to implement different architectures and applications, and hence it provides a flexible platform for building spoken dialogue systems.

To illustrate the system's capability, two typical dialogues are shown in Tables 1 and 2. The first is an example of a straight forward query for which the system is able to provide recommendations immediately. The second is an example where there is no restaurant matching the user's requirements. In this case, the user relaxes his constraints and the system offers him alternatives.

Table 1 An example of a straightforward dialogue

Sys	Hello, welcome to the Cambridge restaurant system? You can ask for restaurants by area, price range or food type. How may I help you?
Usr	Hi, I'm looking for an Indian restaurant in the north.
Sys	There are 3 restaurants in the north of town serving Indian food! Would you like something in the cheap, moderate, or expensive price range?
Usr	I don't mind
Sys	Royal spice is a nice restaurant in the north of town serving Indian food.
Usr	What's the address?
Sys	Sure, It is on Victoria Avenue Chesterton.
Usr	Thanks, bye

Table 2 An example dialogue where the user relaxes their constraints

Sys	Hello, welcome to the Cambridge restaurant system? You can ask for restaurants by area, price range or food type. How may I help you?
Usr	Hi, I'm looking for an expensive Chinese restaurant in the west.
Sys	I am sorry, but there is no expensive restaurant that is in the west part of town and serves Chinese food.
Usr	Do you have anything in the moderate price range?
Sys	Sorry would you like something in the expensive price range or in the moderate price range?
Usr	Moderate.
Sys	I am sorry, but there is no moderately priced restaurant that serves Chinese food and in the west part of town.
Usr	How about the south?
Sys	Would you like Chinese food or you don't care?
Usr	I'd like a Chinese restaurant.
Sys	The Lucky star is a moderately priced Chinese restaurant in the south of town.
Usr	What's the address?
Sys	Sure, It is on Cambridge Leisure Park Clifton Way Cherry Hinton.
Usr	Thank you, goodbye

Next, we describe in more detail each of the basic components of the system.

The **Automatic Speech Recognition (ASR)** module uses the Cambridge HTK/ATK speech recogniser. It accepts live speech input and voice activity detection (VAD) is used to determine the start and end of each user utterance. A tri-gram language model and word-internal triphone HMMs are then applied in a two-pass decoding scheme. In the first pass, a bi-gram language model is used and the resulting lattice is rescored by a tri-gram language model in the second pass. The output is the 10 best hypotheses along with their confidence scores. The pruning thresholds and model size are optimised so that the decoder runs in real time. The acoustic model is trained on data from various sources, including desktop microphone (wide-band data), landline and mobile phone (narrow-band data) and VoIP recording (in-domain narrow-band data), about 130 hours in total. The front-end uses perceptual linear predictor (PLP) features with energy and first, second and third derivatives. A heteroscedastic linear discriminant analysis (HLDA) transform is used to project these features down to 39 dimensions. The language model (LM) is trained using a class-based language modeling technique. The language model training data is about 400K words. The basic idea is to train a generic statistical LM and then expand this generic LM to a specific LM using the information for the specific task in hand. The word list (dictionary) of the generic LM consists of common words used in dialogues (such as good-bye, hello, can ...) and abstract slot labels (such as SLOT_NAME, SLOT_AREA ...). This generic LM is trained on a corpus generated from the 400K-word corpus by substituting abstract slot labels for the actual slot values in our previously collected data. Hence, the generated LM is a special class-based LM consisting of both common words and abstract slot values. We then expand this LM to a specific word-based LM using a class-to-word mapping from the database.

The **Spoken Language Understanding (SLU)** module converts the N-Best ASR output to N-Best dialogue acts with confidence scores. A rule-based hand-crafted Phoenix semantic parser is used in the baseline system (Ward 1991). Various minor modifications of the original system were made to convert the output of this parser into the Cambridge dialogue act format. The Cambridge dialogue format structures the semantic representation as an act type followed by a sequence of items, where each item contains a concept and a value (both of which are optional). Each ASR hypothesis from the N-Best list is parsed into a distinct dialogue act. Identical dialogue acts are then merged and the confidence scores are added together. The resultant multiple dialogue act hypotheses with confidence scores are then fed into the dialogue manager. Table 3 shows the dialogue acts in the Cambridge dialogue semantic representation format for the example shown in Table 1.

Table 3 Example dialogue in the Cambridge dialogue act format

Sys	hello()
Usr	inform(food=indian, area=north, type=restaurant)
Sys	confreq(count="3", type=restaurant, food=indian, area=north, pricerange, option=cheap, option=moderate, option=expensive)
Usr	inform(=dontcare)
Sys	inform(name="Royal spice", type=restaurant, food=Indian, area=north)
Usr	request(addr)
Sys	inform(addr="Victoria Avenue Chesterton")
Usr	bye()

The **Dialogue Management (DM)** component uses the Bayesian Update of Dialogue State (BUDS) framework for updating belief states and a learned policy to select system action (Thomson and Young, 2010). The BUDS DM is an approach within the Partially Observable Markov Decision Process (POMDP) framework. In this approach, the DM maintains a distribution over the dialogue state space, which is referred to as the belief state. The belief state is updated according to the user goal, the observed SLU hypotheses and the dialogue history. The system action is then selected according to the updated belief state. In order to deal with the intractability of both updating belief states and action selection, two techniques of compressing the dialogue state space are employed. The first optimisation is to factorize the user goal and dialogue history further into a series of concepts with concept-level goals (called sub-goals) and concept-level histories.

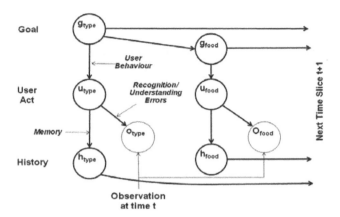

Figure 2 Dependencies in the BUDS dialogue system

Conditional independence assumptions are taken such that the concept-level goal is only dependent on the previous system action, the same concept-level goal of the previous turn and optionally the parent concept-level goal of the previous turn (see figure 2). With the factorization, tractable belief state update can be developed. In the case of the baseline restaurant system, the concepts are name, area, price range, food type, address, post code and signature dish. The second optimisation is used to simplify the action selection process. After updating the belief state, the full state space (also referred to as a master state space) is mapped into a much smaller summary space, in which action selection takes place, resulting in a summary action that gets mapped back into master space using information from the original belief state. The mapping from the (summarized) belief state to system action is referred to as a policy. In the baseline system, a handcrafted policy is used.

The **Natural Language Generation (NLG)** component reads in the dialogue act output by the DM component and generates natural language responses. In the baseline system, we use a template-based NLG. We have also added basic anaphoric references to the NLG to make replies more natural. For instance, when the system suggests a restaurant to the user, the name of that restaurant becomes the topic of the conversation. Instead of repeating the restaurant's name many times, the system refers simply to "it". When the system operates in noisy conditions, it is often the case that the system has to repeat the same dialogue act. In such cases NLG adds variations in the output text in order to appear more natural to the user.

The **Text-To-Speech (TTS)** component gets text from the NLG components and synthesizes speech using a trainable HMM-based speech synthesizer. Text analysis is first performed to convert raw text into phone-level context labels. Context-dependent HMMs are then used to model spectrum, fundamental frequency and duration of the context-dependent phones. During synthesis, the duration, spectrum and fundamental frequency are generated from the HMMs and then converted to speech waveform. As it is a parametric model, HMM-TTS has more flexibility than traditional unit-selection approaches and is especially useful for producing expressive speech (Yu, Mairesse and Young 2010).

3   PROOF-OF-CONCEPT DEPLOYMENT

The Restaurant Information task was selected for the deployment and evaluation of the system described above in an automotive environment. Using a far field hands-free microphone, users can ask for information directly about a restaurant by name but more interestingly, they can search for a restaurant by expressing their preferences for food type, price range and area. The system will then suggest restaurants and when requested provide the user with details such as post-code, address and telephone number. Users can also change their mind or adjust their constraints if no venues match their initial query. It must be noted that the system uses an open microphone architecture, i.e. there is no push-to-talk. The system is always listening and the user can speak at any time.

In a real world scenario the user would be calling the system via mobile phone while driving, for example someone driving to a new town would like to find out a place to eat. The system would have to select the target town either using location based information (e.g. the selected destination in the GPS device), or if the user specifically requests a town. We have integrated our dialogue system with an online restaurant information service provider. However, we currently assume that the town is known in advance. The TopTable (http://www.toptable.com) web-service provides information on around 20,000 restaurants from around the world. Once the town is selected, the system extracts the following information for each venue: the venue's name, area, price range, food type, address, post code and signature dish. The system ontology is updated with a list of possible names and food types and the semantic decoder is also updated to be able to understand the user talking about the restaurants' names and food types. The ASR and TTS pronunciation dictionaries are automatically updated to include new unknown words, such as restaurant names, addresses, dishes etc. This enables our dialogue system to operate for any town, as long as it is available through the web-service. For the evaluation process the city of Cambridge was selected, and a database of about 150 venues was built.

A VoIP interface was implemented so that remote users can use telephones to communicate with the system. SIPGate (www.sipgate.co.uk) was used as the SIP (Session Initiation Protocol) service provider. This service converts incoming PSTN (Public Switched Telephone Network) calls to IP data and streams them to our server. An open source SIP client, PJSIP, is used to manage SIP calls from SIPGate and the low-level audio IO functionality required by the dialogue server.

4   EVALUATION

The evaluation process took place in two stages. In the first stage we used a handcrafted system to collect training data. We then used the collected data to train statistical models for the semantic decoder and the dialogue manager. In the second stage we evaluated the trained dialogue components in contrast to the handcrafted ones. The aim of the experiment was to test the robustness and performance of a spoken dialogue system in a moving car.

The data collection stage of the experiment was performed in a standing car. About 25 subjects were asked to perform a set of 20 randomly generated tasks. The subject was sitting in the front passenger seat while the operator was sitting in the driver's seat. A call was made using an android mobile phone paired via Bluetooth with an OnStar mirror (www.onstar.com). The mirror's built-in microphone and loudspeakers were used as the speech interface. Two recording conditions were tested; for the first one the car's fan was off, while for the second one the fan was set in its maximum speed. For each task, the subject read one of the tasks, while the instructor placed the cal to the system. After, the dialogue completion the subject was asked if the dialogue was successful or not.

A total of 511 dialogues were collected for both conditions. The dialogue success rate was 67.1% (79.6% with the fan off, and 55.5% with the fan on). The collected speech data was automatically transcribed using Amazon Mechanical Turk crowd sourcing (Jurčíček et al 2011). The total audio length of all the dialogues was 2.2 hours (user turns only). The resulting speech database was further split into a training set (~60%) and a test set (~40%). This was done at the speaker level; the word error rate (WER) per speaker was measured from the transcriptions, and then each speaker was assigned to either the train or the test set. The distributions of WERs per speaker, as well as the ratio of male to female speakers were kept the same in both train and test sets. The training set was used to adapt the ASR acoustic model. This resulted in an absolute WER reduction of about 9% (from 25.8% to 16.9%) for the in-car test data set. The LM was not updated since the available data was not sufficient for language model training.

In the second stage of the experiment we trained a POMDP dialogue system and evaluated its performance in comparison with a hand-coded one based on MDP and optimized for the restaurant info task. Henceforth, the two systems are termed as:

- *MDP,* hand-coded dialogue manager based on an MDP architecture
- *POMDP,* trained POMDP dialogue manager

Both systems used the same Phoenix-based semantic decoder and the same retrained speech recognition system.

The experimental setup was similar to the one used for data collection, while taking place in a car driven around the city of Cambridge by a professional driver. The subject sat in the front seat and the experiment instructor in the rear seat. Each subject was asked to complete 14 dialogue tasks (7 for each of the two systems). After the completion of each dialogue, the subject reported to the instructor if he or she considered the task successful or not. A total of 30 subjects took part in the experiment and a total of 399 dialogues were collected. Some of the subjects did not complete the whole set of tasks due to technical difficulties. Table 4 summarises the subjective performance for each of the dialogue systems. For each system, the table shows the total number of dialogues, the number of successful dialogues, the estimated subjective success rates, and the word error rate (WER). The standard error on the success rates was estimated under the assumption that the success rates follow a binomial distribution (Thomson and Young, 2010). The actual word error rates may be lower than the ones reported here since the crowd sourcing transcription process usually introduces additional word errors.

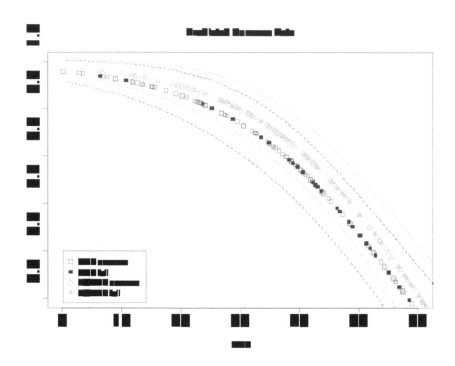

Figure 3 The effect of the WER factor on success rates of the MDP and POMDP systems.

Table 4 Subjective evaluation results

System	# Dialogoues	# Successful	% Success Subjective	WER
MDP	200	130	65.00 ± 3.37	46.83
POMDP	199	141	**70.85 ± 3.22**	43.93

We performed analysis of variance to test the statistical significance of the results. The test shows that the difference in the success rate of the two systems is not statistically significant at the 95% confidence level [p-value = 0.20]. We also tested the statistical significance of the WER factor, which also showed that there is no statistically significant difference between the word error rates of the two systems [p=0.44]. To have a more accurate depiction of the WER factor on the performance of the systems, figure 3 shows the logistic regression of the probability of success against the word error rate for a particular dialogue. The plotted points show the true WER of a given dialogue along with the predicted probability of success according to the logistic regression model. Unfilled markers depict successful dialogues, while filled ones depict unsuccessful dialogues. The dotted lines show one standard error on each side of the predicted probability of success. The POMDP system has a higher success rate on average. Although this difference is not statistically significant, the results strongly suggest that the POMDP system can cope better with ASR errors. We can see, for example, that the MPD system has some failure points even when the speech recognition performs reasonably well (WER<15%), while, on the other hand, the POMDP system starts to break at WER above 25%. This suggests that the improvement in the performance of the POMDP system is not due to better speech recognition performance but rather due to a more robust dialogue manager.

5 CONCLUSIONS

We have investigated a statistical approach to dialogue management for in-car speech-based information systems with the aim of increasing robustness to recognition errors. A Restaurant Information task was selected for this pilot study. A statistical dialogue manager based on POMDPs was trained and evaluated in comparison with a non-statistical one optimized manually for the same task. The evaluation took place in a realistic scenario where human subjects were driven around in an urban environment while talking to the dialogue system in an open microphone manner. Moreover, the system was integrated with an online service providing up to date restaurant information. Although the corpus of 400 dialogues collected for the evaluation is too small to give statistical significance, the results did indicate that the statistical dialogue manager has a higher success rate and is more robust to speech recognition errors. This suggests that in noisy situations, such as the automotive environment, where speech recognition is prone to errors, the statistical approach to dialogue management can improve overall performance.

ACKNOWLEDGMENTS

This work was funded in part by General Motors in the framework of the GM project "Robust Dialogue for Infotainment" (GM Project Number: NV887).

REFERENCES

Jurčíček, F., Keizer, S., Gašić, M., Mairesse, F., Thomson, B., Yu, K., & Young, S. (2011) "Real user evaluation of spoken dialogue systems using Amazon Mechanical Turk". Proc. Interspeech, Florence, Italy.

Mairesse, F., Gasic, M., Jurcicek, F., Keizer, S., Thomson, B., Yu, K., & Young, S. (2009). "Spoken language understanding from unaligned data using discriminative classification models". Proc. ICASSP, Taipei, Taiwan.

Thomson, B. & Young, S. (2010)"Bayesian Update of Dialogue State: A POMDP framework for spoken dialogue systems." Computer Speech and Language 24(4):562-588.

Ward, W.H., (1991), "The Phoenix system: Understanding spontaneous speech." Proc. ICASSP, Toronto, Canada.

Williams, J. & Young, S. (2007). "Partially Observable Markov Decision Processes for Spoken Dialogue Systems." Computer Speech and Language 21(2): 393-422.

Yu, K., Mairesse, F. & Young, S. (2010). "Word-level Emphasis Modelling in HMM-based Speech Synthesis." Proc. ICASSP, Dallas, TX.

CHAPTER 76

How Gender Influences Road User Behaviors: The Bringing-in of Developmental Social Psychology

Marie-Axelle Granié[1], Alexia Abou-Dumontier[2], Ludivine Guého[1,3]

[1]IFSTTAR-MA,
F-13300, Salon de Provence, FRANCE,
[2]Université Paris Ouest Nanterre-La Défense, Laboratoire Psychologie des Processus et Conduites Complexes, FRANCE,
[3]Université de Provence Aix-Marseille 1, Laboratoire de Psychologie Sociale « Influences, Représentations et Pratiques Sociales »,
FRANCE,
marie-axelle.granie@ifsttar.fr
alexia.abou@gmail.com
ludivine.gueho@ifsttar.fr

ABSTRACT

Gender differences are well known in accidentology and manifest themselves very early on in different types of accidents. Boys and men have more frequent and serious accidents than girls and women, and risk exposure does not appear to be the only explanatory variable. This phenomenon can notably be explained by greater risk-taking among boys and men. Numerous psychologists ascribe the male-female difference in risk-taking to gender roles and gender stereotypes. It was not until recently that the effects of sex-stereotype conformity were taken into account in explaining differences between males and females in risky behaviors as driver or pedestrian. In this paper, are exposed our previous studies on gender conformity effect on pedestrian, cyclist and driver behaviors, in a developmental perspective, from childhood to adulthood. Gender stereotype conformity is a better predictor of declared injury-risk behaviors than sex as early as in childhood and that, among adolescents, sensation seeking could be consider as mediator between masculinity, femininity and risk-taking as pedestrian. Furthermore, gender stereotype conformity explains behavioral compliance of children and adults with pedestrian rules, and that rule internalization and perception of intentionality are mediator factors to explain gender influence on injury-risk behavior, as pedestrian, cyclist and driver, from childhood to adulthood. From these results, means to act in education and prevention actions are discussed to minimize gender differences in injury-risk behaviors and in accidentology.

Keywords: gender, compliance, risk-taking, development, psychology

INTRODUCTION

Gender differences are well known in accidentology and manifest themselves very early in different types of accidents, and in particular in traffic injury rates. Men, and particularly young men, are involved in more crashes than young women. The difference between the sexes increases further from childhood, until it reaches a maximum among adults, in France, with 70 to 80% men among traffic fatalities between the ages of 15 and 59 (Assailly, 2001). This difference is not unique to France: in most Western countries, male drivers are 2–3 more likely to die in traffic crashes than female drivers (Hanna et al., 2006, Nell, 2002). This phenomenon can notably be explained by greater risk-taking among males (Byrnes et al., 1999).

In childhood, school-age boys tend to take more risks than girls (Ginsburg and Miller, 1982) and have more frequent and severe injuries than girls (Baker et al., 1992). Boys reported more injuries than girls and perceived them as less severe (Morrongiello, 1997). In adulthood, men reported more traffic violations (Lonczak et al., 2007), showed a lower level of normative motivation to comply with traffic laws (Yagil, 1998), tend to commit more violations, whereas female drivers tend to commit more errors (Aberg and Rimmö, 1998). Males, especially younger ones, have greater self-assessed driving abilities (Farrow and Brissing, 1990, Tronsmoen, 2008, Özkan and Lajunen, 2006), felt safer while driving (Bergdahl, 2007) and used the automobile to enhance self-efficacy more than females (Farrow and Brissing, 1990). Young male drivers reported more driving injury risks behaviors and more traffic offences (Harré et al., 1996), have riskier driving attitudes and behaviors (Harré et al., 2000) than young female drivers.

In this paper, we will show how recent studies on gender conformity effect on pedestrian, cyclist and driver behaviors can contribute to explain gender differences in risk-taking and in rule compliance, in a developmental perspective, from childhood to adulthood.

1. FROM A BIOLOGICAL TO A SOCIAL EXPLANATION OF GENDER DIFFERENCES IN RISK-TAKING

1.1 Gender-role and gender stereotypes

Men's tendency to take more risks has been generally explained in the past by a combination of biological and evolutionary theories (Wilson and Daly, 1985). Only recently has research begun to explore social environment influences on adults' gender differences in risk-taking behavior. Thus, studies have shown that parents contribute to the sex differences found in children's risk taking and unintentional injuries (Hagan and Kuebli, 2007, Morrongiello and Hogg, 2004).

Numerous psychologists indeed ascribe the male–female difference in risk-taking to gender roles (d'Acremont and Van der Linden, 2006), that is expectations about behavior that are generated by the social group and depend upon the gender group to which the individual belongs (Basow, 1992). For instance, boys and girls are treated differently by their parents at a very early age (Fagot, 1995). Girls are encouraged to be nurturing and polite, whereas boys are encouraged to be autonomous, adventuresome and independent (Pomerantz and Ruble, 1998). Gender-role socialization is based on gender stereotypes, which can be defined as the set of beliefs about what it means to be a male or a female in a given society (Deaux and Lewis, 1983). In particular, gender stereotypes about risk-taking characterize it as a typically masculine type of behavior (Bem, 1981), which is consistent with gender norms about risk-taking (Yagil, 1998). For example, by the age of 6, children rate girls as having a greater risk of injury than boys, although boys routinely experience more injuries than girls (Morrongiello et al., 2000). Among adolescents, drinking and driving is seen to be more acceptable for boys than for girls (Rienzi et al., 1996), and male drivers are subject to less supervision than female drivers (Parker et al., 1992). Recognizing certain female-stereotyped traits in oneself, however, does not mean seeing oneself as having all the components of femininity, nor even not recognizing in oneself certain masculine traits (Bem, 1974). Thus, conformity to gender stereotypes can explain why males and females differ in risk-taking, but also help to understand differences in male groups and female groups in risk taking.

1.2. Gender stereotypes and risk-taking

Research now shows that gender stereotypes – through gender role conformity – have an effect on self-reported injury risk behavior in driving behavior (Özkan and Lajunen, 2006, Sibley and Harré, 2009) and in general risk-taking among adolescents (Raithel, 2003). Recent studies showed that this effect of gender stereotypes on risk-taking display very early.

Thus, masculinity score, femininity score, and injury-risk behaviors of 3–6-year old children were measured indirectly in previous study (Granié, 2010). Results show that boys' and girls' injury-risk behaviors are predicted by masculine stereotype conformity and that both girls' masculine behaviors and injury-risk behaviors decline with increasing age.

In another recent study, sex-stereotype conformity and self-reported risky behaviors have been measured among adolescent pedestrians aged 12–16 (Granié, 2009). The results show an effect of sex-stereotype conformity on risk-taking behaviors: masculine stereotype conformity explains greater risk-taking among boys and girls.

Sensation seeking has been often studied as predictor of risk-taking (Abou et al., 2008) and is also linked to gender stereotype conformity. Abou-Dumontier (2012) showed sensation seeking varies according to masculine stereotype conformity. Gender stereotype conformity explains sensation seeking and the both explain risk taking. The more 9-14 years-old boys and girls comply with masculine stereotypes, the more they seek sensations and the more they take risks.

Furthermore, gender stereotypes seem to be more predictive of risky behavior than biological sex (Granié, 2010, Granié, 2009, Abou Dumontier, 2012, Raithel, 2003). The gender stereotype effect has also been studied for effective behavior on driving simulators (Schmid Mast et al., 2008). In all cases studied, research shows masculine stereotype conformity leads to more frequent injury risk behavior than feminine stereotype conformity.

2. FROM RULE COMPLIANCE TO RULE INTERNALIZATION

2.1 Gender stereotypes and rule compliance

Dangerous behaviors and involvement in accidents among adult drivers were shown to be more often due to rule-breaking in males than in females (Harré et al., 1996, Simon and Corbett, 1996). Previous studies have also shown that male pedestrians violate more rules than female pedestrians do (Moyano Diaz, 2002, Rosenbloom et al., 2004).

The results of a study on 162 preschool children aged 3 to 6, showed that girls' behaviors were more compliant than those of boys and that girls had better knowledge of pedestrian safety rules (Granié, 2007).

In a recent study, 400 adult pedestrians (200 men and 200 women) were observed at two signalized and two unsignalized crossroads, using a taxonomic observation grid which detailed 13 behavioral categories before, during and after crossing (Tom and Granié, 2011). The results showed that the temporal crossing compliance rate is lower among male pedestrians but spatial crossing compliance does not differ between genders and was modulated by the crossroad configuration.

In another recent study on drivers aged 18-79, Guého & Granié (2012b) showed that males committed more ordinary violations than females as measured by the driver behavior questionnaire. In a validation of a pedestrian behavior scale on 343 pedestrians (126 males et 217 females) aged 15 to 78 (Granié et al., 2012), it has been shown that pedestrian males committed more violations than pedestrian females.

2.2. Gender differences in rule compliance and internalization

Nevertheless, traffic behaviors are also likely to be influenced by attitudes toward rules. Thus, concerning compliance with traffic rules, Yagil (1998) found that male drivers expressed a lower level of normative motives to comply with traffic laws than did female drivers. Then, females seem to have internalized traffic rules more than males. Internalization is the process by which individuals transform social values and prescriptions acquired from external sources into personal attributes, values and self-regulated behaviors (Grolnick et al., 1997). Moral and prudential rules – relative to one's own well-being and that of others – form a set of internalized rules that are hard for an individual to transgress (Turiel, 1998).

Studies on antisocial behaviors have shown the relationship between delinquent behaviors and the lack of internalization among young people (Tavecchio et al., 1999). Furthermore, research has shown the relationship between risk taking among adolescents and the categorization of social knowledge on drug consumption (Nucci et al., 1991) and on various types of risky behavior (Kuther and Higgins-d'Alessandro, 2000). In both cases, adolescents who are not involved classify these behaviors in the prudential and moral domains, whereas adolescents involved in risky behaviors tend to classify them in the personal or conventional domains. Thus, internalization, through its effect on conformity with rules, can influence risk-taking.

The relationship between gender stereotype conformity and rule internalization as mediator factors to explain gender difference in injury-risk behavior has been explored in several recent studies from childhood to adulthood on pedestrians, cyclists and drivers.

Already in preschool years, girls had better knowledge of rules, and exhibited greater rule internalization than boys (Granié, 2007). These findings suggest that girls and boys have different motives for obeying safety rules. Furthermore, in a study among children aged 9 to 12 (Granié, 2011), it has been shown that girls differ from boys on cycling rules internalization, risk perception, and risk-taking propensity as cyclist. Among adolescents, it has been proved that sex-stereotype conformity has an effect on the internalization of traffic rules and risky behavior (Granié, 2009). Furthermore, the results show an effect of internalizing traffic rules on the risky pedestrian behaviors. Thus, it appears in both studies that, more than biological sex, it is the level of masculinity and the level of rule internalization that explain gender differences in risk taking among child cyclists and adolescent pedestrians.

Among adults, masculinity and femininity levels and normative beliefs towards pedestrian rules were measured in another study (Granié, 2008). Results show than mistaken and unsafe behaviors are more disapproved by men than by women. Furthermore, individual with high conformity to masculine gender role more endorse risk-taking and violations than individuals with high conformity to feminine gender role.

In a recent study on drivers aged 18-79 (Guého and Granié, 2012a), results show that masculinity reinforces ordinary and aggressive violations whereas femininity reinforces positive driver behavior.

3. GENDER STEREOTYPES IN DRIVING BEHAVIORS

These studies are all founded on hypotheses that gender roles and stereotypes influence another neutral behavior, such as driving behavior. However, differentiated social expectations about driving behavior among men and women, that is stereotypes associated with women's and men's driving and the content of these stereotypes has not yet been explored in depth, neither among adults, nor among adolescents, although some research has tended to show that it is used by drivers (Davies and Patel, 2005, Derks et al., 2011). Furthermore, these stereotypes may have effects on driving behavior, as one study showed that the threat of the stereotype of women behind the wheel – that is activating the stereotype that females are poor drivers – leads to a breakdown in driving performances in women driving on a simulator (Yeung and von Hippel, 2008).

Therefore, beliefs about driving by men and women, as well as the effect of the age and gender of the perceiver, were explored using the free association method with preadolescents and adolescents between 10 and 16 years of age (Granié and Pappafava, 2011). The results show that gender stereotypes are indeed associated with driving, from the age of 10. Male drivers are more frequently perceived as not complying with traffic rules, aggressive, being good drivers, and using their vehicle to promote their self-image. Female drivers are more frequently perceived as complying with traffic rules, having either few or a lot of accidents, being bad drivers, and being engaged in a non-feminine activity. While the representation of male drivers is already stable at 10, the negative representation of female drivers appears to develop with age between 10 and 16 and the beginning of driver training.

4. DISCUSSION

These results improve our knowledge of the mechanisms which explain sex differences in risk-taking. They confirm that these differences are not only innate tendencies but are also due to gender-role development and social pressures. This knowledge could have practical implications. Thus, from these results, the question remains how to act in education and prevention actions to minimize gender differences in injury-risk behaviors and in accidentology.

Faced with these sex differences in risk-taking, a twofold response may be given in the present state of the gender stereotypes: risk education should be more differentiated for adolescents, and less differentiated for preschoolers.

Injury-risk behaviors increased as a function of masculinity for boys and girls. Sex differences in risk-taking do not arise from innate temperamental differences between sexes. Rather, gender is a social and cultural construct and gender stereotypes contents reflect perceivers' observations of men's and women's daily life behaviors (Eagly, 1987). This could be used to change the relationship between

760

children's gender-roles and risky behaviors through risk education and media campaigns. In this way, some of the feminine characteristics, which were found to be related to more careful behaviors among adolescents (Granié, 2009), might also be attached to masculine characteristics of role models. Risk education for male adolescents can thus use examples of the numerous male models who do not match with gender stereotypes about risk-taking while being socially recognized as masculine.

In socialization process, role models play a double function of information on gender stereotype (acquisition) and of behavior production (adoption) (Bussey and Bandura, 1999). In risk education, use of feminine characteristics by masculine role models could lead acquisition of modified gender stereotypes and adoption of less risky behaviors among adolescent males. In comparison, Medias' insistence on males' risk-taking can unfortunately strengthen adolescents' and parents' stereotypic beliefs, and therefore reinforce psychological essentialism (Heyman and Giles, 2006) and differential socialization about risk-taking between boys and girls.

For preschool children, injury prevention should be based on a less differentiated risk education. Virtually all the children's socialization agents (parents, peers, teachers, Media) have different expectations and behaviors depending on the children's sex (Bussey and Bandura, 1999). Therefore, socialization agents should be sensitized to their role in building sex differences in injury-risk, before acting on individuals in a sex-differentiated way to prevent risky behaviors.

More broadly, the results of these studies show that, beyond road user behavior itself, the representation of males and females as road users appears to be a field of expression of personal and social identity in a culture where seeking out risk is a part of the manliness construction, one of its outcomes being gender difference in traffic injuries.

REFERENCES

Aberg, L. & Rimmö, P.-A. 1998. Dimensions of aberrant driver behaviour. *Ergonomics,* 41: 39-56.

Abou, A., Granié, M. A. & Mallet, P. 2008. Recherche de sensations, attachement aux parents et prise de risque dans l'espace routier chez l'adolescent piéton. *In:* M. A. Granié & J.-M. Auberlet (eds.) *Le piéton et son environnement: quelles interactions? Quelles adaptations? Actes n°115.* Arcueil: Les collections de l'INRETS.

Abou Dumontier, A. 2012. *La prise de risque dans l'espace routier chez le préadolescent : Implication de l'identité sexuée, la recherche de sensations, l'estime de soi, l'attachement aux parents et la supervision parentale.* PhD, Université Paris Ouest La Défense.

Assailly, J.-P. 2001. *La mortalité chez les jeunes [Mortality among teenagers],* Paris: Que sais-je? P.U.F.

Baker, S. P., O'neill, B. & Ginsburg, M. J. 1992. *The injury fact book,* New York: Oxford University Press.

Basow, S. A. 1992. *Gender stereotypes and roles,* Pacific Grove, CA: Brooks/Cole.

Bem, S. L. 1974. The measurement of psychological androgyny. *Journal of consulting and clinical psychology*, 42: 155-162.

Bem, S. L. 1981. Gender schema theory: a cognitive account of sex-typing. *Psychological Review*, 88: 354-364.

Bergdahl, J. 2007. Ethnic and gender differences in attitudes toward driving. *The Social Science Journal*, 44: 91-97.

Bussey, K. & Bandura, A. 1999. Social Cognitive Theory of Gender Development and Differentiation. *Psychological Review*, 106: 676-713.

Byrnes, J. P., Miller, D. C. & Schafer, W. D. 1999. Gender differences in risk taking: a meta-analysis. *Psychological Bulletin*, 125: 367-383.

D'acremont, M. & Van Der Linden, M. 2006. Gender differences in two decision-making tasks in a community sample of adolescents. *International Journal of Behavioral Development*, 30: 352-358.

Davies, G. M. & Patel, D. 2005. The influence of car and driver stereotypes on attributions of vehicle speed, position on the road and culpability in a road accident scenario. *Legal and Criminological Psychology*, 10: 45-62.

Deaux, K. & Lewis, L. L. 1983. Components of gender stereotypes. *Psychological Documents*, 13: 25.

Derks, B., Scheepers, D., Laar, C. V. & Ellemers, N. 2011. The threat vs. challenge of car parking for women: How self- and group affirmation affect cardiovascular responses. *Journal of Experimental Social Psychology*, 47 178-183.

Eagly, A. H. 1987. *Sex differences in social behavior: A social-role interpretation*, Hillsdale, NJ: Erlbaum.

Fagot, B. I. 1995. Parenting boys and girls. *In:* M. Bornstein (ed.) *Handbook of parenting. Vol. 1. Children and parenting*. Hillsdale, NJ: Erlbaum.

Farrow, J. A. & Brissing, P. 1990. Risk for DWI: A New Look at Gender Differences in Drinking and Driving Influences, Experiences, and Attitudes among New Adolescent Drivers. *Health Education & Behavior*, 17: 213-221.

Ginsburg, H. J. & Miller, S. M. 1982. Sex differences in children's risk-taking behavior. *Child Development*, 53: 426-428.

Granié, M.-A. 2007. Gender differences in preschool children's declared and behavioral compliance with pedestrian rules. *Transportation Research Part F: Traffic Psychology and Behaviour*, 10: 371-382.

Granié, M.-A. 2008. Influence de l'adhésion aux stéréotypes de sexe sur la perception des comportements piétons chez l'adulte. *Recherche - Transports - Sécurité*, 101: 253-264.

Granié, M.-A. 2009. Sex differences, effects of sex-stereotype conformity, age and internalization on risk-taking among pedestrian adolescents. *Safety Science*, 47: 1277-1283.

Granié, M.-A. 2010. Gender stereotype conformity and age as determinants of preschoolers' injury-risk behaviors. *Accident Analysis & Prevention*, 42: 726-733.

Granié, M.-A. 2011. Différences de sexe et rôle de l'internalisation des règles sur la propension des enfants à prendre des risques à vélo. *Recherche - Transports - Sécurité*, 27: 34-41.

Granié, M.-A., Pannetier, M. & Guého, L. 2012. Developping a self-report method to measure pedestrian behaviors: a French validation. *Accident Analysis & Prevention*, submitted.

Granié, M.-A. & Pappafava, E. 2011. Gender stereotypes associated with vehicle driving among French preadolescents and adolescents. *Transportation Research Part F: Traffic Psychology and Behaviour*, 14: 341-353.

762

Grolnick, W. S., Deci, E. L. & Ryan, R. M. 1997. Internalization within the family: the self-determination theory perspective. *In:* J. E. Grusec & L. Kuczynski (eds.) *Parenting and children's internalization of values.* New York: John Wiley.

Guého, L. & Granié, M. A. "Effects of age and gender identity on driving behaviors". Paper presented at ICTTP 2012, August, 29-31 Gronningen (Netherlands). 2012a.

Guého, L. & Granié, M. A. 2012b. French validation of a new version of the Driver Behaviour Questionnaire. *Accident Analysis & Prevention,* submitted.

Hagan, L. K. & Kuebli, J. 2007. Mothers' and fathers' socialization of preschoolers' physical risk taking. *Journal of Applied Developmental Psychology,* 28: 2-14.

Hanna, C. L., Taylor, D. M., Sheppard, M. A. & Laflamme, L. 2006. Fatal crashes involving young unlicensed drivers in the US. *Journal of Safety Research,* 37: 385-393.

Harré, N., Brandt, T. & Dawe, M. 2000. The Development of Risky Driving in Adolescence. *Journal of Safety Research,* 31: 185-194.

Harré, N., Field, J. & Kirkwood, B. 1996. Gender differences and areas of common concern in the driving behaviors and attitudes of adolescents. *Journal of Safety Research,* 27: 163-173.

Heyman, G. D. & Giles, J. W. 2006. Gender and Psychological Essentialism. *Enfance,* 3: 293-310.

Kuther, T. L. & Higgins-D'alessandro, A. 2000. Bridging the gap between moral reasoning and adolescent engagement in risky behavior. *Journal of Adolescence,* 23: 409-422.

Lonczak, H. S., Neighbors, C. & Donovan, D. M. 2007. Predicting risky and angry driving as a function of gender. *Accident Analysis & Prevention,* 39: 536-545.

Morrongiello, B. A. 1997. Children's perspectives on injury and close-call experiences: sex differences in injury-outcome processes. *Journal of Pediatric Psychology,* 22: 499-512.

Morrongiello, B. A. & Hogg, K. 2004. Mother's reactions to children misbehaving in ways that can lead to injury: implications for gender differences in children risk taking and injuries. *Sex Roles,* 50: 103-118.

Morrongiello, B. A., Midgett, C. & Stanton, K.-L. 2000. Gender Biases in Children's Appraisals of Injury Risk and Other Children's Risk-Taking Behaviors. *Journal of Experimental Child Psychology,* 77: 317-336.

Moyano Diaz, E. 2002. Theory of planned behavior and pedestrians' intentions to violate traffic regulations. *Transportation Research Part F: Traffic Psychology and Behaviour,* 5: 169-175.

Nell, V. 2002. Why young men drive dangerously: implications for injury prevention. *Current Directions in Psychological Science,* 11: 75-79.

Nucci, L., Guerra, N. & Lee, J. 1991. Adolescent judgments of the personal, prudential, and normative aspects of drug usage. *Developmental Psychology,* 27: 841-848.

Özkan, T. & Lajunen, T. 2006. What causes the differences in driving between young men and women? The effects of gender roles and sex on young drivers' driving behaviour and self-assessment of skills. *Transportation Research Part F: Traffic Psychology and Behaviour,* 9: 269-277.

Parker, D., Manstead, A. S. R., Stradling, S. G., Reason, J. T. & Baxter, J. S. 1992. Intention to commit driving violations: an application of the theory of planned behavior. *Journal of Applied Psychology,* 77: 94-101.

Pomerantz, E. M. & Ruble, D. N. 1998. The role of maternal control in the development of sex differences in child self-evaluative factors. *Child Development,* 69: 458-478.

Raithel, J. 2003. Risikobezogenes Verhalten und Geschlechtsrollenorientierung im Jugendalter [Risk-taking behavior and gender role orientation in adolescents]. *Zeitschrift für Gesundheitspsychologie,* 11: 21-28.

Rienzi, B. M., Mcmillin, J. D., Dickson, C. L. & Crauthers, D. 1996. Gender differences regarding peer influence and attitude toward substance abuse. *Journal of Drug Education*, 26: 339-347.

Rosenbloom, T., Nemrodov, D. & Barkan, H. 2004. For heaven's sake follow the rules: pedestrians' behavior in an ultra-orthodox and a non-orthodox city. *Transportation Research Part F: Traffic Psychology and Behaviour*, 7: 395-404.

Schmid Mast, M., Sieverding, M., Esslen, M., Graber, K. & Jäncke, L. 2008. Masculinity causes speeding in young men. *Accident Analysis & Prevention*, 40: 840-842.

Sibley, C. G. & Harré, N. 2009. A gender role socialization model of explicit and implicit biases in driving self-enhancement. *Transportation Research Part F: Traffic Psychology and Behaviour*, 12: 452-461.

Simon, F. & Corbett, C. 1996. Road traffic offending, stress, age, and accident history among male and female drivers. *Ergonomics*, 39: 757-780.

Tavecchio, L. W. C., Stams, G.-J. J. M., Brugman, D. & Thomeer-Bouwens, M. a. E. 1999. Moral judgement and delinquency in homeless youth. *Journal of Moral Education*, 28: 63-79.

Tom, A. & Granié, M. A. 2011. Gender Differences in Pedestrian Rule Compliance and Visual Search at Signalized and Unsignalized Crossroads. *Accident Analysis & Prevention*, 43: 1794-1801.

Tronsmoen, T. 2008. Associations between self-assessment of driving ability, driver training and crash involvement among young drivers. *Transportation Research Part F: Traffic Psychology and Behaviour*, 11: 334-346.

Turiel, E. 1998. The development of morality. *In:* N. Eisenberg (ed.) *Handbook of Child Psychology*. New York: Wiley.

Wilson, M. & Daly, M. 1985. Competitiveness, risk taking, and violence: the young male syndrome. *Ethology and Sociobiology*, 6: 59-73.

Yagil, D. 1998. Gender and age-related differences in attitudes toward traffic laws and traffic violations. *Transportation Research Part F: Traffic Psychology and Behaviour*, 1: 123-135.

Yeung, N. & Von Hippel, C. 2008. Stereotype threat increases the likelihood that female drivers in a simulator run over jaywalkers. *Accident Analysis & Prevention*, 40: 667-674.

CHAPTER 77

Integrated Analysis of Driver Behavior by Vehicle Trajectory and Eye Movement

Hidetoshi Nakayasu[1], Tetsuya Miyoshi[2] and Patrick Patterson[3]

[1]Konan University
8-9-1, Okamoto, Higashinada, Kobe, Hyogo, Japan
nakayasu@konan-u.ac.jp

[2]Toyohashi Sozo University
20-1, Mastushita, Ushikawa, Toyohashi, Aichi, Japan
miyoshi@sozo.ac.jp

[3]Texas Tech University
201 IE Building, P.O. Box 43061, Lubbock, TX 79409, USA
pat.patterson@ttu.edu

ABSTRACT

An experimental paradigm is proposed for the measurement of time histories of eye movements and vehicle trajectories during simulator driving. Eye movements, both distance of eye movement and numbers of fixations, along with car log data were recorded as participants maneuvered through an urban road scenario that included turning through an intersection. Eye movements during a traffic event did show how and on what the driver focused their visual attention during specific traffic events, such as right and left turns. Another analysis was conducted linking vehicle trajectory and driver eye movements. Using an integrated analysis of video clips, vehicle trajectory, and time histories of eye movements enables a better determination of the possibility for an accident occurrence.

Keywords: Vehicle trajectories, Driving simulator, Eye tracking system, Traffic safety, Synchronized data analysis.

INTRODUCTION

To maintain a highly safe and reliable driving system, it is important that this system be viewed as a human-machine system, made up not only of the movement of vehicles, but also of human perception, cognition, and responses. A survey of accident reports shows that errors in visual perception are commonly stated causes (ITARDA, 2001).

Visual cognition requires acquiring and judging available information. Human performance is based on simple feedback control, where motor output is a response to an error signal that represents the difference between the actual state and intended state in a time-space environment, with a control signal derived at a specific point in time. An example of these states is eye tracking task. In real life this mode is used rarely and only for slow, very accurate movements, such as a turning task when driving an automobile. This type of human performance is based on feed forward control and depends upon a very flexible and efficient dynamic response based on perceptual information (Rasmussen, 1983). Feed forward control is necessary to explain rapid coordinated movements, for instance, in a steering task when turning an automobile in line with an intended direction of travel.

In this paper, we investigated the relationship between eye movements and driving behavior in situations of familiar and unfamiliar cultural regulations using an integrated approach of a driving simulator and an eye tracking system.

EXPERIMENTS

To investigate a driver's behavior, including eye movement, vehicle operation and vehicle trajectory, a driving simulator with a 6-axis motion base system (Honda Motors; DA-1102) simulated driving an automatic transmission vehicle in conjunction with an EyeLink II eye tracking system. The instruments on the dashboard, steering wheel, gearshift, side brake, accelerator, and brake pedal were positioned in a manner similar to that in a real car as shown (Figure 1). The simulator is controlled by a network of computers, which also generate images from the driver's point of view. Ethernet-linked computers acquired a log of input and

Figure 1 Schematic view of the DS experiment with the eye tracking system

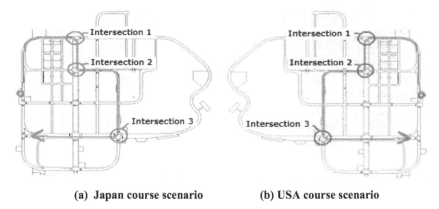

(a) Japan course scenario   (b) USA course scenario

Figure 2   Locations of the traffic incidents on the driving courses scenario

output variables (such as a steering, accelerator, brakes, the position of the car and speed). This data was acquired every 10msec.

Eye movements of driver were recorded by a head mounted eye tracking system (SR Research Ltd.; EyeLink II) at 250 Hz with a spatial resolution of 0.022 degree within a range of ± 30 degrees in the horizontal axis and ± 20 degrees in the vertical axis, controlled by a personal computer which managed the timing of the experiment and collected data. Head movements were monitored, and the system automatically modified the data on eye position with a camera. The horizontal and vertical eye movements of both eyes were recorded. At the same time, the scene from the driver's perspective was recorded by a CCD camera on a head mounted cap (Figure 1), and was capable of recording the view of the driver as video clips, having a spatial resolution of (720, 480) pixels and a frame resolution of 30fps. All eye movement data were analyzed off-line by computer programs that calculated the number of saccades and eye movement distance.

Two scenarios, one Japan and one a USA course, were used in the experiments. The USA course was produced by symmetrical rotation of the Japanese course; therefore the length and dimensions of left turn in USA course were the same as the right turns in the Japan course. Slight modifications to the shape of signal, replacing Japanese letters, and so on were formatted as found in the USA. There are three intersections in the scenario that are corresponding to right, left and right turns in Japan respectively, while to left, right and left turn in USA. (Figure 2)

For the analysis, differences in eye position at successive frames were calculated. These data were analyzed on the basis of velocity and position for each traffic incident to evaluate the mean distance of eye movement, fixation frequency, and duration. This paper defines a saccade as when an eye moves at a velocity of more than 90 deg/sec. An eye moves after the occurrence of a saccade. This paper defines a fixation of the eye as when an eye does not move for at least 48 msec after an eye movement and a saccade. Therefore, the fixation time refers to the time duration from the saccade after one eye movement to the saccade before another eye movement.

Table 1 Details of participants

Subject	Gender	Age	Eye site (R, L)	License (years)	DS Exp. in Japan course (hour/w)	DS Exp. in USA course (hour/w)
A	Male	21	(1.5,1.2)	3	10	5
B	Male	22	(1.2,1.0)	2	6	4
C	Male	21	(1.5,1.5)	3	11	5
D	Male	22	(1.2,1.0)	2	5	5
E	Male	22	(1.0,1.0)	2	4	4
F	Male	22	(1.0,1.2)	2	1	None
G	Male	22	(1.2,1.2)	2	1	None
H	Male	21	(1.0,1.0)	2	1	None
I	Male	22	(1.0,1.0)	3	1	None
J	Male	22	(1.0,1.0)	2	1	None

Ten male students (mean age = 21.9; std.dev = 0.458) participated in this experiment as drivers. The participants had standard Japanese automobile licenses and more than one year's worth of driving experience. Half of the participants received more than 4 hours of training on how to use the driving simulator on the USA course before the experiment (Table 1). Prior to the experiment, the purpose of the study and the procedures were explained to the participants, and the informed consent of all participants was obtained. All participants had normal or corrected-to-normal vision. The following two basic instructions were given to participants to ensure the psychological conditions of the experiment were consistent for all. 1) Do not cause an accident and 2) Follow all traffic laws during the experiment.

RESULTS

The data log of vehicle behavior and driver responses were obtained every 10 msec. Driver eye movements were recorded through the eye tracking system with a 4 msec sampling time. The following instructions were given the participants before the experiment.
   1) Close your eyes and push the brake pedal.
   2) Open your eyes and release the brake pedal simultaneously, then start driving.
This allowed us to synchronize vehicle behavior, driver responses, and eye movement data. The initial point of car log data was located as the point when the record of brake was released (Table 2), and that of eye movement data was located when the eyes opened (Table 3). The eye movements of one driver are summarized in Table 4.

Table 2 Synchronized points in car log data

	A	B	C	D	E	F	G	H
1	No	System	Button	time	SteerAngle	Throttle	Brake	Clutch
851	9633	13:22:33:6	0	96.33	-0.063	0	18.230	100
852	9634	13:22:33:6	0	96.34	-0.063	0	17.238	100
853	9635	13:22:33:6	0	96.35	-0.063	0	15.884	100
854	9636	13:22:33:6	0	96.36	-0.063	0	13.989	100
855	9637	13:22:33:7	0	96.37	-0.063	0	11.823	100
856	9638	13:22:33:7	0	96.38	-0.063	0	9.205	100
857	9639	13:22:33:7	0	96.39	-0.063	0	6.318	100
858	9640	13:22:33:7	0	96.40	-0.063	0	3.430	100
859	9641	13:22:33:7	0	96.41	-0.063	0	0.451	100
860	9642	13:22:33:7	0	96.42	-0.063	0	0.000	100
861	9643	13:22:33:7	0	96.43	-0.063	0	0.000	100
862	9644	13:22:33:7	0	96.44	-0.062	0	0.000	100
863	9645	13:22:33:7	0	96.45	-0.062	0	0.000	100

Table 3 Synchronized points in the eye movement data

1735	8903074	.	.	0	.	.	.	.C.C.
1736	8903078	.	.	0	.	.	.	IC.C.
1737	8903082	.	.	0	.	.	.	.C.C.
1738	8903086	973.4	10.9	133	1165	-17.9	1109	IC.C.
1739	8903090	973.4	10.9	133	1165	-17.9	1109	.C.C.
1740	8903094	1012.9	5.2	151	1160.1	-19.4	1149	I.RC.
1741	8903098	1997.0	1365.7	170	1172.6	74.8	1190	...C.
1742	8903102	1844.9	1301.5	207	1165.5	55.4	1271	...C.
1743	8903106	1720.9	1252	281	1146	26.9	1433	...C.
1744	8903110	1721.3	1253.1	503	1136.1	11.4	1548	...C.
1745	8903114	1618.1	1213.5	580	1125.3	-9.4	1640	...C.
1746	8903118	1531.5	1183.1	641	1117.9	-28.2	1709	...C.
1747	8903122	1420.2	1165.9	1136	1107.8	-50.8	1788	.C.C.
1748	8903126	1409.8	1151	1224	1101.3	-62.1	1860	.C.C.

Table 4 Example of a driver eye movement summary

Number of fixation:8		
	Mean distance of eye movement (degree)	Mean duration of fixation(msec)
1	25.31	520
2	14.34	984
3	40.2	416
4	14.34	488
5	7.03	256
6	23.39	100
7	7.12	160
8	5.49	
Mean distance of eye movement (degree)	12.63	
Mean duration of fixation(msec)	417.71	

Distance of eye movement=SQRT((X_eye1-X_eye2)^2+(Y_eye1-Y_eye2)^2)
Durayion oggicsyion=Time2-Time1

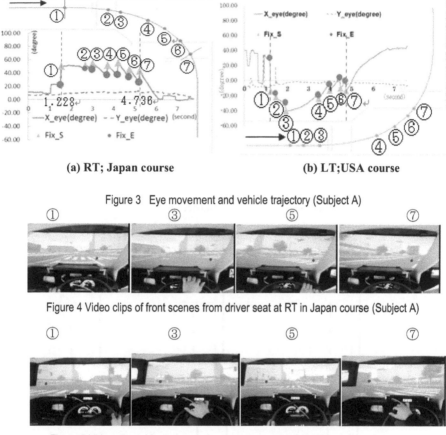

(a) RT; Japan course   **(b) LT;USA course**

Figure 3   Eye movement and vehicle trajectory (Subject A)

Figure 4 Video clips of front scenes from driver seat at RT in Japan course (Subject A)

Figure 5 Video clips of front scenes from driver seat at LT in USA course (Subject A)

For the purpose of comparative consideration, Figures 3, 4, and 5 show the relationships among vehicle trajectories, eye movement of driver, and the video clips respectively. Figure 3(a) and (b) show a track of this driver's eye movement, showing some horizontal movement. Fix_s and Fix_e designate the start and end of fixations as x_eye moves while y_eye does not. This figures also show the trajectory of the vehicle when it turns right (Figure 3(a)) in Japan and in left turns (Figure 3(b)) in USA with the fixation points superimposed. Figures 6 allows us to compare the vehicle trajectories to evaluate the skill of the same driver for the corresponding event, e.g. right turn (RT) in Japan and left turn (LT) in USA. In addition, it is useful in Figure 4 and 5 to compare eye movements where the driver focuses perceptive attention during the turning operation.

DISCUSSIONS

Aoki (Aoki, 2010) divided turning behavior into three phases: introduction, practice confirmation and completion stages. In the introduction stage, the driver

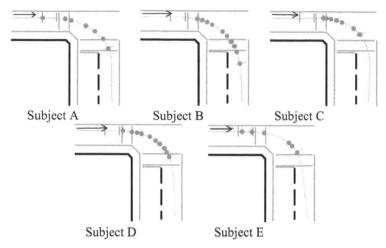

Figure 6 Vehicle trajectories and fixation points (trained driver)

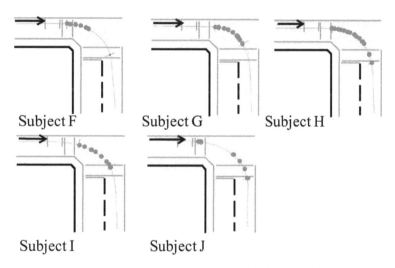

Figure 7 Vehicle trajectories and fixation points (untrained driver)

performs perception and cognition. During the practice confirmation stage, the driver judges, practices and confirms his/her actions. In the completion stage, the driver confirms the completion of the operation and then shifts to the next target, once again moving into the introduction stage.

From these points of view, the trajectories and fixation points during intersection 1 such as right turn in Japan course focused on the behaviors during the practice confirmation stage for both trained and untrained drivers (Figures 6 and 7). From these figures, it was seen that the locations of fixation points on the trajectories of the untrained drivers were biased while those of trained were unbiased. In other words, the points of fixation superimposed on the trajectories during the practice confirmation stage in right turning were different by individual in untrained group, i.e., the focus of subject F was concentrated at the beginning of the practice

Table 5 Eye movements of trained and untrained driver (RT; Japan)

JPN-R1	Number of fixation(times)	Mean distance of eye movement (degree)	Mean duration of fixation(msec)
Trained	8.0	16.48	419.58
Untrained	8.8	31.22	404.92

confirmation stage while the focus of subject J was located only at the end of the practice confirmation stage. On the other hand, there were few differences among trained driver. Another consideration was derived from the summaries of eye movement for the trained and untrained groups (Table 5). The mean eye movement distance for untrained drivers during the practice confirmation stage were larger than that the trained drivers, while there were no differences in the number of fixations or mean duration of fixation. This suggests that the perceptive activity of an untrained driver is more variable as compared to the trained driver, even though the region of perceptive search for untrained drivers was wider than that of trained driver. It is seen that the similar experimental results mentioned the above were obtained at the case of intersection 2 and 3 as well as those at the case of intersection 1. This helps to explain why the distance of eye movement during the practice confirmation stage of turning and the variations of the points of fixation found on the trajectories of untrained drivers were larger than those of trained drivers.

Another point arises from the comparison of vehicle trajectories for the two scenarios, Japan and USA courses. The USA course was produced by symmetrical rotation of the Japanese course; therefore the length and dimensions of left turn in USA course were the same as the right turns on the Japan course. Since the length and dimensions of LT of USA were the same as the RT of Japan, the vehicle trajectories for the USA LT can be imposed on those of Japan RT for comparison study of vehicle behavior.

Figures 8 and 9 show the results of comparisons of the differences amongst vehicle trajectories and fixation points for trained and untrained drivers, where the experimental results of vehicle trajectories of left turns in USA course were superimposed on those of right turns in Japan course to compare with Japan and USA. For trained drivers, half trajectories of LT USA were coincided with the trajectories of RT Japan. On the other hand, for untrained drivers all the trajectories of LT USA were out of alignment to the trajectories of RT Japan. In a Japanese traffic lane an automobile must keep left, while an automobile must keep right in the USA. This meaned that the experimental results by untrained gave warning because of possibility of a head on crash when the trajectory of LT in USA is undershot as compared to its performance in the Japan scenario. From this point of view, it was found that for three cases of untrained drivers the trajectories of LT in the USA course showed the dangerous situations that undershot the required turning trajectory as compared to Japan. On the other hand, all the trajectories of LT in USA of trained drivers were not undershot when compared to their performance on the Japan course. Almost similar tendencies of experimental results at intersection2 and 3 were obtained as well as those at intersection 1 mentioned the above.

This suggests that the vehicle behavior of trained driver was safer since the trained driver focused their visual attention throughout the practice confirmation stage in a LT on USA as well as on a RT in Japan. The mean distance of eye movement of untrained driver on the Japan and USA courses was larger than those of trained driver (Table 6), indicating inappropriate eye movement by the untrained drivers.

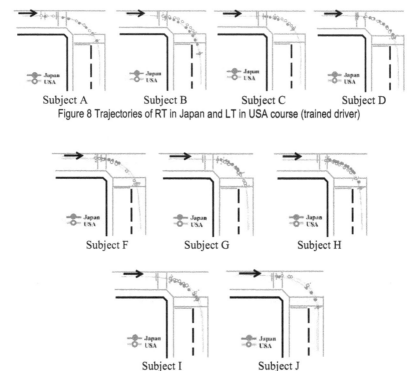

Figure 8 Trajectories of RT in Japan and LT in USA course (trained driver)

Figure 9 Trajectories of RT in Japan and LT in USA course (untrained driver)

Table 6 Comparisons of trained and untrained driver eye movement data

Intersectionon 1 (JPN_R)	Trained			Untrained		
	Number of fixation(times)	Mean distance of eye movement (degree)	Mean duration of fixation(msec)	Number of fixation(times)	Mean distance of eye movement (degree)	Mean duration of fixation(msec)
Japan	8.5	16.43	404.28	8.8	31.22	404.92
USA	7.8	11.74	390.08	7.3	29.73	406.71

CONCLUSIONS

An experimental paradigm to evaluate driver safety is proposed that measures time histories of eye movement with vehicle trajectories during driving in a simulator. Eye movement and car log data were recorded as participants maneuvered through an urban road scenario with intersections configured as a Japan course and a USA course. Our basic findings using this technique:

(1) It was found that the mean eye movement distance of the untrained drivers during the practice confirmation stage was larger than that of trained drivers, while there were no differences in the number of fixations or the mean duration of fixation. This suggests that the perceptive activity of an untrained driver is more variable than the trained driver, even though the region of perceptive search of untrained driver was wider than that of a trained driver. This helps explain why the distance of eye movement during the practice confirmation stage of turning and the variations of the points of fixation in the trajectories of untrained drivers were larger than those of trained drivers.

(2) It is suggested that the vehicle behavior of trained driver were safer as the trained drivers focused their visual attention throughout all of the practice confirmation stage during both LT in USA as well as RT in Japan.

It is hoped that using this style of integrated analysis, using vehicle trajectory and eye movement, will be a powerful tool for evaluating physical behavior and human driving performance in the future.

ACKNOWLEDGMENTS

The authors express great thanks to Mr. M. Kitagawa, T. Kuroki and K. Sasahara for their cooperation to the experiments. In addition, the authors want to appreciate Mr. H. Ono who is an assistant manager of Honda Driving Safety Promotion Center, Honda Motor Co., Ltd., and Mr. I. Koizumi, Monte System Co. Ltd for their helpful support to these experiments.

REFERENCES

Aoki, H., H. Nakayasu, N. Kondo, and T. Miyoshi. 2010. "Cognitive Study on Driver's Behavior by Vehicle Trajectory and Eye Movement in Virtual Environment, " Paper presented at 3rd AHFE International Conference, Miami, 2010.

ITARDA. 2001. Institute for Traffic Accident Research and Data Analysis Information, What Sort of Human Errors Cause Traffic Accidents? -Focus on Mistake Assumption-, No.33. (in Japanese)

The Vision Society of Japan Eds. 2001. Handbook of visual information process, pp.91–92., pp.393–398. (in Japanese)

Rasmussen, J., 1983. "Skills, Rules, and Knowledge; Signal, and Symbols, and Other Distinctions in Human Performance Models," IEEE Trans. on Systems, Man, and Cybernetics, Vol. SMC-13, No.3, pp. 257-266.

Section VIII

Studies of Locomotives, Drivers and Rail Systems

CHAPTER 78

From Flight Data Monitoring to Rail Data Monitoring

Dr Guy Walker and Dr Ailsa Strathie

Institute for Infrastructure and the Environment
Herriot-Watt University
Edinburgh
g.h.walker@hw.ac.uk

ABSTRACT

Flight Data Monitoring (FDM) is the process by which data from on-board recorders, or so-called 'black boxes', is analyzed after every journey in order to detect subtle trends. These trends, in turn, are useful in providing leading indicators of issues relating to fuel economy, maintenance, passenger comfort and safety. An opportunity has been identified to advance the state of the art in FDM processes by coupling recorder data to established Human Factors methodologies so that issues arising from the problematic human/system interface can be better understood and diagnosed. The research has also identified a significantly underused source of recorder-data from the railway industry. This paper, therefore, describes the overall concept of applying Human Factors methods to this new data source, and the knowledge exchange that is currently underway between the rail and aviation sectors.

Keywords: Data Monitoring/Telemetry, HF Methods, Human Error, Risk

1   FLIGHT DATA MONITORING (FDM)

The UK has a long history in the use of Flight Data Recorders (FDR's) and has led the world in important developments in the use of FDR data. As originally conceived it was anticipated that FDR's would provide accident investigators with a vital tool in retrospective accident analysis. This form of analysis focuses on tracing backwards from the accident event in order to diagnose the factors, and their interactions, which led to it. Because these factors are identified after the event they are said to be 'lagging indicators'.

In the 1970's the UK Civil Aviation Authority and British Airways pioneered an alternative approach. Instead of using the data after accidents had already occurred, they instead used the far greater quantity of data collected during routine flights. This approach is premised on the ability to detect trends that might, in future, develop into more serious safety concerns. The process is called Flight Data Monitoring (FDM) and it makes use of performance metrics, or 'leading indicators', which grant the opportunity to pro-actively detect problems before accidents happen. Broadly speaking, FDM comprises five steps:

Table 1 The five step FDM Process as encapsulated in the UK Civil Aviation Authority's CAP739 Guidance Document (CAA, 2003).

	Step	Example
1	Areas of operational risk are identified and safety margins quantified	Rates of rejected take-offs, hard landings, unstable approaches etc.
2	Changes in operational risks are identified and quantified, highlighting when non-standard, unusual or unsafe circumstances occur	Increases in: rejected take-offs, hard landings, unstable approaches, new events and new locations.
3	Assess risks and determine which may become unacceptable if the discovered trend continues	A new procedure has introduced high rates of descent that are approaching the threshold for triggering ground proximity warnings.
4	Put in place appropriate risk mitigation measures	having found high rates of descent the standard operating procedures are changed to improve control of the optimum/maximum rates of descent being used.
5	Confirm the effectiveness of any remedial action by continued monitoring	Confirm that the other measures at the airfield with high rates of descent do not change for the worse after changes in approach procedures (CAA, 2003, p. 2,2).

The FDM process has been developed considerably since the 1970's. It now extends beyond safety management to make a significant contribution to economic and environmental objectives. One area in particular is fuel economy. In the airline industry, changes in operational procedures based on FDM analyses have given rise to savings of 1.5% of the total fuel budget for an airline, a figure equating to several million pounds (Amhered & Halfliger, 2010). Other areas of benefit include optimizing air crew utilization and reducing fatigue, determining the most efficient maintenance regimes based on the actual demands made on a given aircraft (rather than arbitrary 'service intervals'), and analysis of operating procedures in relation to passenger comfort (GAO, 1997). Despite these numerous demonstrable benefits the opportunity to go further remains. As will be described shortly, significant opportunities exist to provide much better diagnosis of problems which reside at the particularly troublesome interface of people and systems.

2   ON-TRAIN MONITORING AND RECORDING (OTMR)

The railway industry, like aviation, is a transport domain wherein serious incidents are comparatively rare, likewise, whilst the frequency of accidents is low, their severity can be extremely high. Compared to the aviation industry data recorders on UK trains are a relatively recent development. Prior to 2002 their use was not mandated by applicable railway group standards. Since then recorder data has played a significant part in accident/incident analyses performed by the UK Rail Accident Investigation Board (RIAB) and in ad-hoc analysis and training activities performed by individual train operating companies. The situation today is that Railway Group Standard GM/RT2472 requires all new and existing trains to be fitted with data recorders. A minimum of fourteen parameters, including variables such as speed, operation of the brake control, operation of safety devices, functioning of train doors and so on, is recorded every 20mS, with every event time stamped and stored in a log. Once recovered, a suite of 'off-train' software tools allows individual (rather than fleet or industry wide) train data to be scrutinized and interpreted.

Reviewing the railway safety literature it is clear that the potential of recorder data is well recognized yet it remains significantly underutilized. In 2003 an international workshop was held into the issue of 'close calls' in which the presence of "mature programs that review flight data recorders after every flight" were noted (Hart, 2003). Likewise, a review of the program of research instituted by the UK's Rail Safety and Standards Board since its inception in 2003 reveals significant interest in quality assurance processes (RSSB, 2004) and a similar interest in identifying and implementing best-practice from the aviation domain (RAeS, 2009). The timeliness of the research described in this paper is highlighted when set against this backdrop. This is because fleet/industry-wide safety margins, operational risks, trends and mitigation techniques are not systematically identified, quantified and/or monitored via on-train recorders, despite virtually every powered rail vehicle now being fitted with one. In other words, an equivalent 'Rail Data Monitoring' process does not currently exist.

3   SHARED HUMAN FACTORS PROBLEMS

Flight Data Monitoring (FDM) represents an innovation for the rail industry. The potential role of Human Factors methods in mature FDM processes represents an innovation for both the aviation and rail industries. The reason for this is that both transport domains are afflicted by a class of accident, or near accident, which resides firmly at the human/system interface. Exemplars in the world of aviation include 'controlled flight into terrain' (Shappell, 2003), 'mode errors' (Endsley & Kiris, 1995) and 'automation surprises' (Sarter & Woods, 1997). The rail industry is afflicted by similar phenomenon like Signals Passed At Danger (SPADS; Wright, Ross & Davies, 2000) and various 'over-speed' events which trigger the Train Protection and Warning System (TPWS). Despite in-depth data logging, and despite even the presence of a mature FDM process, problems such as these remain difficult to detect in advance and resistant to a wide range of conventional safety interventions such as training and additional defenses built into the system. In these cases data logging provides extensive information on the 'what' and 'how' but not on the 'why'. In other words, 'why' were three in-cab warnings acknowledged, and three automatic applications of the emergency brake cancelled by the driver, yet the train still passed a red signal? Likewise, despite a prominent Ground Proximity Warning, 'why' was the fully functional aircraft flown dangerously close to an obstacle?

The question of 'why' is extremely complex, involving in-depth study of human cognition, perception, feedback, expectancy, situation awareness and numerous other issues related to the question of 'what was it about that particular situation and context which made that set of human behaviors make sense to the person carrying them out'. In a context in which the object of study (i.e. the human and their interaction with their environment) is difficult to examine using purely engineering tools, a context that does not lend itself well to assumptions concerning determinism and rationality, Human Factors methodologies provide a unique counterpoint. Herein lies justification for not merely mapping mature FDM process onto the rail industry, but to try and enhance FDM processes themselves.

4   THE CONCEPT

The concept of coupling human factors methods to On-Train Monitoring and Recording (OTMR) data is an exciting one, both theoretically and practically. The issue of Signals Passed at Danger (SPADs) has already been mentioned and a more detailed example of how Human Factors methodologies could contribute to the state of the art in this area can now be provided. In this case Signal Detection Theory (SDT) serves as an applicable human factors method, with recorder data on the driver's responses (and response time) to an in-cab warning system used to drive it.

4.1   The Automatic Warning System (AWS)

The in-cab warning system in question is called AWS (Automatic Warning System). It is a legacy system that dates back to the 1930's and was originally conceived as a means to prevent SPADS (demonstrating that this particular Human Factors problem does indeed have a long history). In-cab AWS alerts and reminders are triggered by an electro-magnetic device placed between the tracks approximately 200 yards prior to the signal, sign or other event to which it refers. Sensors underneath the train detect the presence of a magnetic field and activate AWS accordingly. McLeod, Walker and Moray (2003) describe the purpose of the Automatic Warning System (AWS) thus:

"AWS serves two functions. The first function is to provide an audible alert to direct the driver's attention to an imminent event (such as a signal or a sign). The second function, linked to the first, is to provide an ongoing visual reminder to the driver about the last warning. [AWS] is there to help provide advance notice about the nature of the route ahead, and thus communicate to the driver the need to slow down or stop" (p.4).

The sequence of actions can be simplified thus: on the approach to a lineside signal displaying a green/proceed aspect a bell or simulated chime at 1200Hz will sound in the train cab. No action is required of the driver, and the AWS visual indicator will remain (or switch to) a solid black color. On the approach to a signal displaying an aspect other than green, a horn sound or steady alarm at 800Hz will sound. The driver is required to cancel the audible warning within a limited timeframe in order to avoid an irreversible emergency brake application. To do this the driver presses a cancellation button, upon which the AWS display will switch to a yellow and black state, serving as a reminder that the last AWS indication received was a warning.

4.2   Driver Responses

Several major incidents have seen the use of AWS, and the number of events it now refers to, being extended. Currently AWS provides warnings in six circumstances:

1.   (Certain types of) permanent speed restriction,
2.   All temporary speed restrictions,
3.   (Some) level crossings,
4.   SPAD indicators,
5.   Cancelling boards,
6.   And other locations (such as unsuppressed track magnets, depot test magnets etc.).

Unfortunately, the simple two state warning (bell/horn) and visual reminder (black/yellow display) are unable to discriminate between these six different events. The binary two state decision of whether to cancel an AWS warning is, therefore, performed under conditions of uncertainty. There is □*no single, fixed behavioral response expected of a driver when in receipt of an AWS warning. Many factors specific to the driver, the class of rolling stock involved, the nature of the movement, and the situation at the time the warning occurs will determine how and when an individual driver reacts*□ (2003, p.9). A small sample of OTMR data reveals the extent of this uncertainty (McLeod, Walker & Moray, 2003). In this sample the driver received 21 AWS warnings, with just three of those events being followed by an application of the brakes. Revealingly, a far greater number (5) were followed, quite appropriately for the circumstances, by the train being accelerated.

OTMR does not merely record the nature of the driver's responses and the subsequent behaviors, it also provides data on their reaction time. An incident in 1999 brought the use of OTMR into sharp relief, and the subsequent accident enquiry made extensive use of the insights this data provided. In this situation the OTMR recorded the driver of the train involved in the accident responding to an AWS warning in respect to a signal displaying caution. This was the first signal in a sequence, and the reaction time between hearing the warning and pressing the AWS cancel button was 1.15 seconds. At the next signal, also displaying a cautionary signal aspect, the reaction time shortened by 0.5 seconds to 650ms. The same fast reaction time occurred in respect to the AWS warning at the final signal, which this time was displaying red (i.e. stop) and which, in this situation, was passed in error. In both cases, 650ms is an extremely fast response given the discussion above about the decision to cancel being performed in conditions of uncertainty. It can be understood with reference to insights provided by in-cab observations. These show drivers to be covering the cancelation button with their hand in expectation of an AWS warning. OTMR data also shows numerous occasions where driver's cancelled the AWS horn before it had even started to sound (suggestive of the driver pressing the button a number of times on the approach to the on-track AWS equipment, which is clearly visible from the train cab). The frequency of AWS events in the situation just described would be compatible with this behavior, with four AWS events occurring in the space of 2.5 miles or just three minutes. The dichotomy between fast response times (650ms) and slower response times (1.15s) suggests two distinct 'modes' of driving. "A top-down mode involving conscious deliberation and effort" and a "bottom up mode composed of well-rehearsed perception-action units that enable experienced drivers [to perform tasks] with little or no conscious attention or effort" (Charlton & Starkey, 2011, p. 457). In the latter case, drivers bring to bear knowledge and schemas that enable behavior to be performed in a feed-forward rather than feed-back manner. This may or may not be appropriate given the specific circumstances that exist in a given situation. The critical issue is whether indices such as response time could serve as metrics for switches in 'driving mode' and therefore leading indicators of emerging problems at the human-system interface?

4.3   Leading Indicator

Signal Detection Theory (SDT) is particularly suited to analyzing human responses which are performed in situations of uncertainty. Under SDT driver responses to AWS can be classified into the following taxonomy:

Table 2 – Responses to AWS indications organized into Signal Detection Theory response categories.

Category	Behavior	Category	Behavior
HIT	Receive in-cab warning and press cancel button.	MISS	Receive in-cab warning and do nothing.
CORRECT REJECTION	Receive in-cab bell or chime and do nothing	FALSE ALARM	Receive no in-cab warning/or bell/chime but press cancel button anyway

With a large quantity of data, of the sort that would be derived from a large scale data monitoring program, SDT is capable of providing a leading indicator called 'response bias'. Response bias is when driver(s) would be more likely to respond one way than another. A so-called 'conservative bias' would indicate that drivers are receiving too many in-cab warnings which require little or no change in behavior, thereby increasing the probability that an appropriate behavioral response (when really needed) would not be generated.

Recorder data and the SDT method can be combined and used to generate Receiver Operator Characteristics (ROC) charts. Under an FDM-type process these would provide a probabilistic analysis of how likely such psychological conditions would arise in a given set of external circumstances. For example, certain sites or situations could be monitored and subtle shifts in response bias used to determine interventions (to the infrastructure, procedures, training etc.) aimed at reducing the risk of a SPAD. More specifically, under an FDM-type process the ROC curve would be populated with live data, an exceedance criterion set, and if exceeded would trigger the risk mitigation and monitoring stages of a defined safety management system. Critically, all this would occur before an incident such as a SPAD takes place.

784

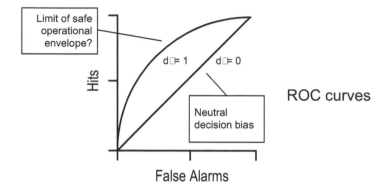

Figure 1   Large scale data from OTMR/Flight Data Recorders enables Receiver Operator Characteristic (ROC) charts to be produced, with response bias providing a leading indicator of a switch in cognitive mode with the potential to increase risk.

5   CONCLUSIONS

The purpose of this paper has been to outline a significant strand of work that is using Human Factors methods to inform best practice and encourage knowledge exchange between transport domains. The example given is merely an illustration of a wider concept: a large number of parameters in both rail and aviation domains are monitored, and a similarly wide range of Human Factors methods could employ this data as an input, offering to the rail and aviation sector much sought after leading indicators of problems and issues that reside at the boundary of people and systems.

ACKNOWLEDGMENTS

The research reported in this paper is funded by the UK Engineering and Physical Sciences Research Council (EPSRC) under grant number: EP/I036222/1.

The authors would also like to acknowledge the involvement and support of the UK Civil Aviation Authority (CAA), Rail Safety and Standards Board (RSSB), the Association of Train Operating Companies (ATOC) and Aerobytes Ltd.

REFERENCES

Amherd, M. and M. Hafliger (2010). The successful implementation of a Fuel Management Information system, in IATA Fuel Efficiency and Conservation Course. Geneva.

Aviation Safety: Efforts to Implement Flight Operational Quality Assurance Programs, in GAO Report to Congressional Requesters. (1997). US Federal Government: Washington, DC.

CAP 739 (2003). Flight Data Monitoring: A Guide to Good Practice. CAA, Safety Regulation Group: Gatwick, UK.

Charlton, S. G. & Starkey, N. J. (2011). Driving without awareness: the effects of practice and automaticity on attention and driving. Transportation Research Part F, 14, 456-471.

Chidester, T.R. (2003). Understanding normal and atypical operations through analysis of flight data. in 12th International Symposium on Aviation Psychology. Dayton, Ohio.

Endsley, M.R. and E.O. Kiris (1995). The out-of-the-loop performance problem and level of control in automation. Human Factors, 37(2): p. 381-394.

Hart, C. (2003). Global Aviation Information Network. in Proceedings of the Human Factors Workshop: Improving Railroad Safety Through Understanding Close Calls. Baltimore, MD.

McLeod, R. W., Walker, G. H., & Moray, N. (2005). Analysing and modelling train driver performance. Applied Ergonomics, 36, 671-680.

RAeS (2009) Recognizing the potential of human factors training in the rail and aviation industries. London, UK: Ashgate.

RSSB (2004). Review of the Efficacy of Assurance Process in Preventing Catastrophic Accidents, in Management. Rail Safety and Standards Board: London.

Sarter, N.B. and D.D. Woods, (1997). Team play with a powerful and independent agent: operational experiences and automation surprises on the Airbus A-320. Human Factors, 39(4): p. 553-569.

Shappell, S.A., A Human Error Analysis of General Aviation Controlled Flight Into Terrain Accidents Occurring Between 1990-1998. (2003). Office of Aerospace Medicine: Washington, DC 20591.

Stanton, N.A., et al. (2005). Human factors methods: A practical guide for engineering and design. Farnham: Ashgate.

Statler, I.C. (2007). Recent advances in mining aviation safety data, in Aerospace Medical Association 78th Annual Scientific Meeting: New Orleans, LA.

Wright, L., A. Ross, and J. Davies (2000). SPAD risk factors: results of a focus group study. ScotRail Railways Ltd.

CHAPTER 79

Professionally Important Qualities of Train Drivers and Risk Prevention of Their Erroneous Work

Valerii SAMSONKIN, Olga LESHCHENKO, Kateryna PERTSEVA,
Oleksandr BUROV

State Scientific-Research Center of Ukrainian Railway Transport
Kyiv, Ukraine
o.burov@iod.gov.ua

ABSTRACT

It was developed and implemented the technique for investigation of psychophysiological important qualities (PPIQ) of rail drivers in conditions of locomotive depots. That technique was realized as a software tool to measure psychophysiological indices that were compared with normative ones. Measure and analysis of rail drivers' PPIQs have revealed reliable differences between drivers and engine-driver's mates on each traction types (diesel and electric locomotives), as well as between drivers of the same traction at different depots. Such a result should be taking into account when designing and use initial professional selection of rail drivers, because could lead to higher risk of unreliable drivers work who do not meet requirements of occupation.

Keywords: rail driver, psychophysiological important qualities, performance

1   INTRODUCTION

Safety and security are growing concerns for businesses, governments and the traveling public. Train accidents continue to occur across West and East Europe despite major improvements in active safety systems and a major refocus on passenger train safety. Policies aimed at reducing transport-related risks depend on state-of-the-art research and technological development, among which human factors issues play increasingly bigger role.

One of the most important activities in the area of rail transport is the machinist's activity. To become a qualified specialist a machinist needs some years for training. The requirements to the occupational selection rise because of the development of this branch, increase of movement speed limits and amount of various irritants.

The labour of machinist becomes more difficult nowadays. On the railways of Ukraine the amount of passenger long component trains which count up to 24 carriages and speed expresses is increased. In freight motion oblong trains, heavy freight trains and coupled trains are introduced. Feeling of personal responsibility increases through monotony of work, features of labour and rest, noise, vibration. During the trip of 600 km long 8-10 thousands of irritants from which only 10 % are important (traffic-lights, moves, pointers of *type*) have an influence on a machinist. At a speed of 80 kilometres per hour 20 alarm irritants in 1 minute have an influence on a machinist, but at a speed of 100 kilometres per hour it is already between 22 and 28 irritants.

The individual qualities of subject of labour are meant under professionally important qualities. They influence on the efficiency of professional activity. Not only psychical but also nonpsychical properties of subject can be professionally important (constitutional, somatic, neurodynamic, etc.). Professionally important qualities are the separate dynamic features of personality, separate psychical and psychomotor characteristics (expressed by the level of development of the proper psychical and psychomotor processes) and also physical qualities which meet a requirement to a person of any profession and facilitate the successful mastering of this profession. On the one hand, professionally important qualities are the pre-condition of professional activity, on the other hand, they are developed during activity, being its new formations; a man changes himself during labour.

Purpose was to reveal reasons of train drivers' (operators) psychophysiological professionally important qualities gap to occupation requirements, their impact on operators professional level and reliability.

2   PPIQs MEASUREMENT

Modern researches of professionally important qualities within the limits of psychology of labour and engineering psychology are conducted on the basis of systemic approach. Activity of machinist is realized on the basis of the system of professionally important qualities. It means: a) the activity requires a certain aggregate of professionally important qualities, b) the activity is not "mechanical" sum of qualities, but it is an organized system. Functional intercommunications and co-operating are set between separate professionally important qualities. A system of professionally important qualities is a certain complex of symptoms of subject characteristic, which is specific for activity of a machinist. It is not set in the complete form, but it can be formed in a worker while he is mastering his activity. Moreover, Some specific subsystems of professionally important qualities are also formed for its basic components (key actions, basic function, etc.).

Therefore, from an internal (actually psychological) side the process of activity shows by itself the dynamic change of integral subsystems of professionally important qualities which provide each of the basic stages (operation, task, function).

There are two major categories of professionally important qualities of a machinist:

- those, which characterized by the most direct relationship with the parameters of activity. They are reflected by the concepts of leading professionally important qualities.

- those, which have the maximal number of internal system relationships with other qualities, that characterized by the most structural weight. They occupy a central place in the whole system of qualities. They are reflected by the concepts of basic professionally important qualities. These qualities can not significantly correlate with the parameters of activity, however they are more important for its realization (Grigor'eva, 2006, p. 65).

Exact basic professionally important qualities are the basis for the forming of professionally important qualities subsystems on the whole. All other qualities of subject, necessary for providing the activity or its basic actions, functions round them and unite, organize themselves on their base. Therefore professionally important qualities are structure-forming for activity of machinist.

Professionally important qualities of mastering the activity of a machinist and its implementation are selected . The first ones are the most essential for effective, high-quality and rapid capture of activity features, the second ones - for its realization at set level. These two groups of professionally important qualities coincide and differ partly.

It is proved that for professional activity of a machinist it is necessary to distinguish those individual qualities which actually "answer" for its performance part, and those which are necessary for perception - they bring professionally meaningful information. In this regard it is accepted to talk about professionally important qualities of implementation and informative professionally important qualities.

For activity of a machinist it is actual to select professionally important qualities which he has in the normal conditions of its implementation, and those which are needed in the complicated or extreme terms (Grigor'eva, 2006, p. 67). For example, extreme conditions require speed of reaction, speed and exactness of motions, good co-ordination of motions, emotional stability. Certainly, these qualities have an important value also in normal conditions, but in extreme conditions the successful resolution of problem situations depends exactly on them.

Driving a train is the most important part of transportation process, where successful implementation depends on successful co-operation of a controller and a machinist, from one hand, and on professional preparation of a machinist and harmony of locomotive brigade's work, from the other hand.

At the modern level of organization of process of transportations it is impossible to be sure, that a machinist will not make an error. For example, the machinists of 1th and 2th classes, which have a sufficient vital experience, normal conditions of life, high professional preparation can make an error. Often machinists can not explain these errors. Obviously it is impossible to provide the absolute rightness of machinist's actions. It is possible only to reduce the certain level of errors, carrying out the special measures (Kozubenko, 1992).

One of important measures to solve the problem of efficiency of professional activity on the railway is a professional selection on the basis of analysis of professionally important qualities which the machinist of locomotive needs. In fact, the lack of these requirements leads to erroneous and untimely actions of the employee.

A professional psychophysiologic selection is a part of professional selection, which contains the complex of measures, directed on the selection of persons, whose activity consists of a locomotive driving, that is directly related to the safety of motion. They also correspond to the requirements of professional activity with their professionally meaningful qualities. In his work the psychologist of locomotive depot uses the automated complex of professional psychophysiologic selection "ADMITTANCE", with the purpose of research of professionally important qualities of machinists. Methods used and professionally important qualities they determine are marked in the table.

Table 1. Methods of determination of psychophysiologic indices for the professional selection of machinists with the recognition of professionally important qualities

BASE METHODS	Professionally important qualities	ADDITIONAL METHODS	Professionally important qualities
Assessment method of sensorimotor response	Simple sensory-motor reaction Complex sensory-motor reaction	Assessment method of fitness to influence of stress	Fitness to stress
Assessment method of response rate on a mobile object	Reaction on a mobile object	Assessment method of ability to make decision and operate in extreme conditions	Readiness to the urgent actions
Assessment method of attention	Firmness of attention	Assessment method of fitness to monotony	Firmness to monotony

Assessment method of visual and auditory memory	Short-term visual memory Short-term auditory memory	Assessment method of fatigue	Individual features of development of fatigue
Assessment method of attention shifting speed	Switching of attention		
Assessment method of emotional firmness and feeling of alarm	Emotional firmness		

The analysis of table shows that the highest level of capacity of machinist is oriented on mobility and stability of work of the perception and analysis systems, stability of emotional sphere and necessary indexes of dynamics of psychomotor and agile functions. Therefore the base methods are used for:

- primary professional psychophysiologic selection;
- periodic psychophysiologic inspection;
- extended psychophysiologic inspection;
- corrective and rehabilitation actions.

3   CONCLUSIONS

The structure of professionally important qualities is a difficult system formation which determines the progress of mastering and implementation of a machinist's performance. The analysis of employee's professionally important qualities is an important measure of erroneous actions prevention.

ACKNOWLEDGMENTS

The authors would like to acknowledge the Ukrainian Railway Transport for financial support of the research.

REFERENCES

Grigor'eva M.V.: Psychology of Labour. Lectures- M.: Higher Education, 2006. - 192 p.
Kozubenko V.G.. Safe management by train: questions and answers – M.: Transport, 1992. – 254 p.
Psychology of Labour. Ed. A.V. Karpov. M.: VLADOS-PRESS. 2005. – 350p.
Samsonkin V.N. System approach as fundamental method of research of ergonomics: essence, application in transport systems // Rail transport of Ukraine. №6. -2008. -p.3-4.
Samsonkin V., Burov O., Burova O. Tools to improve Human Performance Technology // 2nd International Conference on Applied Human Factors and Ergonomics.-14-17 Yuly 2008, Las Vegas, Nevada, USA. – 5p.

CHAPTER 80

The Future of Driver Training: Integrating Non-technical Skills

Kate Bonsall, Ann Mills

RSSB
London, UK
ann.mills@rssb.co.uk

ABSTRACT

Non-technical skills (NTS) have been defined as the cognitive, social and personal resource skills that complement technical skills and contribute to safe and efficient task performance. NTS are more general than technical skills and can be applied to a range of tasks and procedures (Flin, O'Connor & Crichton, 2008). Examples of NTS are conscientiousness, communication, rule compliance and workload management.

In response to growing evidence highlighting the key role of NTS in safe and effective performance, RSSB undertook a research project to develop, pilot and evaluate NTS training courses and other reinforcement activities for front line staff and their managers.

Keywords: Non Technical Skills, Training, Evaluation

1 BACKGROUND

The overall model for train driver training in Great Britain has remained largely the same since the introduction of formalised driver training in the 1970s. Although the model is successful to an extent, it focuses predominantly on development of technical skills and underpinning knowledge. As a result, it tends to be rules-based, trainer-centred and reliant on time as a means of assuring competence. The recent withdrawal of mandatory GB Railway Group Standards and associated guidance relating to training provides an opportunity to review current training models and refocus them in line with safety requirements and business needs.

2   NON-TECHNICAL SKILLS

Analyses of incident and accident reports in the rail industry have shown that the majority of errors that occur relate to people at the front line (63%) rather than to management (30%) or design factors (7%) (RSSB, 2009a). Closer evaluation of incident and accident reports has shown that consistently, non-technical skills (NTS) such as situational awareness and workload management are key contributors to these events. This is supported by other research across safety-critical industries highlighting how important NTS are to safety (Flin, O'Connor and Crichton, 2008). NTS have been defined as the cognitive, social and personal resource skills that complement technical skills and contribute to safe and efficient task performance (Flin, O'Connor & Crichton, 2008). NTS are more general than technical skills and can be applied to a range of tasks and procedures. NTS can be used with technical skills to enhance the way that a task or procedure is carried out and can increase safety by helping to manage threats and errors when they occur. Human error is inevitable but people can develop skills and expertise that can help them to mitigate risks. For example, a driver who shows signs of being conscientious might be more likely to quickly notice threats as they occur and if they are good at managing workload and communicating with others they might effectively mitigate that threat.

While behavioural preferences form part of the train driver selection criteria, there is very little formal coverage of NTS in other areas of the driver competence management system. At present, training programmes for operational staff within the GB rail industry are based largely on rules and traction handling, and ongoing competence development is concerned only with technical skills.

Similar training programmes (known as Rail Resource Management or Crew Resource Management) have already been implemented in rail industries in other countries, and in other safety-critical industries. It can be difficult to outline the exact commercial benefits of training interventions, particularly when accidents occur on an infrequent basis, but the available information regarding the costs and benefits of NTS training are encouraging (Lowe, Hayward & Dalton, 2007). The case has been argued that the application of NTS training to the Rail Industry can be regarded as an investment (Roop et al, 2007). Evaluations of NTS training in a range of safety critical industries show that it generally produces positive reactions among trainees (Salas et al, 2006). Evidence suggests that NTS training has resulted in improvements in knowledge, behaviour and attitudes in a range of industries (e.g. O'Connor, Flin & Fletcher, 2001; Salas et al, 2001; Powell & Hill, 2006). For example, evaluations of non-jeopardy observations in the aviation industry have shown desired changes in behaviour (Helriech & Foushee, 1993). Behavioural improvements of between 6% and 20% have been consistently reported (Salas et al, 1999). The Federal Railroad Administration (FRA) in the USA notes positive safety benefits across a range of safety critical industries including improved communication and situational awareness (Morgan et al, 2006). NTS training is reported to have contributed to a reduction in the number of accidents and incidents caused by human error in rail (Klampher et al, 2007), aviation (Fleming & Lardner, 2000), and shipping (Bydorf, 1998). For example, Canadian Pacific Railway report a 46% decrease in human-caused incidents and the lowest incident rate for Class One Railways in North America and attribute this to the Rail Resource Management Program that they implemented in 2002 (Klampher et al, 2007). Queensland Rail in Australia report that their evaluation of NTS training showed that trainee drivers were more than twice as likely to pass a signal at danger in their first 12 months if they had not completed the NTS training compared to if they had (26.3% compared to 13.1%) (Queensland Rail, 2011).

The decision to make Crew Resource Management training mandatory for pilots (Civil Aviation Authority in the UK) and crew (Federal Aviation Authority in the USA) within the aviation industry, and the increasing adoption of this style of training into other safety critical industries is a clear demonstration that NTS training is considered effective and worthwhile.

When the FRA applied outcomes from the aviation industry to the actual and estimated data from the rail industry, they concluded that NTS training can be expected to have similar benefits in the rail industry and net positive effects at an industry and individual company level (Morgan et al, 2006).

3 DEVELOPMENT OF NTS BEHAVIOURAL TRAINING

In summary, evaluation of the current provision of training within the GB rail industry suggests that there is scope for development in this area based on the evidence emerging from railway incidents and leading training practice. In response to this, RSSB undertook a project to develop, pilot and evaluate a NTS training course. This process involved the development of a complimentary NTS course for managers, and guidance on company policy on the integration of NTS into competence management systems and on the implementation of NTS training.

3.1 Method

3.1.1 Development of Non Technical skils behavioural markers

The first stage was to develop a draft list of NTS applicable to the driver role and corresponding behavioural markers (examples of good and poor behaviour). A thorough approach was taken to the development of this list and the markers. The draft list was based upon a review of existing information including; the selection criteria used for recruitment for train drivers, previous work conducted by RSSB with a train operating company to identify and measure NTS in the simulator (RSSB, 2009b), incidents and accidents recorded on the RSSB Human Factors incident database, research literature on the role of behaviour in safety critical roles (e.g. Flin, O'Connor & Crichton, 2008), a list of NTS compiled by RailCorp and used in task analysis workshops to identify training needs (RailCorp, 2008), and National Occupational Standards for train drivers (GoSkills, 2009). The markers were developed in accordance with the available guidance on behavioural marker development (Daimler-Und, & Benz-Stiftung, 2001).

This draft list was then validated by subject matter experts through a number of workshops. In these workshops attendees were presented with an inventory of all the tasks that drivers are required to carry out as part of their role. Attendees were asked to consider, for each element of the driver role, which (if any) of the NTS were relevant and why. The final NTS list is provided in table 1 below.

Table 1 List of Non Technical Skills

NTS Categories		NTS Skills	
1	Situational Awareness	1.1	Attention to detail
		1.2	Overall awareness
		1.3	Maintain concentration
		1.4	Retain information
		1.5	Anticipation of risk
2	Conscientiousness	2.1	Systematic & thorough approach
		2.2	Checking
		2.3	Positive attitude towards rules & procedures
3	Communication	3.1	Listening (people not stimuli)
		3.2	Clarity
		3.3	Assertiveness
		3.4	Sharing information
4	Decision making & action	4.1	Effective decisions
		4.2	Timely decisions
		4.3	Diagnosing & solving problems
5	Cooperation & working with others	5.1	Considering others' needs
		5.2	Supporting others
		5.3	Treating others with respect
		5.4	Dealing with conflict/ aggressive behaviour
6	Workload management	6.1	Multi-tasking & selective attention
		6.2	Prioritising
		6.3	Calm under pressure
7	Self-management	7.1	Motivation
		7.2	Confidence & initiative
		7.3	Maintain & develop skills & knowledge
		7.4	Prepared and organised

Feedback was also sought on the comprehensiveness of the NTS list and markers, and whether and re-wording or description was necessary. The results of these workshops confirmed the relevance of each NTS on the list, and the clarity and appropriateness of each behavioural marker.

3.1.2   Development & piloting of the NTS training course

The second stage of this work involved developing and piloting a NTS training course. Course development began with a collation of relevant underpinning information for each NTS, for example on the limitations of the human information process system (RSSB, 2008). Training staff from across the industry were then invited to participate in the development of a NTS course for drivers. Working with RSSB, the group produced a training course which combines a collection of learning methods (e.g. group discussion, practice and role play), drawing on relevant industry incidents to illustrate the relevance of NTS to the driver role, the reasons why things can go wrong, and how NTS can be used to anticipate, manage and mitigate these risks.

Pre-course materials, a facilitator guide and delegate workbook were also designed to complement the course delivery. Given the potential relevance of NTS to other safety-critical roles in the industry, the course materials were developed to be generic enough to be adaptable to other roles, and guidance was included on suitable adaptations.

The course materials were then reviewed by more senior training representatives from across the industry, and union members, to ensure that they were supportive of the content and methods used. It is widely agreed that demonstrable support from management (e.g. Predmore, 1999) and reinforcement of principles (e.g. Helmreich, Merritt & Wilhelm, 1999) are vital to the success of NTS training programmes. For this reason, as well as involving senior personnel throughout the project in steering groups and review meetings, a specific course was also developed for the staff who manage the personnel who will receive the NTS training.

This manager course outlined the importance of the role of the manager in promoting the value of NTS, and in developing the competence of drivers from 'competent' to expert. It included practical guidance on how to observe and document NTS, how this related to existing competence management systems, and how to provide meaningful feedback to promote NTS development. Within the course and throughout the project it was made clear that the purpose of the NTS training was to raise awareness of such skills and promote their development, and not to make pass/ fail assessments. The guidance on feedback included on the course was based on a combination of findings from previous RSSB research (RSSB, 2009b) and literature on effective management and coaching (e.g. Lombardo & Eichinger, 2006). The course combines theory, group discussion and the opportunity to practice (e.g. through the observation of pre-recorded driver simulator sessions). Practical resources were developed to complement the course delivery and the managers' role, including quick-reference guides. Finally, the course content was reviewed for suitability by a sample of managers across the industry.

The courses were piloted with a sample of managers and drivers from two train operating companies. Trainer representatives from these two pilot companies were included throughout the project in order to gain their complete understanding and buy-in to the project. These trainers delivered the courses within their respective companies.

3.2 Evaluation of the NTS training course

3.2.1 Evaluation methods

In order to attempt to demonstrate the effectiveness of the NTS training course, methods were put into place to evaluate the course. The method of evaluation of the driver course was designed to assess the Four Levels of Evaluation outlined by Kirkpatrick (1979); reactions, learning, behaviour and results. Reactions were gathered through course feedback sheets distributed after each course module and question sessions were used to check learning during the course. Behaviour change was measured through two methods. Once the managers had received their training, they conducted a pre-course measurement of their drivers' NTS, and repeated this process one month and 6 months following their drivers' completion of the course. Drivers also completed a self-measurement of their NTS at the start of the course, and were encouraged to keep a self-reflective log of their progress.

It is expected that the investment in this training, along with support and reinforcement from managers, will lead to a reduction in incidents and accidents. The companies involved in the piloting of the training programme committed to monitoring their incident and accident rates over time.

3.2.2 Evaluation Results

In analysing the evaluation data, it was expected that improvements would be found for each of the dimensions. It was not possible to analyse this information on an individual-by-individual basis as much of the information had been collected confidentially and some data was missing for the later time points. Instead analyses were conducted at a group level.

3.2.2.1 Manager and driver reactions

Manager reactions to the course were generally very positive. Overall the manager course received an average rating of 4.25 on a one to five scale of 1 = not very useful, and 5 = very useful, a score of 4.25 for relevance (1 = not at all, 5 = very) and a score of 4.42 for how interesting the course was (1 = not at all, 5 = very).

An average rating of 2.23 also indicated that there was a reasonably good balance between theory and practical exercises on the course (1 = too much theory, 5 = too much practical), with a slight lean to too much theory.

As with the manager reactions, driver reactions were generally very positive. Overall the driver course received an average rating of 4 on a one to five scale of 1 = not very useful, and 5 = very useful, a score of 4.17 for relevance (1 = not at all, 5 = very) and a score of 3.83 for how interesting the course was (1 = not at all, 5 = very).

An average rating of 2.97 also indicated that there was a good balance between theory and practical exercises on the course (1 = too much theory, 5 = too much practical).

3.2.2.2 Manager attitudes and behaviour - NTS and KSAs

The analyses showed positive results with significant improvements seen across manager NTS and managerial skills (as perceived by managers themselves). Across the sample (n = 11) the average ratings had shown an improvement for all of the NTS categories and 'Manager Knowledge, Skills and Attitude' (KSA) at sixth months. Significant improvements were reported for 'Maintain concentration' (Z=-2.449, p=0.014), 'Effective decisions' (Z=-2.449, p=0.014), 'Knowledge of safety critical NTS and their relevance to the role' (Z=-2.449, p=0.014) and 'Knowledge of what should be documented and why' (Z=-2.919, p=0.004).

A small average improvement was seen in ratings of attitude to safety (71% total score before the course compared to 72% after the course, n = 12), but these improvements did not reach statistical significance.

3.2.2.3 Driver attitudes and behaviour - NTS and KSAs

Data collected before the course, and then six months after the course was compared. The analyses showed positive results with significant improvements seen across driver NTS (as perceived by managers and drivers).

Using the behavioural markers when observing their drivers, the managers rated their drivers significantly higher on the situational awareness (z = -2.506, p =.012, n = 16), workload management (z = -2.032, p=.042, n = 16) and conscientiousness categories (z = -2.527, p=.012, n = 16), and for a number of sub-skills.

Ratings provided by managers before the driver course, one-month after the course and six months after the course showed that experienced drivers were not rated, on average, any higher or lower than inexperienced drivers in any of the NTS categories, or overall.

In the drivers' own ratings, they rated themselves higher on all NTS at one month after the course than before the course (n = 26), and all ratings further improved at six months (apart from self management which although higher than pre-course levels did drop slightly down from the 1 month measurement). At six months (n = 17), significantly higher ratings were found for 'Situational awareness' (Z = -3.316, p=.001), 'workload management' (Z = -2.926, p =.003), 'Decision making and action' (Z=-2.939, p=.003), 'Conscientiousness' (Z = -3.016, p=.003) and 'Co-operation and Working with Others' (Z = -2.897, p = .004).

A small average improvement was seen in ratings of attitude to safety (69% total score before the course compared to 71% after the course, n = 18), but these improvements did not reach statistical significance.

3.2.2.3 Managers perceptions of safety culture

It was expected that there would be improvements in safety culture, as the training should enable drivers and managers to talk more frankly about the challenges they face and to learn from near-misses.

Improvements were seen in ratings of attitude to safety and safety culture, although these improvements did not reach statistical significance. The safety culture scores were already high before the NTS training (an average score of 80%, raising to 86% after the course, n = 11) and so it is thought that this could by why the improvement did not reach significance.

3.2.2.4 Incident and accident rates

The pilot courses were delivered in spring / summer 2011 and at present it is too early to judge what impact the courses have had on the level of incidents and accidents among the sample from each of the pilot companies who attended the course.

4 CONCLUSIONS

The evaluation data suggests that the course did have a positive impact on the demonstration of NTS on the job. Managers felt their own skills and knowledge had improved, and there were small (but not significant) improvements in safety culture.

Due to the small numbers of drivers and managers involved in the pilot these results should be used only as an indication of the possible impact of NTS training. Also, the results may differ for different companies depending on a number of factors eg the competence of the trainer, how well the training is integrated in NTS strategy, the existing safety culture of the company.

The evaluation process provided some useful feedback that was then used to inform the final refinements to the NTS training course. Changes included a restructure and shortening of the course, and the managers tools were made more usable.

A final report currently being drafted will provide industry with guidance on suitable adaptations that individual companies could make to reflect the knowledge, skills and attitudes of staff within their company, as well as more general recommendations for integrating NTS into company culture, for example through safety briefing days, and incorporating the consideration of NTS into incident investigations. Industry will also be advised to re-visit their training provision and competence assessment at a more general level, in line with another RSSB project to provide an overall review of driver training that has been running in parallel to the non-technical skills project.

REFERENCES

Bydorf, P. (1998). Human factors and crew resource management: An example of successfully applying the experience from CRM programmes in the aviation world to the maritime world. As cited in Health and Safety Executive (2003) *Factoring the human into safety: Translating research into practice. Crew resource management training for offshore operations.*

Daimler-Und, G. & Benz-Stiftung, K. (2001). *Enhancing performance in high risk environments: Recommendations for the use of behavioural markers.* Workshop, Swiss Air training centre, Zurich, 5-6 July 2001. Kolleg Group interaction in High Risk Environments.

Fleming, M., & Lardner, R. (2000). It's all gone pear shaped. *The Chemical Engineer, 6th July* 2000.

Flin, R., O'Connor, P., & Crichton, M (2008). *Safety at the Sharp End: A guide to non-technical skills.* Hampshire: Ashgate Publishing Limited.

GoSkills (2009). National Occupational Standards for Rail Services. Rail services final version approved May 2009. (*http://www.goskills.org/index.php/industries/6/1/22*)

Helmreich, R. L. & Foushee, H. C. (1993). Why crew resource management? Empirical and theoretical bases of human factors training in aviation. In E. Weiner, B., Kanki, & R. Helmreich (Eds.), *Cockpit Resource Management* (pp. 3-45). San Diego, CA: Academic Press.

Helmreich, R. L., Merritt, A. C., & Wilhelm, J. A. (1999). The evolution of crew resource management training in commercial aviation. *International Journal of Aviation Psychology, 9*(1), 19-32.

Kirkpatrick, D. (1979). Techniques for evaluating training programs. *Training and Development Journal, 33*(6),78 – 92.

Klampfer, B., Walsh, C., Quinn, M., Hayward, B., & Pelecanos, S. (2007). The national rail resource management (RRM) project. Launch presentation, Sydney.

Lombardo, M., & Eichinger, R. W. (2006). For your improvement: A guide for development and coaching. Lominger International: A Korn/ Ferry company.

Lowe, A.R., Hayward, B.J., & Dalton, A.L. (2007). Guidelines for rail resource management. *Report prepared by Dédale Asia Pacific for Public Transport Safety Victoria and Independent Transport Safety and Reliability Regulator, NSW.*

Morgan, C., Olson, L. E., Kyte, T. B., Roop, S., & Carlisle, T. D. (2006). Railroad Crew Resource Management (CRM): Survey of Teams in the Railroad Operating Environment and Identification of Available CRM Training Methods. *Report produced by Texas Transportation Institute for the U.S. Department of Transportation, Federal Railroad Administration.*

O'Connor, P., Flin, R., & Fletcher, G. (2001). Methods used to evaluate the effectiveness of CRM training in the aviation industry. *UK Civil Aviation Authority Project 121/SRG/R&AD/1.*

Powell, S. M. & Hill, R. K. (2006). My co-pilot is a nurse – Using crew resource management in the OR. *Official Journal of Association of periOperative Registered Nurses, 83, 179-202.*

Predmore, S. (1999). Managing safe behaviour on the ramp: Delta Airlines' Experience. Presentation at CRM/TRP & Ramp Safety conference, London.

Queensland Rail (2011). Rail Resource Management Evaluation Report.

RailCorp (2008) RBTNA Non-technical skills list. QTMS-FO-58 V3.

Roop, S. S., Morgan, C. A., Kyte, T. B., Arthur, Jr., W., Villado, A. J., & Beneigh, T. (2007). Rail crew resource management (CRM): The business case for CRM training in the railroad industry. *Report produced by Texas Transportation Institute for the U.S. Department of Transportation, Federal Railroad Administration.*

RSSB (2008). Good practice guide on cognitive and individual risk factors. Accessed 21 February 2012. *http://www.rgsonline.co.uk/Railway_Group_Standards/Traffic Operation and Management/RSSB Good Practice Guides/RS232 Iss 1.pdf*

RSSB (2009a). An analysis of formal inquiries and investigations to identify human factors issues: Human factors review of railway incidents. Acessed 21st February 2012. *http://www.rssb.co.uk/SiteCollectionDocuments/pdf/reports/research/T635_HFrpt_final.pdf*

RSSB (2009b). A model for the measurement of non-technical skills and the management of errors on the simulator. Rail Safety and Standards Board.

Salas, E., Burke, C. S., Bowers, C. A., & Wilson, K. A. (2001). Team Training in the Skies: Does Crew Resource Management (CRM) Training Work? *Human Factors,* 41(1), 641-674.

Salas, E., Foulkes, J. E., Stout, R. J., Milanovich, D. M., & Prince, C. (1999). Does CRM training improve teamwork skills in the cockpit? Two evaluation studies. *Human Factors, 41,* 326-343.

Salas, E., Wilson, K. A., Burke, C. S., & Wightman, D. C. (2006). Does crew resource management training work? An update, an extension and some critical needs. *Human Factors, 48(2),* 392-412.

CHAPTER 81

Active Planning Requires the Appropriate Graphical Support Tools: The Case of Scheduling Trains at Stations

Rebecca Charles, Nora Balfe, John Wilson, Sarah Sharples, Theresa Clarke

University of Nottingham and Ergonomics team, Network Rail

ABSTRACT

Network Rail own, maintain and continuously improve the entire UK rail network infrastructure, ensuring over 1 billion rail journeys can be made every year. Keeping the trains running to time is a major focus of the company and regulating the railways and station areas specifically are a vital part of keeping delays to a minimum. Railway signallers have the responsibility of managing these delays. Along with computerised train running systems providing real time train information, some signalling teams currently also use paper based tools (either lists or graphs) to manage delays and plan around them. This paper reports the findings from Critical Decision Method-based interviews and observations carried out at eight sites to investigate the planning strategies and techniques that operators use to manage station areas. Clear differences were observed between the strategies of the operators depending on whether they were using the list-based or graphical tools. The list-based tools required a high degree of interpretation on the part of the operators and mainly allowed them to deal with each problem as it occurred. The graphical representations show the movements within a station area and enable the signaller to see at least four hours of platform occupation at any one time. The signaller is then able to plan any knock-on changes well in advance, keeping delays to a minimum. This proactive management of problems arising in the station area could be a result of any computational offloading taking place due to the 'at a glance view' that the graphical representation provides.

802

Keywords: Decision making, rail human factors, computational offloading, critical decision method, real-time re-planning

1   INTRODUCTION

The organisation and planning of the running of station areas within the UK Rail Network is a complex task. Trains must arrive in their correct platform, at the planned time, so that passengers can travel on the right train. Timetables are worked out months in advance and every train unit has a specific planned 'working timetable', based mainly on maintenance requirements. When disruption occurs, the operator (the station signaller) must plan around the problem and ensure that movements return to the planned timetable as quickly as possible keeping disruption and delays to a minimum.

The movement of trains in the station area involves co-ordination within and between groups. Typically the Train Operating Companies (TOCs) will inform the Shift Signalling Manager (SSMs) of any changes to the timetable. These changes are then fed to the signaller who makes the change and signals the train into the correct platform. Any changes may then have to be communicated to many sources, including TOCs, maintenance staff, station staff or train drivers.

Most signallers use list based representations to handle the station areas, many using two lists: one showing the trains in arrival order and one showing the trains in departure order. Some signallers however have developed a graphical representation of these lists (called 'Dockers') that can be updated with changes to the timetable such as late or cancelled trains (see figure 1).

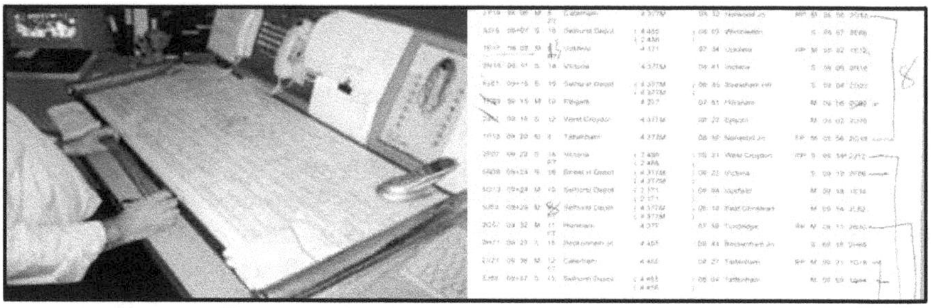

Figure 1 - Docker in use and annotated list

The graphical representations show the movements within a station area and enable the signaller to see at least four hours of platform occupation at any one time. Any changes to the timetable are physically written on to the graph, giving an up-to-date visualisation of the station area at any one time. The signaller is then able to plan any knock-on changes well in advance, keeping delays to a minimum.

This type of representational artefact is not unique to railway signalling. Paper flight strips are one example (Mackay 1999). Still used within Air Traffic Control, the paper flight strip represents a single aircraft and details information about the aircraft such as its speed and filed flight plan. These strips can be annotated by the controllers and are then mounted on a board that provides a schematic of the current situation. In conjunction with radar (a current representation of the state of the network) and voice communications the controller is able to build an accurate picture of the current and future situation (Fields, Wright et al. 1998).

In other domains, the interactivity of hospital status boards has been observed to aid between groups interaction, and in a similar manner to the dockers, were developed by the users (Wears, Perry et al. 2007). Developing the representation in this way enables the users to build on their existing knowledge and lay out the information in a way that further aids planning and problem solving (Reisberg 1997). Most literature however, focuses on the generation of plans and the processes associated with this, rather than considering the execution of the plans as a planning activity in itself. Xiao et al (1997) carried out a study investigating the planning behaviours of anaesthesiologists during complex surgery. They suggest that when a plan is made, it can be rehearsed and examined prior to execution via mental or physical activities. These activities could therefore be a result of events that have occurred, or to the anticipation of future events. The use of the docker could be considered a tool to aid this rehearsal process along with existing knowledge and experience of the operator by making it easier to visualise possible future situations.

As the UK railway continues to modernise and introduce new technologies to reduce delays and increase capacity, there is an opportunity to build on these paper based methods with an electronic tool that has a consistent approach to managing station operations nationwide. This paper presents the findings of interviews and observations carried out regarding the list and graphical based tools with a view to discovering the strategies signallers use alongside these tools in order to manage station areas.

2   METHODS

A semi structured interview technique, based around the Critical Decision Method (CDM) was devised. Although the interest was not in specific 'incidents' the nature of the questioning technique and the probing method seemed suited to exploring the domain further in terms of specific tasks, although the timeline element was not utilised. A total of eight signal boxes were visited at least once (referred to as boxes A – H for the remainder of this paper). Boxes A – F used list dockers and G and H used Graphical ones. A visit typically lasted three hours, but due to the nature of the domain, availability of signallers was opportunistic. SSMs were spoken to at all sites visited, and if signallers were instrumental in the management of disruption at a particular site then they were interviewed. A total of 11 SSMs and 14 signallers were interviewed. All participants were on shift and performing their tasks during the visits, so around 30 hours of general observations were also gathered.

3   RESULTS

Coding the data revealed various tasks carried out by the signallers as well as situations they were carried out in and various limiting factors. Through several short conversations with subject matter experts (three managers based at the stations where the Dockers are used) these were grouped into five key situations (Late Running, Cancelled Service(s), Stock Swaps requested by TOC, Line blockage affecting station area and Platform blockage) and six key strategies (Swap Sets with existing train in station area, Use a spare train from the depot to replace another, Step up sets, Split sets, Join sets or Re-dock train in alternative platform) encountered when re-docking trains. The data for the other six stations was then sorted in the same way to ensure that all the coded data had been accounted for under these 11 groups. The results will be presented in terms of strategies used.

3.1 Swap sets

Sets are typically swapped when the Train Operating Company (TOC) needs a train to meet a specific maintenance schedule. For example, a service will be assigned to a train and this train may be required to be serviced in a certain amount of miles, so the TOC may wish to swap the service to an alternative train in order to meet this requirement. Typically, these are planned in advance and emailed through to the box. If the SSM is responsible for organising these, they will then feed this information to the signaller. At most boxes, the signaller was told which trains to swap. Whether they were instructed on how to achieve the (e.g. what platform to use) varied between boxes. For example, the TOC may instruct them to 'swap the service running on train X to train Y' which may impact on other services and may mean swapping platforms or clearly specify 'swap the service running on train X to train Y and run out of platform 6'. In either instance, the signallers would first have to identify the trains affected. The signaller then locates these trains, and assesses the situation of each. They need to be aware of what time the trains arrive, when they depart, what type of trains they are and also the services around them.

The signallers using the lists were able to find the trains relatively quickly – many of the signallers observed had memorised the timetable and could recall the working timetable without referring to it – but identifying the state of the other trains in their proximity took longer. In contrast, the signallers using the graphical dockers were able to identify the trains and establish the surrounding situations very quickly. The 'list users' typically referred to both lists side by side, marked the affected trains and then looked for other trains with the same platform number. The 'graph users' were able to glance quickly and identify all potentially affected trains.

The 'list users' found the situation particularly difficult when the platforms were not specified by the TOC. One signaller commented that it was so time consuming making a swap that ⊐if the TOC wants to change something, they can sort it out" (Box A). This was not a lone opinion, and many other signallers who used the list based dockers often refused changes as "you can get in a right mess. By the time you have worked out where to put it you end up with more trains delayed. It's not worth it" (box D). The signallers who used the Graphical based dockers however, seemed to have the opposite problem: "I think we are victims of our own success. Because it is so easy to use the docker to make changes and plan, we get asked to do it all the time. That said, it does mean that we save time as if we have sorted the move, we know it's gonna work" (box G).

One of the main issues observed was that the TOCs often had access to the same information, but in a different form, so when the signaller and TOC attempted to devise a plan together it was often a laboured process. One box had overcome this problem by issuing the TOC the same imformation, in the same format which meant that "it's really easy. You just say train such and such, 3rd one down on page 2 and they can find it straight away. Then we work out what to do with them!" (Box B).

3.2 Step up sets

'Step ups' are used primarily to overcome late running, or failed or cancelled services: "The difference between step ups and stock swaps is that step ups are forced upon us through train failure, infrastructure failure and so on." (Box G)

Train sets are 'borrowed' to replace a service whose incoming train has been delayed elsewhere, and then another is used to replace that service and so on, until the original train becomes available again or a replacement can be sought: "Step ups. You are basically robbing Peter to pay Paul. Looking anywhere and everywhere to get a set. Splitting sets, joining sets, anything. We can step up for hours. I think the longest I have done it is about three hours. Ideally an empty set will be brought in eventually to get you back on track." (Box H)

The main reason observed for stepping up sets was to overcome late running services. At most of the boxes visited, the SSM kept an eye on the service using real time train planning systems. If a train was running late, the signaller would assess whether the situation needed attention through various rules of thumb, usually developed by the operators themselves utilising their experience: "If he is 2 minutes late there [points to bridge on screen] then he is ok. If he's 5 minutes late he's not, as other services will usually impact on him, and we will have to hold him outside the station." This was a typical rule used by the operators as the first stage of the decision making process. Many, like this one used static objects such as bridges or crossings to identify how late the train is likely to be when it reaches the station. However, how much the operator could actually do in these situations was different from box to box. This is for a number of reasons including; relationship with TOC, workload of signaller, complexity of station area, density of services, routes in and out of the station and whether the services are local or long distance.

One box for example (Box B): "If you have one a few minutes late, you try and step up sets to help out, but then because the service is so tight, especially round the peak, you end up getting delays coming in. Sometimes you can do it, and not get any delay, but most of the time, if that happens, you're ******."

Knock-on effects from any change were common place in most of the boxes, but again, how efficiently they were minimised differed from box to box. In the boxes where the operators had less input into the changes, knock on effects and delays particularly from late running trains were more common. The process of stepping up trains seemed more laboured: referring to a simplifier, list docker and co-ordinating with the TOC during times of disruption. Boxes where the operators gave input more regularly seemed to be able to handle step ups due to late running trains and plan for any subsequent knock-on effects. This could imply that boxes that had a greater input had a better understanding of the situation from being more involved. It could however imply that the practices and artefacts used at different boxes affected the range of decisions the signallers were able to make.

Boxes that used a graphical docker seemed particularly efficient at handling late running trains. Late runners were quickly identified, and potential moves and step ups were spotted quickly. One particular move observed involved eight step ups to ensure there were no further delays due to one service running late. Each one of these changes involved dialogue with the TOC, station staff and signallers, but took less than three minutes to develop the plan for the eight changes and inform everyone involved. Another box observed who handled similar traffic (locals and long distance) at a terminus station found it difficult to develop plans to handle late runners: (Box B) "Eight step ups? To handle one late runner? Nah, there's no way we could sort that. It would take too long." This was commonplace at busier stations that handled lots of local services and a dense service. Delays were usually accepted and the knock on delays could take a few hours to deal with. The signallers used strategies in this situation to minimise the repercussions of the delays, rather than the delays themselves, for example by minimising delay to high priority trains.

These reactionary strategies were frequently observed. Particularly when the delays were mounting and the signaller was waiting for a gap in the service to start to rectify the situation. As already discussed, many boxes felt unable to plan a strategy to deal with knock on effects from late runners, but could focus easily on each train as an individual object and deal with them one by one, minimising the damage as much as possible.

3.3 Split sets

Sets can only be split if they are already running as two separate units joined together. They can also only be split and utilised if there is a driver available for both sections of the train. Splits are usually timetabled just before the evening peak. Trains run as joined units all day, then split to provide a more frequent service – i.e. a service every 15 minutes instead of 30 minutes. Dealing with these splits appears simple on the face of it; the moves are timetabled and do not require platform moves (as the train remains in one platform as two trains). However, some stations appeared to struggle handling these moves if there was also disruption in the station area, or the trains required stepping up in addition.

The main problem seemed to arise when stations were using list 'dockers'. One list shows the trains in arrival order, the other in departure order. If a train arrives as a double set due to split, it will show as one train arriving on one of the lists. There will then be 2 departure head codes written next to it. On the other list, if a train splits it will be shown as two separate trains in the order they depart. The signaller then goes through the simplifier to find the relevant trains and their arrival /departure times – these could be up to 8 hours in advance or previous. The main method observed for the signallers managing these was for them to annotate the simplifier / docker, drawing lines to represent associations between trains (see figure 1). Many signallers carried this task out at the beginning of their shift, using both of the lists. "If you don't spend time sorting all this out, and getting it straight in your head, you can get in a mess, especially when things start to unravel" (Box E)

Signallers observed using the lists all seemed to struggle with splitting trains as far as initially identifying them and planning. One signaller whilst on the phone to the TOC was observed tracing his finger down the hand drawn line over two pages in order to 'find' a train.

The operators would rarely consider using a split train as a step up or even a stock swap due to the complexity of the information available to them. In contrast, boxes with access to a graphical docker had less difficulty. The graphical format allowed the signallers to see quickly and easily if the train was due to split and when it was departing. This made it easier to use these trains as step ups and swaps. At one box that used a graphical docker, it was commonplace. At box G, a train was available every day that was used in the morning then not used again until the evening peak. This was used a lot to replace sets or step up sets. Using a timetabled split like this was observed at other boxes, however few of them had the luxury of a spare train 'sitting in' all day.

3.4 Join sets

Joining sets is handled in a similar way to splitting sets. The paperwork is annotated in the same way but it is used as a tactic to handle slightly different situations. The trains have to be identical otherwise they will not 'tie up'. This information can be gained from looking at the diagram number (a unique train identifier) which is visible on both the graphical and list dockers.

Ad-hoc joins were often used if trains were blocking other trains: If a train service is cancelled an the sets are lying in the station, joining them with another train and running them as a strengthened service can be an effective way of dealing with the situation. The example below is from Box G where graphical dockers are used. All local services were cancelled due to a problem on the line which meant there were excess sets in the station.

"We have all these trains and only so many platforms and trains are in the platforms so trains can't get in or they can't get out. Basically it's like a car park with no spaces left. That's when you start moving trains around. A lot of the time you just double the sets up. So 3 cars and 4 cars will be doubled up to make 6 cars and 8 cars just to get them out of the station." Signallers without a graphical docker had no way of visualising the station when changes were made, and so were less inclined to make changes;

"You can't plan. You deal with stuff when it happens" (Box B)

This was the response of one signaller when asked if joins were used as a way of dealing with delays. The annotated paperwork was said to be hard to follow, and although changes were logged, it was observed that only two to three changes were made at a time and a great deal of effort was required to do this, with many signallers commenting "I'm sorry, I just have to sort this" when arranging joins, being unable to answer questions simultaneously.

3.5 Use a spare train

The use of a spare train was usually a process that followed stepping up trains as a result of late running or cancelled / failed services. 'Spare' trains are obtained predominantly from two sources: empty trains form the depot or one that is in the stations for a long time (called a long lier); typically a train that is used in the morning peak, then left in a platform until the evening peak.

When step ups are carried out, the process has to eventually end: either by using the train that had to be stepped up in the first place or finding a spare train. Through dialogue with the TOC a solution is sought. If the solution is to bring a train from the depot, the TOC would arrange and notify the box. The main factor to consider was whether the driver had sufficient route knowledge. This, however is left up to the TOC to organise in all cases.

The other option is to use a long lier to replace a service. Two of the boxes visited had a daily long lier that was used primarily for this purpose that can be used until it ir required again; "we have a diesel sprinter in platform 6. Sits there all day, so we just use him as and when we need him."

3.6 Key factors and strategies

Train types are a key factor when swapping or moving trains to another platform. If the train types are not compatible, they cannot be swapped: "There are three types of electric sets, 314, 318 and 334. These are not compatible with one another. One of the problems we have is that if you end up swapping one train type with another, this set may be scheduled to tie up with another set later on in its running, and it won't tie up." All but one of the stations handled a mix of diesel and electric trains and also had certain platforms that were used for certain services – locals, long distance. Also, no two stations are the same in terms of physical attributes. Factors such as number of platforms, length of platforms and availability of platforms all had an effect on the types of moves that were possible. The main consideration was the type of platforms. Four of the stations visited had only terminus platforms, three had terminus and through platforms and one had just through platforms. This in itself led to operators developing and using different strategies to overcome problems.

Terminus platforms are relatively simple to operate. The train arrives, a train may dock on top of it and this one must leave to allow the other train to leave. The only consideration in this case is whether there are sufficient routes into the platform. Most stations had a rule of thumb that if a train was moved you would either try and keep the service running out of its booked platform or as close to it as possible. This not only assisted the station staff and in turn the passengers

(passengers already waiting at a platform will not have to walk far) but will also assist the signaller in terms of routing the train into the platform. Another rule of thumb observed at all stations was a 'top / bottom' split of the station. If a train was coming in on one of the top lines, the operator would try to re-dock it in a top station.

The boxes with access to the graphical dockers were able to see quickly and clearly if there were any gaps and how the move would affect other trains. The boxes using lists found it more difficult, referring to one list, then the other in order to identify the locations of trains.

4   DISCUSSION

Clear differences were observed between the strategies of the operators depending on whether they were using the list based or graphical tools. The list based tools required a high degree of interpretation on the part of the operators and mainly allowed them to deal with each problem as it occurred. The speed at which trains were identified and problems detected was considerably quicker with the users of the graphical dockers. The task of locating a train took a considerable amount of effort for the signallers using the list dockers.

The users of the graphical docker were also able to identify trains more quickly, and could build a clear picture of the potential disruption with relative ease. They provide a simple, reliable method of controlling station areas by providing an external representation of the problem (Zhang 1997), which supports (and is supported by) the signaller's internal understanding and representation of the problem. By physically drawing the new plan on top of the old on the graphical Docker, late running trains can be handled with ease, as the "current" plan is fully visible. By adding this interactivity to a visualisation, more cognitive benefits can be gained (Rogers and Brignull 2003). This ability to try out moves and rehearse the strategies before carrying them out can strengthen the strategy and improve its effectiveness (Xiao, Milgram et al. 1997).

By relying on the list based dockers and using them in a non-interactive way, the signallers internally formulate the solutions, requiring greater computational effort (Larkin and Simon 1987). By marking changes directly onto the Docker, the signaller is kept in the loop, and is able to obtain instant feedback as to whether a change is possible allowing them to concentrate on problem solving.

The users of the list dockers dealt with issues as they appeared and were often unable to deal with more than one issue in more than one area at a time. They were unable to plan around issues effectively and often accrued delay minutes as a result. The graph users however could deal with many varied incidents at the same time and could develop complex plans relatively quickly. These plans could involve many moves and often running up to six hours into the future. This proactive management of problems arising in the station area could be a result of any computational offloading taking place due to the 'at a glance view' that the graphical representation provides. The interactivity of the graphical docker may

810

explain the way it is used as a forward planning tool, to a greater extent than was observed in the stations that were using lists.

In being able to develop complex plans quickly, and visualise their effects instantly, the graphical docker users had increased confidence in their decisions. They could quickly identify if a move was going to work, if this would affect anything else and construct clear time frames and develop concise plans that could then be relayed to the signallers straight away. The moves that they planned were always possible whereas the list based users often made mistakes, such as trying to dock trains in platforms that were already full due to not having the clear visibility. This in turn led to the signallers taking longer to overcome problems and have to check several times using both lists, and often hand written notes before committing to a change.

Currently, the graphical Docker provides a simple, reliable method of controlling station areas by providing an external representation of the problem (Zhang 1997), which supports (and is supported by) the signallers internal understanding and representation of the problem.

5   CONCLUSIONS

The benefits of being able to physically create an updated plan on the graphical docker and 'try out' moves before carrying them out can assist greatly when developing a strategy to overcome disruption. The users are able to manage complex situations and get the stations back to normal operation quickly and clearly has benefits in terms of disruption management and workload. An experiment is currently being carried out in order to measure the cognitive benefits of using the graphical docker compared to the list based one in a controlled setting and with learned knowledge and experience stripped away. This experiment will also aim to assess the benefits of automating some of the key decisions in an electronic version of the graphical tool.

6   REFERENCES

Fields, R. E., P. C. Wright, et al. (1998). Air Traffic Control as a Distributed Cognitive System: a study of external representations. Proceedings of ECCE-9, the 9th European Conference on Cognitive Ergonomics, Roquencourt: France, European Association of Cognitive Ergonomics.

Larkin, J. H. and H. A. Simon (1987). "Why a Diagram is (Sometimes) Worth Ten Thousand Words**." Cognitive Science 11(1): 65-100.

Mackay, W. E. (1999). "Is paper safer? The role of paper flight strips in air traffic control." ACM Trans. Comput.-Hum. Interact.: 6(4) 311-340.

Reisberg, D. (1997). Cognition: exploring the science of the mind. New York, Norton.

Rogers, Y. and H. Brignull (2003). Computational Offloading: Supporting Distributed Team Working Through Visually Augmenting Verbal Communication. 25th Annual Meeting of Cognitive Science Society, Boston.

Wears, R. L., S. J. Perry, et al. (2007). "Emergency department status boards: user-evolved artefacts for inter and intra-group coordination." Cogn Tech Work: 163-70.

Xiao, Y., P. Milgram, et al. (1997). "Planning behavior and its functional role in interactions with complex systems." Systems, Man and Cybernetics, Part A: Systems and Humans, IEEE Transactions on 27(3): 313-324.

Zhang, J. (1997). "The nature of external representations in problem solving." Cognitive Science 21(2): 179-217.

CHAPTER 82

Fatigue and Shiftwork for Freight Locomotive Drivers and Contract Trackworkers

Ann Mills, Sarah Hesketh

RSSB
London, UK
Ann.mills@rssb.co.uk

Karen Robertson, Mick Spencer, Alison McGuffog, Barbara Stone

QinetiQ
Farnborough, UK

ABSTRACT

The paper presents a study to understand the risk associated with current shift patterns of GB freight train drivers and contract trackworkers and to develop strategies for risk reduction and control.

A variety of data collection methods were used including company visits, focus groups and a literature review to elicit background information about the factors influencing the pattern of work and the development of fatigue. A questionnaire and a diary relating to 28 duty periods were completed. An analysis of accident and incident data was collected to identify factors associated with increased risk. An analysis of fatigue countermeasures gathered information from the literature and from the diary and questionnaire study. The main issues relating to fatigue from all of the sources were interpreted in the context of existing knowledge and a number of potential guidelines and fatigue countermeasures are proposed.

The main issues identified relating to fatigue and accident risk were long duty periods, time of day, time without a break, consecutive duties, inadequate recovery time, workload, roster variability and travel time. Useful coping strategies included napping, obtaining sufficient breaks and the careful use of caffeine. Shift designs that take into account the main issues identified in this research programme have the potential to reduce accident risk.

Keywords: Fatigue, Shiftwork

1   BACKGROUND

A serious rail accident at Clapham Junction, in which fatigue of a signal maintainer was found to be one of the underlying causes, lead the British Railways Board to introduce an industry set of maximum limits on working hours (Hidden, 1989). The limits were generic and reflected what was achievable in operational terms at the time, based on expert opinion and agreed good practice. While the limits were not mandatory, companies monitored employees using the limits as a standard, and had systems to identify breaches. The Hidden limits are shown in Table 1.

Table 1 Hidden limits

Hidden Limits
A turn of duty should be no longer than 12 hours in duration
No more than 72 hours should be worked per calendar week (Sunday to Saturday)
There should be a minimum rest period of 12 hours between turns of duty
No more than 13 turns of duty should be worked in any 14 day period

In 1994, the Railways Safety Critical Work regulations introduced the first legal requirement on railway employers to ensure that, so far as is reasonably practicable, ⌐no employee of his undertakes any safety critical work for such number of hours as would be liable to cause him fatigue which could endanger safety; and in determining whether he would be so liable regard shall be had to the length of time between periods on duty⌐ More recent legislation implementing aspects of the European Railway Safety Directive has superseded the Railways Safety Critical Work regulations, and introduced guidance for fatigue risk management systems (ORR, 2012). Despite these advances, much of industry continues to use the ⌐Hidden Limits⌐as a basic framework within which schedules are designed and there remains no guarantee that workers will not continue to experience fatigue.

Fatigue risk is influenced by many job, organisational and individual factors. The ⌐Hidden Limits⌐are generic in that they are not specific to any particular group of safety critical workers. This prompted RSSB to conduct an earlier study of fatigue and the patterns of work of drivers within the passenger train operating community (McGuffog, 2004). The study proposed alternative working limits and good practice to assist duty holders comply with the Railways and Other Guided Transport Systems (Safety) Regulations (2006).

The current study described in this paper was commissioned to research similar issues within the freight operating company (FOC) and infrastructure maintenance contractor (IMC) communities. These two groups differ in their working practices and environment from those of passenger train operating company drivers. For example, there is a greater physical component to many of the tasks in the infrastructure community, most activities are undertaken outdoors, and there may be long periods spent travelling to and from work. In both the FOC and IMC communities there is a greater requirement to work at night. Understanding the contribution of these factors to fatigue, in addition to more well established issues such as shift length, the number of consecutive shifts and reduced rest periods, will ensure the risks of fatigue are also effectively managed within these communities.

2   METHODS

The study involved a range of approaches to provide an in-depth understanding of the culture of shift working practices, methods to monitor fatigue, relationship of accidents and incidents to factors associated with shift work and analysis of data from train drivers

Specifically this included a review of the relevant scientific literature and information and reports on fatigue specifically of interest to the rail community, building on an earlier review undertaken as part of a related research project (McGuffog, 2004). A number of company visits and nine focus groups were undertaken to elicit background information about the factors influencing the pattern of work and the development of fatigue. A questionnaire was used to gather general information about shift working and health and well-being from 362 drivers and 568 contract Trackworkers. A diary was completed by 102 drivers and 105 Trackworkers, this collected complete details of their work pattern, fatigue and sleep covering 28 work days. Accident and incident data was collected and analysed to identify factors associated with increased risk. An analysis of fatigue countermeasures gathered from the diary and questionnaire study was undertaken. The main issues relating to fatigue from all of the sources were interpreted in the context of existing knowledge and a number of potential guidelines and fatigue countermeasures were proposed.

In this paper, we are presenting the results in the form of guidelines for the management and reduction of fatigue.

3   RESULTS

The literature review highlighted the increase in information relating to the risk of accidents among shift workers and in particular, the increased risk of incidents associated with shifts of 12 hours or more. The review also drew attention to the development and increasing use of fatigue risk management systems (FRMSs) in the rail community. FRMSs involve the management of the risk of fatigue through the use of data, policies and procedures designed to gather information about activities in the workplace

Overall, there was a low response rate both from the freight and infrastructure workers to the questionnaire and diary studies. Therefore the findings relate to a relatively small sample of the two populations and the extent to which they apply to staff in general is uncertain. Nevertheless, a large amount of data has been collected from which it has been possible to obtain a clearer picture of the working practices within the FOC and IMC communities, and to identify specific factors related to the development of fatigue. The resulting guidelines have been formulated using information from the questionnaires, diaries, and the accident and incident analysis. In addition, information has been drawn from the scientific literature and other knowledge bases on fatigue and accident risk. It is appreciated that commercial imperatives and other factors may complicate the introduction of the suggested guidelines.

In spite of the large differences in the type of work, the pattern of work and the working environment between the freight drivers and the infrastructure workers, many of the factors contributing to an increased risk of fatigue were similar in the two groups; the main exception was the issue of travel time which applied particularly to infrastructure workers. The key issues identified and the resulting guidelines are summarized below.

3.1   Shifts

Long periods of duty

The Hidden Limits state that shifts should be no more than 12 hours in duration. Data from other studies, such as those of UK signalling staff, have raised concerns about levels of fatigue during 12-hour shifts (Ryan et al 2008) and highlighted the importance of adequate sleep and breaks (Ryan et al 2007). In a review of the relative risk of accidents or injuries, the risk of an incident was shown to increase with increasing shift length over eight hours (Folkard et al 2005). Relative to eight hour shifts, 10-hour shifts were associated with an increased risk of 13% and 12-hour shifts with an increased risk of 27%.

The duties recorded in the diaries used in the current study were generally shorter than 12 hours, with just 1.5% of freight personnel and 3.3% of infrastructure staff working more than 12 hours. However, based on the responses to the questionnaires approximately one in five individuals in both groups reported having undertaken duties of more than 13 hours at least three times in the past year.

The duration of the shift is a key factor influencing fatigue and long shifts have been linked with an increased risk of accidents. There is a strong case for limiting the duration of a shift to 12 hours, with further restrictions on duties, such as nights and early starts, that impinge significantly on the normal hours of sleep. For example, while it may be acceptable to work a 12-hour day shift, lower limits such as 10 hours should be considered where night shifts or early morning start times are planned. This guidance closely follows that given previously for passenger train drivers in the earlier related study (McGuffog, 2004).

Influence of workload and task

In addition to the duration of the shift, the type of work carried out within the shift also made a significant contribution to levels of fatigue. For the drivers, it was the total amount of driving undertaken that was most important, whereas for the infrastructure workers it was the physical component of the work.

The timing and duration of breaks

In a review of the impact of breaks on fatigue and performance (Tucker 2003), the overall conclusion was that rest breaks can be effective in maintaining performance, although direct epidemiological evidence indicating a reduced risk of accidents was lacking. Frequent short breaks were associated with benefits, and better outcomes resulted when the timing of rest was at the discretion of the individual.

In the current study there was a general tendency to perceive breaks as being unsatisfactory, with comments in the diaries about the poor facilities and in the questionnaires, there were reports that there was nowhere to relax and that breaks were interrupted by work activities.

Problems with the timing of some breaks was identified, with a large proportion of individuals taking their breaks either at the beginning or end of the shift and, among freight staff, there were reports that breaks were sometimes rostered at the start or end of a shift. There were also issues with the short duration of some breaks in the infrastructure group.

Among freight drivers there were some reports of both long periods of driving and of inadequate breaks. In addition, there was a high proportion of drivers who would like to have taken a short nap had it been possible to do so (N.B. drivers were not asked why they were unable to nap). More importantly there were long sequences of consecutive night shifts during which the ability to nap may have been beneficial.

For both groups, having inadequate breaks was linked to issues such as health, dissatisfaction with the shift pattern and levels of fatigue. There was also some evidence, from the analysis of Signal Passed at Danger (SPAD) data, of an increased risk of accidents in both groups when the time without a break exceeded seven hours.

To maximise the beneficial effects of breaks greater consideration needs to be given to their timing within the shift. Scheduling breaks at the start or end of a shift reduces any beneficial effects. It is important that they are scheduled to occur at a suitable time with respect to the task activities and ideally towards the middle of the shift. It is also important that breaks provide a genuine opportunity to relax and include access to adequate rest facilities, including food and drink. Previous guidance based on passenger train driver data recommended that a break of 15 minutes should be scheduled after four hours of work (McGuffog, 2004). The current study has also shown that breaks are beneficial but, due to the limited amount of data, it has not been possible to base specific recommendations for breaks on the results of this study taken in isolation. Therefore, it is proposed that a similar break strategy to that proposed in the aforementioned study should be implemented. Additionally for freight drivers, it would be beneficial to implement a napping policy to allow rest to be taken in the cab, assuming that controls are in place to assure safety.

Start time and time of day

As in previous studies of shift workers, the shift start time and time of day that the shift was being undertaken had a strong influence on levels of fatigue. The infrastructure workers reported their highest levels of fatigue on duties starting between 01:00 and 04:00.

A large proportion of FOC staff (81%) and over half of IMC staff (58%) reported that they found the first night shift the most tiring. Both groups of workers indicated that mistakes were more likely to occur during night shifts and particularly on weekend nights for the infrastructure workers. Freight staff also identified early shifts as duties where mistakes were more likely to occur.

3.2   Factors influencing fatigue during a series of duties

Consecutive duties

The risk of accidents has been estimated to increase over successive work days, irrespective of shift type, and there is also evidence that the risk is greater over successive night shifts than successive morning or day shifts (Folkard et al 2005). In other maintenance environments, such UK civil aviation, aircraft maintenance engineers duty hours are limited based on the type and duration of the shift e.g. six consecutive night shifts of up to 8 hours, four shifts of between 8.1 and 10 hours and two shifts of 10.0 hours or longer (CAA, 2003).

The Hidden Limits specify that no more than 13 turns of duty should be completed in 14 days. There are additional limits for signalling and telecom-munication testing staff, which specify a maximum of 23 turns in any two consecutive 14 day periods.

In the infrastructure community, there were a number of reports that staff had worked more than 13 consecutive shifts. Although the incidence of non-compliance with the Hidden Limits was not high, there were reports of some very long shift sequences. This is of some concern because of the association of such long sequences both with sleep problems, based on the results from the questionnaire, and with an increased risk of an incident or accident, based on the analysis of the SPAD data.

There was sometimes considerable variation in the day-to-day timing of consecutive shifts of the same type, which adds to the problems of obtaining sufficient sleep before work. These changes in shift start time applied particularly to the FOCs. In addition, there were examples of freight staff working up to eight consecutive night shifts, eight consecutive day or six consecutive early shifts. Among infrastructure staff there were some examples of longer sequences of the same shift type, with up to 12 consecutive earlies, 11 consecutive day and seven consecutive night shifts.

It is proposed that in normal circumstances, it would be advisable to set a limit of six consecutive duties and a lower limit where duties impinge significantly on the normal hours of sleep such as consecutive night and early shifts. As an example, where night shifts exceed eight hours, consecutive shifts should be limited to a maximum of three. This guidance is similar to that proposed in the earlier passenger train driver study (McGuffog 2004), though it has been modified to address the greater requirement for night-work among IMCs and FOCs.

Duty hours

The Hidden Limits state that no more than 72 hours should be worked per calendar week (Sunday to Saturday). On average, the weekly duty hours for staff in both groups were within these limits. Only 7% of freight staff claimed to work on average more than 50 hours per week, though over the last year 14% said they had worked a maximum of over 70 hours. For IMC staff the average weekly duty hours was 50 per week, although 24% of staff said that on average they had worked more than 60 hours per week. In addition, over the previous 12 months 12% of respondents had worked a maximum of more than 80 hours per week.

Discussions during some of the focus groups indicated that there was wide variation in the attitude towards overtime. Some freight staff reported that they felt pressurized to work overtime due to a shortage of drivers. Among infrastructure workers there were reports that overtime was expected to be completed on a regular basis and other reports of no pressure to work additional hours.

Currently work hours are determined on a weekly basis, which allows a high number of hours to be worked each week. The proposed alternative approach is be to adopt a seven-day rolling period, which is limited to 55 hours before a rest day is scheduled.

Recovery from consecutive duties and days off

Evidence from the current study indicated that where rest periods were reduced to less than 12 hours between consecutive shifts there were implications for fatigue, sleep and health. Therefore, it is recommended to include a minimum rest period of 12 hours between consecutive shifts. In addition, it is recommended that rest periods should be extended to 14 hours between consecutive night shifts. However, this additional provision would in most cases be addressed by limiting the night shift to 10 hours, as outlined in 3.1.

Working on a rest day was associated with an increase in mental tiredness among freight staff and an increase in physical fatigue among infrastructure personnel. Working on scheduled days off was not uncommon, with only 15% of freight and 22% of infrastructure workers reporting that they never worked a rest day. Though staff were not asked specifically why they worked a rest day, questions relating to overtime indicated that most individuals chose to work additional hours to earn extra income. In this situation, staff have some control over duty hours, which was identified as an important issue. This contrasts with the experience that many workers had (over 50% of respondents in both communities), where additional duty hours were due to tasks taking longer than originally anticipated. This suggests that some of the additional hours may have been due to circumstances beyond the control of the workers.

The scheduling of days off following consecutive shifts varied widely and in some circumstances staff were not provided with sufficient time to recover from certain shift sequences, for example night and early shifts. In particular, infrastructure staff tended to work longer night shifts at the weekend than during the week and have only one day off. In addition, it was not uncommon for a single day off to be scheduled after a longer sequence of night shifts. Not surprisingly, a lack of recovery at the start of a shift was an important issue influencing levels of fatigue during the subsequent shift.

Travel time (to and from the workplace)

This was a significant factor increasing levels of fatigue among the infrastructure workers. Though most individuals reported spending less than 90 minutes travelling, in 7% of cases travel time was more than two hours. In some cases, lodgings were used to avoid travelling long distances, though there were differences between companies in the provision and use of lodgings.

Guidance should be given to freight and infrastructure workers on the risks of long periods spent driving, particularly during the drive home after the night shift. Where travel time is anticipated to exceed one and half hours, some alternative arrangements should be made in the form of the provision of transport or lodgings for workers, particularly when long shifts are anticipated.

3.3   Other Issues

Rostering practices

Information elicited from the focus groups indicated that the lack of stability in the rosters and changes in the pattern of work contributed to fatigue, stress, problems with sleep and job dissatisfaction. The results from the questionnaires indicated that there were frequent changes to the shift pattern and many of these changes were reported to be at short notice. This can cause problems for individuals trying to prepare for work, particularly with respect to planning sleep and domestic arrangements. In addition, the way in which some of the shift sequences were put together was not sympathetic to the management of sleep and alertness. For example, many infrastructure staff worked day shifts during the week and night shifts at the weekend. The rapid change from one type of shift to another and back again often did not include sufficient time for recovery and this was highlighted in comments made in the diaries.

Health problems

In the current study, there were correlations between the frequency of working nights and indigestion and upset stomach in both categories of workers. In addition, among freight staff insomnia, stress and anxiety were also correlated with the frequency of night shifts. Furthermore, a lack of control over the shifts worked and the stated working hours was correlated with indigestion, irritability, insomnia, stress and anxiety.

Factors associated with the pattern of work and the way in which duties were rostered was correlated with certain health issues. In both groups, individuals attributed some sleep problems to the pattern of work, and the amount of time off between shifts influenced the severity of the sleep problem. Stress at work was correlated with long shifts and time spent travelling.

3.4   Current worker fatigue management strategies

The most common fatigue management strategies employed by staff at work included getting fresh air, eating and drinking (including caffeinated drinks). Many freight drivers also indicated that they took the opportunity to nap during a break in a shift.

4   CONCLUSIONS

The programme of work has shown that fatigue is an important issue in rail operations and specifically on freight train driver and trackworker performance.

In keeping with the FRMS approach and the move away from prescriptive regulations, it is not proposed to specify fixed limits for the GB rail industry relating to duty hours and roster design. Where specific numbers are mentioned in the sections above, these should be regarded as indicative of best practice rather than as prescriptive limits.

Although each issue was discussed in turn, it should be recognised that there is a considerable interaction between the individual factors and that it is only by considering the roster pattern as a whole that sympathetic rostering can be achieved.

REFERENCES

Hidden, A., 1989. *Investigation into the Clapham Junction Railway Accident*. Her Majesty's Stationary Office, London, UK. ISBN: 0 10 1081029.

HMSO 1994 *The Railways (Safety Critical Work) Regulations 1994*. Her Majesty's Stationary Office. ISBN 0110432991.

McGuffog A, Spencer M, Turner C, Stone B. 2004. *T059 Human Factors study of fatigue and shift work. Appendix 1: Working patterns of train drivers:implications for fatigue and safety.*

Statutory Instrument 2006, Number 599. Health and Safety. *The Railways and Other Guided Transport Systems (Safety) Regulations 2006.* Accessed February 13 2012 : www.opsi.gov.uk/si/si2006/uksi_20060599_en.pdf

Office of Rail Regulation. (2012) *Managing Rail Staff Fatigue* Accessed February 13 2012: http://www.rail-reg.gov.uk/server/show/nav.1521

Ryan B, Wilson JR, Sharples S, Kenvyn F, Clarke T. (2008) Rail signallers□assessments of their satisfaction with different shift work systems. *Ergonomics 51(11); 1656 □1671*

Ryan B, Wilson JR, Sharples S, Marshall E. (2007) Collecting human factors attitudes and opinions from signallers: Development and use of REQUEST (The Railway Ergonomics Questionnaire). In: *People and Rail Systems.* Ed Wilson J, Norris B, Clarke T, Mills A. Ashgate Publishing Ltd, Hampshire, UK, 247 □256.

Folkard S, Lombardi DA, Tucker PT. (2005) Shiftwork: Safety, sleepiness and sleep. *Industrial Health 43; 20-23*

Tucker P.(2003) The impact of rest breaks upon accident risk, fatigue and performance: a review. *Work and Stress 17(2):123-137.*

Civil Aviation Authority, Safety Regulation Group (2003). *Work hours of aircraft maintenance personnel.* CAA Paper 2002/06. Accessed 12 February 2012: http://www.caa.co.uk/docs/33/PAPER2002_6.PDF

CHAPTER 83

Meeting the Challenges of Individual Passenger Information with Personas

Cindy Mayas, Stephan Hörold and Heidi Krömker

Ilmenau University of Technology
Ilmenau, Germany
cindy.mayas@tu-ilmenau.de

ABSTRACT

Personas are an established basic concept in human-centered design process. This paper suggests the adaption of the persona technique within the public transport sector. The developed procedure model combines qualitative and quantitative methods in order to operationalize behavioral variables of the users. It has been applied in a case study in German public transport system.

Keywords: public transport, passenger information, personas, human-computer-interaction

1   INTRODUCTION

Today, passenger information in public transport is becoming more and more indivi-dual, dynamic and mobile (Radermacher, 2011). Therefore, the provided passenger information varies in shape and content. For instance, there may be audito-ry hints, visual texts or icons to indicate changes, departures or service information.

In order to individualize this information, we recommend considering human-centered design in the public transport sector. The human-centered design includes a definite reference to users☐ requirements based on a detailed user description. Within this progress, public transport has to meet the following challenges:

(1) The widespread heterogeneous user group requires a detailed differentiation.

(2) The shape and content of passenger information requires a specified relation to the different attitudes, goals and the context of users.

In regard to these demands, the method of creating personas is adapted to public transport. We reveal generic behavioral variables, identify typical behavioral patterns and exemplify personas according to individual passenger information within the German public transport system.

2   USER MODELING WITH PERSONAS

Personas are hypothetical stereotypes constructed from the characteristics and behavior of real people. A persona description is fictitious, but precise and specific in order to encourage the empathy of the development team (Cooper, 1999). By dropping redundant or unessential information personas keep their stereotypical characters and become clearly distinguishable from each other.

The advantage of personas over traditional user roles or market segments consists in the naturalistic and memorable description of the users. This presentiveness of personas engenders the imagination of all team members for users who might be different from themselves. Furthermore, the description form in common language facilitates an exchange of ideas between engineers, designers and software developers according to the users. (Pruitt and Adlin, 2006)

Thus, personas are a special method supporting the involvement of user requirements in the development process of a product. Constructing personas is a process involving a wide range of qualitative and quantitative data pertaining to the users in a realistic, vivid and illustrated description of a stereotype user. The concrete procedures differ slightly from author to author. The two most prevalent procedure models of Cooper (Cooper, Reimann and Cronin, 2007) and Pruitt (Pruitt and Adlin, 2006) are com-pared with the diversification of Baumann (Baumann, 2010) in table 1. In addition to Cooper, Baumann suggests a ⌐persona and scenario building toolkit□which is based on clustering via card sorting combined with a peer-review process (Baumann, 2010). But in general, all procedure models consist of the three key stages ⌐identify variables and values⌐, ⌐identify patterns□and ⌐describe personas⌐, as shown in figure 1.

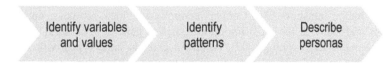

Figure 1 Generalized procedure model to construct personas

As table 1 illustrates, all approaches leave a wide scope of interpretations, especially to identify the variables and values. First of all, all procedures require the specification of basic categories or behavioral variables, which the identification of patterns is based on. For the identification of patterns the authors provide different procedures including several steps of synthesizing and evaluating. Finally, the descriptions of personas are created and prioritized.

Table 1 Persona procedure models of Cooper (Cooper, Reimann and Cronin, 2007), Pruitt (Pruitt and Adlin, 2006) and Baumann (Baumann, 2010)

Artifact	Cooper	Pruitt	Baumann
Variables and values	Identify behavioral variables; Map interview subjects to behavioral variables.	Discuss categories of users; Process data.	In reference to Cooper.
Patterns	Identify significant behavior patterns; Synthesize characteristics and relevant goals; Check for redundancy and completeness.	Identify and create skeletons; Evaluate and prioritize skeletons.	Collect and write on cards persona attributes; Sort the attitudes in clusters of similar properties; Bring the attitudes in a sequence; Repeatedly select a specific subset of attributes; Peer-review by usability and domain experts.
Personas	Expand description of attributes and behaviors; Designate persona types.	Develop skeletons into personas; Validate the personas.	Write a persona description and a scenario; In reference to Cooper.

3 BEHAVIORAL VARIABLES FOR PUBLIC TRANSPORT

Behavioral variables are categories specifying actions and attitudes with respect to the product. In order to adapt the persona technique to the public transport sector behavioral variables have to be operationalized into concrete variables of the behavior in public transport first. Cooper mentions activities, attitudes, aptitudes, motivations and skills as the most important types of generic behavioral variables (Cooper, Reimann and Cronin, 2007). Goodwin extends these types with frustrations, environment and interactions with other people, products and services (Goodwin, 2009). In contrast to Cooper, Pruitt suggests processing data first and building cluster labels in a data assimilation process afterwards (Pruitt and Adlin, 2006).

Baumann has already applied the persona technique to mobility-related services and revealed the behavioral variables personal attributes, context, means of transportation, tickets and on-board services with further subcategories (Baumann, 2010). For instance, the variable personal attributes consists of the three subcategories equipment, carrying and disability. But the purpose of these variables is to create mobile services. The course of actions as well as the attitudes, experiences and knowledge of the users are not considered. Thus, we need to go more into detail and find a procedure to identify specific behavioral variables especially related to passenger information.

The suggested procedure model bases on Hackman's concept of a task performance process and its effects on human behavior (Hackman, 1969). In addition, the concept is combined with qualitative and quantitative data research methods. The procedure model consists of four steps to identify and validate appropriate variables for the construction of personas, see figure 2.

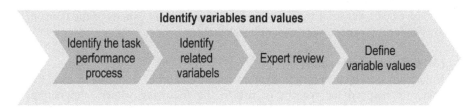

Figure 2 Procedure model to identify behavioral variables and values

In the following, this procedure model is applied to the sector of passenger information in public transport.

(a) Identify the task performance process

From the view of human-centered design the user behavior is mainly driven by the task performance process. The task performance process consists of the steps task redefinition, hypothesis of behavior or strategy, the process sequence and evaluation (Hackman, 1969). According to this model, a generic task performance process for the use of passenger information composes the following elements:

- Task redefinition: identifying information needs.
  Passengers require information for different reasons, e.g. to continue the journey.
- Hypothesis: considering information sources and information types.
  There are different information sources, e.g. timetable or network plan, and information, e.g. destination time or station name, available to the passenger. The passenger has to identify the appropriate information source for the required information type.
- Process sequence: localization, interaction and cognitive orientation.
  Passengers detect an information source with information and digest the information to take a decision for further action.
- Evaluation: assessing the usefulness of information.
  Passengers evaluate, whether the information or the consequent decision was good or bad and draw conclusions about the usefulness of the used information source.

826

(b) Identify related variables

The task performance process is influenced by personal factors of the performer. Hackman mentions the personal factors understanding and acceptance of task, idiosyncratic needs, previous experiences, abilities, performance motivation and level of arousal which have interactions with the task (Hackman, 1969). In order to include these personal factors into persona construction, they are mapped to behavioral variables of the user. For instance, the hypothesis is mainly influenced by the previous experiences of the user. This personal factor can be mapped to the behavioral variables knowledge of a system, knowledge of a place and experiences influencing the consideration of information source and type.

(c) Expert review

The revealed behavioral variables have to be proofed by experts of the domain, e.g. by conducting expert interviews with stakeholders.

(d) Define variable values

Values are specific manifestations of a variable and differ from user to user. For instance, we mapped the reasons of use to the following values: commuting, business trip, schooling trip, errand trip, leisure trip and holiday trip (Verband Deutscher Verkehrsunternehmen, 2011).

These four steps are applied within the German research and standardization project IP-KOM-ÖV to analyze the heterogeneous user requirements on passenger information. The revealed framework of behavioral variables for passenger information is shown in figure 3. In order to provide a better overview, the variables are categorized in classes. Nevertheless, each variable can be applied separately and combined independently.

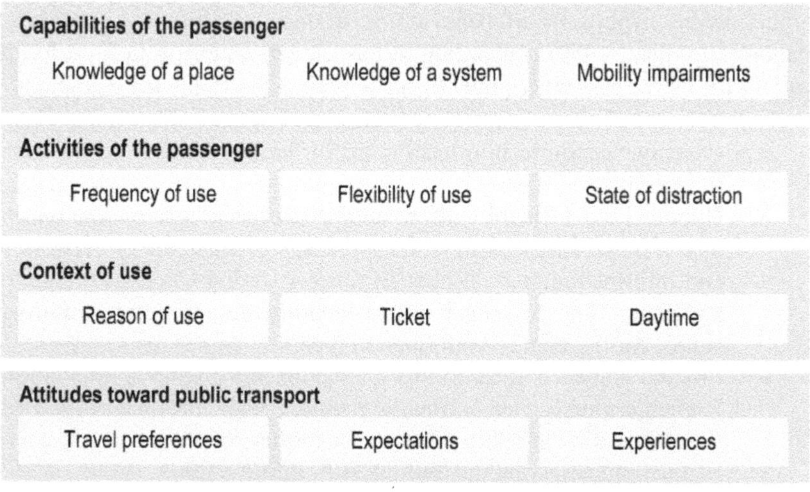

Capabilities of the passenger		
Knowledge of a place	Knowledge of a system	Mobility impairments

Activities of the passenger		
Frequency of use	Flexibility of use	State of distraction

Context of use		
Reason of use	Ticket	Daytime

Attitudes toward public transport		
Travel preferences	Expectations	Experiences

Figure 3 Classified behavioral variables for passenger information in public transport

4   TYPICAL PATTERNS FOR PASSENGER INFORMATION

Based on the behavioral variables, user data is gathered and processed by qualitative and quantitative research methods. In reference to Cooper, the characteristics of each user are mapped to a value of each variable. The most frequent combinations of variable values result in behavioral patterns. These typical behavioral dispositions are the decisive factor for the identification of personas.

In the case study, we combined the results of expert and user interviews with statistical data. The revealed characteristics are mapped to the variables shown in figure 3 and analyzed for behavioral patterns. In the process, the following key variables turned out to identify the most significant patterns:

- Knowledge of a place: includes the knowledge and cognition of streets, directions, points of interest and the relationship between them.
- Knowledge of a system: includes the knowledge and cognition of stations and lines related to city map as well as further system specifications, e.g. color guiding system.
- Flexibility of use: combines limitations regarding route and hours caused by special tickets, appointments at the destination, mobility impairments or travel preferences.

According to these key variables, the characteristics of each user and of typical statistical combinations are mapped to the values ⌐none⌐, ⌐low⌐, ⌐moderate⌐, ⌐high⌐ and ⌐very high⌐ As shown in table 2, we revealed five distinguish key patterns for passenger information.

Table 2 Typical key patterns for passenger information

Pattern name	Knowledge of a place	Knowledge of a system	Flexibility of use
tourist	none	low	low
ad hoc	high	none	moderate
occasional	high	moderate	low
commuter	moderate	high	none
extreme	very high	very high	high

The five key patterns in German public transport system ⌐tourist⌐, ⌐ad-hoc⌐, ⌐occasional⌐, ⌐commuter⌐ and ⌐extreme⌐ are validated in a nationwide survey with 145 subjects.

828

5   PERSONAS IN GERMAN PUBLIC TRANSPORT SYSTEM

With respect to the results of pattern building, one or more personas are deduced from each key pattern, depending on the diversity and priority of additional variable values. In order to achieve a holistic view of the user and to encourage more empathy, the behavioral variables of the personas are enriched by demographic variables. Each persona gets an age, gender and a meaningful name. The characters of personas are created with a family status, a profession, additional hobbies and a place of domicile. Finally several character attributes are topped of according to the behavioral variables.

The case study in German public transport system revealed a variety of seven personas. Each persona belongs to one pattern. Thereby the patterns commuter and extreme are segmented more detailed in respectively two personas, according to their reason of use.

Finally, we created the description of personas. Each persona description consists of the following part:

- name with additional term related to the key behavior,
- a persona illustration and typical statement, representing the most significant requirement,
- an overview of personal information, expectations and the public transport profile in note form,
- a description of a typical daily routine,
- a summarizing description.

An example of the persona description is presented in figure 4. The basic set of seven personas is introduced in the following brief descriptions:

Power user Maria (22 years) is studying musicology in Bonn and uses public transportation daily. Maria is always on the go. She meets friends spontaneously, goes shopping or attends lessons at the university. She does not own a car and relies on the flexibility of public transport. Through her intensive use of public transport, she knows all the ways around the city and the transport network very well.

Daily user Martina (42 years) lives in Stuttgart with her three children and her husband. Currently, she is on parental leave. Her husband uses the car to go to work and Martina uses public transportation for all the daily tasks during the week. She got to know the public transport system and the city on her special ways very well. Due to her children she prefers more time to transfer and needs further information about mother-child-seats.

Commuter Michael (34 years) is a single corporate consultant from Stuttgart, who uses and prefers public transport over the car to get to work. Due to his regular travel, Michael knows his daily routine and the public transport system well. He only uses passenger information in the event that something would not work out as planned.

Commuter Michael

"The main thing is, that I arrive punctually at the destination!"

Michael is a 34 year old, single corporate consultant from Stuttgart, who uses and prefers public transport over the car to get to work. He is punctual, endowed with technical affinities and tries to live ecology-minded. Due to his regular travel, Michael knows his daily routine and the public transport system well. His journey is about 35 minutes long and takes him to the center of Stuttgart. During his journey he has to transfer once and therefore he predominantly uses the street car, which gets him to work and back quickly. Michael does not want any unnecessary information during his daily way to work, which already is familiar to him. He only uses passenger information in the event that something would not work out as planned. In that case, Michael expects that he would be informed about disturbances as soon as possible, ideally before setting off on a journey, so that he is able to avoid the disturbance by using an alternative connection.

... is 34 years old and single.
... lives in Stuttgart.
... is corporate consultant.
... likes biking and gliding.
... is punctual, ecology-minded and technophile.

Michael expects:
- real-time information about service disturbances
- quick alternative connections
- no unnecessary information

Daily use of commuter traffic system by tram;
occasional use for business trips by train;
knowledge of a place: good   **alternatives:** bike, car
knowledge of a system: good   **ticket:**   monthly ticket
restriction: none
preferences: comfort, quietness, work en route

PT profile

Figure 4 Extract of persona description of commuter Michael (Mayas, Hﬁrold and Krﬁmker, 2012)

School commuter Kevin (15 years) is a 9th grade pupil in a middle school in Berlin. He and his friends use the public transport system nearly daily to get to school. Usually they have much to talk about or they are viewing the newest music videos together, so that the travel time goes by quickly. Kevin is used to his daily route very well and does not matter to come late.

Casual user Hildegard (69 years) is widowed and retired. She lives in a small apartment in Wilhelmshaven. Sometimes she manages her errands or travels to her garden via bus. Currently walking independently is a challenge for her. Therefore, she would be edified to have a secured seat and a undisturbed route. She feels very unconfident by using unknown lines and routes.

Ad-hoc user Bernd (51 years) Bernd lives together with his family in a suburb of Düsseldorf and works in the marketing segment. He is a typical car driver who usually avoids public transport at all. But at times Bernd is dependent for business travel on public transport, for instance when his car won't start.

Tourist Carla (29 years) is a nurse from Barcelona. This year she spends her holidays together with her husband in Germany. They have planned to use public transportations mainly for the city tours. Both do not know the country or the public transport system anyway and have difficulties with the German language. Hence, Carla prepares her journeys well and gets information about possible routes beforehand.

6   PUTTING PERSONAS INTO PRACTICE

The suggested generic set of variables and patterns for personas provides an orientation to gather user data and construct personas in public transport sector. In order to construct personas for a special transportation company or region, we recommend checking all variables and values for cultural differences to map the user characteristics correctly.

Moreover, the basic set of personas might be enriched by additional personas for special physical handicaps.

In addition, personas can fulfill the following supporting functions in the entire development process:

- Identifying and describing scenarios in reference to each persona,
- Prioritization of different information needs in different context by personas,
- Deriving individual passenger information in shape and content based on personas. For instance, individual passenger information according to commuter Michael has to include real-time information on deviations from the actual time table and the favorite personal alternative route.

7 CONCLUSION

The suggested procedure model presents a generic approach to identify task-related variables for specific personas. Basic variables and patterns in relation to passenger information in public transport are identified and confirmed in a case study in German public transport system which figured out an exemplary set of seven personas.

The adaptation of the persona technique to public transport emphasizes the relevance of user description and encourages the introduction of human-centered design in public transport. Personas intensify the distinguish consideration of the widespread target group and the transformation of passengers needs into concrete passenger information. Not only the development team but also service staff or marketing staff of the transportation company can use this information to gain a better understanding of their passengers.

ACKNOWLEDGMENTS

Part of this work was funded by the German Federal Ministry of Economy and Technology (BMWi) grant number 19P10003L within the IP-KOM-ÖV project.

REFERENCES

Baumann, K. 2010. Personas as a user-centred design method for mobility related service. In. *Information Design Journal 18(2)*: 157-167.

Cooper, A. 1999. *The inmates are running the asylum. Why High-Tech Poducs Drive Us Crazy and How to Restore the Sanity.* New York: Macmillan.

Cooper, A., R. Reimann and D. Cronin 2007. *About face 3: the essentials of interaction design.* Indianapolis. Ind.: Wiley.

Goodwin, K. 2009. *Designing for the digital age. How to create human-centered products and services.* Indianapolis. Ind.: Wiley.

Hackman, J. R. 1969. Toward understanding the role of tasks in behavioral research. In. *Acta Psychologica 31*: 97-128. Amsterdam: North-Holland Publishing Co.

Krömker, H., C. Mayas, S. Hörold, A. Wehrmann and B. Radermacher 2011. In den Schuhen des Fahrgastes - Entwickler wechseln die Perspektive. In. *DER NAHVERKEHR, 7-8/2011:* 45-49.

Mayas, C., S. Hörold and H. Krömker (eds.) 2012. *Das Begleitheft für den Entwicklungsprozes -. Personas, Szenarien und Anwendungsfälle aus AK2 und AK3 des Projektes IP-KOM-ÖV.* Available: urn:nbn:de:gbv:ilm1-2012200028.

Pruitt, J. and T. Adlin 2006. *The persona lifecycle: keeping people in mind throughout product design.* Amsterdam: Elsevier.

Radermacher, B. 2011. IP based communication in Public Transport - A German Research & Standardisation Project. In. *Travel 2020 London.* presentation (unpublished)

Verband Deutscher Verkehrsunternehmen 2011. *VDV-Schriften 720* 07/11 □ *Kundeninformationen über Abweichungen vom Regelfahrplan.*

CHAPTER 84

A Human Performance Operational Railway Index to Estimate Operators☐ Error Probability

Mr. Miltos Kyriakidis, Dr. Arnab Majumdar, Prof. Washington Y. Ochieng

Centre for Transport Studies, Imperial College London
m.kyriakidis@imperial.ac.uk

ABSTRACT

Over the years, a large number of railway accidents around the world resulting in many injuries, fatalities and a high economic cost have occurred due to operational human degraded performance. The literature shows that it is the train drivers, signallers and controllers (are referred to as operators) that mostly affect the network in terms of safety. Operators' capabilities and attitudes, the provided training, organizational, task or environmental conditions amongst others, influence to some extent human performance within the railway system.

Assessing human performance in the railway system can help to identify the safety limitations before they lead to major accidents. Thus, this chapter introduces a new approach referred to as the Human Performance Railway Operational Index (HuPeROI), which aims not only to estimate the human error probability for railway operations but also to propose mitigation strategies to minimize phenomena such as operators' degraded performance.

Keywords: railway safety, human performance, railways performance shaping factors taxonomy, human performance railway operational index.

1   INTRODUCTION

The railway system is a major component of the economy of most countries, daily transporting millions of passengers as well as millions of dollars worth of goods from origin to destination (Dhillon, 2007). The relevant operational, regulatory and governmental bodies of every country with a rail network aim for a highly reliable, excellent quality and safe railway system (Wilson et al., 2007). The definition of a railway system in this chapter it includes the terms of infrastructure, rolling stock and rail workers. The safety of railway operations within this system depends on several factors including rail traffic rules, infrastructure reliability, organisational safety culture and human factors (Hollnagel, 1998). In recent years, interest in the area of human factors within railway operations has increased significantly (Priestley and Lee, 2008). Whilst there are numerous definitions, the United Kingdom's Health and Safety Executive (HSE) definition of Human Factors (HFs) is adopted, i.e. "environmental, organisational, job factors, and human and individual characteristics, which influence behaviour at work in a way, which can affect health and safety" (Bell and Holroyd, 2009). It is well recognised that a large number of railway accidents occur due to degraded human performance (Dhillon, 2007). Human performance, which can be either positive or negative, can be described as the human capabilities and limitations that have an impact on the safety and efficiency of operations (Maurino, 1998). However, researchers are usually interested in negative human performance. A recent study (Evans, 2011) shows that at least 75% of fatal railway accidents in Europe between 1990-2009 were due to human error. The literature shows that it is the train drivers, signallers and controllers (are referred to as operators) who mostly affect the network in terms of safety (Dhillon, 2007). Several studies have been conducted in the field of HFs and human performance in the railway domain (Wilson et al., 2007). However, most of these studies are based on previous studies in the field of Human Reliability Analysis from other domains, which are not well suited to the rail industry and can be difficult to apply reliably to railway specific operations (RSSB, 2004).

In light of the current limitations, this chapter presents a new approach referred to as the Human Performance Railway Operational Index (HuPeROI). HuPeROI aims not only to estimate the human error probability for railway operations but also to propose mitigation strategies to minimise phenomena such as operators' degraded performance (HuPeROI considers as a railway operation any "train movement from one point to another or during a shunting operation"). HuPeROI is being developed based on a Performance Shaping Factors (PSFs) taxonomy designed for the rail industry. This taxonomy, referred to as the Railway Performance Shaping Factors (R-PSFs), was initially developed based on an extensive literature review in the field of HFs and on a railway operational system architecture, which described the interconnections amongst the rail operators. Maintenance or design personnel are not included in this index and furthermore, accidents or incidents either due to passengers, trespassers or third parties' responsibility, e.g. level crossing accidents.

This chapter introduces HuPEROI and its evolution process. Section 2 describes the current state of PSFs in several domains. Section 3 presents the R-PSFs taxonomy and the underlying theory, while Section 4 introduces HuPeROI and its development process. Finally Section 5 summarizes the findings and charts the future research directions.

2   PERFORMANCE SHAPING FACTORS

Performance Shaping Factors (PSFs) can be described as "all these factors such as age, working conditions, team collaboration, mental and physical health, work experience or training which enhance or degrade human performance" (Boring et al., 2007). Again, researchers are interested mainly in the negative impact of PSFs on human performance.

PSFs can be categorised as internal and external based on their characteristics. The latter are related to situation characteristics, job and task characteristics or environmental circumstances whereas internal PSFs refer to individual characteristics (Kumamoto and Henley, 1996). It is a widely accepted and applied from researchers categorization (Sasou and Reason, 1999). A recent study (Boring et al., 2007) divides PSFs into direct and indirect with the former defined as those that "can be measured directly, whereby there is a one-to-one relationship between the magnitude of the PSF and that which is measured". Conversely, indirect PSFs are defined as those that "cannot be measured directly, whereby the magnitude of the PSF can only be determined multivariately or subjectively" (Boring et al., 2007). The use of direct and indirect PSFs may be inefficient and pose considerable problems to researchers if the criteria to distinguish between them are not explicit. Several PSFs taxonomies are addressed in the literature (Hollnagel, 1998, Rasmussen, 1982, Swain and Guttmann, 1983, Williams, 1988). The different taxonomies can be divided into those that comprise a detailed set of PSFs and those that include a more generic set. Despite this, detailed and generic taxonomies have considerable similarities and include similar PSFs regardless of their domain, for development and application. As the set of the possible PSFs is limited, there is an obvious overlap between the generic and detailed taxonomies, i.e. generic taxonomies are expressed by several categories, which contain individual detailed PSFs. Another issue relates to the definitions given to PSFs in different taxonomies. This happens because researchers describe identical or similar generic factors with different names or characteristics.

The impact of individual PSFs on human performance is a matter of concern for researchers, since they do not affect humans equally. However, the identification of the exact influence of each factor on performance is not an explicit process. Certain techniques, such as CREAM (Hollnagel, 1998) and SLIM (Forester et al., 2006) consider the differences in the influence of PSFs on humans and use simple equations to estimate this (Embrey et al., 1984, Hollnagel, 1998), yet it is not justified whether they provide reliable results (Bell and Holroyd, 2009).

Finally, the relationship between PSFs and time poses considerable problems since many of PSFs are "dynamic" variables, changing continuously even while a task is executed, e.g. weather conditions or fatigue. Because it is too complex to estimate these changes, most of the techniques neither account for this nor do they analyse them separately for every condition. In this study PSFs that do not change while a task is executed e.g. experience or safety culture, are referred to as "static".

3   RAILWAY PERFORMANCE SHAPING FACTORS TAXONOMY

Given the limitations of PSFs (direct-indirect; dynamic-static; individual influence of a PSF on human performance) a new, simple and detailed PSFs taxonomy is proposed which attempts to bridge the limitations described in Section 2. The new taxonomy, which is referred to as the *"Railway Performance Shaping Factors"* (R-PSFs), enables researchers (irrespective of experience in the field of human factors) to better examine and study human performance.

R-PSFs taxonomy has been derived from an extensive literature review in the field of transportation as well as from other domains, such as the nuclear, healthcare and offshore energy exploration and production industries. The R-PSFs taxonomy aims to:

- Identify, define and categorise, based upon their common characteristics, those PSFs that influence human performance on railway operations.
- Investigate and measure interdependencies between PSFs.
- Assess ("weight") PSFs according to each operator's duties (e.g. train driver, signaller and controller) in order to propose mitigation strategies.
- Account for the distinction between dynamic and static PSFs.

The structure and the validation process of the taxonomy are extensively described in (Kyriakidis et al., 2012).

3.1   Structure of R-PSFs taxonomy

Initially 248 PSFs were identified and a mapping amongst them was performed based on their definitions to narrow down the list of factors. The most frequently identified factors are: distraction; training; communication; quality of procedures; work experience; time pressure; task complexity; fatigue or stress. Finally 45 PSFs, divided into 7 categories, were derived for the R-PSFs taxonomy considering the duties of the train drivers, signallers and controllers (Kyriakidis et al., 2012). For the PSFs structure, a simplified model was chosen (Hammerl and Vanderhaegen, 2009) and modified as shown in Figure 1. The model illustrates the interactions between the operator, executed task and railway PSFs.

The 7 R-PSFs categories are distinguished into those contain the dynamic factors, i.e. those that are strongly related to the precise moment of the operation, and those contain the static factors. A sample of R-PSFs is shown in Figure 1, while the specific factors and the categories are described into detail in (Kyriakidis et al., 2012).

A detailed definition and an example for each one of the R-PSFs is provided in (Kyriakidis, 2011) to avoid potential misunderstandings amongst researchers with different backgrounds or biases.

836

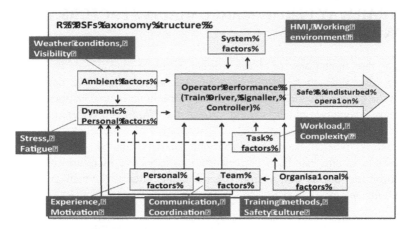

Figure 1 R-PSFs structure

3.2 Validation of R-PSFs taxonomy

To validate the taxonomy, 179 accident and incident railway reports were analysed, which have been collected from several railway stakeholders (Kyriakidis et al., 2012). The reports were analysed and the findings compared to those in the literature review.

The reports contain information on the following: the type of railway, the occurrence type and the associated event, e.g. train derailment, the location and time of the accident, the immediate cause of the accident, the causal factor of the accident, e.g. train driver falls asleep, what PSFs played a role in that occurrence and the severity of its consequences. The sources of the reports represent the majority of the situations experienced in railway operations, e.g. different types of networks, varying geography (Australia and Switzerland), different regulations and different staffing (two drivers on the Australian trains instead of one on European ones). Therefore, the sampled data can be argued to be representative of the population of interest. In order to ensure the reliability of the data the following data quality dimensions were taken into account (PhD thesis submitted to Imperial College London (Dupuy, 2011)), such as: *accessibility* (availability of the data to the users), *accuracy* (data entries are correct), *completeness* (missing values), *consistency* (data values are coherent) , *timeliness* (currentness of the data in terms of the delay in providing the information), *interpretability* (clarity of reported data), *relevance* (suitability of the reported data for a specific domain).

Finally, it was found that 18 PSFs, as shown in Table 1, are responsible for more than 80% of the occurrences (from the 179 reports) (Kyriakidis et al., 2012). However, to enhance this study the authors are currently reviewing 150 additional rail accident and incidents reports.

It is clear that the factors have either a direct or an indirect impact on each other. For instance, adverse weather conditions may affect directly operators' capability to control the train, while some organisational factors such as the level of an organisation's safety culture has an indirect impact on the operator's performance. Therefore, the interdependencies amongst R- PSFs have been investigated by the authors, as shown in Figure 2. The investigation was conducted in three stages.

Table 1 Most Frequently PSFs Met on Railway Operations

PSF	Level	PSF	Level
Distraction □ Concentration	DP	Experience - Familiarity	P
Expectation □ Routine	DP, T	Time pressure (time to respond)	T
Communication	Te	Fit to work □ Health	P, O
Safety culture	O	Workload	T
Training □ Competence	P,O	Visibility	A
Perception	DP	Work Coordination - Supervision	Te, O
Quality of procedures	O	Weather conditions	A
HMI quality	S	Stress	DP
Fatigue (sleep lack, shifts)	DP, O	Risk awareness	P
DP: Dynamic personal P: Personal, T: Task, Te: Team, O: Organisational, A: Ambient, S: System			

Stage 1 involved the identification of the interdependencies based on the operators' duties as specified in railway operational manuals (marked with symbol □ in Figure 2) (Network Rail, 2008a, Network Rail, 2008b, Network Rail, 2008c). Then R-PSFs interdependences that contributed to the occurrence were identified from the reports analysis in the second stage (marked with (a)) and finally an experts elicitation assessment was conducted in collaboration with the Swiss Federal Railways (SBB) (marked with (b)) in stage 3.

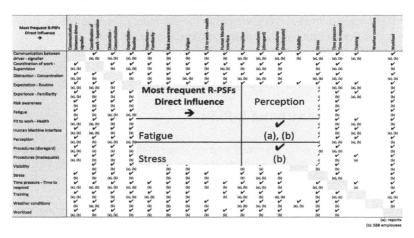

Figure 2 Interdependencies amongst R-PSFs

4   HUMAN PERFORMANCE RAILWAY OPERATIONAL INDEX

The new approach referred to as the Human Performance Railway Operational Index (HuPeROI) is being developed based on the R-PSFs taxonomy, which was described in section 3. The process to develop HuPeROI contains four main steps, as shown in Figure 3, and is described in this section.

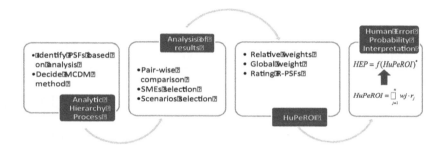

Figure 3 HuPeROI development process

4.1   Step 1: R-PSFs identification and multi-criteria decision methodolgy

Due to the fact that the quality and applicability of HuPeROI depends strongly on the quality of the R-PSFs taxonomy in the first step the accuracy of R-PSFs was assessed and ensured (Kyriakidis et al., 2012). In addition, it was decided the multi-decision theory so as to determine the contribution of each individual R-PSF on the overall quality of the human factors. The Analytic Hierarchy Process (AHP) was selected for this work because:

- It is well known and well reviewed in the literature (Triantaphyllou, 2000).
- It includes an efficient pair-wise comparison process.
- Is simple, intuitive, and yet has mathematical rigour.
- Integrates subjective judgments with numerical data (Kariuki, 2007).
- It is widely used in several domains (Saaty, 2008)

4.2   Step 2: Quantifying R-PSFs

The second step involves the weighting system, which describes the contribution of each R-PSF towards human performance. For instance, for a task A, lighting or training may influence human performance more than communication, while the reverse may be the case for another task B.

These weights will be defined by AHP (Saaty, 1982). The weighting will be based on a set of pair wise comparisons, which eventually will determine the relative impact of each R-PSF on the overall human performance for specific scenarios. The scenarios are identified from the accident-incident reports analysis, as described in (Kyriakidis et al., 2012). Subject matter experts (SMEs) in the field of railway industry and human factors such as operators, safety analysts, regulators and academics will conduct the comparisons.

SMEs will be asked, via questionnaires, to compare elements at the same level (in our case two levels). In the first level SMEs will compare the seven main R-PSFs categories, whereas in the second level the individual R-PSFs which are included in the same category. Finally, a set of 8 matrices will be derived. One matrix will correspond to the first level, while 7 matrices will correspond to the second level.

4.3   Step 3: Global weighting and Rating of R-PSFs

Once the 8 matrices have been originated, the relative weights for each one of the performance shaping factors is calculated, based on AHP (Saaty, 1982). Consequently, the global weights will be derived by multiplying the relative weights of the elements in level I and II. It should be mentioned that the relative weights of factors belonging to the same level and/or category should attain a value of 1.

After having calculated the global weighting for each factor, it is essential to also rate the quality of each shaping factor for the specific scenario. Since the railway organization influences its employees performance SMEs will be asked to rate the current condition of the system being analyzed (Kariuki and Löwe, 2006), as described in (Kariuki, 2007).

4.4   Step 4: Mathematical approach to estimate HuPeROI

A combination of weighting factors and rating factors defines what it was described as human performance railway operational index. HuPeROI is derived using the mathematical expression:

$$HuPeROI = \sum_{j=1}^{n} w_j . r_j$$

(1)

where w is the normalized weighting of each individual R-PSF and r the rating (performance measure) of the R-PSFs category involved in this human factor area.

Finally, the calculated Human Error Probability will be expressed as a function of HuPEROI however, the exact mathematical expression for this calculation has not been defined yet.

5   CONCLUSIONS

Human performance and its contribution to safe operations is of great concern for the railway industry. Therefore, in this chapter a new approach referred to as

Human Performance Railway Operation Index (HuPeROI) is developed for the railway industry.

At this stage of the research, the SMEs of the organizations have been selected and the questionnaires to weight the R-PSFs are being constructed. In addition the mathematical equation to convert HuPEROI to HEP is being investigated.

The introduction of the HuPeROI, and the R-PSFs taxonomy are highly promising. Their implementation is manifold and will enable researchers and analysts to address, deal with and mitigate the impact of degraded performance of railway operators. Firstly, HuPeROI can be used to estimate human error probability in the railway industry. Secondly, both HuPErOI and R-PSFs taxonmy can be applied from the railway orgaisations as a mean to propose mitigation strategies to minimize phenomena such as operators' degraded performance. Thirdly, even non-human factors experts can easily apply them to several railway cases. Finally, railway organisations can implement both of them as a tool for risk management.

ACKNOWLEDGMENTS

The authors are most grateful for the active participation of Swiss Federal Railways (SBB) to this study. Support from the Lloyd's Register Educational Trust in undertaking this study is also gratefully acknowledged.

REFERENCES

Bell, J. & Holroyd, J. (2009) Review of human reliability assessment methods. Norwich, Health & Safety Laboratory.

Boring, R. L., Griffith, C. D. & Joe, J. C. (2007) The Measure of Human Error: Direct and Indirect Performance Shaping Factors, *Proceedings of joint 8th IEEE Conference on Human Factors and Power Plants / 13th Conference on Human Performance, Root Cause and Trending (IEEE HFPP & HPRCT)*,

Dhillon, B. S. (2007) *Human Reliability and Error in Transportation Systems*, Springer.

Dupuy, M.-D. (2011) Framework for the analysis of separation-related incidents in aviation, *Centre for Transport Studies, Department of Civil & Environmental Engineering, Imperial College London*, London.

Embrey, D. E., et al. (1984) SLIM-MAUD: an approach to assessing human error probabilities using structured expert judgement. Washington DC, United States Nuclear Regulatory Commission.

Evans, A. W. (2011) Fatal train accidents on Europe's railways: 1980-2009, *Accident Analysis and Prevention*, 43, 391-401.

Forester, J., et al. (2006) Evaluation of human reliability analysis methods against good practices.

Hammerl, M. & Vanderhaegen, F. (2009) Human Factors in Railway System Safety Analysis Process, *Proceedings of 3rd International Human Factors Conference* Lille, France.

Hollnagel, E. (1998) *Cognitive Reliability and Error Analysis Method*, Elsevier.

Kariuki, S. G. (2007) Integrating human factors into chemical process quantitative risk analysis *Plant and Safety Technology, TU Berlin*, Berlin.

Kariuki, S. G. & Löwe, K. (2006) Increasing Human Reliability In The Chemical Process Industry Using Human Factors Techniques, *Process Safety and Environmental Protection,* 84(B3), 200-207.

Kumamoto, H. & Henley, E. J. (1996) *Probabilistic Risk Assessment and Management for Engineers and Scientists,* New York, IEEE Press.

Kyriakidis, M. (2011) Railway Performance Shaping Factors - Definitions and Examples. Imperial College London.

Kyriakidis, M., et al. (2012) Development and Assessment of Performance Shaping Factors Taxonomy for Railway Operations, *Proceedings of TRB 91st Annual Meeting,* Washington D.C.

Maurino, D. (1998) ICAO supports proactive approach to managing human factors issues related to advanced technology, *ICAO Journal.*

Network Rail (2008a) National operations centre controller job description.

Network Rail (2008b) National operations signaller job description.

Network Rail (2008c) National operations train driver job description.

Priestley, K. & Lee, G. (2008) Human Factors in Railway Operations, *Proceedings of International Conference on Challenges for Railway Transportation in Information Age,* Hong Kong, China.

Rasmussen, J. (1982) Human errors. A taxonomy for describing human malfunction in industrial installations, *Journal of Occupational Accidents,* 4, 311-333.

RSSB (2004) Rail-specific HRA technique for driving tasks. London, Rail Safety and Standards Board.

Saaty, T. L. (1982) *Decision Making for Leaders,* Lifetime Learning Publications.

Saaty, T. L. (2008) Decision making with the analytic hierarchy process, *International Journal of Services Sciences,* 1.

Sasou, K. & Reason, J. (1999) Team errors: definition and taxonomy, *Reliability Engineering & System Safety,* 65.

Swain, A. D. & Guttmann, H. E. (1983) Handbook of human reliability analysis with emphasis on nuclear power plant applications. United States Nuclear Regulatory Commission.

Triantaphyllou, E. (2000) *Multi-Criteria Decision Making Methods: A Comparative Study,* Kluwer Academic Publishers.

Williams, J. C. (1988) A Data-Based Method for Assessing and Reducing Human Error to Improve Operational Procedures, *Proceedings of 4th IEEE Conference: Human Factors and Power Plants,* Monterey, CA , USA

Wilson, J. R., et al. (2007) The railway as a socio-technical system: human factors at the heart of successful rail engineering, *Proceedings of the Institution of Mechanical Engineers, Part F: Journal of Rail and Rapid Transit,* 221, 101-115.

CHAPTER 85

Human Factors at the Interface between Road and Rail Systems

Christian Wullems, Narelle Haworth & Andry Rakotonirainy

Centre for Accident Research and Road Safety □Queensland
Queensland University of Technology
Brisbane, Australia
c.wullems@qut.edu.au

ABSTRACT

Railway level crossings present an arguably unique interface between two transport systems that differ markedly in their performance characteristics, their degrees of regulation and their safety cultures. Railway level crossings also differ dramatically in the importance they represent as safety issues for the two modes. For rail, they are the location of a large proportion of fatalities within the system and are therefore the focus of much safety concern. For the road system, they comprise only a few percent of all fatalities, although the potential for catastrophic outcomes exist. Rail operators and regulators have traditionally required technologies to be failsafe and to demonstrate high levels of reliability. The resultant level of complexity and cost has both limited their extent of application and led to a need to better understand how motorists comprehend and respond to these systems.

Keywords: railway level crossings, human factors

1   INTRODUCTION

According to (Independent Transport Safety Regulator, 2011), collisions between road and rail vehicles account for approximately 30% of rail related fatalities, excluding suicides, and remain one of the biggest safety challenges for rail operators in Australia. In the 10-year period between 2000 and 2009, there were 695 collisions, in which 97 people were fatally injured. Approximately 36% of collisions occurred at public level crossings with passive controls (i.e. give-way or stop signs) and approximately 51% occurred at public level crossings with active controls (i.e. flashing lights only or flashing lights and boom barriers). The remaining 13% of collisions occurred at level crossings on private roads. Despite a significant decrease in collisions over the 10-year period from 2000 to 2009, the number of fatalities remains relatively constant (Independent Transport Safety Regulator, 2011).

Fatalities at level crossings comprise a relatively small proportion of road fatalities (an average of 0.6% over the 10-year period), however, given the significant decrease in road fatalities (Bureau of Infrastructure, 2011), the proportion of road fatalities that occur at level crossings has effectively increased. From the perspective of rail operators, crashes at level crossings are becoming more severe compared to other railway collisions, and from the road safety perspective they are becoming a relatively larger road safety problem.

Differences in perceptions of risk in road and rail drive different approaches to traffic control. Many of the examples in this paper are from Australia, which is largely similar to North America in terms of legislative responsibility for rail and road safety (Williams and Haworth, 2007). Contrasts with European practice are made where appropriate. The scope of traffic considered in this paper is limited to road and rail vehicles; and non-motorized transport is not included (although it is recognized that pedestrian fatalities at level crossings are another important issue).

2 DIFFERENCES IN ECONOMIC EVALUATIONS OF RISK REDUCTION MEASURES IN ROAD AND RAIL CONTEXTS

One of the key differences in perception of risk between a crash at a road intersection and a railway level crossing is the potential for catastrophic consequences. Catastrophic accidents have the potential to fast-track policy changes for controlling risks from what is economically reasonable to prevention of future occurrences of the accident. Society has greater expectations of safety on the railway than it does for the road, due to the high frequency of crashes that occur on the road compared to collisions at railway level crossings. The number of road fatalities does not vary by a large fraction from year to year, unlike smaller numbers such as commercial airline crashes or passenger train crashes which can easily double from one year to the next. The relative stability and predictability of the number of highway fatalities gives an aura of being under control, suggesting there is no crisis to which a response is required.

Much of the lower public profile of road fatalities relates to their scattered nature, in ones and twos across the country (Williams and Haworth, 2007) with little public awareness of the total number of deaths. For example, from 1996 to 2002, ninety-four percent of the US road deaths occurred in crashes where one or two people died (Farmer and Williams, 2005). The contrast is usually made with commercial airline crashes, which effectively capture public attention and concern despite the total number of deaths per day being much less. It is only the worst bus crashes in which large numbers of passengers are killed that seem to capture public and political concern and lead to the introduction of measures that would not have otherwise satisfied economic criteria.

The criteria for determining whether a given risk control is economically viable typically involves a comparison of costs and benefits. The cost of a treatment is compared to benefits such as the monetary value of mitigated safety loss, which includes human costs, avoided property damage, avoided delays and other general costs that are avoided. Human costs are often estimated as a value for preventing a fatality (VPF) and are commonly calculated using one of two approaches: human capital or willingness to pay (WTP). Refer to (Tooth, 2010) and (Tooth and Balmford, 2010) for a discussion of issues in determining the socio-economic costs of road crashes and accidents at railway level crossings. The WTP approach is currently considered good practice for determining a VPF.

A key difference between road and rail is who makes the decision on the implementation of risk controls. In the road environment, the authorities (state departments of transportation) are responsible for these decisions, whereas in rail, accredited rail operators (AROs) are responsible. Consistent with a safety management systems approach to regulation, it is up to the AROs to demonstrate that they have sufficiently controlled risks to safety in their operations and that they meet their obligations under the Rail Safety Act.

In determining applicability of a risk control, road authorities would typically seek a benefit to cost ratio (BCR) of greater than or equal to 1. While a similar approach seems reasonable for AROs, legislative obligations under the Rail Safety Act require that risks are eliminated, and where this is not reasonably practicable, that those risks are reduced so far as is reasonably practicable (SFAIRP) (National Transport Commision, 2006). The Australian National Transport Council provides guidance for interpretation of this requirement (Salter, 2008). The guideline states that in considering the cost of eliminating or reducing the risk, practitioners must demonstrate that the likelihood of the risk eventuating is remote or that the cost is grossly disproportionate to the safety benefit (Salter, 2008). While the likelihood of the risk of a collision at a level crossing eventuating is relatively low, a higher disproportion factor is likely to be required in order to account for societal concerns.

Legislative requirements for safety in the road environment differ significantly from the rail environment. In contrast, road safety legislation typically relates to individuals (i.e. prosecution of individuals for dangerous driving behavior, lack of roadworthiness of road vehicle, etc.) (Tingvall and Haworth, 1999).

3   DIFFERENT APPROACHES TO TRAFFIC CONTROL BY ROAD AND RAIL AUTHORITIES

For several years, appropriateness of road traffic control approaches at railway level crossings has been a topic of discussion as part of a strategy to reduce the costs and in same cases improve the efficacy of warnings at railway level crossings. This section compares different approaches to traffic control at road intersections and railway level crossings within the context of a simple conceptual framework. Human factors issues relating to warning design in both contexts are discussed.

3.1 Traffic Conflict Management at an Intersection

Figure 1 illustrates a hierarchy of controls for managing traffic conflicts at an intersection. Two methods of traffic conflict management have been identified in the hierarchy: spatial separation and time separation. Spatial separation requires that flows of traffic be physically separated, whereas time separation requires that an intersection with multiple flows of traffic is time-shared, such that flows of traffic are separated into stages of non-conflicting phases. Time-sharing can be facilitated by signals or rules of precedence.

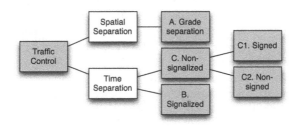

Figure 1. Traffic control hierarchy at an intersection (note that ⌐C1. Signed⌐ includes road markings)

A series of traffic control classes have been defined for the two methods of traffic conflict management. **Table 1** provides a comparison between types of traffic control used at road intersections and railway level crossings, and a brief description of the decision making process required of the user is provided.

Table 1. Traffic control type comparison

Traffic control class	Road intersection	Rail level crossing	Decision making process (road user⌐s perspective)
A. Grade separation	Overpass, underpass	Road passes over or under railway	None required.
B. Signalized	Traffic lights	Level crossing with active warning lights	Attend to warning and decide whether to comply. Decision not to comply requires violation.
B. Signalized	No equivalent at a road intersection	Level crossing with active warning lights and boom-barrier.	Attend to warning and decide whether to comply. Decision not to comply requires gross violation.
C1. Signed (non-signalized)	Stop sign	Level crossing with Stop signs	Attend to warning and situation; decide to come to complete stop.

C1. Signed (non-signalized)	Give-way sign	Level crossing with Give-way signs	Decide to give way to all other vehicles. Decision not to comply requires violation. Attend to warning and situation, decide to slow down and be ready to stop, decide to give way to all other vehicles.
C2. Non-signed (non-signalized)	Road markings only (i.e. no vertical signs)	Uncontrolled level crossing	Attend to situation, recall rules, then look for traffic and proceed.
C2. Non-signed (non-signalized)	Intersection with no vertical or horizontal signs	Uncontrolled level crossing	Attend to situation, recall rules, then look for traffic and proceed.

We expect the paradigms for stop and give-way signs to be similar for road intersections and railway level crossings. Stop signs appear to be more effective at level crossings than Give-way signs, potentially due to road rules that require road users to come to a complete stop. Failure to do so is a punishable violation. Stopping, or at least reducing speed, in theory affords the road user more time to look for trains and determine whether the level crossing is safe to traverse.

An analysis of the performance of level crossing controls in Australia and the U.S. was performed by (Independent Transport Safety Regulator, 2011), comparing flashing lights with half boom barriers, flashing lights only, stop signs and give-way signs. They observed a significantly lower number of collisions at level crossings with stop signs compared to give-way signs. The figures indicate approximately 37% less crashes per million trains and 100 million road vehicles than give-way signs. Of particular interest to us is the comparison of traffic lights and active warnings at railway level crossings in terms of differences in risk perception and human factors issues. The following sections discuss traffic control signals in more depth.

3.2   Traffic control signals

One of the key differences between traffic control signals at road intersections and at railway level crossings is that communication of system state varies significantly between the two. While nominal (correct operation) states and failure states of traffic lights at road intersections are well known, the states of level crossing warnings are not easily differentiated by the road user. For most road users, the train approach warning appears to be identical to the indication of failure. Confusion of what the level crossing warning device is attempting to communicate can potentially encourage the road user to engage in risky behavior, and this behavior can influence road user performance at other railway level crossings, especially if the failure condition occurs frequently or for prolonged periods of time.

This state of confusion is often termed □mode error,□ where an individual applies the operation appropriate for one mode when in fact they are in another (Norman, 1993). This leads to the execution of an inappropriate action. Mode errors occur frequently in systems that do not provide clear feedback of their current state.

Table 2 compares the signal states of traffic lights at road intersections and level crossing warning systems.

Table 2. Traffic signal states from the road user's perspective

Signal state	Road intersection	Rail level crossing
Nominal signal aspects	Traffic lights have the following signal aspects: red (stop), amber (if safe, prepare to stop short of the intersection), green (proceed).	The RX5 flashing light assembly (Standards Australia, 2007) consists of flashing red lights that activate on approach of a train and deactivate once the railway is safe to cross. Some level crossings have boom barriers and audible alarms.
Signal right-side failure (a failure that is detected by the system and results in the system entering a safe state)	If a right-side failure occurs, the signal shows a flashing amber aspect. This indicates that the road user should exercise caution as opposing traffic may enter the intersection.	The right-side failure mode for the RX5 flashing light assembly is equivalent to the train approach warning; however, the flashing rate may be different. Level crossings with boom barriers will close on detection of a failure.
Signal wrong-side failure (a failure that is undetected by the system and results in the system entering an unsafe state)	If a wrong-side failure occurs (e.g. power failure), the road user reverts to road rules, giving way to traffic on their right.	If a wrong-side failure occurs, it is a fundamentally dangerous state, as road users are inadvertently trained to assume that no signal means safe. For crossings with boom barriers, depending on the type of wrong-side failure, the boom barriers may close. A train detection failure will not result in closure of the crossing, whereas boom barriers are designed to close by gravity if there is a power failure.

The following subsections discuss control paradigms and traffic procedures associated with traffic control signals.

3.2.1 Signal Control Paradigms

Vehicle detection is a prerequisite for signal phase control, as it allows the signal's logic controller to optimize traffic flow based on information of vehicle presence and potentially vehicle rate for each traffic flow. Road traffic signals typically use magnetic induction loops to detect vehicles. In contrast, rail level crossings typically use one of the following type approved train detection mechanisms, depending on the performance of the mechanism in the target context and environment:

- Jointed / jointless track circuits
- Axle counters
- Treadles (for low speeds)

In road signal control, vehicle detection is typically used either to change signal stage (i.e. multiple non-conflicting phases) or to extend the period of the current stage, depending on where vehicles are detected. The road signals are configured with a cycle consisting of a sequence of stages. Conflict detection functionality insures that the signal reverts to the failure mode in the case conflicting phases are detected (i.e. green aspect for conflicting flows of traffic). Failure of vehicle detection can result in a delay in passing to the next stage of the cycle. While the condition is not technically unsafe, sufficiently long delays are likely to encourage the road users to engage in risky driving behavior.

In rail level crossing warning controls, train detection is used to activate the warning at least 25 seconds before the train enters the level crossing (Standards Australia, 2007), and deactivate the warning once the train has left the clearance point. The road authority for the particular level crossing defines the exact pre-emptive timing required for the level crossing. This is typically in the range of 25 to 35 seconds and depends on site characteristics such as sighting distance and whether heavy vehicles (e.g. B-doubles) operate on the route. There are several signal failure conditions that can occur as a result of train detection failures:

- *Warning not given when train is approaching level crossing.* This is an unsafe condition, as the road user assumes it is safe to traverse the level crossing (refer to previous discussion on signal states).
- *Insufficient or excessive warning is given when train is approaching level crossing.* Insufficient warning is an unsafe condition. Excessive warning may encourage road users to engage in risky behavior.
- *Warning extinguished before train has passed the level crossing clearance point.* This is an unsafe condition, as in low-visibility conditions, the road user may not be aware that the level crossing is occupied.
- *Warning remains active for excessive time after train has passed the level crossing clearance point (i.e. trail ringing condition).* This can be particularly dangerous in situations where there is a second train. Frequent tail ringing may condition the road user to assume that the level crossing is in a state of failure rather than a second train approaching.

There are significantly higher consequences to failure of train detection at a level crossing compared with failure of vehicle detection at an intersection. This is mainly due to the fact that failed train detection is likely to result in the crossing entering an unsafe state (i.e. a state of wrong-side failure). While there are human factors issues relating to lengthy delays at an intersection, road signals remain in a technically safe state.

3.2.2 Traffic Procedures

Another key difference between traffic signals at road intersections and railway level crossings relates to traffic operating rules. At a road intersection, all vehicles are subject to the same set of rules, whereas at railway level crossings, the rules for trains vary depending on the type of crossing and jurisdiction. For the most part, trains in Australia have right of way, regardless of the level crossing signal state. The level crossing warning signal serves to indicate to the road user when the crossing is clear, rather than timesharing the intersection. **Table 3** compares the warning and decision making process at different types of level crossings from the train driver's perspective.

Table 3. Level crossings from train driver's perspective

Level crossing type	Warning to rail traffic	Decision making process from train driver's perspective
Interlocked	Aspect to proceed is given on the basis that level crossing is protected.	Aspect may be enforced where ATP (automatic train protection) or TPWS (train protection & warning system).
Autonomous with healthy-state indication or side lights	Health state indication or sidelights are visible on approach to the crossing. The indication is a flashing light on the top of a mast.	Health state indication allows a driver to determine whether the level crossing warning is working correctly. It does not necessarily provide sufficient warning to be able to stop before the crossing. Section is assumed clear given that train has authority for section. Train driver decides to inform train controller of failure. Train controller decides to stop or slow down subsequent traffic.
Autonomous	No warning of level crossing state.	None. Section assumed clear given that train has authority for section.

The white or red side lights commonly found on level crossings in Australia are provided to the train driver to indicate that the level crossing warning device is working. Some level crossings provide a healthy state indication. According to the ARTC code of practice (Australian Rail Track Corportation, 2009), the driver must inform the train controller of a failure, who will subsequently advice the signal maintenance technician for the area. The train controller will advise any trains or other rail vehicles in the affected section to approach with caution, and arrange for any further rail traffic from entering the section. (At the time this paper was written, the Australian Railway Industry Safety and Standards board were in the process of developing a national code of practice for Australian Network Rules and Procedures (ANRP). In the absence of national rules, the code of practice from the railway that spans the largest number of jurisdictions has been cited).

To the authors' knowledge, similar operating rules exist in the U.S. Some countries in Europe have installed level crossings that provide an obstacle detection feature, indicating to the train driver the protection state of the level crossing and whether it is clear. The indication must be installed at a distance from the crossing sufficient for a train at line speed to come to a complete stop before the level crossing. In this paradigm, while trains have right of way, they can also be stopped if the crossing is not clear. Stopping trains, however, may have significant economic consequences, especially if the obstacle detection system regularly exhibits false positives.

3.2.3   Summary of Traffic Control Signal Human Factors Issues

Several human factors issues have been discussed in relation to traffic control signals and their meaning to the road user. While the authors acknowledge that there are other factors at play, this discussion has focused on road user interpretation of how signals communicate state and signal control issues. **Table 4** summarizes safety issues identified in this discussion.

In developing new interventions for level crossings, a number of issues discussed in this paper need to be taken into consideration:

- While state communication of level crossing warning devices can be potentially improved, designers or new interventions must be aware that road users have been inadvertently trained to recognize the absence of a signal at the level crossing to mean safe.
- The safety integrity of a level crossing warning system, and the reliability of vehicle detection at level crossings are more safety critical that at road intersections. Making an argument for reliability equivalent to road traffic signals needs to take into consideration the failure modes of the signal and how they are interpreted. There are also significant legal issues in demonstrating that risks have been controlled so far as is reasonably practicable. Less than fail-safe interventions should not be considered in a manner that is inconsistent with duty of care obligations.

Table 4. Summary of safety issues related to signal state communication

Signal Condition	Road intersection	Rail level crossing
Signal wrong-side failure	Mitigated by procedures	Unsafe
Signal right-side failure	Mitigated by procedures	Potentially unsafe
Timing failures	Can lead to unsafe behavior	Can lead to unsafe behavior

The Centre of Accident Research and Road Safety □Queensland (CARRS-Q) is currently leading a project investigating the effects of right-side failures on human behavior at railway level crossings (Gildersleeve and Wullems, 2012). The project will develop human reliability assessment models to quantify human performance at level crossings, and will help inform reliability targets as well as the design of improved state communication methods.

5   CONCLUSIONS

The approaches taken to managing safety in road and rail have been shown to be very different. Some of these differences stem from the divergent performance characteristics of road and rail vehicles, while others reflect historical differences in safety cultures. Recent changes in approaches to road safety appear to be creating more similarities, however. The traffic control hierarchy at intersections (road-road or road-rail) developed and discussed in this paper provides a novel framework for identifying potential human factors issues arising from road user confusion.

REFERENCES

Australian Rail Track Corportation 2009. Working of Level Crossings: Rules 1 to 7. *TA20 - ARTC Code of Practice for the Victorian Main Line Network.*

Bureau of Infrastructure, Transport and Regional Economics 2011. Australian Infrastructure Statistics Yearbook. Canberra, Australia

Farmer, C. M. & Williams, A. F. 2005. Temporal factors in motor vehicle crash deaths. *Injury Prevention,* 11, 18-23.

Gildersleeve, M. & Wullems, C. 2012. A human factors investigation into the unavailability of active warnings at railway level crossings. *ASME/IEEE 2012 Joint Rail Conference.* American Society of Mechanical Engineers.

Independent Transport Safety Regulator 2011. Level crossing accidents in Australia. *Transport Safety Bulletin.*

National Transport Commision 2006. Model Legislation - Rail Safety Bill. *F2006L04074.* Australia: http://www.comlaw.gov.au/Details/F2006L04074/.

Norman, D. A. 1993. Design rules based on analyses of human error. *Communications of the ACM,* 26, 254-258.

Salter, P. 2008. National Guideline for the Meaning of Duty to Ensure Safety So Far As Is Reasonably Practicable. *National Railway Safety Guideline.* National Transportation Commission.

Standards Australia 2007. AS1742.7-2007 Manual of uniform traffic control devices Part 7: Railway crossings.

Tingvall, C. & Haworth, N. 1999. Vision Zero - An ethical approach to safety and mobility. *6th ITE International Conference Road Safety & Traffic Enforcement: Beyond 2000.* Melbourne

Tooth, R. 2010. The cost of road crashes: A review of key issues. LECG.

Tooth, R. & Balmford, M. 2010. Railway Level Crossing Incident Costing Model. Railway Industry Safety and Standards Board (RISSB).

Williams, A. F. & Haworth, N. L. 2007. Overcoming barriers to creating a well-functioning safety culture: A comparison of Australia and the United States. *Improving Traffic Safety Culture in the United States: The Journey Forward.* Washington, DC: AAA Foundation for Traffic Safety.

CHAPTER 86

The On-call System of Work: Fatigue, Well-being, and Stress amongst UK Railway Maintenance Workers

Nuno Cebola, John Wilson, David Golightly, Theresa Clarke

University of Nottingham and Ergonomics team, Network Rail
Nottingham, UK
Nuno.cebola@networkrail.co.uk

ABSTRACT

This paper presents current research to develop understanding of on-call systems of work and explore the relationships on-call work has with fatigue, well-being and performance. In particular the system of work is place for UK rail maintenance staff working for Network Rail (the Great Britain rail infrastructure owner and operator). Initially 71 semi-structured interviews were conducted at five separate maintenance depots across the UK. From the key themes identified in the analysis an on-call questionnaire for managerial staff was developed and data from across the country was collected generated 479 individual responses. Results from both studies indicate that when discussing on-call there are three separate on-call situations; being on-call, receiving calls, and responding to calls; which influence the study variables differently. Working on-call is perceived as a leading cause of stress, and anxiety in particular, poor quality of sleep and fatigue. This is due to the inherent unpredictability of on-call work, which the authors feel is the key differencing factor between on-call work and other types of working-hours systems. Receiving and responding to calls are perceived as detrimental to general well-being both to workers and their families, fatigue, performance. In the next stage of work a diary based study will be conducted to further detail how the timing and amount of call-outs impact on the life, health, and work performance of on-call managers.

Keywords: on-call, fatigue, well-being, stress

1   INTRODUCTION

On-call work is a common form of scheduled work in many of today's industries with 20.2% of us in the European Union (EU) already working under one or another on-call system of work (Eurofound, 2010). On-call work is a cross-sector form of work that is used in a great variety of occupations including doctors, utility workers, ships engineers, media personnel, pilots and many others. The rail industry in the UK is not an exception and many of its workers have to perform duties under this scheduling system across the organisation. Furthermore, for many of these occupations on-call working is not an option but a part of the job itself as it is one of the cheaper and easier ways to provide 24/7 coverage demanded of many organisations in the21st century society. This is not a system of work restricted to one type of occupation and is in fact common to all types of work, even if more common for *high-skilled clerical* workers and *low-skilled manual* workers. Table 1 shows further details.

Table 1 - Results from 2010 EWCS survey question Q37: ☐Does your work include on-call time?☐(Source ☐ Eurofound, 2010)

		2010	
		Yes	N
	High-skilled clerical	23.1%	8116
All 27	Low-skilled clerical	16.9%	15033
EU	High-skilled manual	20.5%	5111
states	Low-skilled manual	23.8%	6717
	Total	20.2%	34977

Surprisingly, despite being such a common system of work on-call has received little research attention apart from the medical industry. This is in sharp contrast with research on other working hours systems, such as shift work and hours of work generally, which have and continue to receive a great deal of attention in research and practical guidelines.

Whilst many industries make use of this type of work there are important differences between *proximal on-call* and *distal on-call*. In most industries on-call work can be called *distal on-call* because the on-call worker is away from the place of work and is called back in case of an emergency. In the medical industry *proximal on-call* is far more common and required that the worker remains in the place of work for the duration of the on-call period. Moreover, in medicine all doctors must go through a period of on-call work in their formation, which might go some way to explain why this sector is the only one that has shown interest in researching the on-call system of work over the past 50 years. The difference between *proximal on-call* and *distal on-call* in medicine and other industries, such as rail, makes generalization of their findings limited and even potentially dangerous.

The bulk of the published work has mainly focused on medical interns, and as such, at *proximal on-call work*, and has concentrated on two areas - health impacts of working on-call and the effects his type of work has on performance. A major review was conducted by Nicol & Botterill (2004) assessing the health impacts of *distal on-call work*. From a total of 16 papers identified up to the year 2000, four health related areas were identified: stress, sleep, mental health, and personal safety. Other studies have found relationships between on-call working and deficits in attention and in working and long term memory, and also an association with feelings of confusion (Wesnes et al., 1997; Sexena & George, 2005).

The reason on-call coverage is the 21st century's preferred scheduling system as that it allows highly trained professionals to be in contact in case of, and respond when required to, critical situations when the normal volume of work does not warrant full shift coverage, i.e. evenings and weekends. Although on-call work makes sense from a cost saving perspective it is not without potential costs to those that are required to work under this system and subsequently to the organisations that choose to use it. It is the overall goal of this research to start clarifying the impacts of this type of work and to be able to provide guidelines for the planning, management, and risk reduction needed. The context of the study was Network Rail (NR), the UK rail infrastructure provider. One of the central activities of NR is to continually maintain and upgrade the infrastructure. Due to the nature of the service most engineering work must be conducted at night when the service is reduced. Both of these require the on-call system of work by both 'frontline' maintenance staff, and at middle and senior management levels.

2   METHODS

2.1   Interviews

A semi-structured interview protocol was developed based on the literature and previous interviews with NR senior managers. The interviews were exploratory and aimed to clarify the processes involved in the on-call system of work, to explore the planning, management, and requirements of on-call work and assess the impact it has on well-being and fatigue of on-call workers. From a total of 40 rail maintenance Delivery Units (DU) the interviews were conducted at DUs in different regions of the United Kingdom with 71 members of staff (65 individual interviews and three in pairs). The interviews covered all four on-call levels (see results section for details on on-call levels) of the on-call system.

These data were analysed through an inductive thematic analysis which wielded an initial 480 proto-themes. These were then re-assessed and re-coded to better fit the data. This left us with a total of around 300 different themes which were organised into high-level themes. In this paper we will focus solely on those related to on-call arrangements and on stress, fatigue, and well-being.

2.2 Questionnaire

Following the interviews a digital format questionnaire was produced to collect data from all 40 DUs across the country. The questionnaire was put together based on the data collected form the interviews and aimed to provide further details into the idiosyncrasies of each factor identified in the analysis such as timing and amount of calls, stress and fatigue. The questionnaire was held online in a NR platform – Ergotools – which made it automatically available to all users in the company. The questionnaire was formatted so that it could be completed both from a computer or a Blackberry phone as Blackberries are widely used.

One early outcome of the interview studies was that the strict enforcement of fatigue management regulations frontline staff was found to lead to fewer issues with the on-call system of work. Managers, who are usually responsible for managing their own working hours, on the other hand had many more issues regarding management of on-call and off-call working hours and as such this study focused solely on manager level on-call workers. Therefore, the target population for this study were all managerial level on-call workers. Invitations were sent out via email to the target population by a senior manager at NR as to elicit a higher response rate and the data collection period was of 6 weeks. At the end of this period 479 completed questionnaires had been returned out of a population of around 800 on-call workers (60% response rate).

3   RESULTS

The results from both studies will be presented together Quotes were typically derived from interview data whereas the majority of statistical results where derived from the survey. For clarity, the source of specific data is noted in the text.
From both studies three separate on-call situations were identified; being on-call, receiving calls, and responding to calls. All results presented are arranged in accordance to each of these three main on-call situations

3.1   On-call arrangements

On-call work is conducted mainly in blocks of weeks, where the most common system is to be on-call for a period of 24h during for seven consecutive days followed by a number of weeks 'off-call'. Survey data revealed that 80% of on-call rosters reported fall into this category and all on-call work is performed in addiction to workers *normal* working hours.

A great variability of on-call rosters was identified throughout the UK with 38 different on-call rosters reported, although, of these 24 were only reported by a single worker. Slow rotating rosters, with a period longer than three weeks between shifts, represent the grand majority of worked rosters (80%). Faster rotation rosters are not considered workable by most interviewed on-call staff considered as they do not allow for enough rest or allow for a good work-life balance e.g. "There is no way we could do week on, week off because it would ruin your – it's all about the balance, the work/life balance".

Nonetheless, there are still a relative high number of workers working under fast rotating rosters e.g. "1 week in 2 weeks", "2 weeks in n weeks", and "1 week in 3" on-call rosters with a total of 109 workers under such rosters or around 23% of all workers. The on-call roster, however, is not the only cause of long work periods for on-call workers as in the event of annual leave, sickness, or any other factors that might cause a worker to be absent from work the on-call period will typically be covered by one of the other on-call workers. This multiplies the on-call period from one week to two or more and 33.4% members of staff surveyed report having worked 14 or more consecutive days on-call in the last year. Not surprisingly, the most common amount of consecutive days working on-call spikes at every multiple of seven with 14 consecutive days being the most common, followed by seven (20%) and then 21 days (10%). Unexpectedly, around 8% of staff (35 workers) state they have worked on-call for 30 days or more in the last year.

During office hours all call-outs are directed to the maintenance depot and are dealt with by the day staff meaning that on-call work effectively only happens outside office hours. There are usually four escalatory levels of on-call staff. Level 1 on-call duties are performed by frontline staff that effectively have a hands-on role and are responsible for the actual physical work involved in attending faults. Levels 2, 3, and 4 are advisory roles and in most cases are not required to physically attend faults but might spend several hours on the telephone giving advice to the lower levels, making decisions, and coordinating efforts with other teams. With most issues do not require the involvement of levels 3 and 4 on-call, it is generally accepted that level 2 workers have the highest number of call-outs as these usually represent the managerial first line. Levels 3 and 4 will only be contacted in case of higher impact incidents such as a derailment or where a fault threatens performance severely.

3.2 Being on-call

3.2.1 Stress, quality of sleep and fatigue

Being on-call can bring feelings of stress reflected in lower sleep quality and in an inability to relax with roughly half the interviewees stating this to be the case. In the survey, to ascertain the magnitude of this feeling we asked participants "How stressed do you feel in general when" on-call and when not on-call. The response scale varied from "1-Not at all" to "5 – Very much" with the average amount of stress when working on-call being higher (M=3.47, SD = 1.22) than when not on-call (M= 2.15, SD = 1.05); t(478) = 23.589, ρ < 0.01.

The key reasons that lead to stress and unrest are the unpredictability of when a call-out will be received (M=3.25, SD= 1.36) e.g. "You're off from work but you can't actually relax because you're always expecting a phone call to come in" and anticipating calls (M=3.26, SD=1.36) due to on-going work e.g. "If there is jobs on obviously I'm at home and I kind of expect phone calls so there's some edginess" or due to weather priming e.g. "When there is bad weather you expect to get more calls, you anticipate it". Additionally, 17 interviewees state this anxiety leads to poorer quality of sleep ("1 – Very poor quality of sleep" to "5 – Very good quality of sleep") when on-call (M=3.04, SD=1.02) when compared to not being on-call (M=4.14, SD=0.73); t(41)=6.613, ρ<0.01 which consequently leads to higher levels of fatigue e.g. "Just the fact that you're on call, I'm normally knackered by the end of an on call week even if I don't get any calls. You don't switch off from work for 7 days". Not surprisingly, fatigue levels ("1 – Not at all" to "5 – A lot") when on-call (M=3.08, SD=0.99) are perceived by survey respondents to be higher than when not on-call (M=2.51, SD=1.22); t(478)=7.905, ρ<0.01

3.2.2   Well-being

On-call workers feel limited in the amount of family and social event that they can plan or attend due to the unpredictability of knowing when one will receive a call is paralysing and many workers decide it is not worth risking being disrupted at a restaurant, the cinema, or any other family or social event as they might need to attend a fault or be on the telephone for long periods of time e.g. "You write it off because you need to be contactable straight away, and the last thing you want is to be half way through a meal in a nice restaurant, your phone goes off". Furthermore for those interviewees with children on-call has a specific impact in the amount of free time they will have to spend with their children and the range of activities available to be performed e.g. "It used to be 2 in 4 but it was too much for my family life, I was missing my kids growing up".

3.2.3   Covering and swapping

Although on-call is seen as very a very disruptive system of work, around one third of interviewees state they still try to maintain their "normal" life and keep planning and attending social events. The most common way to deal with the intrusiveness of on-call work is to swap on-call periods (usually the whole week) with a colleague or to request coverage for a night or even a few hours. Interviewees believe this level of flexibility allows for a better their work-life balance e.g. "We do our best to cover for each other so, say I have a party this weekend and I'm on-call I'll swap with some one to have that shift covered. And at some point the opposite will happen so we cover for each other and even the managers are quite flexible and will do 1st line on-call if needed". This informal swapping and covering system is not solely used to allow for a better work-life balance by allowing workers to attend social and family events but also to relieve a fatigued co-worker as it is recognised that after a few 'bad' nights working on-call one might not be able to deal with the workload anymore e.g. "I covered last Sunday for the guy who was on-call. He had a particularly bad Friday and Saturday, so I had seen he'd done it so I said 'Well, I'll cover you during the day on Sunday'.".

3.3 Receiving and responding to calls

3.3.1 Sleep disruption, sleep loss and fatigue

According to interviewed staff the amount and duration of received calls when on-call and of any work carried out as a consequence vary too much for patterns to be clearly identified. A 'quiet' week could have none or very few calls and a 'bad' week can reach twenty calls plus. Moreover, it is not just call duration and frequency that is problematic as any one single call can prevent rest for several hours due to worrying about the decision made, or the advice given, or simply because one might need to get or give an update every few hours e.g. "when you've gone through it, write everything down, you've made your decision, after putting the phone down you are wide awake".

For most surveyed staff (56%) at least half the call-outs happen after they have gone to bed, therefore, increasing the level of disruption of each call. In addition, respondents claim at least 76% of all call-outs prevent them from falling back to sleep for a period of 30 minutes or more, further disturbing their rest. For at least 66% of the respondents the worst time to be called-out is between the period of time between 1 and 4 hours after falling a sleep. This period corresponds to the period of deeper sleep (Pinel, 1992) and therefore the hardest period of which to awake from. It is also the period when we would expect to see more sleep inertia and deteriorated abilities e.g. "you are impaired, you know, you are not at your best at 3 o'clock in the morning when the phone is ringing", however, only nine interviewees thought this to be the case and surveyed respondents feel their ability ("1 – Not at all" to "5 – Very much") is only affected slightly (M=2.56, SD=1.001).

The average amount of call-outs received on an on-call period by surveyed staff shows a large variation with most respondents (around 48%) receiving on average four calls or less per period but with around 23% of them receiving seven or more per period, or at least one a day for most staff. Worryingly, around 15% state they receive on average 10 or more calls per period (n=69). It is also important to note that there were no respondents stating they usually receive no calls on their on-call period. Furthermore, the average amount of sleep per night drops from an average of around seven hours and 15 minutes when not on-call to an average of around four hours and 30 minutes when on-call and responding to calls, with on-call but not receiving calls averaging around six hours and 40 minutes. As a result, fatigue levels ("1 – Not at all" to "5 – A lot") are perceived as higher when on-call and receiving call-outs (M=3.34, SD=1.35) when compared to an on-call period with no calls (M=2.82, SD=1.010); t(478)=8.797, p<0.01.

3.3.2   Well-being

Receiving call-outs have an unsettling effect on on-call workers and their families independently of the time of the call. As discussed in the previous section out of hours call-outs have a highly disruptive effect on workers sleep. Over half of interviewees declare their partners also get woken up e.g. "My wife is a professional so she has a busy life as well, obviously it disturbs her nights" with at least nine admitting it causes their partners to feel fatigued also e.g. "Sometimes she complains she's tired the next day too". Not surprisingly, on-call was pointed out as a source of conflict with 'better half' by at least 21 interviewees who declared their partners were not happy at all with the on-call system of work e.g. "I don't think any woman would be happy with the on-call. Honestly, I don't think any partner would like getting called out at night". When asked "How happy does your partner feel about you working on-call?" survey respondents believe their partners, on average, are not happy with their on-call system (M=2.26, SD=1.079; "1 = Extremely unhappy" to "5 = "Extremely happy"). Whilst not happy with the system, eleven interviewees also consider that their partners are now 'used' to the system and as such they have also learn to accept it as part of their lives e.g. "She's been doing it for as long as I've been doing it so she's quite used to it now".

Being called-out can result in the interruption of family activities and prevent workers from spending *quality time* with their families' by disrupting their family life (15 interviewees) e.g. "It affects the quality of time you have with your family, you will be there but you're on the phone, you're sort of home and you're not" or by preventing them from playing with their children (6 interviewees) e.g. "It's not good being on the phone when you're trying to play with your children". Survey respondents further confirmed this belief by affirming both their family and social lives are strongly affected by on-call work when compared to not working on-call (Family: M=3.52, SD=1.159 versus M=1.95, SD=1.024; t(478)=23.816, $p<0.01$); Social: M=3.66, SD=1.153 versus M=1.96, SD=1.018; t(478)=25.978, $p<0.01$). Additionally, survey respondents feel on-call work (M=3.16, SD=1.183) moderately affects their mood ("1 – Not at all" to "5 – Very much") with eight interviewees confirming their mood deteriorates when on-call e.g. "I'm never in a good mood when I'm on-call".

3.3.3   Coping

In accordance with current NR fatigue regulations workers must be given 12 hours rest after a call-out. For managers, who are responsible for managing their own hours, this rest many times is not taken as most arrangements and meetings must be dealt with during the day e.g. "If you don't come in who's gonna do your 'day job'? The next day is double". Managers (27 interviewees) may instead choose to have a late start in the morning or leave earlier in the day when possible to help deal with the tiredness e.g. "If I had a particularly bad night I might come in late so I get some rest".

4   DISCUSSION

Week-long on-call rosters are the preferred option when rostering at NR but day-long rosters can in theory be more beneficial to workers. By only working on-call for one day continuous anxiety is removed from the workers' mind and when called-out the workers will be able to recover faster from the disruption as it will be limited to one night. This is an option that, we feel, has not been explored enough at NR but could quickly reduce the levels of stress and fatigue in on-call workers.

It is the idiosyncratic unpredictability associated with on-call work that differentiates it from other types of work schedules. In all other types of scheduling systems workers posses the knowledge of when they are expected to work and for how long which allows them to prepare for their shift and conduct any family and social activities around that schedule. This is not the case when working on-call and the inability to organise their time causes workers to feel anxiety and prevents them from disconnecting from work. Furthermore, this anxiety then affects workers quality of sleep and a result causes their fatigue levels to rise. Similar results have been found in other forms of on-call work (Torsvall & Åkerstedt, 1988).

The unpredictability of when a call-out might be received can be paralyzing, preventing many workers from maintaining their *normal* lives as they might be disturbed and required to spend long period of time on the phone or to attend the fault. What's more, the sleep disruption caused by call-outs during the night and its subsequent impact on fatigue and performance are major concerns regarding the on-call system of work. Although that survey results show that on workers do not perceive their abilities to be severely impaired, research has identified a clear relationship between sleep loss, fatigue, and performance (e.g. Van Dongen et. al, (2003); Kaida et. al, 2007). It is not far fetched to think that in an industry with a long and well-rooted male, and potentially even macho, culture many workers will shy from admitting feeling tired or stressed. In such a culture, even a flexible swapping and covering system as is such as the one available at NR will not be effective if workers will not make use of it when feeling tired, as indeed, only one third of interviewees admitted taking extra rest and starting late or finishing early the next day.

5   CONCLUSIONS

The fact that a large number of people in Network Rail and other industries already work under this system makes this research crucial in order that sensible risk management policies and actions can be put in place. As far as the authors are concerned on-call research to date has yet to produce a set of specific on-call guidance or management recommendations and, as such, the production of said guidelines based in science and research is a key objective of our work. Guidelines based on the finding reported here have already been incorporated to Network Rail's on-going development of a high quality Fatigue Risk Management System.

To further explore the idiosyncrasies of on-call work and its relationships with fatigue, well-being, and stress a diary study is under way with 50 plus on-call managers at NR. We will analyse specifically how the timing of each call-out impacts fatigue and performance. Fatigue accumulation due to successive call-outs and call-outs in successive nights will also be further researched both for day-long and week-long on-call periods.

6   REFERENCES

Eurofound (2010). *European Working Conditions Survey, Does your work include on-call time?   Q37.   Retrieved   online   from :* *http://www.eurofound.europa.eu/surveys/smt/ewcs/ewcs2010_02_12.htm*

Kaida, K., Åkerstedt, T., Kecklund, G., Nilsson, J.P., & Axelsson, J. (2007). Use of subjective and physiological indicators of sleepiness to predict performance during a vigilance task. *Industrial Health*, 45(4), pp. 520-526.

Nicol, A., Botterill, J., S. (2004). On-call work and health: a review. *Environmental Health: A Global Access Science Source*, 3:15.

Pinel, J.P.J. (1992). *Biopsychology*. Needham Heights, MA: Allyn & Bacon.

Saxena, A., & George, C. (2005). Sleep and motor performance in on-call internal medicine residents. *Sleep.* 11, 1, 1386-1391.

Torsvall, L., & Akerstedt, T. (1988). Disturbed sleep while being on-call: An EEG study of ships' engineers. Sleep, 11(1), 35-38.

Van Dongen, H.P.A., Maislin, G., Mullington, J.M., & Dinges, D.F. (2003). The Cumulative Cost of Additional Wakefulness: Dose-Response Effects on Neurobehavioral Functions and Sleep Physiology from Chronic Sleep Restriction and Total Sleep Deprivation. *Sleep*, 26(2), pp. 117-126.

Wesnes, K. A., Walker, M. B., Walker, L. G., Heys, S. D., White, L., Warren, R., et al. (1997). Cognitive performance and mood after a weekend on call in a surgical unit. *British Journal of Surgery,* 84(4), 493-495

Author Index